AF776676

Sensor Fusion Approaches for Positioning, Navigation, and Mapping

IEEE Press
445 Hoes Lane
Piscataway, NJ 08854

IEEE Press Editorial Board
Sarah Spurgeon, *Editor-in-Chief*

Moeness Amin
Jón Atli Benediktsson
Adam Drobot
James Duncan
Ekram Hossain
Brian Johnson
Hai Li
James Lyke
Joydeep Mitra
Desineni Subbaram Naidu
Tony Q. S. Quek
Behzad Razavi
Thomas Robertazzi
Diomidis Spinellis

Sensor Fusion Approaches for Positioning, Navigation, and Mapping

How Autonomous Vehicles and Robots Navigate in the Real World: With MATLAB Examples

Mohamed M. Atia
Department of Systems and Computer Engineering,
Carleton University, Ottawa,
Ontario, Canada

IEEE Press
WILEY

Copyright © 2025 by The Institute of Electrical and Electronics Engineers, Inc. All rights reserved.

Published by John Wiley & Sons, Inc., Hoboken, New Jersey.
Published simultaneously in Canada.

No part of this publication may be reproduced, stored in a retrieval system, or transmitted in any form or by any means, electronic, mechanical, photocopying, recording, scanning, or otherwise, except as permitted under Section 107 or 108 of the 1976 United States Copyright Act, without either the prior written permission of the Publisher, or authorization through payment of the appropriate per-copy fee to the Copyright Clearance Center, Inc., 222 Rosewood Drive, Danvers, MA 01923, (978) 750-8400, fax (978) 750-4470, or on the web at www.copyright.com. Requests to the Publisher for permission should be addressed to the Permissions Department, John Wiley & Sons, Inc., 111 River Street, Hoboken, NJ 07030, (201) 748-6011, fax (201) 748-6008, or online at http://www.wiley.com/go/permission.

The manufacturer's authorized representative according to the EU General Product Safety Regulation is Wiley-VCH GmbH, Boschstr. 12, 69469 Weinheim, Germany, e-mail: Product_Safety@wiley.com.

Trademarks: Wiley and the Wiley logo are trademarks or registered trademarks of John Wiley & Sons, Inc. and/or its affiliates in the United States and other countries and may not be used without written permission. All other trademarks are the property of their respective owners. John Wiley & Sons, Inc. is not associated with any product or vendor mentioned in this book.

Limit of Liability/Disclaimer of Warranty: While the publisher and author have used their best efforts in preparing this book, they make no representations or warranties with respect to the accuracy or completeness of the contents of this book and specifically disclaim any implied warranties of merchantability or fitness for a particular purpose. No warranty may be created or extended by sales representatives or written sales materials. The advice and strategies contained herein may not be suitable for your situation. You should consult with a professional where appropriate. Further, readers should be aware that websites listed in this work may have changed or disappeared between when this work was written and when it is read. Neither the publisher nor authors shall be liable for any loss of profit or any other commercial damages, including but not limited to special, incidental, consequential, or other damages.

For general information on our other products and services or for technical support, please contact our Customer Care Department within the United States at (800) 762-2974, outside the United States at (317) 572-3993 or fax (317) 572-4002.

Wiley also publishes its books in a variety of electronic formats. Some content that appears in print may not be available in electronic formats. For more information about Wiley products, visit our web site at www.wiley.com.

Library of Congress Cataloging-in-Publication Data Applied for:

Hardback ISBN: 9781394155262

Cover Design: Wiley
Cover Images: © xoverian/Shutterstock, © Jason marz/Getty Images

Set in 9.5/12.5pt STIXTwoText by Straive, Pondicherry, India

To my mother. This book is dedicated to your soul, with all my love

Contents

About the Author

Dr. Mohamed M. Atia received his BSc and MSc degrees in computer systems from Ain Shams University in 2000 and 2006, respectively, and a PhD in electrical and computer engineering from Queen's University at Kingston in 2013. Before transitioning to academia, Dr. Atia spent several years in industry, where he developed and implemented advanced algorithms and computer systems utilizing artificial intelligence, machine learning, estimation, and signal processing for diverse applications, including natural language processing, speech recognition, and multi-sensor navigation systems for robotics. He is currently a professor in the Department of Systems and Computer Engineering at Carleton University in Ottawa. Dr. Atia has authored over 100 publications and holds five granted patents.

Preface

Background

The early industrial revolutions of the first (eighteenth century), second (nineteenth century), and third (twentieth century) were marked by the invention of machines, transportation means, and computer/communication systems, respectively. The fourth industrial revolution that is currently evolving and shaping our twenty-first century is marked by two key aspects: autonomy and connectivity. The invention of micro-electromechanical systems (MEMS) sensors, digital imaging sensors, the deployment of global navigation satellite systems (GNSSs), fifth generation (5G) networks, and the advances in semiconductor manufacturing has created an unprecedented opportunity to develop intelligent autonomous systems that can perform complex, challenging tasks in different fields such as agriculture, farming, transportation, construction, health, and services such as snow removal, lawn mowing, cleaning, warehouse operations, package delivery, not to mention space exploration and settlement on the moon or Mars. Sensor fusion algorithms that support full autonomy is the fuel that powers these new emerging technologies and applications. To support full autonomy, positioning, navigation, and mapping are vital features that any robotic platform or intelligent system must have to navigate in the physical world. Although many books address different navigation and mapping systems, most resources address specific sensor technology. A few books address the problem using a sensor fusion approach, where various sensors are integrated into one system. The fusion of multiple sensors has the benefit of maximizing the overall accuracy and mitigating the limitations of each individual sensor. This book explains fundamental sensor fusion concepts for positioning, navigation, and mapping. In addition, the book describes the concept of operations, mathematical models, and the signal processing of navigation and mapping sensors and technologies. The book integrates common navigation sensor technologies and fusion algorithms in one source with examples and sample code projects for maximum clarity.

Scope of the Book

Positioning, navigation, and mapping can be categorized into low level and high level. Low-level positioning means determining objects' location, velocity, and orientation, while low-level navigation refers to estimating objects' trajectories. Low-level mapping refers to the point-based mapping of the environment. High-level positioning, navigation, and mapping include higher level tasks such as object recognition, scene understanding, path planning, and collision avoidance. The book concerns low-level positioning, navigation, and mapping technologies and their integration.

Positioning, navigation, mapping technologies, sensors, and systems are broad and diverse. Each sensor, technology, or system is a sophisticated research area by itself. The book provides introductory-level knowledge about the most used sensors and positioning, navigation, and mapping technologies. Then, the book explains sensor fusion methods and algorithms with applications on different platforms, such as wheeled vehicles, aerial vehicles, and legged robotic platforms. The book collects the critical building blocks in math, physics, estimation, and signal processing that contribute to building accurate positioning, navigation, and mapping systems.

MATLAB Projects' Repository

The book provides MATLAB projects to demonstrate the concepts and explain the implementation details of sensor fusion systems/algorithms for different platforms. The code, data, and instructions of the projects can be found in the following GitHub link:

https://github.com/SensorFusionBook

MATLAB Version Compatibility

The projects and examples in this book were developed and tested using a specific version of MATLAB. While MATLAB strives to maintain backward compatibility, some functions may become obsolete or exhibit different behavior in newer versions. If you encounter any version-related issues or unexpected behavior, please report them to SensorFusionBook@outlook.com. Your feedback is invaluable in addressing potential compatibility concerns and enhancing the usability of the provided resources.

Who Should Read This Book?

This book is designed for undergraduate/graduate students, researchers, and industry engineers working on developing positioning and mapping algorithms for robotics and autonomous vehicles using multiple integrated sensors. Integrating different multi-rate sensors in one system in real time is quite challenging and requires solid knowledge of multidisciplinary topics. Being an expert in all interdisciplinary topics related to sensor technologies and positioning/mapping methods is almost impossible. However, having a minimum level of knowledge in these multidisciplinary topics is essential to developing reliable positioning and mapping algorithms for robotics and autonomous vehicles. Engineers focusing on one sensor technology or algorithmic method would benefit from the multidisciplinary topics discussed in this book. Understanding how other sensors work and how measurements are integrated would enhance collaborative work and highlight essential gaps in individual sensors.

Motivations

Despite the importance of the topic of sensor fusion and the significant impact of precise positioning, navigation and mapping on autonomous vehicles and intelligent robotics technologies, the author could not easily find a single book that combines these many varied topics which constitute sensor fusion for autonomous systems positioning, navigation, and mapping. Although some books cover individual topics in location, positioning, navigation, and mapping technology, most books address a specific sensor or problem domain with little focus on the big picture of sensor fusion and integration. The topic of sensor fusion for positioning, navigation, and mapping is usually overlooked due to the complex details of each sensor or technology. As the number of sensors and technologies has significantly increased, the topic of sensor fusion has become complex and requires a separate textbook to explain its concepts adequately. Here are some additional reasons that motivated the author to write this book:

- There are a few books on sensor fusion that provide a balance between theory and actual implementation of sensor fusion methods. Unfortunately, the books that explain the theoretical foundation usually lack the experience of practical implementation and the books that explain implementation details typically lack a solid theoretical foundation.
- As sensor fusion combines different (but related) topics (linear algebra, signal processing, probability, random variable, stochastic processes, and estimation), researchers usually face significant challenges and struggle to build the necessary mathematical and physical background and systems implementation skills.
- The dominant fusion methods in the literature are based on conventional estimation theory. While there is a significant leap in machine learning and artificial intelligence, the application of these tools in sensor fusion has not been extensively explained in a book.

Book Structure

- **Chapter 1** provides an essential background of critical mathematical concepts and definitions related to Euclidean geometry, such as points, vectors and vector space, matrices, linear algebra, Lie algebra, coordinate frames, and the transformation of frames. The chapter explains the Earth's coordinate frames and the transformation between local and global frames. The chapter also describes the derivations of differential equations of motion of rigid objects in 3D space.
- **Chapter 2** explains the basics of signals and systems, including probability concepts, random processes, Fourier analysis, stochastic linear dynamic systems, differential equations, transfer functions, state space models, linearization of nonlinear systems, and the basics of digital signal processing that are commonly used in the field of positioning, navigation, and mapping.
- **Chapter 3** presents most sensor fusion methods' theoretical and mathematical foundations. Common optimization and filtering algorithms, including Kalman filtering (KF), Extended Kalman filter (EKF), particle filtering (PF), and Graph optimization, are explained. The chapter also includes a brief introductory description of machine learning and artificial intelligence since they became popular tools for performing sensor fusion.
- **Chapter 4** is dedicated to inertial sensors, inertial measurement units (IMU), and inertial navigation systems (INS). The theory of operation and the common errors of inertial sensors (accelerometers and gyroscopes) and magnetic sensors are explained and analyzed using practical data and tools such as Allan variance. Fundamental concepts about 3D position/velocity/orientation modeling using differential equations that process inertial measurement are explained. Concepts are further illustrated using MATLAB code samples.

- **Chapter 5** explains the basic concepts of radio positioning and navigation systems. The fundamental concepts include navigation signal structure, time of arrival estimation, positioning signal acquisition and tracking, phase look loops, frequency look loops, and pseudo-range estimation. The concepts are demonstrated using the popular global satellite navigation system (GNSS) technology, taking the Global Positioning System (GPS) as an example. Standard and advanced GPS positioning modes are explained with MATLAB code samples.
- **Chapter 6** discusses active ranging sensors such as light detection and ranging (LiDAR) and radio detection and ranging (RADAR). The basic principle of operation and signal processing, such as chirp signal generation and echo signal detection using the matched filter, 2D Fourier transform for range and Doppler estimation, range equation, Doppler/speed estimation, and angle estimation. Pulsed and frequency-modulated continuous wave (FMCW) RADAR technologies are explained. Range-based odometry algorithms are presented with MATLAB project examples.
- **Chapter 7** explains vision sensors using a pinhole camera projection model as an example. Key concepts include projection modelling, camera parameters (intrinsic and extrinsic), epipolar geometry, recovery of the camera's 3D motion and the visual features' 3D location from images, monocular camera systems, and stereo camera systems. Camera calibration and vision-based odometry algorithms are explained with MATLAB project examples.
- **Chapter 8** explains the foundation of mapping systems and algorithms, including mapping algorithms and simultaneous localization and mapping (SLAM) algorithms. The derivations of standard mapping methods include occupancy mapping, EKF SLAM, PF SLAM, Rao-Blackwellized SLAM, Fast-SLAM, and Graph-SLAM. Loop closure and bundle adjustment are explained, and both range-based and vision-based SLAM algorithms are elaborated and supported by MATLAB project illustrations.
- **Chapter 9** is a case study that applies sensor fusion methods on 2D wheeled vehicles. The 2D wheeled motion model applies to most ground vehicles. 2D wheeled motion model utilizes inertial sensors and wheel encoders to predict the vehicle's trajectory. Inertial sensors and wheel encoders are fused with GPS and LiDAR measurements in loosely coupled and tightly coupled modes using different fusion methods. Analysis of the convergence and accuracy of the fusion algorithms are demonstrated and supported by MATLAB implementation to illustrate the key concepts further.
- **Chapter 10** is a case study that considers advanced 3D vehicular motion. 3D vehicular motion models apply to aerial vehicles such as drones. Advanced 3D system and error models are developed, and the fusion of inertial sensors with GPS, Vision, and RADAR measurements are explained in detail in loosely and tightly coupled modes. The fusion scenarios are supported and illustrated by MATLAB projects that show the implementation details.
- **Chapter 11** describes two fusion modules used in legged and humanoid platforms. Attitude and heading reference system (AHRS) and pedestrian dead-reckoning (PDR) are two fusion modules. AHRS fuses inertial and magnetic sensor measurements to provide precise and accurate orientation estimation. PDR utilizes pattern recognition techniques to detect the motion patterns of humanoid platforms, such as walking steps. With accurate AHRS, PDR provides robust navigation for legged and humanoid platforms. An example implementation of AHRS using MATLAB is provided.
- **Chapter 12** discusses the application of machine learning (ML) and artificial intelligence (AI) in sensor fusion. ML and AI are being explored as an augmenting or alternative solution that addresses several corner cases that are difficult to handle using conventional estimation and

signal processing methods. This chapter is an introductory brief that discusses the formulation of loss functions and the recent trends and examples of applying ML and AI to solve complex problems that are difficult to solve using filtering and optimization algorithms, such as systems' nonlinearity and 3D dense depth estimation.

Commitment to Accuracy and Continuous Improvement

Despite the best efforts to ensure accuracy and clarity, nothing is perfect, and occasional errors or oversights may exist. Readers are encouraged to share any mistakes or constructive feedback by contacting SensorFusionBook@outlook.com. Such input is invaluable for improving the book for future editions.

Acknowledgment

I express my deepest gratitude to my family for their love, patience, support, and encouragement. I would like to extend my thanks to my colleagues who assisted in reviewing the book; their invaluable insights and thoughtful critiques have greatly enriched this work.

1

Coordinate Systems and Motion Modeling

1.1 Introduction

The field of positioning, navigation, and mapping utilizes geometric representation of the physical world as we perceive it. From a spatial and mathematical point of view, our physical world is a three-dimensional (3D) space defined by the axioms and theories of Euclidean space. Describing objects' motion within Euclidean space requires knowledge of Euclidean geometry, which in turn requires knowledge of linear algebra. This chapter provides an essential background of critical mathematical concepts and definitions related to Euclidean geometry, such as points, vectors and vector space, matrices, linear algebra, Lie algebra, coordinate frames, the transformation of frames, and motion equations as functions of object's rotation and translation rates.

1.2 Points, Vectors, and Matrices

1.2.1 Point, Set, Field, and Space

A set is a group of elements (i.e., numbers or objects). A set is called a space if there are some axioms or general rules that describe relationships (or a structure) between the elements of the set. Any set can have subsets with unique characteristics. A vector is a group of elements of a set. A scalar is a single element of a field. A field is a set where addition, subtraction, multiplication, and division are defined. A set of vectors is called a vector space if it is closed under summation and scalar multiplication, which means that any vector in the set's summation and scalar multiplication belong to the set. For example, the set of real numbers $\mathbb{R}$ is called a space because the addition of any two real numbers is a real number, and the multiplication of any real number by a scalar number is also a real number.

1.2.1.1 Points and Coordinates

A point is an element in a set. For example, $x_1 \in \mathbb{R}$ is a point in the set of real numbers $\mathbb{R}$. It is evident that $\mathbb{R}$ forms a field, and at the same time, it is a one-dimensional (1D) space where a single (scalar) value uniquely represents elements (i.e., points). The 1D real number space is referred to by the symbol $\mathbb{R}^1$. The concept of 1D real number space can be extended to any dimension n. For example, a point P in the 2D real number space $\mathbb{R}^2$ is represented by an ordered pair of numbers $\langle p_x, p_y \rangle$. Similarly, a point P in the 3D real number space $\mathbb{R}^3$ is represented by three-tuple $\langle p_x, p_y, p_z \rangle$. The three-tuple $\langle p_x, p_y, p_z \rangle$ is called a coordinate of the point. There is no limit on dimensions. In other

Sensor Fusion Approaches for Positioning, Navigation, and Mapping: How Autonomous Vehicles and Robots Navigate in the Real World: With MATLAB Examples, First Edition. Mohamed M. Atia.
© 2025 The Institute of Electrical and Electronics Engineers, Inc. Published 2025 by John Wiley & Sons, Inc.

words, we can define a point in the $\mathbb{R}^n$ space with n-tuple coordinates $\langle x_1, x_2, x_3, \ldots, x_n \rangle$. It is obvious that the summation of two points in $\mathbb{R}^n$ is also a point in $\mathbb{R}^n$, and the scalar multiplication of a point in $\mathbb{R}^n$ is also a point in $\mathbb{R}^n$.

1.2.1.2 Euclidean Space

Euclidean space is a space over the real numbers field whose elements are points $P \in \mathbb{R}^n$ that satisfy the five axioms of Euclid, which can be found in [1]. In Euclidean space, a scalar is a real number, and a vector is a group of points that form a line segment that has a length (magnitude) and a direction. Although the physical world can be described by 2D ($\mathbb{R}^2$) or 3D ($\mathbb{R}^3$) Euclidean space, Euclidean space can be extended to any dimension $\mathbb{R}^n$. Euclidean geometry is the set of axioms, definitions, and theorems that describe the geometry of our physical world as perceived by our brains. Concepts such as the distance between two points, a straight line, the direction of a vector, the length of a vector, the definition of parallelism, intersection, angles between vectors, definitions of shapes such as circles and spheres are examples of axioms, definitions, and theorems in the Euclidean space. We use Euclidean space and geometry daily to describe shapes, sizes, orientations, locations, and positions. More information about Euclidean space can be found in [1] and [2].

1.2.1.3 Other Popular Spaces

Euclidean space is a very special space that is used to describe the physical world that we live in. However, in mathematics, physics, and statistics, there are other popular spaces with different elements, definitions, and dimensions. For example, in Hilbert space [3], elements could be complex numbers or even functions or waveforms. In addition, Hilbert space may include infinite-dimensional elements. Given this definition, Euclidean space is a special case of Hilbert space. Hilbert space is important in solving systems that incorporate random signals, and it is used in deep learning algorithms. More information about Hilbert space and its applications can be found in [4].

1.2.2 Vectors

Abstractly, a vector is a quantity that is defined by more than one point in a set. In Euclidean space, a vector has a start point and an endpoint, and consequently, it has a magnitude and a direction. A vector has an infinite number of points. Figure 1.1 shows a vector $\overrightarrow{AB}$ in the $\mathbb{R}^2$ space. In $\mathbb{R}^2$ space, points are defined by a pair of coordinates. This pair of coordinates is determined by two perpendicular axes, X and Y, as seen in Figure 1.1. This coordinate system is the popular 2D X–Y Cartesian coordinate system. We will discuss coordinate systems later in this chapter. A vector in the 2D Cartesian coordinate frame is uniquely defined by two points, as shown in Figure 1.1. An additional illustration of a 3D vector defined in the 3D Cartesian coordinate system is shown in Figure 1.2. Note that the axes of the Cartesian coordinate system are perpendicular (orthogonal) to each other. This orthogonality is a condition of the uniqueness of the point representation in the coordinate frame. Point "O" is the origin of the frame with coordinate $O = (0, 0, 0)$.

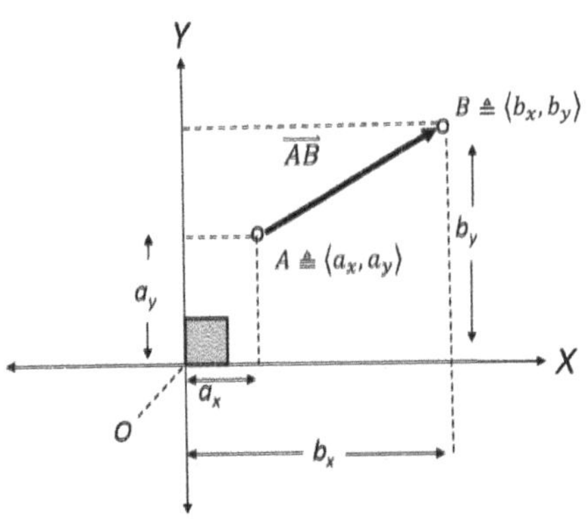

Figure 1.1 A vector defined in the 2D Cartesian coordinate frame.

1.2.2.1 Equivalent Vector

In the previous definition of vectors, we specified a start point and an endpoint. A vector represented in this way is called a located vector $\overrightarrow{AB}$, where A is the start point and B is the end point. Another simpler way to define vectors is by making the origin of the coordinate frame O as the start point of the vector. Any vector that is defined by a pair of points has an equivalent vector that starts at the origin O. This can be shown in Figure 1.3, where vector $\overrightarrow{AB}$ is represented by the equivalent vector $\overrightarrow{OV}$. To guarantee the equivalency of the two vectors, the point V is selected such that $\overrightarrow{AB}$ is parallel to $\overrightarrow{OV}$ and the length of the two vectors is equal.

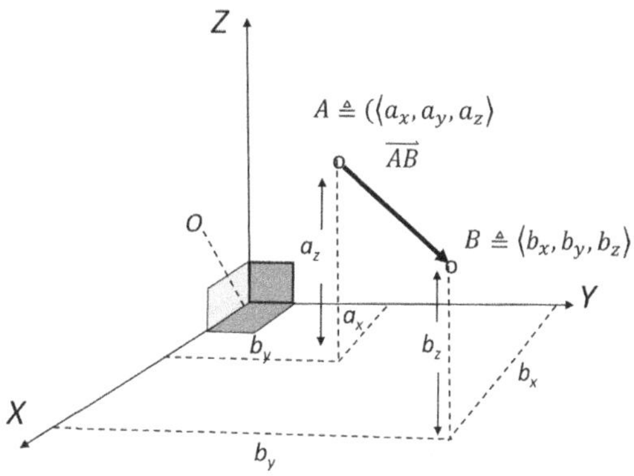

Figure 1.2 A vector defined in the 3D Cartesian coordinate frame.

For simplicity, the vector $\overrightarrow{OV}$ will be referred to by $\vec{V}$. When we mention a vector $\vec{P}$ using only one letter P, this implicitly means that the start point of this vector is the origin O and the endpoint of the vector is P.

1.2.2.2 Vector Space

A set of vectors $\boldsymbol{S}$ is called a linear vector space if the addition of two vectors $\vec{V_1}$, $\vec{V_2} \in \boldsymbol{S}$ is another vector in $\boldsymbol{S}$, and the scalar multiplication of a vector in $\boldsymbol{S}$ is also a vector in $\boldsymbol{S}$. This can be mathematically written as follows:

$$\alpha \vec{V_1} + \beta \vec{V_2} \in \boldsymbol{S} \tag{1-1}$$

where α and β are scalar values. An illustration of the $\mathbb{R}^3$ vector space is shown in Figure 1.4. Vector space satisfies the following properties:

$$\vec{V_1} + \vec{V_2} = \vec{V_2} + \vec{V_1} \quad \forall \ \vec{V_1}, \vec{V_2} \in \boldsymbol{S} \tag{1-2}$$

$$\vec{V_1} + \left(\vec{V_2} + \vec{V_3}\right) = \left(\vec{V_1} + \vec{V_2}\right) + \vec{V_3} \ \forall \ \vec{V_1}, \vec{V_2}, \vec{V_3} \in \boldsymbol{S} \tag{1-3}$$

$$\vec{V_1} + O = O + \vec{V_1} = \vec{V_1} \ \forall \ \vec{V_1} \in \boldsymbol{S} \tag{1-4}$$

$$\vec{V_1} + (-1)\vec{V_1} = O \ \forall \ \vec{V_1} \in \boldsymbol{S} \tag{1-5}$$

Figure 1.3 Equivalent vector $\boldsymbol{V}$ for located vector $\overrightarrow{AB}$.

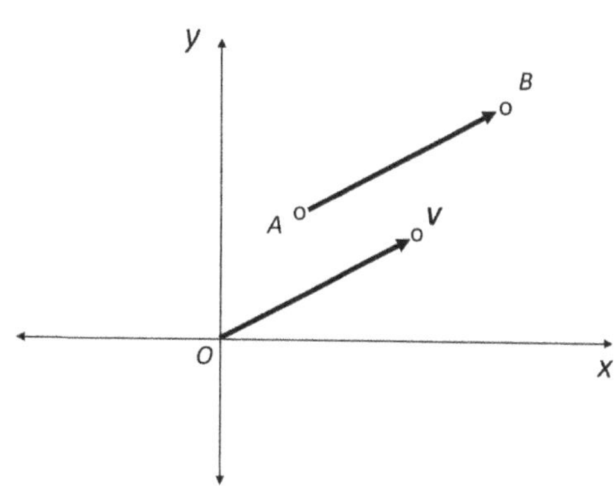

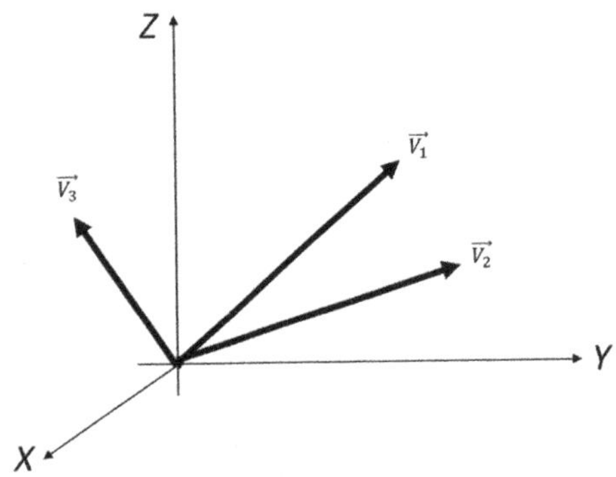

Figure 1.4 Illustration of the $\mathbb{R}^3$ vector space.

$$(\alpha\beta)\,\vec{V_1} = \alpha\left(\beta\,\vec{V_1}\right)\forall\,\vec{V_1}\in \boldsymbol{S} \tag{1-6}$$

$$1\left(\vec{V_1}\right) = \vec{V_1}\ \forall\ \vec{V_1}\in \boldsymbol{S} \tag{1-7}$$

1.2.2.3 Linear Independence, Orthonormality, and Basis Vectors

Basis and Unit Vectors

Abstractly, a set of vectors $\boldsymbol{B}$ where none of the vectors can be represented as a linear combination of other vectors is called a "linearly independent" set. This linear independence can be mathematically expressed as follows:

$$\alpha_1\vec{V}_1+\alpha_2\vec{V}_2+\ldots+\alpha_n\vec{V}_n=0 \text{ if an only if } \alpha_i=0\forall i \tag{1-8}$$

In other words, the vector $\vec{V}_i\in\boldsymbol{B}$ cannot be expressed as a linear combination of other vectors $V_{j\neq i}\in\boldsymbol{B}$. If in a vector space $\boldsymbol{S}$ there is a subspace of vectors $\boldsymbol{B}$ such that (1) all vectors of $\boldsymbol{B}$ are linearly independent and (2) any vector $\vec{V}\in\boldsymbol{S}$ can be expressed as a linear combination of vectors of $\boldsymbol{B}$, then the vector subspace $\boldsymbol{B}$ is called a basis of vector space $\boldsymbol{S}$. A vector space may have an infinite number of basis subsets. For example, in the 2D Cartesian vector space, any pair of perpendicular vectors is a basis set. In Figure 1.5, unit vectors $\vec{I}_x=(1\,0)^T$ and $\vec{I}_y=(0\,1)^T$ are chosen as the basis where any vector $\vec{V}$ can be represented as a linear combination of $\vec{I}_x,\vec{I}_y$ as shown in Figure 1.5 and as demonstrated in the following equation:

$$\vec{V}=\left(v_x\,v_y\right)^T=v_x\vec{I}_x+v_y\vec{I}_y=v_x(1\,0)^T+v_y(0\,1)^T \tag{1-9}$$

An orthogonal unit vector set is called orthonormal. In 2D Cartesian coordinates system, $\vec{I}_x$ and $\vec{I}_y$ are unit vectors along the axes X and Y. From Figure 1.5, we can see that the components of the vector $\vec{V}=\left(v_x\,v_y\right)^T$ are given by:

$$v_x=\left\|\vec{V}\right\|\cos(\theta) \tag{1-10}$$

$$v_y=\left\|\vec{V}\right\|\sin(\theta) \tag{1-11}$$

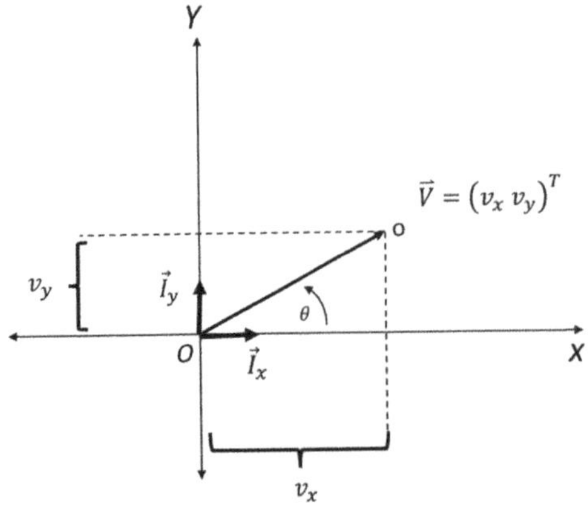

Figure 1.5 Vector $\vec{V}$ represented as a linear combination of basis vectors v_x, v_y.

where

$\left\|\vec{V}\right\|$ is the magnitude (length) of the vector $\vec{V}$
v_x and v_y are the projection of vector $\vec{V}$ on the basis vectors of the 2D Cartesian vector space.

The same concept can be generalized to any n-dimensional vector space. For example, in the 3D Cartesian coordinate system described in Figure 1.2, we used three perpendicular axes X, Y, and Z to uniquely define points and vectors. Figure 1.6 shows a 3D Cartesian coordinate system where a vector $\vec{P}=\left(p_x\,p_y\,p_z\right)^T$ can be written as follows:

$$\vec{P}=\left(p_x\,p_y\,p_z\right)^T=\vec{p}_x+\vec{p}_y+\vec{p}_z \tag{1-12}$$

Figure 1.6 3D Cartesian coordinate space.

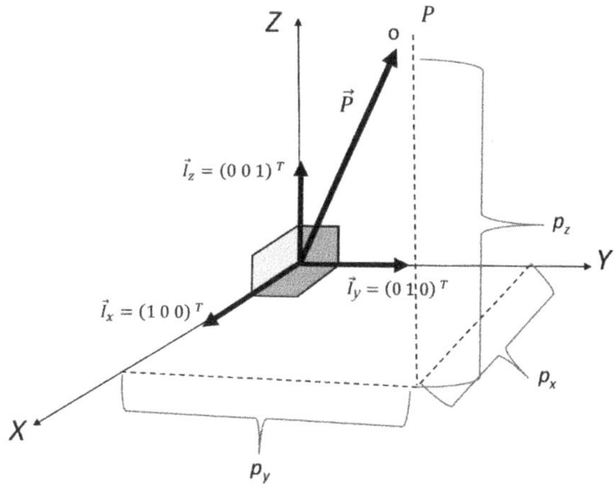

$$\vec{P} = \begin{pmatrix} p_x \\ p_y \\ p_z \end{pmatrix} = \begin{pmatrix} p_x \\ 0 \\ 0 \end{pmatrix} + \begin{pmatrix} 0 \\ p_y \\ 0 \end{pmatrix} + \begin{pmatrix} 0 \\ 0 \\ p_z \end{pmatrix} \tag{1-13}$$

where $\vec{p}_x = (p_x\,0\,0)^T, \vec{p}_y = \left(0\,p_y\,0\right)^T$, and $\vec{p}_z = \left(0\,0\,p_z\right)^T$.

The three vectors $\vec{p}_x, \vec{p}_y, \vec{p}_z$ can be written as follows:

$$\begin{aligned} \vec{p}_x &= p_x \vec{I}_x \\ \vec{p}_y &= p_y \vec{I}_y \\ \vec{p}_z &= p_z \vec{I}_z \end{aligned} \tag{1-14}$$

where p_x, p_y, p_z are scalar values and $\vec{I}_x, \vec{I}_y, \vec{I}_z$ vectors defined as follows:

$$\begin{aligned} \vec{I}_x &= (1\,0\,0)^T \\ \vec{I}_y &= (0\,1\,0)^T \\ \vec{I}_z &= (0\,0\,1)^T \end{aligned} \tag{1-15}$$

The vectors $\vec{I}_x, \vec{I}_y, \vec{I}_z$ are called unit vectors (their magnitude is one). These unit vectors are orthogonal, or their dot products (to be defined shortly) are zero. These three axes are linearly independent. Geometrically, linear independence means that vectors are perpendicular to each other. We can use unit vectors $\vec{I}_x, \vec{I}_y, \vec{I}_z$ to represent any point and vector in this 3D Cartesian coordinate system. This 3D Cartesian coordinate system can be seen as a vector space where vectors are uniquely defined by a linear combination of $\vec{I}_x, \vec{I}_y, \vec{I}_z$ unit vectors.

Linear Span

If any vector in a vector space ***S*** can be expressed as a linear combination of basis vectors ***B***, we say ***B*** spans the entire vector space ***S***. All bases of any vector space have the same number of vectors. For example, all bases in the 2D Cartesian coordinate system have two vectors. The dimension of a vector space is determined by the number of vectors in its basis. For example, the dimension of the 2D

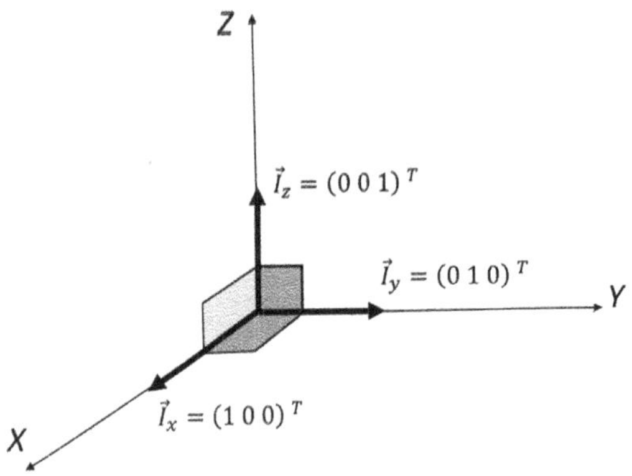

Figure 1.7 Basis vectors of the 3D Cartesian coordinate space.

Cartesian coordinate system is two because we have two vectors in its basis. It is easy to realize that the same concepts of basis and unit vectors apply to any dimension such as 3D space ($\mathbb{R}^3$) or any n-D space ($\mathbb{R}^n$). Figure 1.7 shows the basis of the 3D Cartesian coordinate space.

1.2.2.4 Dot Product and Inner Product

The dot product generates a scalar value; therefore, it is also known as a scalar product. Algebraically, the dot product of vectors $\overrightarrow{X} = (x_1\, x_2\, x_3 \ldots x_n)^T$ and $\overrightarrow{Y} = (y_1\, y_2\, y_3 \cdots y_n)^T$ is defined as follows:

$$\overrightarrow{X} \cdot \overrightarrow{Y} = \sum_{i=1}^{n} x_i y_i \tag{1-16}$$

Geometrically, the dot product of $\overrightarrow{X}$ and $\overrightarrow{Y}$ is defined as follows:

$$\overrightarrow{X} \cdot \overrightarrow{Y} = \left\|\overrightarrow{X}\right\| \left\|\overrightarrow{Y}\right\| \cos(\theta) \tag{1-17}$$

where θ is the angle between the two vectors $\overrightarrow{X}$ and $\overrightarrow{Y}$ as seen in Figure 1.8.

The dot product is a special case of the more general operation "inner product" represented by the operator "< >." The dot product is a key function in Euclidean space because it defines a "metric." As we will see later, this metric will be used to measure the distance between points and the length of a vector. It is also a measure of the angle between vectors. It is easy to observe that the dot product of two orthogonal vectors is zero because the angle between them is 90° and cos (90°) is zero. In addition, the dot product operation can be used to express any vector in a vector space using its basis. For example, a vector $\overrightarrow{V} = (v_x\, v_y\, v_z)^T$ can be represented using the dot product as follows:

$$\begin{aligned} v_x &= \overrightarrow{V} \cdot \overrightarrow{I}_x \\ v_y &= \overrightarrow{V} \cdot \overrightarrow{I}_y \\ v_z &= \overrightarrow{V} \cdot \overrightarrow{I}_z \end{aligned} \tag{1-18}$$

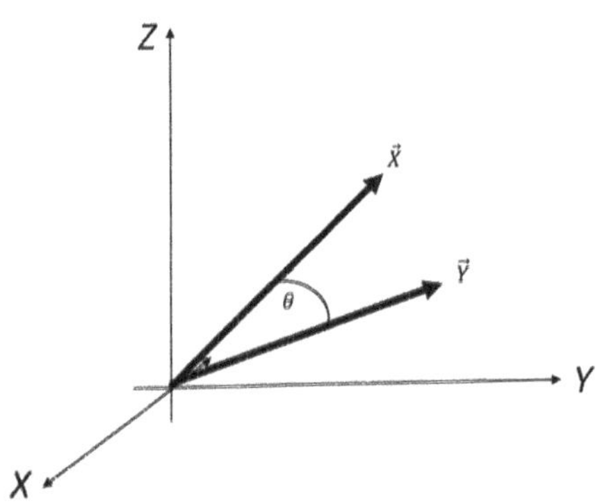

Figure 1.8 Two vectors in the 3D Cartesian coordinates.

1.2.2.5 Euclidean Norm (Magnitude)

Abstractly, a norm is a function that generates a non-negative value regardless of the value of the vector, and it behaves as a distance, length, or proximity measure between the endpoint of the vector and the origin. In Euclidean space, the norm of a vector $\overrightarrow{X} = (x_1\, x_2\, x_3 \ldots x_n)^T$ is defined as follows:

$$\left\|\overrightarrow{X}\right\| = \sqrt{x_1^2 + x_2^2 + \ldots + x_n^2} \tag{1-19}$$

Referring to the dot product definition, it is easy to observe that the magnitude of a vector is related to the dot product of the vector by itself. Geometrically, in Euclidean space, the norm is the magnitude of a vector, which is also the length of the vector. In the 3D Cartesian coordinate system, the length of a vector $\overrightarrow{\boldsymbol{X}} = (x\,y\,z)^T$ is defined as:

$$\left\|\overrightarrow{\boldsymbol{X}}\right\| = \sqrt{x^2 + y^2 + z^2} \tag{1-20}$$

In Cartesian (rectangular) coordinate frames, the magnitude of the vector is the Euclidean distance between the endpoint of the vector and the origin. In n-dimensional Euclidean space, the Euclidean distance between point $P = (p_1\ p_2\ p_3 \dots p_n)$ and point $O = (o_1\ o_2\ o_3 \dots o_n)$ is defined as the norm of the vector between the two points:

$$d = \sqrt{(p_1 - o_1)^2 + (p_2 - o_2)^2 + (p_3 - o_3)^2 + \dots + (p_n - o_n)^2} \tag{1-21}$$

1.2.2.6 Cross Product and the Skew Matrix

In 3D Euclidean space, using the 3D Cartesian coordinate system, the cross product of two vectors $\vec{A} = (x_A, y_A, z_A)$ and $\vec{B} = (x_B, y_B, z_B)$ is another vector $\vec{C} = (x_C, y_C, z_C)$ defined as follows:

$$\begin{aligned} \vec{C} &= \vec{A} \times \vec{B} = (x_C\ \ y_C\ \ z_C)^T \\ x_C &= y_A z_B - z_A y_B \\ y_C &= z_A x_B - x_A z_B \\ z_C &= x_A y_B - y_A x_B \end{aligned} \tag{1-22}$$

Geometrically, the cross product of two vectors is a vector perpendicular to the plane that contains the two vectors and pointing to the direction that completes the right-hand-side rule as can be seen in Figure 1.9. The magnitude of the cross-product output is defined as follows:

$$\left\|\vec{C}\right\| = \left\|\vec{X}\right\| \left\|\vec{Y}\right\| \sin(\theta) \tag{1-23}$$

The cross product $\vec{C} = \vec{A} \times \vec{B}$ is orthogonal to $\vec{A}$ and $\vec{B}$. The direction of cross-product vector $\vec{C}$ depends on the order of the product. The cross product of any two basis vectors in the 3D Cartesian coordinate system equals the third basis vector as follows:

$$\begin{aligned} \vec{I}_x \times \vec{I}_y &= \vec{I}_z \\ \vec{I}_y \times \vec{I}_z &= \vec{I}_x \\ \vec{I}_z \times \vec{I}_x &= \vec{I}_y \end{aligned} \tag{1-24}$$

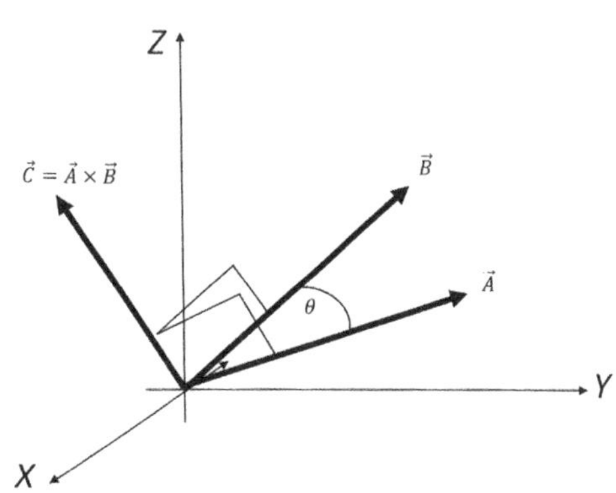

Figure 1.9 Cross product of two vectors A and B.

The cross-product $\vec{C} = \vec{A} \times \vec{B}$ can be mathematically implemented using the skew matrix.

The skew matrix of a vector is an antisymmetric matrix representation of the vector such that the diagonal elements are zeros, and the upper right off-diagonal elements equal the negative of the lower left off-diagonal elements. The skew matrix of a vector $\vec{A}$ is defined as follows:

$$\left(\vec{A}\times\right) = \begin{pmatrix} 0 & -a_z & a_y \\ a_z & 0 & -a_x \\ -a_y & a_x & 0 \end{pmatrix} \quad (1\text{-}25)$$

The cross product of $\vec{C} = \vec{A} \times \vec{B}$ can be defined as matrix-vector multiplication:

$$\vec{C} = \left(\vec{A}\times\right)\vec{B} \quad (1\text{-}26)$$

A skew matrix, in general, is also called skew symmetric matrix. Matrix $A \in \mathbb{R}^{n\times n}$ is skew symmetric matrix if it satisfies the following:

$$A^T = -A \quad (1\text{-}27)$$

1.2.3 Matrices

Matrices are heavily used in the mathematics of positioning, navigation, and mapping. This section will discuss some common concepts, important features, and operations about matrices. Abstractly, an $m \times n$ matrix arranges elements (commonly numbers) in m rows and n columns as follows:

$$A_{m\times n} = \begin{pmatrix} a_{1,1} & \cdots & a_{1,n} \\ \vdots & \ddots & \vdots \\ a_{m,1} & \cdots & a_{m,n} \end{pmatrix} \quad (1\text{-}28)$$

The symbol of a matrix is commonly a big letter with a subscript that shows the dimension of the matrix. If the dimension of the matrix is clearly known, the subscript can be omitted.

1.2.3.1 Range, Rank, and Null Space of a Matrix

In the Euclidean space and linear algebra, matrices can be viewed as a mapping operator that maps vectors from one space to another. A matrix $A \in \mathbb{R}^{m\times n}$ can be seen as a mapping operator from $\mathbb{R}^n$ to $\mathbb{R}^m$. In other words, when we have a vector $\vec{V} \in \mathbb{R}^n$ and multiply it by a matrix $\in\mathbb{R}^{m\times n}$, we map the $\vec{V}$ from $\mathbb{R}^n$ to $\mathbb{R}^m$ so we obtain a new vector with dimension $m \times 1$ vector as follows:

$$A\vec{V} = \vec{U}$$

$$\begin{pmatrix} a_{1,1} & \cdots & a_{1,n} \\ \vdots & \ddots & \vdots \\ a_{m,1} & \cdots & a_{m,n} \end{pmatrix}\begin{pmatrix} v_1 \\ \vdots \\ v_n \end{pmatrix} = \begin{pmatrix} u_1 \\ \vdots \\ u_m \end{pmatrix} \quad (1\text{-}29)$$

The range of a matrix A is defined as all the vectors $\vec{U}$ that can be formed by the multiplication of a matrix A with a vector $\vec{V}$. Note that the set of vectors $\vec{U}$ that can be formed by the multiplication of a matrix A with a vector $\vec{V}$ is a linear combination of the columns of A. Thus, the range of a matrix is the vector space whose basis is the linearly independent columns of the matrix A. The number of the linearly independent columns of the matrix A is called the "rank" of the matrix. The rank can also be defined as the dimension of the smallest submatrix of a matrix A whose determinant is zero. If, for a matrix, A there exists a vector $\vec{V}$ such that:

$$A\vec{V} = 0 \tag{1-30}$$

for a subset of vectors, $\vec{V}$, this subset of vectors is called the null space of the matrix A. Note that the null space of the matrix is the set of vectors that are orthogonal to its rows. The null space is also called the kernel of a matrix. If we have the following matrix equation:

$$A\vec{X} = \vec{B} \tag{1-31}$$

this equation will have a solution if $\vec{B}$ is in the range of the matrix A. The equation will have a unique solution if the null space (the matrix kernel) is empty because if the null space has some vectors, this means that for any $\vec{X}$ that satisfies Eq. (1-31) there will be other solutions $\vec{X} + \vec{V}$ that satisfy the equation if $\vec{V}$ is one of the null space vectors of A. This is also equivalent to saying the Eq. (1-31) has a unique solution if the matrix $A_{n\times n}$ is full rank (its rank is n). If the rank of A is less than n, the null space will not be empty.

1.2.3.2 Eigenvalues, Eigenvectors, and Eigenvalue Decomposition (EVD)

Another important feature of a matrix is Eigenvalues/Eigenvectors. Eigenvalues/Eigenvectors are only defined for square matrices $A \in \mathbb{R}^{n\times n}$. They are special pairs of scalar and vectors $\left\{e, \vec{E}\right\}$ such that the following equation holds:

$$A\vec{E} = e\vec{E} \tag{1-32}$$

Eigenvectors describe a special set of vectors where its multiplication by the matrix A changes only the scale (the length) of the vector. Eigenvectors corresponding to different Eigenvalues are linearly independent. Mathematically, Eigenvalues e are the roots of the characteristic equation of the square matrix A defined as follows:

$$\text{determinant}(eI - A) = 0 \tag{1-33}$$

Eigenvalues do not have to be real numbers. Sometimes, Eigenvalues are complex numbers. Readers can review the concept of complex numbers in [2] and [5]. In this case, the complex conjugate of Eigenvalue is also an Eigenvalue.

Applications of Eigenvectors/Eigenvalues

The concept of Eigenvectors/Eigenvalues is important in 3D modeling and estimation. For example, it can describe the rotation between two frames because the rotation vector between two frames does not change its direction when multiplied by the transformation matrix (transformation between two frames will be explained later in this chapter). This means that if we have a direction cosine matrix between two frames, a rotation vector must be an Eigenvector of this matrix. Another important feature of Eigenvectors is the decomposition of matrices. Eigenvectors/Eigenvalues can be used to factorize a square matrix A to be written as follows:

$$A = MDM^{-1} \tag{1-34}$$

where

D is a diagonal matrix, whose diagonal elements are the Eigenvalues of A
M is a matrix whose columns are Eigenvectors of A.

This is called Eigenvalue decomposition (EVD) and it is illustrated as follows:

$$\begin{pmatrix} a_{1,1} & \cdots & a_{1,n} \\ \vdots & \ddots & \vdots \\ a_{n,1} & \cdots & a_{n,n} \end{pmatrix} = \left(\vec{E}_1 \dots \vec{E}_n\right)\begin{pmatrix} e_1 & 0 & 0 \\ 0 & \ddots & \vdots \\ 0 & \cdots & e_n \end{pmatrix}\left(\vec{E}_1 \quad \cdots \quad \vec{E}_n\right)^{-1} \tag{1-35}$$

1.2.3.3 Matrix Symmetry

A matrix $A \in \mathbb{R}^{n \times n}$ is symmetric if its transpose A^T equals to itself as follows:

$$A^T = A \tag{1-36}$$

Transpose of matrix is simply done by exchanging its rows as its columns and its columns as its rows as illustrated in the following example:

$$\begin{pmatrix} a_{1,1} & \cdots & a_{1,n} \\ \vdots & \ddots & \vdots \\ a_{m,1} & \cdots & a_{m,n} \end{pmatrix}^T = \begin{pmatrix} a_{1,1} & \cdots & a_{m,1} \\ \vdots & \ddots & \vdots \\ a_{1,n} & \cdots & a_{m,n} \end{pmatrix} \tag{1-37}$$

The symmetric matrix A is called positive semidefinite if for any vector $\vec{X} \neq 0$:

$$\vec{X}^T A \vec{X} \geq 0 \tag{1-38}$$

The matrix is called positive definite if for any vector $\vec{X} \neq 0$

$$\vec{X}^T A \vec{X} > 0 \tag{1-39}$$

If A is a real symmetric matrix, all its Eigenvalues are real and its Eigenvectors corresponding to distinguished Eigenvalues are orthogonal. This means that the Eigenvectors of real symmetric matrix $A \in \mathbb{R}^{n \times n}$ forms a basis of the vector space $\mathbb{R}^n$. For any positive semidefinite or definite matrix, all eigenvectors are positive.

1.2.3.4 Matrix Norm

We defined a norm for the vector to be a function that gives a positive value that can be used to express the magnitude or length of a vector. We can also define the same function for matrices. Because the matrix is more complicated than vectors, its norm has many forms. Recall, matrix $A \in \mathbb{R}^{m \times n}$ can be seen as a mapping operator from n-dimensional vector space $\mathbb{R}^n$ to m-dimensional vector space $\mathbb{R}^m$. There is a family of norms called p-norms of a matrix defined as follows: If $p = 1$, then 1-norm of matrix A is the maximum absolute column sum of the matrix. If $p = 2$, then 2-norm is the largest singular value of the matrix. In other words, 2-norm is the maximum value of the singular value decomposition (SVD) of the matrix. SVD of matrix is explained in the next section. If $p = \infty$, then the norm here is the maximum absolute row sum of the matrix.

1.2.3.5 Singular Value Decomposition (SVD)

Recall that the concept of the Eigenvalues and Eigenvectors are only applicable to square matrices. It is useful to find similar decomposition for nonsquare matrices. This is what SVD is used for. Both SVD and EVD are used very frequently in robotics navigation mainly to formulate a problem as matrix multiplied by an unknown vector. If we have a rectangular matrix $A \in \mathbb{R}^{m \times n}$ where $m \geq n$ and the rank of A is p, then A can be represented as follows:

$$A = UDV^T \tag{1-40}$$

where

$U \in \mathbb{R}^{m \times p}$ and $V \in \mathbb{R}^{p \times n}$ are matrices with orthonormal columns
$D \in \mathbb{R}^{p \times p}$ is a diagonal matrix defined as:

$$D = \text{diag}\{\sigma_1, \sigma_2, ...\sigma_p\} = \begin{pmatrix} \sigma_1 & 0... & 0 \\ \vdots & \ddots & 0 \\ 0 & 0 & \sigma_p \end{pmatrix} \tag{1-41}$$

The orthonormal vectors of U and V are called the singular vectors pairs and the diagonal elements of D are called the singular values of the matrix A. The proof of the SVD can be found in [1]. One of the most important applications of SVD is finding the pseudo-inverse of a nonsquare matrix as the inverse of a matrix is only defined for square matrices. It can be shown that if the SVD of matrix $A = UDV^T$, a pseudo-inverse of the matrix A is defined as:

$$A^{\dagger} = VD^{-1}U^T \tag{1-42}$$

where D^{-1} is the inverse of the matrix D defined as:

$$D^{-1} = \text{diag}\left\{\frac{1}{\sigma_1}, \frac{1}{\sigma_2} \cdots \frac{1}{\sigma_p}\right\} = \begin{pmatrix} \frac{1}{\sigma_1} & 0... & 0 \\ \vdots & \ddots & 0 \\ 0 & 0 & \frac{1}{\sigma_p} \end{pmatrix} \tag{1-43}$$

Having the pseudo-inverse of a nonsquare matrix A enables us to solve the following linear system:

$$A\vec{x} = \vec{b} \tag{1-44}$$

The solution of the above linear system is the vector $\vec{x}^*$ that minimizes the following cost function:

$$J(\vec{x}) = \left\|A\vec{x} - \vec{b}\right\| \tag{1-45}$$

$$\vec{x}^* = \arg\min_{\vec{x}} J(\vec{x}) \tag{1-46}$$

$$\vec{x}^* = A^{\dagger}\vec{b} \tag{1-47}$$

1.2.3.6 Differentiations with Respect to Vectors and Matrices

In several mathematical formulations, we need to perform differentiation of scalar, vector, or matrices functions with respect to scalar, vectors, or matrices. Table 1.1 shows the common formulas for such differentiation:

Table 1.1 Differentiation with respect to vectors and matrices.

	Scalar y	Vector $\vec{y}$	Matrix Y
Scalar x	$\frac{dy}{dx}$	$\frac{d\vec{y}}{dx} = \left(\frac{\partial y_1}{\partial x}, \ldots, \frac{\partial y_n}{\partial x}\right)$	$\frac{d\mathbf{Y}}{dx} = \begin{pmatrix} \frac{\partial y_{1,1}}{\partial x} & \cdots & \frac{\partial y_{1,n}}{\partial x} \\ \vdots & \ddots & \vdots \\ \frac{\partial y_{n,1}}{\partial x} & \cdots & \frac{\partial y_{m,n}}{\partial x} \end{pmatrix}$
Vector $\vec{x}$	$\frac{dy}{d\vec{x}} = \left(\frac{\partial y}{\partial x_1}, \ldots, \frac{\partial y}{\partial x_n}\right)$	$\frac{d\vec{y}}{d\vec{x}} = \begin{pmatrix} \frac{\partial y_1}{\partial x_1} & \cdots & \frac{\partial y_1}{\partial x_m} \\ \vdots & \ddots & \vdots \\ \frac{\partial y_n}{\partial x_1} & \cdots & \frac{\partial y_n}{\partial x_m} \end{pmatrix}$	
Matrix $\mathbf{X}$	$\frac{dy}{d\mathbf{X}} = \begin{pmatrix} \frac{\partial y}{\partial x_{1,1}} & \cdots & \frac{\partial y}{\partial x_{1,m}} \\ \vdots & \ddots & \vdots \\ \frac{\partial y}{\partial x_{n,1}} & \cdots & \frac{\partial y}{\partial x_{m,n}} \end{pmatrix}$		

1.3 Coordinate Frames and Systems

Coordinate frames are systems to describe points, vectors, locations, and orientation. Any coordinate frame system must uniquely identify points using its coordinates. Any coordinate system must have a basis vector set that can be used to express any point or vector in the space that the coordinate system describes. A point can be represented as a linear combination of the basis vectors. Orientation of objects is described as a transformation between coordinate frames. Coordinates can be described using different systems. Detailed explanation of coordinate frames and systems is given in this section.

1.3.1 Rectangular Cartesian Coordinate Frames

Cartesian coordinate frame is a system where points can be uniquely determined by the signed distance from a set of perpendicular axes with respect to a definitive origin. These distances are called the coordinates of the point. In the 2D Cartesian coordinate frame, a point P coordinates are defined by two numbers ($P \in \mathbb{R}^2$). In the 3D Cartesian coordinate frame, a point P coordinates are be defined by three numbers ($P \in \mathbb{R}^3$). Figures 1.10 and 1.11 show a point P as described in 2D and 3D Cartesian coordinate frames, respectively. Figures 1.10 and 1.11 show the perpendicular axes and the origin point O.

It is obvious that the 3D Euclidean space can be represented by a 3D Cartesian coordinate frame. In this case, any point in the space is uniquely identified by its coordinates $\mathbb{R}^3$. We already used the 2D and 3D Cartesian coordinate systems as examples in the previous sections in this chapter. Cartesian coordinate systems are very common. They are also called "rectangular" as they use orthogonal axes and basis unit vectors along these orthogonal axes forming a rectangle (in the 2D case) or a rectangular cuboid shape (in the 3D space). The 3D Cartesian coordinate frame follows the right-hand rule such that the cross product of any ordered pair of its basis vectors gives the third basis vector. For example, the 3D Cartesian coordinate system has three orthogonal axes, an origin (0, 0, 0) and three-unit vectors that form the canonical basis of the system; those are $\overrightarrow{I_x}$, $\overrightarrow{I_y}$, and $\overrightarrow{I_z}$. A vector $\overrightarrow{V}$ defined in the 3D space can be represented using the dot product as follows:

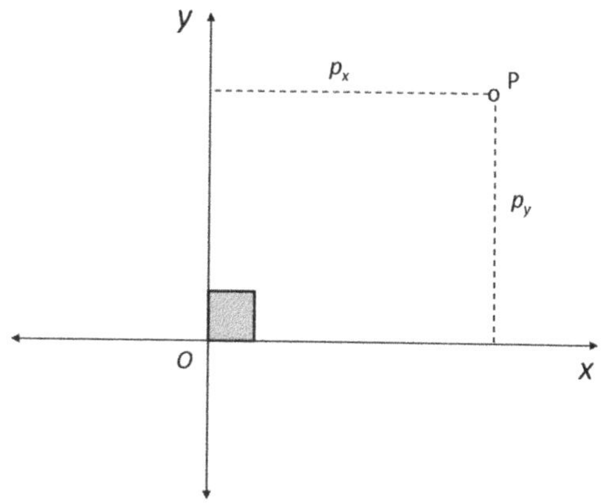

Figure 1.10 A point P described in the 2D Cartesian coordinate frame.

$$\begin{aligned}\overrightarrow{V} &= (v_x \quad v_y \quad v_z)^T\\ v_x &= \overrightarrow{V}\cdot\overrightarrow{I}_x\\ v_y &= \overrightarrow{V}\cdot\overrightarrow{I}_y\\ v_z &= \overrightarrow{V}\cdot\overrightarrow{I}_z\end{aligned} \qquad (1\text{-}48)$$

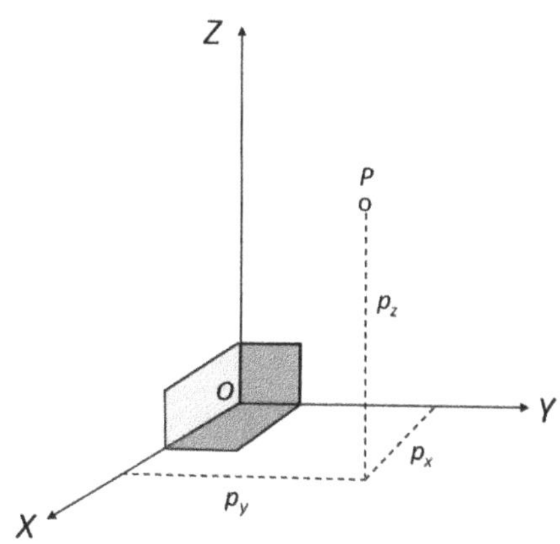

Figure 1.11 A point P described in the 3D Cartesian coordinate frame.

The values v_x, v_y, v_z are called the components of the vector $\overrightarrow{V}$. They are the projection of the vector $\overrightarrow{V}$ on the three axes of the right-handed Cartesian coordinate frame. We can express the vector $\overrightarrow{V}$ as a linear combination of the unit vectors $\overrightarrow{I}_x$, $\overrightarrow{I}_y$, and $\overrightarrow{I}_z$ as follows:

$$\overrightarrow{V} = v_x\overrightarrow{I}_x + v_y\overrightarrow{I}_y + v_z\overrightarrow{I}_z \qquad (1\text{-}49)$$

1.3.2 Polar and Spherical Coordinate Systems

The 2D polar coordinate system identifies a point by an angle θ and a distance from an origin r. In the 2D polar coordinate system, we have two basis vectors; these are the $\vec{r}$ and $\vec{\theta}$. $\vec{r}$ starts from the origin and points outward forming a circle around the origin with a radius $r \in [0, \infty[$. What uniquely identifies the point P on that circle is the angle value of $\theta \in [0, 2\pi]$. The direction of $\vec{r}$ is outward from the origin O, and the direction of $\vec{\theta}$ is perpendicular to $\vec{r}$ and conforms with the right-hand rule as shown in Figure 1.12.

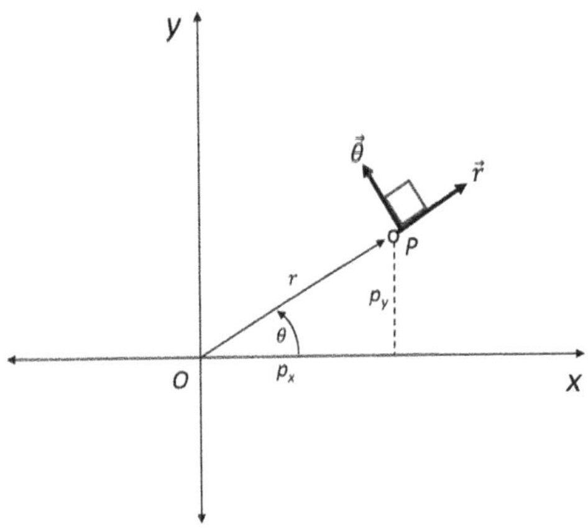

Figure 1.12 2D polar coordinate system.

It can be observed that the 2D Cartesian coordinates components p_x and p_y can be calculated from the polar coordinates as follows:

$$p_x = r\cos(\theta)$$
$$p_y = r\sin(\theta) \tag{1-50}$$

The concept of the 2D polar coordinate frame can be extended to the 3D space leading to what we refer to by the cylindrical (or spherical) coordinate frame by introducing a third basis vector $\vec{\varphi}$ that is orthogonal to $\vec{r}$ and $\vec{\theta}$ as seen in Figure 1.13. In this case, $\vec{r}$ starts from the origin and pointing outward forming a sphere around the origin with a radius $r \in [0, \infty[$. What uniquely identifies the point P on that sphere is the value of $\theta \in [0, 2\pi]$ and $\varphi \in [0, 2\pi]$. The direction of $\vec{r}$ is outward from the origin O. $\vec{r}$, $\vec{\theta}$ and $\vec{\varphi}$ are orthogonal and conforms with the right-hand rule as shown in Figure 1.13. Sometimes the angle φ is known as the Elevation and the angle θ is known as the Azimuth. In the geodetic Earth coordinate system (to be explained later), the angle φ is known as the latitude and the angle θ is known as the longitude.

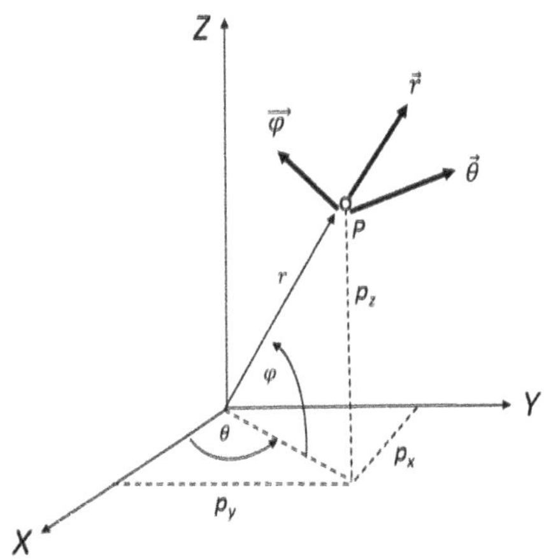

Figure 1.13 3D polar coordinate system.

It can be observed that the 3D Cartesian coordinates components p_x, p_y, and p_z can be calculated from the polar coordinates as follows:

$$p_x = r\cos(\varphi)\cos(\theta)$$
$$p_y = r\cos(\varphi)\sin(\theta) \tag{1-51}$$
$$p_z = r\sin(\varphi)$$

1.3.3 Earth Coordinate Systems and Parameters

The study of Earth coordinate systems, frames, and parameters is essential for describing the global positioning of objects. In small areas, such as a room, a floor, or a building, simple local coordinate systems and frames may suffice. However, in larger-scale environments, like those involving autonomous cars and aerial vehicles, global positioning and navigation coordinate systems and frames become necessary. Additionally, the measurements of certain sensors, such as inertial and magnetic sensors, are influenced by Earth's physical parameters, including gravity, surface geometry, and rotation [5] and [6]. Therefore, these Earth parameters are critical factors in processing data from inertial and magnetic sensors. Furthermore, the Global Navigation Satellite System (GNSS) [7] rely on satellites that are orbiting the Earth to provide global navigation, offering positioning information within global coordinate frames. Due to Earth's movement and rotation, the origin and the axes of coordinate frames change [5] and [8]. Therefore, it is crucial to define a coordinate system that accurately reflects these changes. This section is dedicated to discussing global coordinate systems, frames, and the associated Earth parameters.

1.3.3.1 Earth Shape and Coordinate System

The closest geometric shape of Earth is an ellipsoid as shown in Figure 1.14 with a semimajor axis R_0 and semiminor axis L_0. In Figure 1.14, point S is a point on Earth surface and point B is the body or object of interest. If the Earth is a perfect sphere with radius R_0, the Earth surface equation will be as follows:

$$R_{sx_e}^2 + R_{sy_e}^2 + R_{sz_e}^2 = R_0^2 \tag{1-52}$$

where

$\vec{R}_S = \left(R_{sx_e}\ R_{sy_e}\ R_{sz_e}\right)^T$ is a vector starts from the Earth's center and ends at point $S = \left(R_{sx_e}\ R_{sy_e}\ R_{sz_e}\right)^T$ on the Earth's surface.

The $\vec{R}_s$ is described by a 3D Cartesian coordinate frame with an origin at Earth's center and the frame itself is fixed to Earth. This 3D Cartesian coordinate system is known as Earth Fixed Earth Centered Frame (*ECEF*) or simply *E* frame. One of the *E* frame's axes is coincided with the axis of rotation of the Earth. *E* frame axes can be defined in different ways. One way is illustrated in Figure 1.14, where *Y*-axis coincides with the Earth rotation axis. In other configurations, *Z*-axis is defined is coincided with the Earth rotation axis, which is the definition adopted by the International Earth Rotation Service (IERS) [6]. The *E* frame is rotating with the Earth around its *Y*-axis, as shown in Figure 1.14 at a rate of $\omega^e = 7.292115 \times 10^{-5}$ rad/s [5] and [9]. But because the Earth is not a perfect sphere, it is approximated by an ellipsoid whose equation is as follows [5]:

$$R_{sx_e}^2 + R_{sz_e}^2 + \frac{R_{sy_e}^2}{(1-e)^2} = R_0^2 \tag{1-53}$$

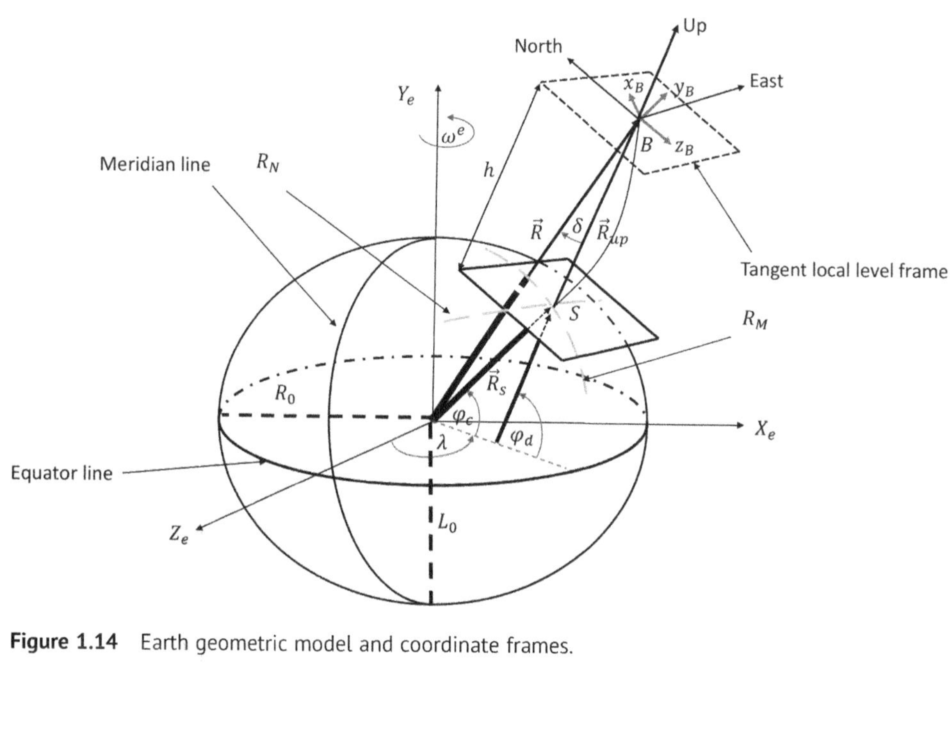

Figure 1.14 Earth geometric model and coordinate frames.

where e is an ellipticity parameter defined as follows:

$$e = 1 - \frac{L_0}{R_0} \tag{1-54}$$

where L_0 is the radius along spinning axis of Earth.

This small ellipticity causes that geodetic latitude φ_d is slightly different from the geocentric latitude angle φ_c as seen in Figure 1.14. Also, the vector $\vec{R}$ connecting the Earth surface to the body (the object that we want to describe its positioning) consists of two vectors as follows:

$$\vec{R} = \vec{R}_s + \vec{R}_{up} \tag{1-55}$$

where

$\vec{R}_s$ is the vector connecting the Earth center to the point S on Earth surface
$\vec{R}_{up} = \left(R_{upx_e}\ R_{upy_e}\ R_{upz_e}\right)^T$ is unit vector starts at point S and pointing upward perpendicular on the ellipsoid surface to point B.

The vector $\vec{R} = \left(R_{x_e}\ R_{y_e}\ R_{z_e}\right)^T$ is the absolute position of the object with respect to Earth (i.e., the ECEF frame). The unit vector $\vec{R}_{up}$ is defined as follows [5] and [7]:

$$\vec{R}_{up} = \begin{pmatrix} R_{sx_e}/R_s' \\ R_{sy_e}/\left(R_s'(1-e)^2\right) \\ R_{sz_e}/R_s' \end{pmatrix} \tag{1-56}$$

where R_s' is defined as follows:

$$R_s' = \sqrt{R_{sx_e}^2 + R_{sz_e}^2 + \frac{R_{sy_e}^2}{(1-e)^4}} \tag{1-57}$$

The magnitude of the position vector $\vec{R}$ is given by the following equation [5]:

$$\left\|\vec{R}\right\| = \sqrt{R_{sx_e}^2 + R_{sy_e}^2 + R_{sz_e}^2 + 2hR_0^2/R_s' + h^2} \tag{1-58}$$

1.3.3.2 Geodetic Coordinates and World Geodetic System (WGS84)

The geodetic latitude and the longitude angles are defined as follows [5] and [7]:

$$\sin\varphi_d = R_{upy_e} = R_{sy_e}/\left(R_s'(1-e)^2\right) \tag{1-59}$$

$$\tan\lambda = \frac{R_{upx_e}}{R_{upz_e}} = \frac{R_{sx_e}}{R_{sz_e}} \tag{1-60}$$

The two angles φ_d and λ determines the horizontal location of the object B and the altitude h is the vertical component. The three parameters φ_d, λ, and h are called the geodetic coordinate of the object B. Any point on Earth has two coordinate values: latitude (φ_d) and longitude (λ). Latitude and longitude measure angles. The third point is the altitude (h) in meters, which measures the distance between the object and the Earth's ellipsoid surface. Points north of the Equator have positive latitude values, while points south have negative values. Latitude values range from −90° to +90°. Lines of latitude are also called parallels because a particular value of latitude forms a circle parallel to the Equator. The meridian, or line of longitude, is formed by a plane that passes through the point and the North and South poles. The longitude value is defined by the angle between that plane and a reference plane. The reference plane is known as the Prime Meridian. The most common prime meridian passes through Greenwich, United Kingdom. Longitude values range from $-180°$ to $+180°$. This geodetic coordinate system is standardized by the World Geodetic System (WGS84). According to the WGS84 standard [9], the ellipsoid parameters are as follows:

- Semimajor axis $a = 6{,}378{,}137$ m
- Semiminor axis $b = 6{,}356{,}752.3142$ m
- Eccentricity $e = \left(\sqrt{a^2 - b^2}\right)/(a) = 0.08181919$

The values of a, b,and e are used to convert positions from the geodetic frame to the ECEF frame which is explained in the next subsection.

1.3.3.3 Earth-Centered-Earth-Fixed (ECEF) Rectangular Coordinate System

The Earth-centered, Earth-fixed coordinate system (ECEF), also known as the geocentric coordinate system, is a Cartesian spatial reference system that represents locations in the vicinity of the Earth as X, Y, and Z measurements from its center of mass. Its origin and the Earth's center of mass are coincided. ECEF frame is fixed at the center of the Earth and rotates along with the rotation of Earth. ECEF axes can be defined in different ways. One way is by taking its y-axis as the rotational axis of the Earth; the z-axis is pointing to the mean Greenwich meridian, while the y-axis is directed to complete a right-handed system as shown in Figure 1.14. Figure 1.14 shows the axes of ECEF, which are three axes named as X_e, Y_e, and Z_e. Other configurations of ECEF that are also common define z-axis to be the Earth rotation axis, x-axis pointing to prime meridian, and y-axis completing the right-hand rule. In the rest of this book, we will adopt the later axes configuration, which defines Z-axis along the Earth rotation axis.

1.3.3.4 Earth-Centered-Inertial (ECI) Frame

Inertial navigation equations are modeled based on Newton's law of motion. For Newton's law to apply, variables must be referenced to a nonaccelerating frame. Therefore, a geocentric-equatorial coordinate system is defined as Earth-Centered Inertial frame (*ECI*). ECI has its origin at the center of the Earth. However, it is not fixed to the Earth (it does not rotate with the Earth). All inertial sensors produce measurements relative to the ECI frame. ECI axes align with ECEF axes at zero longitude. ECEF rotates with respect to ECI around the Earth's rotation axis at a rate of $\omega^e = 7.292115 \times 10^{-5}$ rad/s.

1.3.3.5 Local-level (Tangent) Frames (*L* and *N* Frames)

The tangent local level frame is a Cartesian coordinate system that is used to describe the navigation state in terms of popular north, east, and up (height) components. The origin of the local level frame is moving with the object B. One configuration of the tangent local level frame is East, North, and Up (ENU), where the x-axis points to east, the y-axis points to north, and the z-axis points upward perpendicular to the ellipsoid surface. We will call this configuration N frame. Another configuration is North, East, and Down (NED), where the x-axis points to north, the y-axis points to east, and the z-axis points downward perpendicular to the ellipsoid surface to complete the right-hand axes rule. We will call this configuration L frame. Always keep in mind that the origins of L and N frames are assumed to be coincided with the origin of the body or object frame B. The position of the object is described as the position of the origin of the L or N frame with respect to the ECEF frame. This is the absolute position vector of the object. This absolute position of the object (which is the same as the position of frames L or N) can be described by geodetic latitude, longitude, and height (φ, λ, h) or by $\vec{R}$ vector components $\vec{R} = \left(R_{x_e}\ R_{y_e}\ R_{z_e}\right)^T$. In some configurations, and to provide local positioning as east/north, we may choose a reference latitude (φ_r) and longitude (λ_r) and calculate relative local north and east position components corresponding to an absolute reference latitude (φ_r) and longitude (λ_r) as follows:

$$n = (\varphi_d - \varphi_0)(R_M + h) \tag{1-61}$$

$$e = (\lambda - \lambda_0)(R_N + h)\cos\varphi_d \tag{1-62}$$

where R_M and R_N are the meridian and normal radius of curvature of Earth at the current location.

Unless a wander angle [5] is used, the default common configuration is that the orientation of L and N frames changes with the object' location with respect to Earth (E frame) to keep on of the horizontal axis (x-axis of L frame or y-axis in N frame) always pointing to the geographic north.

1.3.3.6 Body Frame (B)

The body frame B is centered at the center of mass of the object. In the case of perfect alignment between B and L, its x-axis is pointing north, the y-axis is pointing east, and the z-axis is pointing downward. The orientation of the object is described as a rotation of the body frame B with respect to a reference frame. This reference frame can be N or E frame. The body frame B has its axes labeled as x_B, y_B, and z_B as seen in Figure 1.14.

1.3.3.7 Conversion Between Geodetic Coordinates and ECEF Coordinates

The geodetic coordinates can be converted to ECEF coordinates using the following equations: Let X_e, Y_e, Z_e be ECEF coordinates and (φ, λ, h) be geodetic latitude, longitude, and height, respectively. Using the Earth ellipsoid parameters (a, b, e) provided in the previous section, we can obtain the radius of curvature R_M, R_N as follows [5], [7], and [8]:

$$R_M(\varphi) = \frac{a(1-e^2)}{\left(1-e^2\sin^2(\varphi)\right)^{3/2}} \tag{1-63}$$

$$R_N(\varphi) = \frac{a}{\left(1-e^2\sin^2(\varphi)\right)^{1/2}} \tag{1-64}$$

where

$R_M(\varphi)$ is meridian curvature of the ellipsoid model at point S, which is also known as the Earth radius of curvature in the north-south direction

$R_N(\varphi)$ is the normal prime-vertical radius of curvature of Earth in the east-west direction.

The relationship between ECEF coordinate components X_e, Y_e, Z_e and geodetic coordinate components (φ, λ, h) is given as follows:

$$Z_e = (R_N(\varphi) + h)\cos(\varphi)\cos(\lambda) \tag{1-65}$$

$$X_e = (R_N(\varphi) + h)\cos(\varphi)\sin(\lambda) \tag{1-66}$$

$$Y_e = \left(R_N(\varphi)\left(1-e^2\right) + h\right)\sin(\varphi) \tag{1-67}$$

Illustration of $R_M(\varphi)$ and $R_N(\varphi)$ is shown in Figure 1.14. There is also an algorithm to convert from ECEF or E frame to geodetic coordinates, which is given as follows:

$$\lambda = a\ \tan(X_e/Z_e) \tag{1-68}$$

We then calculate φ and h iteratively in a loop as follows:

- Initialization:

$$h = 0, R_N = a, p = \sqrt{Z_e^{\ 2} + X_e^{\ 2}} \tag{1-69}$$

- Iterate until convergence:

$$\sin(\varphi) = \frac{Y_e}{R_N(1-e^2) + h} \tag{1-70}$$

$$\varphi = \tan^{-1}\left(\left(Y_e + R_N e^2 \sin(\varphi)\right)/p\right) \tag{1-71}$$

$$R_N(\varphi) = \frac{a}{\left(1 - e^2 \sin^2(\varphi)\right)^{1/2}} \tag{1-72}$$

$$h = \frac{p}{\cos(\varphi)} - R_N \tag{1-73}$$

1.4 Transformation Between Frames

Transformation is used to describe the orientation and the position of an object with respect to a reference coordinate frame. For example, to describe the orientation of an airplane, we describe how its body coordinate frame is rotated with respect to a reference coordinate frame. We can also describe the location of the airplane by a vector that starts at the origin of the reference coordinate frame and ends at the origin of the airplane body frame. By defining the rotation of a frame with respect to another, we can transform any vector defined in one frame to the other. Therefore, it is important to understand the rotation and transformation between coordinate frames.

1.4.1 Direction Cosine Matrix

Consider a vector $\overrightarrow{V}$ that is defined in a 3D right-handed reference Cartesian coordinate frame L as seen in Figure 1.15. The same vector $\overrightarrow{V}$ is seen as $\overrightarrow{V}^A = \left(v_{xA}\ v_{yA}\ v_{zA}\right)^T$ from frame A with unit vectors $\overrightarrow{I}_{xA}$, $\overrightarrow{I}_{yA}$, and $\overrightarrow{I}_{zA}$. The vector $\overrightarrow{V}$ can be written as follows:

$$\overrightarrow{V} = v_{xA}\overrightarrow{I}_{xA} + v_{yA}\overrightarrow{I}_{yA} + v_{zA}\overrightarrow{I}_{zA} \tag{1-74}$$

The values v_{xA}, v_{yA}, v_{zA} are the components of the vector $\overrightarrow{V}$ in frame A defined as the magnitude of $\overrightarrow{V}$ multiplied by the cosine of angle between $\overrightarrow{V}$ and the unit vectors $\overrightarrow{I}_{xA}$, $\overrightarrow{I}_{yA}$, and $\overrightarrow{I}_{zA}$. The

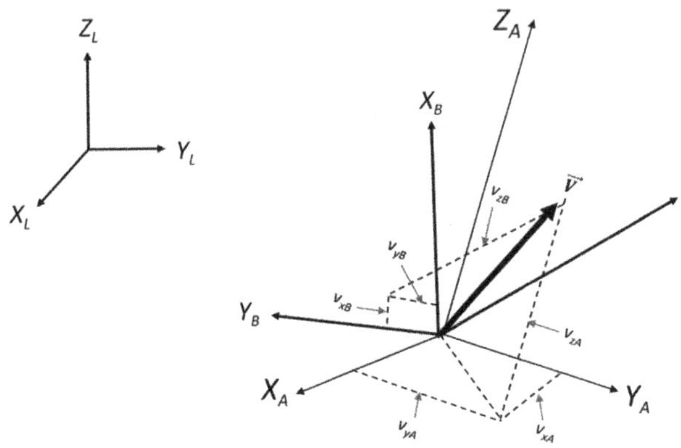

Figure 1.15 Illustration of a vector $\vec{V}$ represented in two frames A and B.

same vector $\vec{V}$ can be seen as $\vec{V}^B = (v_{xB}\ v_{yB}\ v_{zB})^T$ from frame B with unit vectors $\vec{I}_{xB}$, $\vec{I}_{yB}$, and $\vec{I}_{zB}$ as follows:

$$\vec{V} = v_{xB}\vec{I}_{xB} + v_{yB}\vec{I}_{yB} + v_{zB}\vec{I}_{zB} \tag{1-75}$$

The values v_{xB}, v_{yB}, v_{zB} are the components of the vector $\vec{V}$ in frame B defined as the magnitude of $\vec{V}$ multiplied by the cosine of angle between $\vec{V}$ and the unit vectors $\vec{I}_{xB}$, $\vec{I}_{yB}$, and $\vec{I}_{zB}$. Note that the vector $\vec{V}$ and the unit vectors $\vec{I}_{xA}$, $\vec{I}_{yA}$, $\vec{I}_{zA}$. $\vec{I}_{xB}$, $\vec{I}_{yB}$, $\vec{I}_{zB}$ are defined in the reference frame L. However, if frame A is perfectly aligned with frame L, we can consider frame A as the base or reference and frame B is rotated with respect to A (Figure 1.15).

The two representations of $\vec{V}$ are equal since they represent the same vector, so we have the following equation:

$$v_{xA}\vec{I}_{xA} + v_{yA}\vec{I}_{yA} + v_{zA}\vec{I}_{zA} = v_{xB}\vec{I}_{xB} + v_{yB}\vec{I}_{yB} + v_{zB}\vec{I}_{zB} \tag{1-76}$$

If we apply dot product of both sides by $\vec{I}_{xA}$ and recalling that $\vec{I}_{xA} \cdot \vec{I}_{yA} = 0$, $\vec{I}_{xA} \cdot \vec{I}_{zA} = 0$ and $\vec{I}_{xA} \cdot \vec{I}_{xA} = 1$ we obtain:

$$v_{xA} = v_{xB}\vec{I}_{xA} \cdot \vec{I}_{xB} + v_{yB}\vec{I}_{xA} \cdot \vec{I}_{yB} + v_{zB}\vec{I}_{xA} \cdot \vec{I}_{zB} \tag{1-77}$$

Similarly, we can show that v_{yA} and v_{zA} are calculated as follows:

$$v_{yA} = v_{xB}\vec{I}_{yA} \cdot \vec{I}_{xB} + v_{yB}\vec{I}_{yA} \cdot \vec{I}_{yB} + v_{zB}\vec{I}_{yA} \cdot \vec{I}_{zB} \tag{1-78}$$

$$v_{zA} = v_{xB}\vec{I}_{zA} \cdot \vec{I}_{xB} + v_{yB}\vec{I}_{zA} \cdot \vec{I}_{yB} + v_{zB}\vec{I}_{zA} \cdot \vec{I}_{zB} \tag{1-79}$$

We obtained the components of $\vec{V}$ in frame A as a function of the components of $\vec{V}$ in frame B. Similarly, the components of $\vec{V}$ in frame B as a function of the components of $\vec{V}$ in frame A as follows:

$$v_{xB} = v_{xA}\vec{I}_{xB} \cdot \vec{I}_{xA} + v_{yA}\vec{I}_{xB} \cdot \vec{I}_{yA} + v_{zA}\vec{I}_{xB} \cdot \vec{I}_{zA} \tag{1-80}$$

$$v_{yB} = v_{xA}\vec{I}_{yB} \cdot \vec{I}_{xA} + v_{yA}\vec{I}_{yB} \cdot \vec{I}_{yA} + v_{zA}\vec{I}_{yB} \cdot \vec{I}_{zA} \tag{1-81}$$

$$v_{zB} = v_{xA}\vec{I}_{zB} \cdot \vec{I}_{xA} + v_{yA}\vec{I}_{zB} \cdot \vec{I}_{yA} + v_{zA}\vec{I}_{zB} \cdot \vec{I}_{zA} \tag{1-82}$$

We can arrange this set of equations into a matrix form as follows:

$$\begin{pmatrix} v_{xA} \\ v_{yA} \\ v_{zA} \end{pmatrix} = \begin{pmatrix} \vec{I}_{xA} \cdot \vec{I}_{xB} & \vec{I}_{xA} \cdot \vec{I}_{yB} & \vec{I}_{xA} \cdot \vec{I}_{zB} \\ \vec{I}_{yA} \cdot \vec{I}_{xB} & \vec{I}_{yA} \cdot \vec{I}_{yB} & \vec{I}_{yA} \cdot \vec{I}_{zB} \\ \vec{I}_{zA} \cdot \vec{I}_{xB} & \vec{I}_{zA} \cdot \vec{I}_{yB} & \vec{I}_{zA} \cdot \vec{I}_{zB} \end{pmatrix} \begin{pmatrix} v_{xB} \\ v_{yB} \\ v_{zB} \end{pmatrix} \tag{1-83}$$

Recall that matrix multiplication requires inner dimensions to match. Given two matrices A an $m \times n$ matrix and B an $n \times l$ matrix, the product AB is defined as follows:

$$A_{m \times n} B_{n \times l} = C_{m \times l}$$

$$\begin{pmatrix} a_{11} & a_{12} & \dots & a_{1n} \\ a_{21} & a_{22} & \dots & \vdots \\ \vdots & \vdots & \ddots & \vdots \\ a_{m1} & \dots & \dots & a_{mn} \end{pmatrix} \begin{pmatrix} b_{11} & b_{11} & \dots & b_{1l} \\ b_{21} & b_{22} & \dots & \vdots \\ \vdots & \vdots & \ddots & \vdots \\ b_{n1} & \dots & \dots & b_{nl} \end{pmatrix} = \begin{pmatrix} c_{11} & c_{12} & \dots & c_{1l} \\ c_{21} & c_{22} & \dots & \vdots \\ \vdots & \vdots & \ddots & \vdots \\ c_{m1} & \dots & \dots & c_{ml} \end{pmatrix} \tag{1-84}$$

where c_{ij} is defined as follows:

$$c_{ij} = \sum_{k=1}^{n} a_{ik} b_{kj} \tag{1-85}$$

Note that c_{ij} is the dot product of the ith row of matrix A with the jth column of matrix B. The vector $\vec{V}$ can be seen as a matrix of dimension 3×1. Therefore, the multiplication of a 3×3 matrix by a 3×1 vector is another vector of the same dimension 3×1. Note that $\vec{I}_{xA}$ and $\vec{I}_{xB}$ are unit vectors, so the dot product of $\vec{I}_{xA}$ and $\vec{I}_{xB}$ equals to the cosine of the angle between $\vec{I}_{xA}$ and $\vec{I}_{xB}$. We call this cosine the "direction cosine" between $\vec{I}_{xA}$ and $\vec{I}_{xB}$ and we refer to the C_B^A matrix as the Direction Cosine Matrix from frame B to frame B defined as follows:

$$C_B^A = \begin{pmatrix} \vec{I}_{xA} \cdot \vec{I}_{xB} & \vec{I}_{xA} \cdot \vec{I}_{yB} & \vec{I}_{xA} \cdot \vec{I}_{zB} \\ \vec{I}_{yA} \cdot \vec{I}_{xB} & \vec{I}_{yA} \cdot \vec{I}_{yB} & \vec{I}_{yA} \cdot \vec{I}_{zB} \\ \vec{I}_{zA} \cdot \vec{I}_{xB} & \vec{I}_{zA} \cdot \vec{I}_{yB} & \vec{I}_{zA} \cdot \vec{I}_{zB} \end{pmatrix} \tag{1-86}$$

and we can write $\vec{V}^A$ (the vector $\vec{V}$ as projected on frame A) as a function of $\vec{V}^B$ (the vector $\vec{V}$ as projected on frame B) using the following equation:

$$\vec{V}^A = C_B^A \vec{V}^B \tag{1-87}$$

Note that the first row of C_B^A is the projection of the unit vector $\vec{I}_{xA}$ on the three axes of frame B. Similarly, the first column of C_B^A is the projection of the unit vector $\vec{I}_{xB}$ on the three axes of frame A. Using the direction cosine matrix, we can describe the orientation of any frame B with respect to another frame to any frame A. It is easy to observe the following important property of C_B^A:

$$\left(C_B^A\right)^{-1} = \left(C_B^A\right)^T = C_A^B, \det\left(C_B^A\right) = 1 \tag{1-88}$$

Also, we can observe that the direction cosine matrix is an orthonormal matrix which means that the magnitude of any row or column of the direction cosine matrix is 1 (recall that each row or a column in C_B^A is a unit vector projected on a frame, so its magnitude is 1) and the dot product of any two rows or columns equal zero. The direction cosine is one way to describe and perform transformation between two frames. By finding the direction matrix from between frames A and B, any vector in A can be transformed into a vector in B and vice versa.

1.4.2 Conversion from Local Frames to ECEF Frame

Given the direction cosine matrix described Section 1.4.1, we can describe the direction cosine matrix that models the rotation of frame E with respect to frame L using the latitude and longitude coordinates. This rotation matrix consists of two Euler rotations, one anticlockwise rotation around Z by longitude angle λ given by:

$$C_E^{E'} = \begin{pmatrix} \cos(\lambda) & \sin(\lambda) & 0 \\ -\sin(\lambda) & \cos(\lambda) & 0 \\ 0 & 0 & 1 \end{pmatrix} \tag{1-89}$$

and another clockwise rotation around X by 90° plus latitude angle φ, which gives the following rotation matrix:

$$C_{E'}^{L} = \begin{pmatrix} -\sin(\phi) & 0 & \cos(\phi) \\ 0 & 1 & 0 \\ -\cos(\phi) & 0 & -\sin(\phi) \end{pmatrix} \tag{1-90}$$

The overall direction cosine matrix C_E^L is given as follows:

$$C_E^{L} = \begin{pmatrix} -\sin(\phi)\cos(\lambda) & -\sin(\phi)\sin(\lambda) & \cos(\phi) \\ -\sin(\lambda) & \cos(\lambda) & 0 \\ -\cos(\phi)\cos(\lambda) & -\cos(\phi)\sin(\lambda) & -\sin(\phi) \end{pmatrix} \tag{1-91}$$

1.4.3 Transformation Groups

The direction cosine matrix that we explained in the previous section defines a group of transformation. Mathematically, a group G is a set of elements plus a binary operator $\odot$ such that the application of the operator $\odot$ on two elements x, y that belong to the group will generate an output element that belongs to the group. In other words, a group G is a set of elements plus a binary operator $\odot$ that takes two elements of the group and produces a third element that belongs to the group. To form a group, the following conditions must apply [1], [10], and [11]:

- The operator $\odot$ must be associative $\odot(x, \odot(y, z)) = \odot(\odot(x, y), z)$
- An identity element I must exist $\odot(x, I) = x$
- Each element $x \in G$ has an inverse $e \in G\ \forall x \in G\ \exists e \in G \rightarrow \odot(x, e) = I$

A simple example of a group is the set of integers under the addition operation (the operator "$\odot$" is the addition "+"). The identity of the group of integers is the zero and the inverse of any element $x \in G$ in the group is $(-x)$, which also belongs to group of integers. Other groups can be defined using any kind of elements and any operator if they satisfy the above-mentioned conditions. The direction cosine matrices form a group whose elements are matrices, and the operator $\odot$ is multiplication.

1.4.4 General Linear Group

All nonsingular invertible square matrices form a group with respect to matrix multiplication. This group is called the general linear group $GL(n)$ and it consists of all invertible square matrices. If the matrix has a determinant of 1, the group is called the Special Linear Group $SL(n)$, where $n \times n$ is the dimension of the matrix. The direction cosine matrix is a subgroup of the $SL(n)$.

1.4.5 Affine Group A(*n*) and Affine Transformation

The rotation between two frames is only one part of the relative positioning of the two frames with respect to each other. The rotation relationship is represented by $SL(3)$ group. Also, the orientation can be represented as a rotation around a rotation vector $\vec{\phi}_B^L$ as seen in Figure 1.16. To fully describe the positioning of frame B with respect to frame L, we must include the translational displacement between the two frames. Figure 1.16 shows that the origin of B is located at a vector $\vec{T}_B^L$ with respect to frame L. Note that the rotation vector $\vec{\phi}_B^L$ is the same in both frames regardless of the displacement vector $\vec{T}_B^L$.

To transform a vector $\vec{V}^B = (v_{xB}\ v_{xB}\ v_{xB})^T$ from frame B to frame L, we use the following equation:

$$\vec{V}^A = C_B^L \vec{V}^B + \vec{T}_B^L \tag{1-92}$$

This form of transformation is called "affine transformation" and the set of all transformations in this form of dimension n ($n = 3$ in our case) is called the affine group $A(n)$. Affine group $A(n)$ has an $n \times n$ matrix C and vector $\vec{T} \in \mathbb{R}^n$. The affine group transformation is not linear. It is always desirable to have linear transformations which facilitate solving equations as we will see later in

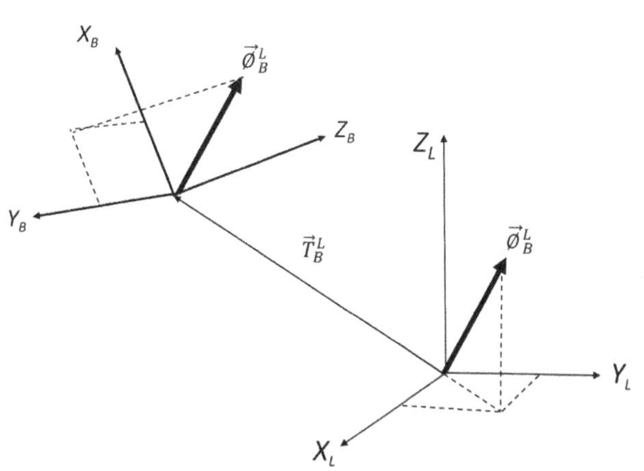

Figure 1.16 Relative positioning of two frames (orientational and translational).

estimation chapter. To make the transformation linear, we introduce a homogeneous coordinate to represent vectors $\vec{V} \in \mathbb{R}^n$ as follows:

$$\begin{pmatrix} \vec{V} \\ 1 \end{pmatrix} \in \mathbb{R}^{n+1} \tag{1-93}$$

In this way, the transformation of a vector $\vec{V}^B$ to vector $\vec{V}^A$ becomes linear as follows:

$$\begin{pmatrix} v_{xA} \\ v_{yA} \\ v_{zA} \\ 1 \end{pmatrix} = \mathrm{T}_{\mathrm{B}}^{\mathrm{L}} \begin{pmatrix} v_{xB} \\ v_{xB} \\ v_{xB} \\ 1 \end{pmatrix} \tag{1-94}$$

where T_B^L is the transformation matrix defined as follows:

$$T_B^L = \begin{pmatrix} C_B^L & \vec{T}_B^L \\ 0 & 1 \end{pmatrix}_{4 \times 4} \tag{1-95}$$

What we have done here is that we included the displacement $\vec{T}_B^L$ in the transformation between the two frames leading to a 4×4 full transformation matrix with two parts, a 3×3 matrix part that includes the direction cosine matrix C_B^L and a 3×1 vector that includes the displacement vector $\vec{T}_B^L$. Any matrix in this form (has $n \times n$ matrix and $n \times 1$ vector) is called an Affine Matrix.

1.4.6 The Orthogonal Group $O(n)$ and Special Orthogonal Group $SO(3)$

We refer to all $n \times n$ orthogonal matrices (i.e., columns and rows are orthonormal vectors), by the Orthogonal Group $O(n)$. It can be observed that the direction cosine matrix belongs to $O(n)$ group. For any matrix $C \in O(n)$, the following relationship holds:

$$CC^T = C^T C = 1 \tag{1-96}$$

If the determinant of the matrix is 1 ($\det(C) = 1$), it is called Special Orthogonal Group $SO(n)$. The direction cosine matrix belongs to the $SO(3)$ matrix group.

1.4.7 The Special Euclidean Group $SE(n)$

In the $A(n)$ group, if the submatrix C is orthogonal ($C \in O(n)$), we get the Euclidean Group $E(n)$. Euclidean means that the transformation group preserves the shape of objects in the transformation (no deformation happens to the transformed vectors and shapes). If the submatrix C is orthogonal and its determinant is 1 ($\det(C) = 1$), which means $C \in SO(n)$, then we get the Special Euclidean Group $SE(3)$. The $SO(3)$ group preserves the norm of vectors and the cross-product of any two vectors. This can be expressed as follows: if frames A and B are related by the transformation matrix $C_B^A \in SO(3)$ then:

$$\left\| \vec{V}^A \right\| = \left\| C_A^B \vec{V}^A \right\| = \left\| \vec{V}^B \right\| \tag{1-97}$$

$$C_A^B \left(\vec{V}^A \times \vec{U}^A \right) = \left(C_A^B \vec{V}^A \right) \times \left(C_A^B \vec{U}^A \right) \tag{1-98}$$

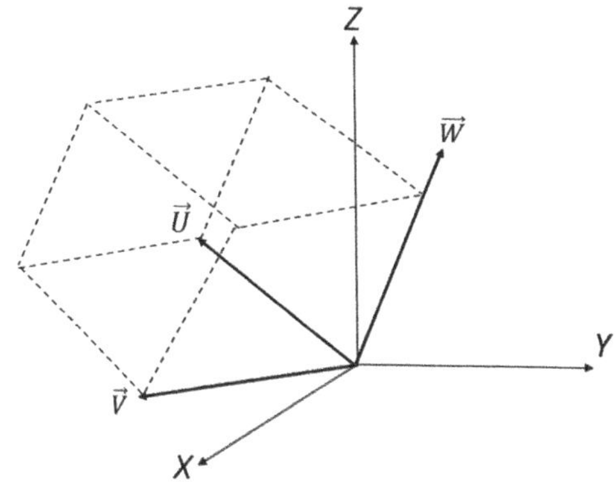

Figure 1.17 Volume spanned by three vectors $\vec{U}$, $\vec{V}$, and $\vec{W}$.

This also implies that the $SO(3)$ transformation preserves the triple scalar product:

$$\left(C_A^B \vec{U}^A \cdot \left(\left(C_A^B \vec{V}^A\right) \times \left(C_A^B \vec{W}^A\right)\right)\right) = \vec{U}^A \cdot \left(\left(\vec{V}^A\right) \times \left(\vec{W}^A\right)\right) \tag{1-99}$$

Note that the triple scalar product of three vectors is the volume spanned by the three vectors as shown in Figure 1.17. In other words, the volume of an object formed by three vectors is the triple product of the three vectors defined by $\vec{a} \cdot \left(\vec{b} \times \vec{c}\right)$.

The above-mentioned three features of the $SO(3)$ group make the $SE(3)$ group transformation represent the rigid body transformation because it preserves both lengths and volumes. The motion of rigid body is either rotational (orientational) or translational or a combination of both. In the rigid body motion, the object does not change its dimensions (lengths of vectors of points that form the object) and its volume. The $SE(3)$ group preserves lengths and orientations. In other words, $SE(3)$ preserves the shape of the object and that is why it used to model rigid bodies motion.

1.4.8 Lie Group

The $SO(3)$ and $SE(3)$ groups belong to a wider group called Lie Group. Lie Group is a group where the operator is the multiplication, and both multiplication and inversion form a continuous smooth mapping of the elements of the group. Another condition to form a Lie Group is that the elements of the group form a continuous differentiable manifold. Away from the complex formal mathematical definition, we can understand the terms "continuous" and "smooth" in the sense that the mapping (multiplication or inversion) of continuous elements in the group generates a continuous set of elements in the group. In other words, there are no spikes or discontinuity in the mapping that the multiplication and inversion cause.

Lie Group definition will take us to the definition of a manifold. A manifold is a space of elements (those elements could be points, vectors, or matrices) with a specific topology (structure) that does not change locally. We can see the topology as structural (or spatial) relationship between points in Euclidean space. For example, a straight line is a topology where all points ordered in a way to form a straight line. A straight line is a manifold because at any point on the line, we see the same structure (the same topology, i.e., a group of points that form a straight line). The space of points on a circle is a manifold because at any point on the circle, you still see connected points that form a curve of the same topology (the spatial relationship between the points is the same, a set of points connected to form a part of a curve of a circle). The eight-shape is not a manifold because at the intersection point, the points do not form a connected curve (they form intersected line segments), which is not the same topology (connected points that form a curve). Both lines and circles are examples of 1D manifold (consists of 1D points). A manifold can be defined as a space that is not necessarily globally Euclidean but can be considered locally Euclidean. The word locally means small changes around an operating point. This concept is very important when we deal with rotation and direction cosine errors. A manifold can be of any dimension. As we will see shortly, the space of rotation matrices forms a manifold because at any rotation matrix value, there is a neighborhood of rotation matrices with the same topology. If the elements of a manifold are continuous, it is a differentiable manifold.

For now, let us keep in mind that $SO(3)$ group and the $SE(3)$ group are Lie Groups. Lie Group has useful mathematical properties that make the mathematical calculations of rotation and transformation linear, which significantly simplifies the solution algorithms that involve rigid body motion parameters in 3D space. In addition, the smoothness and continuity of the Lie Group enables us to describe any rotation as a concatenation of a large number of infinitesimally small rotations (infinitesimal means a very small quantity such that is close to zero than any other real number). Lie Group and manifold are important concepts to understand the mathematics of rigid body motion in 3D space which is explained in the next section.

1.5 The Mathematics of Rigid Body Motion in 3D Space

We discussed fundamental concepts of 3D space representation and Euclidean space, and we introduced the direction cosine matrix and the full transformation matrix that describes both rotation and translational motions of a rigid body in the Euclidean space. In this section, we will study differential equations that describe rigid body motion in 3D space. 3D motion is described by equations that model the rate of change of rotation and transformation matrices. These equations are fundamental tools to estimate objects' location and orientations.

1.5.1 Basic Definitions

We have seen how the $SE(3)$ (the Special Euclidean Group) is defined as a composition of an $SO(3)$ matrix $C_B^L \in SO(3)$ and a 3×1 vector $\vec{T}_B^L$. The matrix part C_B^L describes rotation (orientation of frame B with respect to frame L). The vector $\vec{T}_B^L$ describes translation (the coordinates of the origin of frame B with respect to frame L). Both rotation and translation can be represented in one 4D homogeneous coordinate system as follows:

$$T_B^L = \begin{pmatrix} C_B^L & \vec{T}_B^L \\ 0 & 1 \end{pmatrix} \tag{1-100}$$

In this 4D homogeneous coordinate system, 3×1 vectors $\vec{V} \in \mathbb{R}^3$ is represented by a 4×1 vector as follows:

$$\begin{pmatrix} \vec{V} \\ 1 \end{pmatrix} \in \mathbb{R}^{3+1} \tag{1-101}$$

As mentioned above $SO(3)$ group forms a Lie Group, which means elements form a continuous differentiable geometric structure where Euclidean space rules apply locally around every element, and multiplication/inversion form a continuous smooth mapping. The importance of the Lie Group concept is that at any element in the group, a special mapping can be designed such that this mapping will map the neighboring points to another domain without destroying the smoothness (continuity and differentiability) of the points. For example, at point P on the small curve in Figure 1.18, the neighboring curve points can be mapped to a straight line. The straight line preserves the original structural relationship between points and preserves the smoothness (i.e., continuity and differentiability) of the transformed points. This special mapping in is a straight line that is tangent to the curve at the point P.

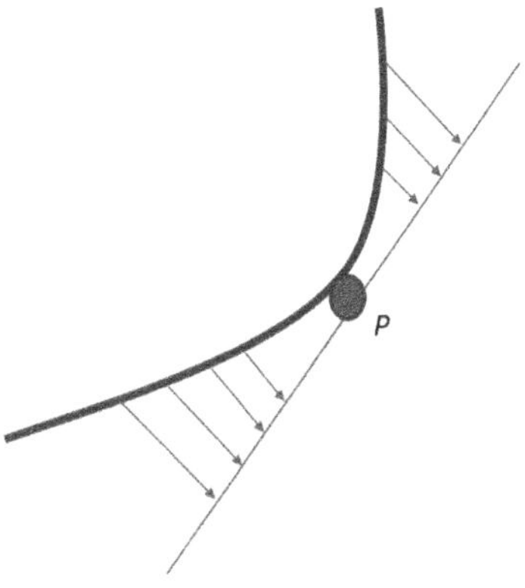

Figure 1.18 Mapping of curve points around point P to a straight line.

The mapping from this curved segment to a straight line makes the operation on the mapped points easier as it is now a linear domain. If the mapped points are close enough, this linear approximation will be accurate. Of course, the further the points apart are from the central point P, the less the linearization approximation accuracy would be. Also, the more aggressive (large slope) the curve is, the less accurate the linearization approximation would be. In the linear domain, several useful mathematical operations, and algorithms work. To apply linear operations, we need to map the rotation and transformation matrices from Lie Group to a linear group. In the next section, we will discuss how to perform this mapping and how this will help us in modeling and estimating rigid body rotational and translational errors and motion differential equations.

1.5.2 Modeling Rotation

1.5.2.1 Rotation Rate Equation

Modeling in this context means formulating a differential equation that describes how a certain quantity changes over time (rate of change). This differential equation will include necessary parameters and relationship between these parameters to describe the quantity. To describe how the rotation rate using a differential equation, we will start from the fundamental transformation equation that models the vector values between rotated frames:

$$\vec{V}^{L} = C_B^L \vec{V}^{B} \tag{1-102}$$

Recall that for all $C_B^L \in SO(3)$, the following condition applies:

$$C_B^L C_B^{L^T} = I \tag{1-103}$$

where I is the identity matrix, which is a constant.

Taking the time derivative of both sides gives us:

$$\frac{d\left(C_B^L C_B^{L^T}\right)}{dt} = \frac{dI}{dt} = 0 \tag{1-104}$$

$$\dot{C}_B^L C_B^{L^T} + \dot{C}_B^{L^T} C_B^L = 0 \tag{1-105}$$

$$\dot{C}_B^L C_B^{L^T} = -\dot{C}_B^{L^T} C_B^L \tag{1-106}$$

$$\dot{C}_B^L C_B^{L^T} = -\left(\dot{C}_B^L C_B^{L^T}\right)^T \tag{1-107}$$

where $\dot{x}$ is the time-derivative operator ($\dot{x} = \frac{dx}{dt}$).

Recall that if a matrix $A \in \mathbb{R}^{n \times n}$ satisfies the condition $A = -A^T$ this means that matrix A is a skew-symmetric matrix that has only three unique parameters. This means that the expression $\dot{C}_B^L C_B^{L^T}$ is a skew symmetric matrix that is characterized by only three parameters (review the definition of skew symmetric matrix in Section 1.2.2.6). This means that there exists a 3×1 vector $\vec{\omega}$ such that:

$$\dot{C}_B^L C_B^{L^T} = -\left(\vec{\omega} \times\right) \tag{1-108}$$

where $\vec{\omega} = (\omega_x \quad \omega_y \quad \omega_z)^T$ is a 3×1 vector and the skew symmetric matrix of $\vec{\omega}$ is defined as follows:

$$\left(\vec{\omega} \times\right) = \begin{pmatrix} 0 & -\omega_z & \omega_y \\ \omega_z & 0 & -\omega_x \\ -\omega_y & \omega_x & 0 \end{pmatrix} \tag{1-109}$$

Multiplying both sides of the equation $\dot{C}_B^L C_B^{L^T} = \left(\vec{\omega} \times\right)$ by C_B^L noting that $C_B^{L^T} = C_B^{L^{-1}}$

$$\dot{C}_B^L(t) = -\left(\vec{\omega}(t) \times\right) C_B^L(t) \tag{1-110}$$

Noting that $\left(\vec{V} \times\right) C = C\left(-\vec{V} \times\right)$, we get the following equation:

$$\dot{C}_B^L(t) = C_B^L(t)\left(\vec{\omega}(t) \times\right) \tag{1-111}$$

This is a very important discovery because it means that we can describe the rate of change of the direction cosine matrix (which is a nine-element matrix) by using only a 3×1 vector $\vec{\omega} \in \mathbb{R}^3$. The first-order approximation of this differential equation can be calculated by:

$$\dot{C}_B^L(t) \cong \frac{C_B^L(t + \Delta T) - C_B^L(t)}{\Delta T} \tag{1-112}$$

where ΔT is a very small period. This gives us the following approximation:

$$\frac{C_B^L(t + \Delta T) - C_B^L(t)}{\Delta T} \cong C_B^L(t)\left(\vec{\omega}(t) \times\right) \tag{1-113}$$

$$C_B^L(t + \Delta T) \cong \left(I + \Delta T\left(\vec{\omega}(t) \times\right)\right) C_B^L(t) \tag{1-114}$$

If $C_B^L(0) = I$ (initially, the two frames B and L are perfectly aligned, which means their direction cosine matrix is identity matrix I), we get the following equation:

$$C_B^L(\Delta T) = \Delta C_B^L \cong I + \left(\vec{\omega}(0) \times\right)\Delta T \tag{1-115}$$

where ΔC_B^L is the direction cosine matrix change corresponding to an infinitesimal small rotational change that happens through a small period ΔT.

This is also an important formula because it shows that ΔC_B^L corresponding to a small (infinitesimal) rotation in time ΔT can be calculated by a 3×3 skew-symmetric matrix space $sk(3)$ modeled by a 3×1 vector $\vec{\omega}$ such that:

$$sk(3) = \left\{ \left(\vec{\omega} \times \right) \mid \vec{\omega} \in \mathbb{R}^3 \right\} \tag{1-116}$$

We will refer to the vector $\vec{\omega}$ as the rotation rate vector. The reason we name it rotation rate is that it is multiplied by time change (ΔT) to get the full rotation change. The equation $C_B^L(\Delta T) = \Delta C_B^L \cong I + \left(\vec{\omega}(0) \times \right) \Delta T$ also means that the vector $\vec{\omega}$ is the rotation rate of frame B around or with respect to frame L. This is also important because as we will see later in inertial sensors chapter, this rotation rate vector can be sensed using three gyroscope sensors attached to the body or the axes of B. This small rotation change is centered around the identity matrix I.

1.5.2.2 Rotation Vector Change

Under the assumption that ΔT is infinitesimally small such that the following approximation formula applies:

$$\Delta C_B^L = C_B^L(\Delta T) \cong I + \left(\vec{\omega}(0) \times \right) \Delta T \tag{1-117}$$

If we multiply the rotation rate vector $\vec{\omega}$ by ΔT it will give a "rotation vector change". This rotation vector change is another simpler way to describe the orientation change of a frame B with respect to frame L. Therefore, we can embed the small-time change ΔT into the vector $\vec{\omega}$ and define a rotation vector $\vec{\Phi}$ as follows:

$$\Delta \vec{\phi} = \vec{\omega} \Delta T \tag{1-118}$$

The rotation vector change $\Delta \vec{\phi}(t)$ describes the rotation between two frames at time t using three parameters instead of nine parameters like the direction cosine matrix. The magnitude of the rotation vector change represents the angle of rotation, and the direction of the rotation vector represents the axis of rotation. The introduction of the rotation vector concept has a great mathematical advantage because it simplified the rotation parameters from a nine-element matrix to a three-element parameter. In addition, as we will see shortly, it maps the rotation from the Lie Group to a linear group, which is the mapping that we need to simplify mathematical operations that involve rotation between different frames.

1.5.2.3 Exponential Mapping

Another way to solve the differential equation $\dot{C}_B^L(t) = C_B^L(t) \left(\vec{\omega}(t) \times \right)$ given that $C_B^L(0) = I$ is the analytical exact solution instead of the first-order approximation solution that we described in the previous section. Given the initial condition $C_B^L(0) = I$, the solution of the differential equation $\dot{C}_B^L(t) = C_B^L(t) \left(\vec{\omega}(t) \times \right)$ given t is initially zero and $C_B^L(0) = I$ is given by the following equation:

$$C_B^L(t) = e^{\left(\vec{\omega}(t) \times \right) t} \tag{1-119}$$

If we embed t inside $\left(\vec{\omega}(t)\times\right)$ matrix, we get the following equation:

$$C_B^L(t) = e^{\left(\vec{\phi}(t)\times\right)} \tag{1-120}$$

The exponential of a matrix is not easy to calculate in a closed form. However, this exponential formula is in the form e^X, where X is a square matrix that can be calculated as a Tylor series as follows:

$$e^X = \sum_{k=0}^{\infty}\frac{1}{k!}X^k \tag{1-121}$$

Applying Tylor series on the function $e^{\left(\vec{\Phi}(t)\times\right)}$ form gives following solution:

$$C_B^L(t) = I + \left(\vec{\omega}(0)\times\right)t + \frac{\left(\left(\vec{\omega}(0)\times\right)t\right)^2}{2!} + \dots \tag{1-122}$$

If the time t is infinitesimally small ($t = \Delta T$), the squared terms and higher order terms will be extremely small compared to the first two terms. This will give us the same first order Tylor approximation of this solution given by

$$\Delta C_B^L = C_B^L(\Delta T) \cong I + \left(\vec{\omega}(0)\times\right)\Delta T \tag{1-123}$$

which is the same equation that we obtained previously in (1-115). Replacing $\vec{\omega}\Delta T$ by the rotation vector $\vec{\Phi}$ we get the following equation:

$$\Delta C_B^L = C_B^L(\Delta T) \cong I + \left(\Delta\vec{\phi}\times\right) \tag{1-124}$$

We can see that the exponential solution $C_B^L(t) = e^{\left(\vec{\phi}(t)\times\right)}$ mapped the rotation matrix $C_B^L(t)$ from the $SO(3)$ group (which is a Lie Group) into a linear vector space (the rotation vector $\vec{\phi}(t)$ space). This exponential mapping of $SO(3)$ to a linear vector space is called Lie Algebra. This Lie Algebra is a linear vector space in the sense that the summation of two rotation vectors, $\vec{\phi}_{AB}$ from frame A to frame B, and $\vec{\phi}_{BC}$ from frame B to frame C is also a rotation $\vec{\phi}_{AC}$ calculated as follows:

$$\vec{\phi}_{AC} = \vec{\phi}_{AB} + \vec{\phi}_{BC} \tag{1-125}$$

The exponential mapping maps the $SO(3)$ domain (Lie Group) to a linear rotation vector space $\vec{\phi}(t)$ (Lie Algebra). This exponential mapping has great mathematical advantage. To calculate multiple rotations, using direction cosine matrix, we need to multiply matrices. Matrix multiplication is very computationally expensive. On the other side, multiplying exponential terms can be easily done by adding the exponents. The addition operation is much more efficient than multiplication. It is easy to observe that in the $SO(3)$ domain, we cannot describe two consecutive rotations as the summation of two direction cosine matrices because the summation of two $SO(3)$ matrices will not generate an $SO(3)$ matrix. On the other side, in the rotation vector space and $sk(3)$ domain, two consecutive rotations can be described as the summation of two rotation rate vectors. This feature simplifies the mathematical equations that describe rigid body motion because vector summation is always easier than matrix multiplication. This concept is illustrated in Figure 1.19.

Figure 1.19 Exponential mapping.

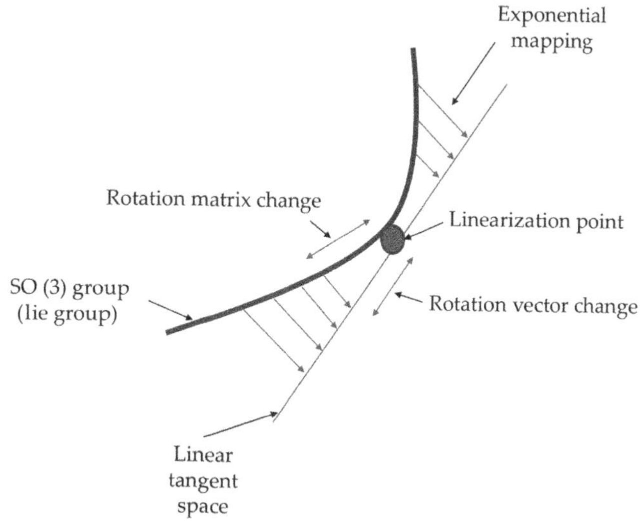

1.5.2.4 Rotation Vector and Rodrigues' Formula

From the above discussions, we realize that full rotation between frames B and L which can be described by 3×3 matrix C_B^L can also be described by a single rotation vector $\vec{\phi}_{LB}$, which is a great simplification and mathematical advantage. However, in the previous discussion, we calculate the change ΔC_B^L corresponding to an infinitesimally small change $\Delta\vec{\phi}_{LB}$ in rotation vector. We can concatenate multiple infinitesimally small rotation changes $\Delta\vec{\phi}_{LB}$ to get a full rotation vector $\vec{\phi}_{LB}$ corresponding to full directional cosine matrix C_B^L. This rotation is visualized in Figure 1.20, where we have two frames B and L initially aligned and then frame B is rotated around the unit rotation vector $\vec{u}_{\Phi_{LB}}$ by an angle ϕ. The figure shows the x-axis unit vector of frame B as projected in frame L defined as $\vec{u}_{xB}^{L}$ rotating around a rotation unit vector $\vec{u}_{\Phi_{LB}}$. This full rotation can be partitioned into multiple infinitesimally small changes $\Delta\vec{\phi}_{LB}$. From the geometry of the rotation, $\vec{u}_{xB}^{L}$ draws an arc around $\vec{u}_{\phi_{LB}}$. The radius of the arc is $\sin\theta$ and the magnitude of this small arc change $\Delta\phi$ cause the following change in $\vec{u}_{\phi_{LB}}$:

$$\left|\Delta\vec{u}_{xB}^{L}\right| = \Delta\phi\sin\theta \tag{1-126}$$

In vector form, this is a cross-product operation between $\vec{u}_{\phi_{LB}}$ and $\vec{u}_{xB}^{L}$ scaled by $\Delta\phi$. In vector form, $\Delta\vec{u}_{xB}^{L}$ can be calculated by the following cross-product:

$$\Delta\vec{u}_{xB}^{L} = \Delta\phi\vec{u}_{\Phi_{LB}} \times \vec{u}_{xB}^{L} \tag{1-127}$$

If we divide both sides by $\Delta\phi$, we get the following differential equation:

$$\frac{d\vec{u}_{xB}^{L}}{d\phi} \cong \frac{\Delta\vec{u}_{xB}^{L}}{\Delta\phi} = \vec{u}_{\phi_{LB}} \times \vec{u}_{xB}^{L} \tag{1-128}$$

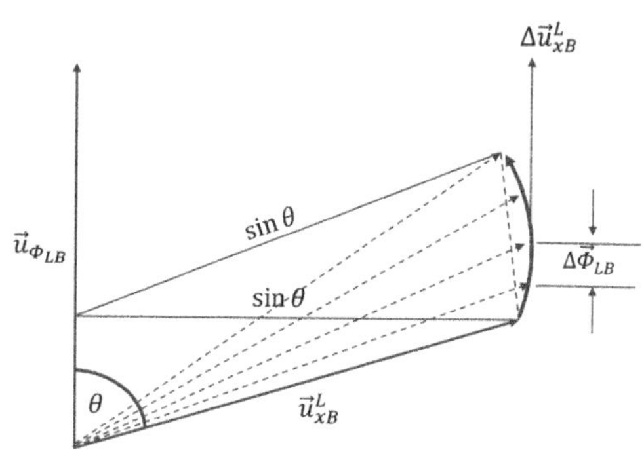

Figure 1.20 Rotation around a rotation vector.

This is a differential equation that relates $\vec{u}_{xB}^{L}$ to $\vec{u}_{\Phi_{LB}}$, which can be solved analytically leading to the following solution:

$$\vec{u}_{xB}^{L} = \left(I + \sin(\phi)\left(\vec{u}_{\phi_{LB}} \times\right) + (1 - \cos(\phi))\left(\vec{u}_{\phi_{LB}} \times\right)\left(\vec{u}_{\phi_{LB}} \times\right)\right)\vec{u}_{xB}^{L}(0) \tag{1-129}$$

The details of the solution of this differential equation can be found in [5]. Note that at $\phi = 0$, the two frames are perfectly aligned, which means $\vec{u}_{xB}^{L}(0) = \vec{u}_{xL}^{L}$ and the solution becomes as follows:

$$\vec{u}_{xB}^{L} = \left(I + \sin(\phi)\left(\vec{u}_{\phi_{LB}} \times\right) + (1 - \cos(\phi))\left(\vec{u}_{\phi_{LB}} \times\right)\left(\vec{u}_{\phi_{LB}} \times\right)\right)\vec{u}_{xL}^{L} \tag{1-130}$$

This is an important relationship that relates the unit vector $\vec{u}_{xB}^{L}$ to the unit vector $\vec{u}_{xL}^{L}$ given that there is a rotation between the two frames by ϕ around a unit rotation vector $\vec{u}_{\Phi_{LB}}$. This relationship can be generalized to all unit vectors as follows:

$$\vec{u}_{yB}^{L} = \left(I + \sin(\phi)\left(\vec{u}_{\phi_{LB}} \times\right) + (1 - \cos(\phi))\left(\vec{u}_{\phi_{LB}} \times\right)\left(\vec{u}_{\phi_{LB}} \times\right)\right)\vec{u}_{yL}^{L} \tag{1-131}$$

$$\vec{u}_{zB}^{L} = \left(I + \sin(\phi)\left(\vec{u}_{\phi_{LB}} \times\right) + (1 - \cos(\phi))\left(\vec{u}_{\phi_{LB}} \times\right)\left(\vec{u}_{\phi_{LB}} \times\right)\right)\vec{u}_{zL}^{L} \tag{1-132}$$

Recall that the direction cosine matrix C_B^L is a relationship between the unit vectors of frames B and L, which is modeled in the above equations as a function of the rotation vector $\vec{\phi}_{LB}$. This gives a relationship between C_B^L and $\vec{\phi}_{LB}$ as follows:

$$C_B^L = I + \frac{\left(\vec{\phi}_{LB} \times\right)}{\left\|\vec{\phi}_{LB}\right\|}\sin\left(\left\|\vec{\phi}_{LB}\right\|\right) + \frac{\left(\vec{\phi}_{LB} \times\right)^2}{\left\|\vec{\phi}_{LB}\right\|^2}\left(1 - \cos\left(\left\|\vec{\phi}_{LB}\right\|\right)\right) \tag{1-133}$$

This relationship is known as Rodrigues' Formula that calculates the direction cosine matrix C_B^L from the rotation vector $\vec{\phi}_{LB}$.

1.5.2.5 Logarithmic Mapping

The exponential mapping maps from $SO(3)$ domain to the rotation vector $sk(3)$ domain. It is also important to calculate the inverse of this exponential mapping. It is straightforward to calculate rotation vector $\vec{\Phi}(t)$ from $C_B^L(t)$ by taking the logarithm as follows:

$$C_B^L(t) = e^{\left(\vec{\phi}_{LB} \times\right)} \tag{1-134}$$

$$\left(\vec{\phi}_{LB}(t) \times\right) = \log\left(C_B^L(t)\right) \tag{1-135}$$

A solution for this equation is a rotation vector $\vec{\phi}_{LB}$ as follows:

$$\left\|\vec{\phi}_{LB}\right\| = \cos^{-1}\left(\frac{trace\left(C_B^L\right) - 1}{2}\right) \tag{1-136}$$

$$\vec{u}_{LB} = \frac{\left\|\vec{\phi}_{LB}\right\|}{\vec{\phi}_{LB}} = \frac{1}{2\sin\left(\left\|\vec{\phi}_{LB}\right\|\right)}\begin{pmatrix} C_B^L(3,2) - C_B^L(2,3) \\ C_B^L(1,3) - C_B^L(3,1) \\ C_B^L(2,1) - C_B^L(1,2) \end{pmatrix} \tag{1-137}$$

The full rotation vector is given by:

$$\vec{\phi}_{LB} = \left\|\vec{\phi}_{LB}\right\|\vec{u}_{LB} = \begin{pmatrix}\phi_x & \phi_y & \phi_z\end{pmatrix}^T \tag{1-138}$$

1.5.2.6 Euler Angles

Another way to describe the rotation transformation between two frames is by rotating one frame in three ordered rotations around its axes until it is aligned with the other frame. Starting from frame A, we will perform the following rotations in sequence as follows: Rotate frame A around its Z-axis by an angle A leading to frame A_1. The direction cosine matrix between A and A_1 is defined as

$$C_{A_1}^{A} = \begin{pmatrix} \cos(A) & -\sin(A) & 0 \\ \sin(A) & \cos(A) & 0 \\ 0 & 0 & 1 \end{pmatrix} \tag{1-139}$$

We obtained $C_{A_1}^{A}$ using the observation that each column in $C_{A_1}^{A}$ is the projection of the unit vectors of A_1 on the basis of A. For example, the first unit vector of A_1 ($\vec{I}_{xA_1}$) projected on the three axes of A is given by (see Figure 1.21).

$$\begin{aligned} \vec{I}_{xA}\cdot\vec{I}_{xA_1} &= \cos(A) \\ \vec{I}_{yA}\cdot\vec{I}_{xA_1} &= \cos\left(\frac{\pi}{2} - A\right) = \sin(A) \\ \vec{I}_{zA}\cdot\vec{I}_{xA_1} &= \cos\left(\frac{\pi}{2}\right) = 0 \end{aligned} \tag{1-140}$$

Now, if we rotated A_1 around its Y-axis by an angle p this will lead to a new frame A_2 where the direction cosine matrix $C_{A_2}^{A_1}$ defined as follows:

$$C_{A_2}^{A_1} = \begin{pmatrix} \cos(p) & 0 & \sin(p) \\ 0 & 1 & 0 \\ -\sin(p) & 0 & \cos(p) \end{pmatrix} \tag{1-141}$$

If we rotate frame A_2 around its X-axis by an angle, we get the target frame B and the direction cosine matrix $C_{A_2}^{B}$ defined by:

$$C_B^{A_2} = \begin{pmatrix} 1 & 0 & 0 \\ 0 & \cos(r) & -\sin(r) \\ 0 & \sin(r) & \cos(r) \end{pmatrix} \tag{1-142}$$

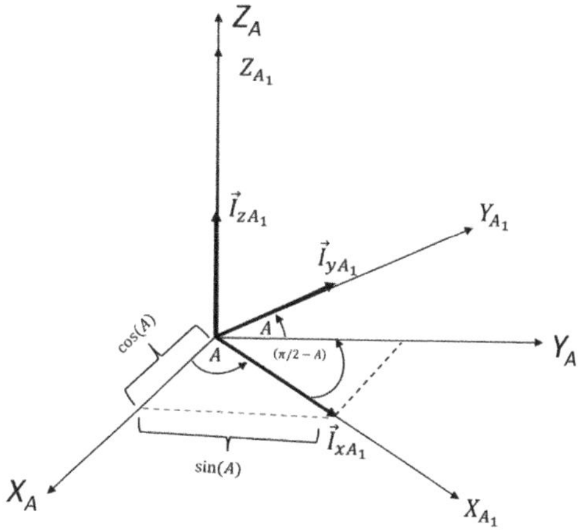

Figure 1.21 Rotation of frame A to frame A_1 around Z_A by angle A. Note the projection of unit vectors of A_1 on the axes of A.

Now, we can obtain the direction cosine matrix C_B^A by multiplying the three cosine matrices as follows:

$$C_B^A = C_{A_1}^A C_{A_2}^{A_1} C_B^{A_2} \tag{1-143}$$

$$C_B^A = \begin{pmatrix} cp\,cA & -cr\,sA + sr\,sp\,cA & sr\,sA + cr\,sp\,cA \\ cp\,sA & cr\,cA + sr\,sp\,sA & -sr\,cA + cr\,sp\,sA \\ -sp & sr\,cp & cr\,cp \end{pmatrix}$$
$$\begin{aligned} cp &= \cos(p) \\ cr &= \cos(r) \\ cA &= \cos(A) \\ sp &= \sin(p) \\ sr &= \sin(r) \\ sA &= \sin(A) \end{aligned} \tag{1-144}$$

The three angles r, p, and A are called *roll*, *pitch*, and *Azimuth* and they uniquely determine the orientation of frame B with respect to frame A and they are called the Euler angles. In some references, Azimuth angle is also called heading. In this book, we will use the two words interchangeably. An interesting fact is that if the Euler angles are infinitesimally small (δr, δp, and δA), the Euler angles form a rotation vector, and the resulting rotation matrix C_B^A will be calculated as follows:

$$\begin{aligned} &\cos(\delta r) \cong 1 \\ &\cos(\delta p) \cong 1 \\ &\cos(\delta A) \cong 1 \\ &\sin(\delta r) \cong r \\ &\sin(\delta p) \cong p \\ &\sin(\delta A) \cong A \\ &\sin(\delta r)\sin(\delta p) \cong 0 \\ &\sin(\delta r)\sin(\delta p)\sin(\delta A) \cong 0 \end{aligned} \tag{1-145}$$

Substituting these approximation formulas into C_B^A gives us:

$$\delta C_B^A \cong \begin{pmatrix} 1 & -\delta A & \delta p \\ \delta A & 1 & -\delta r \\ -\delta p & \delta r & 1 \end{pmatrix} \tag{1-146}$$

$$\delta C_B^A \cong \begin{pmatrix} 1 & 0 & 0 \\ 0 & 1 & 0 \\ 0 & 0 & 1 \end{pmatrix} + \begin{pmatrix} 0 & -\delta A & \delta p \\ \delta A & 0 & -\delta r \\ -\delta p & \delta r & 0 \end{pmatrix} \tag{1-147}$$

$$\delta C_B^A \cong I + \left(\begin{pmatrix} \delta r \\ \delta p \\ \delta A \end{pmatrix} \times \right) \tag{1-148}$$

This interesting note is important when modeling the error and the rate of change of Euler angles.

1.5.2.7 Quaternion

Although the rotation representation using direction cosine matrix and Euler angles are easy to understand and convenient ways to describe rotations and orientations of frames with respect to each other, they introduce several numerical issues in practical implementation on computers due to the inherent singularities in the trigonometric functions. For example, if the pitch angle p reaches 90°, we get a situation called "gimbal lock" where the Azimuth rotation and the roll rotation are no longer distinguishable. Quaternion is another domain where rotation can be defined without singularities, gimbal lock, and with less numerical issues. Quaternion can be seen as another form of transformation between frames. Quaternion is an extension of the complex form representation of vectors [5]. A vector $v = (v_x\, v_y)^T$can be represented as a real part and an imaginary part as follows:

$$v = v_x + jv_y \tag{1-149}$$

$$j \cdot j = j^2 = -1 \tag{1-150}$$

$$\|v\| = \sqrt{{v_x}^2 + {v_y}^2} \tag{1-151}$$

If we have another vector $u = (a\ b)^T$defined as follows:

$$\begin{aligned} u &= a + jb \\ a &= \cos\theta \\ b &= -\sin\theta \end{aligned} \tag{1-152}$$

The complex multiplication of v and u we get the following form:

$$\begin{aligned} w &= uv \\ w &= \left(v_x + jv_y\right)(a + jb) \\ w &= \left(v_x + jv_y\right)(\cos\theta - j\sin\theta) \\ w &= \left(v_x\cos\theta + v_y\sin\theta\right) + j\left(v_y\ \cos\theta - v_x\sin\theta\right) \\ w &= \begin{pmatrix} w_x \\ w_y \end{pmatrix} = \begin{pmatrix} \cos\theta & \sin\theta \\ -\sin\theta & \cos\theta \end{pmatrix} \begin{pmatrix} v_x \\ v_y \end{pmatrix} \end{aligned} \tag{1-153}$$

So, the complex multiplication with vector $u = a + jb$,where $a = \cos\theta$ and $b = -\sin\theta$ has the effect of rotating the vector v by an angle θ around a vector perpendicular to the plane that contains v and u. Here, we represented a single 2D rotation around a single axis by a 2D vector u defined as follows:

$$u = \begin{pmatrix} \cos\theta \\ -\sin\theta \end{pmatrix} \tag{1-154}$$

The quaternion extends this simple concept to the 3D vectors domain representing the rotation by a 4D rotation vector. This is simply the quaternion transformation theory defined as follows: If a frame B is rotating around a rotation vector $\vec{\phi}_{LB} = \left(\phi_x \quad \phi_y \quad \phi_z\right)^T$ with respect to frame L, the transformation between B and L can be represented by a four-component quaternion vector defined as follows:

$$\begin{aligned}
\vec{q}^{\,L}_B &= (a \quad b \quad c \quad d)^T \\
a &= \cos\left(\frac{\left\|\vec{\phi}_{LB}\right\|}{2}\right) \\
b &= \frac{\phi_x}{\left\|\vec{\phi}_{LB}\right\|}\sin\left(\frac{\left\|\vec{\phi}_{LB}\right\|}{2}\right) \\
c &= \frac{\phi_y}{\left\|\vec{\phi}_{LB}\right\|}\sin\left(\frac{\left\|\vec{\phi}_{LB}\right\|}{2}\right) \\
d &= \frac{\phi_z}{\left\|\vec{\phi}_{LB}\right\|}\sin\left(\frac{\left\|\vec{\phi}_{LB}\right\|}{2}\right)
\end{aligned} \tag{1-155}$$

The quaternion transformation vector $\vec{q}^{\,L}_B$ can be used to transform a vector $\vec{V}^{B} = \left(v_{Bx}\, v_{By}\, v_{Bz}\right)^T$ from frame B to frame L by the following transformation formula:

$$\vec{V}^{L} = q^L_B \quad \vec{V}^{B} \quad q^{L*}_B \tag{1-156}$$

The terms of Eq. (1-156) are 4D vectors defined as follows:

$$\begin{aligned}
\vec{V}^{B} &= (0 \quad v_{Bx} \quad v_{By} \quad v_{Bz})^T \\
\vec{V}^{L} &= (0 \quad v_{Lx} \quad v_{Ly} \quad v_{Lz})^T \\
\vec{q}^{\,L}_B &= (a \quad b \quad c \quad d)^T \\
\vec{q}^{\,L*}_B &= (a \quad -b \quad -c \quad -d)^T
\end{aligned} \tag{1-157}$$

The operator $*$ is the complex conjugate and the operator $\circ$ is used to describe the "4D vector" quaternion product operator, which is defined in matrix form as follows:

$$\vec{q}^{\,L}_B \circ \vec{V}^{B} = \begin{pmatrix} a & -b & -c & -d \\ b & a & -d & c \\ c & d & a & -b \\ d & -c & b & a \end{pmatrix} \vec{V}^{B} \tag{1-158}$$

$$\vec{V}^{B} \circ \vec{q}^{\,L*}_B = \begin{pmatrix} a & b & c & d \\ -b & a & -d & c \\ -c & d & a & -b \\ -d & -c & b & a \end{pmatrix} \vec{V}^{B} \tag{1-159}$$

and

$$\vec{V}^{L} = \vec{q}_{B}^{L*}\vec{V}^{B*}\vec{q}_{B}^{L*} = \begin{pmatrix} a & -b & -c & -d \\ b & a & -d & c \\ c & d & a & -b \\ d & -c & b & a \end{pmatrix}\begin{pmatrix} a & b & c & d \\ -b & a & -d & c \\ -c & d & a & -b \\ -d & -c & b & a \end{pmatrix}\vec{V}^{B} \tag{1-160}$$

$$\vec{V}^{L} = \begin{pmatrix} (a^2+b^2+c^2+d^2) & 0 & 0 & 0 \\ 0 & (a^2+b^2-c^2-d^2) & 2(bc-ad) & 2(bd+ac) \\ 0 & 2(bc+ad) & (a^2-b^2+c^2-d^2) & 2(cd-ab) \\ 0 & 2(bd-ac) & 2(cd+ab) & (a^2-b^2-c^2+d^2) \end{pmatrix}\vec{V}^{B} \tag{1-161}$$

which gives the direction cosine matrix from quaternion vector as follows:

$$\begin{aligned} C_B^L(1,1) &= (a^2+b^2-c^2-d^2) \\ C_B^L(1,2) &= 2(bc-ad) \\ C_B^L(1,3) &= 2(bd+ac) \\ C_B^L(2,1) &= 2(bc+ad) \\ C_B^L(2,2) &= (a^2-b^2+c^2-d^2) \\ C_B^L(2,3) &= 2(cd-ab) \\ C_B^L(3,1) &= 2(bd-ac) \\ C_B^L(3,2) &= 2(cd+ab) \\ C_B^L(3,3) &= (a^2-b^2-c^2+d^2) \end{aligned} \tag{1-162}$$

where $C_B^L(i,j)$ is the element located at the ith row and j^{th} column of the matrix C_B^L.

As these formulas show, we can represent rotation using four different ways, direction cosine matrix, rotation vector, Euler angles, and quaternion. Each method has its advantages and limitations.

1.5.3 Modeling Rotation and Translation

To complete the modeling of the full rigid body motion, the translation motion (the displacement between the origins of two frames) must be included. In this section, we show that including the translation motion preserves the Lie Group and Lie Algebra characteristics. The full rigid body motion is described using transformation between frames. The transformation between frames belongs to the Special Euclidean transformations $SE(3)$ defined as follows:

$$SE(3) \equiv \left\{ g = \left(R, \vec{T}\right) \mid R \in SO(3), \vec{T} \in \mathbb{R}^3 \right\} \tag{1-163}$$

Assuming the body has a frame B and we want to describe the motion of the body with respect to a reference frame L. The rotational relationship between the body B and the reference frame L using the rotation matrix (direction cosine matrix) $C_B^L \in SO(3)$. As we have seen in the previous section, the C_B^L can be mapped into a 3×1 rotation vector $\vec{\Phi}_B^L \in \mathbb{R}^3$. The translational part is modeled by a vector $\vec{T}_B^L \in \mathbb{R}^3$ defined as the coordinate of the origin of frame B with respect to frame L as can be seen in Figure 1.22.

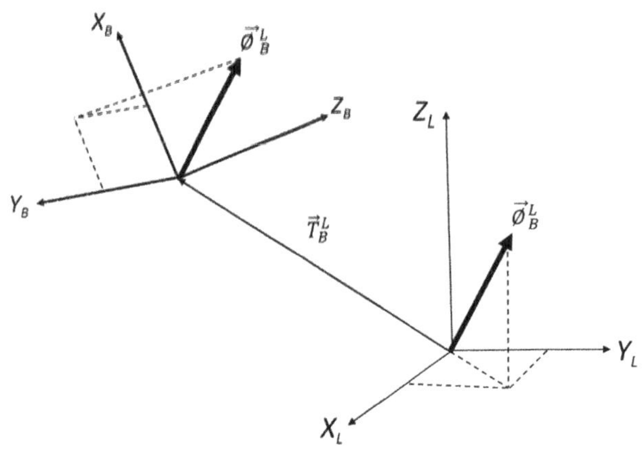

Figure 1.22 Describing the full rigid body motion of frame B with respect to frame L.

The $SE(3)$ transformation between frame B and frame L can be described in a homogeneous form as follows:

$$SE(3) \equiv \left\{ T^L_B = \begin{pmatrix} C^L_B & \vec{T}^L_B \\ 0 & 1 \end{pmatrix}_{4\times4} \mid C^L_B \in SO(3), \vec{T}^L_B \in \mathbb{R}^3 \right\} \tag{1-164}$$

We will follow the same scheme that we followed in the previous section to show that $SE(3)$ is a Lie Group where Lie Algebra is the exponential mapping. We begin by calculating $\dot{T}^L_B(T^L_B)^{-1}$. It can be easily shown that the inverse of T^L_B is defined by:

$$(T^L_B)^{-1} = \begin{pmatrix} (C^L_B)^T & -(C^L_B)^T \vec{T}^L_B \\ 0 & 1 \end{pmatrix}_{4\times4} \tag{1-165}$$

The rate of change of the transformation matrix T^L_B can be calculated as:

$$\dot{T}^L_B = \begin{pmatrix} \dot{C}^L_B & \left(\dot{\vec{T}}^L_B\right) \\ 0 & 0 \end{pmatrix}_{4\times4} \tag{1-166}$$

Multiplying both sides by the inverse of T^L_B we get

$$\dot{T}^L_B(T^L_B)^{-1} = \begin{pmatrix} \dot{C}^L_B(C^L_B)^T & \left(\dot{\vec{T}}^L_B\right) - \dot{C}^L_B(C^L_B)^T \vec{T}^L_B \\ 0 & 0 \end{pmatrix} \tag{1-167}$$

Like what we have done in the previous section, we will replace $\dot{C}^L_B(C^L_B)^T$ by skew symmetric matrix of a rotation rate vector $\left(\vec{\omega}\times\right)$ and velocity vector $\vec{V}$ defined as follows:

$$\vec{V} = \left(\dot{\vec{T}}^L_B\right) - \left(\vec{\omega}\times\right)\vec{T}^L_B \tag{1-168}$$

This relationship is very important. It describes how the translation motion vector $\vec{T}_B^L$ changes with time. It changes with time due to two factors; one factor is its own velocity in its own frame $\left(\dot{\vec{T}}_B^L\right)$ and another factor is the rotation of frame B with respect to frame L. This second factor is expressed by the term $-\left(\vec{\omega}\times\right)\vec{T}_B^L$. The value of $\dot{T}_B^L\left(T_B^L\right)^{-1}$ can then be described as:

$$\dot{T}_B^L\left(T_B^L\right)^{-1} = \begin{pmatrix} \left(\vec{\omega}\times\right) & \vec{V} \\ 0 & 0 \end{pmatrix} = \mathcal{F} \in \mathbb{R}^{4\times4} \tag{1-169}$$

Multiplying both sides by T_B^L gives us:

$$\dot{T}_B^L = \mathcal{F}T_B^L \tag{1-170}$$

This is the same differential equation form that describes the rotation rate. If we set initial condition $T_B^L(0) = I$, we would have the same exponential solution form like in the rotation rate defined as follows:

$$T_B^L = e^{\mathcal{F}t} \tag{1-171}$$

The exponential $e^{\mathcal{F}t}$ can be calculated by:

$$e^{\mathcal{F}t} = \sum_{i=0}^{\infty}\frac{(\mathcal{F}t)^n}{n!} \tag{1-172}$$

which can be approximated for infinitesimally small changes through time $t = \Delta T$ in in the same way the $e^{\left(\vec{\omega}(t)\times\right)t}$ was approximated.

$$T_B^L(\Delta T) \cong I + \mathcal{F}\Delta T \tag{1-173}$$

In conclusion, the transformation matrix T_B^L models the rotational and translational positioning (pose) of frame B with respect to frame L. The transformation matrix T_B^L has two parts, a rotation matrix (direction cosine matrix) C_B^L and a translation vector $\vec{T}_B^L$. The translation vector is the coordinates of the origin point of frame B with respect to frame L as projected in frame L. The group of transformation matrices T_B^L form a Lie Group. The rate of change of the transformation matrix T_B^L can be modeled starting from the identity matrix where the two frames are aligned with each other. Starting from the identity matrix, the rate of change of the rotation matrix of frame B with respect to frame L (C_B^L) and the velocity of the translation vector T_B^L can be modeled (described by a differential equation) that is fully described using six parameters, a 3×1 rotation rate vector $\vec{\omega}$ and a 3×1 translational rate vector $\vec{V}$. $\vec{\omega}$ models the rotation rate of the body frame (B) with respect to a reference frame L as projected in the L frame. $\vec{V}$ models the velocity of the origin point of the body frame B with respect to the reference frame L as projected in the L frame. When the motion is infinitesimally small, the rotation and translation motion of frame B with respect to frame L can be approximated by Lie Algebra. Lie Algebra is the exponential mapping that maps the 4×4 transformation matrix T_B^L into six parameters ($\vec{\omega}$ and $\vec{V}$). If the infinitesimally small change in the transformation matrix T_B^L happens through a small period ΔT, this small change can be approximately calculated using the following first-order approximation equations:

$$C_B^L(\Delta T) \cong I + \left(\vec{\omega} \times\right)\Delta T \tag{1-174}$$

$$\vec{T}_B^L(\Delta T) = \vec{T}_B^L(0) + \vec{V}\Delta T \tag{1-175}$$

The multiplication of $\vec{\omega}$ by ΔT can be used to describe the rotation vector $\vec{\Phi}$ that describes this infinitesimally small rotation as a single rotation by an angle $\left\|\vec{\Phi}\right\|$ around the unit vector $\vec{U}_\Phi = \frac{\vec{\Phi}}{\left\|\vec{\Phi}\right\|}$. Similarly, the multiplication of $\vec{V}\Delta T$ gives the infinitesimally small translation change through period ΔT. It is important to keep in mind that the rotation and translation rate equations (the 3D kinematic models of rigid bodies) are all depending on the assumption that the change in the motion is infinitesimally small to keep the conditions of the Lie Group, which are continuity and smoothness of the multiplication and inversion. This puts a lot of pressure on the processing rate when we implement motion equations on a digital discrete processing computer. The processing rate needs to be high enough to assure that the change in the motion from epoch to epoch is infinitesimally small. If the dynamics of the object are slow, the processing rate can be relaxed. If the dynamics of the object is high or the maneuvering of the object has spikes or sudden sharp changes, this will affect the continuity and smoothness conditions, which make the motion differential equations less accurate. We will analyze this when we study odometry algorithms using inertial sensors and vision sensors.

1.6 Problems

1.6.1 Problem 1

Given the description of the N and L local-level tangent frames, what is the rotation matrix between the L frame and the N frame?

1.6.2 Problem 2

Given an object B that is oriented by a roll angle (0°), pitch angle (0°), and heading angle (−15°) with respect to L frame, calculate the direction cosine matrix C_B^L of the object with respect to the L frame.

1.6.3 Problem 3

Calculate the quaternion vector of the object described in Problem 2.

1.6.4 Problem 4

Given an object located at a latitude/longitude coordinate (−32.830,774°, – 68.792,782°), calculate the rotation matrix that describes the rotation of N frame at this location with respect to Earth Centered Earth Fixed E frame.

1.6.5 Problem 5

Given the following rotation matrix that describes the location of an object with respect to the E frame:

$$C_N^E = \begin{pmatrix} 0.3617 & -0.5053 & -0.7835 \\ 0 & 0.8404 & -0.5420 \\ 0.9323 & 0.1961 & 0.3040 \end{pmatrix}$$

Calculate the latitude/longitude coordinate of the object in degrees.

1.6.6 Problem 6

Consider an object at the following location given in geodetic coordinates:

$$\varphi = -32.8211099^\circ$$
$$\lambda = -68.7948642^\circ$$
$$h = 721 \text{ m}$$

where (φ, λ, h) are geodetic latitude (in degrees), longitude (in degrees), and altitude (in meters), respectively. Given the following reference location:

$$\varphi_r = -32.830774^\circ$$
$$\lambda_r = -68.792782^\circ$$

Calculate the meridian R_M and normal R_N Earth radius of curvature at the object's location and calculate the north and east coordinate of the objects in meters.

1.6.7 Problem 7

Given rotation matrix C_B^L of an object B with respect to L frame:

$$C_B^L = \begin{pmatrix} -0.9628 & -0.2701 & 0.0011 \\ 0.2701 & -0.9628 & -0.0053 \\ 0.0025 & -0.0048 & 1 \end{pmatrix}$$

which is corresponding to the following quaternion vector:

$$\vec{q}_B^{\,L} = \begin{pmatrix} 0.1363 \\ 0.0009 \\ -0.0025 \\ 0.9907 \end{pmatrix}$$

Calculate the rotation vector $\vec{\phi}_{LB} = (\phi_x\ \phi_y\ \phi_z)^T$ corresponding to this rotation and verify that the following relationship between $\vec{\phi}_{LB}$ and $\vec{q}_B^{\,L}$ holds:

$$\vec{q}_B^{\,L}(1) = \cos\left(\frac{\left\|\vec{\phi}_{LB}\right\|}{2}\right)$$

$$\vec{q}_B^{\,L}(2) = \frac{\phi_x}{\left\|\vec{\phi}_{LB}\right\|}\sin\left(\frac{\left\|\vec{\phi}_{LB}\right\|}{2}\right)$$

$$\vec{q}_B^{\,L}(3) = \frac{\phi_y}{\left\|\vec{\phi}_{LB}\right\|}\sin\left(\frac{\left\|\vec{\phi}_{LB}\right\|}{2}\right)$$

$$\vec{q}_B^{\,L}(4) = \frac{\phi_z}{\left\|\vec{\phi}_{LB}\right\|}\sin\left(\frac{\left\|\vec{\phi}_{LB}\right\|}{2}\right)$$

1.6.8 Problem 8

Given an object B rotated with respect to L frame by the following rotation vector:

$$\vec{\phi}_{LB} = \begin{pmatrix} 0.0002 \\ 0.0007 \\ 2.8591 \end{pmatrix}$$

Use Rodrigues' formula to calculate the direction cosine matrix C_B^L.

1.6.9 Problem 9

Calculate the ECEF rectangular coordinate of an object located on Earth at the following geodetic coordinate:

$$\varphi = 43.6538^\circ$$
$$\lambda = -79.3799^\circ$$
$$h = 98.1674 \text{ m}$$

1.6.10 Problem 10

If a traveler was at Greenwich in England at 10:00 a.m. England Time and the traveler sailed toward an unknown city. If the journey took exactly 10 hours and when the traveler arrived, the local time was 15:00 (3:00 p.m.), calculate the longitude of the city the traveler arrived at, noting that Greenwich is the zero-longitude line.

References

1 Y. Ma, S. Soatto, J. Košecká and S. S. Sastry, "*An Invitation to 3-D Vision, From Images to Geometric Models,*" Springer – Science+Business Media, LLC, New York, 2004.

2 M. L. O'Leary, "*Linear Algebra,*" Hoboken, NJ: Wiley, 2021.

3 C. E. Rasmussen and C. K. I. Williams, "*Gaussian Processes for Machine Learning*", Cambridge: MIT Press, 2006.

4 E. Provenzi, "*From Euclidean to Hilbert Spaces: Introduction to Functional Analysis and its Applications,*" Hoboken, NJ: Wiley, 2021.

5 P. Savage, "*Starpdown Analytics,*" Maple Plain, MN: Starpdown Associates Inc., 2007.

6 The International Earth Rotation Service (IERS), "Science background", [Online]. Available: https://www.iers.org/IERS/EN/Science/science.html. [Accessed 30th August 2024].

7 P. D. Groves, "*Principles of GNSS, Inertial, and Multisensor Integrated Navigation Systems,*" Boston, MA: Artech House, 2013.

8 T. Wolfgang, M. Jürgen, M. Jürgen, "*Geodesy,*" Fourth Edition, Berlin, Germany: De Gruyter, 2012.

9 National Geospatial-Intelligence Agency (NGA), "World Geodetic System 1984, Its Definition and Relationships with Local Geodetic Systems," National Geospatial-Intelligence Agency (NGA), Office of Geomatics, NGA.STND.0036_1.0.0_WGS84, 2014-07-08 version 1.0.0.

10 A. Wedel and D. Cremers, "*Stereo Scene Flow for 3D Motion Analysis,*" New York: Springer – Verlag London Limited, 2011.

11 D. Cremers, "Multiple View Geometry, Course Notes," Departments of Informatics Mathematics, Technical University of Munich, 2019.

2

Signals and Systems

2.1 Introduction

Navigation systems integrate measurements from different sensors. Sensor measurements are signals that are propagating through connected subsystems where a series of processing is applied to filter noise, fuse measurements, and extract useful information. In our case, the useful information that we are interested in is the platform's positioning and the map of the environment. Therefore, it is key to understanding the basics of signals and systems which are briefly reviewed in this chapter.

2.2 Probability

Sensor measurements are always contaminated by different errors. Part of these errors is random, leading to stochastic measurements. Another part is systematic or deterministic. Sensor fusion methods and algorithms use deterministic mathematical models to deal with the deterministic components. The random and stochastic components are dealt with using probability concepts. To understand randomness in sensor measurements and how to model it, we need to review basic probability concepts, random variables, and random processes. More detailed information about probability concepts can be found in [1] and [2].

2.2.1 Sample Points, Sample Space, and Events

Random quantities are the outcome of an uncertain process such as flipping a coin. Sample space is the set S of all possible outcomes of a random process (experiment). A possible outcome $s \subset S$ of an experiment is called a sample point. A subspace $e \subseteq S$ is called an "*event*." An elementary event is a single sample point $s \subset S$. For example, if we roll a uniform six-face dice once, the sample space is $S = \{$"1", "2", "3", "4", "5", "6"$\}$. A possible event in this example could be "*observing a prime number*." This event can be defined as a subspace of the sample space $e_p \in S = \{$"2", "3", "5"$\}$. Two events A and B are called mutually exclusive (disjoint) if they cannot occur at the same time. Two events A and B are called independent if the occurrence of one does not affect the possibility of the occurrence of the other.

Sensor Fusion Approaches for Positioning, Navigation, and Mapping: How Autonomous Vehicles and Robots Navigate in the Real World: With MATLAB Examples, First Edition. Mohamed M. Atia.
© 2025 The Institute of Electrical and Electronics Engineers, Inc. Published 2025 by John Wiley & Sons, Inc.

Table 2.1 Probability distribution example. Process: Flipping a coin once and counting the number of heads.

s	*P*(*s*)	Sequence
0	1/4	Tail-tail
1	2/4	Head-tail, tail-head
2	1/4	Head-head

2.2.2 Probability Distribution

Informally, probability is a value in the range [0, 1] that expresses how likely an event may occur. For example, if we roll a dice (uniform six-face dice) once, the sample space is $S = \{$"1", "2", "3", "4", "5", "6"$\}$. One possible event is observing "1." Because the dice is uniform, all faces are equally likely and the probability of observing "1" is simply $P(\{$ "1"$\}) = 1/6$. One possible interpretation of probability is frequency, where the probability of an event is the count of all possible occurrences of the event divided by the sum of counts of occurrences of all disjoint events. We should be careful when we use this definition. In the dice example, events are disjoint, and the number of occurrences of each event is the same, which is 1. In other experiments, it could be more complicated. For example, if we flip a coin twice and count the number of heads, the sample space would be $S = \{0, 1, 2\}$ and the probability of all sample points is shown in Table 2.1. In this example, the probabilities of sample points are not equal because the possible count of observing one head is two (we may get a "*head-tail*" sequence or a "*tail-head*" sequence). The probability of all sample points is called probability distribution. The probability distribution of our example is shown in Figure 2.1.

2.2.3 Discrete vs. Continuous Probability

If the sample space has definite discrete values, the probability distribution is called probability mass function (PMF) and the variable is called a discrete random variable. An example of a discrete random variable is the outcome of the experiment of flipping a coin twice. The PMF of this experiment is shown in Table 2.1 and its PMF is shown in Figure 2.2a.

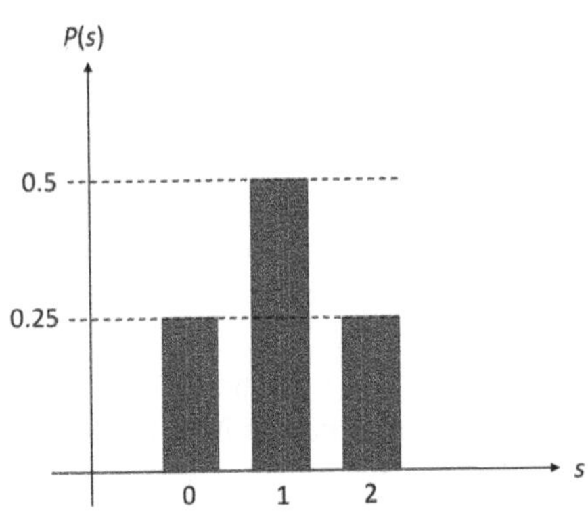

Figure 2.1 Probability distribution of flipping a coin twice and counting the number of heads.

If the sample points are continuous (we have an infinite number of sample points), the random variable is called a continuous random variable and the probability distribution over the sample points is called probability density function (PDF). For example, assume we have range sensors with an error $\pm e$, the output of the sensor will be a continuous random variable whose samples lie somewhere in the range $[d - e,\ d + e]$, where d is the true range. Example PDF $f_X(x)$ of this range sensor is shown in Figure 2.2b. Note that in continuous domain, as the sample point is infinite, probability distribution is defined as a continuous curve where we can only calculate the probability of the random variable in an interval. In the continuous domain, we usually

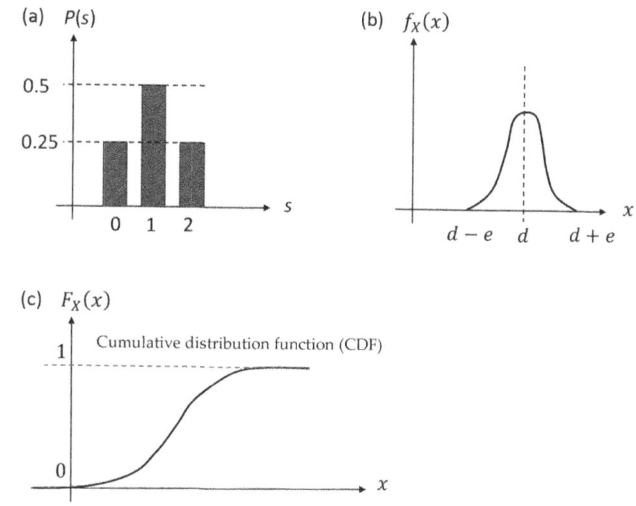

Figure 2.2 (a) PMF of flipping a coin twice and observing the number of heads, (b) PDF of a range sensor with a random error range $\pm e$, (c) CDF (cumulative distribution function).

define the PDF in terms of the probability cumulative distribution function (CDF) (see Figure 2.2c) defined as a function that maps the continuous sample space to the range [0, 1] as follows:

$$F_X(x) = P(X \leq x) \tag{2-1}$$

$$P(a < x < b) = \int_a^b f_X(x)dx = F_X(b) - F_X(a) \tag{2-2}$$

$F_X(x)$ is a monotonically increasing function such that:

$$F_X(-\infty) = 0 \text{ and } F_X(\infty) = 1 \tag{2-3}$$

Mathematically, the PDF assigns a probability that the continuous random variable X will take a value in the infinitesimally small range $[x, x + \delta x]$. PDF is mathematically the derivative of the CDF:

$$f_X(x) = \frac{d}{dx} F_X(x) \tag{2-4}$$

$$F_X(x_1 \leq X \leq x_2) = \int_{x_1}^{x_2} f_X(x)dx \tag{2-5}$$

For an infinitesimally small δx, we have:

$$P(\bar{x} < X < \bar{x} + \delta x) \approx f_X(\bar{x})\delta x \tag{2-6}$$

Note here $f_X(\bar{x})\delta x$ is simply the area under curve $f_X(x)$ at $\bar{x}$, which is approximated as a rectangle of length $f_X(\bar{x})$ and width of δx.

2.2.4 Statistical Moments

Any random variable has statistical properties that can be calculated from its probability distribution. If a random variable is one-dimensional, we refer to it as scalar random variable. In the following discussion, we will consider scalar random variables. Statistical properties of a random variable X are its nth-order moments defined as follows:

$$E\langle(X)^n\rangle = \int_{-\infty}^{\infty} (x)^n P_X(x)dx, \quad P_X(x) \text{ is continous} \tag{2-7}$$

$$E\langle(X)^n\rangle = \sum_{i=1}^{N} (x_i)^n P_X(x_i), \quad P_X(x_i) \text{ is discrete} \tag{2-8}$$

The operator $E\langle.\rangle$ is called the expectation operator. The first-order moment of a random variable is called the mean, and it is calculated as follows:

$$\mu_X = E\langle X\rangle = \int_{-\infty}^{\infty} x f_X(x)dx, \quad \text{if } P_X(x) \text{ is continous} \tag{2-9}$$

$$\mu_X = \sum_{i=1}^{N} x_i P_X(x_i), \qquad \text{if } P_X(x_i) \text{ is discrete} \tag{2-10}$$

The variance of a random variable X is defined by:

$$\sigma_X^2 = E\langle(X-\mu_X)^2\rangle = \int_{-\infty}^{\infty} (x-\mu_X)^2 P_X(x)dx, \quad P_X(x) \text{ is continous} \tag{2-11}$$

$$\sigma_X^2 = \sum_{i=1}^{N} (x_i-\mu_X)^2 P_X(x_i), \quad P_X(x_i) \text{ is discrete} \tag{2-12}$$

The square root of the variance is called the standard deviation. If we have two scalar continuous random variables X and Y, the covariance between them is an indication of how similar the two random variables are, and it is calculated as follows:

$$\text{cov}(X,Y) = \sigma_{XY} = E\langle(X-\mu_X)(Y-\mu_Y)\rangle = \int_{-\infty}^{\infty}\int_{\infty}^{\infty} (x-\mu_X)(y-\mu_Y)P_{XY}(x,y)dxdy \tag{2-13}$$

where $P_{XY}(x, y)$ is called the joint probability, the probability of simultaneously getting $X = x$ and $Y = y$. If X and Y are discrete, their covariance is given by:

$$\text{cov}(X,Y) = \sigma_{XY} = E\langle(X-\mu_X)(Y-\mu_Y)\rangle = \sum_{i=1}^{N}\sum_{j=1}^{M}(x_i-\mu_X)(y_i-\mu_Y)P_{XY}(x_i,x_j) \tag{2-14}$$

Covariance is a measure of the relationship between two random variables. In other words, it indicates how the change in one variable is related to a change in the other variable. Covariance can be positive (an increase in one variable corresponds to an increase in the other), negative (a decrease in one variable corresponds to a decrease in the other), or zero (uncorrelated variables). This measure is frequently used in estimation algorithms that involve random variables. A normalized measure of correlation between two random variables is the correlation, which is calculated as follows:

$$\text{corr}(X,Y) = \frac{\text{cov}(X,Y)}{\sigma_X\sigma_Y} \tag{2-15}$$

If the two variables are independent, their joint probability $P_{XY}(x, y) = P_X(x)P_Y(y)$ and the covariance expression will be reduced to the following equation:

$$\text{cov}(X,Y) = \sigma_{XY} = E\langle(X)(Y)\rangle - \mu_X\mu_Y \tag{2-16}$$

2.2.5 Vector Random Variables

A random variable can be a vector $\vec{X} = (X_1, X_2, ..., X_M)^T$, where each element in the vector is a scalar random variable with its own probability distribution. The mean of a vector random variable is a vector defined as follows:

$$\vec{\mu}_X = \begin{pmatrix} \mu_{X_1} \\ \mu_{X_2} \\ \vdots \\ \mu_{X_M} \end{pmatrix} = \begin{pmatrix} E\langle X_1 \rangle \\ E\langle X_1 \rangle \\ \vdots \\ E\langle X_M \rangle \end{pmatrix} \tag{2-17}$$

The covariance of two random vectors is called the cross-covariance and it is a matrix defined as follows:

$$\operatorname{cov}\left(\vec{X}, \vec{Y}\right) = E\left\langle \left(\vec{X} - \vec{\mu}_X\right)\left(\vec{Y} - \vec{\mu}_Y\right)^T \right\rangle = \begin{pmatrix} \sigma_{X_1Y_1} & \cdots & \sigma_{X_1Y_N} \\ \vdots & \ddots & \vdots \\ \sigma_{X_NY_1} & \cdots & \sigma_{X_NY_M} \end{pmatrix} \tag{2-18}$$

The auto-covariance of a random vector is its covariance with itself as follows:

$$\operatorname{cov}\left(\vec{X}, \vec{X}\right) = E\left\langle \left(\vec{X} - \vec{\mu}_X\right)\left(\vec{X} - \vec{\mu}_X\right)^T \right\rangle = \begin{pmatrix} \sigma_{X_1}^2 & \cdots & \sigma_{X_1Y_M} \\ \vdots & \ddots & \vdots \\ \sigma_{X_1Y_M} & \cdots & \sigma_{X_M}^2 \end{pmatrix} \tag{2-19}$$

The normalized auto-correlation matrix of a random vector is defined as follows:

$$\operatorname{corr}\left(\vec{X}, \vec{X}\right) = \begin{pmatrix} 1 & \cdots & \dfrac{\sigma_{X_1Y_M}}{\sigma_{X_1}\sigma_{Y_M}} \\ \vdots & \ddots & \vdots \\ \dfrac{\sigma_{X_1Y_M}}{\sigma_{X_M}\sigma_{Y_1}} & \cdots & 1 \end{pmatrix} \tag{2-20}$$

We notice here that the diagonal elements of the autocorrelation matrix have the value "1" because each element of the vector is perfectly correlated with itself. If elements of the random vector are independent (they are completely uncorrelated), their covariance and correlation will be zero and we will have a diagonal autocovariance matrix and an identity autocorrelation matrix.

2.2.6 Transformation of Random Variables

If a random variable is an input to a function $Y = g(X)$, it is always important to calculate the probability of the output Y given the input probability X. This feature is used in Monte Carlo simulation sensor fusion algorithms such as Particle Filtering. This process is called transformation, and it can be applied to both scalar and vector random variables. In general, if X is a discrete random variable, the transformed probability distribution of Y is given by:

$$P_Y(y) = \sum_{x:g(x) = y} P_X(x) \tag{2-21}$$

where P_Y and P_X are the PMF of X and Y, respectively. This relationship is intuitive as it states that the probability of the transformed variable $Y = y$ equals to the sum of the probabilities of all $X = x$ that leads to y under the transformation g(X). This is a simple application of the sum rule. If X is continuous random variable, we have to use the CDF as follows:

$$P(x < X < x + \delta x) = P(y < Y < y + \delta y) \tag{2-22}$$

$$P(x < X < x + \delta x) \approx f_X(x)\delta x \tag{2-23}$$

$$P(y < Y < y + \delta y) \approx f_Y(y)\delta y \tag{2-24}$$

$$f_Y(y) = f_X(x)\left|\frac{dX}{dY}\right| \tag{2-25}$$

where f_X and f_Y are the PDF of X and Y, respectively. Note that this formula is only applicable if a derivative of the inverse transformation exists $\left(\frac{dX}{dY}\right)$ exists (see [3] and [4] for more information about transformation of random variables).

2.2.7 Probability Axioms

The probability has three fundamental axioms:

- $p(A) \geq 0$: The probability of any event A is a nonnegative number.
- $p(S) = 1$: The probability of the whole sample space is 1.
- $p(A \cup B) = p(A) + p(B)$: Union probability of two disjoint events.

These axioms are the foundation for several useful rules such as:

- The completement rule: $p(not\ A) = 1 - p(A)$
- monotonicity rule: $p(A) \leq p(B) \rightarrow A \subseteq B$
- probability of the empty set: $p(\emptyset) = 0$

Further rules are to be reviewed in the following subsections.

2.2.8 Bayes' Rule

If we have more than one event, A and B, it is useful to describe the relationship between the two events. The probability of the event that both A and B happen simultaneously is called joint probability. The probability of event A or event B to happen is called union probability. The probability that event A happens, given that event B has already happened, is called conditional probability. We use the following mathematical expressions to express these probabilities:

- **Joint Probability**: The probability of A and B expressed by $p(A \cap B)$
- **Union Probability**: The probability of A or B expressed by $p(A \cup B)$
- **Conditional Probability**: The probability of A given B expressed by $p\ (A \mid B)$

The concepts of conditional and joint probabilities take us to the concept of Bayes' rule, which is a fundamental key rule that is frequently used in estimation and sensor fusion algorithms. Bayes' rule is explained in the next subsection. The relationship between joint, union, and conditional probability is given as follows:

$$p(A \cap B) = p(A \mid B)p(B) = p(B \mid A)p(A) \tag{2-26}$$

$$P(A \cup B) = P(A) + P(B) - p(A \cap B) \tag{2-27}$$

The joint probability equation gives the popular Bayes' rule:

$$p(A \mid B) = \frac{p(B \mid A)p(A)}{p(B)} \tag{2-28}$$

If events A and B are independent, then:

$$p(A \cap B) = p(A)p(B) \tag{2-29}$$

$$P(A \cup B) = P(A) + P(B) \tag{2-30}$$

$$p(A \mid B) = p(A) \tag{2-31}$$

$$p(B \mid A) = p(B) \tag{2-32}$$

Equations (2-26) and (2-29) are called the product rule. They are very useful in estimating the intersection or joint probability of multiple events.

2.2.9 Bayesian Networks

In several real-life problems, events may occur in a network or a sequence structure. For example, the trajectory of an object can be modeled as a sequence of stochastic states where states represent the object's location, direction, and velocity. The probabilistic relationship between events can be represented by a directed graph called Bayesian Networks, where events are illustrated as nodes and the conditional probabilities between events are illustrated by arrows as seen in Figure 2.3. The Bayesian network in Figure 2.3 is a fully connected graph where all pairs of events are connected. The joint probability of a sequence of N events in a fully connected network is:

$$p(X_1X_2 \dots X_N) = p(X_1)p(X_2 \mid X_1)p(X_3 \mid X_2X_1) \dots (X_N \mid X_{N-1}X_{N-2} \dots X_1) \tag{2-33}$$

In the simple example of Figure 2.3, the joint probability of the network is given by the following equation:

$$p(X_1X_2X_3) = p(X_1)p(X_2 \mid X_1)p(X_3 \mid X_2X_1) \tag{2-34}$$

The probability of the sequence X_1X_3 is given by the following expression:

$$p(X_1X_3) = p(X_1)p(X_3 \mid X_1) \tag{2-35}$$

If the network of events is not fully connected (there are missing links between some events), the general rule is given by the following equation:

$$p(X_1X_2 \dots X_N) = \prod_{i=1}^{N} p(X_i \mid \text{All parent events of } X_i) \tag{2-36}$$

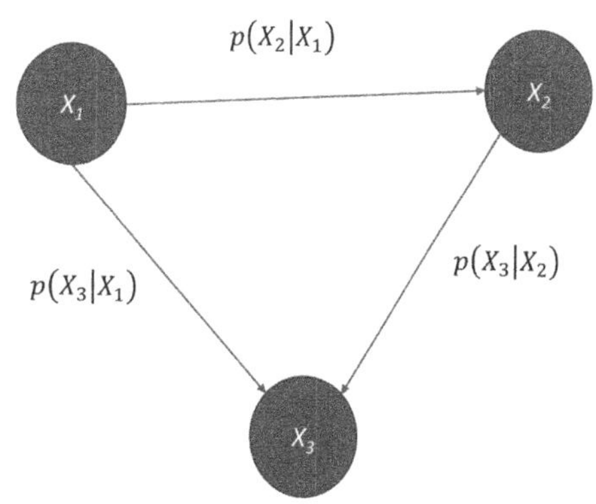

Figure 2.3 Network of events. A network may contain multiple sequences.

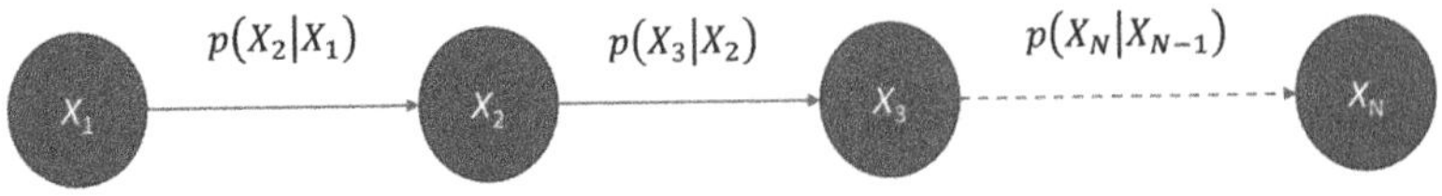

Figure 2.4 Markovian sequence.

2.2.10 Markov Sequences

If the Bayesian network has a single sequence, the probability of an event X_k depends only on the previous event X_{k-1} as seen in Figure 2.4, then the sequence is called "*Markov Sequence*," and the probability of the sequence is given by the following equation:

$$p(X_1X_2...X_N) = p(X_1)\prod_{i=1}^{N} p(X_{i+1} \mid X_i) \tag{2-37}$$

The Markov assumption is widely used in navigation and tracking systems where it is reasonable to assume that the object's location at time t_k depends only on the object location at time t_{k-1} and the sensor measurements u_k. This assumption simplifies the probabilistic estimation algorithms when applied in navigation, positioning, and mapping systems. More details about probabilistic estimation algorithms will be given in the next chapter.

2.2.11 Marginal Probability (the Sum Rule)

If in a Bayesian network, an event B can be reached from N possible prior events A_i $i = 1, 2 ... M$ as seen in Figure 2.5 and we have information about the conditional probability $p(B \mid A_i)$, then, the probability of event B is given as follows:

$$p(B) = \sum_{i=1}^{M} p(B \cap A_i) = \sum_{i=1}^{M} p(B \mid A_i)p(A_i) \tag{2-38}$$

This is called "*marginal probability*," where we marginalize out the variable A from the probability of the variable B. If A_i is a continuous set, the summation becomes integration and $p(B \mid A)$ is given by:

$$p(B) = \int p(B \mid A_i)p(A_i)dA_i \tag{2-39}$$

where dA_i is an integration variable over all possible values of A_i.

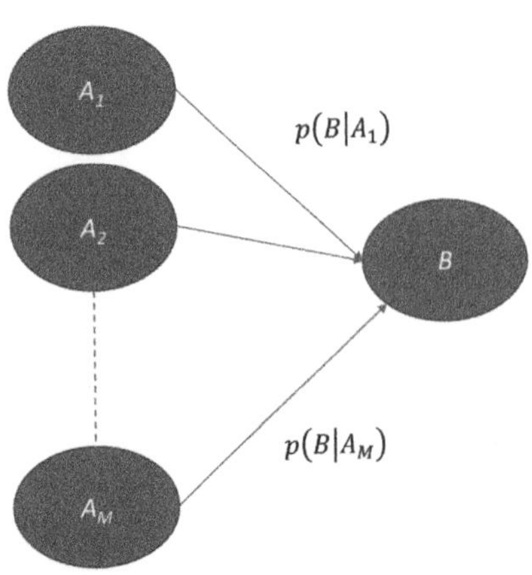

Figure 2.5 Illustration of marginal probability.

2.2.12 Gaussian (Normal) Probability Distribution

One of the most common PDFs that is very convenient and used very frequently in signals and systems is the Gaussian Distribution (also known as the normal distribution). A scalar random variable $X \sim \mathcal{N}(\mu, \sigma^2)$ has a Gaussian PDF function defined as:

$$f_G(x) = \frac{1}{\sqrt{2\pi}\sigma} e^{-\left(\frac{(x-\mu)^2}{2\sigma^2}\right)} \tag{2-40}$$

$f_G(x)$ represents the probability that the random variable X takes the value x. The expression $X \sim \mathcal{N}(\mu, \sigma^2)$ means that X is a scalar Gaussian random variable where mean μ and variance σ^2 are given by the following expressions:

$$E\langle X\rangle = \int_{-\infty}^{\infty} x f_X(x)dx = \int_{-\infty}^{\infty} x \frac{1}{\sqrt{2\pi}\sigma} e^{-\left(\frac{(x-\mu)^2}{2\sigma^2}\right)} dx = \mu \tag{2-41}$$

$$E\langle (X-\mu_X)^2\rangle = \int_{-\infty}^{\infty} (x-\mu)^2 f_X(x)dx = \int_{-\infty}^{\infty} (x-\mu)^2 \frac{1}{\sqrt{2\pi}\sigma} e^{-\left(\frac{(x-\mu)^2}{2\sigma^2}\right)} dx = \sigma^2 \tag{2-42}$$

The Gaussian PDF can effectively model the behavior of many real-life random signals. The intuitive interpretation of Gaussian PDF is that the most likely value of the random variable is μ while the standard deviation σ represents the range within which the value is likely to vary. The statistical interpretation of the Gaussian PDF is usually expressed in terms of the standard deviation σ using common terms such as one-sigma, two-sigma, and three-sigma as seen in Figure 2.6. One sigma (σ) means the probability that the value of the random variable will lie in the range $[\mu - \sigma,\ \mu + \sigma]$, which is 68.27%. Two sigma (2σ) means the probability that the value will lie in the range $[\mu - 2\sigma,\ \mu + 2\sigma]$, which is 95.45%. Three sigma (3σ) means the probability that the value will lie in the range $[\mu - 3\sigma,\ \mu + 3\sigma]$ which is 99.73%.

In practice, there are several other PDFs that can be used to describe the distribution of a random variable. Not all phenomenon that includes randomness can be well described by a Gaussian PDF. However, Gaussian PDF has great mathematical characteristics that make them very attractive in simplifying sensor fusion algorithms as we will see in the next chapter.

2.2.13 Multivariate Normal (MVN) Distribution

If we have Gaussian random variable that is a vector of size n, the distribution is called Multivariate Normal (MVN) distribution. MVN PDF is given by:

$$f_G\left(\vec{x}\right) = \frac{1}{\sqrt{(2\pi)^n}\sqrt{|\Sigma|}} e^{-\frac{1}{2}\left(\vec{x}-\vec{\mu}\right)^T \Sigma^{-1}\left(\vec{x}-\vec{\mu}\right)} \tag{2-43}$$

where vector $\vec{\mu} = (x_1, x_2 ... x_n)^T$, $\vec{\mu} \in R^n$ is the mean of X and Σ is a covariance matrix of size $n \times n$. The diagonal values of the covariance matrix Σ consist of the variances of the individual

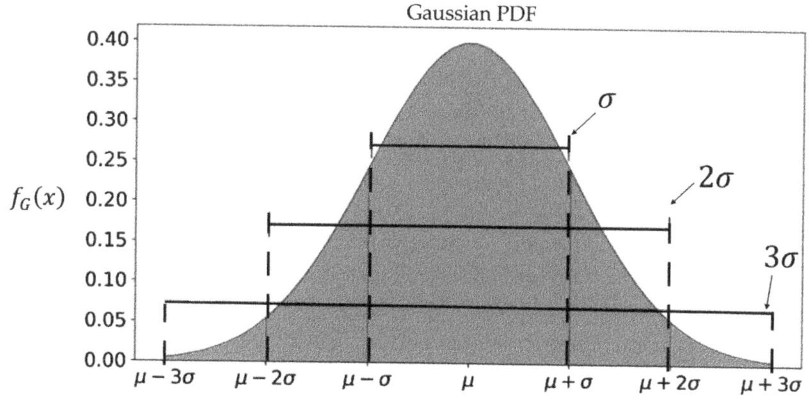

Figure 2.6 Gaussian PDF.

components of the vector $\vec{X}$ and the off-diagonal elements are the covariance between the individual components of the vector $\vec{X}$ as shown below:

$$\Sigma = \begin{pmatrix} \sigma_{x_1}^2 & \sigma_{x_1x_2} & \cdots & \sigma_{x_1}x_n \\ \sigma_{x_2x_1} & \sigma_{x_2}^2 & \cdots & \vdots \\ \vdots & \vdots & \ddots & \vdots \\ \sigma_{x_n}x_1 & \cdots & \cdots & \sigma_{x_n}^2 \end{pmatrix} \tag{2-44}$$

The expression $\left(\vec{X}-\vec{\mu}\right)^T \Sigma^{-1}\left(\vec{X}-\vec{\mu}\right)$ is called the Mahalanobis distance [3, 4] between $\vec{X}$ and $\vec{\mu}$. Mahalanobis distance measures the distance between a point and a distribution, and it is frequently used in several estimation algorithms.

2.2.14 Inference in Multivariate Normal Distribution

One of the key theories in probability is the marginal and conditional probabilities of MVN distribution. Given two MVN random variables (x_1, x_2), where mean and covariances are given by the following equations:

$$\mu = \begin{pmatrix} \mu_1 \\ \mu_2 \end{pmatrix}, \Sigma = \begin{pmatrix} \Sigma_{11} & \Sigma_{12} \\ \Sigma_{21} & \Sigma_{22} \end{pmatrix}, \Lambda = \Sigma^{-1} = \begin{pmatrix} \Lambda_{11} & \Lambda_{12} \\ \Lambda_{21} & \Lambda_{22} \end{pmatrix} \tag{2-45}$$

and their marginal distributions are normal distributions given by:

$$\begin{aligned} p(x_1) &= \mathcal{N}(x_1 \mid \mu_1, \Sigma_{11}) \\ p(x_2) &= \mathcal{N}(x_2 \mid \mu_2, \Sigma_{22}) \end{aligned} \tag{2-46}$$

the conditional posterior distribution $p(x_1 \mid x_2) = \mathcal{N}\left(x_1 \mid \mu_{1|2}, \Sigma_{1|2}\right)$ is given by the following equations:

$$\begin{aligned} \mu_{1|2} &= \mu_1 + \Sigma_{12}\Sigma_{22}^{-1}(x_2 - \mu_2) \\ \Sigma_{1|2} &= \Sigma_{11} - \Sigma_{12}\Sigma_{22}^{-1}\Sigma_{21} = \Lambda_{11}^{-1} \end{aligned} \tag{2-47}$$

The proof of equations (2-47) can be found in [1]. Note that $\Sigma_{1\,|\,2}$ is smaller than Σ_{11} which indicates that the uncertainty in the value of x_1 is reduced when we have some knowledge (mean and variance) about the value of x_2. This theory has a great use in reducing the size of complex estimation problems where the parameters of interest form an MVN distribution. Partial knowledge about some of the parameters enables the interference of the probability distribution of the other parameters given their joint covariances.

2.2.15 Central Limit Theorem

The central limit theorem states that for any probability distribution (not necessarily Gaussian) with mean μ and variance δ^2, if we take n independent identically distributed (IID) samples $\{x_1, x_2 \ldots x_n\}$ and calculate the mathematical average as follows:

$$\overline{x}_n = \frac{\sum_{i=1}^{n} x_i}{n} \tag{2-48}$$

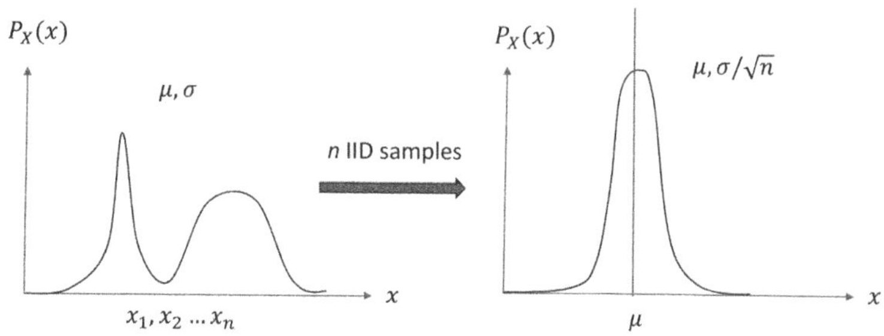

Figure 2.7 Illustration of the central limit theorem.

Then $\overline{x}_n$ will tend to μ as n tends to infinity:

$$\lim_{n \to \infty} \overline{x}_n = \mu \tag{2-49}$$

If we repeat the experiment (taking n IID samples) and calculate the sample mathematical average $\overline{x}_n$, we will get a Gaussian distribution with mean μ and variance δ^2/n and the resulting distribution of the sample averages will be $\mathcal{N}\left(\mu, \frac{\sigma^2}{n}\right)$. The central limit is an important theorem because it means that the tools and methods applied to Gaussian distribution can be applied to non-Gaussian distributions if IID samples are taken. An illustrative figure of the central limit theorem is shown in Figure 2.7. It is worth noting that converting a non-Gaussian distribution to a Gaussian one can be risky when systematic errors are present, as this can cause the states to converge to incorrect values. Statistical tools can be used to verify the accuracy and precision of the measurements before and after applying the central limit theorem.

2.3 Systems

In navigation systems, sensor signals go through different subsystems for processing. Therefore, reviewing fundamentals of systems theory is important. Systems theory models a physical system using a mathematical model. This mathematical model describes the system as a set of states $x(t)$. States are internal variables that describe the system at a certain time. The system states $x(t)$ evolves with time according to a system model as shown in Figure 2.8. The delayed feedback of $x(t)$ is interpreted as the system future state depends on current and/or previous states. In some cases, the

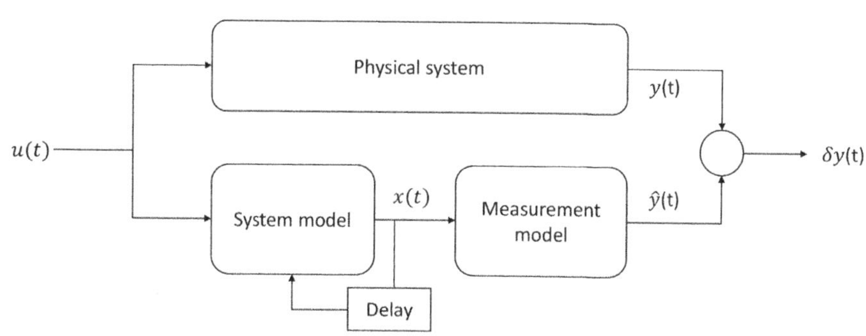

Figure 2.8 System modeling.

system may also accept a control input signal $u(t)$. Any system has observables or measurables $y(t)$, which are the visible or measurable outputs of the system. These are the variables that we can measure using sensors. System states are mapped to system measurables by a measurement model. The general system modeling is shown in Figure 2.8. A physical system can be modeled by a system model and a measurement model. The accuracy of the model can be evaluated by calculating the error $\delta y(t)$ between the output of the model $\hat{y}(t)$ and the observed/measured output $y(t)$. This error is commonly known as the residuals. A residual close to zero suggests that the mathematical model has effectively captured the physical behavior.

Mathematically, the system model, which describes how system states evolve with time is commonly represented in three different ways [5, 6]:

- Differential equations
- State space model
- Transfer function

We will discuss each technique in a separate subsection. Before we discuss each technique, we would like to choose a physical system as an example. We will use two examples to demonstrate each technique. The two systems are illustrated in Figure 2.9. The system in Figure 2.9a is a mass-spring system (a linear system) and the system in Figure 2.9b is a 2D car kinematics system (a non-linear system). These are simple systems examples that we will use for illustrating main concepts related to systems modeling.

2.3.1 Differential Equations

A system with a state variable $x(t)$ and an optional control signal $u(t)$ can be modeled by the following general ordinary differential equation:

$$x^{(n)}(t) = f\left(x^{(n-1)}(t) \ldots x(t), u^{(n-1)}(t) \ldots u(t)\right) \tag{2-50}$$

where $x(t)$ is the state of the system, $u(t)$ is an optional control input signal, and $x^{(n)}(t)$ is the nth time derivative of the state $x(t)$. The first time derivative $x^{(1)}(t)$ is denoted by $\dot{x}(t)$ and $x^{(0)}(t)$ means simply $x(t)$. Sometimes we represent first-, second-, and third-order time-derivatives using superscript dots as follows:

$$x^{(1)}(t) = \dot{x}(t) \tag{2-51}$$

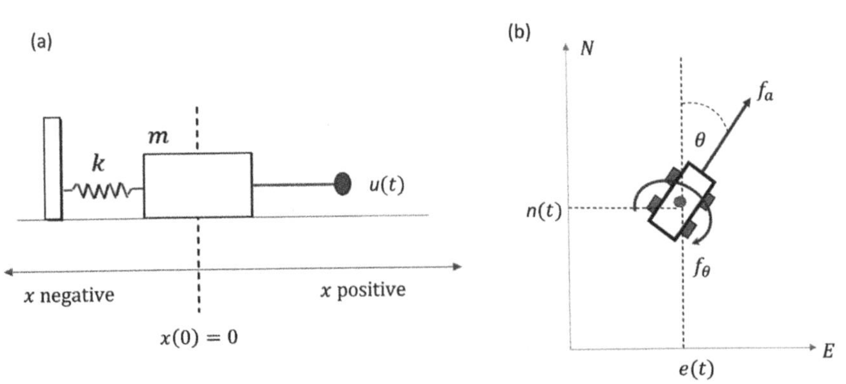

Figure 2.9 (a) Mass-spring system example (linear system), (b) 2D vehicle kinematic system example (nonlinear system).

$$x^{(2)}(t) = \ddot{x}(t) \tag{2-52}$$

$$x^{(3)}(t) = \dddot{x}(t) \tag{2-53}$$

The word ordinary means that the states and their derivatives are functions of a single variable, which is time t in our case. If the system is linear and time-invariant (LTI system), the general differential equation will be given as follows:

$$x^{(n)}(t) + a_1 x^{(n-1)}(t) + a_2 x^{(n-2)}(t) + \ldots + a_n x(t) = b_1 u^{(n-1)}(t) + b_2 u^{(n-2)}(t) + \ldots + b_n u(t) \tag{2-54}$$

A linear system satisfies the superposition principle [5] described as follows:

$$\begin{aligned} &\text{If } y_1(t) = f(x_1(t)) \text{ and } y_2(t) = f(x_2(t)) \text{then} \\ &f(\alpha x_1(t) + \beta x_2(t)) = \alpha y_1(t) + \beta y_2(t) \end{aligned} \tag{2-55}$$

A linear system is a system where operations on the states are linear operators such as scaling, time shifting, and addition.

2.3.1.1 Linear System Example

The mass-spring system example in Figure 2.9a is a linear system. We can verify this by deriving its differential equations. Physics laws tell us that the applied force $f(t)$ equals to the total forces of the system. The total forces of the system come from acceleration, friction with the surface, and the strength of the spring. Without going into so many details about laws of physics, which can be found in many other resources, the differential equation for this mass-spring system example is given by:

$$m(t)\ddot{x}(t) + b(t)\dot{x}(t) + k(t)x(t) = u(t) \tag{2-56}$$

$$\ddot{x}(t) + \frac{b(t)}{m(t)}\dot{x}(t) + \frac{k(t)}{m(t)}x(t) = \frac{1}{m(t)}u(t) \tag{2-57}$$

where $x(t)$ is the position of the mass, $m(t)$ is the mass of the object, $b(t)$ is the friction factor between the mass and the surface, $k(t)$ is the spring factor, and $u(t)$ is the force applied to the mass. System parameters $b(t)$, $k(t)$,and $m(t)$ are written as variables in time; however, they can be constant, making the system linear time-invariant (LTI) system. This is a linear differential equation that describes how the system state (mass position $x(t)$) evolves with time driven by the control signal (the force $u(t)$), the force applied to the mass. The linearity of the system can be easily determined by the superposition test. If we define the input to the system as the applied force $u(t)$ and the output as the displacement $y(t) = x(t)$, we can verify the superposition applies as follows:

$$\ddot{x}_1(t) + \frac{b(t)}{m(t)}\dot{x}_1(t) + \frac{k(t)}{m(t)}x_1(t) = \frac{1}{m(t)}u_1(t) \tag{2-58}$$

$$\ddot{x}_2(t) + \frac{b(t)}{m(t)}\dot{x}_2(t) + \frac{k(t)}{m(t)}x_2(t) = \frac{1}{m(t)}u_2(t) \tag{2-59}$$

Multiplying the first equation by α and the second equation by β and adding the two new equations gives the following:

$$(\alpha\ddot{x}_1(t) + \beta\ddot{x}_2(t)) + \frac{b(t)}{m(t)}(\alpha\dot{x}_1(t) + \beta\dot{x}_2(t)) + \frac{k(t)}{m(t)}(\alpha x_1(t) + \beta x_2(t)) = \frac{1}{m(t)}(\alpha u_1(t) + \beta u_2(t)) \tag{2-60}$$

which is the same system equation (2-57), but for the combined output $x(t) = \alpha x_1(t) + \beta x_2(t)$ corresponding to the composed input $\alpha u_1(t) + \beta u_2(t)$. This shows that the application of two scaled control signals separately is equivalent to applying the sum of the two scaled control signals simultaneously. This is simply the superposition rule.

2.3.1.2 Nonlinear System Example

The 2D vehicle kinematic system in Figure 2.9b is a nonlinear system. We can verify this by deriving its differential equations as follows:

$$\ddot{e}(t) = f_a(t)\sin(\theta(t)) \tag{2-61}$$

$$\ddot{n}(t) = f_a(t)\cos(\theta(t)) \tag{2-62}$$

$$\dot{\theta}(t) = f_\theta(t) \tag{2-63}$$

where $e(t)$ is the east position, $n(t)$ is the north position, $\theta(t)$ is the heading angle, $f_a(t)$ is the acceleration control signal and $f_\theta(t)$ is the steering angular velocity control signal. The system state here is a vector defined as follows:

$$\vec{x}(t) = \begin{pmatrix} e(t) \\ n(t) \\ \theta(t) \end{pmatrix} \tag{2-64}$$

and the control signal is also a vector defined as follows:

$$\vec{u}(t) = \begin{pmatrix} f_a(t) \\ f_\theta(t) \end{pmatrix} \tag{2-65}$$

As $\theta(t)$ is part of the system states, the differential equations of $\vec{x}(t)$ are nonlinear due to the embedding of $\theta(t)$ into a cosine and sine functions, which are not linear operators. The system equations do not satisfy the superposition test in the sense that the application of two acceleration control signals separately is not equivalent to applying the sum of the two acceleration control signals.

2.3.2 State Space Model

2.3.2.1 System Model

A more systematic representation of systems is the state space model. The higher-order differential equations in the general differential equation form are usually complex to solve either analytically or numerically. It would be much easier to deal with the system model if all equations were first-order differential equations. First-order differential equations are easier to solve analytically and numerically using a digital computer. To represent the system using the state space model, we need to define the state variables such that we have a first-order differential equation for all state variables. We will define a state vector $\vec{x} = \{x_1, x_2, ..., x_n\}$ such that:

$$\begin{aligned} x_1 &= x \\ x_2 &= x^{(1)} \\ x_3 &= x^{(2)} \\ &\dots \\ x_n &= x^{(n-1)} \end{aligned} \tag{2-66}$$

We can also do the same for the control input $\vec{u} = \{u_1, u_2, ..., u_n\}$. Then we define the first-order differential equations of x_i as follows

$$\begin{aligned}\dot{x}_1 &= x_2\\ \dot{x}_2 &= x_3\\ &\dots\\ \dot{x}_n &= f(x_n \dots x_1, u_n \dots u_1)\end{aligned} \tag{2-67}$$

The state variables $x_1 \dots x_n$ form a state vector $\vec{x}$ and the same for $u_1 \dots u_n$ that form $\vec{u}$ so we obtain a system in the general form:

$$\dot{\vec{x}} = f\left(\vec{x}, \vec{u}\right) \tag{2-68}$$

If the system model is linear, the state space model will take the form:

$$\dot{\vec{x}}(t) = A(t)\vec{x}(t) + G(t)\vec{u}(t) \tag{2-69}$$

where $A(t)$ is called the transition matrix or the system matrix and the matrix $G(t)$ is called the input shaping matrix. In the mass-spring example, if we define $v(t)$ as the speed of the object, and $p(t)$ as the position of the object, the state space model can be defined as:

$$\dot{p}(t) = v(t) \tag{2-70}$$

$$\dot{v}(t) = \frac{1}{m(t)}(u(t) - b(t)v(t) - k(t)p(t)) \tag{2-71}$$

In matrix form, the state space model can be written as:

$$\begin{pmatrix}\dot{p}(t)\\ \dot{v}(t)\end{pmatrix} = \begin{pmatrix}0 & 1\\ \frac{-k(t)}{m(t)} & \frac{-b(t)}{m(t)}\end{pmatrix}\begin{pmatrix}p(t)\\ v(t)\end{pmatrix} + \frac{1}{m(t)}u(t) \tag{2-72}$$

which is a linear system of the following form:

$$\dot{\vec{x}}(t) = A(t)\vec{x}(t) + G(t)\vec{u}(t) \tag{2-73}$$

$$\vec{x}(t) = \begin{pmatrix}p(t)\\ v(t)\end{pmatrix} \tag{2-74}$$

$$A(t) = \begin{pmatrix}0 & 1\\ \frac{-k(t)}{m(t)} & \frac{-b(t)}{m(t)}\end{pmatrix} \tag{2-75}$$

$$G(t) = \begin{pmatrix}\frac{-1}{m(t)}\end{pmatrix} \tag{2-76}$$

$$\vec{u}(t) = (f(t)) \tag{2-77}$$

If the system is an LTI system, all matrices will be constants ($A(t) = A$, $G(t) = G$). In the 2D car kinematics system example, we can define vehicle's position states: $e(t)$ and $n(t)$, vehicle's velocity states: $v_e(t)$ and $v_n(t)$, and vehicle's heading angle state: $\theta(t)$ so the state vector is as follows:

$$\vec{x}(t) = \begin{pmatrix}e(t)\\ p(t)\\ v_e(t)\\ v_n(t)\\ \theta(t)\end{pmatrix} \tag{2-78}$$

The state space model can be written as:

$$\dot{e}(t) = v_e(t) \tag{2-79}$$

$$\dot{n}(t) = v_n(t) \tag{2-80}$$

$$\dot{v}_e(t) = f_a(t)\sin(\theta(t)) \tag{2-81}$$

$$\dot{v}_n(t) = f_a(t)\cos(\theta(t)) \tag{2-82}$$

$$\dot{\theta}(t) = f_\theta(t) \tag{2-83}$$

These differential equations resemble the general form $\dot{\vec{x}} = f\left(\vec{x}, \vec{u}\right)$. This 2-D vehicle's kinematic system cannot be arranged in a linear matrix form due to the sine/cosine terms so, it is a nonlinear system.

2.3.2.2 Measurement Model

Measurement model maps system observables to the system internal states. If the measurement model is linear, it will take the following linear form:

$$\vec{y}(t) = H(t)\vec{x}(t) \tag{2-84}$$

where $H(t)$ is called the design matrix. If the system is an LTI system, the design matrix will be constant ($H(t) = H$). If the measurement model is nonlinear, it takes the following general form:

$$\vec{y}(t) = h\left(\vec{x}(t)\right) \tag{2-85}$$

where $h(.)$ is a general nonlinear function of the system states. In the mass-spring system example, if the observables are the position $x(t)$ and velocity $v(t)$ of the object, the measurement model will be given as follows:

$$\vec{y}(t) = \begin{pmatrix} 1 & 0 \\ 0 & 1 \end{pmatrix} \begin{pmatrix} p(t) \\ v(t) \end{pmatrix} \tag{2-86}$$

For the 2D vehicle's kinematic system, if the measurement is the position of the car, we will have the following linear measurement model:

$$\vec{y}(t) = \begin{pmatrix} 1 & 0 & 0 & 0 & 0 \\ 0 & 1 & 0 & 0 & 0 \end{pmatrix} \begin{pmatrix} e(t) \\ p(t) \\ v_e(t) \\ v_n(t) \\ \theta(t) \end{pmatrix} \tag{2-87}$$

which is a linear relationship in the state vector $\vec{x}(t)$. If the measurement of the system in the 2-D vehicle's kinematic example is the forward speed of the car, the measurement model would be given by:

$$y(t) = h\left(\vec{x}(t)\right) = h\begin{pmatrix} e(t) \\ p(t) \\ v_e(t) \\ v_n(t) \\ \theta(t) \end{pmatrix} = \sqrt{v_e^2(t) + v_n^2(t)} \tag{2-88}$$

which is a nonlinear relationship due to the squares and square root operators.

2.3.3 Transfer Function

The transfer function is a useful tool to represent LTI systems as sometimes, dealing with linear differential equations in the time domain is complex. The transfer function technique uses the Laplace transform to map an LTI from the time domain to another domain called the S-domain. The greatest benefit of Laplace transform is that when it is applied to linear ordinary differential equations, it makes the solution much easier, as we will see in the next subsection.

2.3.3.1 Laplace Transform

Laplace transform of a time-domain function $f(t)$ is defined as follows [7]:

$$F(s) = \mathcal{L}(f(t)) = \int_{-\infty}^{\infty} f(t)e^{-st}dt \tag{2-89}$$

where s is a complex plane with two parts, real and imaginary as follows:

$$s = \sigma + j\omega \tag{2-90}$$

The popular Fourier transform is a special case of Laplace transform where $s = j\omega$ where ω is the angular frequency. The frequency domain is an important tool to analyze signals and we will briefly discuss it in the next sections. The Laplace transform of a function $f(t)$ may exist for only specific values (regions) of the real part of s. This region is where the Laplace transform integral will have a value. This region is called the convergence region of Laplace transform of the function $f(t)$ commonly referred to by the symbol c. What is special about Laplace transform is that it simplifies the solution of ordinary linear differential equations due to the following differentiation property of Laplace transform:

If

$$\mathcal{L}(f(t)) = F(s) \tag{2-91}$$

then

$$\mathcal{L}\left(f^{(n)}(t)\right) = s^n F(s) \tag{2-92}$$

where $f^{(n)}(t)$ is the nth-order time-derivative of $f(t)$. This feature makes the ordinary linear differential equation a linear algebraic equation that does not include differentiations. When we solve a system equation in the S-domain, we apply the inverse Laplace transform to get the solution in the time-domain:

$$f(t) = \int_{c-j\infty}^{c+j\infty} F(s)e^{st}ds \tag{2-93}$$

where c is the region, where the Laplace transform of $f(t)$ is defined.

2.3.3.2 Impulse Response

To understand why Laplace transform is a useful tool to solve linear systems, we need to review the concept of the impulse response. An impulse (also known as "*Dirac delta function*") is an infinitesimally small pulse as seen in Figure 2.10. An impulse is mathematically defined as follows:

$$\int_{-\infty}^{\infty} P(t)dt = 1 \tag{2-94}$$

$$\delta(t) = \lim_{\epsilon \to 0} P(t) \tag{2-95}$$

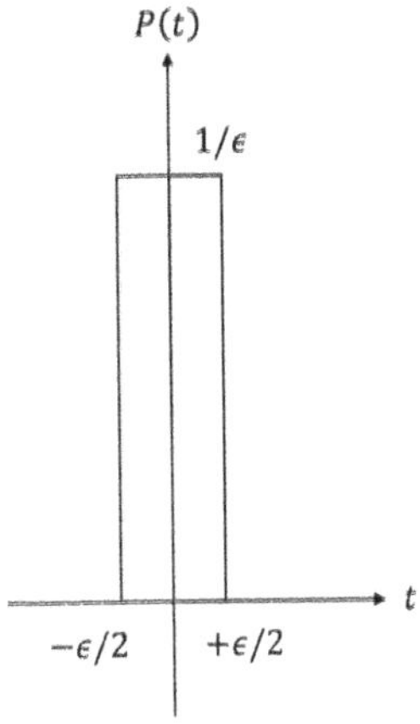

Figure 2.10 Pulse with infinitesimally small period.

The key feature of the impulse function is that it can be used to represent any signal as a pulse train as follows:

$$s(\tau) = \int_{-\infty}^{\infty} \delta(t-\tau)s(t)dt \tag{2-96}$$

This means that any signal $s(t)$ can be represented as continuous train or sum of impulses where each impulse at time τ is modulated by $s(\tau)$ as seen in Figure 2.11.

Given an LTI system, if we applied the Dirac delta function $\delta(t)$, we get an output response $h(t)$ that is called the impulse response of the system as seen in Figure 2.12.

Since the system is linear (the superposition rule applies) and any input signal can be represented as a sum of impulses at different times modulated by $x(t)$, the system output for any signal $x(t)$ can be calculated as the sum of the impulse response modulated by $x(t)$. The system output can be described by:

$$y(t) = \int_{-\infty}^{\infty} x(\tau)h(t-\tau)d\tau \tag{2-97}$$

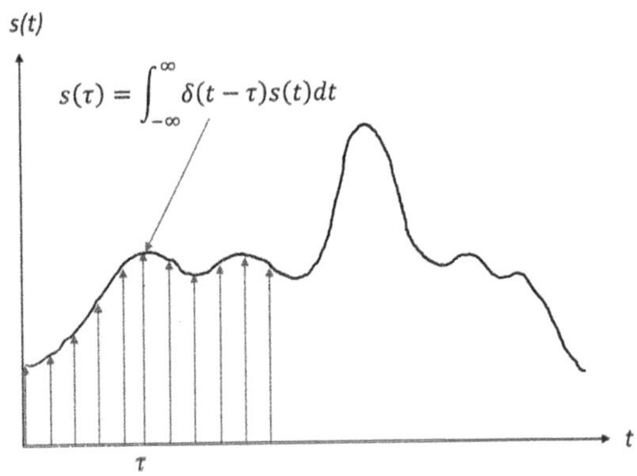

Figure 2.11 Representing a signal as a train of impulses modulated by the signal.

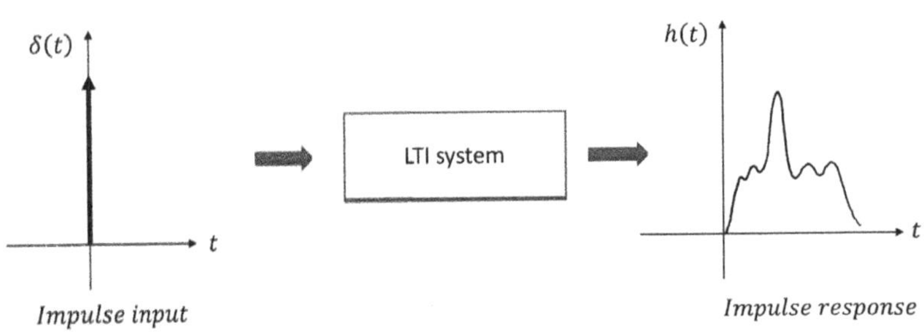

Figure 2.12 Impulse response of LTI system.

This formula is called the convolution operator (written as $x(t) * h(t)$), which is a very common technique to study systems. If we apply Laplace Transform to both sides of the convolution formula and recall that multiplying exponential is simply done by adding exponents, we will get the following useful algebraic formula:

$$Y(s) = H(s)X(s) \tag{2-98}$$

$$H(s) = \frac{Y(s)}{X(s)} \tag{2-99}$$

where $Y(s)$ is the Laplace transform of the output, $X(s)$ is the Laplace transform of the input, and $H(s)$ is the Laplace transform of the impulse response which turns out to be the transfer function of the system.

2.3.3.3 Transfer Function from Differential Equation

The transfer function can be derived from the system differential equation. If the system is linear, it is represented by an ordinary linear differential equation. If we took the Laplace transform of both sides, we would have the following:

$$\mathcal{L}\left(x^{(n)}(t) + a_1 x^{(n-1)}(t) + a_2 x^{(n-2)}(t) + \ldots + a_n x(t)\right) = \mathcal{L}\left(b_1 u^{(n-1)}(t) + b_2 u^{(n-2)}(t) + \ldots + b_n u(t)\right) \tag{2-100}$$

$$s^n X(s) + a_1 s^{n-1} X(s) + a_2 s^{n-2} X(s) + \ldots + a_n X(s) = b_1 s^{n-1} U(s) + b_2 s^{n-2} U(s) + \ldots + b_n U(s) \tag{2-101}$$

By rearranging the terms, the transfer function $H(s) = \frac{X(s)}{U(s)}$ can be given by

$$H(s) = \frac{X(s)}{U(s)} = \frac{b_1 s^{n-1} + b_2 s^{n-2} + \ldots + b_n}{s^n + a_1 s^{n-1} + a_2 s^{n-2} + \ldots + a_n} \tag{2-102}$$

Once we get the system transfer function, $H(s)$, we can solve the system under any given input (i.e., calculating $x(t)$ given $u(t)$), by calculating $\mathcal{L}(u(t)) = U(s)$ and substitute in the transfer function as follows:

$$X(s) = H(s)U(s) \tag{2-103}$$

This is a simple algebraic equation without differentiation. Once we calculate $X(s)$, we apply inverse Laplace transform to get $x(t)$ as follows:

$$x(t) = \mathcal{L}^{-1}(X(s)) \tag{2-104}$$

2.3.3.4 Transfer Function Example

Given the differential equation of the mass-spring system example, its transfer function is given by the following formula:

$$H(s) = \frac{X(s)}{U(s)} = \frac{1/m}{s^2 + \frac{b}{m}s + \frac{k}{m}} \tag{2-105}$$

Now assume the input force $u(t)$ is a unit step given by the following formula:

$$u(t) = \begin{cases} 1 & t \geq 0 \\ 0 & t < 0 \end{cases} \tag{2-106}$$

The Laplace transform of a unit step function is given by:

$$\mathcal{L}(f(t)) = \frac{1}{s} \tag{2-107}$$

The solution $X(s)$ would be given by the following formula:

$$X(s) = \frac{1/m}{\left(s^2 + \frac{b}{m}s + \frac{k}{m}\right)} \frac{1}{s} \tag{2-108}$$

If we calculate the inverse Laplace transform of $X(s)$, assuming $b = m = k = 1$, it will give us the following time-domain solution:

$$x(t) = -e^{-\frac{t}{2}} \cos\left(\frac{\sqrt{3}t}{2}\right) - \frac{1}{\sqrt{3}}\left(e^{-\frac{t}{2}} \sin\left(\frac{\sqrt{3}t}{2}\right)\right) + u(t) \tag{2-109}$$

Even if it is difficult to calculate the $x(t)$ from $U(s)$, it turns out that the study of the transfer function provides useful information about the solution $x(t)$. For example, $X(s)$ has no zeros (there are no values of s that make $X(s) = 0$), and it has poles at $s = 0$ and $s = \frac{-\frac{b}{m} \pm \sqrt{\left(\frac{b}{m}\right)^2 - \frac{4k}{m}}}{2} = \frac{-1 \pm \sqrt{-3}}{2} = \frac{-1}{2} \pm j\frac{\sqrt{3}}{2}$. $X(s)$ has poles at the imaginary axis and real axis. Any pole on the imaginary axis indicates a sinusoidal term and any pole on the negative real axis indicates a damping exponential term, which is clearly shown in the $p(t) = x(t)$ solution formula above. The solution plot is given in Figure 2.13.

2.3.4 Stochastic Systems

In the previous discussions, we assumed that the system inputs and outputs are deterministic signals. In real-life applications, we always deal with random signals. Therefore, it is important to study the effect of random inputs on systems. We will limit our review to linear systems as they are the most widely used models in navigation systems. Even nonlinear systems can be linearized allowing the application of linear tools and algorithms. In most engineering applications, random noises are modeled by a zero-mean Gaussian distribution. Therefore, we will study linear dynamic

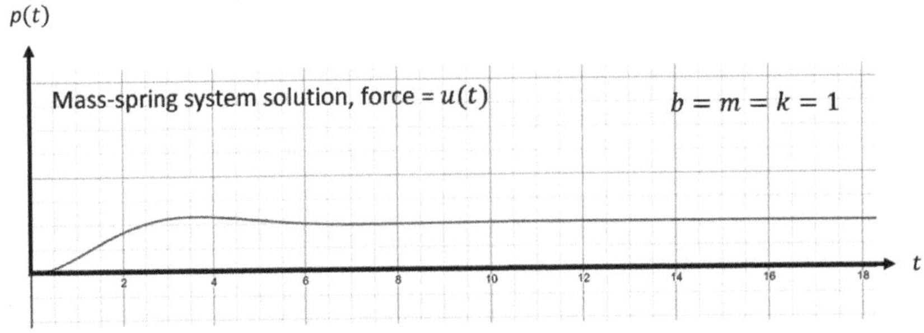

Figure 2.13 Analytical solution of the mass-spring system.

systems under zero-mean Gaussian inputs. Systems with random input are called stochastic systems. We refer to linear dynamic stochastic system with inputs contaminated by random zero-mean Gaussian noise by Gauss–Markov processes.

2.3.4.1 Gauss-Markov Processes

We want to study the effect of random inputs on the system output. We assume that the stochastic system has a deterministic part that can be described using state space model and stochastic random part. The stochastic random part is assumed to be an additive zero-mean Gaussian noise that is simply added to the deterministic part of the system. A linear system with these noise assumptions is called a Gauss–Markov process. Therefore, in our analysis, we will assume that the linear system has the following stochastic model [5, 6]:

$$\dot{\vec{x}}(t) = F(t)\vec{x}(t) + G(t)\vec{u}(t) + \vec{w}(t) \tag{2-110}$$

where $\vec{w}(t)$ is random noise. In some configurations, $\vec{u}(t)$ can be embedded as part of the states like the case in error dynamics systems model. In this configuration, the linear stochastic system is given by the following equation:

$$\dot{\vec{x}}(t) = F(t)\vec{x}(t) + G(t)\vec{w}(t) \tag{2-111}$$

$$\vec{y}(t) = H(t)\vec{x}(t) + \vec{v}(t) \tag{2-112}$$

$$\vec{w} \sim \mathcal{N}(0, Q(t)) \tag{2-113}$$

$$\vec{v} \sim \mathcal{N}(0, R(t)) \tag{2-114}$$

$$\text{cov}(v(t), w(t)) = 0 \tag{2-115}$$

$$\text{cov}(x(t), w(t)) = 0 \tag{2-116}$$

$$\text{cov}(x(t), v(t)) = 0 \tag{2-117}$$

We will refer to $G(t)$ as the noise shaping matrix, $w(t)$ as the process or system noise and the $v(t)$ as the measurement noise. Both noise vectors are zero-mean Gaussian, and both are uncorrelated to each other and to the system states. The goal is to answer the following question: given a linear dynamic system with additive zero-mean Gaussian noise, what are the mean and covariance of the system states? To answer this question, we will study the mean and covariance propagation through the linear system. Having the system linear and the random part additive independent zero-mean Gaussian greatly simplifies this analysis as we will see. To further simplify the analysis, we will consider the discrete version of the system model given by:

$$\vec{x}_{k+1} = \emptyset_k \vec{x}_k + G_k \vec{u}_k + \vec{w}_k \tag{2-118}$$

where $\emptyset_k$ and G_k are the discrete equivalent of the transition matrix $F(t)$ and the noise shaping matrix $G(t)$, respectively. In other configurations where the control input is embedded into the state vector, the model will be given by the following equation:

$$\vec{x}_{k+1} = \emptyset_k \vec{x}_k + \vec{w}_k \tag{2-119}$$

$$\vec{y}_{k+1} = H_k \vec{x}_k + \vec{v}_k \tag{2-120}$$

$\vec{w}_k$ and $\vec{v}_k$ are the discrete equivalent of $\vec{w}(t)$ and $\vec{v}(t)$ given as follows:

$$\vec{w}_k \sim \mathcal{N}(0, Q_k) \tag{2-121}$$

$$\vec{v}_k \sim \mathcal{N}(0, R_k) \tag{2-122}$$

2.3.4.2 Propagation of the Mean

The mean of the state can be calculated by taking the expectation of both sides of the system model equation in (2-118) as follows:

$$E\left\langle \vec{x}_{k+1} \right\rangle = E\left\langle \emptyset_k \vec{x}_k + G_k \vec{u}_k + \vec{w}_k \right\rangle \tag{2-123}$$

$$E\left\langle \vec{x}_{k+1} \right\rangle = E\left\langle \emptyset_k \vec{x}_k \right\rangle + E\left\langle G_k \vec{u}_k \right\rangle + E\left\langle \vec{w}_k \right\rangle \tag{2-124}$$

Since $\emptyset_k$, $\vec{u}_k$, and G_k are deterministic, they can be taken out of the expectation operator $E\langle . \rangle$ and we assume $E\left\langle \vec{w}_k \right\rangle = 0$, which gives the following equations:

$$E\left\langle \vec{x}_{k+1} \right\rangle = \emptyset_k E\left\langle \vec{x}_k \right\rangle + G_k \vec{u}_k \tag{2-125}$$

$$\vec{\mu}_{x_{k+1}} = \emptyset_k \vec{\mu}_{x_k} + G_k \vec{u}_k \tag{2-126}$$

Therefore, the mean of the state vector $\vec{\mu}_x$ is not affected by the random noise. It propagates normally through the system, which means the noisy output of the system will have the same mean as the noise-free output.

2.3.4.3 Propagation of the Covariance

The state covariance matrix of the system state is defined as follows:

$$\operatorname{cov}\left(\vec{x}_k\right) = P_k = E\left\langle \left(\vec{x}_k - \vec{\mu}_{x_k}\right)\left(\vec{x}_k - \vec{\mu}_{x_k}\right)^T \right\rangle \tag{2-127}$$

$$P_{k+1} = E\left\langle \left(\vec{x}_{k+1} - \vec{\mu}_{x_{k+1}}\right)\left(\vec{x}_{k+1} - \vec{\mu}_{x_{k+1}}\right)^T \right\rangle \tag{2-128}$$

$$P_{k+1} = E\left\langle \left(\emptyset_k \vec{x}_k + G_k \vec{u}_k + \vec{w}_k - \emptyset_k \vec{\mu}_{x_k} - G_k \vec{u}_k\right)\left(\emptyset_k \vec{x}_k + G_k \vec{u}_k + \vec{w}_k - \emptyset_k \vec{\mu}_{x_k} - G_k \vec{u}_k\right)^T \right\rangle \tag{2-129}$$

$$P_{k+1} = E\left\langle \emptyset_k \left(\vec{x}_k - \vec{\mu}_{x_k} + \vec{w}_k\right)\left(\emptyset_k \left(\vec{x}_k - \vec{\mu}_{x_k} + \vec{w}_k\right)\right)^T \right\rangle \tag{2-130}$$

$$P_{k+1} = E\left\langle \emptyset_k \left(\vec{x}_k - \vec{\mu}_{x_k} + \vec{w}_k\right)\left(\emptyset_k \left(\vec{x}_k - \vec{\mu}_{x_k} + \vec{w}_k\right)\right)^T \right\rangle \tag{2-131}$$

$$P_{k+1} = \emptyset_k E\left\langle \left(\vec{x}_k - \vec{\mu}_{x_k}\right)\left(\left(\vec{x}_k - \vec{\mu}_{x_k}\right)\right)^T \right\rangle \emptyset_k^T + E\left\langle \vec{w}_k \vec{w}_k^T \right\rangle \tag{2-132}$$

$$E\left\langle \vec{w}_k \vec{w}_k^T \right\rangle = Q_k \tag{2-133}$$

$$P_{k+1} = \emptyset_k P_k \emptyset_k^T + Q_k \tag{2-134}$$

Similarly, if we have a linear measurement model in the following form:

$$\begin{aligned} \vec{y}_k &= H_k \vec{x}_k \\ \vec{x}_k &\sim \mathcal{N}\left(\vec{\mu}_{x_k}, P_k\right) \end{aligned} \tag{2-135}$$

the predicted measurements mean is given by:

$$\vec{\mu}_{y_k} = H_k \vec{\mu}_{x_k} \tag{2-136}$$

and the covariance of the predicted measurements will be given by:

$$\text{cov}\left(\vec{y}_k\right) = H_k P_k H_k{}^T \tag{2-137}$$

The covariance equation shows that the covariance of the state vector and the predicted measurements are increasing by two factors, the transition matrix $\emptyset_k$ and the system noise Q_k. This means that when applying Gaussian noisy inputs to a linear dynamic system, the state vector mean is not affected, and it just follows the deterministic part of the dynamic system model. However, the state vector covariance (which is the deviation from the mean) is increasing. As the system output is a linear combination of the state, the output covariance will also increase, which means that the system-predicted output becomes nosier and nosier with time as illustrated in Figure 2.14, which shows the response of a linear system to an additive zero-mean Gaussian noise input with zero initial conditions $x(0) = 0$. If there is no noise, the system state and output should stay zero according to the deterministic system and measurement models. Under additive zero-mean Gaussian noise, the state covariance and hence the output covariance increases with time.

2.3.5 Control Systems

In navigation systems, we usually encounter problems where we want to control or steer the system state to follow a certain desired value. An example of this control loop is the phase lock loop (PLL) [3, 8] that is used in many radio ranging and navigation systems. Control systems theory is concerned about the methods and techniques that maintain a system output at a desired value. Closed-loop feedback control is commonly applied where the control system constantly monitors the error between the measured system output $\hat{y}(t)$ and the desired system output $y_d(t)$ and issues a control signal $u(t)$ to steer the system output toward the desired $y_d(t)$ as seen in Figure 2.15. Control systems are used in many navigation systems where signals with specific characteristics are to be

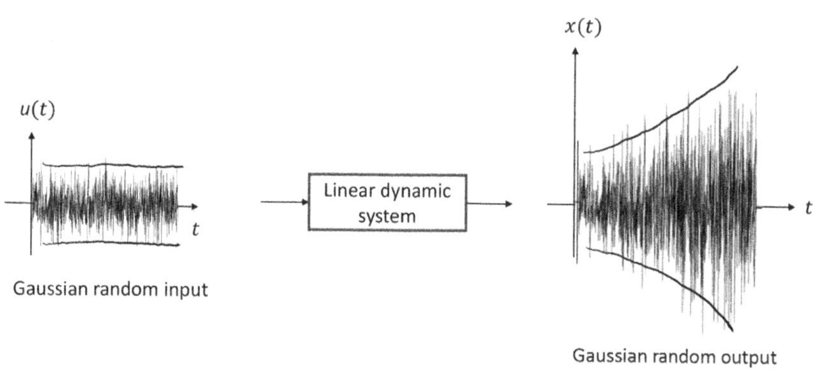

Figure 2.14 Effect of additive zero-mean Gaussian noise on the output of LTI system.

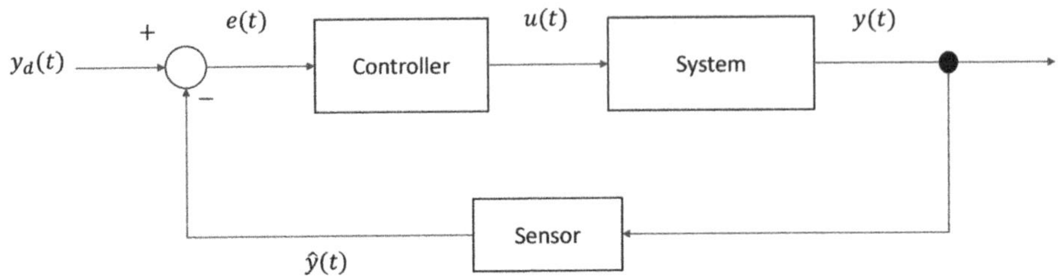

Figure 2.15 Closed-loop feedback control system.

detected and tracked. This signal tracking control system can be found in GPS, Radar, LiDAR, and other navigation systems/sensors.

Figure 2.15 shows the general diagram of closed-loop control system. For linear systems, control methods are usually studied in the S-domain to avoid the complexity of the differential equations. In the S-domain, the closed-loop control system is illustrated in Figure 2.16.

The general transfer function of the overall system is given by:

$$\frac{Y(s)}{Y_d(s)} = \frac{F(s)C(s)}{1 + F(s)C(s)} \tag{2-138}$$

where $C(s)$ is the transfer function of the controller and the $F(s)$ is the transfer function of the LTI system. Given a desired output $Y_d(s)$, we can use the transfer function to study the response of the system output $Y(s)$ under controller $C(s)$. The most common control method for LTI systems is PID controllers [4]. PID stands for proportional, integral, differential controller. Although PID controllers have been studied for decades, they are still used in many navigation systems due to their efficiency and simplicity. The transfer functions of PID controllers are shown in Table 2.2.

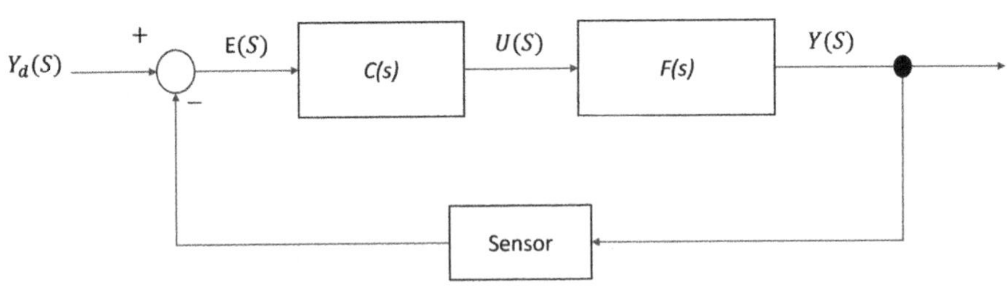

Figure 2.16 Closed-loop feedback control of LTI system.

Table 2.2 Transfer functions of PID controllers.

P controller	PI controller	PD controller	PID controller
$C(s) = K_p$	$C(s) = \frac{K_p s + K_I}{s}$	$C(s) = K_p + K_D s$	$C(s) = \frac{K_D s^2 + K_p s + K_I}{s}$

Being a closed system, stability becomes a great concern. There are several methods to study the stability of control systems, most of which are based on the S-domain, such as Root Locus and Bode-Plots [4]. The design of PID controllers depends on the system being controlled. The design decisions include which component will be used (P, I, or D) and the best tuning of the controller constants K_p, K_D, and K_I. The control theory and the details of control systems analysis are out of the scope of this book. It is mentioned briefly here because, as we will see later, a control loop is used in PLLs in radio and light ranging and navigation sensors/systems. More information about control systems and PID controller design can be found in [9] and [4].

2.3.6 Linearization

Linear systems facilitate the analysis and solutions to systems. If the system model is not linear (which is the case in most navigation systems), linearization is commonly used to enable us to apply the same analysis/solutions tools that we apply with linear systems. In this discussion, we will assume that the system state, measurement, and control variables are all scalar. However, all concepts are applicable to vectors. The general model for nonlinear systems is given by:

$$\dot{x}(t) = f(x(t), u(t)) + \epsilon(t) \tag{2-139}$$

$$y(t) = h(x(t)) + v(t) \tag{2-140}$$

where $\epsilon(t)$ and $v(t)$ are random components that model the effect of input and measurements noises. For now, we cannot assume Gaussian noise because the system is nonlinear. A common technique to deal with nonlinear systems is linearization by applying Tylor series expansion of the nonlinear models. Tylor series expansion of any function $f(x)$ around an operating point $\overline{x}$ is given by the following series:

$$\begin{aligned} f(x,u) = {} & f(\overline{x},\overline{u}) + \left.\frac{\partial f(x,u)}{1!\partial x}\right|_{x=\overline{x}} (x-\overline{x}) + \left.\frac{\partial^2 \partial f(x,u)}{2!\partial x^2}\right|_{x=\overline{x}} (x-\overline{x})^2 + \ldots \left.\frac{\partial f(x,u)}{1!\partial u}\right|_{u=\overline{u}} (u-\overline{u}) \\ & + \left.\frac{\partial^2 \partial f(x,u)}{2!\partial x^2}\right|_{u=\overline{u}} (u-\overline{u})^2 + \ldots \end{aligned} \tag{2-141}$$

$$h(x(t)) = h(\overline{x},\overline{u}) + \left.\frac{\partial h(x)}{1!\partial x}\right|_{x=\overline{x}} (x-\overline{x}) + \ldots \tag{2-142}$$

Tylor series estimates the value of $f(x, u)$ given a known operating point $\{\overline{x},\overline{u}\}$ and the derivatives of $f(x, u)$ at this operating point. The more terms we use, the more accurate the series output will be. However, if the differences $\delta x = (x-\overline{x})$ and $\delta u = (u-\overline{u})$ are small, the power of δx and δu becomes significantly small, which makes the first-order Tylor series expansion a reasonable approximation of the value of $f(x, u)$ in the region close and around the operating point $\{\overline{x},\overline{u}\}$. An illustration of first-order Tylor series expansion on a 1-D function $f(x)$ is shown in Figure 2.17.

Applying first-order Tylor series expansion on the nonlinear models of the system around an operating point will lead to the following linear models.

$$\delta\dot{x}(t) = F(t)\delta x(t) + G(t)\delta u(t) + w(t) \tag{2-143}$$

$$\delta y(t) = H(t)\delta x(t) \tag{2-144}$$

$$F(t) = \left.\frac{\partial f(x(t),u(t))}{\partial x}\right]_{x=\overline{x}} \tag{2-145}$$

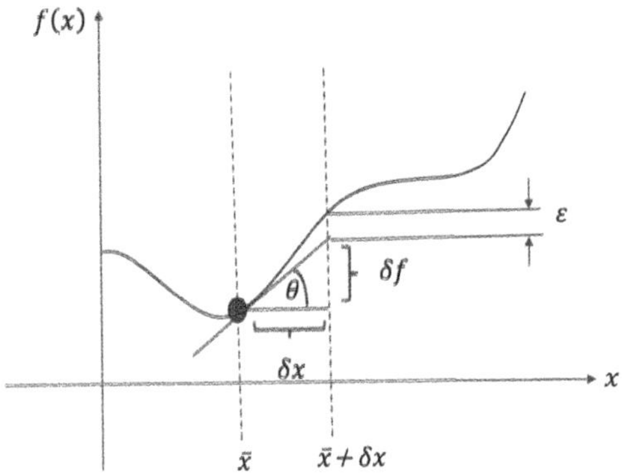

Figure 2.17 First-order Tylor series approximation.

$$G(t) = \left.\frac{\partial f(x(t), u(t))}{\partial u}\right]_{u=\overline{u}} \tag{2-146}$$

$$H(t) = \left.\frac{\partial h(x(t))}{\partial x}\right|_{x=\overline{x}} \tag{2-147}$$

where $w(t)$ is zero-mean Gaussian noise. In many configurations, we can embed the error in the control signal $\delta u(t)$ into the error state vector $\delta x(t)$. In this case, the linearized error model is given by the following simpler equations:

$$\delta\dot{x}(t) = F(t)\delta x(t) + w(t) \tag{2-148}$$

$$\delta y(t) = H(t)\delta x(t) + v(t) \tag{2-149}$$

The above linear models are functions in the error of the system states, not the state themselves. That is the reason we usually refer to the linear system models as error models. The linearized system enables us to apply linear analysis and solution techniques. The matrices $F(t)$, $G(t)$, and $H(t)$ represent the linearized system model at time t around an operating point $\{\overline{x}, \overline{u}\}$. In discrete form, the model will be given by the following equations:

$$\vec{x}_{k+1} = \emptyset_k \vec{x}_k + \vec{w}_k \tag{2-150}$$

$$\vec{y}_{k+1} = H_k \vec{x}_k + \vec{v}_k \tag{2-151}$$

$$\vec{w}_k \sim \mathcal{N}(0, Q_d) \tag{2-152}$$

$$\vec{v}_k \sim \mathcal{N}(0, R) \tag{2-153}$$

2.3.7 Numerical Solution

A system with a deterministic analytical input signal can be solved by solving its differential equations or by the transfer function technique as discussed earlier or numerically. The first-order time derivative of a signal $x(t)$ is given by:

$$\dot{x}(t) = \lim_{\delta t \to 0} \frac{x(t + \delta t) - x(t)}{\delta t} \tag{2-154}$$

For small period, the numerical solution approximates the time-derivative using an approximation formula such as Euler first-order zero-hold formula given by:

$$\dot{x}(t + \delta t) \cong \frac{x(t + \delta t) - x(t)}{\delta t} \tag{2-155}$$

If the system model is given by:

$$\dot{x}(t) = f(x(t), u(t)) \tag{2-156}$$

then, the first-order approximate numerical solution is given by:

$$x(t + \delta t) \cong x(t) + f(x(t), u(t))\, \delta t \tag{2-157}$$

where δt is a small-time increment called the sampling period. Given an initial condition $x(t - T_k)$, we can approximate $x(t)$ using this simple first-order Euler integration method. There are more sophisticated numerical integration methods such as Runge–Kutta. Unfortunately, we cannot use analytical solutions or numerical solutions in a straightforward way if the input to the system (either the initial condition or the control input) is random or has a random part. To solve systems with random inputs, we use estimation methods. Estimation will be discussed in the next chapter.

2.4 Signals

A signal can be seen as a vector of values given as a function of a parameter (mostly time). Most sensors signals are function of time (domain is time). Being a function means at any instant of time t where the signal is defined, the signal has one single value. The value of the signal could be real or complex. A signal could be a function of parameters other than time. For example, images can be represented as a function of time and pixel location. In the following subsection, we will discuss common characteristics of signals. More detailed information about signals can be found in [6] and [7].

2.4.1 Continuous vs. Discrete Signals

Mathematically, continuous signals have values at infinitesimally small instants of time, which means that the signal domain t is infinite. If the derivative of the signal function (the slope of the signal curve) is defined at any point of time, the signal is called differentiable. A differentiable signal is also called a smooth signal. Note that continuity does not imply smoothness or differentiability. For example, the signal in Figure 2.18a is continuous and differentiable at the shown duration. However, the signal in Figure 2.18b is continuous but not differentiable at time t_1.

Most signals in nature are continuous, including electrical signals (voltage or current). However, in digital systems, we can only measure and process the signals at discrete instants of time even if the signal source is continuous. The discrete version of the signal is a finite set of samples of the signal taken at specific instants of time as seen in Figure 2.19b. Note that by taking discrete samples of the signal we are losing some information about the signal because the value of the signal in-between samples is not known. We will discuss information loss due to sampling when we discuss the analog-to-digital conversion.

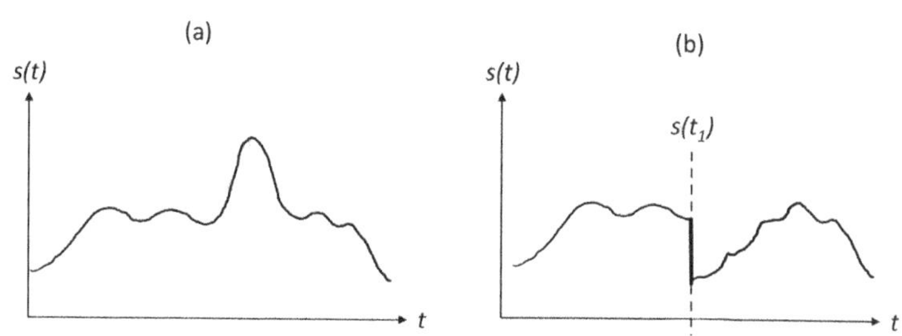

Figure 2.18 (a) Continuous differentiable signal and (b) continuous but not differentiable at t_1.

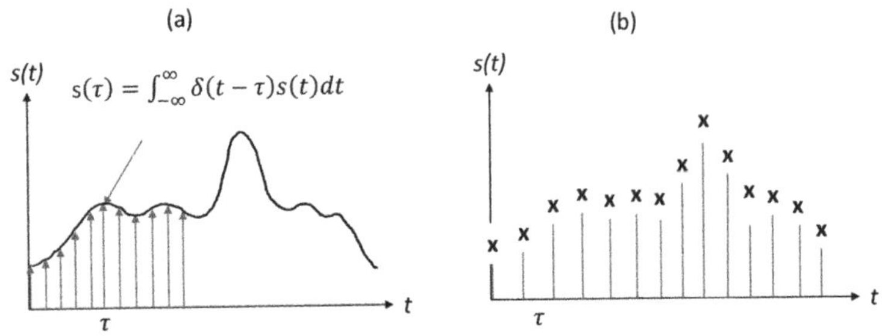

Figure 2.19 (a) Continuous signal and (b) discrete signal.

2.4.2 Analog vs. Digital Signals

Signals in nature are mostly analog signals like the example shown in Figure 2.20a. The term analog means that the signal can assume any value. On the other hand, digital signals' values are restricted to specific discrete values (levels) like the one shown in Figure 2.20b. The digitized signal takes a constant value during the sampling period. Digitized signals can be processed by a digital computer.

The constant definite levels that the signal can take in the discrete domain enable the digital representation of the signal values in a computer system where numbers are stored and processed in binary numbering system.

2.4.3 Periodic vs. Nonperiodic Signals

Signals can be categorized into periodic and nonperiodic signals. Periodic signals repeat every specific period T as follows:

$$s(t) = s(t - T) \tag{2-158}$$

Periodic signals, especially sinusoidal signals, are of great importance in radio and laser ranging, navigation, mapping, imaging, and communication systems. A sinusoidal signal of period T measured in seconds, and phase φ measured in radians, is an example of a periodic signal that can be seen in Figure 2.21. The length of a period is called wavelength λ. The frequency f of the signal represents the number of cycles per seconds and it is measured in Hertz (Hz). If this periodic sinusoidal signal travels a distance Δd in Δt time with speed s, the relationship between its wavelength λ, period T, frequency f and phase φ are given by:

$$\lambda = s.T \tag{2-159}$$

$$T = 1/f \tag{2-160}$$

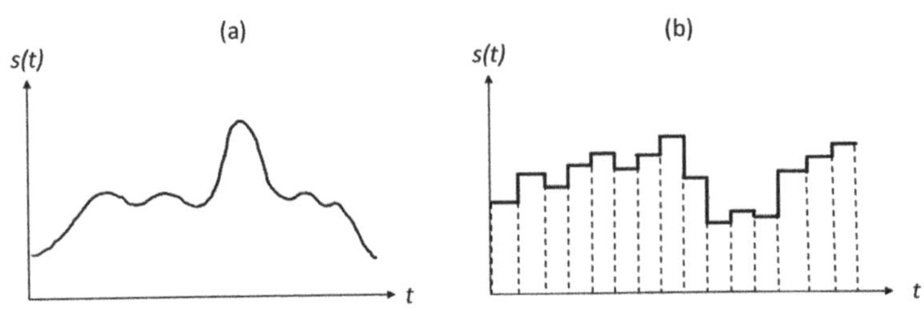

Figure 2.20 (a) Analog continuous signal and (b) digital signal.

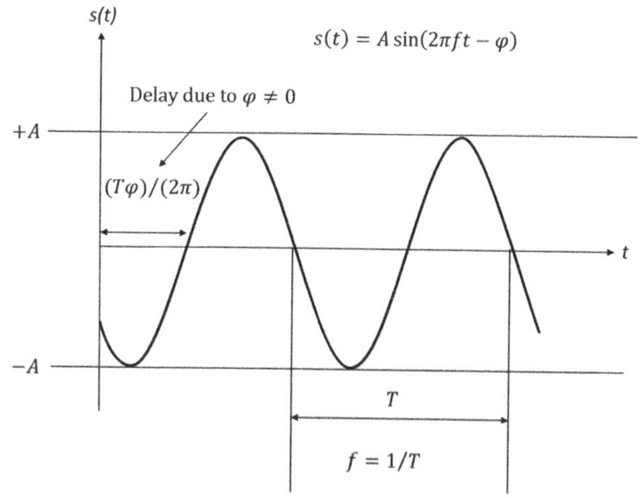

Figure 2.21 Sinusoidal signal.

$$s = f.\lambda \tag{2-161}$$

$$\Delta d = s.\Delta t \tag{2-162}$$

$$\Delta\varphi = 2\pi f\Delta t \tag{2-163}$$

$$\Delta t = T(\Delta\varphi/2\pi) \tag{2-164}$$

$$\Delta\varphi = 2\pi f\left(\frac{\Delta d}{\lambda}\right) \tag{2-165}$$

$$f = \frac{1}{2\pi}\frac{d\varphi}{dt} \tag{2-166}$$

$$\varphi(t) = 2\pi\int f(t)dt = 2\pi ft + \varphi(0) \tag{2-167}$$

Note here that phase $\varphi(t)$ is an integration of frequency $f(t)$. This relationship is important to realize as it will be used in PLL analysis used in radio and laser ranging, detection, navigation, imaging, and mapping systems.

2.4.3.1 Phasor Form of Sinusoidal Signals

Sinusoidal signals have special importance in engineering. For reasons that will be clear later, it is mathematically useful to express sinusoidal signal in exponential complex form called Phasor. The complex exponential form has several mathematical advantages, and it has been a very useful tool in simplifying mathematical notations. A common way to represent sinusoidal signals using exponential form is Euler formula:

$$e^{j\emptyset} = \cos(\emptyset) + j\sin(\emptyset) \tag{2-168}$$

$$A\cos(\emptyset) = \frac{A}{2}\left(e^{j\emptyset} + e^{-j\emptyset}\right) \tag{2-169}$$

This is a complex formula that has an imaginary part and real part. The real-valued sinusoidal signals are commonly represented by the following exponential formula:

$$A\cos(2\pi ft + \phi) = \Re e\left\{Ae^{j(2\pi ft + \phi)}\right\} = \Re e\left\{Ae^{j\phi}e^{j2\pi ft}\right\} \tag{2-170}$$

This part $Ae^{j\phi}$ is called the "*phasor*" of the sinusoidal signal. Here, the signal $A\cos(2\pi ft + \phi)$ is a cosine signal leading by a phase of ϕ. If ϕ is positive, this means the cosine signal is leading by ϕ. If ϕ is negative, this means that the cosine signal is delayed by $\frac{\phi}{2\pi f}$ or $\frac{\phi}{2\pi}T$, where f is the frequency and T is the period of the signal.

2.4.3.2 In-Phase/Quadrature (I/Q) Form of Sinusoidal Signals

Transmitting a sinusoidal with a specific desired phase and amplitude can be easily done using the in-phase/quadrature representation of sinusoidal signals. A sinusoidal signal can be represented as two sinusoidal signals that are 90° apart in phase. A cosine signal delayed by a phase of $\phi(t)$ can be represented as follows:

$$x(t) = A(t)\cos(2\pi ft - \phi(t)) \tag{2-171}$$

$$x(t) = A(t).\cos(2\pi ft)\cos(\phi(t)) + A(t).\sin(2\pi ft)\sin(\phi(t)) \tag{2-172}$$

$$x(t) = I_x(t)\cos(2\pi ft) + Q_x(t)\sin(2\pi ft) \tag{2-173}$$

$$I_x(t) = A(t).\cos(\phi(t)) \tag{2-174}$$

$$Q_x(t) = A(t).\sin(\phi(t)) \tag{2-175}$$

$$x(t) = \sqrt{I^2 + Q^2}\cos\left(2\pi ft - \tan^{-1}\left(\frac{Q_x(t)}{I_x(t)}\right)\right) \tag{2-176}$$

"I" is called the in-phase component/branch and "Q" is called the quadrature component/branch. Any sinusoidal can be represented as I/Q data, which is the common way to store signal samples in many signal processing software tools, signal processing, and communication algorithms. Any signal is composed as I/Q data using numerically controlled oscillator (NCO) and decomposed as I/Q branches as shown in Figure 2.22. By controlling I/Q values, we control the

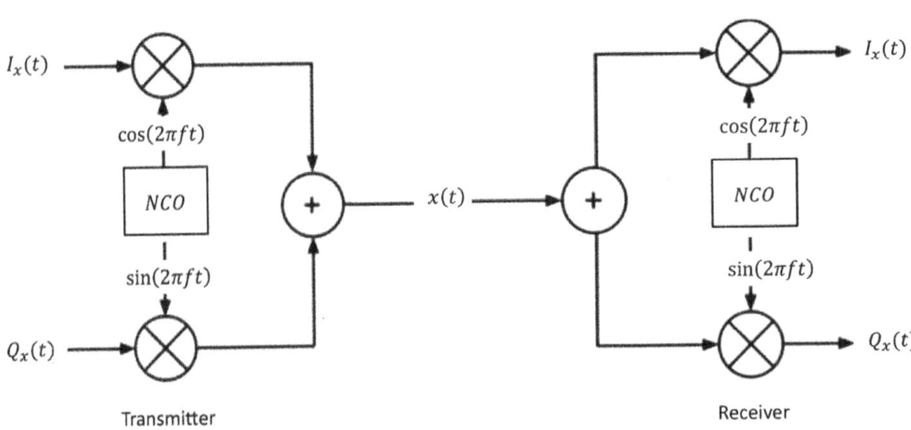

Figure 2.22 I/Q representation of sinusoidal signals.

phase of the generated signal, so this method can be used to implement phase modulation. In fact, using *IQ* form of the sinusoidal signal enables us to implement amplitude, frequency, and phase modulation.

2.4.4 Deterministic vs. Random/Pseudorandom Signals

Deterministic signals have specific values at any specific time. It is completely predictable. The signal in Figure 2.21 is a deterministic signal that is generated by a known function (sinusoidal function) with deterministic parameters. In real-life systems, signals are always contaminated by random noise. A random signal is a signal whose values cannot be certainly predicted like the one in Figure 2.23a. In engineering systems, sometimes we encounter signals with mixed characteristics where a deterministic part is mixed with a random part like the one in Figure 2.23b. The signal in Figure 2.23b can be mathematically described by:

$$s(t) = A\ \sin\left(2\pi\frac{1}{T}t - \varphi\right) + n(t) \tag{2-177}$$

where $n(t)$ is a random signal. Although a random signal cannot be predicted, its values can be described using probability distributions. Random noise comes from different sources. The very common source is the random variations in electrical current/voltage in electrical/electronic circuits. No matter the accuracy of electrical/electronic circuits, there are always small random fluctuations that add noise to signals. These fluctuations can be affected by temperature or other environmental factors. For example, a radar that transmits a perfect deterministic sinusoidal signal like the in Figure 2.21, the actual transmitted signal may look like the one in Figure 2.23b. The reflected signal from an object may look noisier due to signal scattering and fading.

In engineering systems, we can generate a pseudorandom signal that has the characteristics of a random signal but has been generated by a purposefully engineered random signal generator with a known formula and hence can be predicted and repeated as seen in Figure 2.24. Pseudorandom signals have several applications in engineering including satellite navigation systems.

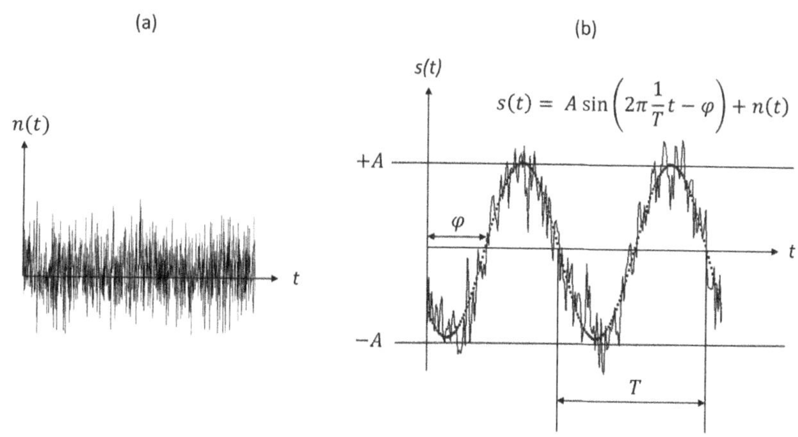

Figure 2.23 (a) Random noise and (b) sinusoidal signal contaminated by random noise.

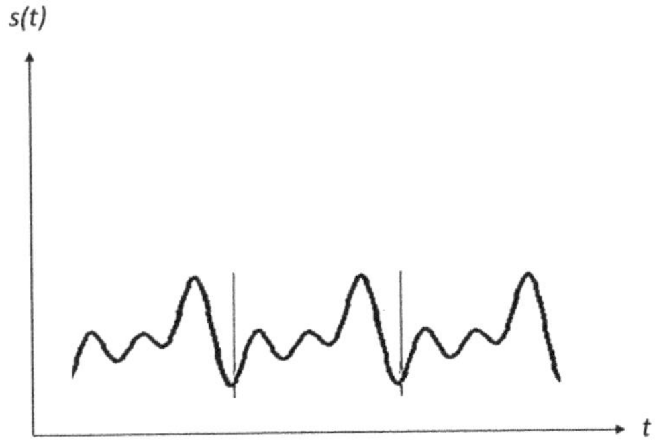

Figure 2.24 A pseudorandom signal repeated three times.

2.5 Random Processes

A random variable that is a function of time is called a stochastic or random process. Most sensors will generate a random process due to random errors that contaminate the sensor measurements. In random processes, the sample space is functions of time not scalar or vectors. Each sample is a function of time with a specific probability. The function space $X(t) \sim N(0, \sigma_n^2)$ is a random process where each sample point is a function $x_i(t)$ as shown in Figure 2.25. At each time value t_i, $X(t_i)$ is not a specific value but a random value that follows a specific probability distribution (Gaussian in this example).

A random process may have a deterministic part and a random part. For example, the following random process $S(t)$ is a stochastic process that has two parts, a deterministic sine wave, and an additive Gaussian random noise part:

$$S(t) = A \sin\left(2\pi \frac{1}{T} t - \varphi\right) + n(t), \quad n(t) \sim N(0, \sigma_n^2) \tag{2-178}$$

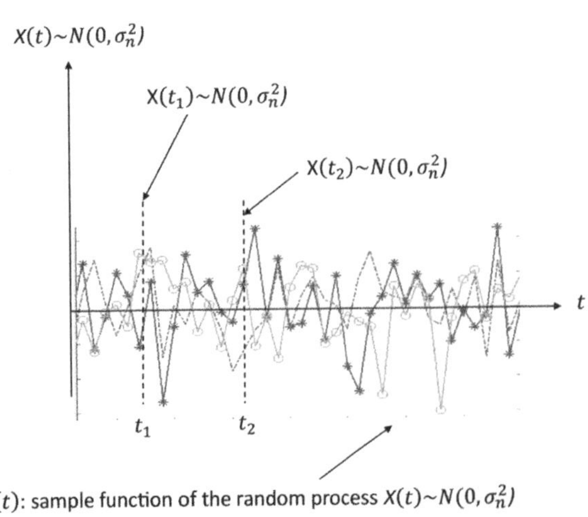

Figure 2.25 Gaussian random process.

where T and φ are deterministic values. At any time t_i, $s(t_i)$ is a scalar Gaussian random variable with mean $\mu(t_1) = A\ \sin\left(2\pi\frac{1}{T}t_1 - \varphi\right)$ and a variance $\sigma^2(t_1) = \sigma_n^2$. Given $n(t)$ is Gaussian, we can mathematically describe the probability distribution of $s(t)$ as follow:

$$f_S(s(t)) = \frac{1}{\sqrt{2\pi}\sigma_n}e^{-\left(\frac{\left[s(t) - A\ \sin\left(2\pi\frac{1}{T}t - \varphi\right)\right]^2}{2\sigma_n^2}\right)} \tag{2-179}$$

where $f_S(s(t))$ is the probability that the random process takes the value $s(t)$ at time t. $f_S(s(t))$ is Gaussian distribution centered around the deterministic part of the process, which is the mean $\mu = A\ \sin\left(2\pi\frac{1}{T}t_1 - \varphi\right)$, with a constant variance of σ_n^2 as illustrated in Figure 2.26. $S(t)$ can also be written as:

$$S(t) \sim N\left(A\ \sin\left(2\pi\frac{1}{T}t - \varphi\right), \sigma_n^2\right) \tag{2-180}$$

More complicated random processes can be formed if the random part is not additive. For example, the random part can be embedded within a function such as the following sinusoidal random process:

$$S(t) = A\ \cos\left(2\pi\frac{1}{T}t + \varphi(t)\right), \quad \varphi(t) \sim N(-\pi, \pi) \tag{2-181}$$

In this example, the random part here is the phase $\varphi(t)$ of the signal and everything else is deterministic. One of the common random processes that are used to model sensor error states is Gauss–Markov random processes [3, 5, 10] defined as follows:

$$\dot{x}(t) = -\beta_x x(t) + \sqrt{2\sigma_x{}^2\beta_x}w_u(t) \tag{2-182}$$

where β is the reciprocal of time constant and σ is the standard deviation and $w_u(t)$ is a unit-variance Gaussian noise. In the following subsections, we will discuss some key aspects of random processes.

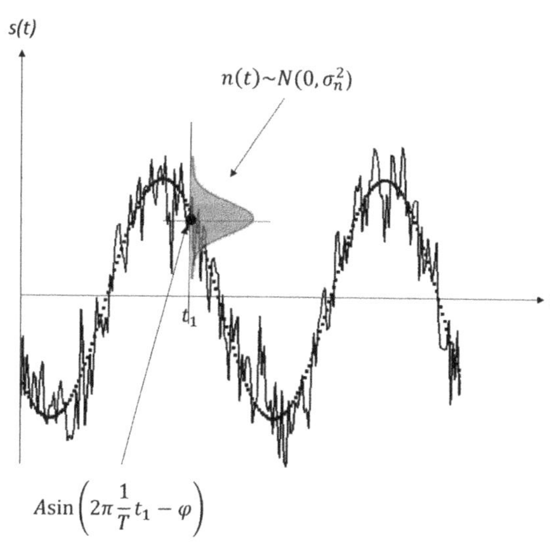

Figure 2.26 A stochastic process: a deterministic sinusoidal part plus a Gaussian random part.

2.5.1 Statistical Properties of Random Processes

As random processes cannot be determined precisely, they can be described using their statistical properties like any random variable. We review the most used statistical properties to describe a random process. The mean of the random process (also known as the first order moment) is generally a function of time calculated by:

$$\mu_x(t) = E\langle X(t)\rangle = \int_{-\infty}^{+\infty} x(t) f_X(x(t))dx(t) \tag{2-183}$$

Note here the integration variable is $dx(t)$ means all possible values of x at time t. The variance of the random process (also known as the second-order moment) is generally a function of time given by:

$$\sigma^2{}_x(t) = E\left\langle (x(t) - \mu_x(t))^2 \right\rangle = \int_{-\infty}^{+\infty} (x(t) - \mu_x(t))^2 f_X(x(t))dx(t) \tag{2-184}$$

The nth-order moment of random process is generally a function of time given by:

$$E\langle (X(t))^n\rangle = \int_{-\infty}^{+\infty} x^n(t) f_X(x(t))dx(t) \tag{2-185}$$

The cross-correlation is a measure of similarity between two random processes:

$$r_{xy}(t_1, t_2) = E\langle x(t_1)y(t_2)\rangle = \int_{-\infty}^{+\infty}\int_{-\infty}^{+\infty} x(t_1)y(t_2)f_{XY}(x(t_1), y(t_2))dx(t_1)dy(t_2) \tag{2-186}$$

Note here we used the joint probability distribution $f_{XY}(x(t_1), y(t_2))$. The autocorrelation is a measure of similarity between the signal and a delayed version of itself:

$$r_{xx}(t_1, t_2) = E\langle x(t_1)x(t_2)\rangle = \int_{-\infty}^{+\infty}\int_{-\infty}^{+\infty} x(t_1)x(t_2)f_{XY}(x(t_1), x(t_2))dx(t_1)dx(t_2) \tag{2-187}$$

A normalized version of cross-correlation and autocorrelation are cross-covariance and autocovariance. Cross-covariance of two random processes is a normalized similarity measure between the two processes defined by:

$$\mathrm{cov}(x(t_1), y(t_2)) = E\langle x(t_1)y(t_2)\rangle - E\langle x(t_1)\rangle E\langle y(t_2)\rangle \tag{2-188}$$

The autocovariance of a random process is a normalized measure between the signal and a delayed version of itself defined by:

$$\mathrm{cov}(x(t_1), x(t_2)) = E\langle x(t_1)x(t_2)\rangle - E\langle x(t_1)\rangle E\langle x(t_2)\rangle \tag{2-189}$$

2.5.2 Stationary Random Processes

If the statistical properties (moments) of the random process do not change with time, we refer to this random process as a strict-sense stationary (SSS) process. This implies that all statistical moments (i.e., mean, variance, and higher order moments) of the random process do not depend on time. If only the mean and variance of a random process are independent of time, we refer to this random process as wide-sense-stationary (WSS). Examples of nonstationary, SSS, and WSS random processes are shown in Figure 2.27.

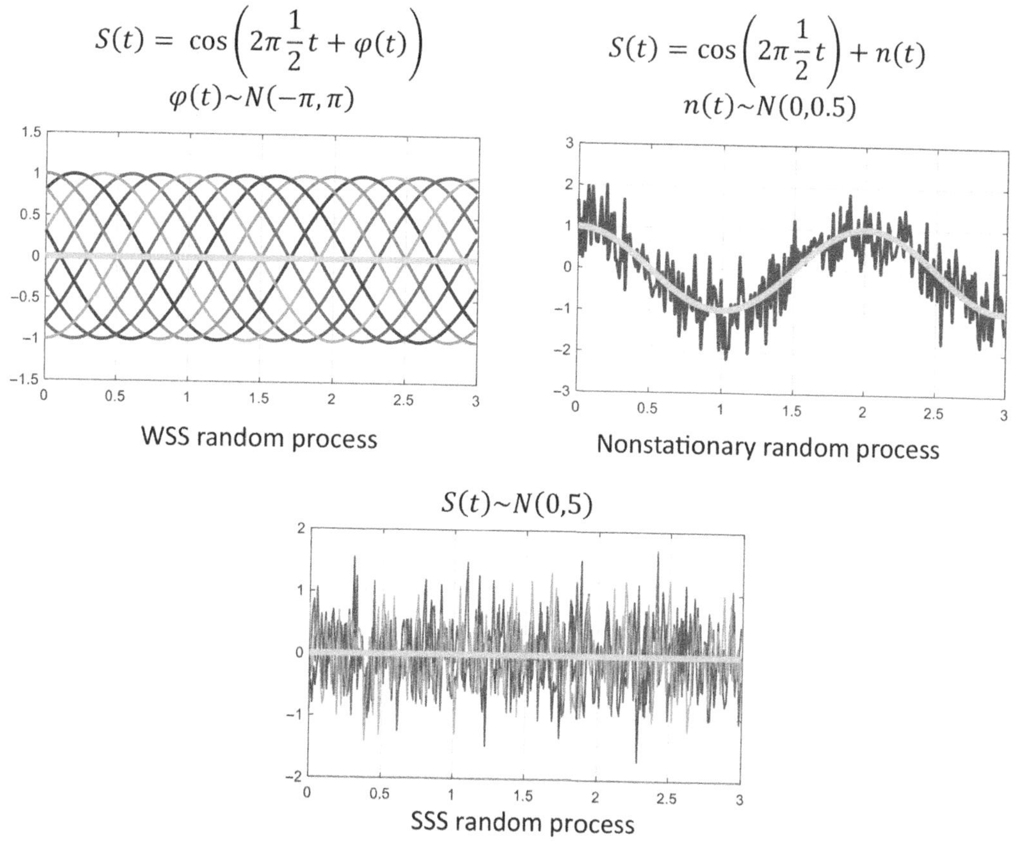

Figure 2.27 Examples of random processes.

If the random process is SSS or WSS, this means that its expected value does not depend on time, and its autocovariance/autocorrelation does not depend on the absolute value of time but instead depends on the time difference. These features can be expressed mathematically as follows:

$$E\langle X(t_1)\rangle = E\langle X(t_2)\rangle \tag{2-190}$$

$$R_{xx}(t_1, t_2) = E\langle x(t_1)x(t_2)\rangle = R_{xx}(t_1 - t_2) = R_{xx}(\tau) \tag{2-191}$$

2.5.3 Ensemble Average vs. Temporal Average

Averaging is an approximate method to estimate a random variable's mean or expected value. Assume we have a set of values $X_s = \{x_1, x_2, x_3 \ldots x_N\}$ drawn randomly from a random variable X with probability distribution $f_X(x)$. This set of values will have a set of unique values $Y_s = \{y_1, y_2, y_3 \ldots y_M\}$ where $M \leq N$. It is expected that the frequency of occurrences of a certain value y_i is proportional to its probability $f_X(x = y_i)$, which is mathematically expressed as follows:

$$\frac{1}{N}\sum_{i=1}^{N} x_i \cong \sum_{j=1}^{M} y_j f_X(y_j) \tag{2-192}$$

If the sample size is large enough, the value of $\sum_{j=1}^{M} y_i f_X(y_i)$ will approach the mean of the random, which gives:

$$\sum_{j=1}^{M} y_i f_X(y_i) \cong E\langle X\rangle \tag{2-193}$$

$$E\langle X\rangle = \lim_{N\to\infty}\left(\frac{1}{N}\sum_{i=1}^{N} x_i\right) \tag{2-194}$$

This is consistent with the central limit theory discussed in Section 2.2.15. The same concept applies to random processes. For example, if we took several samples (time sequences) of the noisy sinusoidal signal of Figure 2.26 and averaged them, this will give us the "*ensemble average.*" As the PDF of this random process is Gaussian, the ensemble average will converge to the mean of the signal, which is the deterministic sinusoidal part in this particular example. Now, let's focus only on the Gaussian noise part $n(t) \sim N(0, \sigma_n^2)$, which is by itself a random process. If we draw three samples (three time sequences) of (t), and call them $n_1(t)$, $n_2(t)$, and $n_3(t)$ and plot them, we will see something similar to three plots in Figure 2.28. The approximate ensemble average at t_1 can be calculated by:

$$\mu(t_1) = \frac{n_1(t_1) + n_2(t_1) + n_3(t_1)}{3} \tag{2-195}$$

This ensemble average will tend to the PDF mean (which is zero in this example) if we took more samples. Theoretically, if we took infinite number of samples and calculated the ensemble average at each instant of time, we would have the exact mean of the random process $E\langle n(t)\rangle$.

$$E\langle n(t)\rangle = \lim_{N\to\infty}\left(\frac{1}{N}\sum_{i=1}^{N} n_i(t)\right) \tag{2-196}$$

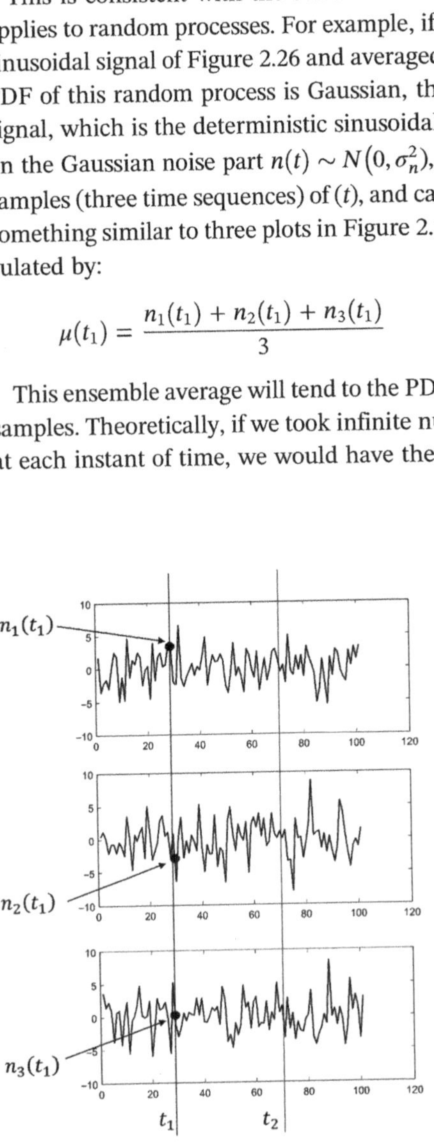

Figure 2.28 Three time-sequence samples of zero-mean Gaussian signal.

The "*temporal average,*" on the other hand, is the average of a single sample of the random process (one time sequence). For example, the temporal average of a sample $n_i(t)$ drawn from $n(t)$ distribution can be given by:

$$\mu_1 = \frac{1}{T}\int_0^T n_1(t) \tag{2-197}$$

It is easy to notice that for zero-mean Gaussian PDF, the temporal average equals the ensemble average if T is large. This can be mathematically expressed by:

$$E\langle n(t)\rangle = \lim_{N\to\infty}\left(\frac{1}{N}\sum_{i=1}^{N} n_i(t)\right) = \lim_{T\to\infty}\left(\frac{1}{T}\int_0^T n_s(t)\right) \tag{2-198}$$

where $n_s(t)$ is any single sample (single time sequence) of the random process $n(t)$. If the ensemble average of a random process equals its temporal average, this random process is called "*ergodic.*" This is a very useful observation because, in practice, it is too restrictive to obtain a large number of samples of $n(t)$ and average them to get the ensemble average. Instead, we obtain one sample of $n(t)$ and calculate the temporal average.

2.5.4 Examples of Random Processes

Random processes are chosen to model the behavior of a stochastic measurement. The choice and the design of the random process are based on the system under consideration. In navigation systems, random walk and Gauss-Markov processes are widely used. The equations and description of random walk and Gauss–Markov random processes are given in this subsection.

2.5.4.1 Random Walk

$$\dot{x}(t) = \sigma_x w(t) \tag{2-199}$$

where $w(t)$ is a unity variance Gaussian random variable. σ_x is an indication of how large the variations of $x(t)$ is expected. Example of samples of a random walk process is shown in Figure 2.29.

2.5.4.2 First-order Gauss–Markov

$$\dot{x}(t) = -\frac{1}{\tau_x}x(t) + \sqrt{\frac{2\sigma_x^2}{\tau_x}}w(t) \tag{2-200}$$

where $w(t)$ is a unity variance Gaussian random variable. σ_x is an indication of how large the variations of $x(t)$ are expected and τ_x is an indication of how fast the changes of $x(t)$ is expected. Examples of first-order Gauss–Markov process are shown in Figure 2.30.

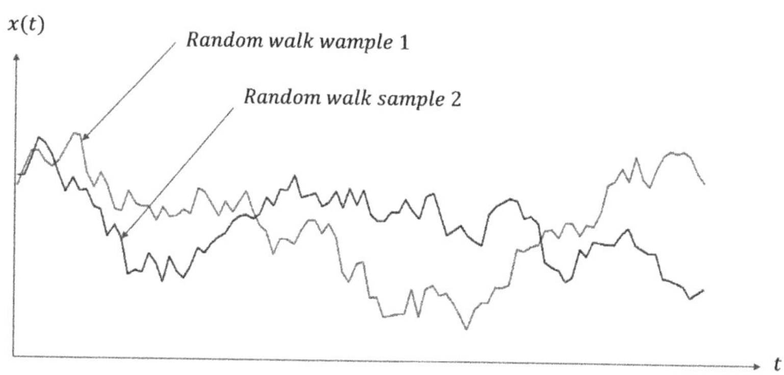

Figure 2.29 Random walk samples.

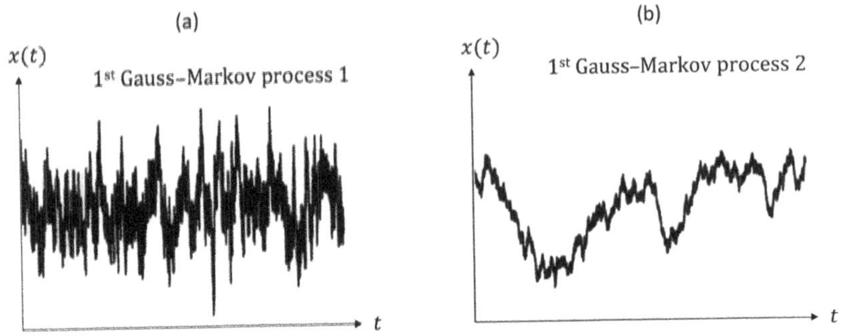

Figure 2.30 First-order Gauss–Markov process. (a) Large and rapid variations; (b) small and slow variations.

2.6 Signal Processing Basics

Signal processing plays a major role in navigation systems. Signal processing is a sophisticated comprehensive research field that is difficult to summarize within a chapter. Therefore, in this section, we review a few key signal-processing concepts that are commonly used in navigation systems. These concepts are important to understand how systems such as global navigation satellite systems, radio detection and ranging sensors, laser detection and ranging scanners, and sound navigation systems work. This section starts by reviewing the Fourier transform and then discusses common digital signal processing concepts, including digital filters. We will conclude with a review of the Fourier analysis of random processes. More details and comprehensive studies about signal processing can be found in [1] and [9]. In addition, scientific computing software tools, such as MATLAB, have a rich signal-processing toolbox with examples.

2.6.1 Fourier Analysis

Dealing with noisy stochastic signals in time domain is challenging due to two main reasons: (i) it is difficult to extract useful information from noisy stochastic signal in the time domain and (ii) if we have a system with impulse response $h(t)$, the convolution operator $\left(y(t) = \int_{-\infty}^{\infty} x(\tau)h(t-\tau)d\tau\right)$ that is used to calculate the system response $y(t)$ to a noisy input $x(t)$ is computationally expensive and mathematically complex. Therefore, most signal processing algorithms transform the time-domain signal into a frequency-domain using Fourier analysis [6, 7]. Fourier analysis uses trigonometric functions as a basis to represent any time-domain signal. Fourier theory states that any infinite periodic signal can be represented as linear combinations of trigonometric functions. The reason to choose trigonometric functions (i.e., sine and cosine) is that they are orthogonal such that:

$$\int_{-\infty}^{\infty} \sin(\omega_i t)\sin(\omega_j t)dt = 0 \quad \forall i \neq j \tag{2-201}$$

Trigonometric functions for infinite periodic time signals are like vector basis for vector space. As any vector in a vector space can be represented uniquely by a linear combination of orthogonal unit vectors. Likewise, any infinite periodic signal $s(t)$ can be represented as a linear combination of trigonometric function.

2.6.1.1 Fourier Series

If the signal is infinite periodic signal with period T, it can be represented by a Fourier series expansion as follows:

$$s(t) = \frac{1}{2}a_0 + \sum_{n=1}^{\infty} a_n \cos\left(\frac{2\pi n}{T}t\right) + \sum_{n=1}^{\infty} b_n \sin\left(\frac{2\pi n}{T}t\right) \quad (2\text{-}202)$$

where T is the signal period, and a_0, a_n, and b_n are coefficients calculated as follows:

$$a_0 = \frac{2}{T}\int_0^T s(t)dt \quad (2\text{-}203)$$

$$a_n = \frac{2}{T}\int_0^T s(t)\cos\left(\frac{2\pi n}{T}t\right)dt \quad (2\text{-}204)$$

$$b_n = \frac{2}{T}\int_0^T s(t)\sin\left(\frac{2\pi n}{T}t\right)dt \quad (2\text{-}205)$$

If the periodic signal is sinusoidal itself, $s(t) = A\cos(2\pi f_1 t)$, then Fourier coefficients will be all zeros (due to the orthogonality of the sinusoidal basis functions) except a_1 which is the coefficient corresponding to frequency f_1. So, in the frequency domain, the infinite periodic function $s(t) = A\cos(2\pi f_1 t)$ is mapped to a single value function $F(f) = A\,\delta(f - f_1)$, where $\delta(.)$ is the Dirac delta function as seen in Figure 2.31. The square periodic pulse in Figure 2.32a can be represented by the following coefficients:

$$a_0 = a_0 = \frac{2}{T}\int_0^T s(t)dt = \frac{2}{T}\left[\int_0^{\frac{T}{2}}(1)dt + \int_{\frac{T}{2}}^{T}(-1)dt\right] = \frac{2}{T}[T/2 - T/2] = 0 \quad (2\text{-}206)$$

$$a_n = \frac{2}{T}\left[\left[\frac{\sin(2\pi(n/T)t)}{2\pi(n/T)}\right]\Bigg|_0^{T/2} - \left[\frac{\sin(2\pi(n/T)t)}{2\pi(n/T)}\right]\Bigg|_{T/2}^{T}\right] = 0 \quad (2\text{-}207)$$

$$b_n = \frac{2}{T}\left[\left[\frac{-\cos(2\pi(n/T)t)}{2\pi(n/T)}\right]\Bigg|_0^{T/2} - \left[\frac{-\cos(2\pi(n/T)t)}{2\pi(n/T)}\right]\Bigg|_{T/2}^{T}\right] = \begin{cases} 0 & n \text{ is even} \\ 4/(n\pi) & n \text{ is odd} \end{cases} \quad (2\text{-}208)$$

Based on the results of Eqs. (2-207) and (2-208), only b_n has values at odd values of n. If we took the first nonzero coefficient, $b_1 = 4/\pi$, we will be approximating the square wave by a sine wave

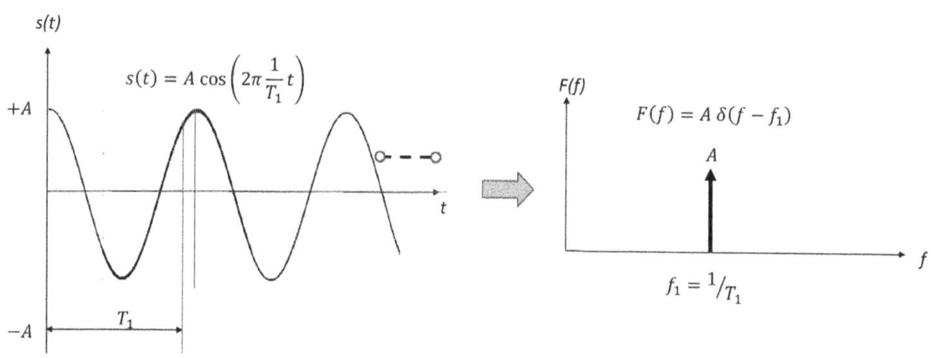

Figure 2.31 Fourier series representation of a sinusoidal periodic signal.

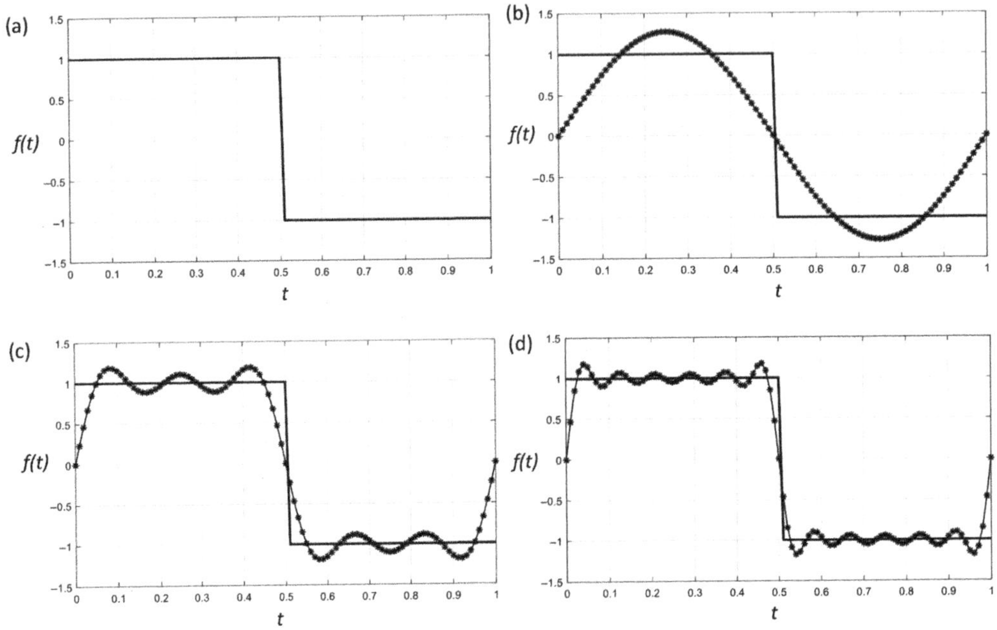

Figure 2.32 Fourier series approximation of a periodic pulse. (a) Original pulse, (b) Fourier approximation with 1 coefficient, (c) Fourier approximation with 3 coefficient, (d) Fourier approximation with 7 coefficient.

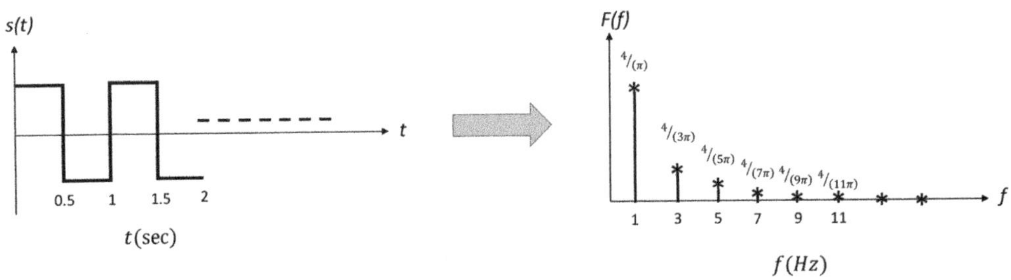

Figure 2.33 Fourier series approximation of a periodic square pulse.

(Figure 2.32b). If we took more coefficients, our approximation would become closer to the original signal. Therefore, we can represent this periodic square signal by a few coefficients (five coefficients would capture most of the signal spectrum as seen in Figure 2.33).

Fourier series coefficient formulas can be simplified into one unified formula using complex representations of sine and cosine. A complex number x is represented by two components, real part $a = \Re e(x)$ and imaginary $b = \Im(x)$ and it is written in the following form:

$$x = a + jb \tag{2-209}$$

where $j^2 = -1$. A complex number has a magnitude A_x and phase $\emptyset_x$ that are calculated as follows:

$$A_x = \|x\| = \sqrt{a^2 + b^2} = \sqrt{\Re e(x)^2 + \Im(x)^2} \tag{2-210}$$

$$\emptyset_x = \tan^{-1}\left(\frac{\Im(x)}{\Re e(x)}\right) \tag{2-211}$$

x can be represented using its amplitude and phase as follows:

$$x = A_x e^{j\emptyset_x} \tag{2-212}$$

$$e^{j\emptyset_x} = \cos\emptyset_x + j\sin\emptyset_x \tag{2-213}$$

The exponential form in (2-213) is called the Euler formula. Using this complex exponential form, the Fourier series of periodic signals $s(t)$ with a period T, $s(t)$ can be represented in a more compact and unified formula as follows:

$$s(t) = \sum_{n=-\infty}^{n=\infty} a_n e^{j\left(2\pi n\frac{1}{T}\right)t} \tag{2-214}$$

$$a_n = \frac{1}{T}\int_0^T s(t)e^{-j\left(2\pi n\frac{1}{T}\right)t}dt \tag{2-215}$$

If we replace $e^{j\left(2\pi n\frac{1}{T}\right)t}$ by $\cos\left(2\pi n\frac{1}{T}\right)t + j\sin\left(2\pi n\frac{1}{T}\right)t$ and substitute in the integral formula, and take the real part of the resulting expression, we will end up having the same real-valued Fourier series formulas. A real sinusoidal signal can be represented in complex exponential form as follows:

$$s(t) = \cos(2\pi ft) = \Re e\left(e^{j2\pi ft}\right) \tag{2-216}$$

$$s(t) = \sin(2\pi ft) = \Im\left(e^{j2\pi ft}\right) \tag{2-217}$$

2.6.1.2 Fourier Transform

If the signal is not an infinite periodic signal, we can still perform Fourier analysis using Fourier transform. Fourier transform is the more general Fourier analysis that is applicable to nonperiodic signals. We will present Fourier transform in its complex form as this is the preferred common form that is widely adopted in the scientific community for several reasons, which will be discussed later when we explain the core navigation and mapping algorithms in the rest of the book. If we replace the discrete summation with a continuous integration and calculate the limit of the a_n as the period T approaches ∞, we will get the Fourier transform as follows:

$$F(f) = \int_{-\infty}^{\infty} s(t)e^{-j(2\pi f)t}dt \tag{2-218}$$

If we have Fourier transform $F(f)$, the time-domain signal can be reconstructed using inverse Fourier transform defined as follows:

$$s(t) = \int_{-\infty}^{\infty} F(f)e^{j(2\pi)t}df \tag{2-219}$$

The term $F(f)$ represents the signal frequency spectrum at frequency f. Some references represent the frequency in angular form as follows:

$$\omega = 2\pi f \tag{2-220}$$

In this case, the Fourier transform, and its inverse will be given by:

$$H(\omega) = \int_{-\infty}^{\infty} s(t)e^{-j\omega t}dt \tag{2-221}$$

$$s(t) = \frac{1}{2\pi}\int_{-\infty}^{\infty} H(\omega)e^{j\omega t}d\omega \tag{2-222}$$

Note that $H(\omega)$ is a complex value, so it will have a magnitude and a phase. The magnitude $\|H(\omega)\|$ represents the amplitude of the original signal at sinusoidal component of frequency f (angular frequency $2\pi f$). The phase $\angle H(\omega)$ represents the phase shift of the sinusoidal component of frequency f (angular frequency $2\pi f$). It is also important to realize that $H(\omega)$ is a symmetric function:

$$H(-\omega) = H(+\omega) \tag{2-223}$$

$$F(-f) = F(f) \tag{2-224}$$

An important property of the Fourier transform is that time shift is translated into exponential scaling in frequency domain as follows: if $s(t) \Leftrightarrow F(f)$ are Fourier transform pairs, the time shift in $s(t)$ gives the following Fourier transform:

$$s(t-\tau) \Leftrightarrow e^{-2\pi f\tau}F(f) \tag{2-225}$$

To verify that Fourier transform is equivalent to Fourier series if the input signal is an infinite periodic sinusoidal $s(t) = A\cos(2\pi f_1 t)$, let's calculate its Fourier transform as follows:

$$F(f) = \int_{-\infty}^{\infty} A\cos(2\pi f_1 t)e^{-j(2\pi f)t}dt \tag{2-226}$$

we can replace $A\cos(2\pi f_1 t)$ by its complex formula as follows:

$$A\cos(2\pi f_1 t) = \frac{1}{2}\left(e^{(2\pi f_1)t} + e^{-j(2\pi f_1)t}\right) \tag{2-227}$$

$$F(f) = \int_{-\infty}^{\infty} \frac{A}{2}\left(e^{(2\pi f_1)t} + e^{-j(2\pi f_1)t}\right)e^{-j(2\pi f)t}dt \tag{2-228}$$

$$F(f) = \int_{-\infty}^{\infty} \frac{A}{2}\left(e^{j(2\pi f_1)t} + e^{-j(2\pi f_1)t}\right)e^{-j(2\pi f)t}dt \tag{2-229}$$

$$F(f) = \frac{A}{2}\int_{-\infty}^{\infty} e^{j(2\pi f_1)t}e^{-j(2\pi f)t}dt + \frac{A}{2}\int_{-\infty}^{\infty} e^{-j(2\pi f_1)t}e^{-j(2\pi f)t}dt \tag{2-230}$$

$$F(f) = \frac{A}{2}\int_{-\infty}^{\infty} e^{-j2\pi(f-f_1)t}dt + \frac{A}{2}\int_{-\infty}^{\infty} e^{-j2\pi(f+f_1)t}dt \tag{2-231}$$

the integration $\int_{-\infty}^{\infty} e^{a(f-f_1)t}dt$ equals zero at frequencies $f \neq f_1$ and equals one at $f = f_1$. The same for the integral $\int_{-\infty}^{\infty} e^{a(f+f_1)t}dt$, it equals zero at frequencies $f \neq f_1$ and equals one at $= -f_1$. Therefore, the $F(f)$ is given by:

$$F(f) = \frac{A}{2}\delta(f-f_1) + \frac{A}{2}\delta(f+f_1) \tag{2-232}$$

which is consistent with the Fourier series with only one difference: the Fourier series of sinusoidal signal is a single-sided function with frequencies greater than or equal zero. Fourier transform has a double-sided spectrum, which can be seen in Figure 2.34.

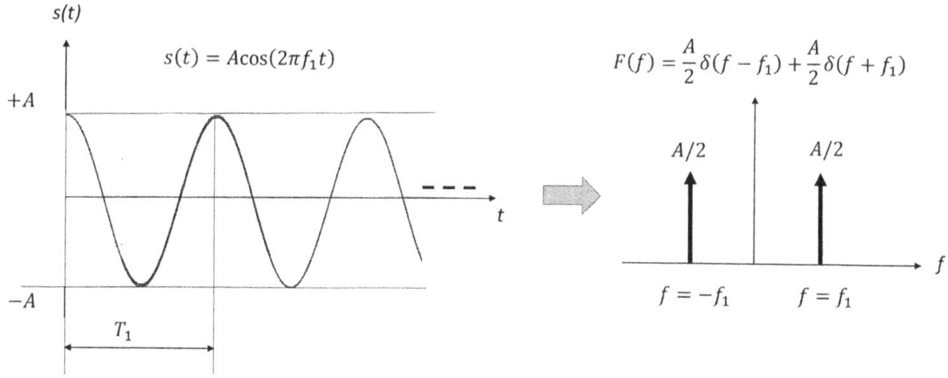

Figure 2.34 Double-sided Fourier transform of sinusoidal signal.

This analysis and this figure can be reached if we represent this pure sinusoidal signal by its complex phasor form. To understand that, let us see the complex phasor form of a pure real sinusoidal signal $f(t) = A\cos(2\pi f_1 t + \varphi)$, which can be represented by:

$$\Re e\{Ae^{j\varphi}e^{j2\pi f_1 t}\} = \frac{A}{2}\left(e^{j\varphi}e^{j2\pi f_1 t} + e^{-j\varphi}e^{-j2\pi f_1 t}\right) \tag{2-233}$$

which shows clearly that the signal has mirrored amplitudes at frequencies $\pm f_1$ as seen in Figure 2.34. The only difference is that in Figure 2.34, the phase angle φ is set to zero. If we contaminate the signal with noise and perform Fourier transform, we will get a "*spectrum*" of frequencies that describes the energy distribution over all frequencies contained in the signal. In practice, a signal has a spectrum of frequencies that can be analyzed and identified using Fourier transform. For example, consider the signal $S(t) = \cos(2\pi 10t) + \cos(2\pi 20t)$ shown in Figure 2.35a. This signal obviously has two frequency components, 10 and 20 Hz. The Fourier transform of the pure noise-free signal will have two peaks at frequencies 10 and 20 Hz as shown in Figure 2.35c. If the signal is contaminated by a Gaussian noise such that:

$$x(t) = \cos(2\pi 10t) + \cos(2\pi 20t) + n(t) \tag{2-234}$$

$$n(t) \sim N\left(0, \sigma_n^2\right) \tag{2-235}$$

we will obtain the noisy signal shown in Figure 2.35b. The Fourier transform of the noisy signal is shown in Figure 2.35d, which still shows slightly distorted peaks at frequencies 10 and 20 Hz in addition to small spectrum components spread over all frequencies. These spread spectrum values are the frequency components of the Gaussian noise. A perfect Gaussian signal has a flat frequency spectrum (approximately constant spectrum component at all frequencies). Because we never get a perfect Gaussian in practice, we get these little, small peaks spread over the whole frequency spectrum. Figure 2.35d also shows that the dominant components are the frequency components at frequencies 10 and 20 Hz. This is because the original signal energy is larger than the noise energy which is reflected in the Fourier amplitude spectrum that shows large peaks at the original signal frequencies and smaller peaks at the noise signal frequencies.

Figures 2.34 and 2.35 show the amplitude spectrum of the Fourier transform. The amplitude spectrum shows how much energy is in each frequency component in the signal. As we mentioned, Fourier transform also has a phase spectrum. The phase spectrum shows the phase angle of each

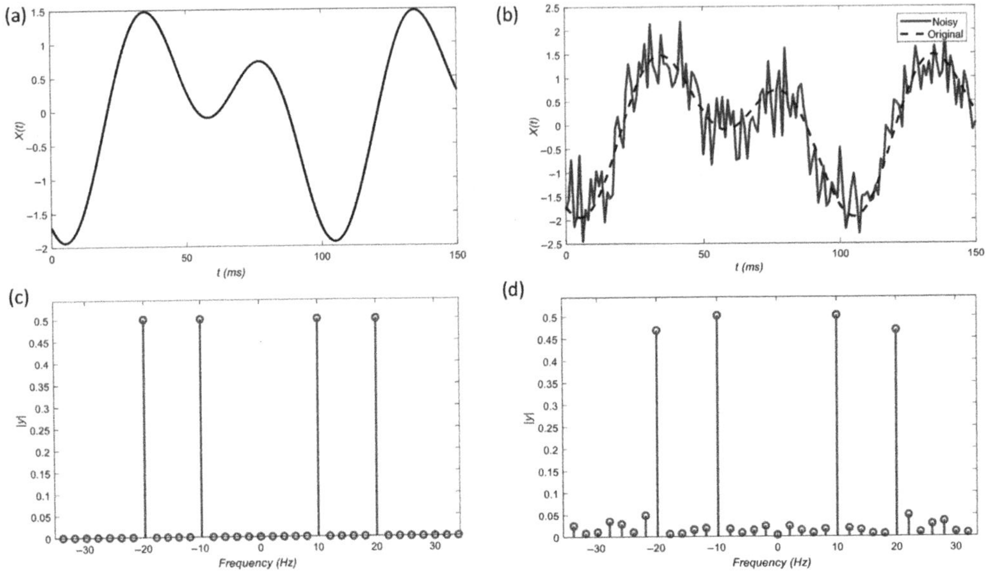

Figure 2.35 Amplitude spectrum of signals *S*(*t*) and *X*(*t*). (a) Original noise-free signal *S*(*t*), (b) signal corrupted with zero-mean random Gaussian noise, (c) double-sized amplitude spectrum of noise-free *S*(*t*), and (d) double-sized amplitude spectrum of noise-signal *X*(*t*).

frequency component. The phase spectrum shows how each harmonic in the signal is shifted from zero phase. An example of the same signal $x(t)$ with phase shifts is given by:

$$x(t) = \cos(2\pi 10t + 0.5\pi) + \cos(2\pi 20t + 0.75\pi) + n(t) \tag{2-236}$$

The double-sided phase spectrum of $x(t)$ is shown in Figure 2.36 with and without Gaussian noise. The phase spectrum tells us what the phase shift of each harmonic (frequency) component in the signal is. A phase shift is also called distortion. It is clear that adding Gaussian noise distorts all frequencies.

Applying the Fourier transform on a noisy sinusoidal wave $s(t) = A.\cos(2\pi ft + \phi)$ generates complex numbers from which we can generate two plots, amplitude spectrum, and phase spectrum. At each frequency, we get a complex number $n = x + jy$ with the following amplitude A_n and phase $\emptyset_n$:

$$A_n = \sqrt{x^2 + y^2} \tag{2-237}$$

$$\emptyset_n = \tan^{-1}\left(\frac{y}{x}\right) \tag{2-238}$$

Even with noise, the amplitude plot will show a sharp peak at the sinusoidal' s frequency f where the peak value is the amplitude of the component of this frequency. If the signal is noise-free (which is unrealistic), its phase plot will also show a sharp peak at the frequency f, where the peak value is the phase shift of the sinusoidal component at this frequency. If the signal is noisy (which is the realistic case), although the amplitude plot will show a clear peak, the phase plot is always noisier than the amplitude plot. Another way to calculate the phase of a signal is to use the I/Q form.

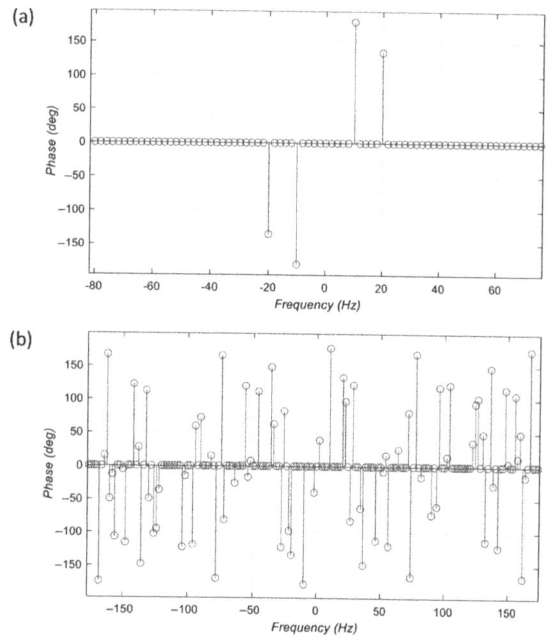

Figure 2.36 Phase spectrum (double sided) of $x(t)$ (a) without Gaussian noise, (b) with Gaussian noise.

The phase can be calculated by multiplying $s(t)$ by $\cos(2\pi ft)$ to get the I branch and $\sin(2\pi ft)$ to get the Q branch and use the following simple formula to calculate the phase:

$$\phi = \tan^{-1}\left(\frac{Q}{I}\right) \tag{2-239}$$

This method is used in PLLs to steer the local oscillator frequency and phase to track an incoming noisy sinusoidal signal (i.e., carrier signal). PLL will be explained in the radio positioning systems chapter (Chapter 5).

2.6.1.3 Power Spectral Density (PSD)

Fourier Transform tells us how a certain frequency is contributing to a signal. When applying Fourier transform on a sample of a random process, the resulting frequency spectrum is not indicative of the signal properties in the frequency domain. A better useful tool to study stochastic signals in the frequency domain is the Power Spectral Density (PSD) [5]. PSD is a measure of the energy concentration (density) at a certain frequency. It is more meaningful than Fourier transform and it is commonly used in navigation systems to assess the noise of sensors. The term density refers to the fact that PSD gives energy per unit frequency (Hz). It is like a probability density function to calculate the absolute energy, we need to integrate the PSD over the desired frequency range of band. The formula of PSD came from the energy formula of a signal $x(t)$, which is given by the following equation:

$$E_x = E\langle x^2(t)\rangle = \int_{-\infty}^{\infty} x^2(t)\, dt = R_{xx}(0) \tag{2-240}$$

where $R_{xx}(\tau)$ is the autocorrelation function of $x(t)$. If $x(t)$ is an ergodic random process (its ensemble average equals to its time average), its autocorrelation function is given by the following equation:

$$R_{xx}(\tau) = \lim_{T\to\infty} \frac{1}{2T}\int_{-T}^{T} x(t)x(t-\tau)dt \tag{2-241}$$

If we have the Fourier transform of the autocorrelation function defined as follows:

$$S_x(f) = \int_{-\infty}^{\infty} R_{xx}(\tau)e^{-j2\pi f\tau}d\tau \tag{2-242}$$

we can obtain $R_{xx}(\tau)$ by the inverse Fourier transform as follows:

$$R_{xx}(\tau) = \int_{-\infty}^{\infty} S_x(f)e^{j2\pi f\tau}df \tag{2-243}$$

Setting $\tau = 0$ will get $R_{xx}(0)$, the energy of the signal as follows:

$$R_{xx}(0) = \int_{-\infty}^{\infty} S_x(f)df \tag{2-244}$$

which shows that $S_x(f)$ is a power density function. The power density function $S_x(f)$ is called the power spectral density and it gives the energy distribution over frequencies per unit frequency. So, if the units of $x(t)$ is V, the power $P = V^2$, the power spectral density units will be V^2/Hz, which is a variance unit. It is common to use the standard deviation of the power spectral density, which is the square root of the power spectral density given in $V/\sqrt{\text{Hz}}$. We can easily realize that $R_{xx}(\tau)$ is the convolution of $x(t)$ with itself and we know from Fourier transformation properties that the convolution in the time domain is equivalent to multiplication in the frequency domain. Therefore, the Fourier transform of $R_{xx}(\tau)$ is proportional to the square of the Fourier transform of $x(t)$:

$$S_x(f) \triangleq |F_x(f)|^2 \tag{2-245}$$

In signal processing literature, the standard power unit that is commonly used is the dB, which is related to the original signal power unit as follows:

$$\text{dB} = 10\log_{10}(P) \tag{2-246}$$

The plots in Figure 2.37 show the power spectral density in dB/Hz for a noisy signal that has two dominant frequencies, 10 and 20 Hz. Note that the dB unit is negative, but the absolute power value

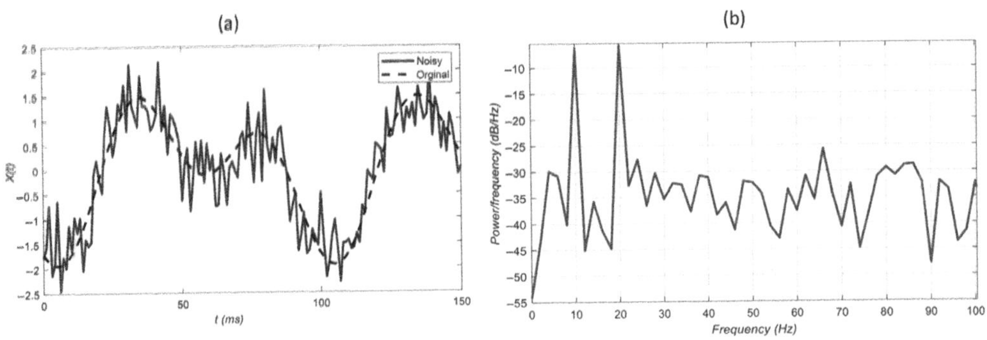

Figure 2.37 Power spectral density of a noisy signal that has two dominant frequencies, 10 and 20 Hz. (a) original signal and (b) power spectral density.

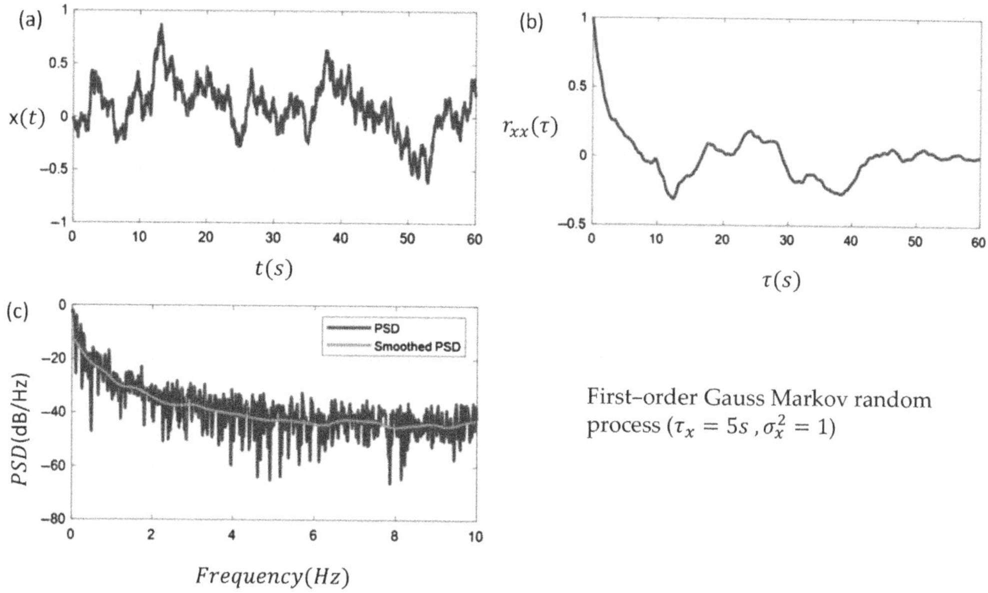

Figure 2.38 PSD for First-order Gauss–Markov random process sample. (a) Gauss–Markov process sample, (b) autocorrelation of Gauss–Markov process sample, and (c) power spectral density of Gauss–Markov process.

P is positive. Since autocorrelation $R_{xx}(\tau)$ is a useful statistical property of random signals, PSD is always preferred to study random processes in the frequency domain. Figure 2.38 shows a sample of a first-order Gauss-Markov process with its autocorrelation and PSD plots.

2.6.1.4 Frequency Bands

The range of frequencies in a signal is called the frequency band. Based on the frequency band, a signal can lie in one of the following common frequency bands: baseband, narrowband, broadband, and ultrawideband (UWB). Noise occupies the whole spectrum, so its frequency band spans the whole spectrum as seen in Figure 2.39, which shows a theoretical illustration of PSD of different frequency bands.

2.6.2 Digital Signal Processing

2.6.2.1 Analog Digital Conversion

Most modern systems use digital computers/digital circuits to process signals. Digital systems represent signals as streams of binary codes at discrete time intervals. In the discrete-time domain, we deal with a sample of the signal at specific intervals. If $x(t)$ is the original continuous signal, discrete time $x_d[n]$ is defined by:

$$x_d[n] = x(nT_s) \qquad n \in \{0, 1, 2...\} \tag{2-247}$$

where T_s is a small interval of time called the sampling period and sampling frequency $F_s = \frac{1}{T_s}$. In between these samples, the signal value is constant, which is known as sample-and-hold. The discrete-domain version of the signal is obtained by the sampling-and-hold circuit as seen in Figure 2.40. The coder will convert the discrete values into a digital binary code that can be processed by a computer. The discretization and coding process is called analog/digital conversion.

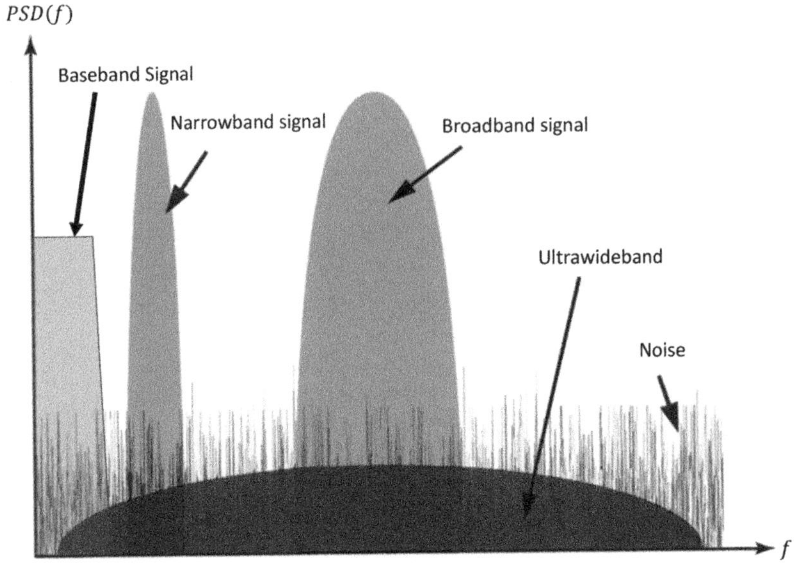

Figure 2.39 Theoretical single-sided illustration of PSD for different frequency bands.

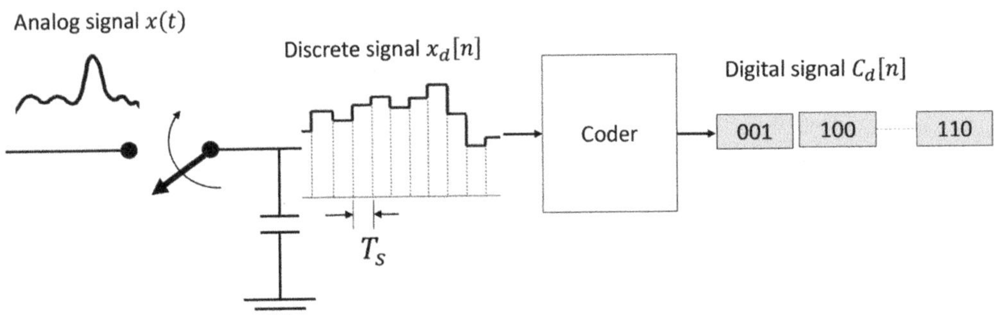

Figure 2.40 Simplified analog to digital converter diagram.

2.6.2.2 Sampling and Nyquist Theorem

It is intuitive to realize that when converting a continuous-time signal to a discrete-domain signal, the sampling frequency $F_s = \frac{1}{T_s}$ should be large enough to preserve the original signal. Nyquist has proved that the minimum sampling frequency to sample a signal whose maximum frequency component is F_{max} is as follows:

$$F_s > 2F_{max} \tag{2-248}$$

To prove the Nyquist theorem, we will study the effect of sampling a sinusoidal signal and then generalize it to any signal. For any real-valued sinusoidal signal:

$$x(t) = A\cos(2\pi Ft + \emptyset) \tag{2-249}$$

the discrete signal is given by:

$$x_d[n] = x(nT_s) = A\cos(2\pi FnT_s + \emptyset) = A\cos\left(2\pi\frac{F}{F_s}n + \emptyset\right) \tag{2-250}$$

We define the following important quantities:

- Sample number n
- Sampling period T_s (s)
- Continuous signal cycle period T (s)
- Continuous frequency F (cycles/s)
- Continuous radial frequency $\Omega = 2\pi F$ (radians/s)
- Sampling frequency F_s (samples/s)
- Normalized radial frequency $\omega = 2\pi F/F_s$ (radians/sample)

When we apply frequency analysis (Fourier-transform and Fourier series) in the discrete domain, we always use the normalized radial frequency $\omega = 2\pi F/F_s$ as the x-axis, so it is important to understand the limits of this x-axis. In the continuous-domain analysis, if continuous frequency f is used as x-axis, frequency ranges between $-\infty$ and $+\infty$ and the units are Hz (cycles/seconds). If continuous radial frequency Ω is used as x-axis, the values range from $-\infty$ and $+\infty$ and the units are radians/seconds. Both F and Ω are not normalized values (they can take any arbitrary value), which leads to a lack of standard framework to study different signals. In digital systems engineering, when we use a sampled version of the signal, everything becomes dependent on the sampling frequency F_s, and in this case, the normalized radial frequency ω is given by:

$$\omega = 2\pi F/F_s \tag{2-251}$$

In practice, we do not allow F_s to be smaller than $2F_{max}$ where F_{max} is the maximum frequency in the signal and $2F_{max}$ is the Nyquist frequency. We can comfortably assume F_s will range from $2F_{max}$ to ∞. If F_s goes to ∞, ω will go to 0 and if F_s approaches $2F_{max}$, ω will be π. Any $F_s > 2F_{max}$ will lead to $\omega < \pi$. In complex frequency analysis, we have double-sided spectrum function where both negative and positive frequencies are plotted in a symmetric plot; therefore, theoretically, the range of normalized radial frequency ω will be:

$$\omega \in [-\pi, +\pi] \tag{2-252}$$

Any normalized radial frequency outside this range $[-\pi, +\pi]$ means that we have too slow sampling frequency that is lower than Nyquist frequency ($2F_{max}$). This normalized radial frequency is useful because it maps all frequencies from $[-\infty, \infty]$ range to $[-\pi, +\pi]$ range.

To further elaborate the effect of sampling the signal, we study the spectrum of the discrete signal. Consider an arbitrary analog signal $x(t)$. Typical signals in communication systems have a high-frequency component F_c (e.g., a sinusoidal high-frequency carrier in a radio system) and a smaller frequency signal (e.g., voice signal or data). Assume the low-frequency signal has a frequency range $[0, B]$ Hz. We call this low-frequency signal "*baseband*" signal with a bandwidth *BHz*. When the low-frequency signal is modulated (e.g., multiplied by) the high-frequency signal, the resulting signal becomes a signal centered around F_c plus or minus B. as shown in Figure 2.41. Note that in Figure 2.41, we plot the double-sided spectrum against the continuous frequency Ω. Note also that the multiplication in the time domain is a convolution in the frequency domain. That is why we see these two peaks centered around F_c.

The digitization (discretization) of a signal is very similar to the modulation example above with one exception, instead of multiplying the signal by a carrier sinusoidal signal, we multiply it with an impulse train with of a frequency F_s sampling frequency as seen in Figure 2.42. In Figure 2.42, to be able to reconstruct the original continuous signal, the triangular peaks in the bottom figure (the frequency spectrum of the digitized signal $Y(\Omega)$) must be separable. If they overlap, the original

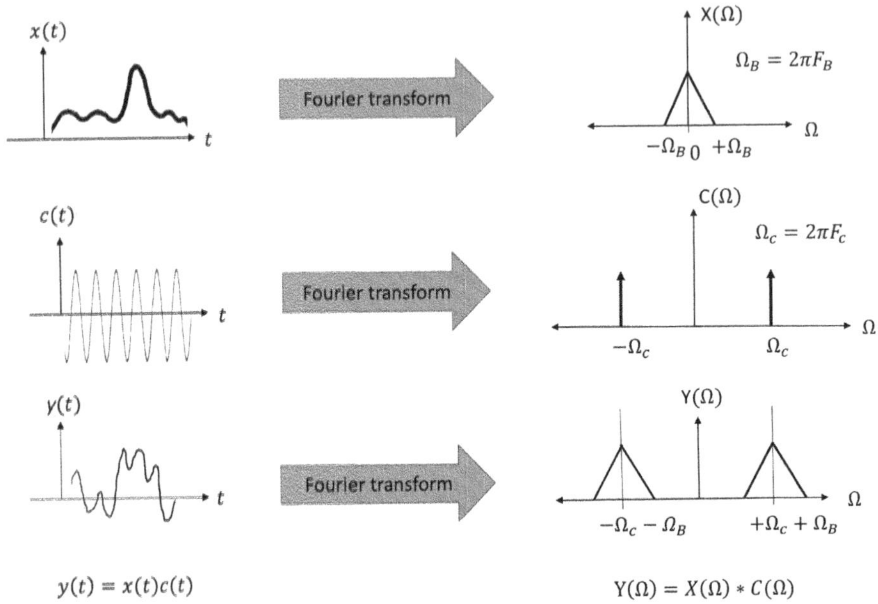

Figure 2.41 Illustrative example of a carrier/signal modulation. The signal and carrier in both time domain and frequency domain are shown on the left. The spectrum of the modulated overall signal is on the right.

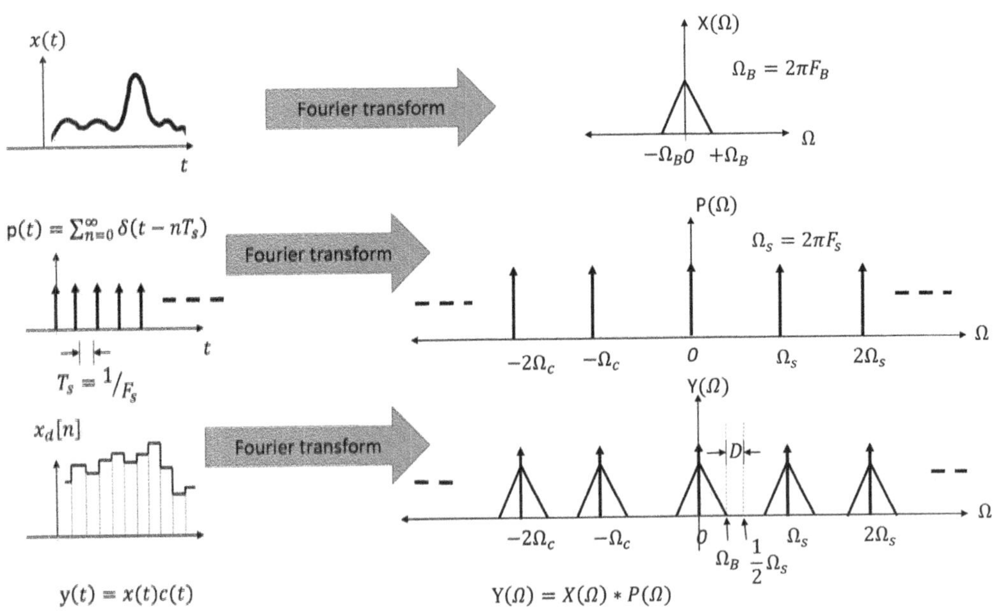

Figure 2.42 Spectrum of digitized (discretized) signal.

signal will be lost. To keep the triangular peaks separable, the frequency difference D shown in the $Y(\Omega)$ spectrum in Figure 2.42 must be greater than zero, which yields the following condition:

$$D = \left(\frac{1}{2}\Omega_s - \Omega_B\right) > 0 \tag{2-253}$$

which means

$$\Omega_s > 2\Omega_B \tag{2-254}$$

Or

$$F_s > 2F_B \tag{2-255}$$

which proves Nyquist sampling theorem.

2.6.2.3 Discrete Fourier Analysis

Recall the continuous-time Fourier transform is given by:

$$H(\omega) = \int_{-\infty}^{\infty} s(t)e^{-j\omega t}dt \quad \omega \in [-\infty, \infty] \tag{2-256}$$

$$s(t) = \frac{1}{2\pi}\int_{-\infty}^{\infty} H(\omega)e^{j\omega t}d\omega \tag{2-257}$$

The discrete-time Fourier transform (DTFT) of a discrete signal $x_d[n]$ is given by:

$$F(\omega) = \sum_{n=-\infty}^{\infty} x_d[n]e^{-j\omega n} \qquad \in [-\pi, \pi] \tag{2-258}$$

$$x_d[n] = \frac{1}{2\pi}\int_{-\pi}^{\pi} F(\omega)e^{j\omega n}d\omega \tag{2-259}$$

The integration usually has a closed formula; however, summation does not have a closed formula. Given that n has an infinity range, it is impossible to compute DTFT in a finite time. In addition, in practice, we rarely know $x_d[n]$ at all values of n. We only know a finite sequence of $x_d[n]$ values. The inverse DTFT is also difficult to calculate for the same reasons. Therefore, instead of calculating DTFT, there is an approximate version of Fourier transform for discrete domain. This is the discrete Fourier transform (DFT). Like the continuous Fourier analysis, we have discrete Fourier series (DFS) for infinite periodic discrete signal. Both DFS and DFT are described in the next subsections.

Discrete Fourier Series (DFS)

For an infinite periodic discrete signal with a period N, DFS is defined as follows:

$$a_k = \sum_{n=0}^{N-1} x_d[n]e^{-j\omega_k n} \qquad k = 0, 1 \ldots N-1 \tag{2-260}$$

$$x_d[n] = \frac{1}{N}\sum_{k=0}^{N-1} a_k e^{+j\omega_k n} \tag{2-261}$$

$$\omega_k = 2\pi\frac{k}{N} \qquad \omega_k \in [-\pi, \pi] \tag{2-262}$$

Discrete Fourier Transform (DFT)

If $x_d[n]$ is nonperiodic and has a finite length N, we still can approximate its Fourier transform by assuming that $x_d[n]$ is one sequence of a periodic signal and uses the DFS to calculate the coefficients a_k. In this case, one important selection is the sequence period. We can pad the $x_d[n]$

sequence by any number of zeros to increase the accuracy of the transform. The more zeros padded into the sequence, the more accurately the DFT coefficients approximate the DTFS.

Fast Fourier Transform (FFT)

Fast Fourier transform (FFT) is a method that performs Fourier transform on discrete vectors of data. FFT algorithm approximates the discrete Fourier transform as bank of digital filters where the signal goes through to calculate the Fourier coefficients. The details of FFT algorithm can be found in several DSP references such as [6] and [7]. FFT is a built-in function in most computing and programming platforms such as MATLAB. When FFT is applied on a discrete vector $x_d[n]$ with length N and sampling frequency F_s, the following key items apply:

- FTT will return a vector of N complex numbers.
- The FFT vector is the double-sided FFT.
- Each complex number has an amplitude and a phase.
- The single-sided FFT will be in the first $N/2$ elements of the FFT vector.
- Maximum detectable frequency is $F_s/2$ (Hz).
- Single-side x-axis frequencies $f \in \left[0, \frac{F_s}{N}, \frac{2F_s}{N}, \ldots \frac{F_s}{2}\right]$ (Hz).
- Frequency resolution is $\frac{F_s}{N}$ (minimum detectable frequency difference).

Discrete Power Spectral Density

Discrete power spectral density can be calculated by calculating the discrete Fourier transform of the discrete autocorrelation of the signal. Given a finite-length instance of a discrete signal $x_d[n]$ given by:

$$x_N[n] = \begin{cases} x_d[n] & -N \leq n \leq N \\ 0 & \text{otherwise} \end{cases} \tag{2-263}$$

Its autocorrelation is given by:

$$R_{xx}[m] = \frac{1}{2N+1} \sum_{n=-N}^{N} x_d[n+m] x_d^*[n] \tag{2-264}$$

The discrete Fourier transform of $R_{xx}[m]$ is given by:

$$S_x(\omega) = \sum_{m=-\infty}^{\infty} R_{xx}[m] e^{-j\omega m}, \quad \omega \in [-\pi, \pi] \tag{2-265}$$

From Eqs. (2-64) and (2-65), the power spectral density S_x is given by:

$$S(\omega) = \frac{1}{2N+1} |F(\omega)|^2, \quad \omega \in [-\pi, \pi] \tag{2-266}$$

where $F(\omega)$ is the DTFT of $x_d[n]$ which can be approximated by FFT of $x_N[n]$. Power spectral density is important to predict the noise we get when we sample a random signal at a certain sampling frequency. Commonly, the noise is a high-frequency component, so we get more noise when we sample at a high frequency. However, we risk losing information if we sample at a very low frequency. Therefore, we need to know the energy distribution over frequencies so when we can sample the signal at a suitable frequency to get the desired data or information without allowing significant noise to pass.

2.6.3 Digital Filters

Digital filters are LTI Systems that filter out specific frequencies from an input discrete-time signal. Filters are used extensively in navigation systems, especially in radio positioning systems and RADAR signal processing. Understanding filters in the frequency domain is easier because, in the time domain, the LTI system output is convolution, while in the frequency domain, the LTI system is a simple multiplication. Filters can be categorized based on the range of frequencies they allow to pass as seen in Figure 2.43. There are common technical terms that are frequently used to describe a filter. For example, a bandpass filter may be characterized by its insertion loss, which is the damping that happens to the desired frequency, which ideally should be zero, but no filter is perfect, so -1.0 dB is a common insertion loss. Recall, -1 dB is corresponding to $10^{-(1/10)}$ gain which is a gain of 0.7943 (the input signal power is multiplied by 0.7943). Zero dB corresponds to 1 (unity gain). Another characteristic is the bandwidth, which is usually described by frequency boundaries at which the filter gain reaches -3 dB threshold as seen in Figure 2.44. About -3 dB corresponds to approximately 50% gain. Figure 2.44 shows the difference between the idealistic filter and the practical filter.

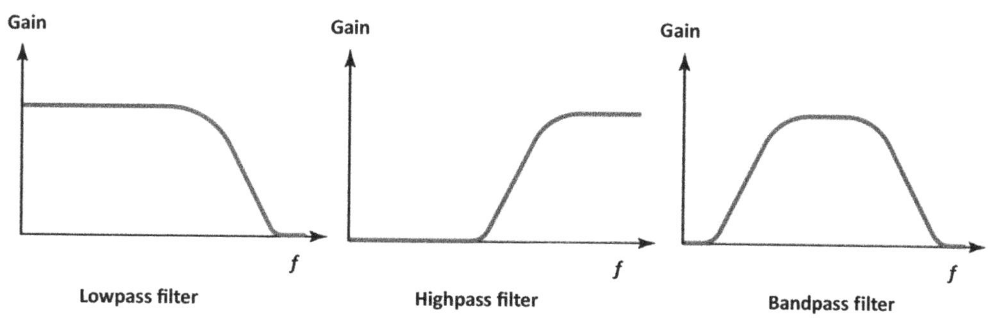

Figure 2.43 Types of filters.

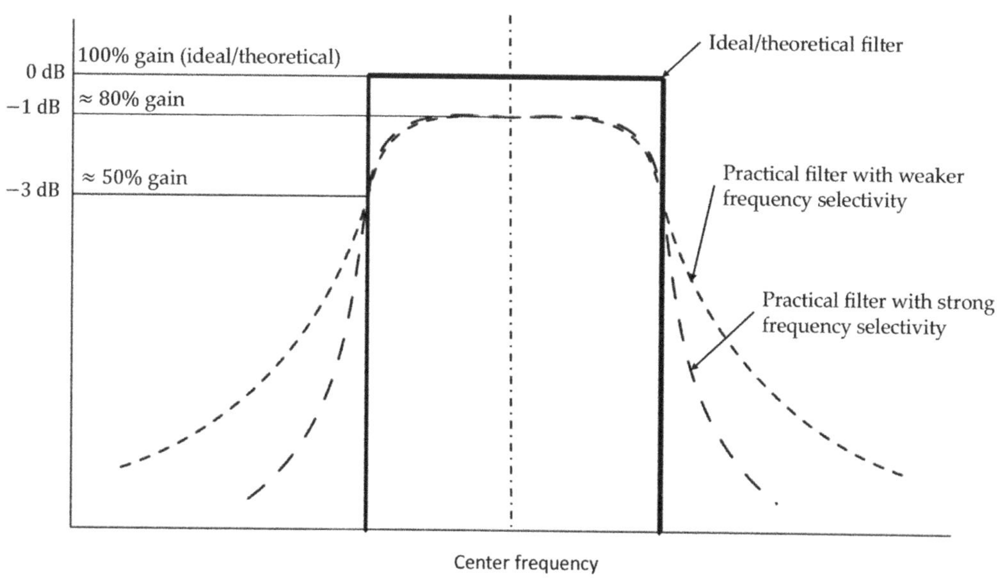

Figure 2.44 Bandpass filter illustration.

In the continuous-time domain, the general differential equation of LTI filters is given by the following equation:

$$y(t) + a_1 y^{(1)}(t) + a_2 y^{(2)}(t) + \ldots + a_N y^{(N)} = b_0 x(t) + b_1 x^{(1)}(t) + \ldots + b_M x^{(M)} \quad (2\text{-}267)$$

This is an LTI system with a transfer function:

$$H(s) = \frac{Y(s)}{X(s)} = \frac{b_0 + b_1 s^1 + \ldots + b_M s^M}{1 - a_1 s - a_1 s^2 - \ldots - a_N s^N} \quad (2\text{-}268)$$

In the discrete-time domain, the differential equation becomes the difference equation given as follows:

$$y[n] + a_1 y[n-1] + a_2 y[n-2] + \ldots + a_n y[n-N] = b_0 x(t) + b_1 x[n-1] + b_2 x[n-2] \ldots + b_M x[n-M] \quad (2\text{-}269)$$

This LTI system has a Z-transform transfer function given by:

$$H(z) = \frac{Y(z)}{X(z)} = \frac{b_0 + b_1 z^{-1} + \ldots + b_M z^{-M}}{1 - a_1 z^{-1} - a_2 z^{-2} - \ldots - a_N z^{-N}} \quad (2\text{-}270)$$

The Z-transform is the equivalent of the Laplace Transform in the discrete-time domain and it is given by:

$$X(z) = \sum_{-\infty}^{\infty} x[n] z^{-n} \quad (2\text{-}271)$$

Note that Z-transform is equivalent Fourier transform at $z = e^{j\omega}$ and Laplace transform at $z = s$. By selecting the coefficients a_i and b_i, we can adjust the transfer function of the LTI system to pass or reject certain range of frequencies. If all a_i coefficients are zeros, the filter is called finite impulse response (FIR) filter as $H(z)$, in this case, will always have a bounded value. If a_i coefficients are nonzero, this means that the filter may have poles at the frequencies at which the polynomial $1 - a_1 z - a_2 z^{-2} - \ldots - a_N z^{-N}$ equals zero. This filter is called the infinite impulse response (IIR) filter.

Digital filters are frequently used in navigation systems for different purposes. For example, digital filters are used extensively in radio positioning and ranging systems such as GPS and RADAR sensors. One of the common processes in GPS and RADAR systems is the mixing of two signals to shift the center frequency of a signal or to remove a high-frequency carrier from a received signal. If we have an incoming signal with f'_c center frequency multiplied by a signal of frequency f_o, we get the following signal:

$$s(t) = A\cos(2\pi f'_c t + \varphi)\cos(2\pi f_o t) = \frac{1}{2}\cos(2\pi[f'_c - f_o]t + \varphi) + \frac{1}{2}\cos(2\pi[f'_c + f_o]t + \varphi) \quad (2\text{-}272)$$

The resulting signal has two components: a low-frequency component and a high-frequency one. If $f'_c = f_o$, the low-frequency component will be a baseband signal centered around zero frequency. This baseband signal is the signal that contains the information of interest. This process is also known as demodulation. An example of the mixing process is shown in Figures 2.45–2.47. Figure 2.45 shows a signal received by a GPS receiver after being shifted to an intermediate carrier frequency of approximately 10 MHz. To remove the carrier, the signal is mixed with a local

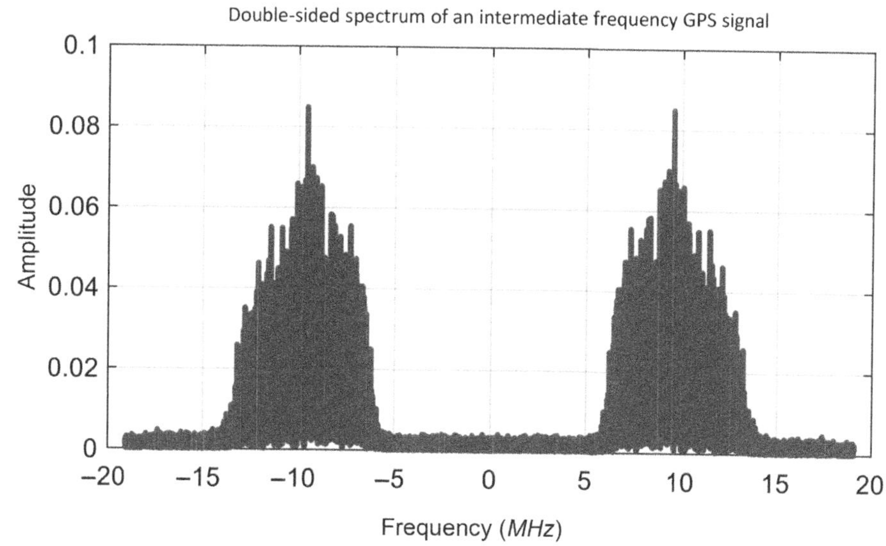

Figure 2.45 Frequency spectrum of incoming signal.

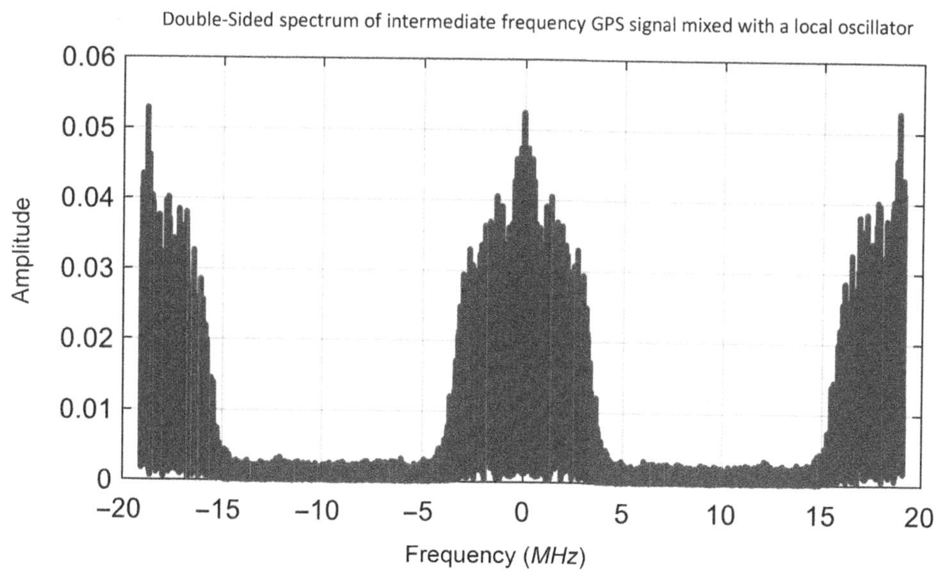

Figure 2.46 Frequency spectrum of mixed signal.

oscillator signal. The mixed signal is shown in Figure 2.46 and the output is low pass filtered and shown in Figure 2.47. The filter used in this example has the following transfer function (z-transform):

$$H(z) = \frac{0.0192 + 0.0486z^{-1} + 0.0651z^{-2} + 0.0486z^{-3} + 0.0192z^{-4}}{1 + 2.0557z^{-1} - 2.093z^{-2} + 1.0676z^{-3} - 0.2333z^{-4}} \tag{2-273}$$

The PSD of the raw input, mixed, and filtered signals are shown in Figure 2.48.

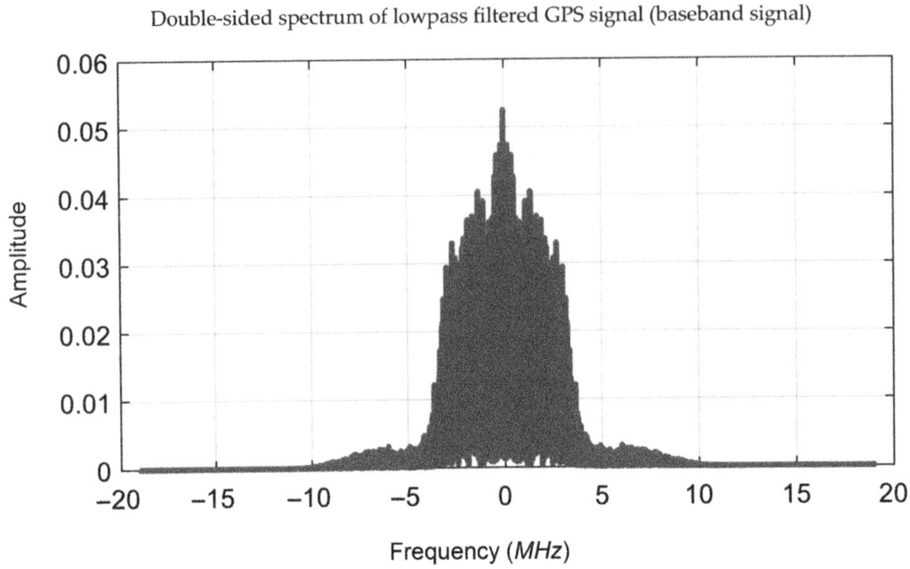

Figure 2.47 Frequency spectrum of lowpass filtered signal (baseband signal).

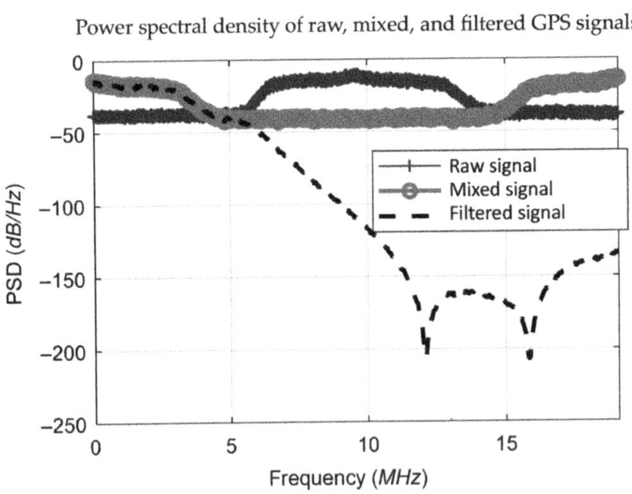

Figure 2.48 Power spectral density of raw, mixed, and filtered signals.

2.7 Problems

2.7.1 Problem 1

Consider a train moving on a straight line as seen in Figure 2.49. Train motion within a limited period can be modeled by a simple 1D constant acceleration motion that uses initial position $y(0)$, initial velocity $v(0)$, and constant acceleration a, to calculate $y(t)$, the position of the train at any time t. Write down the position of the train as a function of initial position $y(0)$, initial velocity $v(0)$, and constant acceleration a.

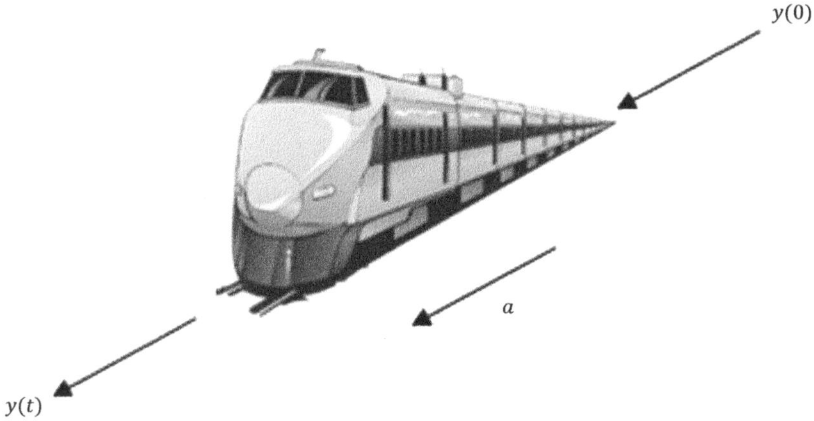

Figure 2.49 1-D dynamic system example.

2.7.2 Problem 2

Several position sensors are placed at different points along the track of the train of Problem 1. Each sensor k records the train position y_k and the measurement time t_k.

A) What is the minimum number of sensors needed to estimate the train's initial position $y(0)$, initial velocity $v(0)$, and constant acceleration a ?

B) Sensors' measurements are arranged into a vector $\vec{y} = (y_1 \quad y_2 \quad \dots \quad y_N)^T$ and the time measurements $\vec{t} = (t_1 \quad t_2 \quad \dots \quad t_N)^T$ are recorded. Derive a measurement model in the form $\vec{y} = H\vec{x}$ where $\vec{x}$ is the train's initial position $y(0)$, initial velocity $v(0)$, and constant acceleration a defined as follows:

$$\vec{x} = \begin{pmatrix} y(0) \\ v(0) \\ a \end{pmatrix}$$

2.7.3 Problem 3

The differential equation of the mass-spring system shown in Figure 2.50 is given as follows:

$$m\ddot{p}(t) = -b\ddot{p}(t) - kp(t) + f(t)$$

Figure 2.50 Mass-spring force system.

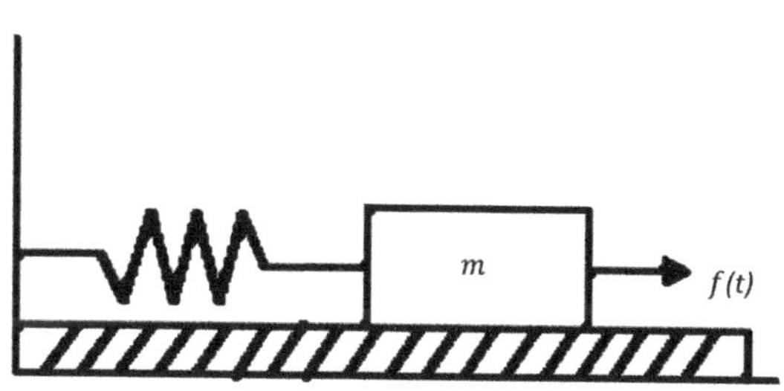

where $p(t)$ is the position of the mass, $f(t)$ is the applied force, b, k are constant parameters. If the force $f(t)$ is the system input and $p(t)$ is the system output, write the transfer function of the system $H(s) = \frac{P(s)}{F(s)}$

2.7.4 Problem 4

What are the values of the in-phase and quadrature arms that generate the following signal:

$$x(\mathrm{t}) = A(t)\cos(2\pi ft - \pi/4)$$

2.7.5 Problem 5

Given MVN random vector of two variables x_1, x_2 with the following mean and covariance matrix:

$$\mu = \begin{pmatrix} 0.2 \\ 0.5 \end{pmatrix}, \Sigma = \begin{pmatrix} 0.32 & 0.01 \\ 0.01 & 0.10 \end{pmatrix}$$

If a sample from the variable $x_2 = 0.5$, calculate the conditional probability $p(x_1 \mid x_2)$.

References

1 J. G. Proakis and D. G. Manolakis, "*Digital Signal Processing, Principles, Algorithms, and Applications*," New Jersey: Pearson, Prentice Hall, 2007.
2 G. Mutambara, "*Design and Analysis of Control Systems*," New York, Washington, DC: CRC Press, 1999.
3 K. P. Murphy, "*Machine Learning: A Probabilistic Perspective*," London, England: The MIT Press, Cambridge, Massachusetts, 2012.
4 C. M. Bishop, "*Pattern Recognition and Machine Learning*," Springer Science+Business Media, LLC, 2006.
5 K. Borre, D. M. Akos, N. Bertelsen, P. Rinder and S. H. Jensen, "*A Software-Defined GPS and Galileo Receiver, A Single-Frequency Approach*," Boston, MA: Birkhauser, 2007.
6 J. A. Farrell, "*Aided Navigation, GPS with High Rate Sensors*," New York: McGraw Hill, 2008.
7 P. D. Groves, "*Principles of GNSS, Inertial, and Multisensor Integrated Navigation Systems*," Artech House, 2013.
8 P. G. Savage, "*Strapdown Analytics*," Maple Plain, MN: Strapdown Associates Inc., 2007.
9 C. L. Phillips, J. M. Parr and E. A. Riskin, "*Signals, Systems, and Transforms*," Fifth Edition, Upper Saddle River, NJ: Pearson, 2014.
10 L. Wang, "*PID Control System Design and Automatic Tuning using MATLAB/Simulink*," John Wiley & Sons, Incorporated, 2020.

3

Sensor Fusion Methods and Algorithms

3.1 Introduction

In the previous chapters, we studied 3D space and rigid body motion modeling. We have also discussed how navigation systems can be modeled using different forms, such as differential equations, state space models, and transfer functions. We have also reviewed basic probability concepts, random signals, and signal processing basics. In this chapter, we study how navigation systems and algorithms estimate objects' locations, orientation, speed, and the surrounding map from multiple noisy sensors using sensor fusion methods. Sensor fusion is based on the art of state estimation, and recently, it has been augmented (and sometimes replaced) by artificial intelligence (AI) and machine learning (ML). Therefore, in this chapter, we study the basics of state estimation, AI, and ML. State estimation can be defined as the art of accurately deducing useful information from noisy data. The foundation of state estimation art is based on optimization techniques. State estimation has been significantly enhanced with the advances in AI and ML and the availability of powerful computing hardware. Although the focus of the chapter will be on models related to navigation and mapping systems, the concepts of state estimation, ML, and AI can be applied in other fields.

3.2 Estimation Philosophy

The philosophy of estimation is based on building a mathematical model of the system under consideration where the state is the parameter of interest (e.g., position, angle, or map point location). Starting from an initial guess of the value of the state, we predict the state using a system model, and we predict the external measurements the system should observe given the current state estimate using a measurement model (see Figure 3.1). We then observe the actual measurements using some sensors and calculate the difference between our predicted measurements and the observed measurements to correct the estimate of the state. We repeat this process as needed to reach the optimal state estimation. To make sure that we approach the optimal estimate of the state, our change of the state must be guided in the right direction, the direction that brings our predicted measurements closer to the observed measurements. In fact, this simple philosophy is the foundation of all estimation and ML algorithms. Figure 3.1 shows an overall diagram that illustrates this process.

Sensor Fusion Approaches for Positioning, Navigation, and Mapping: How Autonomous Vehicles and Robots Navigate in the Real World: With MATLAB Examples, First Edition. Mohamed M. Atia.
© 2025 The Institute of Electrical and Electronics Engineers, Inc. Published 2025 by John Wiley & Sons, Inc.

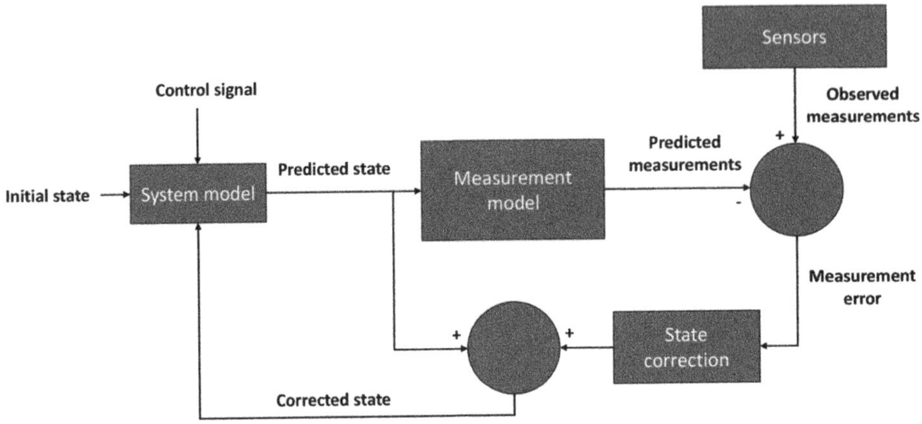

Figure 3.1 General state estimation diagram.

3.3 Gauss-Markov Process Model

Without loss of generality, most real-life navigation systems can be described as first-order Markov processes [1, 2] as shown in Figure 3.2. The system has a control input $\vec{u}(t)$, state vector $\vec{x}(t)$ and observables $\vec{y}(t)$. Due to noisy inputs and sensors' imperfection, states are contaminated by noise $\vec{w}(t)$ and observations are contaminated by noise $\vec{v}(t)$. In first-order Markov models, the state at any time depends on the previous state and the current control input. The following state-space model gives the general mathematical model that describes this Markov random process:

$$\dot{\vec{x}}(t) = f\left(\vec{x}(t), \vec{u}(t)\right) + \vec{w}(t) \tag{3-1}$$

$$\vec{y}(t) = h\left(\vec{x}(t)\right) + \vec{v}(t) \tag{3-2}$$

The vector symbol "→" is used to show that states and the observables are, in general, vectors. Here we assumed additive random noise. First-order models can be used to model complex higher-order dynamics using the state space model. In discrete form, the Markov process is given by:

$$\vec{x}_{k+1} = f\left(\vec{x}_k, u_k\right) + \vec{w}_k \tag{3-3}$$

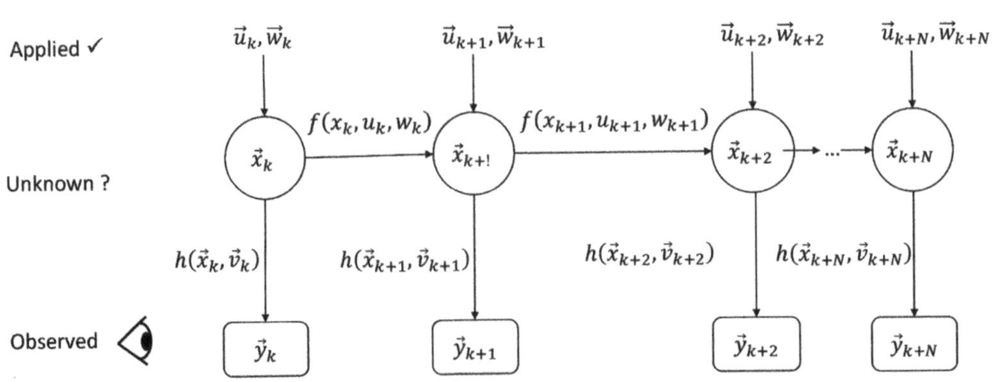

Figure 3.2 First-order Markov process.

$$\vec{y}_k = h\left(\vec{x}_k\right) + \vec{v}_k \tag{3-4}$$

If the system is linear, it will take the following form:

$$\vec{x}_{k+1} = \emptyset_k \vec{x}_k + G_k \vec{u}_k + \vec{w}_k \tag{3-5}$$

$$\vec{y}_{k+1} = H_k \vec{x}_k + \vec{v}_k \tag{3-6}$$

In some models, the control signals $\vec{u}_k$ is embedded into the state vector $\vec{x}_k$ as state variable. In this case, the system will take the following form:

$$\vec{x}_{k+1} = \emptyset_k \vec{x}_k + G_k \vec{w}_k \tag{3-7}$$

$$\vec{y}_{k+1} = H_k \vec{x}_k + \vec{v}_k \tag{3-8}$$

In some problems, there may be no dynamic system, and we have only a measurement model in the form $y_k = h(x_k) + v_k$ where v_k is measurement noise and $h(x_k)$ is a general measurement function. In this Markov model, if the system noise $\vec{w}_k$ and measurement model $\vec{v}_k$ are Gaussian, we call this model Gauss-Markov process which is a special case of the Bayesian network model that we discussed in the previous chapter. Given this system model, we will discuss the estimation techniques used to estimate the system state $\vec{x}_k$ given the measurements $\vec{y}_k$. We will also discuss how AI and ML can be applied to support conventional state estimation techniques or to perform the state estimation entirely.

3.4 State Estimation

State estimation can be approached from a probabilistic perspective or optimization perspective. Instead of considering the state vector value, we consider the probability distribution in probabilistic forms. Under the first-order Gaussian-Markov assumption, the probabilistic formula of the estimation problem is given by the following conditional probability:

$$\vec{x}_k \triangleq p\left(\vec{x}_k \middle| \vec{y}_{1:k}\right) \tag{3-9}$$

which is interpreted as the probability of the system state x_k given all measurements since the beginning of time. The initial knowledge about the system's states is modeled by the distribution $p\left(\vec{x}_0\right)$. At any given time k, the prior probability distribution before k^{th} measurement becomes available is given by:

$$p\left(\vec{x}_k\right) \triangleq p\left(\vec{x}_k \middle| \vec{y}_{1:k-1}\right) \tag{3-10}$$

Posterior probability distribution represents the updated knowledge about the system's states after a kth measurement becomes available:

$$p\left(\vec{x}_k \middle| \vec{y}_k\right) \triangleq p\left(\vec{x}_k \middle| \vec{y}_{1:k}\right) \tag{3-11}$$

To further illustrate these definitions, assuming that states and observations are scalars and all distributions are discrete, the initial, prior, and posterior distributions are illustrated in Figure 3.3, where sample space of x_k is represented as nodes $x_{k,\,j}$, where k is time index $k = 1, 2...N$ and j is sample index $j \in \{1, 2...M_k\}$ and M_k is the size of the sample space of state probability at time k.

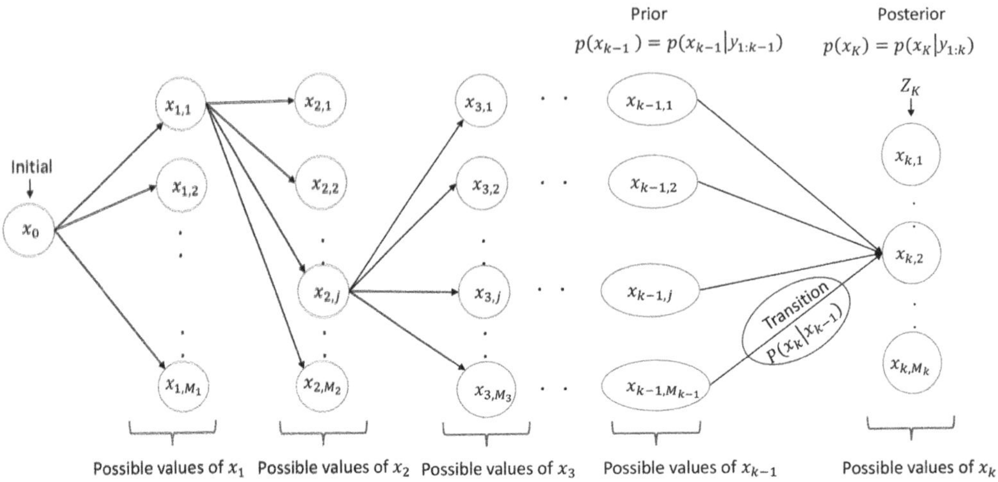

Figure 3.3 Dynamic Bayesian network (DBN).

The diagram in Figure 3.3 can be viewed as a dynamic Bayesian network that estimates the probability of any node in the network given the previous nodes information. To find the posterior probability at time k, we use the three sources of information: the prior $p\left(\vec{x}_{k-1}\right)$, the transitional conditional probability $p\left(\vec{x}_k \mid \vec{x}_{k-1}\right)$, and the measurement likelihood $p\left(\vec{y}_k \mid \vec{x}_k\right)$. Given these three sources of information, the posterior distribution of the system's states is given by Bayes' rule as follows:

$$p\left(\vec{x}_k \middle| \vec{y}_k\right) = \frac{p\left(\vec{y}_k \middle| \vec{x}_k\right) p\left(\vec{x}_k\right)}{p\left(\vec{y}_k\right)} \tag{3-12}$$

The state $\vec{x}_k$ is reached from $\vec{x}_{k-1}$ through the transition probability $p\left(\vec{x}_k \middle| \vec{x}_{k-1}\right)$. Therefore, the prior distribution $p\left(\vec{x}_k\right)$ is calculated by the following marginal probability (the sum rule):

$$p\left(\vec{x}_k\right) = \int p\left(\vec{x}_k \middle| \vec{x}_{k-1}\right) p\left(\vec{x}_{k-1}\right) d\vec{x}_{k-1} \tag{3-13}$$

We used integral formula assuming that the sample space is continuous. The integration variable $d\vec{x}_{k-1}$ means all possible values of $\vec{x}_{k-1}$ (sample space at time $k-1$). If the sample space is discrete (we have a definite list of possible values of the state at each time index k), the marginal probability of $\vec{x}_{k,j_k}$ is given by:

$$p\left(\vec{x}_{k,j}\right) = \sum_{i=1}^{M_{k-1}} p\left(\vec{x}_{k,j} \middle| \vec{x}_{k-1,i}\right) p\left(\vec{x}_{k-1,i}\right) \tag{3-14}$$

The term $p(\vec{y}_k)$ in (3-12) is a normalization factor to make sure that the sum of the posterior probability over all possible values of $\vec{x}_k$ is one. The $p(\vec{y}_k)$ is called the support and it is given by the following formula:

$$p(\vec{y}_k) = \int p(\vec{y}_k \mid x_k) p(\vec{x}_k) d\vec{x}_k \tag{3-15}$$

However, the support does not change the shape of the posterior distribution of $\vec{x}_k$ so we can write the posterior probability $p(\vec{x}_k | \vec{y}_k)$ as follows:

$$p(\vec{x}_k | \vec{y}_k) \propto p(\vec{y}_k | \vec{x}_k) p(\vec{x}_k) \tag{3-16}$$

If we are interested only in the state $\vec{x}_k$ it is enough to find $p(\vec{y}_k | \vec{x}_k) p(\vec{x}_k)$ as the normalization factor (support) $p(\vec{y}_k)$ will not affect the statistical characteristics of $p(\vec{x}_k | \vec{y}_k)$, so the mean and variance of $\vec{x}_k$ will be the same with or without the support $p(\vec{y}_k)$. Note that transitional conditional probability $p(\vec{x}_k | \vec{x}_{k-1})$ is defined by the system model and measurement likelihood $p(\vec{y}_k \mid \vec{x}_k)$ is defined by the measurement model. If the probability distributions are Gaussian and the system model and the measurement model are linear, the integral equations to calculate $p(\vec{x}_k | \vec{y}_k)$ can be analytically solved and we get a closed-form analytical solution for $p(\vec{x}_k | \vec{y}_k)$. If the probability distributions are non-Gaussian and/or the system model and/or the measurement model are nonlinear, the integrals become intractable and there is no closed-form analytical solution to calculate the posterior probability $p(\vec{x}_k | \vec{y}_k)$.

There are two main approaches to solving this estimation problem: the Bayesian approach and the maximum likelihood or optimization approach. In the Bayesian approach, we calculate the conditional probability distribution $p(\vec{x}_k | \vec{y}_k)$ using Bayes' rule. Note that we are always interested in finding a final single value of the system state, not the full distribution. When we report a probability distribution as a solution, we either provide a mean $\vec{\mu}_x$ and a standard deviation $\vec{\sigma}_x$ of a Gaussian distribution or we report a range of values of the state vector $\vec{x}$ with a probability with each value of $\vec{x}$ (probability mass function). We use the mean of this probability as a final best estimate $\vec{x}$ and the variance of this probability distribution as an uncertainty measure that reflects the confidence in this estimate. We will refer to the final best estimate of $\vec{x}$ by $\vec{x}^{\,+}$. Maximum Likelihood approach can be used to calculate the best estimate of $\vec{x}$ without calculating the full distribution of $\vec{x}_k$ as follows:

$$\vec{x}_k^{\,+} = \arg\max_{\vec{x}_k} p(\vec{x}_k | \vec{y}_k) \tag{3-17}$$

The probability values are usually very small which leads to problems in solving the optimization problem in (3-17). A better approach is to take the log of the probability, which transforms the numerical range of the problem to a more stable region. Taking the negative log makes the problem a minimization problem as follows:

$$\vec{x}_k^{\,+} = \underset{\vec{x}_k}{\arg\min}\left(-\ln\left(p\left(\vec{x}_k \middle| \vec{y}_k\right)\right)\right) \tag{3-18}$$

The optimization approach tackles the state estimation problem by formulating $p\left(\vec{x}_k \middle| \vec{y}_k\right)$ as a cost function $J\left(\vec{x}_k\right) = -\ln\left(p\left(\vec{x}_k \middle| \vec{y}_k\right)\right)$ and tries to find the optimal state $\vec{x}_k^{\,+}$ that minimizes this cost function as follows:

$$\vec{x}_k^{\,+} = \underset{\vec{x}_k}{\arg\min}\left(J\left(\vec{x}_k\right)\right) \tag{3-19}$$

3.4.1 Linear Estimation

Linear estimation considers linear systems modeled by following equations:

$$\vec{x}_{k+1} = \emptyset_k \vec{x}_k + G_k \vec{u}_k + \vec{w}_k \tag{3-20}$$

$$\vec{y}_{k+1} = H_k \vec{x}_k + \vec{v}_k \tag{3-21}$$

Or, if the control signal is embedded into the state vector:

$$\vec{x}_{k+1} = \emptyset_k \vec{x}_k + \vec{w}_k \tag{3-22}$$

$$\vec{y}_k = H_k \vec{x}_k + \vec{v}_k \tag{3-23}$$

This linear form has several mathematical advantages that enable us to obtain closed-form solutions, as we will see shortly in the next subsections.

3.4.2 Least Squares

If we assume Gaussian noise, ignored the state dynamics, and assumed that we know a Gaussian prior distribution of the parameters $p\left(\vec{x}_k\right)$ and the measurement likelihood $\left(\vec{y}_k | \vec{x}_k\right)$, we will have the following definitions of $p\left(\vec{x}_k\right)$ and $p\left(\vec{y}_k | \vec{x}_k\right)$:

$$p\left(\vec{x}_k\right) \sim \mathcal{N}\left(\vec{\mu}_x, P_x\right) \tag{3-24}$$

$$p\left(\vec{y}_k | \vec{x}_k\right) \sim N\left(H_k \vec{\mu}_x, R_k\right) \tag{3-25}$$

where P_x and R_k are covariance matrices of the noise $\vec{w}_k$ and $\vec{v}_k$, respectively. The posterior distribution will be given by:

$$p\left(\vec{x}_k \middle| \vec{y}_k\right) \propto p\left(\vec{y}_k | \vec{x}_k\right) p\left(\vec{x}_k\right) \tag{3-26}$$

$$p\left(\vec{x}_k \middle| \vec{y}_k\right) \sim \mathcal{N}\left(\vec{\mu}_{x|y}, P_{x|y}\right) \tag{3-27}$$

$$\vec{\mu}_{x|y} = P_{x|y}\left(H_k^{\;T} R_k^{-1}\left(\vec{y}_k\right) + P_x^{\;-1} \vec{\mu}_x\right) \tag{3-28}$$

$$P_{x|y}^{-1} = P_x^{-1} + H_k^{\;T} R_k^{-1} H_k \tag{3-29}$$

where $\vec{\mu}_{x|y}$ is the updated state mean and $P_{x|y}$ is the updated state covariance matrix. The proof of Eqs. (3-28) and (3-29) can be found in [2] and [3]. Note that this simple and closed-form solution is possible because of the mathematical characteristics of the exponential formula of Gaussian distribution and the linear form of the measurement model. The proof of this closed-form solution is left as an exercise. If we have no prior information about the states, we can assume that the prior distribution is a zero-mean Gaussian with infinity covariance:

$$\mu_x = 0 \tag{3-30}$$

$$P_x \to \infty \tag{3-31}$$

Substituting $\mu_x = 0$ and $P_x \to \infty$ will give us the following solution:

$$P_{x|y}{}^{-1} = H_k{}^T R_k^{-1} H_k \tag{3-32}$$

$$\vec{\mu}_{x|y} = \left(H_k{}^T R_k^{-1} H_k\right)^{-1} H_k{}^T R_k^{-1} \vec{y}_k \tag{3-33}$$

which is the popular weighted least-squares solution formula where the weights are set to R_k^{-1}. If all measurements are independent and equally weighted, $R_k = I$, which leads to the standard least-squares solution:

$$\vec{\mu}_{x|y} = \left(H_k{}^T H_k\right)^{-1} H_k{}^T \vec{y}_k \tag{3-34}$$

The same formula can be reached using maximum likelihood and optimization approaches. In the maximum likelihood approach, we want to find the system state vector $\vec{x}_k^{\,+}$ that maximizes the following measurement likelihood:

$$\vec{x}_k^{\,+} = \underset{x_k}{\operatorname{argmax}} p\left(\vec{y}_k | \vec{x}_k\right) \tag{3-35}$$

$$p\left(\vec{y}_k | \vec{x}_k\right) = \frac{1}{\sqrt{(2\pi)^n}\sqrt{|R_k|}} e^{-\frac{1}{2}\left(\vec{y}_k - H_k \vec{x}_k\right)^T R_k{}^{-1}\left(\vec{y}_k - H_k \vec{x}_k\right)} \tag{3-36}$$

where n is the number of measurements (y_k is $n \times 1$ vector). Taking the negative log of both sides will give the following equation:

$$-\ln p\left(\vec{y}_k | \vec{x}_k\right) = -\left(\ln(R_k) + \frac{1}{2}\ln((2\pi)^n) - \frac{1}{2}\left(\vec{y}_k - H_k \vec{x}_k\right)^T R_k{}^{-1}\left(\vec{y}_k - H_k \vec{x}_k\right)\right) \tag{3-37}$$

The term $-\ln p(y_k|x_k)$ is called the negative log-likelihood. Maximizing $p\left(\vec{y}_k | \vec{x}_k\right)$ is equivalent to minimizing the negative log-likelihood. The first two terms of the negative log-likelihood do not depend on $\vec{x}_k$ so the negative log-likelihood will be given by the following equation:

$$-\ln p\left(\vec{y}_k | \vec{x}_k\right) = \text{const.} + \frac{1}{2}\left(\vec{y}_k - H_k \vec{x}_k\right)^T R_k{}^{-1}\left(\vec{y}_k - H_k \vec{x}_k\right) \tag{3-38}$$

The best solution $\vec{x}_k^{\,+}$ is the solution that minimizes $J\left(\vec{x}_k\right)$, where $J\left(\vec{x}_k\right)$ is a cost function defined as follows:

$$\vec{x}_k^{\,+} = \underset{x_k}{\operatorname{argmin}} \left(J\left(\vec{x}_k\right)\right) \tag{3-39}$$

$$J\left(\vec{x}_k\right) = \frac{1}{2}\left(\vec{y}_k - H_k\vec{x}_k\right)^T R_k^{-1}\left(\vec{y}_k - H_k\vec{x}_k\right) \tag{3-40}$$

Taking the derivative of $J\left(\vec{x}_k\right)$ with respect to $\vec{x}_k$ and setting the result to zero gives the following solution:

$$\frac{\partial}{\partial x}J\left(\vec{x}_k^{\,+}\right) = 0 \xrightarrow{\text{yields}} \vec{x}_k^{\,+} = \left(H_k^T R_k^{-1} H_k\right)^{-1} H_k^T R_k^{-1}\vec{y}_k \tag{3-41}$$

which is the same solution obtained by Bayesian approach. The same solution can be obtained by optimization approach if we define the squared error as a function of the state vector $\vec{x}_k$. The error vector is simply the difference between the measurements $\vec{y}_k$ and our model prediction $H_k\vec{x}_k$ as follows:

$$\vec{e}_k = \left(\vec{y}_k - H_k\vec{x}_k\right) \tag{3-42}$$

The weighted squared error equation in vector-form is given by:

$$J\left(\vec{x}_k\right) = \frac{1}{2}\left(\vec{y}_k - H_k\vec{x}_k\right)^T R_k^{-1}\left(\vec{y}_k - H_k\vec{x}_k\right) \tag{3-43}$$

$J\left(\vec{x}_k\right)$ is the weighted sum of squared errors. The best estimate of $\vec{x}_k$ is the one that minimizes $J\left(\vec{x}_k\right)$, which can be obtained by taking the derivative of $J\left(\vec{x}_k\right)$ with respect to $\vec{x}_k$ and setting the result to zero:

$$\frac{\partial}{\partial x}J\left(\vec{x}_k^{\,+}\right) = 0 \xrightarrow{\text{yields}} \vec{x}_k^{\,+} = \left(H_k^T R_k^{-1} H_k\right)^{-1} H_k^T R_k^{-1}\vec{y}_k \tag{3-44}$$

which is the same solution as Bayesian and Maximum likelihood approaches.

3.4.3 Integer Least-squares Algorithm

In some problems, the state vector that we need to estimate $\vec{x}$ has two parts, an integer part $\vec{a}$ and a float part $\vec{b}$ such that $\vec{x} = \begin{pmatrix}\vec{a}\\ \vec{b}\end{pmatrix}$. This problem requires an integer least square (ILS) method. In this case, the linear measurement model is given in the following form:

$$\begin{aligned}\vec{y} &= H_a\vec{a} + H_b\vec{b} + \vec{v}\\ &= \underbrace{\left(H_a \;\; H_b\right)}_{H}\underbrace{\begin{pmatrix}\vec{a}\\ \vec{b}\end{pmatrix}}_{\vec{x}} + \vec{v}\\ \vec{v} &\sim \mathcal{N}(0, R)\end{aligned} \tag{3-45}$$

The measurement model in (3-45) adds an additional constraint to the estimation problem where the values of $\vec{a}$ must be integers. To find the best state estimation of $\vec{x}$ and at the same time ensure that the values of $\vec{a}$ are integer, a special least-square algorithm is applied called linear least-squares (LLS) algorithm. The steps of the LLS algorithm can be summarized as follows:

1) The float solution is calculated using regular least-squares formula as follows:

$$\vec{x}^{f+} = \begin{pmatrix} \vec{a}^f \\ \vec{b}^f \end{pmatrix} = \left(H^T R^{-1} H\right)^{-1} H^T R^{-1} \vec{y} \tag{3-46}$$

This solution has a covariance matrix Q that is calculated as follows:

$$Q = \left(H^T R^{-1} H\right)^{-1} = \begin{pmatrix} Q_{aa} & Q_{ab} \\ Q_{ba} & Q_{bb} \end{pmatrix} \tag{3-47}$$

In practical problems, the correlation between the values of the elements of vector $\vec{a}^f$ is high making the search for best integer values difficult and time consuming. It is better to map $\vec{a}^f$ to an orthogonal space where the covariance matrix is diagonal. We do this by the following orthogonalization step using the LAMBDA method.

2) Given $\vec{a}^f$ and Q_{aa}, the LAMBDA (least-squares ambiguity decorrelation adjustment) is applied to estimate the integer values $\vec{a}^I$. The details of the LAMBDA method can be found in [4] and [5]. The LAMBDA method decomposes the covariance matrix Q_{aa} into lower diagonal matrix decomposition as follows:

$$Q_{aa} = LL^T \tag{3-48}$$

The lower diagonal matrix L is reduced into a near diagonal matrix L_{red} using the following transformation:

$$L_{red} = ZL \tag{3-49}$$

To find the transformation matrix Z that performs this reduction step, an algorithm like Lenstra–Lenstra–Lovász (LLL) algorithm can be used (see [4], [5], and [6]).

3) Both $\vec{a}^f$ and Q_{aa} are transformed by the Z matrix as follows:

$$\begin{aligned} Q_{zz} &= ZQ_{aa}Z^T \\ \vec{a}_z^f &= Z^{-1}\vec{a}^f \end{aligned} \tag{3-50}$$

4) Search for the best integer values $\vec{a}_z^I$ that minimize the following function:

$$\vec{a}_z^I = \arg\min_{\vec{a}_z^I \in \mathbb{Z}^n} \left(\vec{a}_z^f - \vec{a}_z^I\right)^T Q_{zz}^{-1} \left(\vec{a}_z^f - \vec{a}_z^I\right) \tag{3-51}$$

The search for integer values $\vec{a}_z^I$ can be done be done using any search algorithm such as best-first or depth-first [4],[5], and [6].

5) The best integer values $\vec{a}^f$ is then calculated as follows:

$$\vec{a}^I = Z\vec{a}_z^I \tag{3-52}$$

6) Once the integer values of $\vec{a}$ is adjusted, the float values of $\vec{b}$ is updated using the Bayesian update rule as follows:

$$\vec{b}^f = \vec{b}^f - (H_b{}^T H_b)^{-1} H_b{}^T \left(y - H_a\vec{a}^I\right) \tag{3-53}$$

3.4.4 Kalman Filter

The least squares algorithm described above considers only the measurement model, which is the mapping between the observables $\vec{y}_k$ and the states $\vec{x}_k$. The measurement model is one source of information. Another source of information is how the states $\vec{x}_k$ evolves with time, which we describe using the system model. Kalman filter is an optimal linear estimation algorithm that fuses both the measurement model and the system mode if the noise is zero-mean Gaussian. Given the following system and measurement models:

$$\vec{x}_{k+1} = \emptyset_k \vec{x}_k + \vec{w}_k \tag{3-54}$$

$$\vec{y}_{k+1} = H_k \vec{x}_k + \vec{v}_k \tag{3-55}$$

where $\vec{w}_k$ and $\vec{v}_k$ are zero-mean Gaussian noise with the following covariances:

$$\vec{w}_k \sim \mathcal{N}\left(\vec{0}, Q_k\right) \tag{3-56}$$

$$\vec{v}_k \sim \mathcal{N}\left(\vec{0}, R_k\right) \tag{3-57}$$

Starting from a prior distribution of the states, Kalman filter performs two steps, prediction, and update. Let the prior state distribution at time epoch k be defined as follows:

$$p\left(\vec{x}_{k-1}{}^+\right) \sim \mathcal{N}\left(\vec{\mu}_{k-1}, P_{k-1}\right) \tag{3-58}$$

In the next subsection, we will discuss how Bayesian probabilistic rules are used to derive the prediction and the update equations of the Kalman filter.

3.4.4.1 Prediction

Given the initial state probability $p\left(\vec{x}_{k-1}{}^+\right)$, we calculate a predicted probability distribution $p\left(\vec{x}_k{}^-\right)$ using the marginalization (sum) rule as follows:

$$p\left(\vec{x}_k{}^-\right) = \int p\left(\vec{x}_k{}^- \middle| \vec{x}_{k-1}{}^+\right) p\left(\vec{x}_{k-1}{}^+\right) dx_{k-1}{}^+ \tag{3-59}$$

If the system model is linear, and all probabilities (system noise and state prior distributions) are Gaussian, the mean and covariance propagation equations (review linear stochastic systems in Chapter 2) this integral form will reduce to the following Gaussian distribution:

$$p\left(\vec{x}_k^{\,-}\right) \sim \mathcal{N}\left(\vec{\mu}_k^{\,-}, P_k^-\right) \tag{3-60}$$

$$\vec{\mu}_k^{\,-} = \emptyset_k \vec{\mu}_{k-1} \tag{3-61}$$

$$P_k^- = \emptyset_k P_{k-1} \emptyset_k^{\,T} + Q_k \tag{3-62}$$

where $\vec{\mu}_k^{\,-}$ is the predicted mean and P_k^- is the predicted covariance. We have demonstrated in Chapter 2 the derivations of the Gaussian mean/covariance propagation analysis through linear systems which lead to the predicted mean and covariance equations given above. The derivation used the sum rule and the linear transformation rule of Gaussian variables.

3.4.4.2 Update

When measurements become available, we update the system state probability distribution using the Bayesian conditional probability rule given the measurement likelihood. The measurement likelihood is defined as follows:

$$p\left(\vec{y}_k | x_k\right) \sim \mathcal{N}\left(H_k \vec{\mu}_x, R_k\right) \tag{3-63}$$

Now, we calculate the new state distribution as follows:

$$p\left(\vec{x}_k^{\,+}\right) \sim \mathcal{N}\left(\vec{\mu}_k^{\,+}, P_k^+\right) \tag{3-64}$$

$$\left(P_k^+\right)^{-1} = \left(P_k^-\right)^{-1} + H_k^{\,T} R_k^{-1} H_k \tag{3-65}$$

$$\vec{\mu}_k^{\,+} = \vec{\mu}_k^{\,-} + P_k^- H^T / \left(H_k P_k^- H_k^{\,T} + R_k\right)\left(\vec{y}_k - H_k \vec{\mu}_k^{\,-}\right) \tag{3-66}$$

The term $P_k^- H^T/(H_k P_k^- H_k^{\,T} + R_k)$ is multiplied by the difference between predicted measurements $H_k \mu_k^-$ and the actual observed measurements y_k, which means it is a correction term. In the literature, this correction term is called Kalman gain defined by the following equation:

$$K_k = P_k^- H^T / \left(H_k P_k^- H_k^{\,T} + R_k\right) \tag{3-67}$$

The details of the proof of the Kalman gain formula and the state posterior probability distribution can be found in [2], [3], and [1]. In [2], Kalman gain is proved using the Bayesian approach by applying Bayesian rule on linear Gaussian systems. In [1], Kalman gain is proved using maximum likelihood approach.

3.4.5 Nonlinear Estimation

The previous sections discussed linear estimation, where all models are linear. The linearity of the system model guarantees that the states propagating through the system stay Gaussian as long as inputs and noise signals are Gaussian. If the system model is nonlinear, the system output is not guaranteed to be Gaussian. The most common approach to estimating nonlinear system states is to linearize the system around a known state called the operating point. The linearization process generates an approximate linear version of the system that models the errors in the state vector, not the state vector itself. This linearization is performed by considering the first-order term of the Tylor series expansion terms. Linearized models are linear models of state errors; therefore, we call them error models. By having linear error models, we can apply all linear estimation techniques, such as least-squares and Kalman filter. Extended Kalman filter (to be explained in the next subsection) is a Kalman filter applied to a linearized approximation of a nonlinear system. Another approach to nonlinear estimation is using nonlinear solutions that directly work on the nonlinear system models. An example of an estimation technique that can deal with nonlinear non-Gaussian systems is the particle filter, which is explained shortly in a later subsection.

3.4.6 Nonlinear Least-Squares

Given the following nonlinear measurement model:

$$\vec{y} = h(\vec{x}) + \vec{v} \tag{3-68}$$

and an initial guess of the state $\vec{x}(0) = \vec{x}^-$, the error in $\vec{y}$ is given by:

$$\delta\vec{y} = \vec{y} - h(\vec{x}^-) \tag{3-69}$$

$\delta\vec{y}$ can be defined by linearizing the measurement model using Tylor series expansion around $\vec{x}^-$ as follows:

$$\delta\vec{y} = h(\vec{x}) - h(\vec{x}^-) = \left.\frac{\partial h(\vec{x})}{\partial \vec{x}}\right|_{\vec{x}=\vec{x}^-} \frac{(\vec{x}-\vec{x}^-)}{1!} + \left.\frac{\partial^2 h(\vec{x})}{\partial \vec{x}^2}\right|_{\vec{x}=\vec{x}^-} \frac{(\vec{x}-\vec{x}^-)^{.2}}{2!} + \ldots \tag{3-70}$$

where $(\vec{x}-\vec{x}^-)^{.n}$ is element-wise n^{th} power of $(\vec{x}-\vec{x}^-)$. By ignoring the second and higher order terms because $\delta\vec{x} = (\vec{x}-\vec{x}^-)$ is presumably small, we get the following approximated linear model:

$$\delta\vec{y} \cong H(\vec{x}^-)\delta\vec{x} \tag{3-71}$$

$$H(\vec{x}^-) \cong \left.\frac{\partial h(\vec{x})}{\partial \vec{x}}\right|_{\vec{x}=\vec{x}^-} \tag{3-72}$$

This system can be solved using the linear least squares solution illustrated in Eq. (3-33) as follows:

$$\delta\vec{x} = \left(H(\vec{x}^-)^T H(\vec{x}^-)\right)^{-1} H(\vec{x}^-)^T \delta\vec{y} \tag{3-73}$$

We then use the error $\delta\vec{x}$ to improve our initial guess as follows:

$$\vec{x}^+ = \vec{x}^- + \delta\vec{x} \tag{3-74}$$

We then repeat the same steps starting from the new estimate $\vec{x}^+$ by setting $\vec{x}^- = \vec{x}^+$ and implementing Eqs. (3-69)–(3-74). We keep repeating the calculations in Eqs. (3-69)–(3-74) until $\delta\vec{x}$ converges below a small threshold.

3.4.7 Newton Method

Another approach to solve the nonlinear measurement model $\vec{y} = h(\vec{x}) + \vec{v}$ is by defining a cost function as follows:

$$F(\vec{x}) = \frac{\left\|\vec{y} - h(\vec{x})\right\|^2}{2} = \frac{\left\|f(\vec{x})\right\|^2}{2} \tag{3-75}$$

Starting from initial guess $\vec{x}^-$, and considering second-order Tylor series expansion, $F(\vec{x}^- + \Delta\vec{x})$ can be approximated using Tylor series expansion as follows:

$$F(\vec{x}^- + \Delta\vec{x}) \approx F(\vec{x}^-) + \mathrm{J}_F(\vec{x}^-)^T \Delta\vec{x} + \frac{1}{2}\Delta\vec{x}_k^T \mathrm{H}_F(\vec{x}^-)\Delta\vec{x} \tag{3-76}$$

$$\mathrm{J}_F(\vec{x}^-) = \frac{\partial F(\vec{x}^-)}{\partial \vec{x}} \tag{3-77}$$

$$\mathrm{H}_F(\vec{x}^-) = \frac{\partial^2 F(\vec{x}^-)}{\partial \vec{x}^2} \tag{3-78}$$

$\mathrm{J}_F(\vec{x}^-)$ is called Jacobian matrix (first derivative) of $F(\vec{x})$ and $\mathrm{H}_F(\vec{x}^-)$ is the hessian matrix (second derivative) of $F(\vec{x})$. Optimal state error $\Delta\vec{x}^+$ is given by:

$$\Delta\vec{x}^+ = \arg\min_{\Delta\vec{x}} \left(F(\vec{x}^-) + \mathrm{J}(\vec{x}^-)^T \Delta\vec{x} + \frac{1}{2}\Delta\vec{x}_k^T \mathrm{H}_F(\vec{x}^-)\Delta\vec{x}\right) \tag{3-79}$$

Differentiating the term $\left(F(\vec{x}^-) + \mathrm{J}_F(\vec{x}^-)^T \Delta\vec{x} + \frac{1}{2}\Delta\vec{x}_k^T \mathrm{H}_F(\vec{x}^-)\Delta\vec{x}\right)$ with respect to $\Delta\vec{x}$ and setting the result to zero gives the following optimal solution of $\Delta\vec{x}$ as follows:

$$\mathrm{J}_F(\vec{x}^-) + \mathrm{H}_F(\vec{x}^-)\Delta\vec{x}^+ = 0 \tag{3-80}$$

$$\mathrm{H}_F(\vec{x}^-)\Delta\vec{x}^+ = -\mathrm{J}_F(\vec{x}^-) \tag{3-81}$$

which can be solved using the singular value decomposition (SVD) of $\mathrm{H}_F(\vec{x})$ (review Chapter 1 for an explanation of SVD matrix solution approach). Once we calculate $\Delta\vec{x}^+$, we update our initial guess and repeat the steps until $\Delta\vec{x}$ converges under a small threshold. An enhanced version of Newton's method is Quasi-Newton method [7], which approximates the calculation of the Hessian matrix. More information about Newton's method and other optimization methods can be found in [7] and [8].

3.4.8 Gauss–Newton Method

To avoid calculating the hessian matrix, the Gauss–Newton method was proposed based on the following approximation of the error function $\vec{y} - h(\vec{x})$ defined as follows:

$$f(\vec{x}) = \vec{y} - h(\vec{x}) \tag{3-82}$$

$$f\left(\vec{x}^- + \Delta\vec{x}\right) \approx f\left(\vec{x}^-\right) + J_f\left(\vec{x}^-\right)^T \Delta\vec{x} \tag{3-83}$$

To minimize the square of the error function $\left\|f\left(\vec{x}\right)\right\|^2$, we want to find $\Delta\vec{x}^+$ that minimizes the approximate formula of $\left\|f\left(\vec{x}\right)\right\|^2$ as follows:

$$\Delta\vec{x}^+ = \arg\min_{\Delta\vec{x}} \frac{1}{2}\left\|f\left(\vec{x}^-\right) + J_f\left(\vec{x}^-\right)^T \Delta\vec{x}\right\|^2 \tag{3-84}$$

where $J_f\left(\vec{x}^-\right)$ is the Jacobian of $f\left(\vec{x}\right)$. Differentiating $\left(\frac{1}{2}\left\|F\left(\vec{x}^-\right) + J_f\left(\vec{x}^-\right)^T \Delta\vec{x}\right\|^2\right)$ with respect to $\Delta\vec{x}$ and setting the result to zero gives the following optimal solution:

$$J_f\left(\vec{x}^-\right)F\left(\vec{x}^-\right) + J_f\left(\vec{x}^-\right)J_f\left(\vec{x}^-\right)^T \Delta\vec{x}^+ = 0 \tag{3-85}$$

$$J_f\left(\vec{x}^-\right)J_f\left(\vec{x}^-\right)^T \Delta\vec{x}^+ = -J_f\left(\vec{x}^-\right)F\left(\vec{x}^-\right) \tag{3-86}$$

which can be solved using SVD without the need to calculate the Hessian matrix. This method only needs the Jacobian of $f\left(\vec{x}\right)$, which is simpler to calculate compared to the Jacobian and Hessian of $F\left(\vec{x}\right)$. Once we calculate $\Delta\vec{x}^+$, we update our initial guess and repeat the steps until $\Delta\vec{x}$ converges under a small threshold. More information about Gauss–Newton's method can be found in [7] and [8].

3.4.9 Levernberg–Marquatdt Method

Linearization around an initial guess may give poor results if the function to be optimized is highly nonlinear and changes rapidly, which causes wrong estimation of $\Delta\vec{x}^+$. Therefore, it is beneficial to know the trusted range of $\Delta\vec{x}^+$ to consider in our calculations. Given an initial guess $\vec{x}^-$, Levernberg–Marquatdt method compares the actual difference in $f\left(\vec{x}^-\right)$ to the approximate difference calculated by the Jacobian $J_f\left(\vec{x}^-\right)$ as follows:

$$\rho = \frac{f\left(\vec{x}^- + \Delta\vec{x}\right) - f\left(\vec{x}^-\right)}{J\left(\vec{x}^-\right)^T \Delta\vec{x}} \tag{3-87}$$

The closer the parameter ρ to 1, the better the approximation and the more confidence we trust the estimated $\Delta\vec{x}$. We can use the parameter ρ to control the size of $\Delta\vec{x}$ and use it as additional constraints in our solution of $\Delta\vec{x}^+$. More information about Levernberg–Marquatdt method can be found in [7] and [8].

3.4.10 Extended Kalman Filter

Extended Kalman filter (EKF) is a method to apply Kalman filter estimation on nonlinear dynamic systems. In the previous chapter, we have seen that a nonlinear stochastic dynamic system can be given in the following form:

$$\dot{\vec{x}}(t) = f\left(\vec{x}(t), \vec{u}(t)\right) + \vec{w}(t) \tag{3-88}$$

$$\vec{y}(t) = h\left(\vec{x}(t)\right) + \vec{v}(t) \tag{3-89}$$

where $\vec{w}(t)$ and $\vec{v}(t)$ are random noise signals. Assuming that the control signal is embedded into the state vector, and starting from an operating known state $\vec{\bar{x}}$, and control input $\vec{\bar{u}}\ (t)$, applying first-order Tylor series expansion gives the following linear models:

$$\delta\dot{\vec{x}}(t) = F(t)\delta\vec{x}(t) + G(t)\delta\vec{u}(t) + \vec{w}(t) \tag{3-90}$$

$$F(t) = \left.\frac{\partial f\left(\vec{x}(t), \vec{u}(t)\right)}{\partial x}\right|_{\vec{x} = \vec{\bar{x}}, \vec{u} = \vec{\bar{u}}}$$
$$G(t) = \left.\frac{\partial f\left(\vec{x}(t), \vec{u}(t)\right)}{\partial \vec{u}}\right|_{\vec{x} = \vec{\bar{x}}, \vec{u} = \vec{\bar{u}}} \tag{3-91}$$

Or, if we embed the control signal error into the error state vector, we get:

$$\delta\dot{\vec{x}}(t) = F(t)\delta\vec{x}(t) + \vec{w}(t) \tag{3-92}$$

$$\delta\vec{y}(t) = H(t)\delta\vec{x}(t) + \vec{v}(t) \tag{3-93}$$

$$F(t) = \left.\frac{\partial f\left(\vec{x}(t), \vec{u}(t)\right)}{\partial \vec{x}}\right|_{\vec{x} = \vec{\bar{x}}, \vec{u} = \vec{\bar{u}}} \tag{3-94}$$

$$H(t) = \left.\frac{\partial h\left(\vec{x}(t)\right)}{\partial \vec{x}}\right|_{\vec{x} = \vec{\bar{x}}, \vec{u} = \vec{\bar{u}}} \tag{3-95}$$

which is a first-order dynamic linear system in the state error $\delta\vec{x}(t)$. Optimal error can be estimated given the observation error $\delta\vec{y}(t)$. As we can observe, EKF is estimating system state error and not the state directly. Once the error is estimated, the state vector can be corrected as follows:

$$\vec{x}(t) = \vec{\bar{x}}\ (t) + \delta\vec{x}(t) \tag{3-96}$$

In discrete form, we will have the following linear error model:

$$\delta\vec{x}_{k+1} = \emptyset_k\delta\vec{x}_k + \vec{w}_k \tag{3-97}$$

$$\delta\vec{y}_{k+1} = H_k\delta\vec{x}_k + \vec{v}_k \tag{3-98}$$

$$\vec{w}_k \sim \mathcal{N}\left(\vec{0}, Q_k\right) \tag{3-99}$$

$$\vec{v}_k \sim \mathcal{N}\left(\vec{0}, R_k\right) \tag{3-100}$$

where $\emptyset_k$ is the discrete version of the dynamic transition matrix. There are different ways to calculate $\emptyset_k$ from $F(k)$ as described in [1]. A simple first-order approximation of the time derivative gives the following formula of $\emptyset_k$:

$$\emptyset_k \cong I + F_k \Delta T \tag{3-101}$$

$$F_k = F(k\Delta T) \tag{3-102}$$

where ΔT is the sampling period. To fuse both the system model and the measurement model, EKF performs prediction and update steps. Both steps are explained in the following subsections.

3.4.10.1 Prediction

We always assume that the error is zero in the beginning of the processing ($\delta\vec{x}_0 = 0$) and we always correct the state vector $\vec{x}_k$ for error $\delta\vec{x}_k$ at each measurement update step. Therefore, during the whole prediction epochs, δx_k is always zero. What we predict is the system state x_{k+1}^- using the nonlinear system model and error covariance matrix P_{k+1}^- using the linearized model as follows:

$$\vec{x}_k^+ \sim \mathcal{N}\left(\vec{\mu}_k^+, P_k^+\right) \tag{3-103}$$

$$\emptyset_k = I + F_k \Delta T \tag{3-104}$$

$$\vec{\mu}_{k+1}^- = f\left(\vec{\mu}_k^+, \vec{u}_k\right) \tag{3-105}$$

$$P_{k+1}^- = \emptyset_k P_k^+ \emptyset_k^T + Q_k \tag{3-106}$$

The prediction probability distributions of the state are given as follows:

$$\vec{x}_{k+1}^- \sim \mathcal{N}\left(\mu_{k+1}^-, P_{k+1}^-\right) \tag{3-107}$$

3.4.10.2 Update

The update step calculates the posterior conditional probability of the state given the measurements. The measurement model is the probability of the observed error given the state error, which is a Gaussian probability whose mean is given by:

$$p\left(\delta\vec{y}_k | \delta\vec{x}_k\right) \sim \mathcal{N}\left(\vec{y}_k - h\left(\vec{x}_k\right), R_k\right) \tag{3-108}$$

Now, we calculate the posterior probability as follows:

$$p\left(\vec{x}_k^+\right) \sim \mathcal{N}\left(\vec{\mu}_k^+, P_k^+\right) \tag{3-109}$$

$$(P_k^+)^{-1} = (P_k^-)^{-1} + H_k^T R_k^{-1} H_k \tag{3-110}$$

$$\vec{\mu}_k^+ = \vec{\mu}_k^- + K_k\left(\left(\vec{y}_k - h\left(\vec{\mu}_k^-\right)\right) - H_k \delta\vec{x}_k\right) \tag{3-111}$$

$$K_k = P_k^- H^T / \left(H_k P_k^- H_k^T + R_k\right) \tag{3-112}$$

Note that in the above equation, we assumed that there is a predicted error state $\delta\vec{x}_k$. However, as explained earlier, we assume that $\delta\vec{x}_k$ is zero because we correct the states whenever we have an update (correction makes the error, theoretically, zero), and we always start from initial conditions $\delta\vec{x}_{k-1} = 0$, so the corrected state mean is given by the following equation:

$$\vec{\mu}_k^{\,+} = \vec{\mu}_k^{\,-} + K_k\left(\left(\vec{y}_k - h\left(\vec{\mu}_k^{\,-}\right)\right)\right) \tag{3-113}$$

3.4.11 Particle Filter

3.4.11.1 Motivation

Another approach to tackle nonlinear systems is to avoid linearization altogether and work directly on the nonlinear models. This will take us to another limitation, which is the Gaussian assumption. A Gaussian input to a nonlinear system will not output a Gaussian signal. Therefore, we need a method that does not assume Gaussian distribution. Recall the Bayesian probabilistic formulation of the state estimation problem using the Bayesian rule and the marginalization, which are given as follows:

$$p\left(\vec{x}_k \middle| \vec{y}_k\right) = \frac{p\left(\vec{y}_k \middle| \vec{x}_k\right) p\left(\vec{x}_k\right)}{p\left(\vec{y}_k\right)} \tag{3-114}$$

$$p\left(\vec{x}_k\right) = \int p\left(\vec{x}_k \middle| \vec{x}_{k-1}\right) p\left(\vec{x}_{k-1}\right) d\vec{x}_{k-1} \tag{3-115}$$

where $p\left(\vec{y}_k \middle| \vec{x}_k\right)$ represents the measurement model (measurement likelihood), $p\left(\vec{x}_k \middle| \vec{x}_{k-1}\right)$ represents the system model, and $p\left(\vec{x}_{k-1}\right)$ is the state prior probability distribution. If the models are nonlinear and/or the distributions are not Gaussian, the Bayesian rule and the marginalization rule, which include integral formulas, will be difficult, if not impossible, to solve analytically. These are the main motivations for introducing particle filters. More information can be found in [9].

3.4.11.2 Particles Cloud

The particle filter approximates any probability distribution by a number of sample points called a particle cloud of the sample space drawn from the desired distribution. Each sample point is represented in the sample cloud such that its frequency is proportional to its probability. The higher the probability of a sample point, the higher the number of its occurrence in the particle cloud. This is called an "unweighted" particle cloud. In the following discussion, and for the simplicity of the notation, we will consider scalar states and measurements. Assume we have a sample space $S = \{x_1, x_2, x_3 \ldots x_N\}$ and arbitrary probability distribution $p(x_j) \in [0, 1]$, $p(x)$ can be approximated by the following unweighted particle cloud:

$$p(x) \Longrightarrow C_x = \left\{x_1^{(m_1)}, x_2^{(m_2)}, x_3^{(m_3)} \ldots x_N^{(m_N)}\right\} \tag{3-116}$$

where the superscript "(m_j)" is the multiplicity of x_j (the number of occurrences of x_j in C_x). C_x can also be represented as a "weighted" particle cloud as follows:

$$C_x = \left\{x_j, w_j\right\} \tag{3-117}$$

$$w_j \propto p\left(x_j\right) \tag{3-118}$$

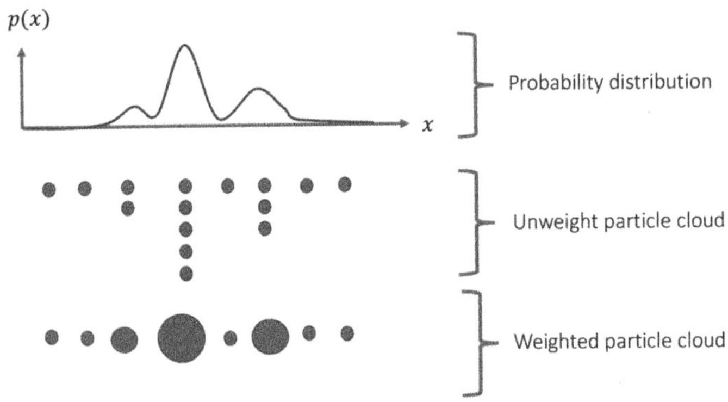

Figure 3.4 Point cloud approximation of probability distribution.

The mean of the distribution $p(x)$ can be approximated by the following form:

$$\mu_x = \sum_{j=1}^{S} w_j x_j \tag{3-119}$$

An illustration of approximating a probability distribution by weighted and unweighted particle cloud is shown in Figure 3.4.

3.4.11.3 Sampling Methods

To approximate $p(x)$ by a particle cloud, for each sample point x_j, we need to generate several samples m_j that is proportional to $p(x_j)$. Sampling from uniform distributions is easy as all sample values are equally probable, so all sample points x_j have the same multiplicity. Most scientific computing software such as MATLAB have a built-in uniform distribution random number generator $u{\sim}U(0, 1)$. However, sampling from an arbitrary probability distribution is not an easy straightforward task. Many sampling methods generate uniformly distributed samples and then map them to the desired target distribution. Recall, the cumulative distribution function (CDF) of $p(x)$ is defined as follows:

$$F(x) = \int_{-\infty}^{x} p(x)dx \tag{3-120}$$

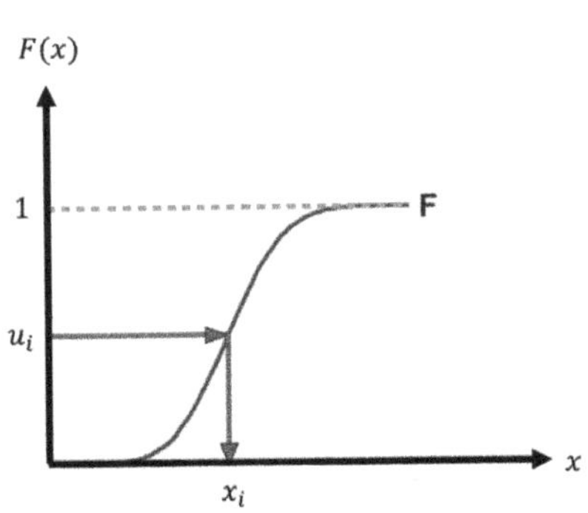

Figure 3.5 Sampling from standard distributions.

If the target distribution has a closed-form formula of F^{-1}, the inverse CDF, we can map any uniformly distributed random sample $u_j \in u{\sim}U(0, 1)$ to our target distribution $p(x_j)$ using the following mapping:

$$x_j = F^{-1}(u_j) \tag{3-121}$$

We will refer to this sampling algorithm by inverse CDF sampling [2]. This process is visually illustrated in Figure 3.5. We keep generating uniformly distributed numbers u_j and map them to our target distribution using $x_j = F^{-1}(u_j)$ until we have the desired number of samples in the particle cloud.

If the distribution is Gaussian, we can use the Box-Muller method [2, 3] to generate samples. Box-Muller method generates two uniform samples u_1, $u_2 \sim U(0, 1)$

and rejects the values that lie outside the unit circle $u_1^2 + u_2^2 > 1$. This will give us uniformly distributed points inside the unit circle. Then, we calculate the corresponding Gaussian samples x_1, x_2 by:

$$x_1 = u_1\sqrt{\left(\frac{-2\ln(u_1^2 + u_2^2)}{u_1^2 + u_2^2}\right)} \tag{3-122}$$

$$x_2 = u_2\sqrt{\left(\frac{-2\ln(u_1^2 + u_2^2)}{u_1^2 + u_2^2}\right)} \tag{3-123}$$

We repeat the process until we reach the required particle cloud size. For the proof of this method, see section 23.2 in [2]. Other advanced sampling methods can be found in several references (see, e.g., section 23.2 in [2]). If the target distribution is not Gaussian and/or it is difficult to obtain the inverse CDF, the rejection sampling method [2, 3] can be used. Rejection sampling approximates the target distribution $p(x)$ with a proposal distribution $q(x)$ that is either Gaussian (so we can sample from it using Box–Muller method) or a distribution with a known inverse CDF (so we can use the inverse CDF sampling method). Then, we reject the samples that do not belong to our target distribution. To implement this method, the proposal distribution $q(x)$ must be an envelope for $p(x)$ as seen in Figure 3.6. This can be achieved by the following condition:

$$M\,.\,q(x) \geq c\,.\,p(x) \tag{3-124}$$

where M and c are constants. Once we have the proposal distribution, we sample from the proposal $x_j \sim q(x)$ (using inverse CDF or Box-Muller method) and then generate a uniform sample from the uniform distribution $u_j \sim U(0, 1)$. We then accept x_j if u_i satisfies the following condition:

$$u_i \leq (c\,.\,p(x))/(M\,.\,q(x)) \tag{3-125}$$

This procedure is visually illustrated in Figure 3.6. As can be seen in Figure 3.6, the accepted samples are the ones that fall within the target distribution $p(x)$.

3.4.11.4 Prediction Using Importance Sampling

Having samples from probability distributions is the first step in particle filter. The second step is to apply calculation on these distribution samples to estimate the state. For example, the predictive distribution $p(x_k)$ given the transition probability $p(x_k|x_{k-1})$ and the prior $p(x_{k-1})$ is given by the following equation:

$$p(x_k) = \int p(x_{k-1})p(x_k|x_{k-1})dx_{k-1} \tag{3-126}$$

The predicted probability is calculated as an integral of the multiplication of the prior probability $p(x_{k-1})$ approximated by the particle cloud $C_x(k-1)$ and the transition probability $p(x_k|x_{k-1})$ modeled by the dynamic system function $f(x_k, u_k)$. If, for example, we assumed Gaussian transition probability (which is not necessary, but we use it here for illustration) as follows:

$$p(x_k|x_{k-1}) \sim \mathcal{N}\left(f(x_k, u_k), \sigma_k^2\right) \tag{3-127}$$

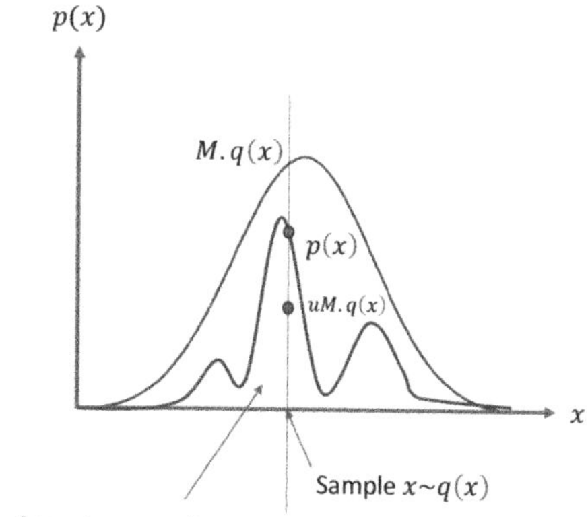

Figure 3.6 Rejection sampling method.

$\mathcal{N}(f(x_k, u_k), \sigma_k^2)$ is a Gaussian probability distribution whose mean is $f(x_k, u_k)$ and variance σ_k^2. The predicted mean is then can be calculated as follows:

$$\mu_{x_k} = \int f(x_{k-1}, u_k)p(x_{k-1})\, dx_{k-1} \tag{3-128}$$

This formula brings us to the concept of importance sampling (IS). IS aims at calculating the expectation of a random variable x that is passing through a function $f(x)$ as follows:

$$\mu_x = \int f(x)p(x)\, dx \tag{3-129}$$

Since $p(x)$ here is arbitrary, we will sample from it using a method that is similar to the rejection sampling method, where we defined a proposal distribution $q(x)$ that is easy to sample from and at the same time includes or envelopes $p(x)$. We can achieve this by choosing $q(x)$ that satisfies the following condition:

$$q(x) > 0\ \mathrm{IF}\, f(x)p(x) \neq 0 \tag{3-130}$$

In this case, the mean μ_x can be calculated as the averaging of samples (or particle cloud) of $q(x)$ that satisfy the above condition. If we refer to this particle cloud of $q(x)$ that satisfies the above condition by: $C_q(x)$, the predicted mean can be given by:

$$\mu_x \cong \frac{1}{|C_q(x)|} \sum\nolimits_{x_j \in C_q(x)} x_j \tag{3-131}$$

In the practical implementation of particle filtering, importance sampling is simply implemented as follows: $p(x_{k-1})$ is approximated by the sample cloud $C_x(k-1)$, and $p(x_k|x_{k-1})$ is replaced by the dynamic system model function $f(.)$, and the integral (or sum) formula is: approximated by passing the particle cloud $C_x(k-1)$ through the dynamic system model function $f(.)$ to get the predictive distribution $p(x_k)$ as follows:

$$p(x_k) \approx C_x(k) \triangleq f(C_x(k-1), u_k) \tag{3-132}$$

Then the predicted state mean is calculated as follows:

$$\mu_{x_k} \approx \frac{1}{|C_x(k)|} \sum\nolimits_{x_j \in C_x(k)} x_j \tag{3-133}$$

This process is illustrated in Figure 3.7.

3.4.11.5 Measurement Update

The update step is performed to calculate the posterior probability $p(x_k \mid y_k)$ when a measurement y_k becomes available. The posterior probability is calculated as a conditional probability using the following formula:

$$p(x_k|y_k) \propto p(y_k|x_k)p(x_k) \tag{3-134}$$

where $p(x_k)$ is the predicted probability modeled by particle cloud $C_x(k)$ and the measurement likelihood, which is centered around the measurement function $h(x_k)$ with a noise variable v_k. This conditional probability can be calculated in a way similar to the idea of importance sampling in the sense that we want to calculate the probability of x_k after passing through the measurement model function $h(x_k)$. The importance sampling method can be applied in the Bayesian conditional probability equation that calculates the posterior distribution of x_k given the prior distribution $p(x_k)$

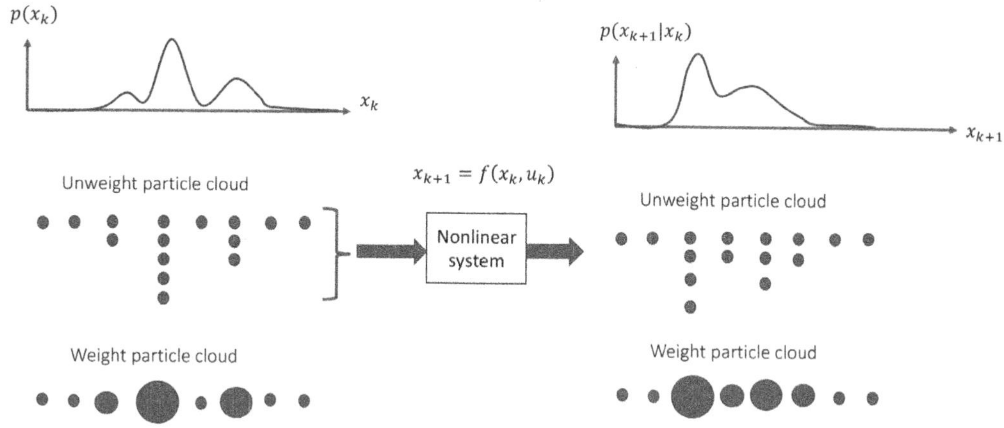

Figure 3.7 Propagating particle cloud through a nonlinear system model.

and the measurement likelihood $p(x_k|y_k)$. In the practical implementation of particle filter, the importance sampling method can be approximated by updating the weights of the prior samples based on the proximity between the predicted particle measurement $h(x_j(k))$ and the actual observed measurement as follows:

$$p(x_k|y_k) \triangleq w^+{}_j(k) \propto w_j(k) \frac{1}{\sqrt{(2\pi)^m}\sqrt{|R|}} e^{-\frac{1}{2}\left(y_k - h\left(x_j(k)\right)\right)^T R^{-1}\left(y_k - h\left(x_j(k)\right)\right)} \tag{3-135}$$

where $w^+{}_j(k)$ is the updated weight, $w_j(k)$ is the prior weight (the weight that represent $p(x_k)$), R is the measurement noise and m is the size of the measurement vector. The resulting particle cloud after the measurement update is now a weighted particle cloud. To repeat the cycle for new iterations, we must resample the particle cloud to generate an unweighted particle cloud. This process is called resampling, and it can be simply done by randomly selecting particles that survive to the next iteration such that each particle is selected several times that is proportional to its weight. The resampling step is illustrated in Figure 3.8. More detailed information about Particle filtering theory and implementation can be found in [9].

3.4.12 Unscented Kalman Filter

Another version of Kalman filter that can deal with nonlinear systems is unscented Kalman filter (UKF) [10]. UKF is like particle filtering in the sense that it uses a cloud of points to represent the state probability. However, it assumes Gaussian probabilities. The advantage of UKF is that it uses a smaller number of sample points. UKF approximates the state distribution by a particle cloud called sigma-points. The sigma-points are passed through the nonlinear system model $f(.)$ to generate predicted sigma-points. Once measurements become available, the sigma-points go through the

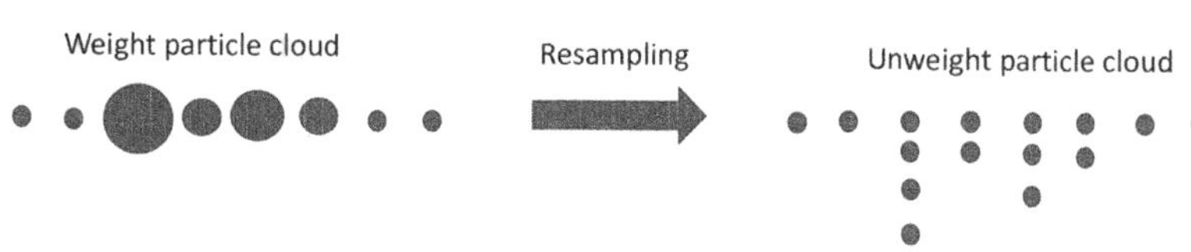

Figure 3.8 Illustration of resampling step.

measurement model $h(.)$ to generate the posterior sigma-points. Sigma-points can be fitted to a Gaussian distribution and a mean and a covariance of the state can be provided. The prediction and update steps of UKF are described in the following subsections.

3.4.12.1 UKF Prediction

Given prior state Gaussian distribution $p\left(\vec{x}_k^{\,+}\right) \sim \mathcal{N}\left(\vec{\mu}_{xk}^{\,+}, \Sigma_{xk}^{+}\right)$, we want to estimate the predictive distribution $p\left(\vec{x}_{k+1}^{\,-}\right) \sim \mathcal{N}\left(\vec{\mu}_{x_{k+1}}^{\,-}, \Sigma_{x_{k+1}}^{-}\right)$ by passing $p\left(\vec{x}_k^{\,+}\right)$ through the nonlinear system model function $\vec{x}_{k+1}^{\,-} = f\left(\vec{x}_k^{\,+}, u_k\right)$. We approximate the distribution $p\left(\vec{x}_k^{\,+}\right)$ by $2d+1$ sigma points $S_{xk}^{+} := \{s_{xk,0}, s_{xk,1}, \ldots, s_{xk,2d}\}$, where d is the dimension of the state vector $\vec{x}_k$ using the following formula:

$$S_{xk}^{+} := \left\{\vec{\mu}_{xk}^{\,+}, \vec{\mu}_{xk}^{\,+} \pm \sqrt{(d+\lambda)\Sigma_{xk}^{+}(:,i)}\right\}, \quad i = 1, 2 \ldots d \tag{3-136}$$

where $\Sigma_{xk}^{+}(:, i)$ is the ith column of the covariance matrix Σ_{xk}^{+}. When the sigma points are passed through the system model function, we get new sigma-points:

$$S_{x_{k+1}}^{-} := \left\{s_{x_{k+1,i}} = f\left(s_{x_{k,i}}, u_k\right)\right\}, \quad i = 1, 2 \ldots d \tag{3-137}$$

We can then calculate the predicted distribution $\mathcal{N}\left(\vec{\mu}_{x_{k+1}}^{\,-}, \Sigma_{x_{k+1}}^{-}\right)$ by:

$$\vec{\mu}_{x_{k+1}}^{\,-} = \sum_{i=0}^{2d} \omega_m^i s_{x_{k+1,i}} \tag{3-138}$$

$$\Sigma_{x_{k+1}}^{-} = \sum_{i=0}^{2d} \omega_c^i \left(s_{x_{k+1,i}} - \vec{\mu}_{x_{k+1}}^{\,-}\right)\left(s_{x_{k+1,i}} - \vec{\mu}_{x_{k+1}}^{\,-}\right)^T + Q_{k+1} \tag{3-139}$$

where Q_{k+1} is the system noise covariance, the weights ω_m^i and ω_c^i are given by:

$$\omega_m^i = \frac{\lambda}{d+\lambda} \tag{3-140}$$

$$\omega_c^i = \frac{1}{2(d+\lambda)} \tag{3-141}$$

The selection of λ is a design parameter called scaling parameter that can be selected based on the specific problem.

3.4.12.2 UKF Update

To integrate the measurements $\vec{y}_{k+1}$, we pass the predictive state distribution $p\left(\vec{x}_{k+1}^{\,-}\right) \sim \mathcal{N}\left(\vec{\mu}_{x_{k+1}}^{\,-}, \Sigma_{x_{k+1}}^{-}\right)$ through the measurement model $h\left(\vec{x}_{k+1}^{\,-}\right)$ to get predictive measurement distribution $p\left(\vec{y}_{k+1}^{\,-}\right) \sim \mathcal{N}\left(\vec{\mu}_{y_{k+1}}^{\,-}, \Sigma_{y_{k+1}}^{-}\right)$. In the prediction step, we already generated the sigma-points that represent $p\left(\vec{x}_{k+1}^{\,-}\right) \sim \mathcal{N}\left(\vec{\mu}_{x_{k+1}}^{\,-}, \Sigma_{x_{k+1}}^{-}\right)$ which is referred to by $S_{x_{k+1}}^{-} = \{s_{x_{k+1,0}}, s_{x_{k+1,1}}, \ldots s_{x_{k+1,2d}}\}$. If we only have $\vec{\mu}_{x_{k+1}}^{\,-}$ and $\Sigma_{x_{k+1}}^{-}$, we can generate the predictive state sigma-points by:

$$S_{x_{k+1}}^{-} := \left\{\vec{\mu}_{x_{k+1}}^{\,-}, \vec{\mu}_{x_{k+1}}^{\,-} \pm \sqrt{(d+\lambda)\Sigma_{x_{k+1}}^{-}(:,i)}\right\}, \quad i = 1, 2 \ldots d \tag{3-142}$$

Then we pass the sigma-points $S_{x_{k+1}}^{-}$ through the measurement model $h(.)$ to get the predicted measurements sigma-points:

$$S_{y_{k+1}}^{-} := \left\{ s_{y_{k+1,i}} = h\left(s_{x_{k+1,i}}\right) \right\}, \quad i = 1, 2 \ldots d \tag{3-143}$$

The predicted measurements mean, and covariance is given by:

$$\vec{\mu}_{y_{k+1}}^{-} = \sum_{i=0}^{2d} \omega_m^i s_{y_{k+1,i}} \tag{3-144}$$

$$\Sigma_{y_{k+1}}^{-} = \sum_{i=0}^{2d} \omega_c^i \left(s_{y_{k+1,i}} - \vec{\mu}_{y_{k+1}}^{-} \right) \left(s_{y_{k+1,i}} - \vec{\mu}_{y_{k+1}}^{-} \right)^T + R_{k+1} \tag{3-145}$$

where R_{k+1} is the measurement noise covariance. Finally, we calculate the posterior state mean and covariance as follows:

$$\Sigma_{x_{k+1}} = \sum_{i=0}^{2d} \omega_c^i \left(s_{x_{k+1,i}} - \vec{\mu}_{x_{k+1}}^{-} \right) \left(s_{x_{k+1,i}} - \vec{\mu}_{x_{k+1}}^{-} \right)^T \tag{3-146}$$

$$K = \Sigma_{x_{k+1}} \left(\Sigma_{y_{k+1}}^{-} \right)^{-1} \tag{3-147}$$

$$\vec{\mu}_{x_{k+1}}^{+} = \vec{\mu}_{x_{k+1}}^{-} + K\left(\vec{y}_{k+1} - \vec{\mu}_{y_{k+1}}^{-} \right) \tag{3-148}$$

$$\Sigma_{x_{k+1}}^{+} = \Sigma_{x_{k+1}}^{-} - K \Sigma_{x_{k+1}} K^T \tag{3-149}$$

The mean $\vec{\mu}_{x_{k+1}}^{+}$ and covariance $\Sigma_{x_{k+1}}^{+}$ represent the posterior state distribution.

3.4.13 Hybrid Filtering (Combining PF and KF: Rao-Blackwellized Particle Filter)

Particle filtering has the advantage of avoiding linearization and matrices inversion while suffering from dimensionality problems. On the other hand, Kalman filtering is more efficient but suffers from linearization and matrices inversion problems, especially if the dimension of the state vector is large. Fortunately, in some cases, we can combine the benefits of both filters if the state can be partitioned into two types, independent states, and inferred states. Inferred states are states that can be inferred if the independent states are known. This applies to the problem of simultaneous localization and mapping (SLAM) problem, where the independent states are the vehicle or robot's pose, and the inferred states are the map (environment features' location). This kind of hybrid filtering is called "Rao-Blackwellized Particle Filter" [2, 3]. We will discuss Rao–Blackwellized particle filter and SLAM algorithm in a future chapter dedicated to mapping systems.

3.4.14 Graph Optimization

Another batch estimation method that is commonly used in mapping and SLAM systems is Graph-based optimization. The Gauss-Markov model in Figure 3.9 can be seen as a graph model. Starting from an initial state x_0, the overall probability of any state batch or sequence $x_{1:N}$ is given by the following joint probability:

$$p\left(\vec{x}_{1:N} \mid \vec{y}_{1:N}\right) = p\left(\vec{x}_{1:N}\right) p\left(\vec{y}_{1:N} \mid \vec{x}_{1:N}\right) \tag{3-150}$$

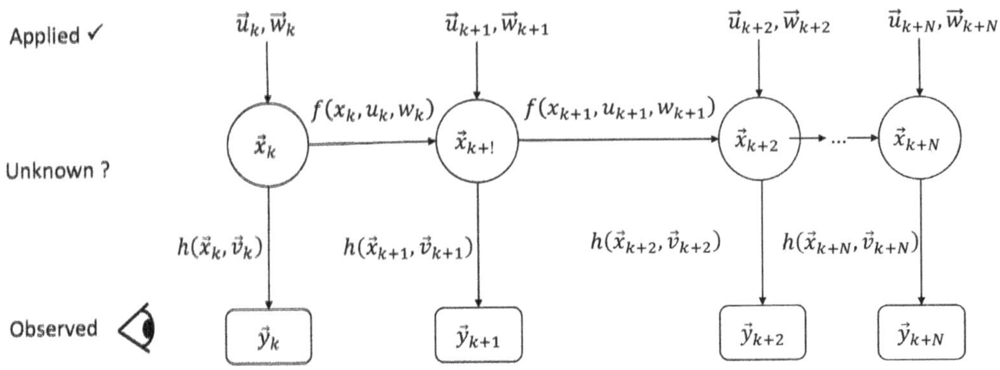

Figure 3.9 Graph model.

where $p\left(\vec{x}_{1:N}\right)$ and $p\left(\vec{y}_{1:N} | \vec{x}_{1:N}\right)$ are given by:

$$p\left(\vec{x}_{1:N}\right) = p\left(\vec{x}_0\right) \prod_{k=1}^{N} p\left(\vec{x}_k | \vec{x}_{k-1}\right) \tag{3-151}$$

$$p\left(\vec{y}_{1:N} | x_{1:N}\right) = \prod_{k=1}^{N} p\left(\vec{y}_k | \vec{x}_k\right) \tag{3-152}$$

which combines the system model and measurement model. The transition probability $p\left(\vec{x}_k | \vec{x}_{k-1}\right)$ and measurement likelihood $p\left(\vec{x}_k | \vec{x}_k\right)$ can be modeled as Gaussian probability centered around the system model as follows:

$$p\left(\vec{x}_k | \vec{x}_{k-1}\right) \sim \mathcal{N}\left(f\left(\vec{x}_{k-1}, \vec{u}_k\right), Q_k\right) \tag{3-153}$$

$$p\left(\vec{y}_k | \vec{x}_k\right) \sim \mathcal{N}\left(h\left(\vec{x}_k\right), R_k\right) \tag{3-154}$$

where $f(.)$ and $h(.)$ are nonlinear stochastic system model and measurement model given by the following general formulas:

$$\vec{x}_{k+1} = f\left(\vec{x}_k, u_k\right) + w_k \tag{3-155}$$

$$\vec{y}_k = h\left(\vec{x}_k\right) + v_k \tag{3-156}$$

where w_k and v_k are random noise signals modeled as follows:

$$\vec{w}_k \sim \mathcal{N}(0, Q_k) \tag{3-157}$$

$$\vec{v}_k \sim \mathcal{N}(0, R_k) \tag{3-158}$$

Being Gaussian, transition probability and measurement likelihood can be written as exponential formulas as follows:

$$p\left(\vec{x}_k | \vec{x}_{k-1}\right) \propto e^{-\frac{1}{2}\left(\vec{x}_k - f\left(\vec{x}_{k-1}, \vec{u}_k\right)\right)^T Q_k^{-1}\left(\vec{x}_k - f\left(\vec{x}_{k-1}, \vec{u}_k\right)\right)} \tag{3-159}$$

$$p\left(\vec{y}_k \mid \vec{x}_k\right) \propto e^{-\frac{1}{2}\left(\vec{y}_k - h\left(\vec{x}_k\right)\right)^T R_k^{-1}\left(\vec{y}_k - h\left(\vec{x}_k\right)\right)} \tag{3-160}$$

Substituting these exponential terms into the equation of the overall probability of any state sequence $\vec{x}_{1:N}$ gives the following general formula:

$$p\left(\vec{x}_{1:N} \mid y_{1:N}\right) \propto e^{-J\left(\vec{x}_{1:N}, y_{1:N}\right)} \tag{3-161}$$

$$J\left(\vec{x}_{1:N}, y_{1:N}\right) = \left(\sum_{k=1}^{N}\left(\vec{x}_k - f\left(\vec{x}_{k-1}, \vec{u}_k\right)\right)^T \frac{Q_k^{-1}}{2}\left(\vec{x}_k - f\left(\vec{x}_{k-1}, \vec{u}_k\right)\right)\right.$$
$$\left. + \sum_{k=1}^{N}\left(\vec{y}_k - h\left(\vec{x}_k\right)\right)^T \frac{R_k^{-1}}{2}\left(\vec{y}_k - h\left(\vec{x}_k\right)\right)\right) \tag{3-162}$$

If we took the negative log of both sides, we would have the following form:

$$-\log\left(p\left(\vec{x}_{1:N} \mid \vec{y}_{1:N}\right)\right) \propto \frac{1}{2}\left(\sum_{k=1}^{N}\left(\vec{x}_k - f\left(\vec{x}_{k-1}, \vec{u}_k\right)\right)^T Q_k^{-1}\left(\vec{x}_k - f\left(\vec{x}_{k-1}, \vec{u}_k\right)\right)\right.$$
$$\left. + \sum_{k=1}^{N}\left(\vec{y}_k - h\left(\vec{x}_k\right)\right)^T R_k^{-1}\left(\vec{y}_k - h\left(\vec{x}_k\right)\right)\right) \tag{3-163}$$

Maximizing $p(x_{1:N}, y_{1:N})$ is equivalent to minimize the above negative log likelihood formula. What is given in this formula is: the initial state x_0, system model function $f(.)$, measurement model function $h(.)$, observations/measurements y_k, system noise covariance Q_k and measurement noise covariance R_k and the unknowns are the state sequence $x_{1:N}$ and cost function $J\left(\vec{x}_{1:N}\right)$ can be written as follows:

$$J\left(\vec{x}_{1:N}\right) = \sum_{k=1}^{N}\frac{1}{2}\left(\vec{x}_k - f\left(\vec{x}_{k-1}, \vec{u}_k\right)\right)^T Q_k^{-1}\left(\vec{x}_k - f\left(\vec{x}_{k-1}, \vec{u}_k\right)\right)$$
$$+ \sum_{k=1}^{N}\frac{1}{2}\left(\vec{y}_k - h\left(\vec{x}_k\right)\right)^T R_k^{-1}\left(\vec{y}_k - h\left(\vec{x}_k\right)\right) \tag{3-164}$$

If we have uncertainty about the initial state guess x_0, the cost function $J(x_{0:N})$ will be given as follows:

$$J\left(\vec{x}_{0:N}\right) = \vec{x}_0^T Q_0^{-1}\vec{x}_0 + \sum_{k=1}^{N}\frac{1}{2}\left(\vec{x}_k - f\left(\vec{x}_{k-1}, \vec{u}_k\right)\right)^T Q_k^{-1}\left(\vec{x}_k - f\left(\vec{x}_{k-1}, \vec{u}_k\right)\right)$$
$$+ \sum_{k=1}^{N}\frac{1}{2}\left(\vec{y}_k - h\left(\vec{x}_k\right)\right)^T R_k^{-1}\left(\vec{y}_k - h\left(\vec{x}_k\right)\right) \tag{3-165}$$

One approach to solve this problem is by using nonlinear optimization methods such as Newton, Gauss-Newton, or Levernberg–Marquatdt methods described earlier in this chapter. This will require the linearization of system model function (.), measurement model function $h(.)$ around an initial guess of the state sequence $\vec{x}_{0:N}(0)$. The initial guess of the state sequence can be simply calculated by applying the system model. Starting from $\vec{x}_0$ we keep calculating

$\vec{x}_{k+1} = f\left(\vec{x}_k, u_k\right), k = 1 : N-1$. By predicting a full-state sequence $\vec{x}_{0:N}(0)$, we can linearize the system model function $f(.)$, measurement model function $h(.)$ around this initial state prediction and then apply Newton, Gauss–Newton, or Levernberg–Marquatdt methods interactively. What Newton, Gauss–Newton, or Levernberg–Marquatdt methods would do is the optimal estimation of the state error $\Delta\vec{x}_{0:N}$, so we can enhance the initial state sequence guess as follows:

$$\vec{x}_{0:N}(n) = \vec{x}_{0:N}(n-1) + \Delta\vec{x}_{0:N}(n) \tag{3-166}$$

This approach is feasible when we have a small number of state elements. The problem becomes too computationally expensive when we have a large state vector size, like in mapping systems where map features are included in the state vector. Graph-based optimization has an additional step that reduces the computational complexity of this problem in mapping and SLAM frameworks. This additional step is the elimination of multiple measurements that are common between multiple states through multiple epochs and replacing them with direct links between the states. This reduction method works nicely in mapping and SLAM systems because most of the measurements are environment features that are observed by the sensors multiple times. If multiple states see the same feature at multiple epochs, the measurement links are eliminated and replaced by direct links between the states. This reduction step is like the variable elimination step in solving multiple simultaneous linear equations. More details about graph-based optimization and graph-based SLAM can be found in [11] and [12].

3.4.15 Hidden Markov Models and Viterbi Algorithm

3.4.15.1 First-Order Markov Process

We have seen how filtering methods estimate the state of a first-order Markov process. Filtering solutions infer linear Gaussian random processes by calculating state probability (x_k) online recursively, which saves time and memory because all that we need to keep is the parameters of $p(x_k)$ at time epoch k. In some scenarios, we may have the option to keep a record of system state predictions and measurements, which enables us to perform batch, buffered, or offline estimation instead of online estimation. Starting from an initial state x_0, we can calculate the probability of any given state sequence $x_{1:N}$ using the following conditional probability:

$$p(x_{1:N} \mid y_{1:N}) = p(x_{1:N})p(y_{1:N} \mid x_{1:N}) \tag{3-167}$$

where:

$$p(x_{1:N}) = p(x_0)\prod_{k=1}^{N} p(x_k \mid x_{k-1}) \tag{3-168}$$

$$p(y_{1:N} \mid x_{1:N}) = \prod_{k=1}^{N} p(y_k \mid x_k) \tag{3-169}$$

which gives the following probability equation:

$$p(x_{1:N} \mid y_{1:N}) = p(x_0)\prod_{k=1}^{N} p(x_k \mid x_{k-1})p(y_k \mid x_k) \tag{3-170}$$

If we can calculate the probability of a complete sequence, we can then search for the sequence that maximizes the joint probability $p(x_{1:N}, y_{1:N})$. This is called offline, batch, or buffered processing and sometimes referred to as "Bundle Adjustment." Batch processing usually gives better and more smooth estimation. However, if the probable states are huge, we will have a huge number of probable state sequences and searching in this huge space will be computationally difficult. This offline approach would be feasible if we may have a definite set of probable states at each time epoch. In other words, system states at time k can take a definite set of values $x_{k,\,j}$ where $j = 1, 2 \ldots M_k$. The state transitions form a Bayesian dynamic network starts from an initial state $x_{0,\,0}$ and evolve through time. This network is called Markov model. If the unknown in this Markov model are the system states given some observations, we call this network a hidden Markov model (HMM) [2, 3]. Given an N-length network, the objective in HMM is to estimate the most probable sequence of states x_{k,j_k} where $k = 1, 2 \ldots N$ and $j_k \in \{1, 2 \ldots M_k\}$ that best explains the observations $y_{1:N} = y_1 y_2 \ldots y_N$. The probability of any sequence of states $x_{1:N} = x_{0,0} x_{1,j_1} x_{2,j_2} \ldots x_{N,j_N}$ is given by the following joint probability equation:

$$p(x_{1:N}) = p(x_{0,0}) \prod_{k=1}^{N} p\left(x_{k,j_k} \mid x_{k-1,j_{k-1}}\right) \tag{3-171}$$

We are interested in finding the most probable sequence $x^*_{1:N}$ that explains the observed measurements $y_{1:N}$. This is equivalent to finding the sequence $x^*_{1:N}$ that maximizes $p(x_{1:N} \mid y_{1:N})$, the joint probability of $x_{1:N}$ and the observations $y_{1:N}$ which can be written as:

$$x^*_{1:N} = \underset{x_{1:N}}{\operatorname{argmax}}\, p(x_{1:N}, y_{1:N}) \tag{3-172}$$

To calculate $p(x_{1:N}, y_{1:N})$, we use two sources of information, the transition probabilities $p\left(x_{k,j_k} \mid x_{k-1,j_{k-1}}\right)$ and the measurement likelihood $p\left(y_k \mid x_{k,j_k}\right)$ formulated as follows:

$$p(x_{1:N} \mid y_{1:N}) = p(x_{1:N}) p(y_{1:N} \mid x_{1:N}) \tag{3-173}$$

$$p(x_{1:N} \mid y_{1:N}) \propto p(x_{0,0}) \prod_{k=1}^{N} p\left(x_{k,j_k} \mid x_{k-1,j_{k-1}}\right) p\left(y_k \mid x_{k,j_k}\right) \tag{3-174}$$

If we know the transition probabilities $p\left(x_{k,j_k} \mid x_{k-1,j_{k-1}}\right)$ (which is shaped by the system model) and the measurement likelihood $p\left(y_k \mid x_{k,j_k}\right)$ (which is shaped by the measurement model), we can use the above formula to calculate the probabilities of all probable sequences and select the sequence that maximizes $p(x_{1:N} \mid y_{1:N})$. However, this will be time-consuming as the network growth in length (N) or in depth ($\max(M_k)$, $k = 1, 2 \ldots N$). A more efficient algorithm that calculates the most probable sequence without having to go through all possible sequences is Viterbi Algorithm explained below.

3.4.15.2 Forward-Backward Algorithm

We will deal with the HMM as a matrix where columns represent time epochs and rows represent probable state values. Viterbi algorithm implements a forward step and backward step to calculate the most probable state sequence as follows:

- **Forward-step (online phase):** at each node x_{k,j_k} at column k, we calculate forward probability or predictive probability which is the probability to arrive at node x_{k,j_k} from any of the previous nodes at column $k-1$ (system model probability) times the probability of observation y_k given state x_{k,j_k} (measurement likelihood) which is given by the following equation:

$$p\left(x_{k,j_k} \mid y_k\right) \propto p\left(y_k \mid x_{k,j_k}\right) \sum_{j_{k-1}=1}^{M_{k-1}} p\left(x_{k,j_k} \mid x_{k-1,j_{k-1}}\right) p\left(x_{k-1,j_{k-1}}\right) \tag{3-175}$$

The forward probability is called the filtered distribution, which is equivalent to Kalman filter in linear Gaussian dynamic systems. In fact, Kalman filter gives us the most probable state at each time epoch k. That is why the forward step is called the online phase. Let's define the following parameters:

$$\alpha(j_k, k) = p\left(x_{k,j_k} \mid y_k\right) \tag{3-176}$$

$$\Psi(j_{k-1}, j_k) = p\left(x_{k,j_k} \mid x_{k-1,j_{k-1}}\right) \tag{3-177}$$

$$\psi(j_k) = p\left(y_k \mid x_{k,j_k}\right) \tag{3-178}$$

where rows j_k represent the j_k^{th} state at column k and each column k represents time epoch k. It is clear that:

$$\alpha(j_k, k) = \psi(j_k) \sum_{j_{k-1}=1}^{M_{k-1}} \Psi(j_{k-1}, j_k)\alpha(j_k, k-1) \tag{3-179}$$

- **Backward-step (offline phase)**: In the offline phase, we calculate the most probable sequence by moving backward. The probability of the state x_{k,j_k} given all measurements $y_{1:N}$ can be decomposed into two parts, a past part that starts from 1 to k and a future part that starts at $k+1$ to N. The probability $p\left(x_{k,j_k} \mid y_{1:N}\right)$ can then be written as:

$$p\left(x_{k,j_k} \mid y_{1:N}\right) \propto p\left(x_{k,j_k} \mid y_{1:k}\right) p\left(y_{k+1:N} \mid x_{k,j_k}\right) \tag{3-180}$$

We have already calculated $p\left(x_{k,j_k} \mid y_{1:k}\right)$, which is the same as $p\left(x_{k,j_k} \mid y_k\right)$, which is calculated and saved in the matrix $\alpha(j_k, k)$. The part $p\left(y_{k+1:N} \mid x_{k,j_k}\right)$ is calculated as follows:

$$p\left(y_{k+1:N} \mid x_{k,j_k}\right) = \sum_{j_{k+1}=1}^{M_{k+1}} p\left(y_{k+2:N} \mid x_{k+1,j_{k+1}}\right) p\left(y_{k+1} \mid x_{k+1,j_{k+1}}\right) p\left(x_{k+1,j_{k+1}} \mid x_{k,j_k}\right) \tag{3-181}$$

By setting $\beta(j_k, k) = p\left(y_{k+1:N} \mid x_{k,j_k}\right)$, we get the following recursive formula:

$$\beta(j_{k-1}, k-1) = \sum_{j_k=1}^{M_k} \beta(j_{k-1}, k-1)\psi(j_k)\Psi(j_{k-1}, j_k) \tag{3-182}$$

It is clear that $\beta(j_N, N) = 1$, so we can calculate $\beta(j_{k-1}, k-1)$ recursively. If we have $\beta(j_k, k)$ and $\alpha(j_k, k)$, we can calculate the smooth posterior probability for all states x_{k,j_k} in all time epochs given all measurements $y_{1:N}$ by:

$$p\left(x_{k,j_k} \mid y_{1:N}\right) = \beta(j_k, k)\alpha(j_k, k) \tag{3-183}$$

The most probable sequence is calculated by selecting the state x^*_{k,j_k} where $p\left(x_{k,j_k} \mid y_{1:N}\right)$ is maximum.

3.4.15.3 Viterbi Algorithm

The forward-backward algorithm is robust and globally optimum but not computationally efficient. A more efficient way to find the most probable sequence is by calculating the probability of ending up to state j_k at time k given that the most probable sequence path $1:k-1$ is selected as follows:

$$\delta(k, j) = \max_{j_{k-1}} \delta(k-1, j_{k-1})\Psi(j_{k-1}, j_k)p\left(y_k \mid x_{k,j_k}\right) \tag{3-184}$$

$$\delta(1,j) = p\left(x_{1,j_1} \mid x_{0,0}\right) p\left(y_1 \mid x_{1,j_1}\right) \tag{3-185}$$

$$a(k,j) = \underset{j_{k-1}}{\operatorname{argmax}}\, \delta(j_{k-1}) \Psi(j_{k-1}, j_k) p\left(y_k \mid x_{k,j_k}\right) \tag{3-186}$$

Then we start from the end node and move backward as follows:

$$x^*_{N,j_N} = \max_{j_N} \delta(N, j_{N-1}) \tag{3-187}$$

$$j^*_N = \underset{j_N}{\operatorname{argmax}}\, \delta(N, j_{N-1}) \tag{3-188}$$

$$x^*_{k-1,j_{k-1}} = a\left(k, j^*_k\right) \tag{3-189}$$

3.5 Machine Learning and Artificial Intelligence

3.5.1 Motivation

The state estimation problem is all about estimating a state variable given some measurements. In the previous sections, we assumed that we knew the general relationship between the state variables and the measurements, and we wanted to estimate the state given the measurements. This approach works well if our knowledge about the relationship between states (system model) and measurements (measurement model) is accurate. If this is not the case, we may need different approaches to state estimation. This need has motivated scientists and researchers to develop new estimation techniques that can autonomously learn complex relationships and patterns from data. This trend has become very popular and successful due to the advancement in computing hardware and the availability of massive amounts of data. This gave rise to ML as an advanced state estimation approach. ML is always confused with AI, and sometimes the two terms are used interchangeably. However, in this section, we will distinguish between ML and AI.

ML- and AI-based systems can perform high-level recognition tasks that can improve sensor fusion performance, especially in vision-based and range-based sensor fusion-based navigation, positioning, and mapping systems. Therefore, this section will provide a brief discussion about ML and AI. It is important to highlight that ML and AI are wide and complex research areas that cannot be grasped in a section of a chapter or even a single book. The goal of this brief discussion here is to introduce the basic concepts and to link these concepts to their potential applications in positioning, navigation, and mapping fields. The material discussed in this section is of an introductory level. For detailed and advanced information and extensive materials about ML and AI topics, readers are encouraged to read dedicated textbooks in these areas such as [13], [14], [15], [2], [16], and [17].

3.5.2 Basic Definitions

ML can be defined as a generic function approximation model that can be trained given training data. Thanks to the advancement in computer hardware (speed and memory), ML has evolved dramatically to become advanced computing tools and data structures that can be tweaked in different ways by designers to solve complex problems. AI can be defined as computer algorithms that mimic human reasoning in solving complex problems. The fundamental difference between ML and AI is that ML models are mostly self-learning models that heavily rely on data to learn patterns, while AI systems rely more on rule-based and computer engineering of human knowledge. Human-crafted

AI systems are computer systems that allow human experts to feed their knowledge in a computerized way to a computer system, which can then apply this knowledge to solve a complex problem. Knowledge-engineered AI systems can be as simple as a set of IF-THEN rules. The main limitation of knowledge-engineered AI systems is that they are difficult to build due to the intense human involvement in the process. As ML models are more data-driven and self-learning, they are currently the preferred models in many applications where big data volumes are available.

ML can be seen as a universal nonlinear multi-input/multi-output approximation function that can be trained on massive amounts of data to generate a desired output given an unseen input. ML can be categorized into the general categories shown in Figure 3.10 with some basic illustrations are shown in Figure 3.11. ML can be used in regression and classification problems. Regression models work on numerical data, while classification models work on labeled classes. Supervised learning is

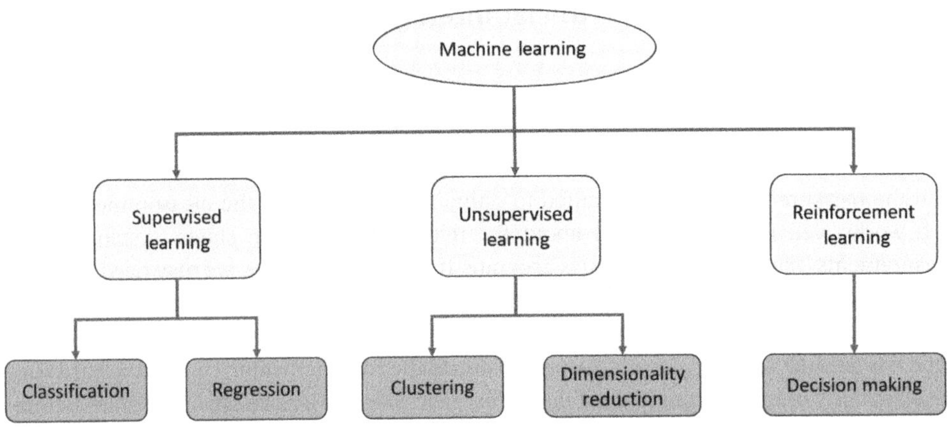

Figure 3.10 Types of machine learning.

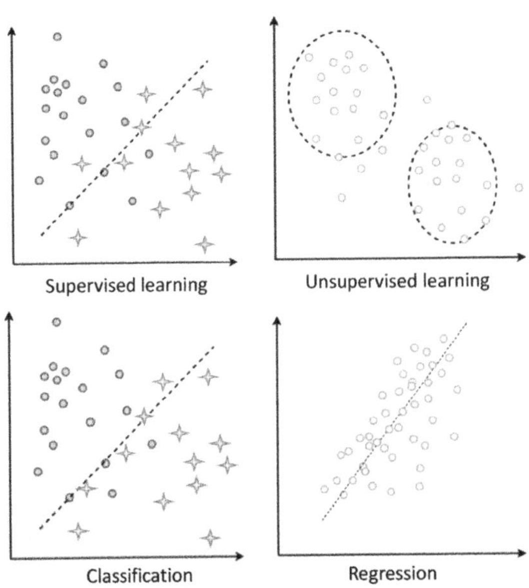

Figure 3.11 Illustration of types of machine learning.

a learning configuration that uses labeled data (ground-truth input/output data) to train ML models. The training is performed by minimizing a loss or a cost function. The loss/cost function is commonly a squared error function that measures the difference between the ML model's output and a desired output. In an unsupervised configuration, the ML model can be trained in an unsupervised mode that does not need ground-truth labeled data. Clustering and dimensionality reduction (data compression) are examples of unsupervised learning. Clustering and dimensionality reduction do not need labeled data, but they can be formulated as a loss or cost function. For example, clustering is basically a grouping of similar data points in definite groups called clusters. This similarity can be formulated as a loss or cost function and an ML model can be designed to minimize the cost function. In both modes, any ML-based system will need a loss/cost function to minimize. In fact, the loss/cost function plays a key role in designing successful ML systems. In Chapter 12, we will demonstrate use-case scenarios in which ML is applied to perform sensor fusion for positioning, navigation, and mapping.

3.5.3 Risks of ML and AI

While AI and ML are powerful tools, they come with significant risks. First, due to the flexibility of AI and ML tools and data structures, too complicated structures and operations may end up solving a specific problem in a kind of a lookup table or an exhaustive search approach that is overly fitted to a specific scenario or a pattern. Because the system is mostly a very complex network that has a tremendous amount of processing nodes, connections, and weights, it may be extremely difficult to explain, analyze, or predict its behavior. Most of the analysis and explanation of AI and ML success is based on testing the developed models on massive amounts of data without complete mathematical justification or error/stability analysis. Therefore, researchers and engineers need to be careful when applying AI and ML to a specific problem, and they need to understand the limits and challenges that come with the application of ML and AI in sensor fusion systems for positioning, navigation, and mapping.

3.5.4 Mathematical Foundation

ML is mainly concerned with mapping certain inputs to desired outputs. Given training data $\{X, Y\}$ and a new unseen input x^*, ML aims to predict output y^*. It is up to the designer how to choose the training data, which variables should be inputs and which variables should be outputs. Since we are always dealing with noisy data, we discuss ML from a probabilistic perspective, which can be described as maximizing the following conditional probability:

$$y^* = \underset{y}{\operatorname{argmax}}\, p(y \mid x^*, X, Y) \tag{3-190}$$

where $p(y \mid x^*, X, Y)$ is the probability distribution of the output given unseen input x^* and the training data $\{X, Y\}$. This general mapping problem can be used to perform the two main learning tasks: regression and classification. Regression is: related to fitting numerical data like curve-fitting, while classification is related to mapping data to higher-level information like object detection. Regression can be used to approximate numerical linear or nonlinear functions. Classification can be used to detect features, patterns, and objects. In the next sections, we will review the basic concepts of ML methods for regression and classification.

3.5.4.1 Regression

The general regression problem can be described using the following equation:

$$y = f\left(\vec{x}\right) + \varepsilon \tag{3-191}$$

where $\vec{x}$ is input vector, y is scalar output, $f(.)$ is generally a nonlinear function, and ε is a random noise. The mapping function $f(.)$ can be generalized to multi-input/multi-output mapping. If multiple inputs are applied at once (like a 2D image, for example), the output becomes a vector. If multiple functions are employed, the output can be a matrix like a 2D image or a more complex multidimensional structure. However, for simplicity, we will assume that the input is a vector and the output is a scalar. If random noise is assumed to be Gaussian, the mapping becomes a Gaussian random process. The Gaussian process has been used as the base for several ML methods, including regression and classification. In the next subsection, we will discuss how the Gaussian process can be used to develop a linear and nonlinear data-driven regression model that predicts the output corresponding to unseen input given a noisy set of input/output data points.

Linear Regression

A Gaussian linear regression model can be written as follows:

$$y_j = f\left(\vec{x}_j\right) = \left(\vec{x}_j\right)^T \vec{w} + \varepsilon \tag{3-192}$$

Which is illustrated in Figure 3.12. If multiple input vectors $\left(\vec{x}_j, j = 1, 2...n\right)$ are arranged as columns into matrix $= \left(\vec{x}_1\, \vec{x}_2...\vec{x}_n\right)$, the model can be written as follows:

$$\vec{y} = (X)^T \vec{w} + \vec{\varepsilon} \tag{3-193}$$

$$\vec{\varepsilon} \sim \mathcal{N}\left(\vec{0}, \Sigma_n\right) \tag{3-194}$$

where $\vec{y}$ is output vector, $\vec{w}$ is a weight vector, and $\vec{\varepsilon}$ is a zero-mean Gaussian noise vector with covariance matrix Σ_n. Our aim is to calculate the best weights $\vec{w}$ given data $D = \left\{X, \vec{y}\right\}$ and use it to predict the output for unseen inputs $p\left(y^* \mid \vec{x}^*, X, \vec{y}\right)$ using Bayesian rule. Given training.

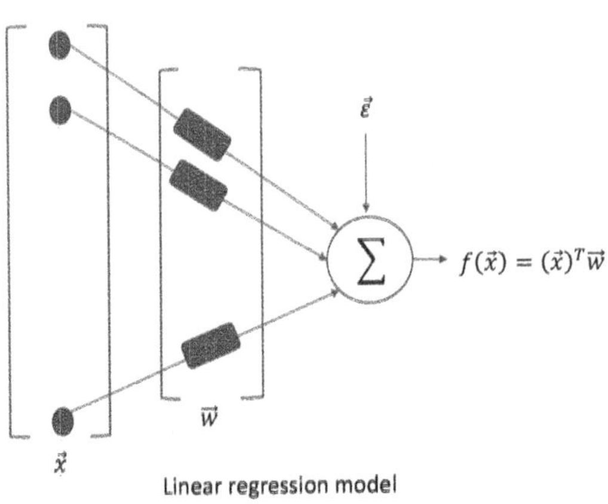

Figure 3.12 Linear regression model.

The probability of the data $D = \left\{X, \vec{y}\right\}$ is the probability of $\vec{y}$ given X and weights $\vec{w}$, which is a Gaussian probability distribution as follows:

$$p\left(\vec{y} \mid X, \vec{w}\right) \sim \mathcal{N}\left(X^T\vec{w}, \Sigma_n\right) \tag{3-195}$$

If the prior distribution of the weights $\vec{w}$ is Gaussian $p\left(\vec{w}\right) \sim \mathcal{N}(0, \Sigma_p)$, using Bayesian rule, we can calculate the posterior distribution of the weights $p\left(\vec{w} \mid \vec{y}, X\right)$ as follows:

$$p\left(\vec{w} \mid \vec{y}, X\right) = \frac{p\left(\vec{y} \mid X, \vec{w}\right)p\left(\vec{w}\right)}{p\left(\vec{y} \mid X\right)} \tag{3-196}$$

As the term $p\left(\vec{y} \mid X\right)$ is a normalization factor that does not affect the shape of $p\left(\vec{w} \mid \vec{y}, X\right)$, we can write it in the form:

$$p\left(\vec{w} \mid \vec{y}, X\right) \propto p\left(\vec{y} \mid X, \vec{w}\right)p\left(\vec{w}\right) \tag{3-197}$$

$$p\left(\vec{w} \mid \vec{y}, X\right) \propto e^{-\frac{1}{2}\left(\vec{y} - X^T\vec{w}\right)^T \Sigma_n^{-1}\left(\vec{y} - X^T\vec{w}\right)} e^{-\frac{1}{2}\left(\vec{w}\right)^T \Sigma_p^{-1}\left(\vec{w}\right)} \tag{3-198}$$

After rearranging the terms in the exponent term, we can show that $p\left(\vec{w} \mid \vec{y}, X\right)$ is a Gaussian distribution defined as follows:

$$p\left(\vec{w} \mid \vec{y}, X\right) \sim \mathcal{N}\left(\vec{\mu}_w, \Sigma_w\right) \tag{3-199}$$

$$\Sigma_w^{-1} = \Sigma_p^{-1} + X\Sigma_n^{-1}X^T \tag{3-200}$$

$$\vec{\mu}_w = \Sigma_w X \Sigma_n^{-1}\vec{y} \tag{3-201}$$

If we do not have any prior information about the weights, their covariance will tend to infinitely ($\Sigma_p \to \infty$). Assuming that $\Sigma_n = I$ (observations are independent and are equally probable), the weight vector will be given by:

$$p\left(\vec{w} \mid \vec{y}, X\right) \sim \mathcal{N}\left(\left(XX^T\right)^{-1}X\vec{y}, XX^T\right) \tag{3-202}$$

which is the standard unweighted least-square solution. If this problem is handled by a least-square approach, we will end up having the following linear equation:

$$\left(XX^T\right)\vec{w} = X\vec{y} \tag{3-203}$$

In Bayesian learning literature, the vector $\left(X\vec{y}\right)$ is called the information vector, (XX^T) is called the information matrix, and $\vec{w}$ is the unknown vector. We will use these terms when we discuss the graph-based SLAM framework in a later chapter. Given the best estimate of weights $\vec{\mu}_w = \Sigma_w X \Sigma_n^{-1}\vec{y}$, we can calculate the probability distribution of a predicted output y^* corresponding to an unseen input $\vec{x}^*$ by marginalizing out $\vec{w}$ as follows:

$$p\left(y^* \mid \vec{x}^*, X, \vec{y}\right) = \int p\left(y^* \mid \vec{x}^*, \vec{w}\right)p\left(\vec{w} \mid \vec{y}, X\right)dw \tag{3-204}$$

Since both $p\left(y^* \mid \vec{x}^*, \vec{w}\right)$ and $p\left(\vec{w} \mid \vec{y}, X\right)$ are Gaussian, this integral can be calculated recalling the lemma that states that: if we have a linear Gaussian model in the form $y = Hx$ where $\sim \mathcal{N}(\bar{x}, \Sigma_x)$, the distribution y is Gaussian given by $y \sim \mathcal{N}\left(H\bar{x}, H\Sigma_x H^T\right)$. The probability of the output given the unseen input and the training data X, $\vec{y}$ is given by:

$$p\left(y^* \mid \vec{x}^*, X, \vec{y}\right) \sim \mathcal{N}\left(\left(\vec{x}^*\right)^T \vec{\mu}_w, \left(\vec{x}^*\right)^T \Sigma_w \left(\vec{x}^*\right)\right) \tag{3-205}$$

Nonlinear Regression

If the mapping between $\vec{y}$ and X is not linear, it will be given as follows:

$$\vec{y} = f(X) + \vec{\varepsilon} \tag{3-206}$$

If $\vec{\varepsilon}$ is zero-mean Gaussian noise with covariance Σ_n, we can still use Gaussian process regression to predict output y^* corresponding to unseen input $\vec{x}^*$ given the data $D = \left\{X, \vec{y}\right\}$. We can do that by mapping the input vectors $\vec{x} \in X$ to another space $\phi(X)$, where we can express the output as a linear function of the mapped input using weight vector $\vec{w}$ as follows:

$$f(X) = \phi(X)\vec{w} \tag{3-207}$$

This nonlinear regression model is illustrated in Figure 3.13.

The space mapping $\phi\left(\vec{x}\right)$ is called the support and it maps the relationship between the input vector and the output vector into a linear space where we can apply the Gaussian linear regression formula discussed above. Under these Gaussian assumptions, the posterior predictive output

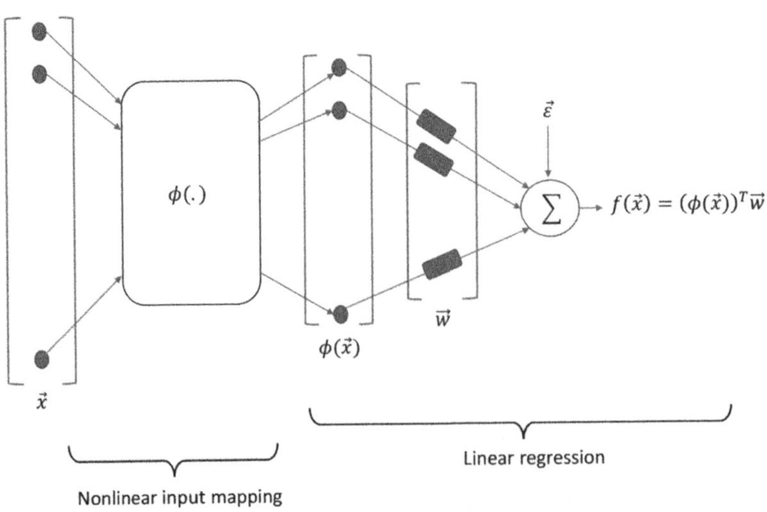

Figure 3.13 Nonlinear regression model.

probability distribution of an unseen input x^* given training data $D = \{X, \vec{y}\}$ is a Gaussian distribution with the following mean and variance:

$$\vec{\mu}_{y^*} = \frac{K(\vec{x}^*, X)}{(K(X,X) + \Sigma_n)} \vec{y} \tag{3-208}$$

$$\sigma_{y^*}^2 = K(\vec{x}^*, \vec{x}^*) - K^T(\vec{x}^*, X)(K(X,X) + \Sigma_n)^{-1} K(\vec{x}^*, X) \tag{3-209}$$

$$K(\vec{x}_i, \vec{x}_j) = \phi^T(\vec{x}_i) \sum\nolimits_p \phi(\vec{x}_j) \tag{3-210}$$

$K(.)$ can be seen as a "kernel" function that represents the covariance (i.e., similarity) between two random inputs. The input space mapping function $\phi(X)$ can be seen as a basis function and the term $\phi^T(\vec{x}_i)\sum_p \phi(\vec{x}_j)$ is an inner product of the mapped input called the kernel. The definition of the basis function $\phi(X)$ and the kernel K are the primary goal of the learning. Once the best kernel is learned, the remaining work will be the linear regression part. In fact, the definition of the basis and the kernel is the foundation for several ML methods such as artificial neural networks, k-nearest, and support vector machines. For example, the input and hidden layers in an artificial neural network are simply a basis function that maps the original input to another space (commonly known as feature space) where the output can be calculated as a linear combination of the features. Given that the kernel is an inner product which measures the similarity between random vectors, the interpretation of y^* is that if x^* is close to one of the observations in X, the prediction mean will be close to the corresponding output in $\vec{y}$. In fact, this is the general rule of several ML and neural network methods, which is "similar inputs give similar outputs." The interpretation of the variance $\sigma_{y^*}^2$ is that the closer the unseen input x^* to one of the inputs in the training data X, the smaller the variance of the prediction. One of the commonly used kernels used in literature is the exponential mapping kernel defined as follows:

$$K(\vec{x}_i, \vec{x}_j) = \sigma_n^2 \delta_{ij} + \sigma_f^2 e^{-\frac{1}{2}\left((\vec{x}_i - \vec{x}_j)^T M (\vec{x}_i - \vec{x}_j)\right)} \tag{3-211}$$

$$\delta_{ij} = \begin{cases} 1 & i = j \\ 0 & i \neq j \end{cases} \tag{3-212}$$

where $\vec{x}_i, \vec{x}_j \in X$ and $y_i, y_j \in \vec{y}$ and δ_{ij} is the Dirac delta function. The parameters σ_n^2, σ_f^2, and M are called the hyper-parameters and they can be tuned to fit a given input/output data set. The intuitive interpretation of σ_n^2 is how much confidence we have in the accuracy of the training data, and σ_f^2 is how much confidence we have in the accuracy of the unseen input $\vec{x}^*$ and M is how fast or slow the output is changing when the input change. If we have multi-output system, we can model each output separately and merge the overall system as shown in Figure 3.14.

3.5.4.2 Classification

Classification assigns a class label c_j to an input vector $\vec{x}_i$. The output in the classification problem is a label, not a numerical value like regression. Given a set of classes $C = (c_1\ c_2 ... c_m)$, input vectors or sample points $X = (\vec{x}_1\ \vec{x}_2 ... \vec{x}_n)$, and classification output $Y = (\vec{y}_1\ \vec{y}_2 ... \vec{y}_n)$ where $\vec{y}_i = (y_{i,1}\ y_{i,2} ... y_{i,m})^T$, $y_{i,j} \in \{+1, -1\}$ where $y_{i,j} = +1$ means $\vec{x}_i \in c_j$ and $y_{i,j} = -1$ means $\vec{x}_i \notin c_j$

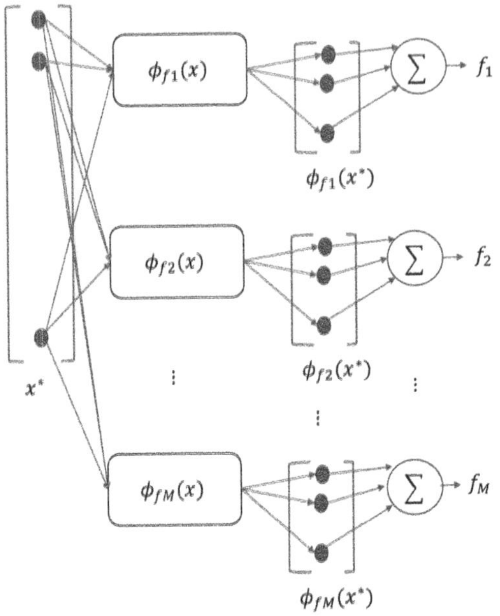

Figure 3.14 Nonlinear multi-output regression model.

The classification of an unseen input vector $\vec{x}$ to an output class c_j, given the training data can be described as a probability distribution as follows:

$$p\left(\vec{x} \in c_j \mid X, Y\right) = p\left(y_j = +1 \mid \vec{x}, X, Y\right) \in [0,1] \tag{3-213}$$

As for any given input $\vec{x}$, we have a probability for each class; the final class assignment j^* for an input vector $\vec{x}$ is the class with the maximum probability.

$$j^* = \arg\max_j p\left(y_j = +1 \mid \vec{x}, X, Y\right) \tag{3-214}$$

Classification Using Linear Logistic Regression

Classification can be modeled as regression where output is a probability value in the region [0, 1]. By limiting the output to the region [0, 1], we can use regression methods to design classification models. Let $\vec{x}$ be an unseen input vector and we want to model the probability that $\vec{x} \in c_j$, given training data X, Y. This probability can be modeled as follows:

$$p\left(y_j = +1 \mid \vec{x}, X, Y, \vec{w}_j\right) = s\left(\vec{x}^T \vec{w}_j\right) = \frac{1}{1 + e^{-\vec{x}^T \vec{w}_j}} \tag{3-215}$$

where $\vec{w}_j$ is a weight vector of class c_j. The weight vectors of all classes can be arranged into a matrix $W = \left(\vec{w}_1\, \vec{w}_2 ... \vec{w}_m\right)$. The conditional probability $p\left(y_j = +1 \mid \vec{x}, X, Y, \vec{w}_j\right)$ is conditioned on the $\vec{w}_j$ weight vector for class c_j. The function $s(.)$ is a sigmoid function shown in Figure 3.15. The sigmoid function $s(.)$ assures that the output will be a value in the range [0, 1]. The sigmoid function $s(.)$ belongs to a family of S-shaped functions called "logistic functions." That is the reason this type of classification is called "linear logistic regression."

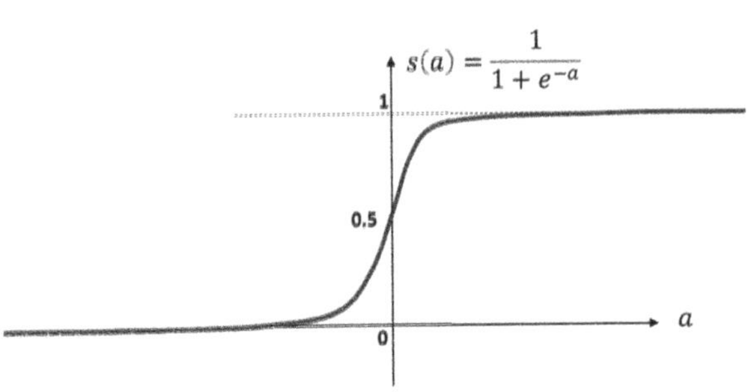

Figure 3.15 Logistic function $s(a)$.

Figure 3.16 Linear separation example in 2D.

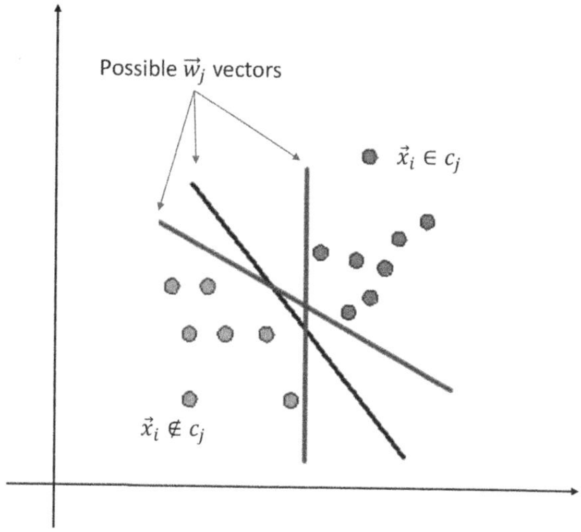

By performing linear logistic regression, we assume that the input vectors or input data points can be separated into separate regions (classes) by straight lines. We want a weight vector $\vec{w}_j$ that divides the input vector space into two regions; one region belongs to class c_j and another region that does not belong to class c_j. This linear separation is shown in Figure 3.16.

From probability axioms, we can write the following constraints:

$$p\left(y_{i,j} = -1 \mid \vec{x}_i, X, Y, \vec{w}_j\right) = 1 - p\left(y_{i,j} = +1 \mid \vec{x}_i, X, Y, \vec{w}_j\right), \tag{3-216}$$

If $y_{i,\ j} = +1$, we have $p\left(y_{i,j} = +1 \mid \vec{x}_i, X, Y, \vec{w}_j\right) = s\left(\vec{x}_i \vec{w}_j\right)$ and if $y_{i,\ j} = -1$ we have $p\left(y_{i,j} = -1 \mid \vec{x}_i, X, Y, \vec{w}_j\right) = 1 - s\left(\vec{x}_i \vec{w}_j\right)$. Noting that $s(-a) = 1 - s(a)$, we can write the probability of any data point $\left\{\vec{x}_i, y_{i,j}\right\}$ given the weight vector $\vec{w}_j$ as follows:

$$p\left(y_{i,j} \mid \vec{x}_i, \vec{w}_j\right) = s\left(y_{i,j} \vec{x}_i^{\,T} \vec{w}_j\right) \tag{3-217}$$

The total probability of the whole training data set for class c_j is given by:

$$p\left(Y \mid X, \vec{w}_j\right) = \prod_{i=1}^{n} s\left(y_{i,j} \vec{x}_i^{\,T} \vec{w}_j\right) \tag{3-218}$$

If we have a prior distribution for the weights $p\left(\vec{w}_j\right)$, we can calculate the posterior probability of the weight given the training data as follows:

$$p\left(\vec{w}_j \mid \vec{x}_i, y_{i,j}\right) \propto p\left(y_{i,j} \mid \vec{x}_i, \vec{w}_j\right) p\left(\vec{w}_j\right) \tag{3-219}$$

$$p\left(\vec{w}_j \mid X, Y\right) \propto p\left(\vec{w}_j\right) \prod_{i=1}^{n} p\left(y_{i,j} \mid \vec{x}_i, \vec{w}_j\right) \tag{3-220}$$

$$p\left(\vec{w}_j \mid X, Y\right) \propto p\left(\vec{w}_j\right) \prod_{i=1}^{n} s\left(y_{i,j} \vec{x}_i^{\,T} \vec{w}_j\right) \tag{3-221}$$

Having the weight probability given the training data, the predictive probability of unseen input $\vec{x}^*$ is given by:

$$p\left(y_j^* \mid \vec{x}^*, X, Y, \vec{w}_j\right) = \int p\left(y_j^* \mid \vec{x}^*, \vec{w}_j\right) p\left(\vec{w}_j \mid X, Y\right) d\vec{w}_j \tag{3-222}$$

Recall that in the regression case, when all distributions were Gaussian, we were able to find a closed-form Gaussian analytical solution for this integral. This integral would be easily evaluated into a Gaussian solution if both $p\left(y_j^* \mid \vec{x}^*, \vec{w}_j\right)$ and $p\left(\vec{w}_j \mid X, Y\right)$ are Gaussian; however, this is not the case in the classification problem because the distributions are not Gaussian due to the sigmoid function $s(a)$, which is not Gaussian. One solution is to use optimization methods such as Gauss-Newton method to find the best $\vec{w}_j^*$ weight vector. The best weight vector $\vec{w}_j^*$ that fits the given data will be given by:

$$\vec{w}_j^* = \underset{w_C}{\operatorname{argmax}}\, p\left(\vec{w}_j \mid X, Y\right) \tag{3-223}$$

If we assume Gaussian prior on the weight $\vec{w}_j \sim \mathcal{N}(0, \Sigma_p)$, we will have the following formula for $p\left(\vec{w}_j \mid X, Y\right)$:

$$p\left(\vec{w}_j \mid X, Y\right) \propto e^{-\frac{1}{2}\left(\vec{w}_j\right)^T \Sigma_p^{-1}\left(\vec{w}_j\right)} \prod_{i=1}^{n} s\left(y_j x_i^T \vec{w}_j\right) \tag{3-224}$$

Taking the negative log of both sides, we get the following formula:

$$J\left(\vec{w}_j\right) = -\log\left(p\left(\vec{w}_j \mid X, Y\right)\right) \propto \frac{1}{2}\left(\vec{w}_j\right)^T \Sigma_p^{-1}\left(\vec{w}_j\right) - \sum_{i=1}^{n} \log\left(s\left(y_j x_i^T \vec{w}_j\right)\right) \tag{3-225}$$

The best weights $\vec{w}_j$ can be obtained by minimizing $J\left(\vec{w}_j\right)$ using any optimization method such as Gauss–Newton method. Once we have the best weights, we can classify any unseen input $\vec{x}^*$ as follows:

$$p\left(y_j^* = +1 \mid \vec{x}^*, X, Y, \vec{w}_j^*\right) = p\left(y_j^* \mid \vec{x}^*, X, Y, \vec{w}_j^*\right) = s\left(\vec{x}^{*T} \vec{w}_j^*\right) \tag{3-226}$$

Linear logistic regression is: illustrated in Figure 3.17.

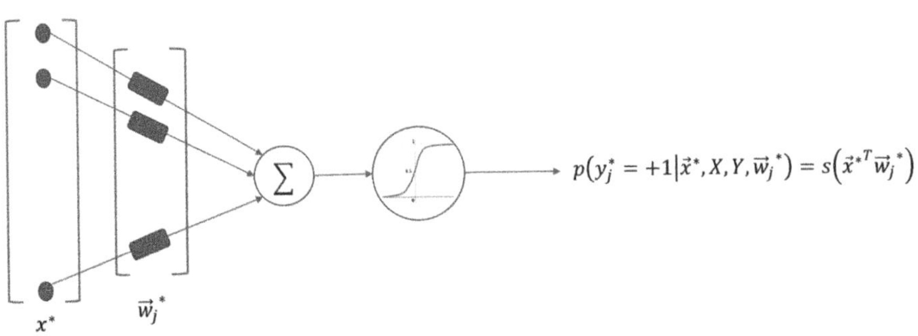

Figure 3.17 Linear logistic regression.

Classification Using Nonlinear Logistic Regression

If the class data points are not linearly separable (there is no straight line that separates the data points into two regions where one region belongs to the class and the other region does not belong to the class), we can use the same trick that we have used in the nonlinear Gaussian process regression. We can use a mapping function (basis function) $\phi\left(\vec{x}\right)$ that maps the input vector into a linearly separable space. In this case, we can write the probability that input vector x belongs to class c as follows:

$$p\left(y_j = +1 \mid \vec{x}, X, Y, \vec{w}_j\right) = s\left(\phi\left(\vec{x}\right)^T \vec{w}_j\right) \tag{3-227}$$

we can then apply the same linear regression technique but in the feature space $\phi\left(\vec{x}\right)$. If we assume Gaussian prior on the weight $w_C \sim \mathcal{N}(0, \Sigma_p)$, we will have the following weight probability formula:

$$p\left(\vec{w}_j \mid X, Y\right) \propto e^{-\frac{1}{2}\left(\vec{w}_j\right)^T \Sigma_p{}^{-1}\left(\vec{w}_j\right)} \prod_{i=1}^{n} s\left(y_j \phi\left(\vec{x}\right)^T \vec{w}_j\right) \tag{3-228}$$

Taking the negative log of both sides, we get:

$$J\left(\vec{w}_j\right) = -\log\left(p\left(\vec{w}_j \mid X, Y\right)\right) \propto \frac{1}{2}\left(\vec{w}_j\right)^T \Sigma_p{}^{-1}\left(\vec{w}_j\right) - \sum_{i=1}^{n} \log\left(s\left(y_j \phi\left(\vec{x}\right)^T \vec{w}_j\right)\right) \tag{3-229}$$

This function of $\vec{w}_j$ can be minimized using any optimization method such as Gauss–Newton method. Once we have the best weights, we can classify any unseen input x^* as follows (see Figure 3.17):

$$p\left(y_j^* = +1 \mid \vec{x}^*, X, Y, \vec{w}_j\right) = p\left(y_j^* \mid \vec{x}^*, X, Y, \vec{w}_j\right) = s\left(\phi\left(\vec{x}^*\right)^T \vec{w}_j^*\right) \tag{3-230}$$

Like the regression case, in the classification case, the idea of the basis function $\phi\left(\vec{x}\right)$ that maps the original input space to another feature space that can be linearly separated is the same idea that many ML methods apply such as artificial neural network and support vector machines. In fact, $\phi\left(\vec{x}\right)$ acts as the input and hidden layers of a neural network that maps the input into a useful feature that can be linearly separated by a linear logistic regression (Figure 3.18).

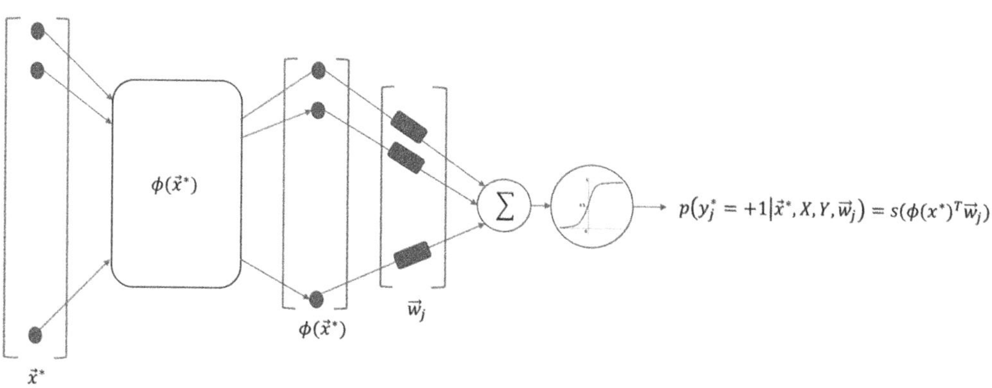

Figure 3.18 Nonlinear logistic regression.

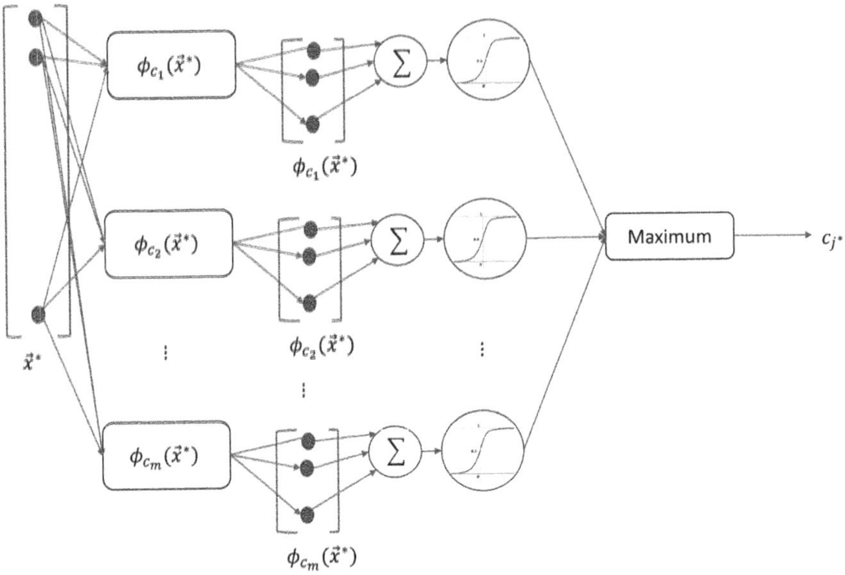

Figure 3.19 Nonlinear multiclassification logistic regression model.

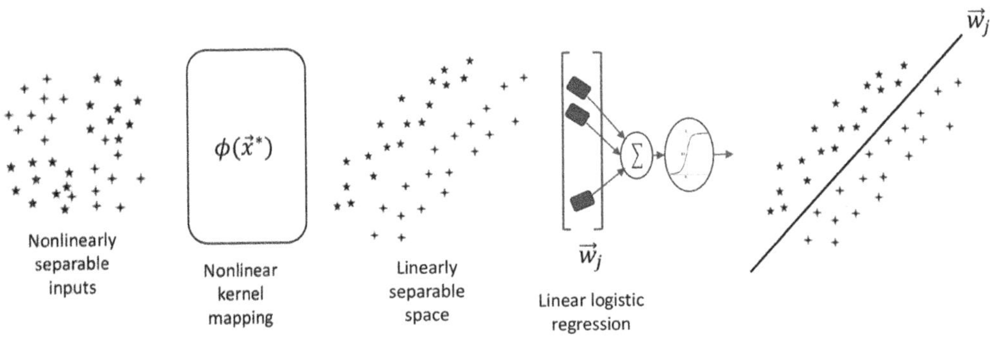

Figure 3.20 Transformation of input to a linear space.

If we have multiple classes, we can build a model for each class and train the weights of each class separately. To classify an unseen input $\vec{x}^*$, we pass the input into all models and select the class with the maximum output probability as shown in Figures 3.19 and 3.20.

3.5.5 Artificial Neural Networks (ANN)

3.5.5.1 Basic Definitions

In the previous sections, we have discussed the idea of the basis function and kernel that maps input vectors $\vec{x}$ into a linear space $\emptyset\left(\vec{x}\right)$ to facilitate modeling nonlinear mapping as follows:

$$y = \left(\emptyset\left(\vec{x}\right)\right)^T \vec{w} \quad \text{for regression} \tag{3-231}$$

$$y = s\left(\emptyset\left(\vec{x}\right)\right)^T \vec{w} \quad \text{for classification} \tag{3-232}$$

This mapping is linear in the weight vector $\vec{w}$ and hence linear regression methods can be applied to estimate the best weights and consequently predict y^* for unseen input $\vec{x}^*$. The remaining part is how to choose good basis function kernel $\emptyset\left(\vec{x}\right)$ that maps the inputs into a linear space (for regression) or linearly separable space (for classification). The idea of the basis function $\emptyset\left(\vec{x}\right)$ gave rise to the concept of artificial neural networks (ANN), where the basis function $\emptyset\left(\vec{x}\right)$ is replaced by multiple layers of processing units. These processing units can take several forms. One simple common form is weighted sum of inputs like the case in the regression models or it can be a logistic function of weighted sum of inputs like the case in the classification models. This processing unit is like neuron cells in the brain and that is the reason it is called a "neuron." Basic neuron model is shown in Figure 3.21. The output of a neuron is given by:

$$y = f\left(\vec{x}^T \vec{w}\right) \tag{3-233}$$

This model is known as McCulloch and Pitts Neuron Model. The function $f\left(\vec{x}^T \vec{w}\right)$ is called the firing function and it can take several forms as shown in Figure 3.22.

ANN consists of connected layers of neurons as shown in Figure 3.23 (left). The connections could be modularized as shown in Figure 3.23 (right). The network may have several complex hidden layers as seen in Figure 3.24. This type of ANN is known as deep ANN (DANN). ML models that are based on DANN are called Deep Learning (DL). Figure 3.25 shows the general multi-input/multi-output ANN structure where we have raw input layer, multiple hidden layers, and an output layer. The hidden layers work as the kernel basis. Multiple hidden layers increase the modeling power of the network, but they also increase the network depth and complexity. Training complex multi-layer ANNs requires massive computational power (i.e., computers' memory/storage and processors). Figures 3.23 and 3.25 show the standard ANN structure where

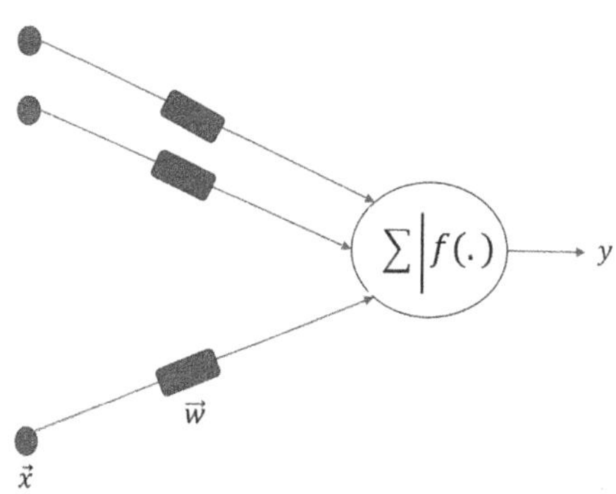

Figure 3.21 Neuron model.

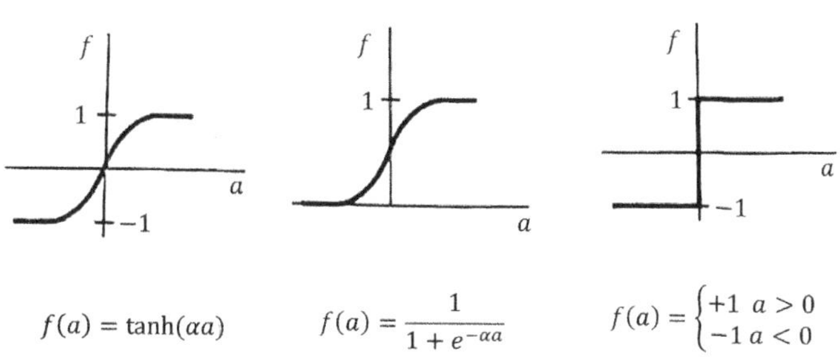

Figure 3.22 Common neuron activation functions.

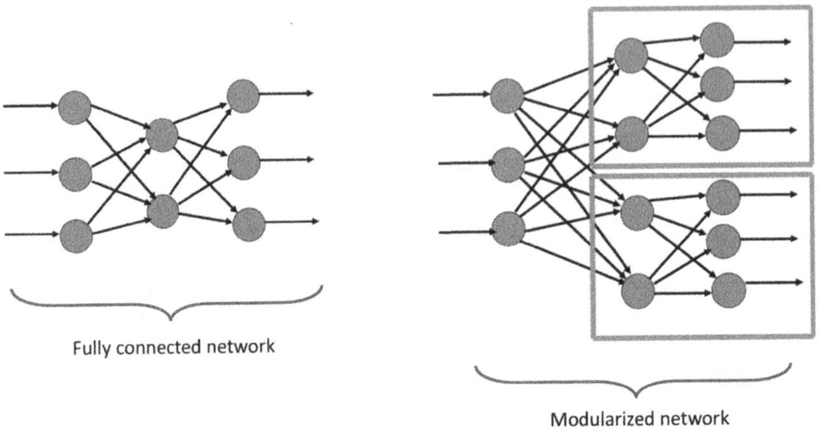

Figure 3.23 Feedforward ANN structure. (Left) fully connected, (right) modularized.

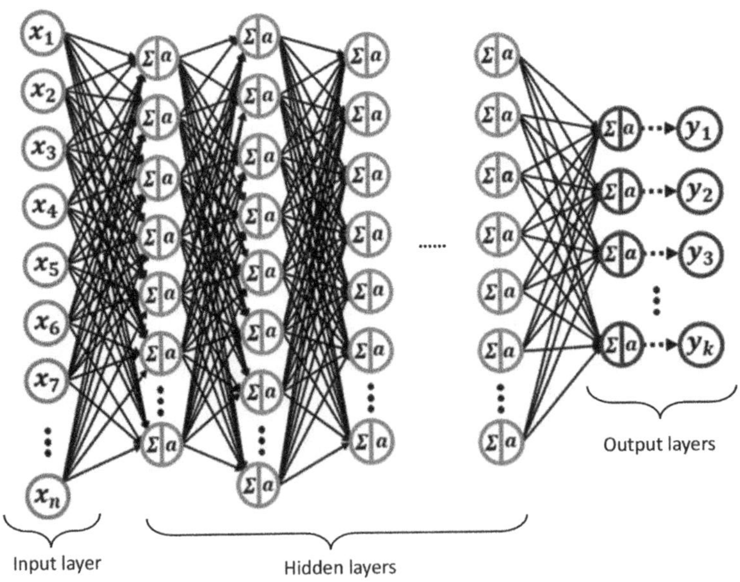

Figure 3.24 Example of a deep neural network.

information flow is one-directional from input to output. This standard structure is called feedforward ANN. Feedforward ANN is suitable for learning static mapping between inputs and outputs. However, feedforward ANN does not have any memory, so it is difficult to use to model dynamic systems.

3.5.5.2 Training Neural Networks

The key algorithm that makes ANN work is the training algorithm that finds the best weights that fit the mapping between the training input and desired output. In the Gaussian regression method, we explained the concept on a single processing unit (one neuron) and we assumed a Gaussian

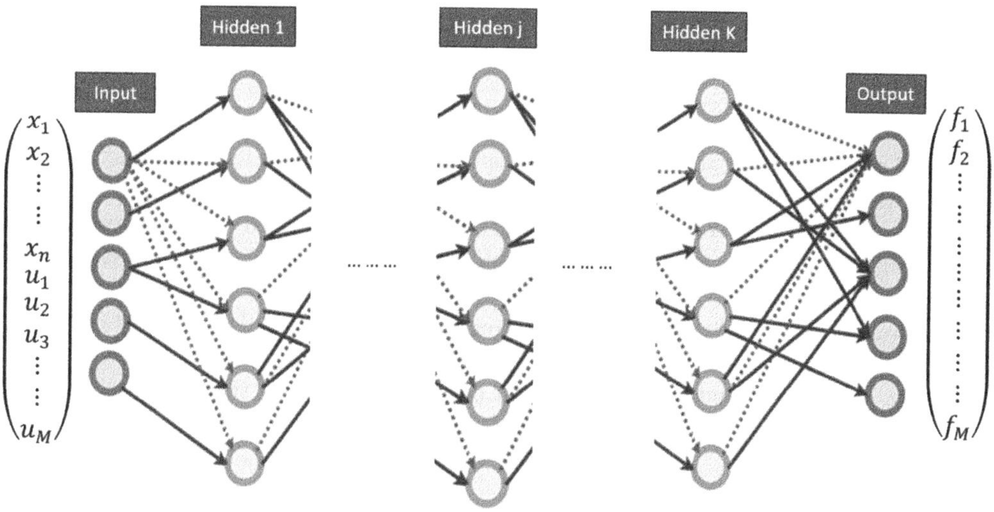

Figure 3.25 Basic general structure of ANN.

prior on the weight and a Gaussian likelihood on the data (given the weight) and then we calculated the weight as a probability distribution using Bayes' rule as follows:

$$p\left(\vec{w} \mid \vec{y}, X\right) \propto p\left(\vec{y} \mid X, \vec{w}\right) p\left(\vec{w}\right) \tag{3-234}$$

where $p\left(\vec{w}\right)$ is the prior distribution of the weights, $p\left(\vec{w} \mid \vec{y}, X\right)$ is the posterior distribution of the weights given the data, and $p\left(\vec{y} \mid X, \vec{w}\right)$ is the likelihood of the data given the weights. $p\left(\vec{w} \mid \vec{y}, X\right)$ means we have a range of weight vectors values, and each weight vector value has a probability. Therefore, the predicted output y^* given unseen input $\vec{x}^*$ is given by the marginalization rule as follows:

$$p\left(y^* \mid \vec{x}^*, X, \vec{y}\right) = \int p\left(y^* \mid \vec{x}^*, \vec{w}\right) p\left(\vec{w} \mid \vec{y}, X\right) dw \tag{3-235}$$

This probabilistic approach was suitable for regression because all distributions were Gaussian which makes the integration analytically tractable, and the solution closed form. Even if the input/output relationship is nonlinear, we have seen how we can use basis functions to transform the raw input into another feature-space that makes the relationship linear. In the classification problem, the likelihood was not Gaussian and we either do the integration numerically and approximating the probability distributions by particle cloud (Monte Carlo Simulation) or we try to find the best value of the weight vector by optimization methods such as Gauss–Newton method. In the ANN case, the output is a sigmoid function which affects the Gaussian assumption. Also, in the ANN case, the basis function is replaced by complex connected layers of processing units (neurons). This makes the probabilistic approach to train ANN and calculating the probability distribution of the weights very difficult. Therefore, ANN is usually trained using optimization approach or loss-function minimization methods using gradient-descent approach. The good thing about optimization method is that it can be solved numerically if a closed-form solution is difficult or does not exist. The most used ANN training algorithm is the backpropagation algorithm. The objective of the learning is to find a formula to update the weight vector iteratively such that the error between the training data observations and the ANN network prediction is minimized.

3.5.5.3 Neuron Learning

The general rule to optimize the weights of ANN is to adjust the weights iteratively such that the ANN output comes closer to the desired output or minimizes a specific loss function in each iteration. In other words, the learning algorithm must adjust the weights such that the error between the ANN output and a desired output decrease. This learning can be implemented using different algorithms such as Gauss-Newton or Levenberg-Marquatdt or any gradient-descent method. In the following discussion, we will explain a basic learning algorithm based on the first-order derivative of the loss function. For simplicity, the loss function is assumed to be the squared difference between the ANN output and a desired output. Consider a single-layer single-neuron case (see Figure 3.21). The error is given by:

$$e = d - y \tag{3-236}$$

where d is the desired output and y is the network prediction given by the neuron output as follows:

$$y = f(a) \tag{3-237}$$

$$a = \vec{x}^T \vec{w} \tag{3-238}$$

The loss function is the total squared error as a function of weights is given by:

$$J\left(\vec{w}\right) = \frac{1}{2}(d - y)^2 \tag{3-239}$$

Because the function $f(a)$ is nonlinear, we cannot simply set the derivative of $J\left(\vec{w}\right)$ with respect to $\vec{w}$ and set the result to zero to find the best weight. We cannot simply separate the weight vector $\vec{w}$ in one side and the other quantities/variables on the other side. The solution to this problem is to iteratively adjust the weight in the direction that minimizes the error, which is simply the negative of the derivative $\frac{\partial J(\vec{w})}{\partial \vec{w}}$ as follows:

$$\Delta\vec{w} = -\eta \frac{\partial J\left(\vec{w}\right)}{\partial \vec{w}} \tag{3-240}$$

where η is a learning parameter that represents how fast or how large the weights change in each iteration. The derivative $\frac{\partial J(\vec{w})}{\partial \vec{w}}$ can be calculated using chain rule as follows:

$$\frac{\partial J\left(\vec{w}\right)}{\partial \vec{w}} = \frac{\partial J\left(\vec{w}\right)}{\partial y} \frac{\partial \hat{y}}{\partial a} \frac{\partial a}{\partial \vec{w}} \tag{3-241}$$

$$\frac{\partial J\left(\vec{w}\right)}{\partial y} = -(d - y) \tag{3-242}$$

$$\frac{\partial y}{\partial a} = f'(a) \tag{3-243}$$

$$\frac{\partial a}{\partial \vec{w}} = \vec{x} \tag{3-244}$$

$$\frac{\partial J(\vec{w})}{\partial \vec{w}} = -(d-y)(f'(a))\vec{x} \tag{3-245}$$

Let us define δ^o as the output gradient as follows:

$$\delta^o = -\frac{\partial J(\vec{w})}{\partial y}(f'(a)) \tag{3-246}$$

$$\delta^o = (d-y)(f'(a)) \tag{3-247}$$

then the weight is then updated using the following equation:

$$\vec{w} = \vec{w} + \eta\delta^o\vec{x} \tag{3-248}$$

For sigmoid function, the derivative $f'(a)$ is given by:

$$f'(a) = f(a)(1-f(a)) \tag{3-249}$$

Therefore, we have all the values to calculate the change in the weights in the direction that minimizes the error or the loss function. If we are using a different firing function $f(a)$, we will need to calculate its derivative $f'(a)$ to perform the learning.

3.5.5.4 Backpropagation Learning

This neuron's learning rule can be generalized to the weights of all neurons in all the layers of the ANN. We will consider the case where we have one hidden layer and an output layer as seen in Figure 3.26. We will derive the learning algorithm per neuron, but we can generalize to all neurons of all layers easily as we will see. The learning rule for k^{th} weight of the jth neuron in the output layer is given by:

$$\Delta w_{jk} = -\eta\frac{\partial J(w_{jk})}{\partial w_{jk}} \tag{3-250}$$

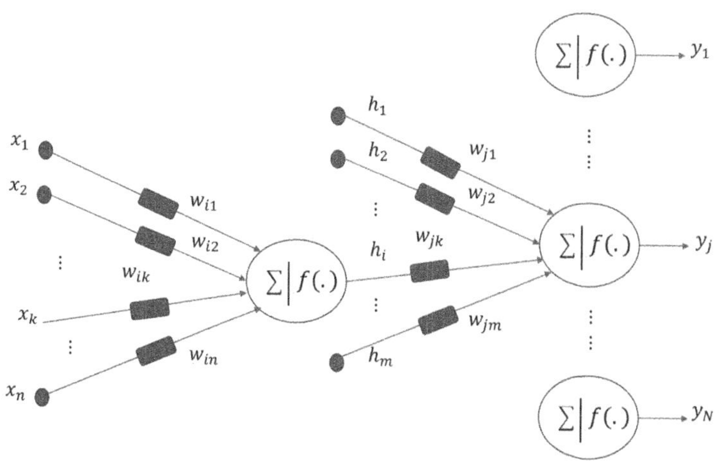

Figure 3.26 Two neurons. One in a hidden layer and one in an output layer.

$$J(w_{jk}) = \frac{1}{2}\sum_{j=1}^{N}\left(d_j - y_j\right)^2 \tag{3-251}$$

$$y_j = f(a_j) \tag{3-252}$$

$$a_j = \sum_{l=1}^{m} w_{jl}h_l \tag{3-253}$$

$$\frac{\partial J(w_{jk})}{\partial w_{jk}} = \frac{\partial J(w_{jk})}{\partial y_j}\frac{\partial y_j}{\partial a_j}\frac{\partial a_j}{\partial w_{jk}} \tag{3-254}$$

which gives:

$$\delta_j^o = \frac{\partial J(w_{jk})}{\partial y_j}\frac{\partial y_j}{\partial a_j} = \left(d_j - y_j\right)f'(a_j) \tag{3-255}$$

$$\Delta w_{jk} = \eta\delta_j^o h_k \tag{3-256}$$

where δ_j^o is the gradient of the jth neuron of the output layer and d_j is desired output for j^{th} neuron in the output layer. Now, consider the learning rule for the kth weight of ith neuron in the input layer. The overall error as a function of the hidden layer weight w_{ik} is given by:

$$h_i = f(a_i) \tag{3-257}$$

$$a_i = \sum_{l=1}^{n} w_{il}x_l \tag{3-258}$$

$$\frac{\partial J(w_{ik})}{\partial w_{ik}} = \sum_{j=1}^{N}\left(\frac{\partial J(w_{ik})}{\partial y_j}\right)\left(\frac{\partial y_j}{\partial a_j}\right)\left(\frac{\partial a_j}{\partial h_i}\right)\left(\frac{\partial h_i}{\partial a_i}\right)\left(\frac{\partial a_i}{\partial w_{ik}}\right) \tag{3-259}$$

The terms $\left(\frac{\partial h_i}{\partial a_i}\right)\left(\frac{\partial a_i}{\partial w_{ik}}\right)$ do not depend on j so it is taken out of the summation.

$$\left(\frac{\partial h_i}{\partial a_i}\right)\left(\frac{\partial a_i}{\partial w_{ik}}\right) = f'(a_i)x_k \tag{3-260}$$

$$\sum_{j=1}^{N}\left(\frac{\partial J(w_{ik})}{\partial y_j}\right)\left(\frac{\partial y_j}{\partial a_j}\right)\left(\frac{\partial a_j}{\partial h_i}\right) = -\sum_{j=1}^{N}\left(d_j - y_j\right)f'(a_j)w_{ji} = -\sum_{j=1}^{N}\delta_j^o w_{ji} \tag{3-261}$$

Similarly, we can define the gradient for the ith neuron of the hth layer as follows:

$$\delta_i^h = f'(a_i)\sum_{j=1}^{N}\delta_j^o w_{ji} \tag{3-262}$$

and the update for the kth weight of the ith neuron will be given by:

$$\Delta w_{ik} = x_k\sum_{j=1}^{N}\delta_j^o w_{ji} \tag{3-263}$$

$$\Delta w_{ik} = \eta\delta_i^h x_k \tag{3-264}$$

We can generalize the backpropagation algorithm to any number of hidden layers. This is the basic foundational concept of ANN. ANN research has dramatically advanced and several variations and modifications of both ANN structures and learning algorithms have been suggested. However, the backpropagation algorithm is still the core foundational concept of ANN.

3.5.5.5 Other Learning Methods

The objective of any learning method is how to adapt the weights of the network to achieve a specific goal. In this context, it should be noted that the backpropagation is just one learning method among others. Backpropagation is the most used learning algorithm in neural networks. However, there are other learning algorithms such as Newton, Newton–Gauss, Conjugate gradient, and Levenberg–Marquardt algorithms. All learning methods are based on the objective of minimizing the loss function $J(.)$. The gradient of the jth neuron of the output layer δ_j^o depends on the derivative of the loss function $J(.)$ and the neuron activation function $f(.)$. The gradient for the ith neuron of the first hidden layer depends on the gradient of the neurons of the output layer and so on until all hidden and input layers are trained. Therefore, the loss function $J(.)$ and the neuron activation function $f(.)$ play the key roles in the network training and convergence.

3.5.5.6 Local Minima and Overfitting

Despite the flexibility of ANN, it has several issues and limitations. A common problem is local minima when weights are stuck in a local minimum. Several algorithms have been suggested to escape from local minima such as simulated annealing. Also, overfitting is a common problem in ANN. It happens when the weights are perfectly tuned on a certain input–output dataset and do not work well if slight modifications or minor changes are added to the data. This makes the ANN sensitive to noise. Also, overfitting limits the generalization power of the ANN. The common solution to the generalizability problem is to train the network over large diverse data sets. Another common issue of ANN is interpretability and predictability. Due to the complexity of the ANN structure, especially if we have large number of hidden layers, sometimes it is complex to understand or predict the ANN behavior in all scenarios, which impose serious risks of failure. Additionally, complex ANN takes so long to train. Although the development of high-performance computers like GPUs (Graphical Processing Units), ANN training still takes considerable amount of time.

3.5.5.7 Types of ANN

The ANN concepts discussed above explain the canonical standard structure and loss function of ANN. To solve a specific problem using ANN, researchers and engineers must customize the structure and the loss function to achieve a specific task. In practice, there is no ANN structure that can fit or solve all problems. Therefore, designing specific ANN framework to solve a specific problem is an active research area and an open-ended design problem. Some ANN designs/structures may work very well on a specific problem under a specific loss function and fail under a different loss function or a different structure. Other types of ANN include recurrent ANN (RANN), Deep ANN (deep learning "DL"), radial basis function (RBF) ANN, and others. More information about types of neural networks can be found in other resources such as [13] and [17]. In the next sections, we will discuss some special structures of ANNs that have been used recently in sensor fusion for positioning, navigation, and mapping. Specific learning-based approaches will be discussed in more details in Chapter 12.

3.5.5.8 Radial Basis Function ANN (RBF-ANN)

RBF-ANN is an ANN designed for function approximation and it depends on the assumption that similar inputs produce similar outputs. It is very close to the concept of Gaussian process regression, where all data points in the training dataset contribute to the prediction of the output given unseen input. The general concept of RBF-ANN is that given a dataset $D = \{x_i, y_i\} : i = 1, 2, ..., N$, the output for an unseen input x^* is a weighted combinations of outputs as follows:

$$y^* = \frac{\sum_{j=1}^{N} r(\|x^* - x_j\|)y_j}{\sum_{j=1}^{N} r(\|x^* - x_j\|)} \tag{3-265}$$

where $r(a)$ is a radial basis function that increases if the a is small and decreases if a is large. The predicted output formula can be written as the following weighted sum:

$$y^* = \sum_{j=1}^{N} r(\|x^* - x_j\|)w_j \tag{3-266}$$

If we have a large data set, this weighted sum will be computationally expensive. In addition, the data points that are very far away from x^* will have a very small weight, so it can be ignored. One approach to optimize this weighted sum is to select key data points c_j, where $j = 1, 2, ..., M$ and $M \ll N$. The best c_j is ideally the centers of clusters of the inputs x_i. The output is then approximated as follows:

$$y^* = \sum_{j=1}^{M} r(\|x^* - c_j\|)w_j \tag{3-267}$$

We can realize this weighted combination using one input layer, one hidden layer, and one output layer as seen in Figure 3.27.

The weight update can be formulated as a function of both cluster centers c_j and weights w_j. Training the network on the dataset should calculate the best values for cluster centers c_j and

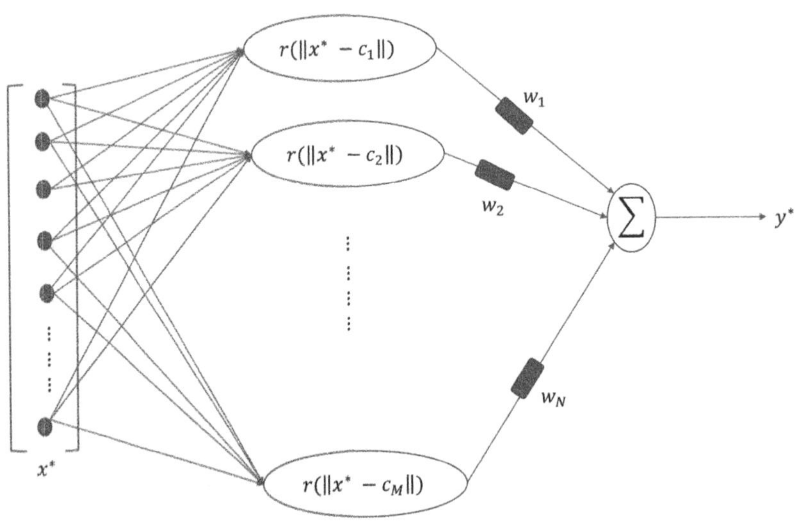

Figure 3.27 Basic RBF-ANN concept.

weights w_j simultaneously. Alternatively, we can find c_j by clustering the data using any clustering technique and then train the weights of the network.

3.5.5.9 Recurrent ANN (RANN) and LSTM Networks

The standard feedforward ANN is suitable for regression and classification where the relationship between input and output is static. In other words, the output depends only on the current input and the training points are independent (not a time-series or a sequence). The standard feedforward structure does not store information. Therefore, standard ANN is not suitable for dynamic systems where the input is a sequence such as a time-series and the output depends on both the input and the current state of the network. Navigation systems are example use case scenario of dynamic systems. In this case, we want the ANN to have a memory and to store state information. This type of ANN can be used to replace conventional methods to model dynamic systems when conventional methods fail or when it is too complex to use conventional methods such as differential equations, transfer function, or state space models. RANN is a modified ANN structure that can model dynamic behavior of sequential data. The output of a RANN neuron can affect its input therefore there is a cycle in the network structure. The general model of a nonlinear dynamic system is given by the following equation:

$$y_k = f(x_k, x_{k-1}, x_{k-2} \dots x_{k-M}) \tag{3-268}$$

RANN can be trained to model the behavior of such system using the ANN structure shown in Figure 3.28. The learning of the RANN can be done using backpropagation algorithm. To model more complex nonlinear dynamic systems, more hidden layers with recurrent connections can be used. However, the problem with the structure in Figure 3.28 will lead to a problem called Vanishing gradient. With the increased depth of the network, the derivative of the error function that is used to train the network weights in the backpropagation algorithm becomes smaller, which is known as "vanishing gradient" problem.

To solve this problem, the long short-term memory (LSTM) ANN was proposed in [14]. LSTM is an enhanced RANN with improved learning stability that is designed to overcome the vanishing gradient problem of conventional RANN so it can deal with long-term dependencies. The LSTM consists of cells. Each LSTM cell has the structure shown in Figure 3.29. LSTM cell has four ports input (*I*), forget (*F*), update (*U*) and output (*O*) ports. The forget gate takes the input $x(t)$, and the

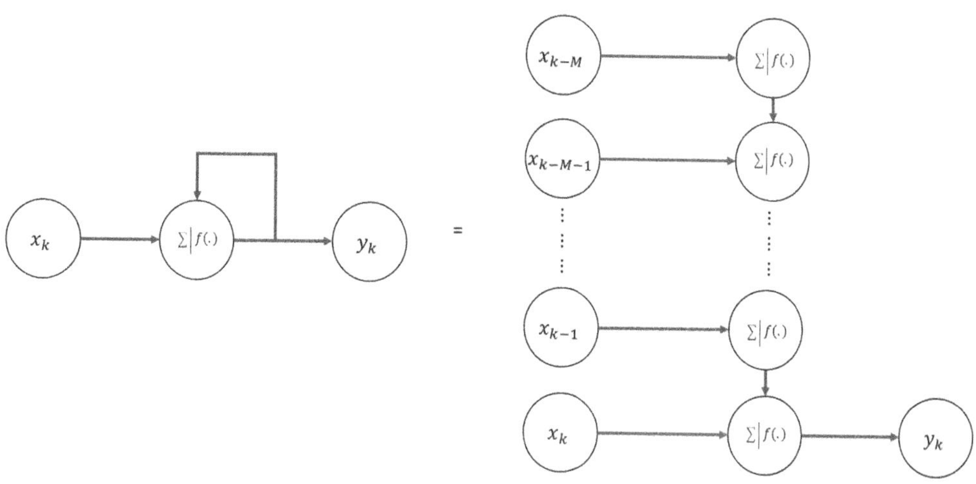

Figure 3.28 Recurrent ANN for dynamic system modeling.

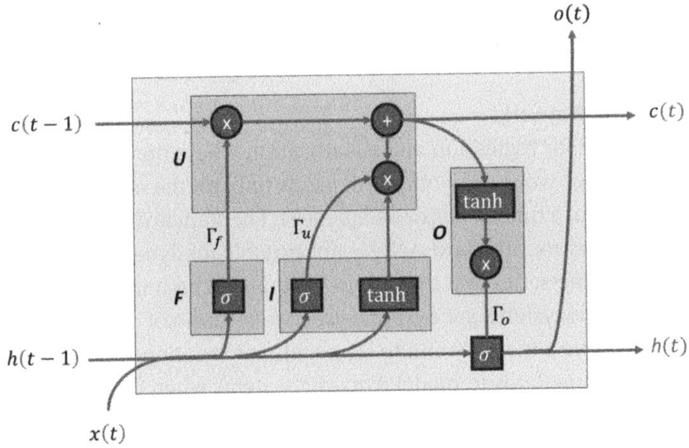

Figure 3.29 LSTM network cell structure.

incoming hidden state $h(t-1)$, as inputs and determines which elements of the incoming cell state $c(t-1)$, and to what extent it should be discarded. The input gate receives the input $x(t)$ and the incoming hidden state $h(t-1)$ and determines which elements of the incoming cell state $c(t-1)$ and to what extent it should be updated by the new information. The forgetting and updating processes happen in the update gate. Eventually, the output gate, receiving the input $x(t)$ and the incoming hidden state $h(t-1)$ and the outgoing cell states $c(t)$, determine the LSTM cell output $o(t)$ and the outgoing hidden state $h(t)$. The function $\sigma(.)$ is the sigmoid function shown in Figure 3.15 and the tan$h$(.) function is the hyperbolic tangent function shown in Figure 3.30 defined as follows:

$$\tanh(x) = \frac{e^{2x} - 1}{e^{2x} + 1} \tag{3-269}$$

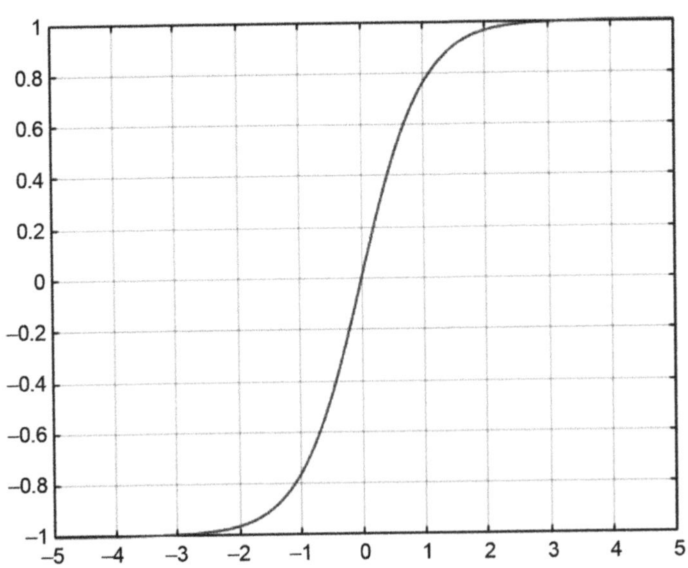

Figure 3.30 Tanh activation function.

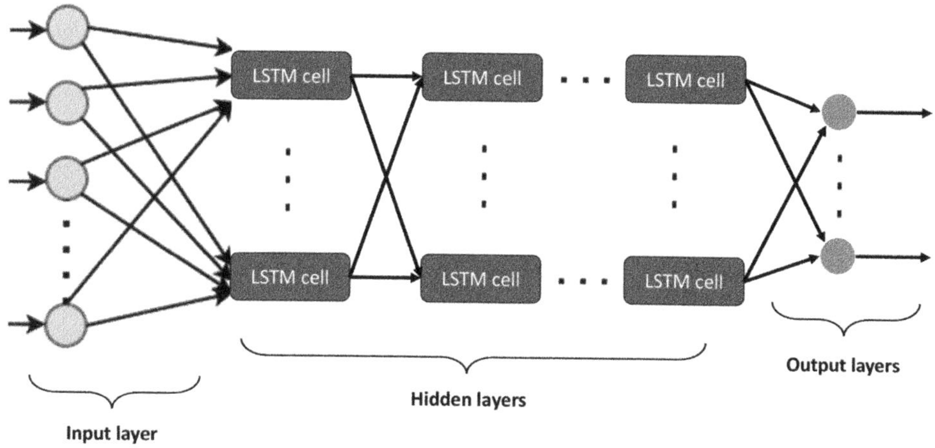

Figure 3.31 Example LSTM network with multiple hidden layers.

An LSTM network consists of layers of LSTM cells where the input to the hidden layer is the output of the input layer, and the output of the network is the output of the output layer. The width of the input layer depends on the length of the input sequence. Example LSTM Network with multiple hidden layers is shown in Figure 3.31. Example application of LSTM can be found in [18].

3.5.5.10 Convolutional Neural Networks

One of the important ANN types in the field of signal processing and computer vision is convolutional neural networks (CNN). CNN belongs to the deep learning models where the number of layers of the ANN is large, and the architecture is modular and complex. The first successful application of CNN was handwritten recognition. CNN are deep feedforward ANN designed to process multidimensional inputs (like images) to extract useful features such as edges, textures, segments, and to recognize higher-level knowledge and semantics such as object types, semantic segmentation, and scene dynamics. The general structure of CNN is shown in Figure 3.32. CNN processing

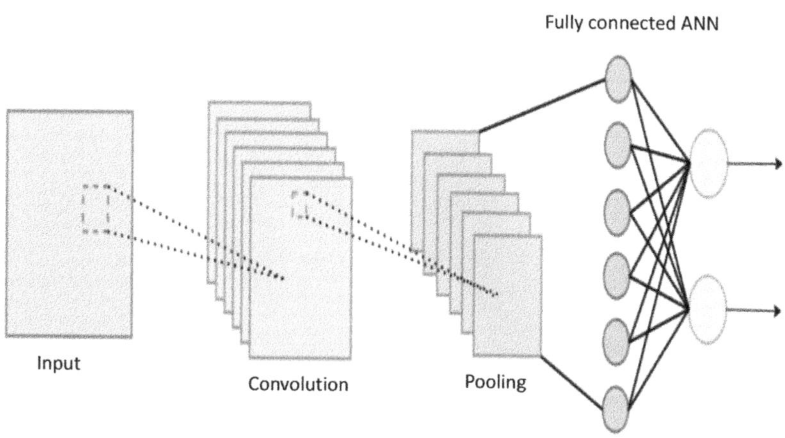

Figure 3.32 General structure of CNN.

starts with the convolution layer. Convolution is the key operation in CNN where a dot-product is applied between a sliding grid of weights and a similar grid of input data (e.g., image pixels). Mathematically, 2D convolution can be defined as follows:

$$C(j,k) = \sum_{l}\sum_{m} I(l,m)\, h(j-l+1, k-m+1) \tag{3-270}$$

Figure 3.33 shows an illustration of how a filter (kernel, or convolution operator) is applied on an input. The input can be an image, or any data arranged in a known structure. The filter is a small window that slides over all the input images, applying the convolution mathematical operation as described in the equation above and stores the output into an output structure, which is also called a tensor. We can apply more than one filter, and, in this case, we will have multilayer output as seen in Figure 3.33. We can add more layers by applying more convolution operators on the output. The new convolution layer will generate another layer of multi-layer outputs of smaller dimensions. However, before we apply another convolution layer, we may apply a pooling layer that converts the multilayer output into a reduced or even a single-layer output where each cell is assigned one value from the multiple filters' outputs. This output could be the maximum (Max Pooling), the sum (Sum Pooling), or the average (Average Pooling). There are several types of pooling operators. The final fully connected layer is a regular ANN that can be trained to produce the final desired output. CNN weights are learned using the backpropagation algorithm. The complex structure and the nonlinear activation function of the neurons in a CNN make it capable of learning complex patterns in the data.

A series of convolutions that is followed by a pooling layer is called an encoder. An encoder shrinks the size of the raw input to a reduced dimension where features are stored. This acts as the nonlinear mapping kernel described in nonlinear functions modeling. The reduced dimension block is called a bottleneck. Sometimes the encoding process is called downsampling because we

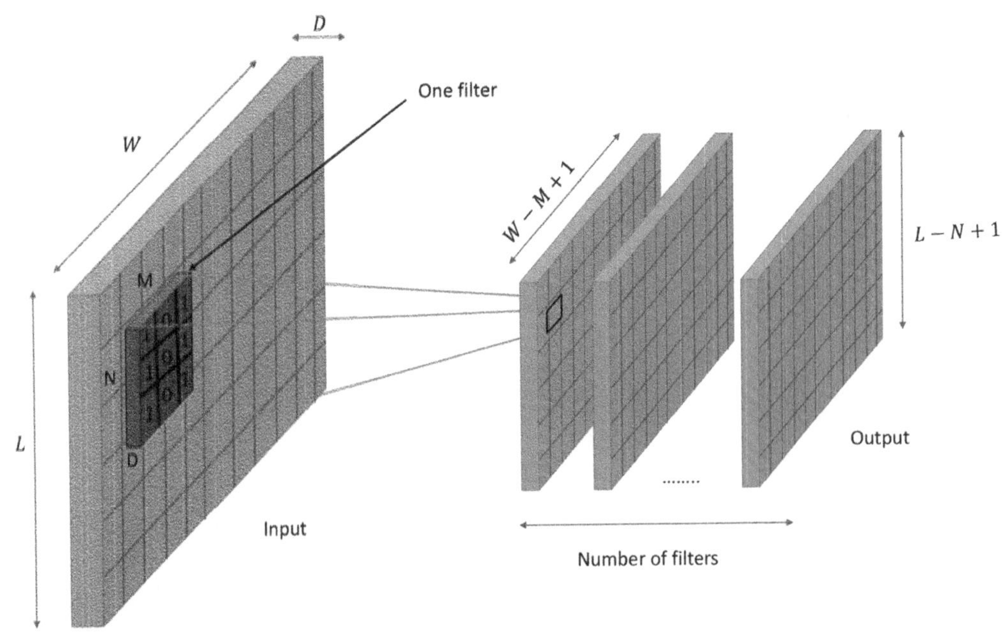

Figure 3.33 Illustration of convolution operation.

reduce the number of data samples with each convolution/pooling layer. To restore the original dimension, an up-sampling of the bottleneck is applied. This process is called decoding. In computer vision, the decoder brings the compressed features back into a form of an image to be visualized by users. The decoding process could be as simple as upsampling operation. We can apply convolutions on the upsampling output with padding to keep the upsampling size. The final network output can be generated by any type of ANN depending on the output type if it is a classification or a regression. An illustration of the processing flow is shown in Figure 3.34.

The training of CNN is based on the backpropagation algorithm. However, the architecture and the loss function of the network are major design factors that influence the network accuracy and performance. The architecture and the loss function are designed and formulated by researchers and engineers to solve a particular problem. One of the challenges of CNN and ML in general is that there are no clear systematic approaches to design the architecture of the networks (i.e., number of layers, convolution filters, pooling operators, ...) to solve a specific problem. However, researchers and designers usually gain extensive understanding and experience by trying different designs. With experience accumulation designers and engineers become familiar with specific types of networks models and architectures that are suitable for specific problems. The loss function plays an important role in the network performance. The loss function determines the error on which the backpropagation algorithm uses to adapt the network weights. In Chapter 12, we will demonstrate some ML network structures and loss functions that are used to solve problems related to positioning and navigation such as optical flow, relative pose, absolute pose, and depth estimation from multiple sensors such as camera, LiDAR, and inertial sensors.

As DL and CNN networks have complex architectures, the simple scalar, vector, and matrices data structures become unsuitable for the representation and the operations applied to train or test the network. Therefore, alternative advanced data structures called tensors have been introduced. The most used complex data structure used in deep ML is tensors. Tensors are advanced multidimensional data structures that are more complicated than simple vectors, 2D, and 3D matrices. In

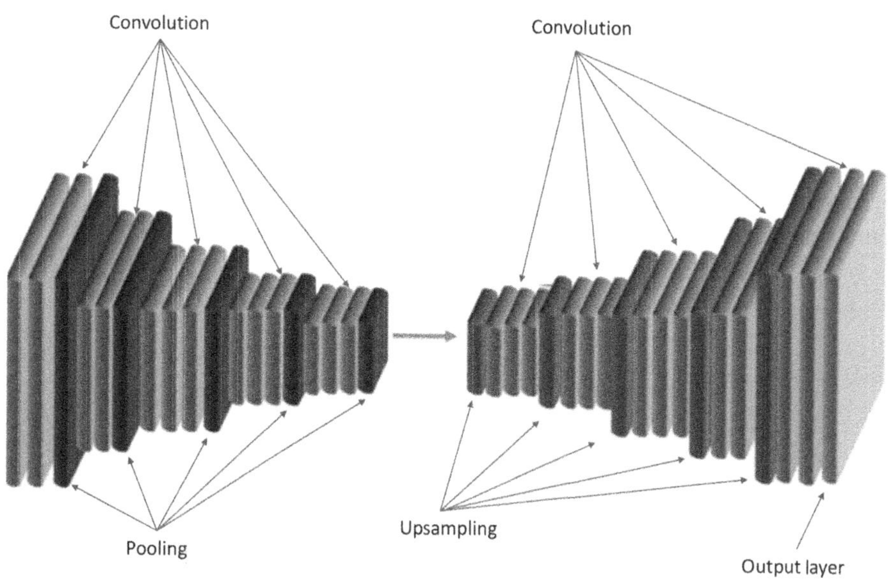

Figure 3.34 Illustration of processing flow in CNN.

fact, scalars, vectors, and matrices are special tensors. A tensor can be thought of as *N*D array. Mathematical operations such as addition, and multiplication, which are required to train deep neural networks, are defined. More information about tensors and operations on tensors can be found in [16].

The purpose of this brief introduction to DL and CNN is to illustrate basic concepts that are needed to understand how ML and AI can be applied in positioning, navigation, and mapping applications. For detailed information about DL and CNN, readers are encouraged to seek specialized textbooks in these fields such as [3], [14], [13], and [17]. In Chapter 12, we will demonstrate use case scenarios and example ANN types and architectures that are used to fuse data from multiple sensors to perform several tasks related to positioning, navigation, and mapping such as optical flow estimation, relative pose estimation, absolute pose estimation (relocalization), and 3D dense depth estimation. Depending on the use case scenario and the objective of the network framework, the basic design factors in designing such networks are the network type, the network architecture, and the loss function.

3.5.6 Reinforcement Learning

We have seen how RNN and LSTM ANN are special networks designed to handle dynamic systems where the output depends on current and past values of the input and the network stores a state. Another useful ML technique that is very suitable for dynamic environments is reinforcement learning. RL learns dynamically by rewarding an agent when performing good actions and penalizing it when taking poor actions. Good and poor here are used to refer to a performance metric such as an error cost function that we want to minimize. Agents in RL learn dynamically by accumulating experience while acting in an environment. For example, a robot can learn to maneuver around an object, avoiding collision by being letting the robot interacting with objects and defining a policy that rewards the robot when it moves at a safe distance from objects and penalizes it whenever it comes too close to an object. In the beginning, the robot may take random actions with several mistakes. However, the accumulated experience expressed in terms of rewards and penalties will make the robot learn the best policy that maximizes the rewards and minimizes the penalties. This "policy" can be learned and stored in an ANN, and it can be updated whenever the environment or the rules change. The general framework of RL is shown in Figure 3.35. The definition and the structure of the agent, environment, state, state transition, rewards, penalties, and possible actions are very flexible and can be designed to suit a specific problem.

Quality Learning (Q-learning) is the most common form of RL algorithms. It uses a Q-table to keep track of state, action, reward and maps the state-action pair to a Q-value. This Q-table represents the accumulated experience or learning that the agent learns overtime. It uses this Q-table to

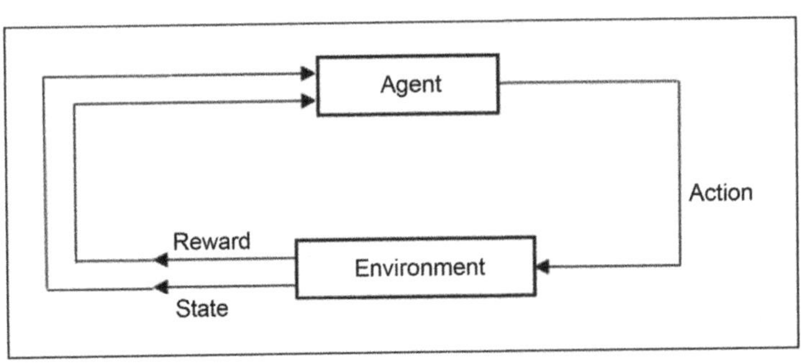

Figure 3.35 The general framework of RL image (*Source:* Adapted from [13]).

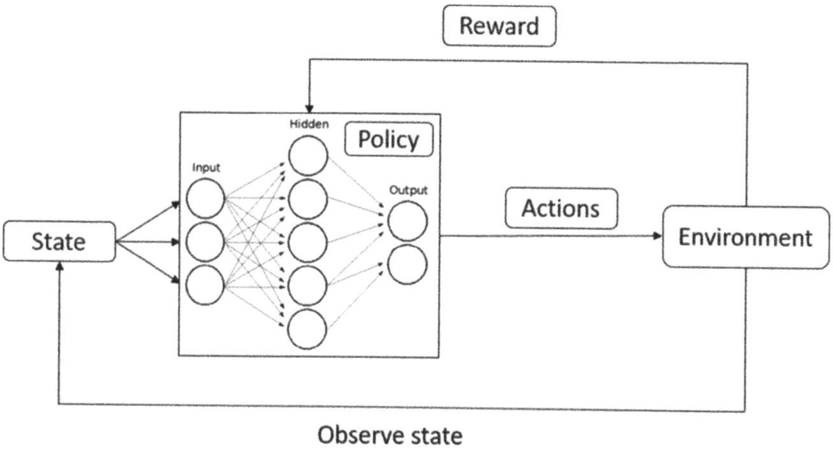

Figure 3.36 Deep Q-reinforcement learning.

generate the best action given the current state and environment. The main limitation of Q-learning is that it keeps the Q-table to decide future actions, which may not be feasible of the size of Q-table becomes too large. To avoid this problem, Deep-Q-learning was proposed to replace the Q-table by an ANN with multiple layers as shown in Figure 3.36.

3.5.7 Decision Trees

Decision tree is a ML approach that uses tree structure to learn patterns from data. Starting from a parent node, the decision tree splits the data by applying a boundary check on a specific feature into multiple sets (children's nodes). The split operation continues until a final output is concluded in the leaf node. Decision trees are most used in classification, but they can also be applied in regression. Decision trees are illustrated in Figure 3.37. The core learning module in decision trees is the splitting operation. Key decisions involve the feature to be split and the splitting rule or boundaries. For classification, the most used splitting method is based on the information gain such that the splitting that leads to the maximum information gain is the desired splitting rule. Information gain can be measured using the entropy of the data. Entropy is a measure of purity of the dataset. For example, the splitting on the left side of Figure 3.38 is an ideal splitting with maximum entropy (i.e., minimum variance or maximum purity). In regression, the most used splitting rule is the minimum variance rule demonstrated in Figure 3.38. Decision trees can be used to build lookup-table-like modules to predict the output of an unknown input based on training datasets. Decision trees are suitable for complex, highly nonlinear input/output mapping. Advanced structures that combine multiple decision trees can be used to build complex models. This multiple decision tree integration approach is called ensemble methods such as voting, bagging, and boosting. More information about decision tree-based ML methods can be found in [15].

3.5.8 Genetic Algorithms

Genetic algorithms are a branch of ML and AI that is inspired by biological evolution theory. A problem can be defined as the fitness of a population of species, and possible solutions can be seen as population individuals. The population is evolving while reproducing new generations. Evolution happens either by mixing the features of the ancestors or by mutations that could happen during reproduction. The new population will have some individuals with healthy features (good

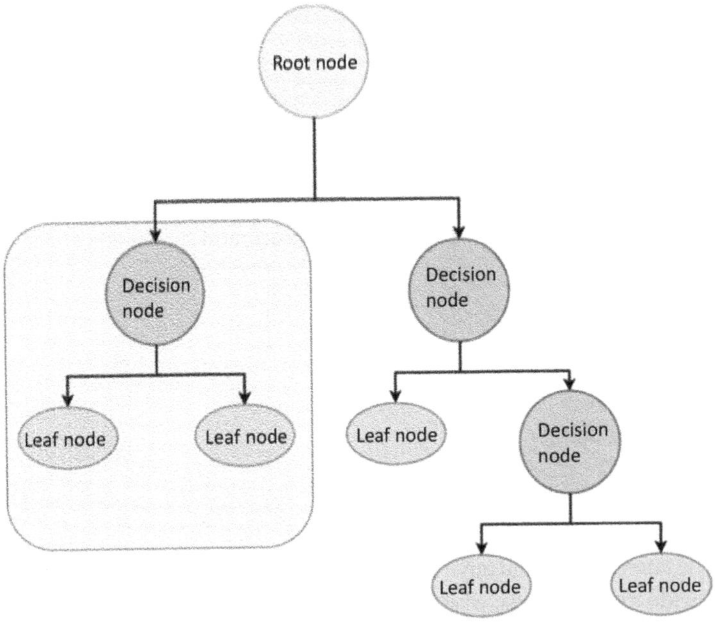

Figure 3.37 Decision tree.

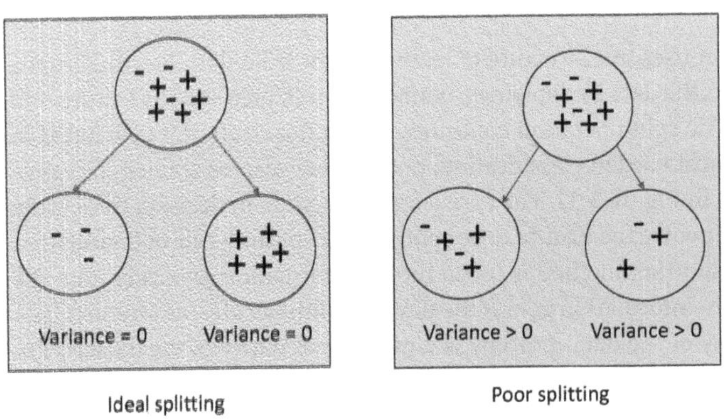

Figure 3.38 Splitting based on minimum variance rule.

solutions to a problem) and some individuals with weak features (poor solutions). Good solutions are naturally selected to survive for the next generation, and poor solutions tend to decay over the generations. This process emulates the natural selection process that happens in nature. The definition and formulation of good and poor solutions depend on the problem and the solution objective, which could be a performance metric such as an error cost function $J(x)$ that we want to find the best solution x^* that minimizes $J(x)$. The general flowchart of the genetic algorithm is shown in Figure 3.39. We briefly explain each step in the following steps:

- **Solution Encoding:** the possible solution x of the problem under consideration must be encoded in a suitable form. The most used encoding form in GA is binary encoding. A possible solution is

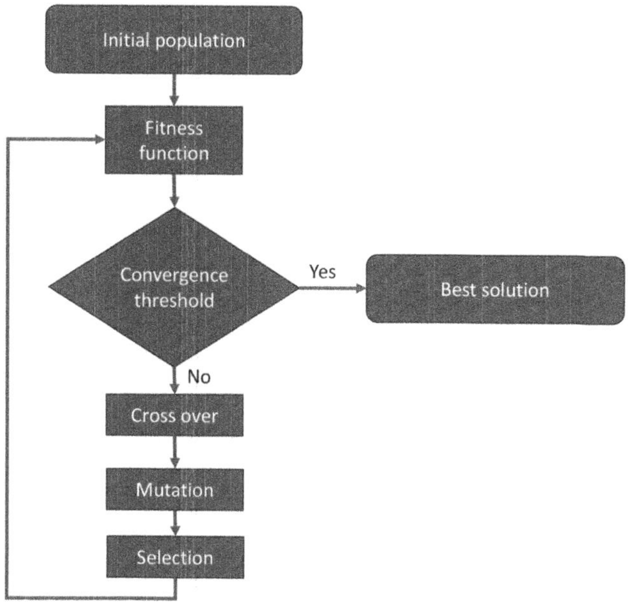

Figure 3.39 General GA flowchart.

called a candidate, and its code is called a chromosome. A chromosome consists of a sequence of genes. If binary encoding is chosen, a chromosome will be a binary number consisting of a sequence of genes where each gene is a binary digit "0" or "1."

- **Fitness Function:** the fitness function $F(x)$ assigns a score between 0 and 100% that indicates how healthy the solution is. A healthy solution is a solution that significantly minimizes the cost function $J(x)$.
- **GA Parameters**: the common parameters of GA include the population size, the crossover rate, the mutation rate, and the maximum number of iterations. Population is a cloud of possible solutions that evolves in each iteration. Evolution happens by crossover and mutation operators. A crossover operator generates new solutions by mixing two existing solutions. The crossover rate is the percentage of the population selected for crossover. The crossover candidates are selected randomly from the population. A mutation operator randomly selects several candidates and randomly mutates or changes some genes in its chromosomes (code). The mutation rate is the percentage of the population selected from mutation. In each iteration, crossover and mutation are applied and new solutions are added to the population. A selection process is applied to bring the population size back to its original size. This selection process selects candidates randomly such that the probability of selecting a candidate to survive to the next iteration is proportional to its fitness. The initial population is generated randomly under the constraint that the generated candidate is a valid solution to the problem under consideration.
- **Convergence Strategy**: The evolution process is stopped if the best solution is obtained or the maximum number of iterations has been reached. The best solution is determined by a threshold such that the problem is considered satisfactorily solved if the fitness of the solution is larger than a threshold. Other convergence strategies can be designed. For example, a convergence policy may state that the problem is considered satisfactory solved if a certain number of candidates achieve a minimum fitness score.

3.5.9 Fuzzy Systems

Most, if not all, sensors have imperfections that cause random errors. Dealing with randomness requires flexibility in the interpretation of the data. Probability theory is a strong tool to deal with randomness. However, sometimes it is complex and difficult to calculate due to numerical issues. Fuzzy theory is an alternative method to deal with randomness, which provides more abstraction to the data. Abstracting sensor data leads to flexible and robust processing that is less sensitive to noise. Instead of dealing with crisp numerical data, Fuzzy systems represent information in a linguistic, logical domain that is less sensitive to noise and more human-understandable. For example, when defining a pressure sensor output into abstract symbols such as "very low", "low," "level," "high," and "very high" and assign each symbol to a range of numerical values, the interpretation of the sensor data will be less prone to noise as small random noise will unlikely lead to misassignment of the sensor reading to the correct symbol. More flexibility is provided by making the abstraction overlap such that some values could be assigned multiple abstract symbols with different probabilities. When modeling the sensor reading as a fuzzy value, we move from the numerical domain to a symbolic domain where randomness can be better represented and processed. The set of axioms, definitions, and algorithms that deal with this fuzzy representation of data is known as fuzzy theory and systems that model the data as fuzzy sets are called Fuzzy Systems. The following are the key concepts in Fuzzy Systems theory:

- **Fuzzy Set:** a fuzzy set $\mathcal{F}$ is an uncertain set of elements x where each element has a degree of belonging to $\mathcal{F}$ in the range [0, 1] that indicates how certain the element x belongs to $\mathcal{F}$. A membership function $\mu_{\mathcal{F}}(x) \in [0,1]$ assigns an element x a membership value. For example, the fuzzy set of small real numbers $S_{\mathbb{R}^1}$ may have a membership function as shown in Figure 3.40. In this way, any randomness in the value of x can be represented as a membership value in the range [0, 1]. We can have more than one fuzzy set and an element can belong to more than one set with different degrees of membership. For example, the random sensor measurement number x_1 in Figure 3.41 belongs to small real numbers $S_{\mathbb{R}^1}$, large positive numbers $P_{\mathbb{R}^1}$, and large negative numbers $N_{\mathbb{R}^1}$ fuzzy sets. It is also clear that $\mu_{S_{\mathbb{R}^1}}(x_1) > \mu_{P_{\mathbb{R}^1}}(x_1) > \mu_{N_{\mathbb{R}^1}}(x_1)$ which expresses that the measurement x_1 is highly likely a small number, there is s small possibility that it is large positive number, but it is highly unlikely a negative number. This fuzzy interpretation of the measurement x_1 transform the random value of x_1 to a symbolic logical domain (small, large, positive, negative, etc.). Experiments have shown that inference in this symbolic logical domain is more robust against random noise and it also enables AI systems such as expert systems to process noisy measurements and model complex systems.

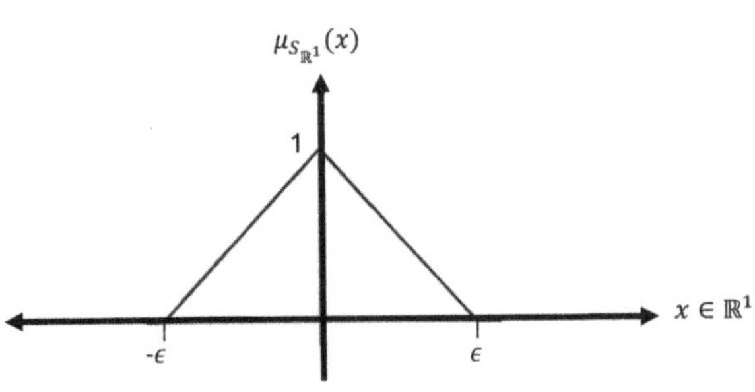

Figure 3.40 Membership function for fuzzy set of small real numbers.

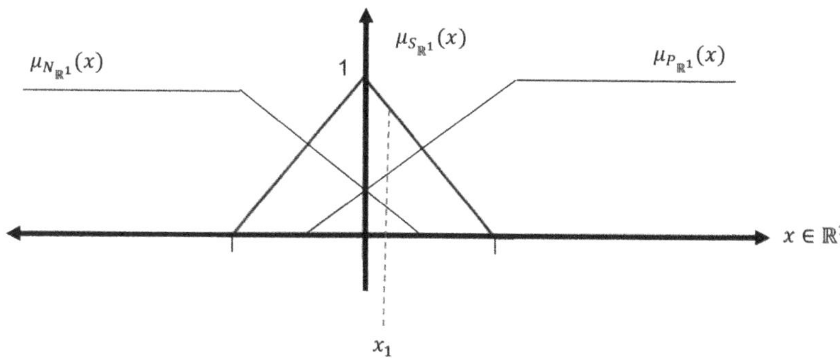

Figure 3.41 Overlapped fuzzy membership function. x_1 belongs to three fuzzy sets: small numbers, positive large numbers, and negative large numbers.

- **Fuzzy Logic**: reasoning in fuzzy measurements is performed using fuzzy logic. Classical logic is a set of rules and operators that work on Boolean variables that take the values {0, 1}. Fuzzy logic is like classical logic, but it is applied to fuzzy variables that take membership values in the range [0, 1] instead of Boolean values {0, 1}. Formally, fuzzy logic can be defined as follows: Given a fuzzy set $\mathcal{F}$ and a membership function $\mu_{\mathcal{F}}(x) \in [0, 1]$, for any $a, b \in \mathcal{F}$ we have:

$$\mu_{\mathcal{F}}(a \wedge b) = \mu_{\mathcal{F}}(a) \wedge \mu_{\mathcal{F}}(b) = \min\{\mu_{\mathcal{F}}(a), \mu_{\mathcal{F}}(b)\} \tag{3-271}$$

$$\mu_{\mathcal{F}}(a \vee b) = \mu_{\mathcal{F}}(a) \vee \mu_{\mathcal{F}}(b) = \max\{\mu_{\mathcal{F}}(a), \mu_{\mathcal{F}}(b)\} \tag{3-272}$$

$$\mu_{\mathcal{F}}(\bar{a}) = 1 - \mu_{\mathcal{F}}(a) \tag{3-273}$$

$$\mu_{\mathcal{F}}(a \Rightarrow b) = \mu_{\mathcal{F}}(a) \Rightarrow \mu_{\mathcal{F}}(b) = \min\{1, 1 + \mu_{\mathcal{F}}(b) - \mu_{\mathcal{F}}(a)\} \tag{3-274}$$

$$\mu_{\mathcal{F}}(a \Longleftrightarrow \mathrm{b}) = \mu_{\mathcal{F}}(a) \Longleftrightarrow \mu_{\mathcal{F}}(\mathrm{b}) = 1 - |\mu_{\mathcal{F}}(\mathrm{a}) - \mu_{\mathcal{F}}(\mathrm{b})| \tag{3-275}$$

  where $\wedge$ is the AND operator, $\vee$ is the OR operator, $\bar{\cdot}$ is the NOT operator, $\Rightarrow$ is the IMPLY operator, and $\Longleftrightarrow$ is the EQUIVELENT operator. A fuzzy relation is: a joint membership $\mu_{A\times B}(a, b)$ that defines a relationship between two fuzzy sets A, B. For example, shows the fuzzy relation "a is slightly larger than b" [19].

- **Fuzzy Rules:** A fuzzy rule is a statement with fuzzy inputs and fuzzy output defined using the IF-THEN style. The general fuzzy IF-THEN rule takes the following form:

  IF a_1 *is* A_1 AND....AND a_n *is* A_n THEN b *is* B

  where $A_1, ...A_n, B$ are **fuzzy** sets with membership functions $\mu_{A_i}(a_i)$ and $\mu_B(b)$. The same rule can be written as follows:

$$\mu_{A_1}(a_1) \wedge \mu_{A_2}(a_2)...\mu_{A_n}(a_n) \Rightarrow \mu_B(b) \tag{3-276}$$

- **Fuzzy Modeling:** Classical models describe systems using mathematical equations. Fuzzy modeling describes a system using fuzzy logic and fuzzy rules. A system can be described by a set of fuzzy rules called fuzzy rule base. Fuzzy modeling can be used to overcome some of the limitations in conventional mathematical modeling such as dealing with nonlinearity and non-Gaussian noise. Fuzzy modeling is a useful modeling tool if the system is too difficult to model using conventional methods such as differential equations or transfer functions. For example, the

Figure 3.42 Unknown system.

unknown system in Figure 3.42 can be modeled using a fuzzy rule base in the following general form:

$$\begin{aligned}&\text{IF } x_1 \textit{ is } X_1 \text{ AND....AND } x_n \textit{ is } X_n\\&\text{THEN } y_1 \textit{ is } Y_1 \text{ AND....AND } y_m \textit{ is } Y_m\end{aligned} \tag{3-277}$$

where X_i and Y_i are input and output values, respectively. Note that the statement "x_i *is* X_i" is interpreted as follows: x_i belongs to the fuzzy subset X_i with a degree of membership equals to $\mu_{X_i}(x_i)$.

- **Fuzzification and Defuzzification:** For example, consider the problem of a n-degree polynomial curve-fitting system, as experts, we have the following knowledge about the input/output relationship:

$$\begin{aligned}&\text{IF } x_1 \textit{ is } X_1 \text{ AND....AND } x_n \textit{ is } X_n\\&\text{THEN } y = a_0 + a_1x_1 + ...a_nx_n\end{aligned} \tag{3-278}$$

where a_0, $a_1...a_n$ are constants. Given numerical input x_1, x_2, ...x_n, we convert these numerical inputs into fuzzy membership values $\mu_{X_1}(x_1), \mu_{X_2}(x_2)...\mu_{X_n}(x_n)$. This process is called fuzzification. We then apply the fuzzy rule, which states that $y = a_0 + a_1x_1 + ...a_nx_n$ noting that the output y is a fuzzy variable with the following membership:

$$\mu_Y(y) = \sup_{y = a_0 + a_1x_1 + ...a_nx_n} \{\mu_{X_1}(x_1) \wedge \mu_{X_2}(x_2)... \wedge \mu_{X_n}(x_n)\} \tag{3-279}$$

If we have more than one rule that generate different values for the output (the output has more than fuzzy set $\mu_{Y_i}(y_i)$), we apply all rules and calculate the final numerical value of y using the center-of-gravity formula given by:

$$y = \frac{\sum_{i=1}^{m} \mu_{Y_i}(y_i)y_i}{\sum_{i=1}^{m} \mu_{Y_i}(y_i)} \tag{3-280}$$

This process is known as defuzzification. Fuzzy modeling of any unknown system includes the definition of the fuzzy sets of the inputs and outputs, the valid range of input/output values, definitions of membership functions and their parameters, and the fuzzy rule base. All these definitions depend on the characteristics of the system. Fuzzy modeling can be either handcrafted by experts in the field or they can be learned by ML methods such as ANN or GA if enough input/output datasets are available. The overall structure of the fuzzy system is shown in Figure 3.43.

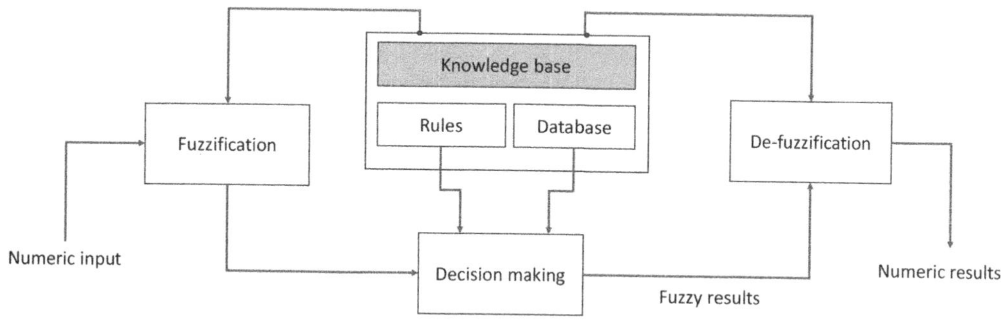

Figure 3.43 Basic structure of fuzzy system.

3.5.10 Hybrid ML/AI Systems

Hybrid AI and ML systems combine the rule-based design and the data-driven learning approach to build intelligent systems. ML can be combined with rule-based AI in a hybrid system that integrates the power of autonomous learning abilities of ANN and the flexibility and human-interpretability of rule-based AI. One of the limitations of rule-based AI is the rules building. Building databases of rules developed by human experts is very time consuming and expensive. A solution to this problem is to learn the rules from the data using ML. These types of systems can be described as hybrid ML/AI systems.

A great example of a hybrid ML/AI system is the Adaptive Neuro-Fuzzy-Inference System (ANFIS). ANFIS has been applied to several complex estimation and control problems [20] and [21]. ANFIS models can be described as ANNs that work on fuzzy inputs, not crisp numeric value numeric inputs. The ANN, in this case, represents the rules in fuzzy systems. In this design, the rules are learned using backpropagation least-squares algorithm. The ANFIS structure has five layers. The general structure of ANFIS is shown in Figure 3.44. The ANFIS structure has five layers.

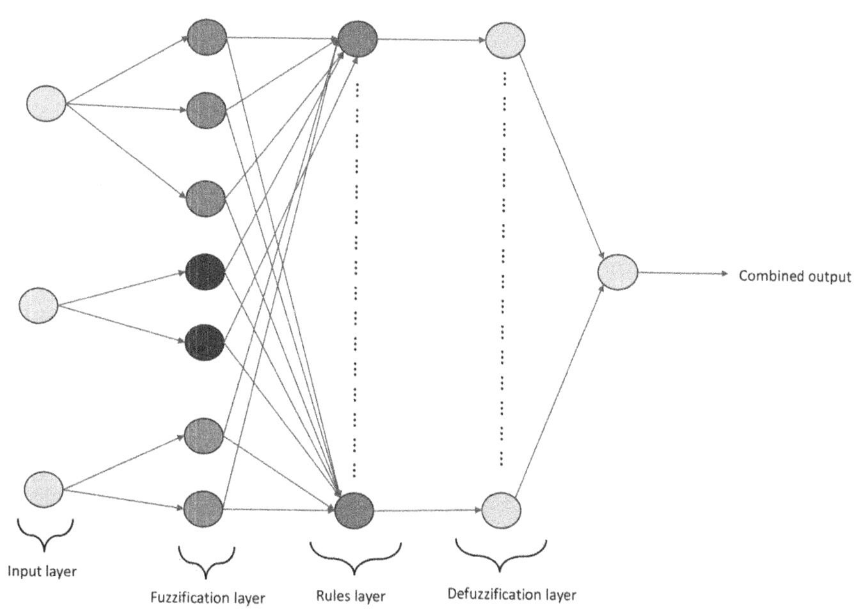

Figure 3.44 Fuzzy model mapped into a neural network.

The first layer of ANFIS is a fuzzification layer that converts numerical input x to a membership value $\mu_A(x)$ where A is an abstract logical linguistic label or symbolic variable. The second layer multiplies the output of the first layer to get a weight for each rule. The third layer normalizes the second layer's weights. The fourth layer multiplies the normalized weights by the output of the second layer. The fifth layer sums the output of the fourth layer to calculate the overall final output.

3.6 Multisensor Temporal and Spatial Extrinsic Calibration

In any multi-sensor system, transformation of measurements from one sensor's frame to a global frame or to another sensor's frame is necessary. This process is known as spatial extrinsic calibration. Accurate sensor fusion also requires precise time synchronization of the measurements of sensors. This time synchronization is called temporal calibration. Temporal and spatial extrinsic calibration can be performed offline in the laboratory, or it can also be performed online in the field. Temporal and spatial extrinsic calibration between two sensors s_r and s_t can be seen as an estimation problem where the unknown parameters are the relative rotation $R_{S_r}^{S_t}$ and translation $T_{S_r}^{S_t}$ and the time-shifts τ_{S_r,S_t} between the measurements of sensors s_r and s_t. The general idea of temporal and spatial extrinsic calibration is to take of the sensors as a reference sensor (s_r) and the other as a target (s_t). Then, we measure the same quantity $\vec{y}$ (e.g., object's pose or features' location) using s_r and s_t. Assume the quantity measured by s_r is $\vec{y}_r$ and the same quantity measured by s_t is $\vec{y}_t$. We then calculate $T\left(\vec{y}_r\right)$, the transform $\vec{y}_r$ to the frame of $\vec{y}_r$ to the target sensor's frame. The final step is to find $R_{S_r}^{S_t}$, $T_{S_r}^{S_t}$, and τ_{S_r,S_t} that minimize the difference between the transformed $T\left(\vec{y}_r\right)$ and $\vec{y}_t$. In case of temporal calibration, the transform $T\left(\vec{y}_r\right)$ is a function of the time-shifts τ_{S_r,S_t}. In case of spatial extrinsic calibration, the transform $T\left(\vec{y}_r\right)$ is a function of $R_{S_r}^{S_t}$, $T_{S_r}^{S_t}$. The general form of the cost function of the temporal and spatial extrinsic calibration is given as follows:

$$J\left(R_{S_r}^{S_t}, T_{S_r}^{S_t}, \tau_{S_r,S_t}\right) = \frac{1}{2}\left\|\vec{y}_t - T\left(\vec{y}_r\right)\right\|^2 \tag{3-281}$$

Spatial transformation can be formulated as follows:

$$T\left(\vec{y}_r\right) = R_{S_r}^{S_t}\vec{y}_r + T_{S_r}^{S_t} \tag{3-282}$$

Temporal transformation can be formulated as follows:

$$T\left(\vec{y}_r\right) = \vec{y}_r(t - \tau_{S_r,S_t}) \tag{3-283}$$

Combined temporal and spatial transformation can be formulated as follows:

$$T\left(\vec{y}_r\right) = R_{S_r}^{S_t}\vec{y}_r(t - \tau_{S_r,S_t}) + T_{S_r}^{S_t} \tag{3-284}$$

The minimization of the cost function in Eq. (3-281) can be performed using least-square algorithm or any optimization method such as Gauss–Newton or Levenberg–Marquardt methods. The specific measurements, the mathematical formulation of the transformation, and the cost function definition varies depending on the type of sensor, the system model, and the measurements model of each sensor. An important caution that must be considered in formulating the cost function is parametrization of the rotation. The parametrization of the rotation using rotation matrix is not ideal due to the redundancy, nonlinearity, and the singularity issues of rotation matrices. In addition, rotation matrices do not form a Euclidean space because addition of rotation matrices is not defined. Rotation vectors and the vector part of quaternion are preferred parameters. Both form a minimal coordinate that is suitable for estimation. The concept of minimal coordinate is further illustrated in Section 10.2.1.4. In Chapter 10, we will discuss Camera-IMU calibration as an example of multisensor spatial calibration.

3.7 Problems

3.7.1 Problem 1

Given the following linear dynamic system:

$$\vec{x}_{k+1} = \emptyset_k \vec{x}_k + G_k \vec{w}_k$$

$$\vec{y}_{k+1} = H_k \vec{x}_k + \vec{v}_k$$

Given measurements at time epochs $k = 1, 2...M$, derive the observability matrix Γ_M that relates the state vector at epoch 1 ($\vec{x}_1$) to the measurement vector $\vec{y} = (y_1 \quad y_2 \quad \dots \quad y_M)^T$ in the following form:

$$\vec{y}_{1:M} = \Gamma_{1:M} \vec{x}_1$$

Determine what is the condition of the measurements to ensure the estimation of the state vector $\vec{x}_1$ at time epoch $k = 1$ (after how many measurements an estimator like Kalman filter will be able to estimate the state vector $\vec{x}_1$?)

3.7.2 Problem 2

Show that the location of the person in Figure 3.45 can be estimated using nonlinear least square solution. Show the first two iterations of the iterative least square solution if $R1 = 3m$, $R2 = 4m$, and $R3 = 5m$ and the location of the three flags are (3, 3), (3, – 2), and (–2, 1) respectively.

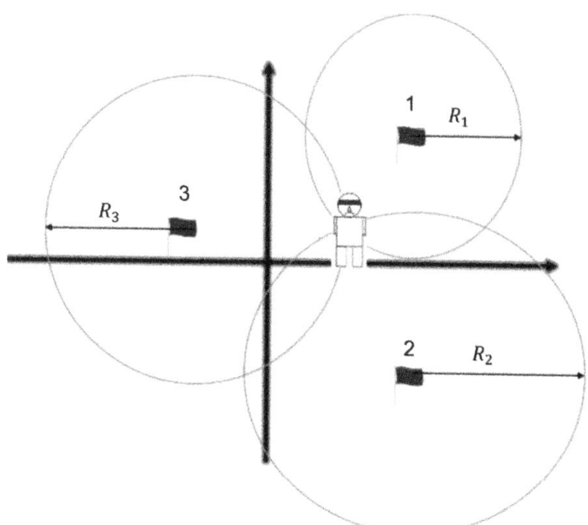

Figure 3.45 Localization using three landmarks.

3.7.3 Problem 3

The 1D straight-line train motion example has the following solution equation:

$$y_k = y_0 + v_0 t_k + \frac{1}{2} a t_k^2$$

where y_0, v_0, and a are train's initial position, initial velocity, and constant acceleration, respectively. Given the following measurements, estimate the parameter vector $\vec{x}$ defined as follows:

$$\vec{x} = \begin{pmatrix} y_0 \\ v_0 \\ a \end{pmatrix}$$

The measurements are given as follows:

$$\vec{y} = (6.7283 \quad 36.7985 \quad 66.9046 \quad 81.2343 \quad 82.2648 \quad 77.9797)^T$$

$$\vec{t} = (0 \quad 0.9000 \quad 1.9000 \quad 2.9000 \quad 3.9000 \quad 4.9000)^T$$

3.7.4 Problem 4

Repeat the solution of Problem 3 if the measurements noise covariance of $\vec{y}$ is:

$$R = \begin{pmatrix} 0.1 & 0 & 0 & 0 & 0 \\ 0 & 0.5 & 0 & 0 & 0 \\ 0 & 0 & 0.2 & 0 & 0 \\ 0 & 0 & 0 & 0.1 & 0 \\ 0 & 0 & 0 & 0 & 0.3 \end{pmatrix}$$

References

1 J. Farrel, "*Aided Navigation: GPS with High Rate Sensors*," 1st edition, McGraw Hill, 2008.
2 K. P. Murphy, "*Machine Learning: A Probabilistic Perspective*," London, England: MIT Press, Cambridge, Massachusetts, 2012.
3 C. M. Bishop, "*Pattern Recognition and Machine Learning*," Springer Science+Business Media, LLC, 2006.
4 J. J. van Aardt and Y. Pailhas, "Integer Least Squares Techniques in Remote Sensing Applications: Review and Practical Considerations," *ISPRS Journal of Photogrammetry and Remote Sensing*, vol. 58, no. 3–4, pp. 337–354, 2004. https://doi.org/10.1016/j.isprsjprs.2004.07.001
5 A. N. Bishop, "Mixed-Integer Least Squares for Simultaneous Estimation of Integer and Real Parameters," *Journal of the Operational Research Society*, vol. 66, no. 1, pp. 51–60, 2015. https://doi.org/10.1057/jors.2014.31.
6 B. Li, S. Verhagen and P. Teunissen, "GNSS Integer Ambiguity Estimation and Evaluation: LAMBDA and Ps-LAMBDA," in *China Satellite Navigation Conference (CSNC) 2013 Proceedings, Springer, Lecture Notes in Electrical Engineering*, vol. 244, p. 291, 2013.

7 N. N. Schraudolph, J. Yu and S. Günter, "A Stochastic Quasi-Newton Method for Online Convex Optimization Online Convex Optimization," in *Proceedings of the Eleventh International Conference on Artificial Intelligence and Statistics*, 2007.
8 X. Gao and T. Zhang, "*Introduction to Visual SLAM*," Springer, 2021.
9 J. V. Candy, "*Bayesian Signal Processing: Classical, Modern, and Particle Filtering Methods*," John Wiley & Sons, Incorporated, 2016.
10 W. Li, H. Leung and Y. Zhou, "Space-Time Registration of Radar and ESM Using Unscented Kalman Filter," *IEEE Transactions on Aerospace and Electronic Systems*, vol. 40, no. 3, pp. 824–836, 2004.
11 S. Thrun and M. Montemerlo, "The Graph SLAM Algorithm with Applications to Large-Scale Mapping of Urban Structures," *The International Journal of Robotics Research*, vol. 25, no. 5-6, pp. 403–s429, 2006.
12 G. Grisetti, R. Kummerle, C. Stachniss and W. Burgard, "A Tutorial on Graph-Based SLAM," *IEEE Intelligent Transportation Systems Magazine*, vol. 2, no. 4, pp. 31–43, 2010.
13 C. C. Aggarwal, "*Neural Networks and Deep Learning*," Springer International Publishing AG, part of Springer Nature, 2018.
14 S. Hochreiter and J. Schmidhuber, "Long Short-term Memory (LSTM)," *Neural Computation*, vol. 9, no. 8, pp. 1735–1780, 1997.
15 M. Jena and S. Dehuri, "Decision Tree for Classification and Regression: A State-of-the Art Review," *Informatica*, vol. 44, no. 4, pp. 405–420, 2020.
16 Y. Toleubay and A. P. James, "Getting Started with TensorFlow Deep Learning," in *Deep Learning Classifiers with Memristive Networks*, Springer, pp. 57–67, 2019.
17 A. Burkov, "*The Hundred-Page Machine Learning Book*," Andriy Burkov, 2019.
18 S. Malakar, S. Goswami, B. Ganguli, A. Chakrabarti, S. S. Roy, K. Boopathi and A. G. Rangaraj, "Designing a Long Short-Term Network for Short-Term Forecasting of Global Horizontal Irradiance," *SN Applied Sciences*, vol. 3, no. 477, 2021. https://doi.org/10.1007/s42452-021-04421-x.
19 G. Chen and T. T. Pham, "*Fuzzy Sets, Fuzzy Logic, and Fuzzy Control Systems*," Washington, DC: CRC Press, 2001.
20 A. E. Mahdi, A. Azouz, A. E. Abdalla and A. Abosekeen, "A Machine Learning Approach for an Improved Inertial Navigation System Solution," *Sensors (MDPI)*, vol. 22, p. 1687, 2022.
21 S. Sremac, D. R. Stojčić and I. Tanackov, "ANFIS Model for Determining the Economic Order Quantity," *Decision Making: Applications in Management and Engineering*, vol. 1, no. 2, p. 81, 2018.

4

Inertial Sensors and Inertial Navigation Systems

4.1 Introduction

Inertial sensors measure the force applied on an object by measuring the acceleration using Newton's second law of motion, which states that force applied to an object equals the acceleration times the object's mass. Depending on the configuration of the mass and its suspension system, inertial sensors can measure rotational motion and translational motion. Inertial sensors are one of the earliest and most used sensors in navigation systems. Inertial sensors are self-contained sensors that can be attached to any object and measure the rate of change of the object's motion in 3D. The most advantageous characteristic of inertial sensors is that they can work independently without the need for external infrastructure or environmental features. In this chapter, we will study the key concepts related to inertial sensors and inertial navigation systems.

4.2 Inertial Measurements

The concept of inertial sensors can be illustrated by the proof-mass-spring system as shown in Figure 4.1. Figure 4.1a shows a simple single-axis accelerometer that senses linear acceleration $a(t)$ along the x-axis. Figure 4.1b shows a simple single-axis gyroscope that senses angular velocity $\omega(t)$ around the z-axis. The axis that we measure the motion on is called the sensitive axis. This proof-mass-spring system is used in MEMS (Micro-Electro-Mechanical-Systems) sensors technology, which has gained significant popularity in recent years due to their low-cost/low-power features, which enabled them to find their way in consumer electronics such as smartphones, tablets, smart glasses, virtual/augmented reality, and gaming sets. It is important to understand that the diagram in Figure 4.1 is for illustrative purposes to explain the physical concept of inertial sensors. Practical design, implementation, and manufacturing are much more complicated and can be found in other references like [1] and [2]. In typical inertial sensors, there are three sensors in three orthogonal axes as seen in Figure 4.2.

4.2.1 Accelerometers

For the accelerometer, at rest equilibrium (see Figure 4.1), the position of the mass m will be at zero position when the forces on the two springs are equal. If there is an acceleration along the sensitive axis, the forces on the two springs will not be equal and therefore, the mass will make a

Sensor Fusion Approaches for Positioning, Navigation, and Mapping: How Autonomous Vehicles and Robots Navigate in the Real World: With MATLAB Examples, First Edition. Mohamed M. Atia.
© 2025 The Institute of Electrical and Electronics Engineers, Inc. Published 2025 by John Wiley & Sons, Inc.

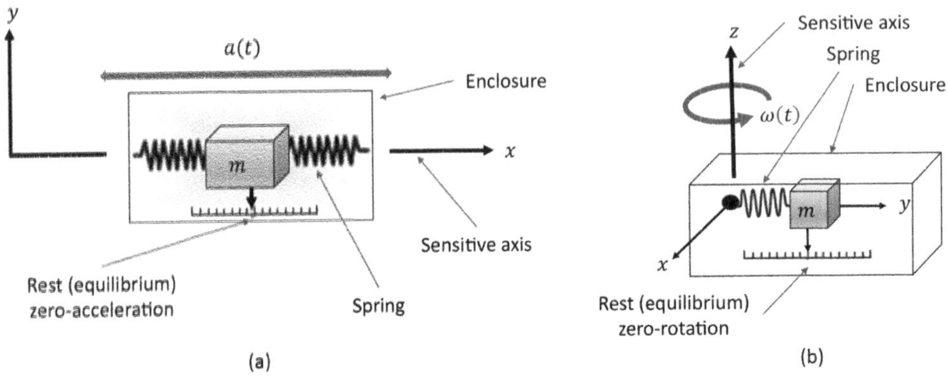

Figure 4.1 Simple illustration of MEMS inertial sensor (a) accelerometer and (b) gyroscope.

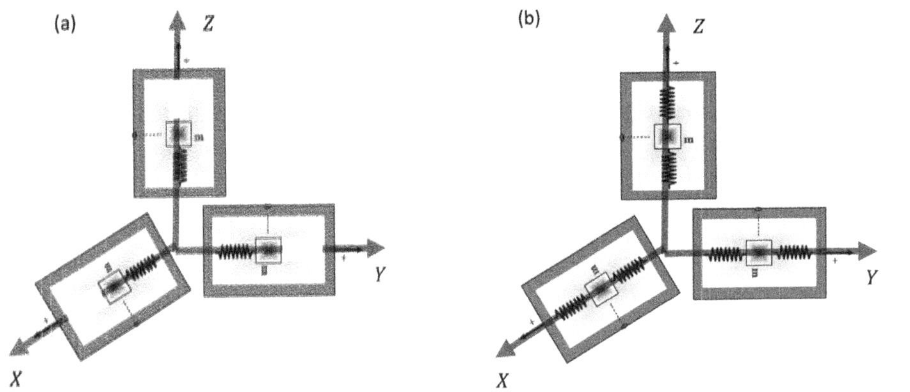

Figure 4.2 Simple illustration three-axis proof-mass-spring accelerometer (a) and gyroscope (b).

displacement proportional to the acceleration. This simple accelerometer is overly simplified to explain the main idea. Practical accelerometers have different designs and configurations, but they use the same concept illustrated in Figure 4.1a. More details about the practical techniques of accelerometer design can be found in [1] and [2]. Typically, accelerometers will have three perpendicular sensitive axes like the one shown in Figure 4.2. In this way, the accelerometer will measure the linear acceleration in 3D. The three perpendicular sensitive axes will form a coordinate frame that is usually referred to as a "body frame." We will see later in this chapter how 3D acceleration measurements in the body frame are used to calculate the object's velocity and position with respect to a reference frame. Considering the simple 1D example in Figure 4.1, the linear position of an object is a displacement $x(t)$ along a sensitive axis (x-axis in Figure 4.1a) starting from a reference "0" point. Let $a(t)$ be the linear acceleration of the object along x-axis. Linear position $x(t)$ and linear velocity $v(t)$ are defined as follows:

$$v(t) = \int_0^t a(t)dt \tag{4-1}$$

$$x(t) = \int_0^t v(t)dt = \int_0^t \int_0^t a(t)dt \tag{4-2}$$

4.2.2 Gyroscopes

Gyroscopes measure angular motion or the rate of rotation. The MEMS gyroscope uses the mass-spring concept illustrated in Figure 4.1 to measure angular rotation rate using the Coriolis effect. Coriolis effect is a force that is generated on an object that is rotating within a frame of reference. The displacement of the mass caused by the Coriolis force is known as the Coriolis effect, and it is proportional to the rotation rate. The illustration in Figure 4.1 is overly simplified. Practical gyroscopes have sophisticated designs. In addition, gyroscopes can be designed using technologies different from the proof-mass system, such as optical technology. An optical gyroscope is another common technique to sense rotation. Optical gyroscopes use the "Sagnac effect" to sense rotation. The Sagnac effect is a change in the speed of light waves when passing through rotating media. If two light signals are passed through two parallel waveguides, such as fiber-optics channels, and there is no rotation, the two light signals coming out from the two channels will arrive at the same time. If there is a rotation around an axis perpendicular to the plane that contains the two light channels, the two light signals coming out from the two channels will arrive at different times that can be measured as phase difference as seen in Figure 4.3. The difference in time is proportional to the rotation rate. Fiber-optics gyros (FOGs) are very accurate, but they are also extremely expensive as working with light waves requires extremely accurate electronic circuits compared to MEMS sensors. More details about the practical techniques and technologies of gyroscope sensor design and manufacturing can be found in [1] and [2]. Like accelerometers, typical gyroscopes will have three sensors mounted on the body-frame axes as seen in Figure 4.2. In this way, the gyroscope will measure the rotation rate in 3-D. We will see later in this chapter how the 3D rotation rate in the body frame is used to calculate the object's orientation with respect to a reference frame.

In the simple 1D example of Figure 4.1b, the angle (i.e., orientation) of the object can be defined as an angle of rotation around the sensitive axis (z-axis), starting from a reference angle. Let $\omega(t)$ be the rotation of the object where $\omega(t)$ is positive if the rotation is counterclockwise around the

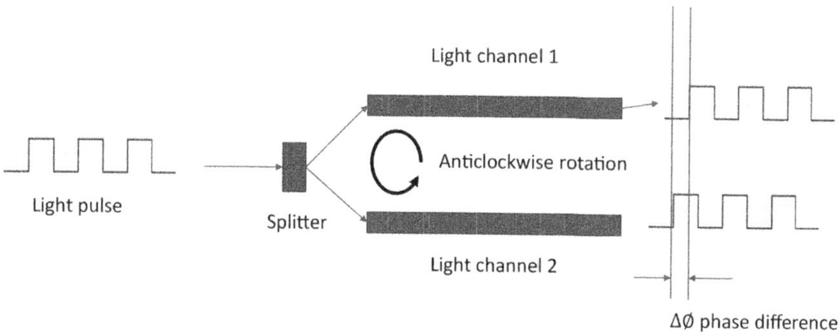

Figure 4.3 Simple illustration of the Sagnac effect.

sensitive axis and negative otherwise. The angle $\theta(t)$ for a single-axis rotation can be calculated as follows:

$$\theta(t) = \int_0^t \omega(t)dt \tag{4-3}$$

4.2.3 Magnetometer

Another common sensor that is usually added to accelerometers and gyroscopes in one device is the magnetometer [1, 2]. The magnetometer measures the magnitude of the magnetic field of Earth at the current object location. The Earth is a huge magnet that has a magnetic field along the meridian lines, as seen in Figure 4.4. The magnetometer sensor senses the Earth's magnetic field vector $\vec{M}$ as projected along three perpendicular axes of the body frame. $\vec{M}$ is always pointing toward the magnetic north pole with an inclination angle depending on the object's location within the Earth's gravity field. At the Earth's magnetic poles, the magnetic field $\vec{M}$ is completely vertical to the ground (has an inclination angle of 90°), whereas the Earth's magnetic field is completely horizontal at the equator (has an inclination angle of 0°). The magnetic field vector $\vec{M}$ always points to the magnetic north pole, which is different from the geographic north. The difference between the geographic north and magnetic north is called the declination angle. Earth's magnetic field magnitude is calculated using physics models that can be found in many references, such as the National Centers for Environmental Information [3]. The magnetometer sensor will measure the magnetic field $\vec{M}$ projected on the sensor's three axes as follows:

$$\vec{m} = \begin{pmatrix} m_x \\ m_y \\ m_z \end{pmatrix} \tag{4-4}$$

$$\sqrt{m_x^2 + m_y^2 + m_z^2} = \left\|\vec{M}\right\| \tag{4-5}$$

A perfect magnetometer will measure a constant magnitude $\left\|\vec{M}\right\|$, no matter what the orientation of the sensor is. Assuming a perfect magnetometer, and given the magnetic measurements

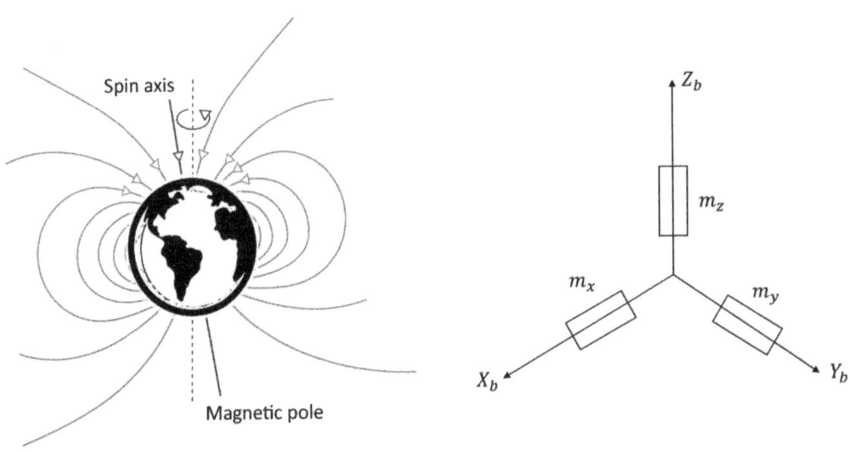

Figure 4.4 Earth magnetic field.

$\vec{m} = (m_x\, m_y\, m_z)^T$, the heading angle of the magnetic sensor A_m is calculated by the following equation:

$$m_E = (-\, m_y \cos(r) + m_z \sin(r)) \tag{4-6}$$

$$m_N = (m_x \cos(p) m_y \sin(r) \sin(p) + m_z \cos(r) \sin(p)) \tag{4-7}$$

$$A_m = \tan^{-1}\left(\frac{m_E}{m_N}\right) + D(\varphi, \lambda) \tag{4-8}$$

where $D(\varphi,\lambda)$ is the declination angle which can be calculated from any reference table in sources such as [3] or Geomagnetism Canada [4]. The magnitude of the magnetic sensor measurements $\|\vec{m}\|$ should be the same regardless of the orientation of the sensor. However, in most scenarios, magnetic field measurements are distorted due to local magnetic disturbances, especially from electrical or magnetic equipment. Therefore, a calibration process is always needed to make sure that $\vec{m} = (m_x\, m_y\, m_z)^T$ always has the same magnitude $\|\vec{M}\|$, no matter the orientation of the sensor. To make sure that $\vec{m} = (m_x\, m_y\, m_z)^T$ always has the same magnitude $\|\vec{M}\|$, a calibration process is performed where the sensor is rotated in different directions and magnetic measurements are measured at different orientations. As the measurements are distorted, the magnitude $\|\vec{m}\|$ will not be the same at different orientations and the results will not form a perfect sphere. Instead, it will form a distorted shifted ellipse. A calibration model is then calculated to transform this distorted shifted ellipse into a perfect sphere. The details of this calibration process can be found in [1]. An example figure that shows the uncalibrated and calibrated magnetic sensor can be seen in Figure 4.5. This figure has been generated from data recorded from an MPU 9250 IMU and calibrated using the open-source code available in [5].

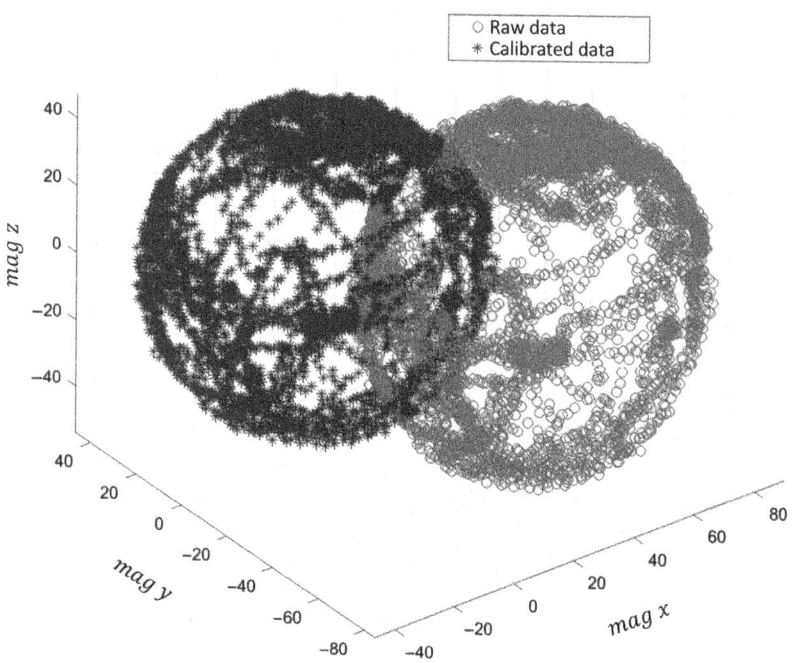

Figure 4.5 Magnetic sensors measurements in different directions. noncalibrated vs. calibrated measurements.

4.2.4 Units of Measurements

In this book, we will use meters (m) as position units, and radians (r) and/or degrees (°) as angle units. Linear velocity is measured in meters per second (m/s). Angular velocity is the rate of angles (rotation rate), which is measured either in degrees per second (°/s) or radians per second (r/s) or (rad/s). Linear acceleration is the rate of linear velocity, so it is measured as distance per second-squared such as m/s^2. Commonly, commercial accelerometer sensors provide measurements in units of gravity. Earth's gravity vector points toward the center of the Earth perpendicular to the tangent plane where the object is located. Standard gravity has an acceleration of $g = 9.80665$ m/s^2. However, precise gravity measurement depends on the latitude and height of objects, so most accelerometer sensors will provide measurements in Earth field gravity g. In Figure 4.1, only 1D motion is explained. However, the same definitions can be generalized into 2D and 3D. In the 3D case, the position, linear velocity, linear acceleration, and angular velocity will be vectors $(\vec{x}(t), \vec{v}(t), \vec{a}(t), \vec{\omega}(t))$ and angular position (orientation) will be a 3D orientation vector such as rotation vector, Euler angles, or quaternion vector, as explained in Chapter 1. The speed of an object is a scalar value defined as the magnitude of its velocity vector $v = \left\|\vec{v}\right\|$. Magnetic measurements are measured in gauss [1, 3, 6]. We will use the following mathematical notation to refer to the accelerometer, gyroscope, and magnetometer measurements:

- $\vec{a}^B_{SF} \triangleq$ acceleration vector as measured in the body-frame B. The "SF" stands for specific force which is the force that the accelerometer senses through its springs after being calibrated for the mass.
- $\vec{\omega}^B_{IB} \triangleq$ rate of the rotation of the body-frame B with respect to the inertial frame as observed in the body-frame B.
- $\vec{\omega}^B_{IB} \triangleq$ rate of the rotation of the body-frame B with respect to the inertial frame as observed in the body-frame B.
- $\vec{m}^B \triangleq$ magnetic field as projected in the body-frame B.

In the subsequent chapters, we will use the superscript to refer to the frame at which the measurement is referenced to or measured at. When we mention rotation, we use the subscript like "LB" to refer that the described rotation is the rotation of frame B around, or with respect to frame L. Reversing the order of the subscript letters like "BL" means the rotation of frame L with respect to frame B. It is worth to mention that $\vec{\omega}^B_{LB} = -\vec{\omega}^B_{BL}$.

4.3 Earth Gravity/Rotation/Curvature Effects on Inertial Measurements

Earth's shape and dynamics influence inertial sensor measurements. Although this influence is small and sometimes it can be neglected, especially with low-precision applications or low-cost MEMS sensors, it is important to understand these effects. Earth can be modeled as an ellipsoid shape, as discussed in Chapter 1. Linear motion on a curved surface (such as Earth's ellipsoid shape) will generate an angular rate that will be sensed by a gyroscope even if the object's motion is linear due to the curvature of Earth surface. This effect is known as Transport Rate. As Earth is rotating, a stationary gyroscope will sense a small rotation even if the sensor is stationary. This effect is called Earth Rate. Earth rotation also induces a centroid force called Coriolis force, which is sensed by the accelerometer measurements. As the object is moving near the Earth's surface on an ellipsoid curve, the transport rate will add additional Coriolis force, which is measured by the accelerometer. Even if the object does not move, the accelerometer will still sense an acceleration due to the Earth's gravity. Before we can use inertial sensors to calculate an object's position, velocity, and orientation, we must compensate gyroscope and accelerometer measurements from all effects that do not actually contribute to the object's motion with respect to the frame of reference of interest (e.g., local-level navigation frame). In the following sections, we will explain each effect and how we can remove it from inertial measurements.

4.3.1 Gravity Effect on Accelerometers

An accelerometer that is free-falling will measure zero in all its axes. For a free-falling accelerometer, it is obvious that it has a velocity in the direction of the center of the Earth, and its position is changing. However, if the accelerometer measurements are zero ($\vec{a}(t) = 0$), how come we have a nonzero velocity and position? This is because the accelerometer senses the specific force (specific acceleration $\vec{a}^B_{SF}$) applied on the mass and measured through the springs and not the total acceleration $\vec{a}(t)$. This specific force is nongravitational force. Before we integrate the accelerometer measurement, we must add the gravity acceleration to its raw measurements. Accelerometer measurements can be modeled as follows:

$$\vec{a}^B_{SF} = \vec{a}^B - \vec{g}^B \tag{4-9}$$

where $\vec{a}^B$ is the total acceleration measurement vector, $\vec{a}^B_{SF}$ is the specific force acceleration, and $\vec{g}^B$ is the gravity vector. The superscript "*B*" means all the three vectors $\vec{a}^B$, $\vec{a}^B_{SF}$, and $\vec{g}^B$ are defined in the body-frame "*B*." Therefore, if we placed an accelerometer on a perfect horizontal flat table such that its *z*-axis points exactly upwards, as shown in Figure 4.6, ideally, the total acceleration $\vec{a}^B$ is zero because the accelerometer does not move. Thus, the accelerometer measurements, according to Eq. (4-9) $\vec{a}^B_{SF}$ will be given by:

$$\vec{a}^B_{SF} = \vec{g}^B \tag{4-10}$$

As the gravity force is pointing downward, and the *z*-axis of the body frame is pointing up, the gravity vector in the body frame is given by:

$$\vec{g}^B = \begin{pmatrix} 0 \\ 0 \\ -g \end{pmatrix} \tag{4-11}$$

where g is the gravity magnitude at the sensor's location. Standard Earth gravity scalar value is $g = 9.80665\ \text{m/s}^2$ according to National Institute of Standards and Technology (NIST), but precise gravity value varies depending on the latitude and the height of the object and there are several complicated models to calculate gravity vector precisely at any given location such as the models in [1] and [7]. The specific force $\vec{a}^B_{SF}$ in the case shown in Figure 4.6 will be:

$$\vec{a}^B_{SF} = \begin{pmatrix} 0 \\ 0 \\ +g \end{pmatrix} \tag{4-12}$$

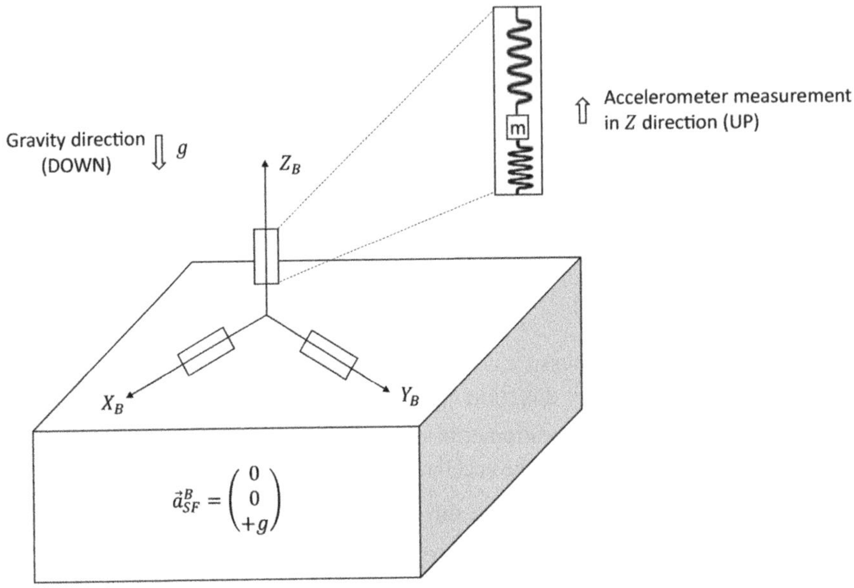

Figure 4.6 An accelerometer placed on a flat horizontal table.

In any other placement orientation, $\vec{a}_{SF}^{B}$ will have nonzero values on the three axes. The output measurement $\vec{a}_{SF}^{B}$ will continues to be the negative of the gravity in any stationary placement orientation such that the total gravity $\vec{a}^{B}$ will be zero, indicating that the object is not moving. Any deviation of $\vec{a}_{SF}^{B}$ from the negative of gravity acceleration means that the object is accelerating.

4.3.2 Earth Rotation Rate

A stationary gyroscope should measure zero angular rate. However, due to Earth's rotation, a stationary gyroscope will measure a small angular rate. In Chapter 1, we have discussed the common coordinate frames used to describe objects' locations and orientations with respect to the Earth. Figure 4.7 shows the common Earth coordinate frames. We will refer to the Earth-Centered-Inertial (ECI) frame by the letter "*I*," the Earth-Fixed-Earth-Centered (ECEF) by the letter "*E*," and to the local-level navigation frame by the letter "*L*." The Earth is rotating around its axis of rotation, which is the *z*-axis of the ECEF frame at a constant rate ω^e. Earth rotation is a rotation of *E* frame with respect to the *I* frame. As the rotation of E frame with respect to to I frame is completely around the *z*-axis of E so, in vector form, the Earth's angular rate is defined as follows:

$$\vec{\omega}_{IE}^{E} = (0 \ \ 0 \ \ \omega^e)^T \tag{4-13}$$

The superscript "*E*" in $\vec{\omega}_{EI}^{E}$ means the vector is observed in *E* frame. The subscript "*IE*" means the rotation is a rotation of *E* frame with respect to *I* frame. Given the sensor's location in Geodetic coordinates form (latitude φ, longitude λ, and altitude h), the rotation rate in the *L*-frame is given by:

$$\vec{\omega}_{IE}^{L} = (\,\omega^e \cos(\varphi) \quad 0 \quad -\omega^e \sin(\varphi)\,)^T \tag{4-14}$$

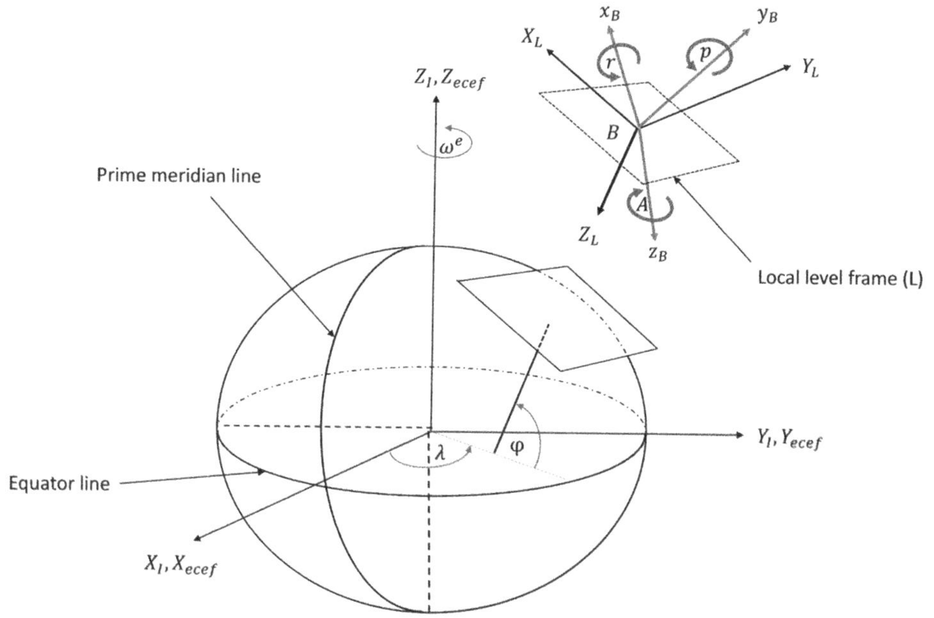

Figure 4.7 Approximate illustration of Earth frames of reference (body "B," local-level "L," Earth-Centered-Earth-Fixed "ECEF," and inertial "I" frames.

which is the projection of $\vec{\omega}_{IE}^{E}$ on the north-east-down components of local level frame (L). If we have an ideal stationary gyroscope attached to a body frame B, the gyroscope will measure the Earth rotation as projected on the body-frame given by:

$$\vec{\omega}_{IE}^{B} = C_L^B \vec{\omega}_{IE}^{L} \tag{4-15}$$

where C_L^B is the direction cosine matrix that transforms matrices from L frame to B frame. If the gyroscope sensor is perfectly aligned with the L frame (its x-axis is pointing to the north and its z-axis is pointing downward), it will measure exactly $\vec{\omega}_{IE}^{L}$. The ω^e is simply the Earth's rotation rate in radians per second, which is approximately 7.2921150×10^{-5} r/s. As the value of ω^e is very small, only extremely accurate and precise gyroscopes can sense this small angular rate.

4.3.3 Gyrocompassing

This phenomenon is used to build a compass that calculates the angle between an object and the north which is called gyrocompass. Given the sensor's latitude, we can calculate the angular rate that the sensor should measure in stationary state in the L frame, which is $\vec{\omega}_{IE}^{L} = (\omega^e \cos(\varphi) \quad 0 \quad -\omega^e \sin(\varphi))^T$. Given the actual gyroscope measurement in body frame $\vec{\omega}_{IE}^{B}$, we know that $\vec{\omega}_{IE}^{B} = C_L^B \vec{\omega}_{IE}^{L}$. The C_L^B matrix is a function of the Euler angles roll, pitch, and Azimuth. If we keep the sensor perfectly horizontal such that roll and pitch angles are zeros, and record the gyroscope measurement, the only unknown in the equation will be the Azimuth, which is the angle between the x-axis of the body frame and the north. If we direct the sensor to measure exactly $(\omega^e \cos(\varphi) \quad 0 \quad -\omega^e \sin(\varphi))^T$, this means that the Azimuth angle is zero and the two frames (B and L) are perfectly aligned ($C_L^B = I$) and the x-axis of the body frame is pointing toward the north. This mechanism is known as gyrocompassing, and it requires highly accurate and precise gyroscopes.

4.3.4 Transport Rate

Linear motion on or near the Earth's surface induces an angular rate due to the ellipsoid curvature of Earth. This linear motion will cause the object to rotate with respect to the E frame at a rate called transport rate. The gyroscope will measure this rotation rate even if the orientation of the body-frame B with respect to L frame does not change. Therefore, as the rotation induced from the transport rate will not contribute to a change in the object's orientation with respect to the L frame so it must be calculated and compensated from gyroscope measurements. Given the object's velocity as observed in L frame $(v_n \; v_e \; v_d)^T$, the transport rate is calculated as follows (see [1] and [7]):

$$\vec{\omega}^{L}_{EL} = \left(\frac{v_e}{(R_N + h)} \quad \frac{-v_n}{(R_M + h)} \quad \frac{-v_e}{(R_N + h)} \tan(\varphi) \right)^T \tag{4-16}$$

where $\vec{\omega}^{L}_{EL}$ is the angular rate of rotation of L frame with respect to E frame represented in the L frame, v_e is the east velocity, v_n is the north velocity, φ is the latitude, h is the altitude, R_N is the normal radius of curvature of Earth at the object's location, and R_M is the meridian radius of curvature of Earth at the object's location. For more details about R_N and R_M and how they can be calculated, review the Earth ellipsoid model in Chapter 1. As we will discuss later in this chapter, when the body moves, the L frame is constantly moving with it such that its north axis always points to geographic north. That is the reason the transport rate is given as a rotation rate of the L frame with respect to E frame. This transport rate vector can be projected in the body frame B and removed from the raw gyroscope measurement.

4.3.5 Coriolis Effect

The Earth is rotating with respect to the I frame, and the L frame is also rotating with respect to the E frame. This means that L frame is rotating with respect to I frame due to both Earth's rotation and transport rate. The combined rotation due to Earth's rotation and transport rotation rate will induce a linear force that is proportional to the object's velocity. This linear force will be sensed by the accelerometer even if the object is not accelerating (the body frame is moving at a constant speed). This linear force is known as the Coriolis effect, and it must be compensated by accelerometer measurements. The acceleration due to the Coriolis effect is calculated by the following cross-product equation:

$$\vec{a}^{L}_{Cr} = \left(\vec{\omega}^{L}_{EL} + 2\vec{\omega}^{L}_{IE} \right) \mathrm{x} \, \vec{v}^{L} \tag{4-17}$$

Proof of Eq. (4-17) can be found in see [7] and [1].

4.4 Inertial Measurement Unit

Raw inertial measurements require several preprocessing, such as prefiltering and noise cancellation, in addition to calibration and unit conversion, before they can be used for navigation. In addition, the sensor measurements need to be made available to external components (typically a digital computer) that run the navigation algorithms. Therefore, inertial sensors come in a unit that integrates the raw inertial sensor, calibration modules, preprocessing modules, power modules, and communication ports. This unit is called an inertial measurement unit (IMU). A general block diagram of an IMU is shown in Figure 4.8. ADC stands for analog to digital converter. Unit conversion

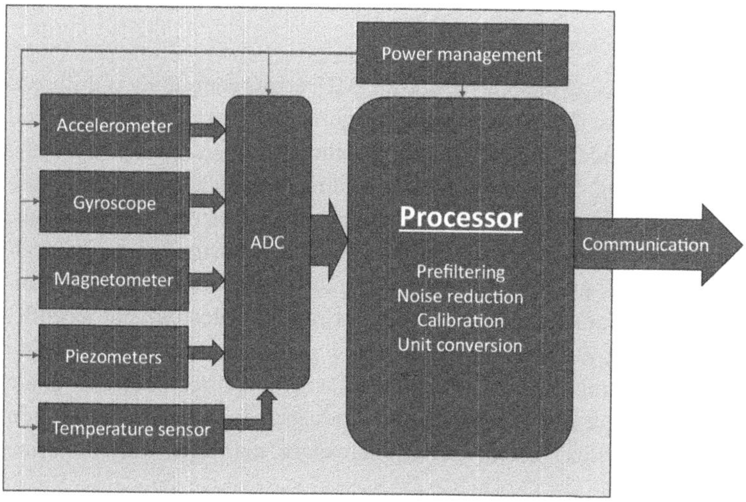

Figure 4.8 General block diagram of an IMU.

converts the inertial measurement from an electrical current or voltage into standard inertial units such as m/s^2 for acceleration and r/s for angular rate. Calibration is the process of removing known error parameters from the measurements before sending them as an output. For example, biases, scale factors, and axes-misalignment errors need to be compensated. Some calibration parameters are temperature-dependent (vary with temperature). Therefore, IMU usually includes a temperature sensor. In the factory, IMU is calibrated at different temperatures to build a temperature calibration model that properly compensates IMU output for these common errors. Factory calibration is not always perfect. Therefore, in practical navigation systems, sensor fusion methods are used to continuously calibrate the IMU using external measurements from other sensors such as Global Navigation Satellite Systems (GNSS). Razor IMU is a commercially available IMU developed by SparkFun and can be seen in Figure 4.9. It has a microcontroller and MPU9250 inertial sensor. In addition, it features USB communication and storage on an SD memory card.

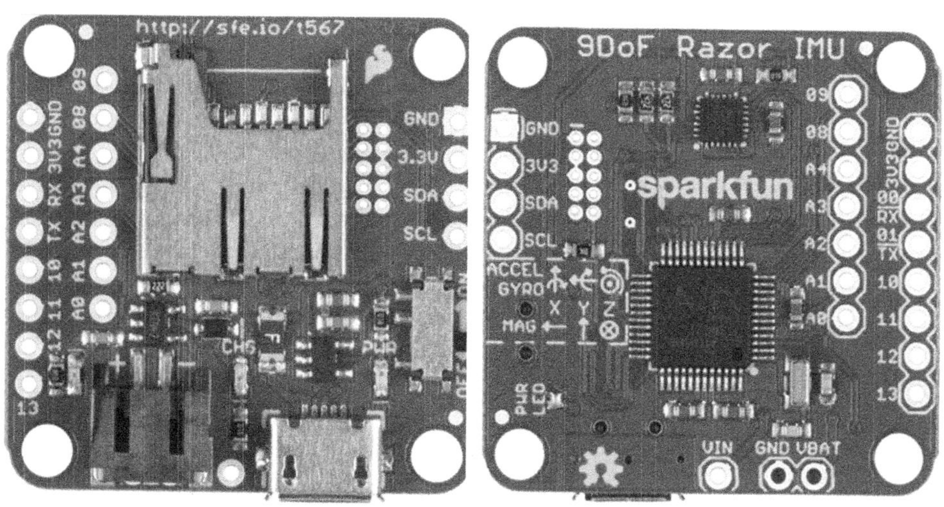

Figure 4.9 Razor IMU by SparkFun. It has a microprocessor and MPU-9250 Sensor. *Source:* SparkFun Electronics.

4.5 Inertial Sensor Errors

Even with the calibration and preprocessing modules inside an IMU, errors commonly continue to exist, especially in low-cost MEMS IMUs. IMU errors can be categorized into systematic errors and stochastic errors. Systematic errors include Bias, Scale factor, Nonlinearity, Sign Asymmetry, Dead Zone, Quantization, and Cross-coupling. Stochastic or Random Errors include Bias Drift, Noise Density, Random Walk noise, and Correlated noise. Depending on the errors' specifications, IMU sensors can be categorized into four categories: consumer-grade, industrial-grade, tactical-grade, and navigation-grade. Consumer-grade IMUs are used in smartphones, tablets, and consumer-electronics products. Consumer grades are mainly used for motion detection, but recently, with the advances in MEMS technology, consumer-grade IMU's performance has been greatly enhanced to the extent that they are being used for non-critical navigation tasks. Industrial-grade IMUs are used in industrial automation and the automotive industry. Tactical and navigation grades are used in safety and mission-critical systems such as aerospace, maritime, and submarine applications.

4.5.1 Systematic Errors

Bias is a deviation from the true measurements. The scale factor is a sloping error that maps the true input (acceleration or rotation) to the sensor output. Ideally, this slope should be exactly one. Scale factor error is a deviation in the slope from the ideal slope value ("1"). Nonlinearity is a skew that happens to the slope that maps the true input (acceleration or rotation) to the sensor output. Common systematic IMU errors are illustrated in Figure 4.10. Other errors can exist from the nonorthogonality of the IMU axes. Nonorthogonality means that the axes are not perfectly aligned at 90° angles. Nonorthogonality errors are like axes misalignment when the IMU axes are not perfectly aligned with the enclosure frame that packages the IMU sensor.

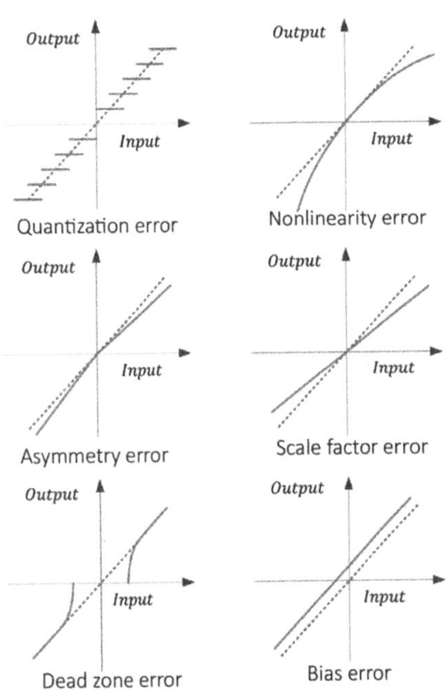

Figure 4.10 Systematic IMU errors.

4.5.2 Stochastic Errors

Stochastic errors include random noise and bias drifts. Random noise is a high-frequency Gaussian noise that contaminates the data. The integration of random Gaussian noise leads to error accumulation called Random Walk, as illustrated in Figure 4.11. Figure 4.11 shows random walk integration for a unity variance Gaussian noise. If we repeat the integration over and over many times and plot the integrated output, we will get the curves in Figure 4.11b. If, at any given time, we take the standard deviation (STDV) of the integrated outputs and plot the STDV against integration time, we will get the curve in Figure 4.11c. If we took the square root of the integration time, we would get the linear relationship in Figure 4.11d, which shows that the STDV of the random walk is proportional to the square root of the integration time. That is the reason the random walk

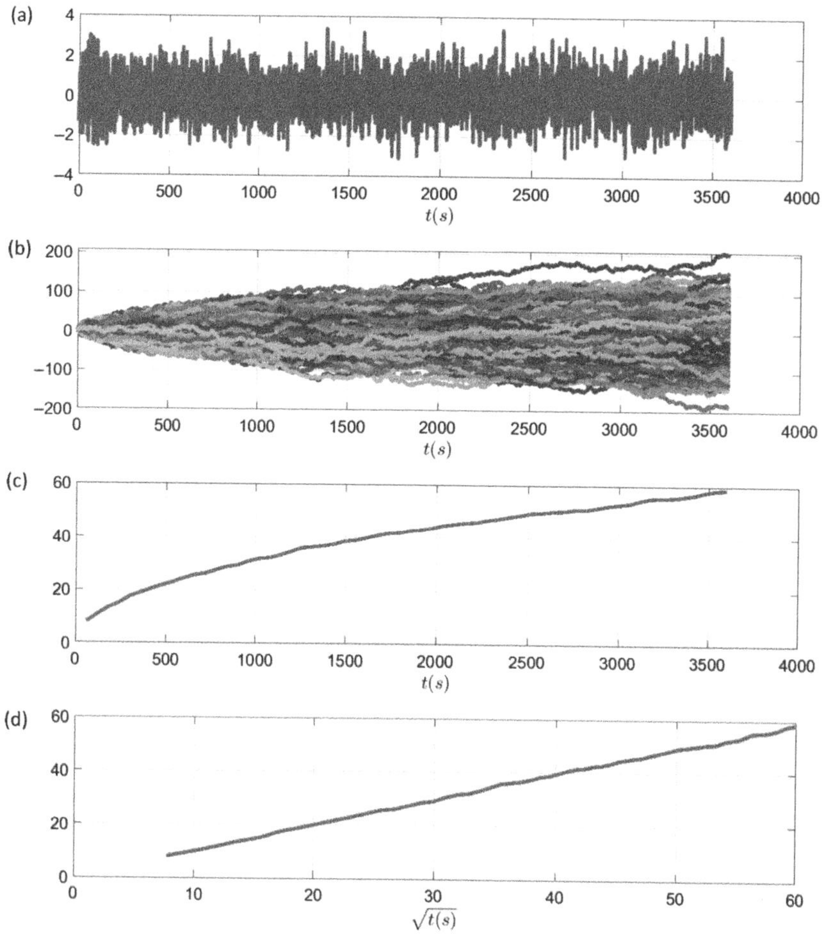

Figure 4.11 Integration of random noise (random walk). (a) Gaussian random noise, (b) random walk (integration of random noise), (c) STDV of random walk vs. integration time, (d) STDV of random walk vs. sqrt of integration time.

is given in units of $^{\circ}/\sqrt{s}$ or $^{\circ}/\sqrt{\text{hr}}$ in case of gyroscope and in units of $\text{m/s}/\sqrt{\text{s}}$ or $\text{m/s}/\sqrt{hr}$ in case of accelerometer. In the case of a gyroscope, it is called an angular random walk, and in the case of an accelerometer, it is called a velocity random walk. Random walk interpretation is that if you integrate the signal over the square root of a second or an hour, you will get an error equal to the random walk value standard deviation.

Noise density is another way to describe random Gaussian noise in terms of its standard deviation. Recall that the power spectral density (PSD) of a signal is the Fourier transform of the autocorrelation of the signal. Fourier transform of the autocorrelation of an idealistic Gaussian noise signal is a flat curve that is equal to the variance of the signal. The PSD of the Gaussian noise signal of Figure 4.11 is shown in Figure 4.12. Note that the smoothed PSD shows that the PSD is zero dB. The zero dB means the PSD is one which is the variance of our Gaussian test signal. Recall that $\text{dB} = 10\log_{10}(P)$ where P is the square of the signal. Our Gaussian test signal has unity variance which is equivalent to a power $P = 1$ which is zero in dB units as Figure 4.12 confirms.

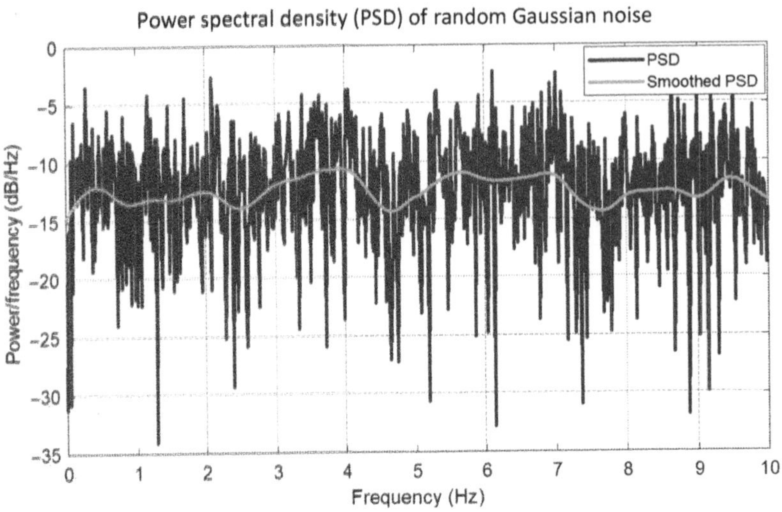

Figure 4.12 PSD of unity variance Gaussian noise sample.

Noise density is another popular measure of Gaussian random noise. Noise Density is commonly given as the square root of the PSD. The PSD unit is the squared units of the input signal per Hz. The square root of the Noise Density is the unit of the input signal per $\sqrt{\text{Hz}}$. This makes sense as the sampling frequency affects how much noise is passed through during the sampling process. Recall that the higher the sampling frequency, the higher the noise passing from the input signal to the digital output. In fact, sampling works as a low pass filter where the cut-off frequency depends on the sampling frequency. That is the reason that the units of noise density are $^{\circ}/\text{s}/\sqrt{\text{Hz}}$ in case of gyroscope and $\text{m/s}^2/\sqrt{\text{Hz}}$ in case of accelerometer. Bias drifts are random changes in the bias over-time. It is a significant factor that is used to categorize IMU grades. Bias drifts are measure of bias instability that shows how much the bias may change from time to time and it is measured in $^{\circ}/\text{s}$ in case of gyroscope and m/s^2 in case of accelerometer.

4.5.3 IMU Error Analysis using Allan Variance

4.5.3.1 Allan Variance

Allan variance [8, 9] is a statistical analysis tool for identifying various noise types that exist in a stochastic signal. Allan variance was used to measure the frequency stability of precision oscillators. Later, this technique was applied to identify the stochastic errors that affect IMU sensors. Allan variance performs overlapped window averaging of the signal using different window sizes and plots the results in a log-log plot. The concept behind averaging is that Gaussian noise can generally be removed by averaging over M values. However, this technique fails when there is a drift over time. In this case, Allan variance plots show the minimum averaging window size required to reduce the drift. The Allan variance plot not only shows the averaging time that minimizes the noise variance, but it also gives useful information about the different types of random processes that contribute to the drift. Usually, Allan deviation is used in the plots and Allan deviation can be calculated by taking the square root of Allan variance. The Allan variance analysis of a time domain signal $s(t)$ consists of computing its root Allan variance or Allan deviation as a function of different averaging times τ and then analyzing the characteristic regions in a log-log plot. The slopes of the

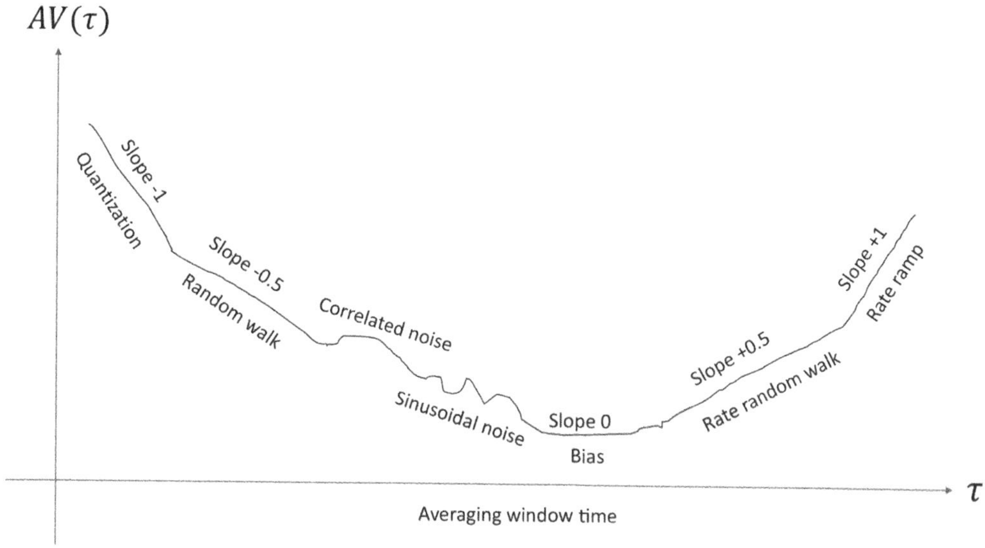

Figure 4.13 General Allen deviation curve.

Allan deviation curve are used to identify the different noise modes existing in the signal. An illustrative Allan deviation plot can be seen in Figure 4.13 [8, 9]. The horizontal axis τ can take different time units such as seconds, minutes, or hours. The unit of the vertical axis is rad/s, °/s or °/h for gyroscopes and g/s^2 or m/s^2 for accelerometers.

The original signal $s(t)$ is collected as N samples with a sampling period τ_s and we then try different window sizes for averaging and split the entire data into clusters of finite duration that is equal to $\tau = m\tau_s, \;\; m = 1, 2 \ldots \frac{N}{2} - 1$. Once all clusters are formed, the output angles (for gyroscope) or the output velocity (for accelerometer) are calculated for all the clusters as follows:

$$y_k = s(k\tau_s)\tau_s \tag{4-18}$$

Allan variance for specific cluster time τ (window size m) can be calculated using the following equation:

$$\sigma^2(\tau) = \frac{1}{2\tau^2(N - 2m)} \sum_{k=1}^{N-2m} (y_{k+2m} - 2y_{k+m} + y_k)^2 \tag{4-19}$$

We then repeat the steps for multiple values of τ. Allan variance is related to the PSD of the signal as follows [10]:

$$\sigma^2(\tau) = 4\int_0^{\infty} S(f)\frac{\sin^4(\pi f \tau)}{(\pi f \tau)^2}df \tag{4-20}$$

Random walk value σ_{rw}^2 is related to the Gaussian random part of the IMU errors. The value of the random walk for a Gaussian noise is simply the variance, which is the same as the value of the PSD. By setting $S(f) = \sigma_{rw}^2$ and performing the integration we get:

$$\sigma^2(\tau) = \frac{\sigma_{rw}^2}{\tau} \tag{4-21}$$

Taking the log of each side:

$$2\log(\sigma(\tau)) = -\sigma_{rw}^2 \log(\tau)$$
$$\log(\sigma(\tau)) = -0.5\,\sigma_{rw}^2 \log(\tau) \tag{4-22}$$

On a log-log plot, Eq. (4-22) is a line with a slope −0.5, which means the random walk σ_{rw}^2 can be calculated by setting $\tau = 1$ and going up vertically until we intersect with the Allan deviation curve at slope = −0.5. Other errors can be estimated from Allan variance as described in [10] and [11].

4.5.3.2 Allan Variance Example of Different IMU Grades

This section demonstrates an example to show how Allan deviation can be used to analyze different IMUs. The example can be found in the projects' repository of the book. This example uses MATLAB to simulate noisy IMU of different grades (consumer, industrial, tactical, and navigation grades). In this example, we used MATLAB imuSensor object (see Figure 4.14) to generate simulated IMU stationary measurements to test Allan variance on different IIMU grades. The values used in the imuSensor object creation have been collected from the datasheets of real IMUs of different grades. The raw gyroscope measurements of the *x*-axis of the four IMU grades are shown in Figure 4.15. The Allan deviation was performed on the data using MATLAB script shown in Figure 4.16. The Allan deviation for the four IMU grades for gyroscope and accelerometer data are shown in Figures 4.17 and 4.18. Figures 4.17 and 4.18 show the expectation and the differences between IMU grades.

To verify the simulation results, we run our MATLAB Allan variance program on real data collected from two IMUs. MTi-100 IMU from Xsens and MPU 9250 from InvenseSense. The results of these two real datasets are shown in Figures 4.19 and 4.20. The results on the real data sets show that the gyroscopes have a similar quality, but the MTi-100 IMU has a better accelerometer. The specific error values (random walk, biases, etc.) can be extracted from the Allan deviation curve as illustrated in Figure 4.13 and explained in [11]. For example, the angular random walk of the MTi-100 gyroscope is calculated from Allan variance by setting $\tau = 1$ and going up until we interest with Allan deviation curve at slope −0.5 as shown in Figure 4.21. The project files including MATLAB code, IMU datasheets, raw IMU data, and instructions are given under "*AllanVarianceProject*" project in the projects' repository of the book.

```
ConsumerGradeIMU = imuSensor('accel-gyro');

ConsumerGradeIMU.Accelerometer = accelparams( ...
    'MeasurementRange',200, ...              % m/s2
    'Resolution',200/2^16, ...               % m/s2
    'ConstantBias',[0.1 0.2 0.3], ...        % m/s2
    'NoiseDensity',300*10^-6*g, ...          % m/s2/sqrt(Hz)
    'RandomWalk', 10*10^-6*g,...             % m/s2*sqrt(Hz)
    'BiasInstability',15*10^-6*9.8);         % m/s2

ConsumerGradeIMU.Gyroscope = gyroparams( ...
    'MeasurementRange',450*D2R, ...                % rad/s
    'Resolution',450*D2R/(2^16), ...               % rad/s
    'ConstantBias',[0.01,0.02,0.03]*D2R,...        % rad/s
    'NoiseDensity',0.01*D2R, ...                   % rad/s/sqrt(Hz)
    'RandomWalk', 0.01*D2R/100,...                 % rad/s*sqrt(Hz)
    'BiasInstability',0.0028*D2R);                 % rad/s
```

Figure 4.14 Using MATLAB imuSensor object to simulate Consumer Grade IMU.

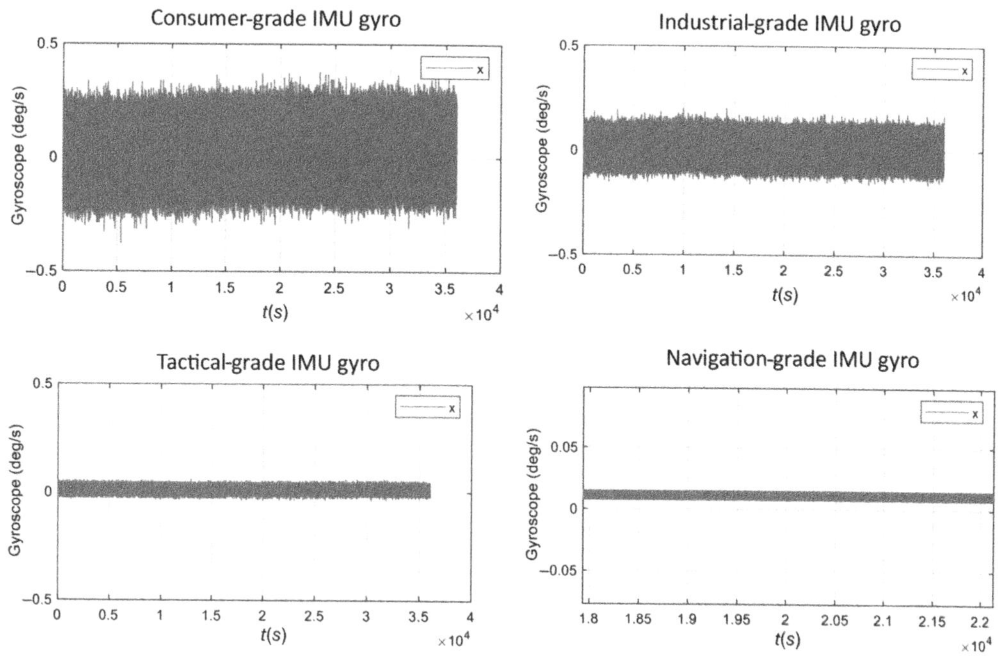

Figure 4.15 Raw gyroscope measurements of *x*-axis of different IMU grades.

```
figure;title('Allan Deviation - GyroX(deg/sec)')
[avar,tau] = allanvar(ConsumerGradeIMU_Gyr(:,1)*R2D,'octave',Fs);
loglog(tau,sqrt(avar),'r','LineWidth',2);hold on;
xlabel('\tau(s)');ylabel('\sigma(\tau) (deg/s)');grid on;
```

Figure 4.16 Using MATLAB allanvar(.) function to calculate and plot Allan deviation.

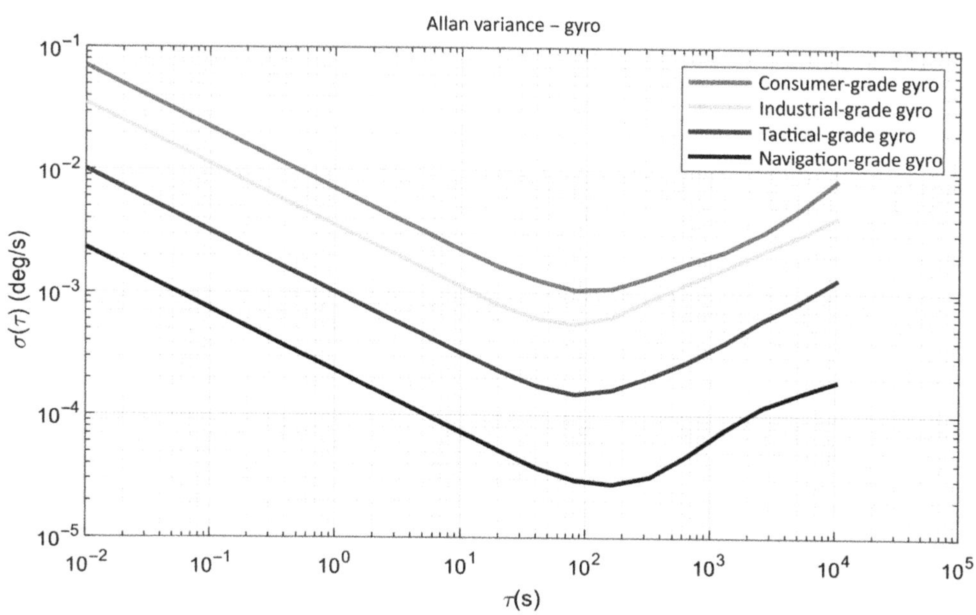

Figure 4.17 Allan deviation for different IMU grades – Gyro Data.

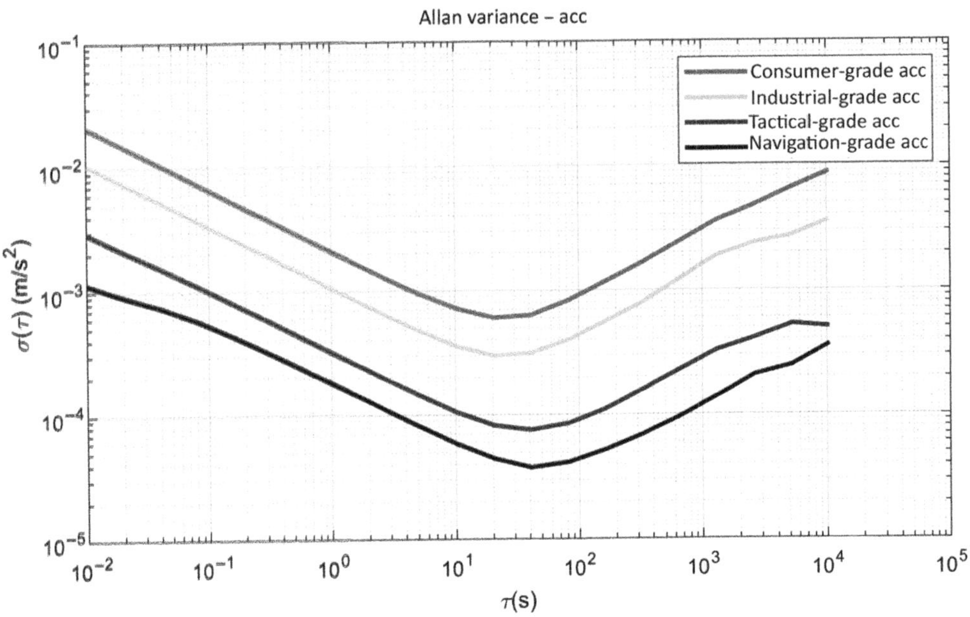

Figure 4.18 Allan deviation for different IMU grades – Accel Data.

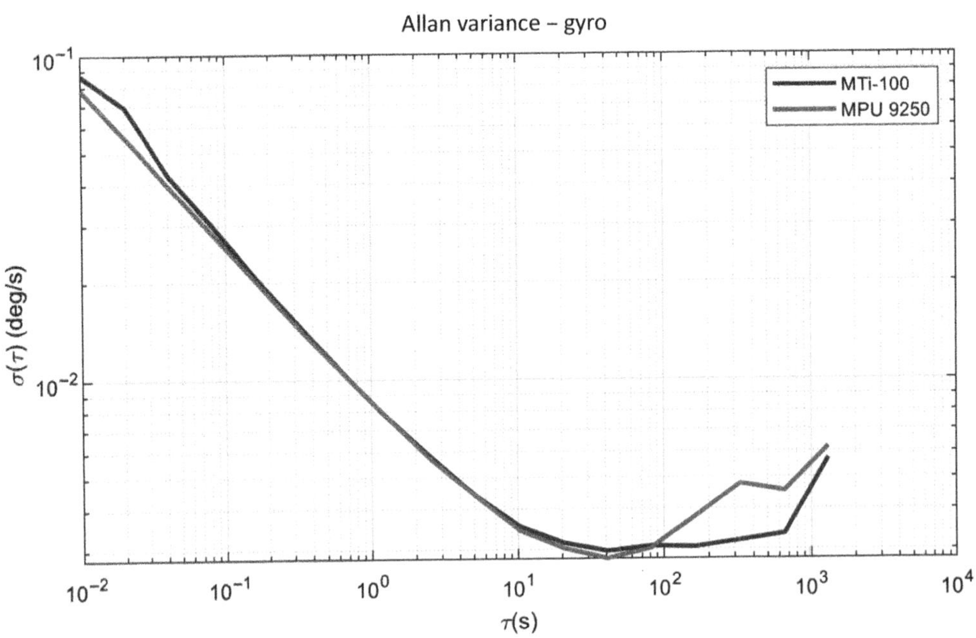

Figure 4.19 Allan deviation for real gyro data (MTi-300 and MPU 9250 IMUs).

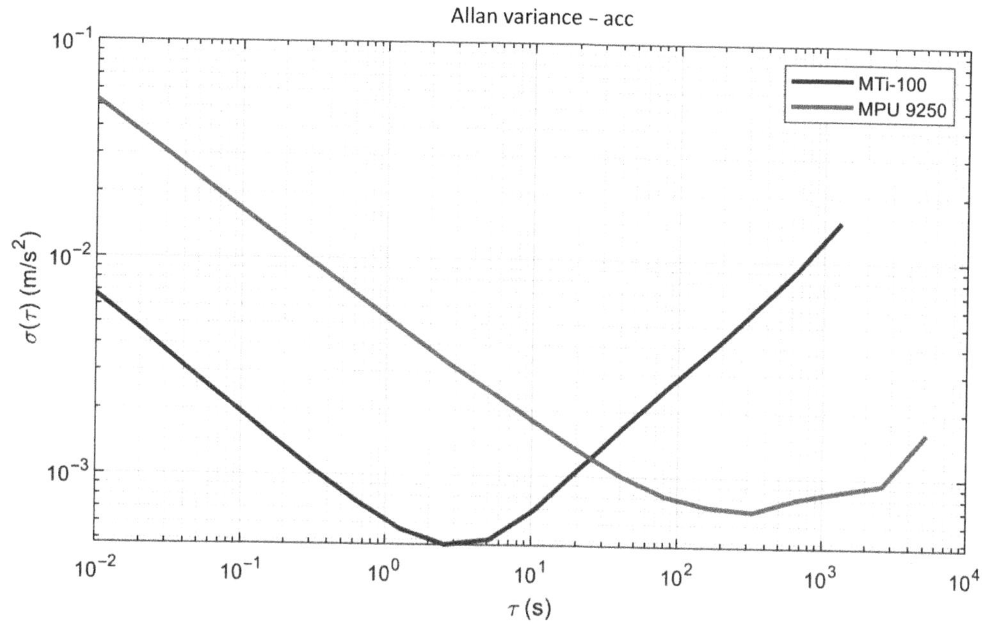

Figure 4.20 Allan deviation for real accel data (MTi-300 and MPU 9250 IMUs).

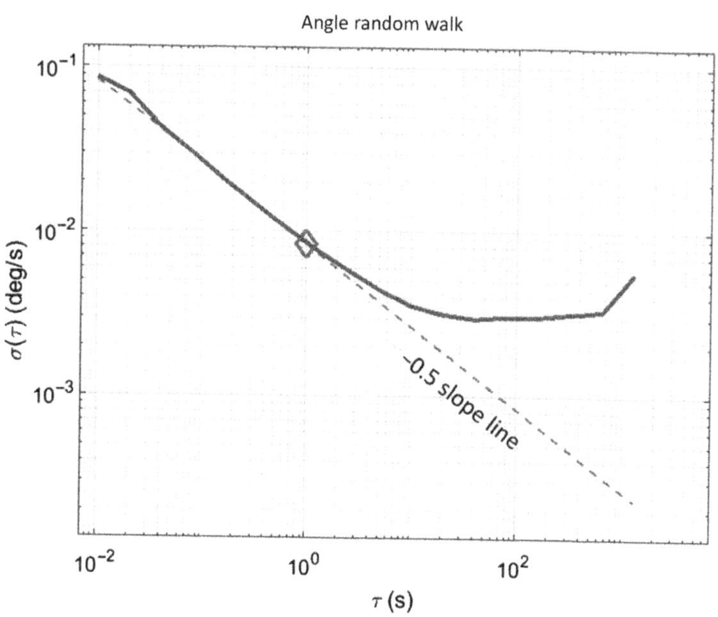

Figure 4.21 Random walk noise from Allan deviation for MTi-100 Gyroscope.

4.6 Inertial Navigation Systems (INS)

Inertial navigation systems (INS) use inertial sensor measurements or IMU data to estimate the object's position, velocity, and attitude (PVA). Attitude is also known as orientation which can be represented in different forms such as direction cosine matrix, Euler angles, rotation vector, and quaternion vector (review Chapter 1 for the definitions of orientation representation). INS equations of motion take the IMU output as input and calculate the sensor's PVA as an output. INS equations can be derived in any frame of reference. Common INS systems prefer the L-frame since it is easier to interpret. In addition, it turned out that implementing the INS equations in the L-frame has technical benefits such as the link between velocity error correction and the pitch/roll angles correction. Another benefit is the Schuller effect, where the pitch/roll angles error due to IMU errors has a cyclic bounded nature. More information about INS equations of motion in L frame and in other frames (e.g., ECEF and ECI) can be found in [1, 6, 7].

4.6.1 Forward Kinematics Equations

INS motion equations are differential equations that describe the time rate of change of position, velocity, and orientation as a function of gyroscope and accelerometer measurements. INS motion equations are also called forward kinematics, and in some literature, they are also called mechanization equations.

4.6.1.1 Orientation Rate Equation

From Chapter 1, the direction cosine matrix rate equation of a body-frame B rotating with respect to a local level navigation frame L is given by:

$$\dot{C}_B^L = C_B^L\left(\vec{\omega}_{IB}^B \times\right) - \left(\vec{\omega}_{IL}^L \times\right)C_B^L \tag{4-23}$$

The term $\left(\vec{\omega}_{IL}^L \times\right)C_B^L$ is included to compensate for the Earth's rotation rate and transport rate. Recall the value of $\vec{\omega}_{IL}^L$ is given by:

$$\vec{\omega}_{IL}^L = \vec{\omega}_{IE}^L + \vec{\omega}_{EL}^L \tag{4-24}$$

where $\vec{\omega}_{EL}^L$ is the Earth rate and $\vec{\omega}_{IE}^L$ is the transport rate. This matrix equation has several problems. For example, C_B^L uses nine equations (nine matrix elements), while 3D orientation requires only three parameters. In addition, trigonometric functions (sine and cosine) cause singularities. For example, when an angle θ approaches 90°, tan(θ) will approach infinity, causing numerical instability. Also, this singularity may lead to gimbal lock where pitch angle approaches 90°, causing ambiguities between roll and heading angles. For these reasons, rotation vector and quaternion are preferred as they provide more numerical stability. Another reason we prefer to use a rotation vector is the error analysis. As we will see later in this chapter, the linearized error model of the INS dynamic system is needed to perform the fusion with other sensors/measurements. If we used the error in the rotation matrix δC_B^L as a state, we would face a problem because the rotation matrix space is a Lie Group where addition is not defined. To be able to add errors, errors must be in the Euclidean space. In Chapter 1, we saw how the Lie Group is mapped to a Euclidean manifold using exponential mapping. The exponent in this mapping was the rotation vector. This is a strong reason to represent the rotation error using a rotation vector. As the quaternion has a redundant fourth component, its vector part can be used as an error state. The redundant fourth component in quaternion is a constraint that makes calculation numerically more stable as we can always assure the stability of the calculation by forcing this constraint after each iteration of calculations. The quaternion equivalent of the rotation matrix rate equation is the quaternion vector rate equation, which is given by:

$$\dot{\vec{q}}^{L}_{B} = \frac{1}{2}\left((W^{B}_{IB}\times)\vec{q}^{L}_{B} - (W^{L}_{IL}\times)\vec{q}^{L}_{B}\right) \tag{4-25}$$

where $(W^{B}_{IB}\times)$ is the 4×4 skew matrix cross-product operator defined by:

$$(W^{B}_{IB}\times) = \begin{pmatrix} 0 & -\omega^{B}_{IB_X} & -\omega^{B}_{IB_Y} & -\omega^{B}_{IB_Z} \\ \omega^{B}_{IB_X} & 0 & \omega^{B}_{IB_Z} & -\omega^{B}_{IB_Y} \\ \omega^{B}_{IB_Y} & -\omega^{B}_{IB_Z} & 0 & \omega^{B}_{IB_X} \\ \omega^{B}_{IB_Z} & \omega^{B}_{IB_Y} & -\omega^{B}_{IB_X} & 0 \end{pmatrix} \tag{4-26}$$

$$(W^{L}_{IL}\times) = \begin{pmatrix} 0 & -\omega^{L}_{IL_X} & -\omega^{L}_{IL_Y} & -\omega^{L}_{IL_Z} \\ \omega^{L}_{IL_X} & 0 & \omega^{L}_{IL_Z} & -\omega^{L}_{IL_Y} \\ \omega^{L}_{IL_Y} & -\omega^{L}_{IL_Z} & 0 & \omega^{L}_{IL_X} \\ \omega^{L}_{IL_Z} & \omega^{L}_{IL_Y} & -\omega^{L}_{IL_X} & 0 \end{pmatrix} \tag{4-27}$$

where $\vec{\omega}^{B}_{IB} = \left(\omega^{B}_{IB_X}\ \ \omega^{B}_{IB_y}\ \ \omega^{B}_{IB_z}\right)^{T}$ is the gyroscope measurements in the body-frame and $\vec{\omega}^{L}_{IL} = \left(\omega^{L}_{IL_X}\ \ \omega^{L}_{IL_Y}\ \ \omega^{L}_{IL_Z}\right)^{T}$ is the transport rate and Earth rate combined. Another way to compensate the raw gyroscope measurements for transport rate and Earth rate is to convert $\vec{\omega}^{L}_{IL}$ from L frame to B frame and remove it from $\vec{\omega}^{B}_{IB}$. This will lead to a rotation rate vector $\vec{\omega}^{B}_{LB}$ that represents the rotation of B frame with respect to frame L directly. In addition, we can also remove the known errors such as biases and scale factors. The compensated gyroscope measurements vector $\vec{\omega}^{B}_{LB}$ can be written as follows:

$$\vec{\omega}^{B}_{IB} = \left(\vec{\omega}^{B}_{LB} + \vec{\omega}^{B}_{IL}\right)\left(1 + \vec{s}_{g}\right) + \vec{b}_{g} \tag{4-28}$$

$$\left(\vec{\omega}^{B}_{LB} + \vec{\omega}^{B}_{IL}\right) = \frac{\vec{\omega}^{B}_{IB}}{\left(1 + \vec{s}_{g}\right)} - \vec{b}_{g} \tag{4-29}$$

$$\vec{\omega}^{B}_{LB} = \frac{\vec{\omega}^{B}_{IB}}{\left(1 + \vec{s}_{g}\right)} - \vec{b}_{g} - \vec{\omega}^{B}_{IL} \tag{4-30}$$

where $\vec{\omega}^{B}_{IB}$ is measured angular rate, $\vec{\omega}^{B}_{LB}$ is compensated angular rate, $\vec{s}_{g}$ is gyro scale factor, $\vec{b}_{g}$ is gyro bias, $\vec{\omega}^{L}_{IL}$ is the Earth + transportation rates combined. The quaternion rate equation using the compensated gyroscope measurements $\vec{\omega}^{B}_{LB}$ is given by the following equation:

$$\dot{\vec{q}}^{L}_{B} = \frac{1}{2}(W^{B}_{LB}\times)\vec{q}^{L}_{B} \tag{4-31}$$

4.6.1.2 Velocity and Position Rate Equations

After calculating the orientation, the velocity is calculated. Velocity is estimated by simply projecting the accelerometer measurements $\vec{a}^{B}_{SF} = \begin{pmatrix} f_x & f_y & f_z \end{pmatrix}^T$ from body B frame to local level navigation frame L and integrate the results once after compensating for gravity force and Coriolis force. The Coriolis force is the force the accelerometer will measure due to the rotation of L frame on the Earth's ellipsoid surface. The velocity equation is calculated as follows:

$$\dot{\vec{v}}^{L} = C^{L}_{B}\vec{a}^{B}_{SF} + \vec{g}^{L} - \left(\vec{\omega}^{L}_{EL} + 2\vec{\omega}^{L}_{IE}\right) \times \vec{v}^{L} \tag{4-32}$$

The operator "×" is the cross-product and $\vec{a}^{B}_{SF}$ is the accelerometer measurements in the body-frame. Once the velocity is calculated, Geodetic position can be calculated as follows:

$$\dot{\varphi} = \frac{v_n}{(R_m + h)} \tag{4-33}$$

$$\dot{\lambda} = \frac{v_e}{(R_n + h)\cos(\varphi)} \tag{4-34}$$

$$\dot{h} = -v_d \tag{4-35}$$

We can also directly calculate the position $\vec{r}^{L}$ as north, east, and down coordinates with respect to the L frame as follows:

$$\dot{\vec{r}}^{L} = \begin{pmatrix} \dot{N} \\ \dot{E} \\ \dot{D} \end{pmatrix} = \begin{pmatrix} v_n \\ v_e \\ v_d \end{pmatrix} - \left(\vec{\omega}^{L}_{EL} \times\right)\vec{r}^{L} \tag{4-36}$$

The term $\left(\vec{\omega}^{L}_{EL} \times\right)\vec{r}^{L}$ accounts for the rotation of L frame with respect to E frame (the transport rate). The velocity of the L frame $\vec{r}^{L}$ has two components; one component is related to the motion of the object and another component is related to the rotation of the L-frame with respect to E frame. For more details about the origin of the $\vec{r}^{L}$ equation, review the velocity vector in rotating frames in Chapter 1. More information about deriving motion equations in rotating frames can be found in [1, 6, 7].

4.6.1.3 MATLAB Implementation of 3D Inertial Navigation System

The projects' repository of the book includes a MATLAB implementation of INS forward kinematics with testing datasets to help readers to get hands-on experience in inertial navigation. The project is under the folder "*InertialNavSystemProject*." A snapshot of the forward kinematics script is shown in Figure 4.22. The function takes initial position in L frame, velocity in N frame, orientation in quaternion vector, current latitude, and the sampling period. The given code implements simple first-order numerical integration. More advanced integration methods can be used. The project comes with a main script named "*demo_INS.m*" that loads datasets and apply forward kinematics on the data and display the results. The dataset comes with reference ground truth solution, clean raw IMU, and noisy version of the raw IMU, position, and velocity. The readers can modify the main script to add custom noise to the clean raw IMU and process the data to compare the output against the ground truth to study the effect of different IMU errors on the resulting position,

```
function [r_out,v_out,q_out] = forward_kinematics(acc,gyr,r_in,v_in,q_in,lat,sampling_period_sec)
    C_NL  = [0 1 0;1 0 0;0 0 -1];                                              %DCM from L to N frame
    C_LN  = C_NL';                                                              %DCM from N to L frame
    we    = 2*pi/(24*60*60);                                                    %Earth rotation rate
    earth_a = 6378137;earth_f = 1/298.257223563;                                %Earth Shape Params
    earth_b = earth_a*(1-earth_f);earth_e2 = 1-(earth_b^2)/(earth_a^2);         %Earth Shape Params

    %--> Calculate Earth Carvature Rn and Rm
    Rn = earth_a/sqrt(1-earth_e2*sin(lat)*sin(lat));
    Rm = earth_a*(1-earth_e2)/((1-earth_e2*sin(lat)*sin(lat))^(1.5));

    %--> Calculate transportation rate and earth rotation rate vectors
    vn   = v_in(2); ve   = v_in(1);
    w_N_EN_value = [-vn/(Rm + r_in(3)); ve/(Rn + r_in(3)); (ve*tan(lat))/(Rn + r_in(3))];
    w_N_IE_value = [0; we*cos(lat); we*sin(lat)];
    w_L_IL_value = C_LN*(w_N_EN_value + w_N_IE_value);

    %--> Prepare quaternion vectors and matrices
    a = q_in(1); b = q_in(2); c = q_in(3); d = q_in(4);

    att_quat_matrix  = [a -b -c -d; b a -d c; c d a -b; d -c b a];

    gyro_quat_vector = [0;gyr];

    w_il_x = w_L_IL_value(1); w_il_y = w_L_IL_value(2); w_il_z = w_L_IL_value(3);

    w_il_quat_matrix = [0          -w_il_x     -w_il_y     -w_il_z;...
                        w_il_x      0          -w_il_z      w_il_y;...
                        w_il_y      w_il_z      0          -w_il_x;...
                        w_il_z     -w_il_y      w_il_x      0   ];

    quat_vector = [a;b;c;d];

    attitude_quat_dot = 0.5*att_quat_matrix*gyro_quat_vector - 0.5*w_il_quat_matrix*quat_vector;

    %--> Advance quaternion
    q_out = q_in + attitude_quat_dot*sampling_period_sec;

    %--> Normalize the quaternion
    q_out = q_out./norm(q_out);

    %--> Calculate Euler angles from Quaternion
    [A, p, r] = quat2angle(q_out','ZYX');

    %-->Reset quaternion in case -180/180 boundaries crossed
    q_out = angle2quat(A,p,r,'ZYX');
    q_out = q_out./norm(q_out);

    %--> Advance velocity
    C_BL_value = angle2dcm(A,p,r,'ZYX');
    C_LB_value = C_BL_value';

    %--> Calculate gravity. For simplicity it is fixed in this example
    g_N = [0; 0; -9.8];

    V_N_dot_value = C_NL * C_LB_value * acc + g_N - cross((w_N_EN_value + 2*w_N_IE_value),v_in);
    v_out = v_in + V_N_dot_value*sampling_period_sec;

    %--> Advance position
    pos_dot = v_in - cross(w_N_EN_value,r_in);
    r_out = r_in + pos_dot*sampling_period_sec;
end
```

Figure 4.22 MATLAB implementation of INS forward kinematics equations.

```
gyro_rw_stdv = 0.1; %deg/s
gyro_bias = 0.25; %deg/s
ems_data.gyro_noisy = ems_data.gyro_clean;
ems_data.accel_noisy = ems_data.accel_clean;
ems_data.gyro_noisy(:,3) = ems_data.gyro_clean(:,3) + ...
                           gyro_bias*D2R + ...
                           gyro_rw_stdv*D2R*randn(ems_data.kn,1);
```

Figure 4.23 Resetting the default noise and adding custom noise to raw IMU data.

velocity, and orientation. For example, the script in Figure 4.23 resets the default noise and adds a random bias with a mean of 0.25°/s and a standard deviation of 0.1°/s to the z-axis gyroscope measurements. Therefore, all raw IMU data are clean except for the z-axis gyroscope. The effect of this gyroscope random noise can be seen in Figures 4.24 and 4.25.

4.6.2 INS System Error Model

The INS error model is a key component that is used to integrate inertial sensors with other measurements using Kalman filtering. The INS motion equations constitute a nonlinear dynamic system in the form:

$$\dot{\vec{x}}(t) = f\left(\vec{x}(t), \vec{u}(t)\right) \tag{4-37}$$

where $\vec{x}(t)$ is the navigation state (position, velocity, and orientation) and $\vec{u}(t)$ is the triggering input signal. In our case, $\vec{u}(t)$ is the IMU raw data. It is important to note that the IMU data is a noisy, imprecise measurement of motion. In other systems, the input could be the thrust input of an engine or the gas pedal and steering wheel angle of a car. In navigation systems, we do not consider the root signal that triggers the motion. Instead, we consider the inertial measurement resulting from these root control signals. In the navigation systems study, we consider the inertial

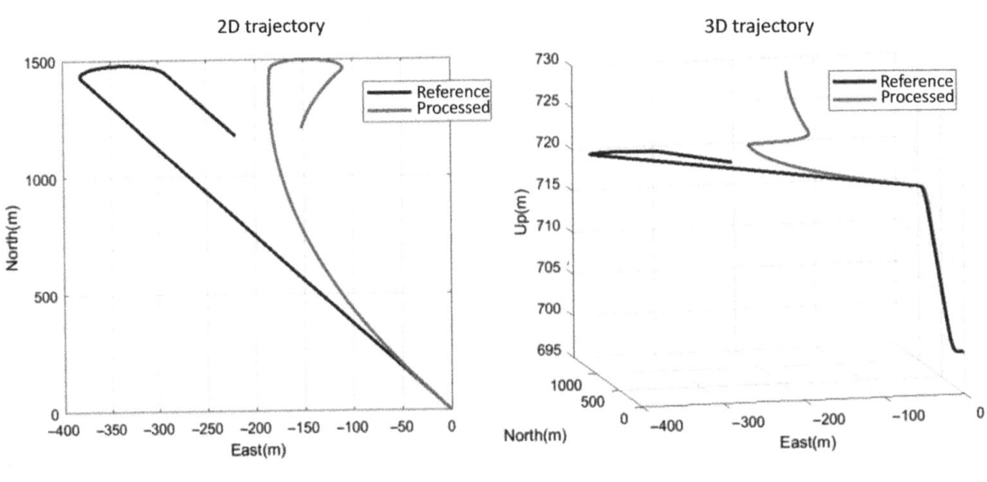

Figure 4.24 Position plots that show the effect of adding random noise to gyroscope z-axis.

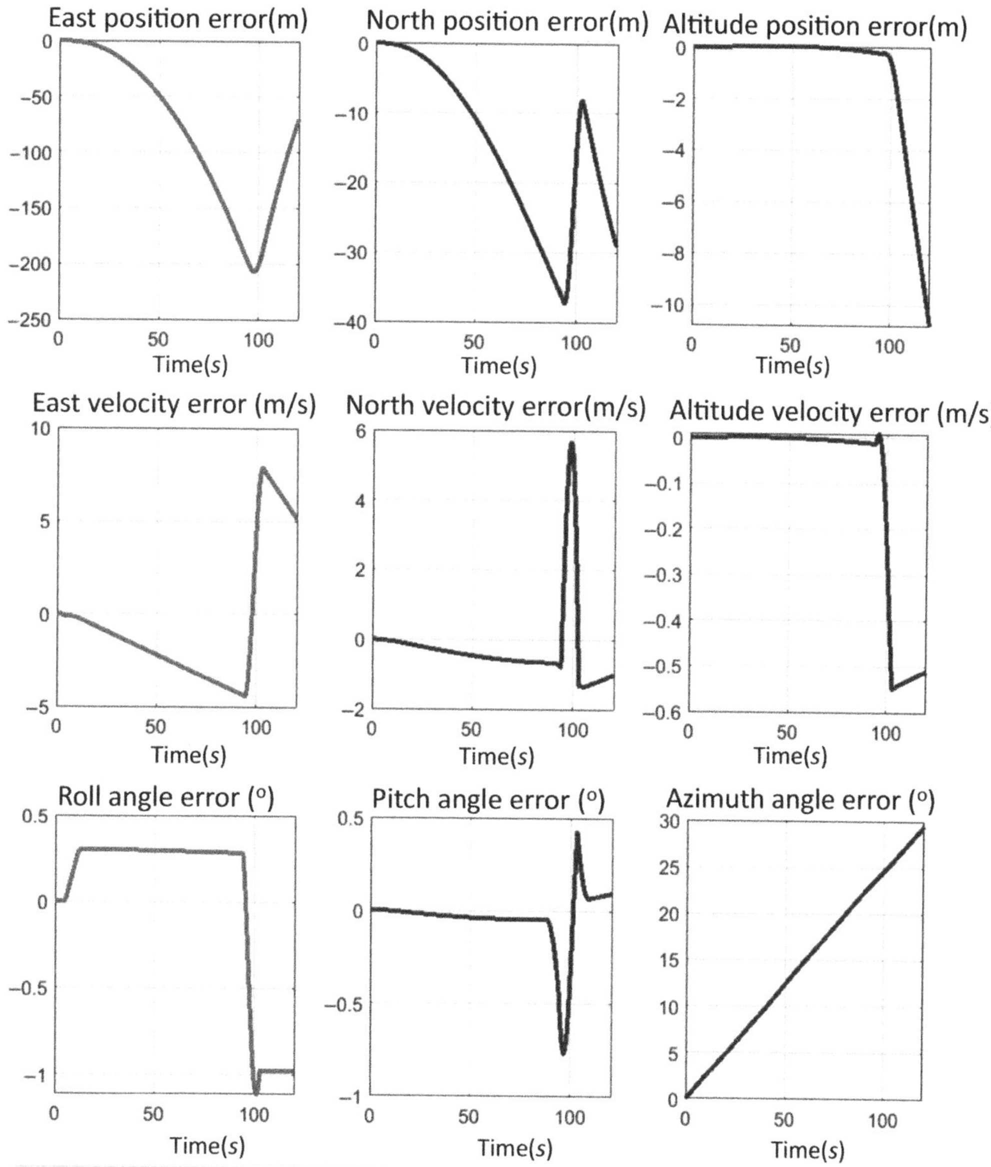

Figure 4.25 Effect of gyroscope *z*-axis noise on position, velocity, and orientation errors.

measurements as the input control signal to our dynamic system. It is important to study the effect of inertial sensor errors (which contaminate the input signal $\vec{u}(t)$) on the error in the navigation state $\vec{x}(t)$. Because the system is a recursive dynamic system given in the state-space model, an error in $\vec{u}(t)$ will lead to an error in $\vec{x}(t)$, which will lead to an error in future values of $\vec{x}(t)$ so the error is a compound effect of both current error in $\vec{u}(t)$ and initial errors in $\vec{x}(t)$. In the coming subsection, we analyze the dynamics of the errors in INS motion equations. The following analysis is a brief discussion based on the detailed derivations that can be found in [7] and other sources such as [8], [1], and [6].

4.6.2.1 INS Error States and Error Model

The INS error equations can be obtained by linearizing the INS motion equations around an operating point (position, velocity, and orientation or attitude). Before doing so, we need to agree on the error states or state vector $\vec{x}(t)$, which will turn into an error state vector $\delta\vec{x}(t)$ when we linearize the INS motion model. To be complete, states should include position, velocity, attitude, and sensors error parameters. Being stochastic and complex, modeling inertial sensor errors is a challenging active research topic. In our analysis, we will consider the most used sensor errors and models.

INS Error States

There are different representations of position, velocity, attitude, and sensor error parameters. In this book, and for reasons that will be clear shortly, we will use the following states:

$$\vec{x}(t) = \begin{pmatrix} \vec{r} \\ \vec{v}^L \\ \vec{\varnothing}_B^L \\ \vec{b}_\omega \\ \vec{b}_a \end{pmatrix} \tag{4-38}$$

The definition of each item of the state vector is defined in Table 4.1.

To complete the dynamic model $\dot{\vec{x}}(t) = f\left(\vec{x}(t), \vec{u}(t)\right)$, we need to have a differential equation for the bias states. A common random process that is always suitable to describe the bias states dynamic behavior is Gauss–Markov process [1, 6, 7].

$$\dot{\vec{b}}_\omega(t) = -\beta_b \vec{b}_\omega(t) + \sqrt{2\vec{\sigma}_b{}^2 \vec{\beta}_b} w_u(t) \tag{4-39}$$

$$\dot{\vec{b}}_a(t) = -\beta_a \vec{b}_a(t) + \sqrt{2\vec{\sigma}_a{}^2 \vec{\beta}_a} w_u(t) \tag{4-40}$$

where β is the reciprocal of time constant and σ is the standard deviation and $w_u(t)$ is a unit-variance Gaussian noise.

Rotation Error Modeling

Rotation matrix error can be described as follows [7]:

$$\hat{C}_B^L = (I + \delta C_B^L) C_B^L \tag{4-41}$$

$$\hat{C}_B^L = C_B^L + C_B^L \delta C_B^L \tag{4-42}$$

where $\hat{C}_B^L$ is erroneous direction cosine matrix, C_B^L is error-free direction cosine matrix, and δC_B^L is an error matrix. δC_B^L can be decomposed into two parts, symmetric and skew-symmetric parts as follows:

$$\delta C_B^L = \delta C_{Bs}^L + \delta C_{Bss}^L \tag{4-43}$$

where δC_{Bs}^L and δC_{Bss}^L are defined as follows:

$$\delta C_{Bs}^L = \frac{1}{2}\left(\delta C_B^L + \left(\delta C_B^L\right)^T\right) \tag{4-44}$$

Table 4.1 Definitions of navigation states.

$\vec{r} \in \mathbb{R}^3$	Position vector can be defined as follows: $\vec{r}^{geo} = \begin{pmatrix} \varphi \\ \lambda \\ h \end{pmatrix}$ φ is latitude, λ is longitude, h is altitude. $\vec{r}^{L} = \begin{pmatrix} n \\ e \\ d \end{pmatrix}$ north, east, down (NED) position. $\vec{r}^{N} = \begin{pmatrix} e \\ n \\ u \end{pmatrix}$ east, north, up (ENU) position.
$\vec{v} \in \mathbb{R}^3$	Velocity vector can be defined as follows: $\vec{v}^{L} = \begin{pmatrix} v_n \\ v_e \\ v_d \end{pmatrix}$ north, east, and down (NED) velocities. $\vec{v}^{N} = \begin{pmatrix} v_e \\ v_n \\ v_u \end{pmatrix}$ east, north, and up (ENU) velocities.
$\vec{\emptyset} \in \mathbb{R}^3$	Rotation vector can be defined as follows: $\vec{\emptyset}_B^{L} = \begin{pmatrix} \emptyset_{Bx}^L \\ \emptyset_{By}^L \\ \emptyset_{Bz}^L \end{pmatrix}$ rotation vector of frame B with respect to frame L. $\vec{\emptyset}_B^{N} = \begin{pmatrix} \emptyset_{Bx}^N \\ \emptyset_{By}^N \\ \emptyset_{Bz}^N \end{pmatrix}$ rotation vector of frame B with respect to frame N.
$\vec{b}_\omega \in \mathbb{R}^3$	$\vec{b}_\omega = \delta\vec{\omega}_{IB}^{B} = \begin{pmatrix} b_{\omega x} \\ b_{\omega y} \\ b_{\omega z} \end{pmatrix}$ gyroscope sensor bias.
$\vec{b}_a \in \mathbb{R}^3$	$\vec{b}_a = \delta\vec{a}_{SF}^{B} = \begin{pmatrix} b_{ax} \\ b_{ay} \\ b_{az} \end{pmatrix}$ accelerometer sensor bias

$$\delta C_{Bss}^{L} = \frac{1}{2}\left(\delta C_B^L - \left(\delta C_B^L\right)^T\right) \tag{4-45}$$

Recall the small change in direction cosine matrix related to a small change in rotation vector is approximately by:

$$\delta C_B^L \cong I + \left(\vec{\gamma}\times\right) \tag{4-46}$$

where $\vec{\gamma}$ is the rotation vector error and $(\vec{\gamma}\times)$ is a skew-symmetric matrix. This means that skew-symmetric part of error δC^L_{Bss} is related to rotation vector error between frame B and L frames. The rotation vector error between frames B and L represents the "misalignment" between them. We will simply refer to this type of error as "misalignment." On the other hand, we can see that:

$$\delta C^L_{Bs} = \frac{1}{2}\left(\hat{C}^L_B\left(\hat{C}^L_B\right)^T - I\right) \tag{4-47}$$

which clearly shows that δC^L_{Bs} is related to orthonormality of the direction cosine matrix $\hat{C}^L_B$. Given the differential equation of the direction cosine matrix:

$$\dot{\hat{C}}^L_B = \hat{C}^L_B(\omega^B_{IB}\times) - (\omega^L_{IL}\times)\hat{C}^L_B \tag{4-48}$$

we can find $\delta\dot{C}^L_{Bs}$ and $\delta\dot{C}^L_{Bs}$ as follows:

$$\delta\dot{C}^L_{Bs} = \delta C^L_{Bs}(\omega^L_{IL}\times) - (\omega^L_{IL}\times)\delta C^L_{Bs} \tag{4-49}$$

$$\delta\dot{C}^L_{Bss} = ((C^A_B\delta\omega^B_{IB})\times) - (\delta\omega^L_{IL}\times) + \delta C^L_{Bss}(\omega^L_{IL}\times) - (\omega^L_{IL}\times)\delta C^L_{Bss} \tag{4-50}$$

The detailed derivation of Eqs. (4-49) and (4-50) can be found in [7]. These equations show that δC^L_{Bs} is not affected by sensors errors and can be avoided if initial values of δC^L_{Bs} are zeros. The more problematic part is δC^L_{Bss} which grows with the error in gyroscope sensor measurements $\delta\omega^B_{IB}$ and errors in transport and Earth rate $\delta\omega^L_{IL}$ which is related to errors in position and velocity. Based on this brief analysis, we will focus on the skew-symmetric part of error δC^L_{Bs} which is related to the misalignment between frame B and L frames. The misalignment can be modeled as a misalignment of L that causes C^L_B to be erroneous or a misalignment of B that causes C^L_B to be erroneous. After ignoring the orthonormality error part, and assuming that L is the misaligned frame (representing the rotation error as a misalignment of L frame), the erroneous direction cosine matrix is given by:

$$\hat{C}^L_B = \hat{C}^{L'}_L C^L_B \tag{4-51}$$

where $\hat{C}^{L'}_L$ is a small misalignment of frame L (we see it L' instead of the correct frame L). In this case, the error in the direction cosine matrix is given by:

$$\delta C^L_B = \hat{C}^L_B - C^L_B \tag{4-52}$$

$$\delta C^L_B = \hat{C}^{L'}_L C^L_B - C^L_B = \left(\hat{C}^{L'}_L - I\right)C^L_B \tag{4-53}$$

The misalignment $\hat{C}^{L'}_L$ can be calculated using the rotation vector error as follows:

$$\hat{C}^{L'}_L \cong I + (\vec{\gamma}\times) \tag{4-54}$$

which, after rearrangements and substitution, gives the following important relationship between δC^L_B and rotation vector error $\vec{\gamma}$:

$$\delta C^L_B = -(\vec{\gamma}\times)C^L_B \tag{4-55}$$

Differentiating δC^L_B with respect to time will get the following:

$$\delta\dot{C}^L_B = -\left(\dot{\vec{\gamma}}\times\right)C^L_B - (\vec{\gamma}\times)\dot{C}^L_B \tag{4-56}$$

substituting $\dot{C}_B^L$ from the differential equations of rotation matrix:

$$\dot{C}_B^L = C_B^L(\omega_{IB}^B \times) - (\omega_{IL}^L \times)C_B^L \quad (4\text{-}57)$$

and rearranging the terms of the equation to separate $\dot{\vec{\gamma}}$ in one side gives:

$$\dot{\vec{\gamma}} = -C_B^L \delta\omega_{IB}^B - (\omega_{IL}^L \times)\vec{\gamma} + \delta\omega_{IL}^L \quad (4\text{-}58)$$

The error term $\delta\omega_{IL}^L$ is obtained by the following equation:

$$\delta\omega_{IL}^L = \delta\omega_{IE}^L + \delta\omega_{EL}^L \quad (4\text{-}59)$$

where $\delta\omega_{IE}^L$ is the error in Earth rate vector, which is a function of latitude error and $\delta\omega_{EL}^L$ is the transport rate error, which is a function of east/north velocity errors, latitude error, and altitude error. Hence, the differential equation of the rotation vector error between frames B and L assuming L is the misaligned frame $\dot{\vec{\gamma}}$ is a function of gyroscope sensor error ω_{IB}^B, initial rotation vector error $\vec{\gamma}$, and velocity/position error states. Since Earth's rotation rate is a small value (when measured in radians per second, it equals approximately 7.2921150×10^{-5}), $\delta\omega_{IE}^L$ is small. Also, since the transport rate error is velocity/position errors divided by Earth's radius of curvature, which is large (approximately 6 371 230m), dividing by this large value makes the error in $\delta\omega_{EL}^L$ small. Therefore, $\delta\omega_{IL}^L$ can be ignored and the rotation vector error equation becomes as follows:

$$\dot{\vec{\gamma}} = -C_B^L \delta\omega_{IB}^B - (\omega_{IL}^L \times)\vec{\gamma} \quad (4\text{-}60)$$

Equation (4-60) relates the rate of change of rotation vector error to the error in the gyroscope measurements. It is good practice to model rotation error in terms of rotation vector error to avoid singularity, nonlinearity, and gimbal locks. Modeling rotation error in direction cosine matrix is not ideal due to the use of trigonometric functions (sine and cosine), which cause singularity, nonlinearity, and gimbal lock issues. The same issue exists with Euler angles. Modeling rotation error using the complete quaternion vector is also not perfect due to the redundancy in the quaternion vector components. This redundancy causes covariance matrix rank problem when we use Kalman filtering. The ideal parameter to model rotation error is the rotation vector as it is the minimal set of parameters, and it also satisfies the exponential mapping from Lie algebra (where errors cannot be simply added) to a linear algebra where errors can be added.

Velocity and Position Error Modeling

If we use geodetic coordinates (latitude, longitude, and altitude) to describe the position, and because latitude and longitude are angles that represent orientation of L (or N) frame with respect to E (ECEF) frame, as explained in [7], it is a good idea to use a rotation vector parameter to describe the error in the horizontal position part (latitude and longitude). The direction cosine matrix C_L^E between L frame and E frame is defined as follows:

$$\dot{C}_L^E = C_L^E(\omega_{EN}^N \times) \quad (4\text{-}61)$$

Like what we have done in modeling the error in C_B^L as a misalignment in L frame, we define the error δC_L^E as a function of a rotation vector error $(\vec{\varepsilon}^L)$ that represents the misalignment of frame L with respect to frame E as follows:

$$\delta C_L^E = -(\vec{\varepsilon}^L \times)C_L^E \quad (4\text{-}62)$$

The two equations above can be used to derive $\vec{\varepsilon}$, which gives the following form:

$$\dot{\varepsilon}^L = -\omega_{EL}^L \times \varepsilon^L + \delta\omega_{EL}^L \tag{4-63}$$

The rate of change of error in altitude is simply given by the following equation:

$$\delta\dot{h} = -\delta v_n = \delta v_u \tag{4-64}$$

This concludes the position error model if we took the geodetic coordinates as position parameters. If we took the NED coordinates to describe the position, we will simply linearize the following NED position differential equation:

$$\dot{r}^L = v^L - \left(\omega_{EL}^L \times\right) r^L \tag{4-65}$$

Since ω_{EL}^L term includes division by Earth's radius of curvature, which is a large number, the differentiation of $\left(\omega_{EL}^L \times\right)$ can be ignored, which gives the following approximated NED position error differential equation:

$$\delta\dot{r}^L \cong \delta v^L - \left(\omega_{EL}^L \times\right) \delta r^L \tag{4-66}$$

Applying the first-order Tylor series approximation to the velocity equation gives the differential equation of the velocity error. By differentiating the following velocity rate equation with respect to state vector $\vec{x}(t) = \begin{pmatrix} \vec{r} & \vec{v}^L & \vec{\emptyset}_B^L & \vec{b}_\omega & \vec{b}_a \end{pmatrix}^T$

$$\dot{v}^L = C_B^L \vec{a}^B + \vec{g}^L - \left(\vec{\omega}_{EL}^L + 2\vec{\omega}_{IE}^L\right) \times \vec{v}^L \tag{4-67}$$

The differentiation will give the following velocity error model (for detailed derivation of this model, see reference [7]):

$$\begin{aligned} \delta\dot{v}^L = {} & C_B^L \delta a_{SF}^B + C_B^L a_{SF}^B \times \vec{\gamma}^L + v^L \times \left(\delta\omega_{EN}^N + 2\omega_{IE}^L \times \vec{\varepsilon}^L\right) \\ & - \left(\omega_{EL}^L + 2\omega_{IE}^L\right) \times \delta v^L + f(h)\frac{g}{R}\begin{pmatrix} 0 \\ 0 \\ \delta h \end{pmatrix} \end{aligned} \tag{4-68}$$

Another form of velocity error rate equation that does not contain the positional rotational rate $\vec{\varepsilon}^L$ can be obtained from the error analysis in the ECEF frame, which gives the following form for $\delta\dot{v}^L$

$$\delta\dot{v}^L = C_B^L \delta a_{SF}^B + C_B^L a_{SF}^B \times \vec{\Psi}^L - \left(\omega_{EL}^L + \omega_{IE}^L\right) \times \delta v^L - \frac{g}{R}\begin{pmatrix} \delta e \\ \delta n \\ 0 \end{pmatrix} + f(h)\frac{g}{R}\begin{pmatrix} 0 \\ 0 \\ \delta h \end{pmatrix} \tag{4-69}$$

The terms $\frac{g}{R}\begin{pmatrix} \delta e \\ \delta n \\ 0 \end{pmatrix}$ and $f(h)\frac{g}{R}\begin{pmatrix} 0 \\ 0 \\ \delta h \end{pmatrix}$ came from the gravity model because an error in the position will cause an error in the gravity which, in turn, will cause an error in the velocity. R here is the magnitude of the position vector connecting the center of Earth to the center of frame L and $f(h) = 2$ for positive altitude and $f(h) = 1$ for negative altitude. The term $\frac{g}{R}$ came from the gravity error. As mentioned earlier, gravity is not just a constant vector, it is a function of latitude and altitude therefore, so any error in latitude and altitude causes error in gravity, which explains the last term of the velocity error differential equation.

Schuler Effect

Schuler [5, 12, 13] (German mathematician in early 1900s) imagined a hypothetical pendulum of a length that is equal to Earth's radius and mounted on Earth's surface and pointing toward Earth's center. Schuler explained that because of gravity, the mounting platform of this hypothetical pendulum will continue pointing toward Earth's center regardless of the horizontal motion of the mounting platform on the Earth's surface. This hypothetical pendulum has a period of $2\pi\sqrt{\frac{g}{R}} = 84.4$ minutes where g is the gravity and R is the Earth's radius. The explanation of this theory is that the linear horizontal motion of the mounting platform on a curved surface would generate an angular motion that keeps the platform vertical at all times. In addition, if a linear displacement horizontal error is applied to the mounting platform, this error will oscillate at a period of 84.4 minutes.

Schuler's observation can be used to study the behavior of inertial navigation system errors for platforms that are moving on or near Earth's surface. To demonstrate the analogy, we will review the velocity error Eq. (4-69). Part of the velocity error is related to the orientation error. If, for example, the platform is stationary, and we have a perfect accelerometer and we ignore the errors that are divided by Earth's radius of curvature, and the errors that are multiplied by Earth's rotation rate, and we consider only the orientation error, we will have the following velocity errors in the horizontal channel:

$$\delta \dot{v}_e \cong g\delta r \quad (4\text{-}70)$$

$$\delta \dot{v}_n \cong -g\delta p \quad (4\text{-}71)$$

From orientation error, assuming we have a perfect gyroscope, and ignoring Earth's rotation part, we get the following approximate equation:

$$\delta \dot{r} \cong \frac{-1}{R+h}\delta v_e \quad (4\text{-}72)$$

$$\delta \dot{p} \cong \frac{-1}{R+h}\delta v_n \quad (4\text{-}73)$$

For most scenarios near Earth's surface, altitude h is very small compared to Earth's radius of curvature R; we get the following further approximation:

$$\delta \dot{r} \cong \frac{-1}{R}\delta v_e \quad (4\text{-}74)$$

$$\delta \dot{p} \cong \frac{-1}{R}\delta v_n \quad (4\text{-}75)$$

From Eqs. (4-74) to (4-75), we notice two interesting observations: (a) there is coupling between roll/pitch angles error and east/north velocities, (b) if we differentiate the horizontal velocity errors, and substitute $\delta\dot{r}$ and $\delta\dot{p}$ by their expressions in Eqs. (4-74)–(4-75), we get the following undamped oscillatory equations:

$$\delta \ddot{v}_e + g\frac{1}{R}\delta v_e = 0 \quad (4\text{-}76)$$

$$\delta \ddot{v}_n + g\frac{1}{R}\delta v_n = 0 \quad (4\text{-}77)$$

The frequency of the oscillation of the above equations is $\sqrt{\frac{g}{R}}$ Hz, which gives a period of approximately 84.4 minutes, which is the same as Schuler's period [7, 12]. This coupling between roll/pitch angles error and east/north velocities can be demonstrated visually in Figure 4.26. The coupling between roll/pitch angles error and east/north velocities can be observed by processing raw IMU data using the MATLAB project "*InertialNavSystemProject*" that we discussed in Section 4.6.1.3. We used the MATLAB environment in the "*InertialNavSystemProject*" folder to process noisy IMU data and plot the errors in Figure 4.27, which shows the coupling.

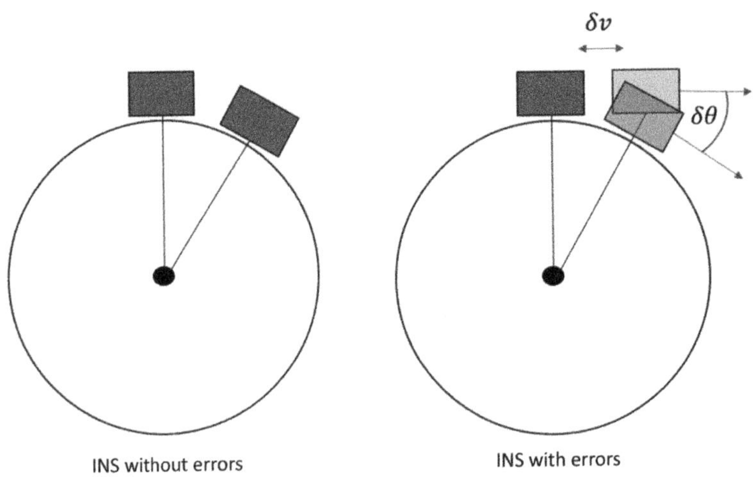

Figure 4.26 Application of Schuler's period on INS errors.

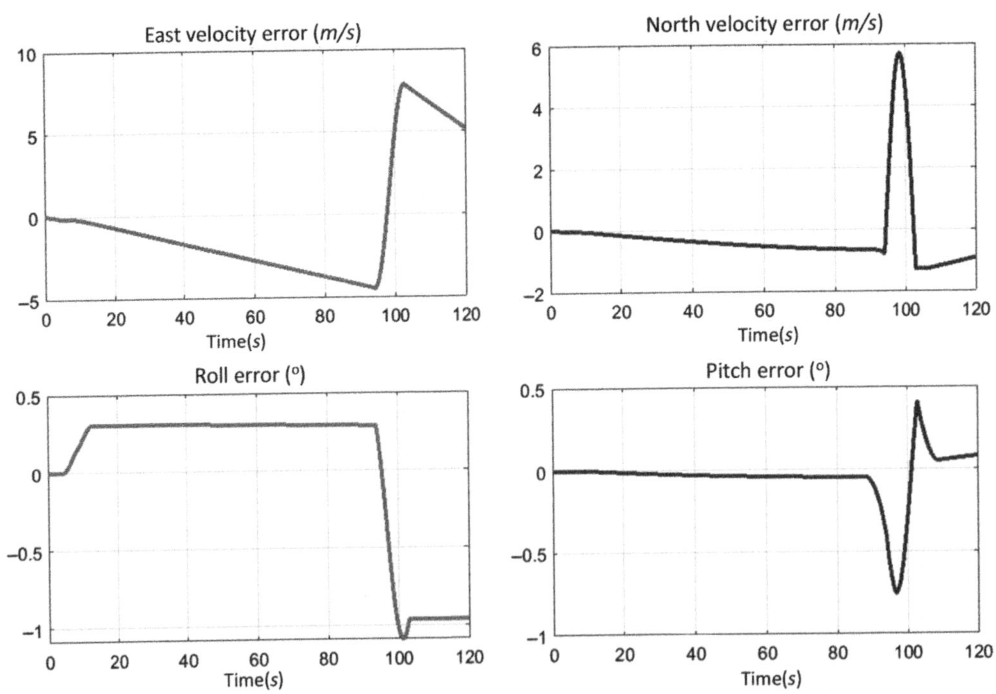

Figure 4.27 Coupling between roll/pitch angles error and east/north velocities.

4.7 Problems

4.7.1 Problem 1

An object located at latitude of −32.8308°. Calculate the Earth rotation rate vector as represented in L frame ($\vec{\omega}_{IE}^{L}$) in degrees per second, given that Earth is rotating around its axis at a rate of approximately 7.2921150×10^{-5} r/s.

4.7.2 Problem 2

If a gyroscope is attached to the object (body frame B) in Problem 1, and the object is stationary, the orientation of the gyroscope with respect to the L frame is given by the following Euler angles:

$$r = 30.7918^{\circ}$$

$$p = 1.5565^{\circ}$$

$$A = 84.8672^{\circ}$$

calculate the gyroscope measurement due to Earth's rotation in degrees per second.

4.7.3 Problem 3

Given a stationary 3-D accelerometer that is positioned at a roll angle (29.7656°), pitch angle (1.8616°), and heading angle (44.6779°) with respect to L at geodetic latitude of −32.8176° and longitude of −68.7967° and altitude of 721. Calculate the accelerometer measurement (in meters per squared seconds) as represented in the accelerometer body frame B.

4.7.4 Problem 4

The direction cosine matrix of an object B with respect to local frame L at time k and $k+1$, respectively, are given as follows:

$$C_B^L(k) = \begin{pmatrix} -0.0202 & -0.8639 & 0.5033 \\ 0.9998 & -0.0176 & 0.0099 \\ 0.0002 & 0.5034 & 0.8641 \end{pmatrix}$$

$$C_B^L(k+1) = \begin{pmatrix} -0.0242 & -0.8639 & 0.5031 \\ 0.9997 & -0.0209 & 0.0123 \\ -0.0001 & 0.5032 & 0.8642 \end{pmatrix}$$

Calculate the combined rotation rate $\vec{\omega}_{LB}^{B}(k)$ of frame B with respect to frame L that led to this change of orientation if the sampling time is 10 ms.

4.7.5 Problem 5

If the velocity of a vehicle in the N frame (east, north, and up) is (3.3776,1.0917, − 0.0279) m/s and the current geodetic location (latitude, longitude, and altitude) of the vehicle is (43.6546 ° , −79.3768 ° , 99.4 m), calculate the Coriolis force that is applied on the vehicle due to its velocity and Earth rotation.

References

1 P. D. Groves, "*Principles of GNSS, Inertial, and Multisensor Integrated Navigation Systems,*" Artech House, 2013.
2 W. Lang, "*Sensors and Measurement Systems,*" River Publishers, 2021.
3 National Centers for Environmental Information (NCEI), "Magnetic field calculators," NOAA, 2023. [Online]. Available: https://www.ngdc.noaa.gov/geomag/calculators/magcalc.shtml.
4 Government of Canada, "Geomagnetism Canada," Natural Resources Canada, 2023. [Online]. Available: https://www.geomag.nrcan.gc.ca/index-en.php.
5 P. Bartz, "Razor-AHRS – magnetometer_calibration," github, 2013. [Online]. Available: https://github.com/Razor-AHRS/razor-9dof-ahrs/tree/master/Matlab/magnetometer_calibration.
6 J. A. Farrell, "*Aided Navigation, GPS with High Rate Sensors,*" New York: McGraw Hill, 2008.
7 P. G. Savage, "*Strapdown Analytics, Maple Plain,*" Minnesota: Strapdown Associates, Inc., 2007.
8 H. C. Lefevre, "*The Fiber-Optic Gyroscope,*" Third Edition, Artech House, 2022.
9 V. F. Kroupa, "*Frequency Stability: Introduction and Applications,*" Wiley-IEEE Press, 2012.
10 IEEE, "IEEE Std. 647-2006 IEEE Standard Specification Format Guide and Test Procedure for Single-Axis Laser Gyros," IEEE, 2006.
11 Mathworks Inc., "Inertial Sensor Noise Analysis Using Allan Variance," Mathworks Inc. [Online]. Available: https://www.mathworks.com/help/nav/ug/inertial-sensor-noise-analysis-using-allan-variance.html.
12 C. Jekeli "*Inertial Navigation Systems with Geodetic Applications,*" De Gruyter, Inc., 2000.
13 P. G. Savage, "Schuler Oscillations," SAI-WBN-14003, Strapdown Associates, Inc, [Online]. Available: www.strapdownassociates.com.

5

Radio Positioning Systems

5.1 Introduction

Radio positioning systems use electromagnetic waves in the radio spectrum to build positioning systems. The radio spectrum is a subsection of the electromagnetic spectrum that is used for communication. A radio wave travels at the speed of light. If the time-of-flight (ToF) of a radio wave is measured, the distance between the transmitter and the receiver can be estimated by multiplying the ToF by the speed of light. If the distances between a target and multiple known-location anchors are known, the position of the object can be calculated using trilateration algorithm (to be explained later in this chapter). One of the very popular radio positioning technology realizations is the global navigation satellite system (GNSS). GNSS technology is one of the dominant navigation systems that are used in civil and military applications. Illustration of the GNSS technology is shown in Figure 5.1. The popular GPS system (the Global Positioning System) is one of the early realizations of GNSS technology developed by the United States Department of Defense. GNSS technology utilizes the upper top section of the radio waves spectrum known as the microwave spectrum. In addition to the main concepts of radio positioning systems, this chapter will explain GNSS technology and its different realizations, taking GPS as an example.

5.2 Concept of Operation

A simple illustration of radio positioning system components is shown in Figures 5.2 and 5.3. A key item in radio positioning systems is the navigation signal. The navigation radio signal is designed to enable target equipment to measure the time of flight of the signal and to estimate the range between the anchors and the target. We will discuss the navigation signal structure in a later section. Using multiple range measurements from different anchors, target location can be estimated using least-squares algorithms. In this chapter, we will explain in detail the least-square algorithms that estimate a target's positioning from radio signal measurements. We can also measure the angle-of-arrival (AoA) of a radio wave by either directing the wave in a certain direction or by using an array of antennas to estimate the angle of arrival of the wave. AoA can be used with ToF to calculate positions using a smaller number of anchors. Other Time-based radio positioning techniques include time-of-arrival (ToA) or time-difference-of-arrival (TDoA) to calculate targets' position. Time-based radio positioning is one of many techniques that can be used to estimate targets' positions from radio wave measurements. Other position-depending measurements can

Sensor Fusion Approaches for Positioning, Navigation, and Mapping: How Autonomous Vehicles and Robots Navigate in the Real World: With MATLAB Examples, First Edition. Mohamed M. Atia.
© 2025 The Institute of Electrical and Electronics Engineers, Inc. Published 2025 by John Wiley & Sons, Inc.

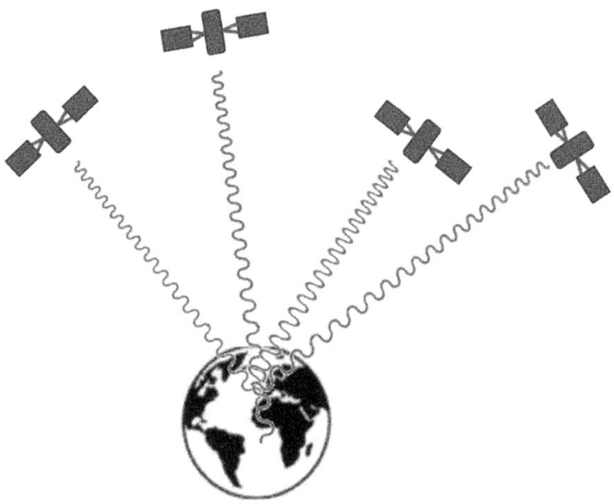

Figure 5.1 Simple illustration of global navigation satellite system (GNSS).

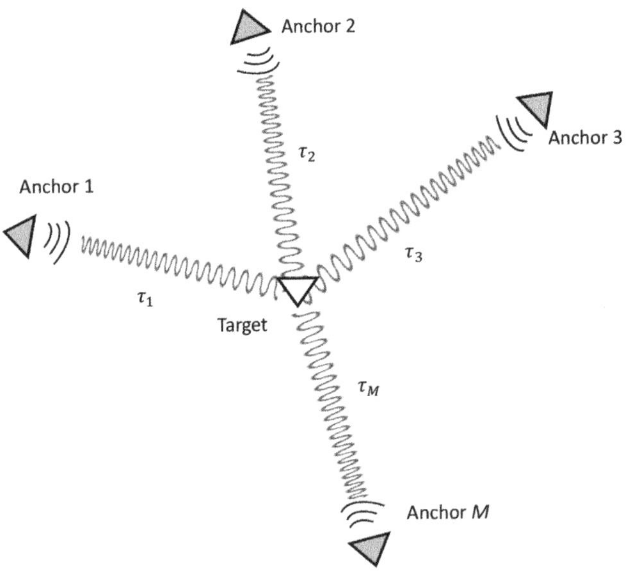

Figure 5.2 Radio positioning system illustration.

be used such as received signal strength (RSS) or radio channel state information "CSI." RSS and CSI can be used to build models that estimate distances between anchors and targets and use the estimated distances to estimate positions. RSS and CSI can also be used to build models that directly map RSS or CSI measurements to a position and save this mapping into a database called a radio map. Targets can then be localized by matching measured RSS and/or CSI against the radio map and estimating the target targets. The rest of the chapter will explain the basic navigation radio signals, receiver design, signal processing, and common radio positioning schemes and configurations. In this chapter, we will study time-based radio positioning systems with GNSS/GPS as an example.

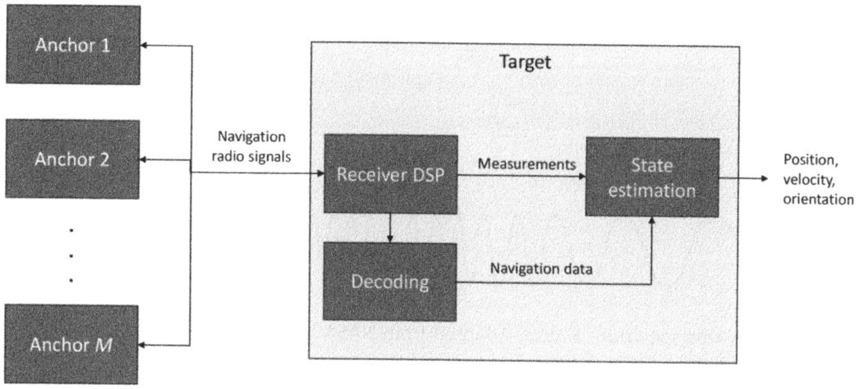

Figure 5.3 Basic illustration of radio positioning system components.

5.3 The Electromagnetic Waves and Radio Spectrum

Electromagnetic waves are generated from the oscillatory motion of electrons through a metal conductor (e.g., an antenna) [1]. This oscillatory motion of electrons can be induced by applying an alternating (oscillating) voltage source through a simple RLC circuit as shown in Figure 5.4. The circuit in Figure 5.4 is overly simplified to illustrate the concept. When electromagnetic waves hit a receiver antenna, they are converted back to an electrical current/voltage that can be digitized and processed by a computer. One of the key advantageous features of electromagnetic waves is that they travel through space at a known speed. Although the signal power is attenuated when traveling long distances and/or going through objects and obstacles, their speed remains constant. This is the fundamental feature that makes electromagnetic waves useful in ranging, mapping, and positioning systems.

Electromagnetic waves are divided into spectrum ranges based on the wave frequency or wavelength as seen in Figure 5.5. An electromagnetic wave has frequency f, wavelength λ, and speed of travel c. Frequency is defined as the number of full wave cycles generated per second. Units of

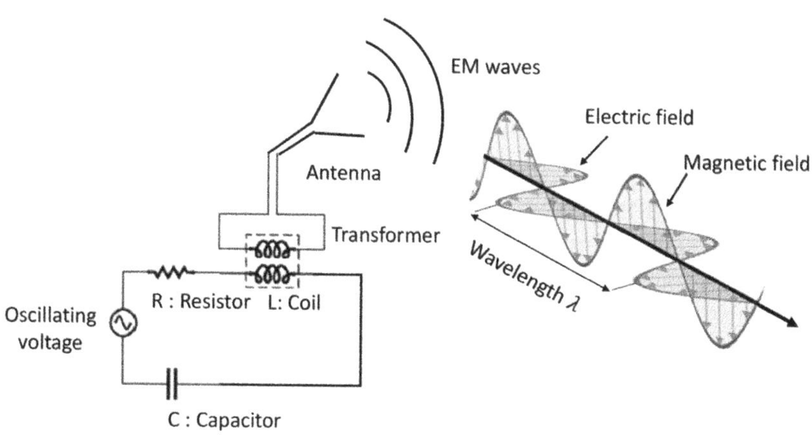

Figure 5.4 Simple illustration of electromagnetic waves generation and transmission.

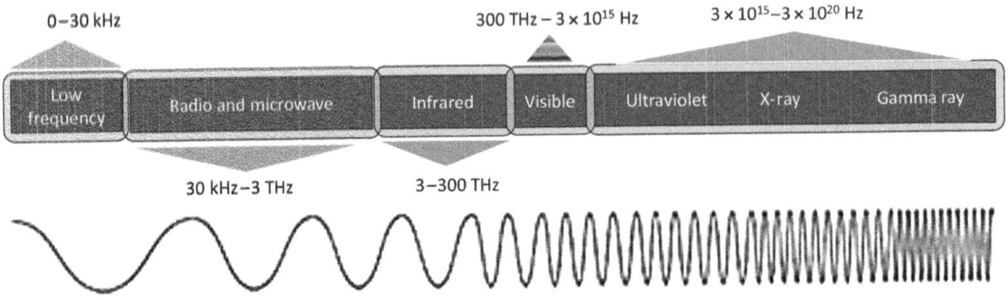

Figure 5.5 Electromagnetic waves spectrum. *Source:* Adapted from NASA [2].

frequency is Hertz (Hz). Wavelength λ is the length of a full wave cycle measured in distance units such as meters (m). Speed of travel c is measured in unit distance per unit of time and meters/seconds (m/s) is mostly used. The time of travel of one wavelength is called the wave period T measured in unit time (seconds "s"). Electromagnetic waves travel at the speed of light, which is estimated at approximately 300,000,000 m/s. The relationships between frequency, period T, speed c and wavelength λ are given by the following equations:

$$f = \frac{1}{T} \tag{5-1}$$

$$f\lambda = c \tag{5-2}$$

Figure 5.5 shows the electromagnetic frequency spectrum. The radio spectrum ranges from 30 kHz to 3 THz. GNSS technology uses the higher part of the radio spectrum called "microwave range" [3]. Microwaves range from 3 MHz to 300 GHz. When electromagnetic waves hit an antenna equipped with necessary circuitry, the antenna induces an electrical voltage which converts back the electromagnetic wave into an electrical charge in the form of electric current or voltage (signal). The selection of microwave spectrum for radio positioning systems is based on the penetration and power characteristics of microwave signals. Microwaves require reasonably low power to generate. In addition, microwave signals can travel very long distances into space, and they can penetrate through the atmosphere and ionosphere layers [1]. Therefore, microwave spectrum is the preferred spectrum for space and satellite communication and radio navigation and radar systems.

5.4 Radio Positioning Transmitter

Radio positioning systems use signal generators and transmitter circuits to transmit navigation radio signals. The transmitted signal consists of three main components: carrier, data, and code. A simple block diagram of the transmitter is shown in Figure 5.6. The practical and detailed implementation of transmitters varies from one system to another, but they all share the general design shown in Figure 5.6. We will discuss later in this chapter how the global positioning system (GPS) implements signal generation and transmission. We will refer to the device that generates and transmits positioning radio signals as the anchor.

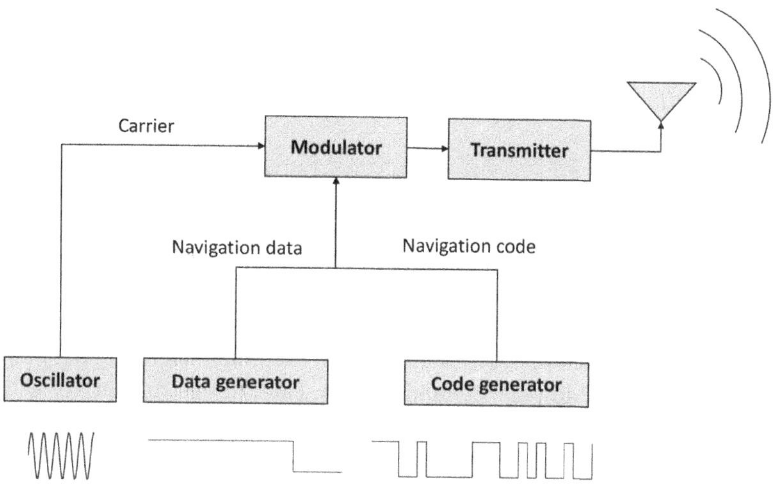

Figure 5.6 Basic block diagram of transmitter in radio positioning systems.

5.5 Positioning Signal Structure

Positioning signals consists of three main parts, carrier wave, positioning code, and navigation data. The three parts and the construction of the positioning signal is explained in this section.

5.5.1 Carrier Wave

The carrier signal is a sinusoidal signal generated by electronic oscillators at a desired frequency, as seen in Figure 5.7. Carrier frequency is selected to be within the radio spectrum, and it is the highest frequency component of the transmitted signal. From the name, the carrier signal carries the desired information (navigation messages and positioning code). We cannot send navigation

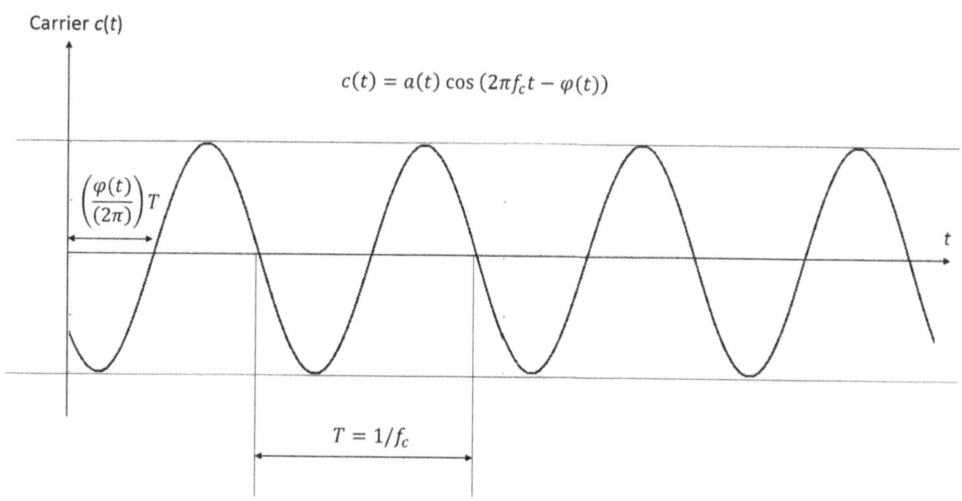

Figure 5.7 Carrier signal.

messages and code directly as they are at a much lower frequency. The general mathematical formula of the carrier signal is given by:

$$c(t) = a(t)\cos(2\pi f_c t - \varphi(t)) \tag{5-3}$$

where

$a(t)$ is the amplitude
$\varphi(t)$ is the phase
Frequency f_c is the carrier frequency.

This general form can be used to carry information on the carrier. Carrying information (i.e., modulation) can be performed by representing the information in the amplitude, the phase, or the frequency. Modulation is a wide topic in radio communication. Later in this chapter, we will explain the modulation scheme used by GPS as an example. But for now, you can think about modulation as simply embedding the required information (in our case navigation data and code) onto the high-frequency carrier signal.

5.5.2 Navigation Data

Navigation data includes key parameters that are needed by the target to calculate the range and position of the received radio signal. For example, in satellite radio positioning systems, navigation data include orbital parameters and transmission time of the signal that can be used to calculate the satellite's position and speed at a given time. As we will explain later, the receiver needs to know the satellites' position and velocity to calculate its own position and velocity. Electronic clocks that are used by transmitters and receivers to track time [4] are core components. In addition, electronic clocks synchronize and control the operations of all components of the circuits, including the oscillators that generate the carrier signals. Electronic clocks are not perfect, and they may have biases and drifts. The navigation data may include clock corrections. In GNSS technology, satellite clocks are managed by control stations that constantly estimate the clocks' biases and drifts. GNSS navigation data includes satellite clock biases and drifts so the target receiver can use this information to calculate the correct transmission time of the signal. Typically, navigation data are partitioned into frames. Frames are identified by known headers or preambles so the target can detect the beginning and the length of a data frame. Navigation data is typically transmitted as a sequence of bits at lower frequencies. More details about navigation data are discussed in the GNSS technology section.

5.5.3 Positioning Code

Code is a unique predefined sequence that identifies the anchor's signal. It is like a fingerprint embedded into the signal that uniquely identifies the anchor. In modern digital positioning systems, code is a binary sequence that looks like the one shown in Figure 5.8.

Code is always chosen to have strong correlation properties. Mathematically, code signals are designed to be orthogonal as follows:

$$r(k) = \frac{1}{m}\sum_{0}^{m-1} c_i(n)\cdot c_j(n+k) = \begin{cases} \cong 0 & \text{if } i \neq j \\ \cong 0 & \text{if } k \neq 0 \\ > 0 & \text{iff } k = 0 \text{ and } i = j \end{cases} \tag{5-4}$$

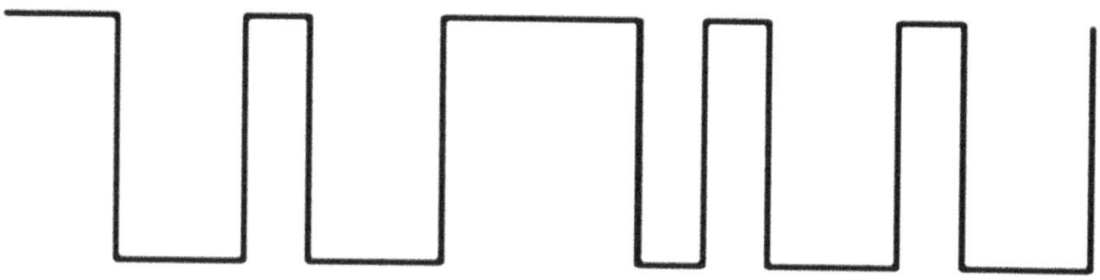

Figure 5.8 Example of a code sequence.

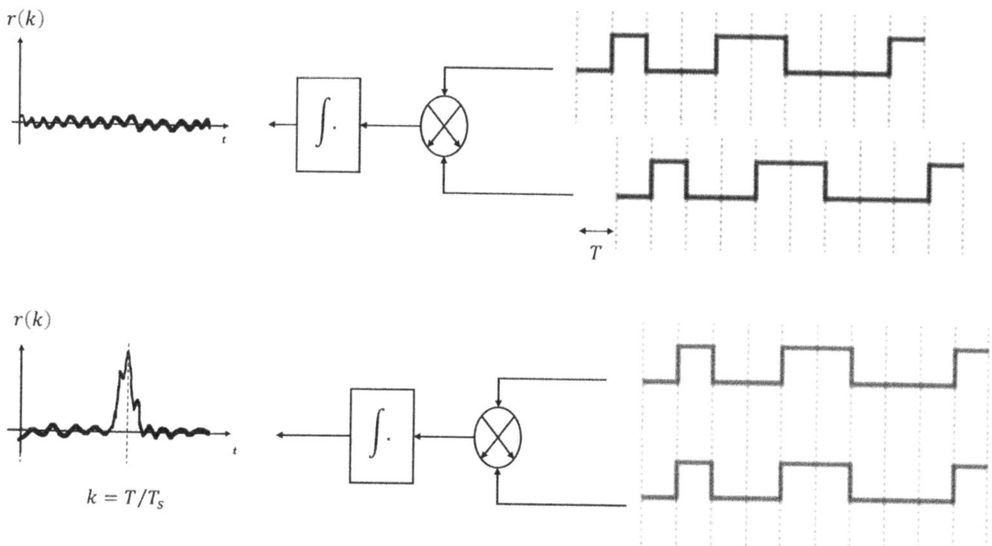

Figure 5.9 Illustration of the correlation properties of the code signal.

If we have two copies of the same code but they are shifted by a delay, T, the correlation $r(k)$ will have a peak at delay $k = T/T_S$ as shown in Figure 5.9.

If we have multiple anchors transmitting at the same carrier frequency (which is the case in GPS technology that uses Code Division Multiple Access (CDMA) communication protocol [5] [6]), we want a way to select a specific anchor and reject other anchors to process its signal. The navigation code is also used for this purpose. To detect the presence of a specific anchor and to cancel out or attenuate the signals of other anchors, the receiver correlates the received signal with an internal copy of the code of interest and searches for a correlation peak. This correlation process is called a matched filter, and it is used in GPA and RADAR systems. Assume the code signal is $c(t)$, the receiver will receive a noisy attenuated delayed copy of $c(t)$ defined as follows:

$$\hat{c}(t) = ac(t - T) + n(t) \tag{5-5}$$

where

a is an attenuation factor

T is the time delay of the signal, which is equivalent to the time shift between the transmitted signal and the received signal

$n(t)$ is a random noise commonly assumed to be white Gaussian noise.

If the signal reflected or refracted from obstacles and objects in the environment, the receiver will receive multiple attenuated copies of the code as follows:

$$\hat{c}(t) = \sum_{j=1}^{m} a_j c(t - T_j) + n(t) \tag{5-6}$$

This scenario is called the "multipath" scenario, which is one of the most challenging problems in radio positioning systems. To explain the main concept, we will ignore the multipath effect. To detect the presence of a specific code from a specific anchor, a correlation is performed between the received code $\hat{c}(t)$ and an internal exact copy of the code $c(t)$ as follows:

$$r_{c\hat{c}}(\tau) = \int \hat{c}(t)c(t - \tau)\, dt \tag{5-7}$$

where τ is the correlation delay in seconds.

In discrete form, and assuming a sampling period T_s and code length of N samples, the correlation is given by:

$$r_{c\hat{c}}(nT_s) = \sum_{k} \hat{c}(kT_s)c(kT_s - nT_s), k = 0, 1, 2, \ldots, N \tag{5-8}$$

where n is the correlation delay in samples.

This correlation will be maximum when $n = T/T_s$ (or $\tau = T$ in the continuous time domain) as illustrated in Figures 5.10 and 5.11. Figures 5.10 and 5.11 show binary digital code signals which is commonly used in most radio positioning systems.

To find unique maximum peaks that are easily distinguishable, the code signal must have strong correlation characteristics. GPS technology uses codes that are called golden sequences. They are pseudo-random sequences (PRN). Being pseudo-random maximizes the correlation characteristics. This is known from random signal properties. The extreme case of this maximum correlation

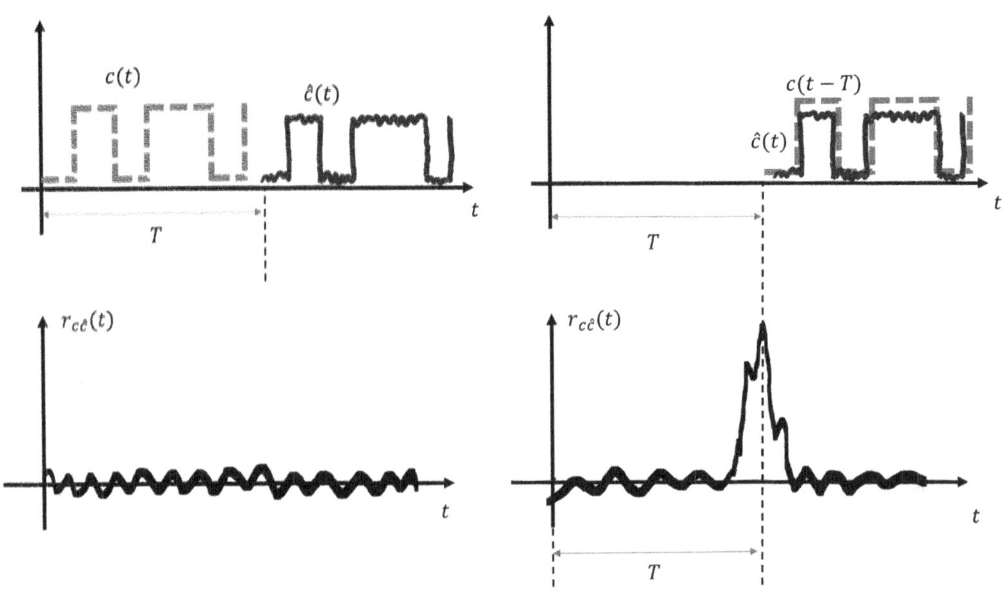

Figure 5.10 Correlation illustration between $c(t)$ and $\hat{c}(t)$.

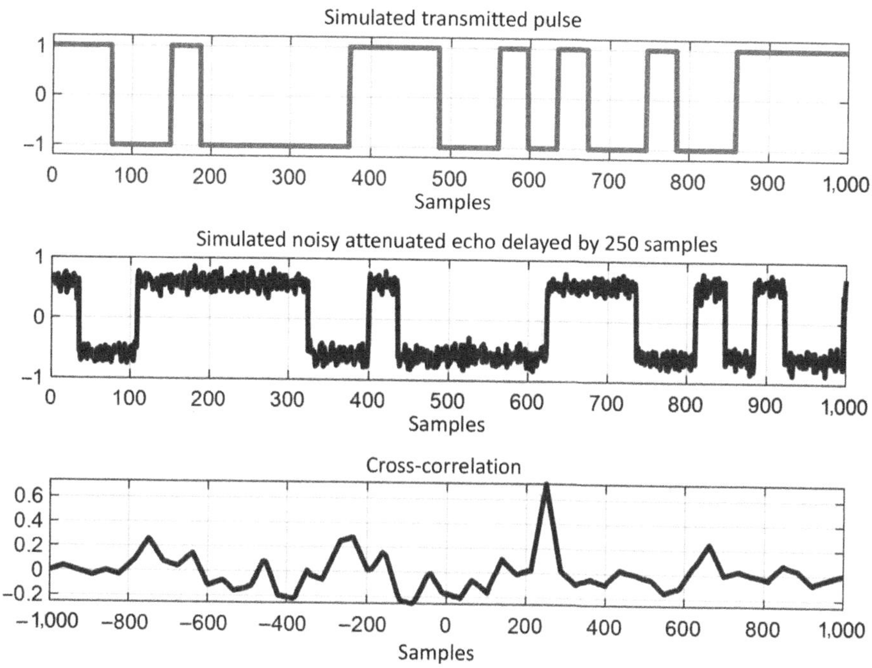

Figure 5.11 Correlation between an ideal replica $c(t)$ and a noisy delayed signal $\hat{c}(t)$.

happens with Gaussian signals. An idealized Gaussian random signal has zero correlation with any small, shifted version of itself (when the two signals are not perfectly aligned) and maximum correlation with itself (when the two signals are perfectly aligned). Figure 5.12 shows the correlation between GPS PRN code 1 and a delayed version of itself. The delay is 0.2 ms. As we can see in Figure 5.12, the correlation has a distinguishable peak at 0.2 ms delay and approximately zero elsewhere.

One key characteristic of a code signal is its frequency. It must have a frequency that is high enough to find the correlation peak with a high resolution. The correlation of low-frequency signals will have a wide peak (see Figure 5.12 bottom plot) that causes large uncertainty in calculating the time of arrival of the signal. The correlation of higher frequency signals will have a narrow sharp peak (see Figure 5.12 top) that provides a high-resolution estimation of the time of arrival.

There are some important definitions of code parameters that need to be clear before discussing range estimation and position calculation in detail. Common key definitions include code sequence, code chip, code phase, and code delay. In most modern radio positioning systems, code is a binary digital sequence of bits in the form shown in Figure 5.13. A code chip is a single bit in the code sequence. The first bit of the code is Chip 0, the second bit is Chip 1, and so on. The code has a length N_c (chips) a duration T_c (s), and chip rate f_{ch} (in chips/s). Chip duration T_{ch} is the chip period, which can be calculated by dividing code duration by code length. In GPS C/A Code, for example, $N_c = 1023$ chips, $T_c = 10^{-3}$ s, $f_{ch} = 1.023$ million chips per second. In the target receiver, all processing is digital, so the incoming signal is converted into discrete samples at a specific sampling frequency. We will use f_s to refer to the sampling frequency. If we used a sampling frequency f_s, this means the code sequence has the following number of samples:

$$N_s = f_s \cdot T_c \tag{5-9}$$

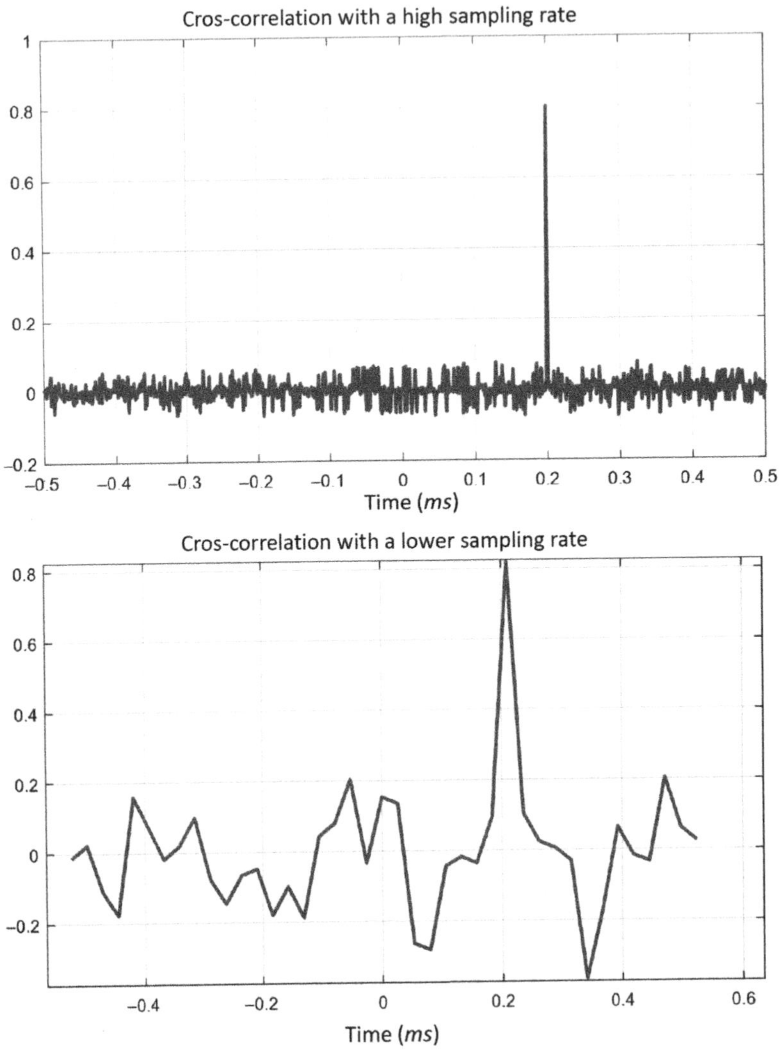

Figure 5.12 GPS RPN1 code correlation with a delay of 0.2 ms. Top plot used 38.192 MHz sampling frequency while bottom plot uses 38.192 kHz.

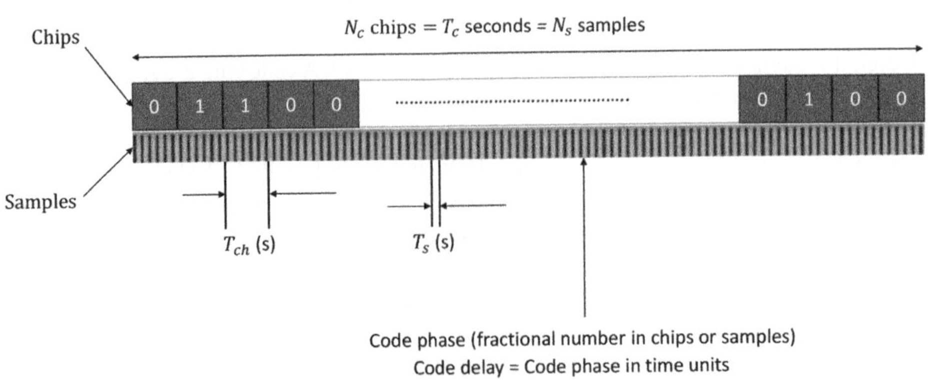

Figure 5.13 Navigation code structure and definitions.

Code phase tells us which chip $[0, N_c - 1]$ or sample $[0, N_s - 1]$ the code signal hit the receiver antenna. Code phase is a fractional number and if multiplied by chip period T_{ch} (or sampling duration T_s if sample number are used as a measurement), this will give us the code delay in seconds. We will refer to the code delay by τ_D and to the code phase (in number of samples) by n_D and we will use these variables when we discuss the acquisition and tracking modules.

Both radio positioning and radar systems use code signals. The main difference between radar and positioning systems is that radar transmits a code and listens to the echo while positioning transmits a continuous stream of codes without listening. In radar, code delay is an indication of distance or range, but in positioning systems, code delay expresses how much time the code sequence signal takes when it hits the receiver's antenna. This time shift is used to calculate the time of arrival of the signal precisely using the internal receiver's clock. In both cases, we are going to apply the matched filter to detect this time shift.

5.5.4 Modulated Signal

The code and data are modulated on the carrier signal and transmitted. The modulation can be done in different ways. For example, in GPS technology, binary phase shift keying (BPSK [4]) modulation is performed. BPSK modulation is a simple modulation that is commonly used to modulate binary data. In GPS L1 signal, Code and Data are combined using XOR operation if the information is coded by 0, 1 bits or by ordinary multiplication if the information is coded by −1, +1 bits. Given an anchor k, that transmits a code signal $C^k(t)$ and data signal $D^k(t)$ using the carrier frequency f_{L1}, the BPSK modulated signal will be generated as follows:

$$S^k(t) = C^k(t) \bigoplus D^k(t) C_{L1}(t) \tag{5-10}$$

This modulation is called binary phase shift keying because the carrier phase is shifted by 180° when a flip in the amplitude envelops $C^k(t) \oplus D^k(t)$ happens. This modulation is shown in Figure 5.14. It is important to note that this is only one part of the actual GPS signals. GPS technology uses multiple codes and different frequencies. This example here is just for the illustration of the concept. More detailed information about different codes and modulation can be found in [3], [7].

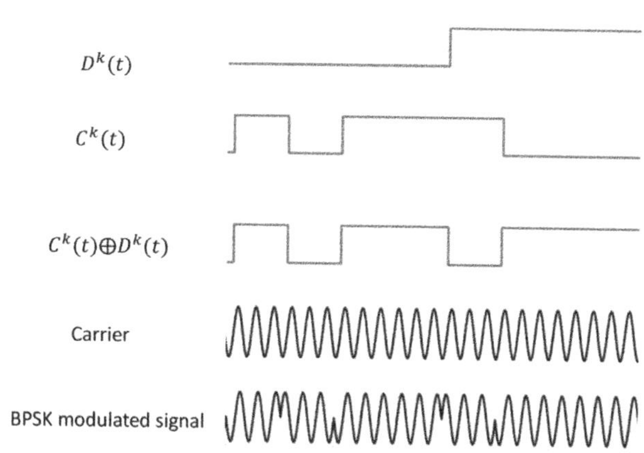

Figure 5.14 BPSK modulated positioning signal with code, data, and carrier combined.

5.6 Receiver Signal Processing

The general block diagram of the receiver is shown in Figure 5.15. Targets receive signals using a module that is called front-end. The front-end is the first module that the signal hits in the receiver. The front-end delivers a digitized down-converted copy of the received signal to the next module, which is the acquisition module. The acquisition module estimates the received signal carrier frequency and code delay. The next module is the tracking module that refines and tracks the changes in carrier frequency and code delay to keep in synchronization (locked) with the incoming signal. Estimating the carrier frequency and code delay enables the receiver to demodulate the signal, extract the navigation data, and at the same time estimate the time of arrival of the incoming signal with high accuracy. This process is called demodulation. We will explain each block in the following subsections.

5.6.1 Front-End Module

The front-end is the first subsystem that receives the anchors' transmitted electromagnetic waves. The front-end's antenna will receive any electromagnetic wave in the space, but the front-end components filter out (or attenuate) waves based on signal frequencies.

5.6.1.1 Main Components of Front-End

The general block diagram of a front-end is given in Figure 5.16. It is important to realize that Figure 5.16 is a very basic overly simplified diagram just to highlight the main concepts and blocks of front-end module. We will see a more detailed diagram of a GPS front-end as an example when we discuss GNSS technology. As Figure 5.16 shows, the front-end starts with an antenna that has an internal band pass filter (BPF). This BPF will be centered around the carrier frequency that the anchor transmits on. The signal is then amplified and serialized to the front-end circuit. The front-end circuit further filters and amplifies the incoming signal. Then, the incoming signal is mixed with a local oscillator and the output is then band-pass filtered to generate the intermediate frequency (IF) signal that is further amplified and digitized using the analog to digital converter (ADC) and serialized as digital samples to the receiver processor. The concept of the intermediate

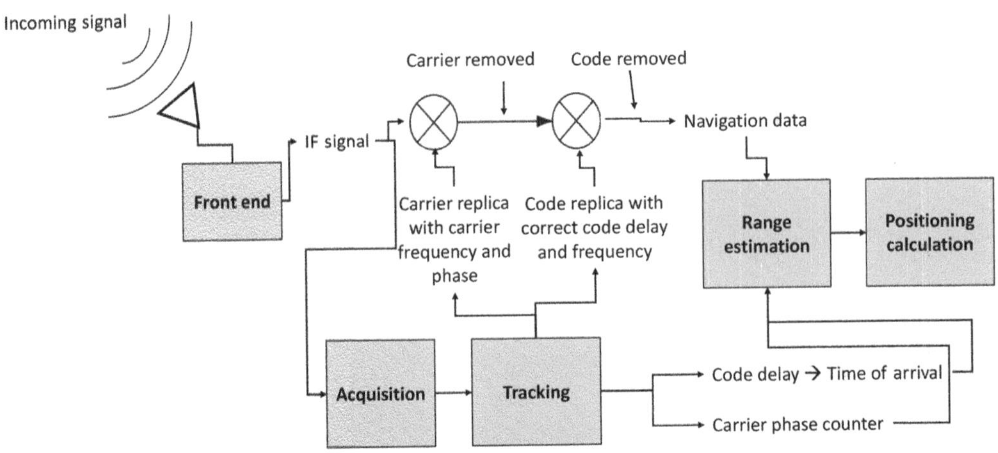

Figure 5.15 General block diagram of the receiver.

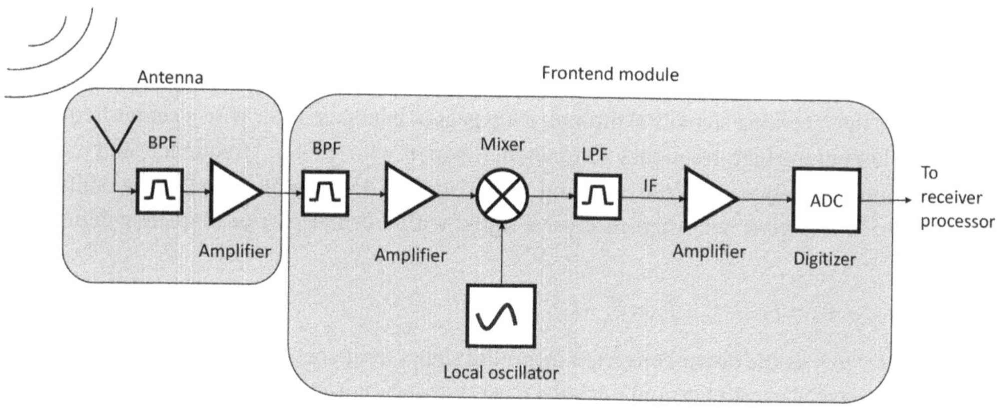

Figure 5.16 General front-end block diagram.

frequency is very important to understand. The mathematical detail of the intermediate frequency is explained in the next subsection.

5.6.1.2 Frequency Down Conversion and Doppler Frequency

The transmitter transmits at a nominal carrier frequency f_c. What the front-end is doing is down-converting the received signal from its high-frequency f_c to a lower frequency so the ADC can digitize the signal and the receiver processor can process the signal. Working directly on the nominal transmitter frequency is not possible due to the very high frequency, which requires an extremely large sampling frequency, which is not practical. The IF frequency is the difference between the nominal carrier frequency f_c and the mixer local oscillator frequency f_{LO}. The concept of the mixer is simple. If we have an incoming signal with f'_c frequency multiplied by a signal of f_{LO}, we get the following signal:

$$s(t) = A\cos(2\pi f'_c t + \varphi)\cos(2\pi f_{LO}t) = \frac{A}{2}\cos(2\pi[f'_c - f_{LO}]t + \varphi) + \frac{A}{2}\cos(2\pi[f'_c + f_{LO}]t + \varphi) \quad (5\text{-}11)$$

If we pass by a low-pass filter, the resulting *IF* signal will be

$$s_{IF}(t) = \frac{A}{2}\cos(2\pi[f'_c - f_{LO}]t + \varphi) \quad (5\text{-}12)$$

The down-converted signal $s_{IF}(t)$ is the one that will go to the acquisition and tracking modules. In the next subsections, we will refer to $s_{IF}(t)$ as the incoming IF signal, but the reader should recognize that this is a down-converted version of the original transmitted signal.

Note that we used f'_c instead f_c because due to the relative motion between the receiver and the transmitter, the receiver receives a signal with a frequency different from the nominal transmitter frequency. The difference between the nominal transmitter frequency f_c and the actual received frequency f'_c is known as Doppler shift. The Doppler shift is used in radio ranging and positioning systems to estimate target's velocity. Therefore, f'_c is shifted from f_c by the Doppler shift. f'_c can be defined as follows:

$$f'_c = f_c + f_D \quad (5\text{-}13)$$

If we define the intermediate frequency as follows:

$$f_{IF} = f_c - f_{LO} \tag{5-14}$$

and if we pass the incoming signal $s(t)$ through a lowpass or bandpass filter with a center frequency f_{IF}, we will filter out the high-frequency component (the part with $[f'_c + f_o]$ frequency) and only the f_{IF} and surrounding frequencies (the code and the navigation data signals plus Doppler shift) will pass through. The receiver will then receiver a signal with a center carrier frequency defined as follows:

$$f_r = f_{IF} + f_D = f_{IF} + f'_c - f_c = f'_c - f_{LO} \tag{5-15}$$

We will refer to f_r as the down-converted incoming signal frequency. As we will see soon in this chapter, the receiver can estimate and track the f_r using acquisition and tracking modules. Knowing f_{IF} and estimating f_r enables us to estimate f_D

$$f_D = f_r - f_{IF} \tag{5-16}$$

The relative speed between the transmitter and the receiver $\dot{\rho}$ given the Doppler frequency f_D and speed of signal c and nominal transmission frequency f_c is given by the following equation [5]:

$$\dot{\rho} = \frac{-cf_D}{f_c} = -\lambda f_D$$

$$f_D = -\frac{1}{\lambda}\dot{\rho} \tag{5-17}$$

The Doppler frequency f_D is an indication of the rate of change of the range between the transmitter and the receiver which is used to calculate the receiver's speed.

5.6.2 Acquisition Module

The goal of the acquisition is to find the code phase (or code delay) and the carrier frequency f_r, which are the basic input to the tracking module. The code phase/delay and carrier frequency are coarse values that are refined by the tracking module.

5.6.2.1 Code Phase/Delay Search

Serial Code Search

The digitized stream of the translated down-converted signal is fed to the acquisition module. The acquisition module has a channel for each anchor identified by the anchor's code. If all anchors are using the same carrier frequency (like the case in CDMA systems), the receiver wants to separate the signal of each anchor. This is done by convoluting the incoming signal with the desired code. As the code is pseudorandom, the multiplication and summation (i.e., correlation) will attenuate other signals from other anchors and leave only the desired anchor's signal. The incoming signal is correlated with a local copy (replica) of the anchor's code signal to achieve this selectivity. To get a peak in the correlation, the local replica must be perfectly aligned with the incoming code signal. In other words, both the incoming code and the internal replica must have the same code phase to detect a peak in the correlation. Otherwise, a peak will not be detected, and the correlation will be zero. This process is illustrated in Figure 5.17 with a sample code {1, 2, 3, 4, 5, 6}. When the receiver starts working, the first code chip that hit the antenna was code chip number 3 (code phase is 2 because code phase starts at zero). To calculate this code phase, the receiver tries different shifted codes. This shifting is called a circular shift as seen in Figure 5.17. The receiver tries different circular shifts, and

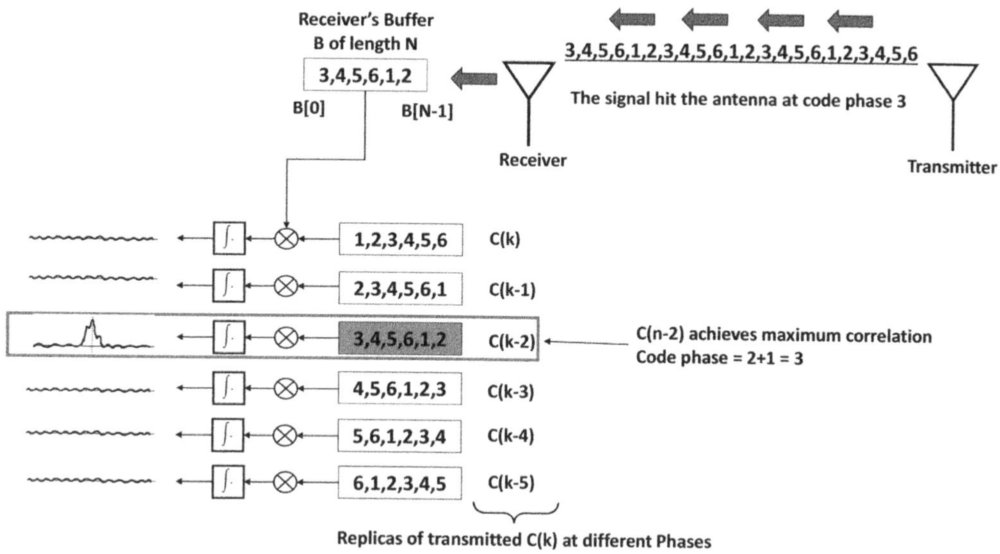

Figure 5.17 Illustration of the correlation with an incoming code signal.

the shift that makes the maximum correlation with the incoming signal is the code phase that we are looking for. The traditional way to do this correlation is to keep shifting the internal copy of the code by small shifts (fractions of a chip) and do the correlation until we detect a peak. If no peak is detected, this means either a signal from this anchor does not exist (this anchor is not visible) or the code phase of the incoming signal and the local replica is not the same. However, this way is very time-consuming. Imagine shifting the GPA C/A code that has 1,023 chips by a half chip and doing the correlation. This means that we will have to do the correlation $2 * 1{,}023$ times, which is not efficient. This way is called serial code search.

Parallel Code Search

A better way to do this search is by implementing a matched filter in the frequency domain. In frequency domain, the convolution operation is a simple multiplication, so we apply discrete Fast Fourier Transform (FFT) on the incoming signal $\hat{c}[n]$ and a reversed internal replica of the code ($h_m[n]$) and apply inverse FFT on the output. The definition of $h_m[n]$ is given by the following formula:

$$h_m[n] = c[-n]. \tag{5-18}$$

The discrete Fourier transform of the received code signal $\hat{c}[n], n = 0, 1, \ldots, N$ is given by the following formula:

$$C[k] = \sum_{n=0}^{N-1} \hat{c}[n]e^{-j2\pi\frac{k}{N}n}\ k = 0, 1, \ldots, N-1 \tag{5-19}$$

The index k here is corresponding to a frequency. As discussed in Chapter 2, the FFT will generate the frequency spectrum at frequencies in the range $f \in \left[0, \frac{F_S}{N}, \frac{2F_S}{N}, \ldots, \frac{F_S}{2}\right]$ (Hz) where F_s is the sampling frequency. The discrete Fourier transform of the reversed internal replica of the code $h_m[k]$ is given by the following formula:

$$H[k] = \sum_{n=0}^{N-1} h_m[n] e^{-j2\pi\frac{k}{N}n} \; k = 0, 1, \dots, N-1 \tag{5-20}$$

We then multiply $C[k]$ and $H[k]$:

$$Y[k] = [k]H[k] \tag{5-21}$$

Inverse Fourier transform is applied on $Y[k]$ as follows:

$$y_{c\hat{c}}[n] = \frac{1}{N}\sum_{k=0}^{N-1} Y[k] e^{+j2\pi\frac{k}{N}n} \tag{5-22}$$

The resulting output $y_{c\hat{c}}[n]$ will have a peak at the code phase. If the code phase is calculated in samples and multiplied by the sampling period, it will give the code delay in seconds. If there is no detected peak, this means that a signal from this anchor (e.g., GPS satellite) does not exist or is too weak/noisy to detect.

5.6.2.2 Matched Filter

The correlation operation is digitally implemented using a digital filter called matched filter. A matched filter is a filter whose impulse response is a reversed version of the original signal to be detected. So, in discrete form, the impulse response of the matched filter is given by:

$$h_m[n] = c[-n] \tag{5-23}$$

The incoming signal $\hat{c}[k]$ is convoluted with $h_m[k]$ as follows:

$$r_{c\hat{c}}[n] = y[n] = \sum_{k=0}^{N} \hat{c}[k] h_m[n-k] \tag{5-24}$$

$h_m[n-k]$ is a reversed delayed version of $c[k]$ (delayed by n samples).

This convolution output $y[n]$ will be maximum when n equals to the time shift between $c[k]$ and $\hat{c}[k]$. Figure 5.18 shows how convolution with a matched filter is nothing but a correlation between the incoming signal and the reversed filter impulse response.

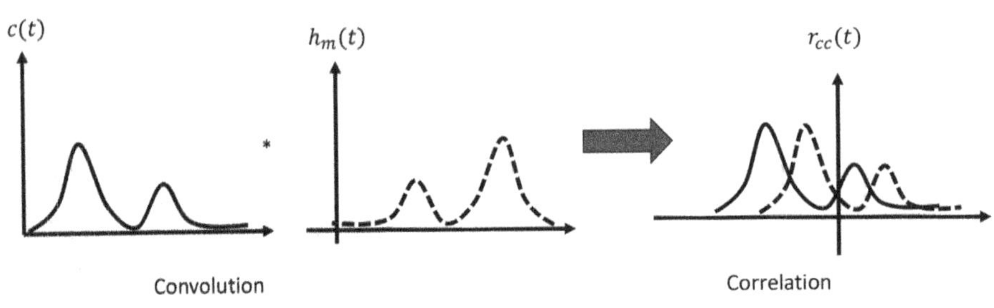

Figure 5.18 Illustration of convolution and correlation.

5.6.2.3 Carrier Frequency Search

In the above discussion, we assumed that the incoming signal has the code signal only which is not true. The incoming IF signal has the down-converted carrier signal and data components. To perfectly correlate with the incoming code, the down-converted carrier signal must be removed from the incoming IF signal. This process is called carrier demodulation. The incoming IF signal is multiplied by an exact local copy (replica), and the output is low-pass-filtered to wipe off the IF carrier signal. Recall that multiplying two exact sinusoidals will generate two signal components, one with double the frequency and one with zero frequency. If the IF signal has a center frequency f_r plus other lower frequency components (the code and data components) and it is multiplied by a sinusoidal that is exactly oscillating at f_r, the output will have two components, one centered around $2f_r$ and one centered around zero. A low-pass filter will remove the one centered around $2f_r$. The remaining part will have only the code and the data. This remaining part is centered around zero frequency, so we call this part a baseband signal. Baseband signal is simply the signal after the down-converted carrier is removed. To find the right down-converted carrier frequency, we try different values from a range of potential frequencies around the intermediate frequency (to account for the Doppler shift). The one that leads to the maximum correlation peak in the code search is the desired one.

After removing the carrier part, the remaining signal has a code and data parts. Therefore, we must still remove the data part to do the code search. However, we cannot separate the data part without removing the code. To work around this problem, radio positioning systems transmit data at a much lower rate than the code and carrier. For example, the data rate in GPS is 50 bits/s, so one bit of data contains 20 codes (code is 1 ms). If our processing buffer is the same length as the code, the data bit value will remain constant within the processing buffer. The code correlation process (the matched filter) will not be affected if the data bit does not flip within the data buffer. It is like multiplying the correlation result by a constant, which does not affect the correlation peak location.

In the above discussion, we did not discuss the carrier signal phase. We just focused on the frequency of the carrier signal. The carrier signal may hit the antenna at a phase that is anywhere between 0 and 2π as Figure 5.19 shows. It is insufficient to match the carrier's frequency to remove the carrier from the signal. We must have matched phase to perfectly remove the carrier signal.

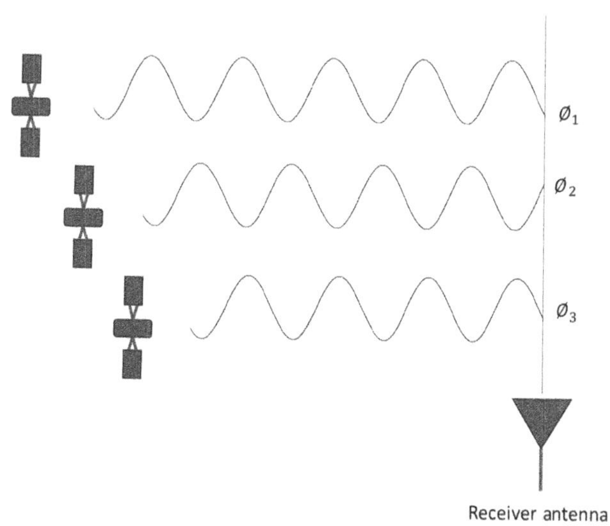

Figure 5.19 Phase of the incoming carrier wave signal.

To do so and to make the carrier frequency search neutral to the carrier phase, we process two branches of the incoming signal, the In-phase (I) branch and the Quadratic (Q) branch. This process is illustrated in Figure 5.20, which shows the general diagram of the acquisition module. To resonate with this incoming signal and wipe off the carrier signal regardless of its phase, we process two branches of the signal, I and Q. The I and Q branches are calculated as follows:

$$I = A\cos(2\pi f_r t + \varphi_r)\cos(2\pi f_{nco} t + \varphi_{nco}) \tag{5-25}$$

$$Q = A\cos(2\pi f_r t + \varphi_r)\sin(2\pi f_{nco} t + \varphi_{nco}) \tag{5-26}$$

where f_{nco} is the local oscillator frequency.

We want f_{nco} to be the same as f_r. The I and Q branches, which are then combined into a complex signal as follows:

$$s(t) = I + jQ \tag{5-27}$$

$s(t)$ is then fed to an FFT module (fast Fourier transform).

FFT is also applied to the local replica of the code and multiplied by the FFT of $s(t)$. The result of the multiplication is converted back to the time domain and a search for a peak is done. If the local carrier replica frequency f_{nco} and phase φ_{nco} equal the incoming signal frequency f_r and the phase φ_r, we will have a correlation peak at the code phase. The peak will occur at the code phase of the incoming signal. Note that we do not have to apply a lowpass filter to the I and Q branches as the high-frequency component does not affect the code correlation peak position. If there is no peak, the system will try different values of f_{nco} until a peak is found. The range of values of f_{nco} is simply the intermediate frequency plus and minus a frequency range that covers possible Doppler shift.

Acquisition is performed in multiple parallel channels where each channel searches for a specific code. The reason we want to know the code phase is to know the time of arrival of the incoming signal. This time calculation is illustrated in Figure 5.21. The code phase/delay is used to calculate the arrival time of the signal. The arrival time of the signal is calculated using the internal clock of the receiver. In Figure 5.21, the arrival times of the M anchors' signals are illustrated as τ_1, τ_2, ..., τ_M. It is important to remember that these times of arrivals are just coarse values that are not precise enough to calculate accurate range. These values will be refined in the tracking module.

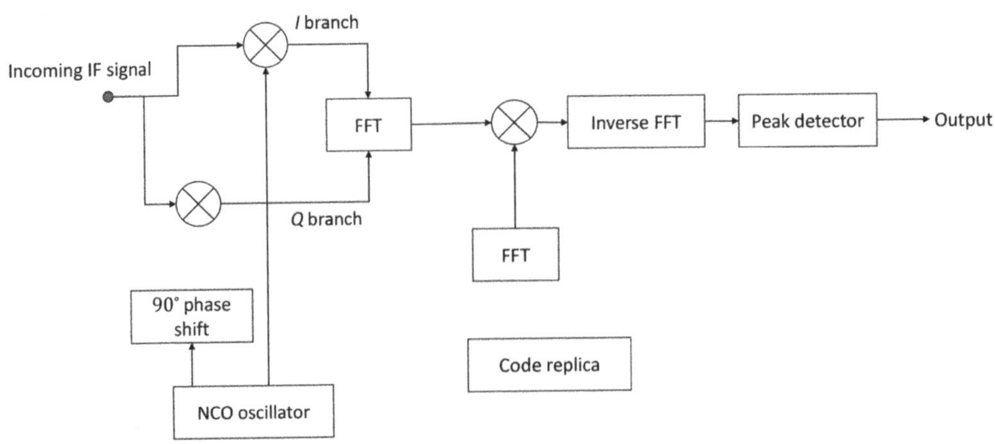

Figure 5.20 General acquisition module.

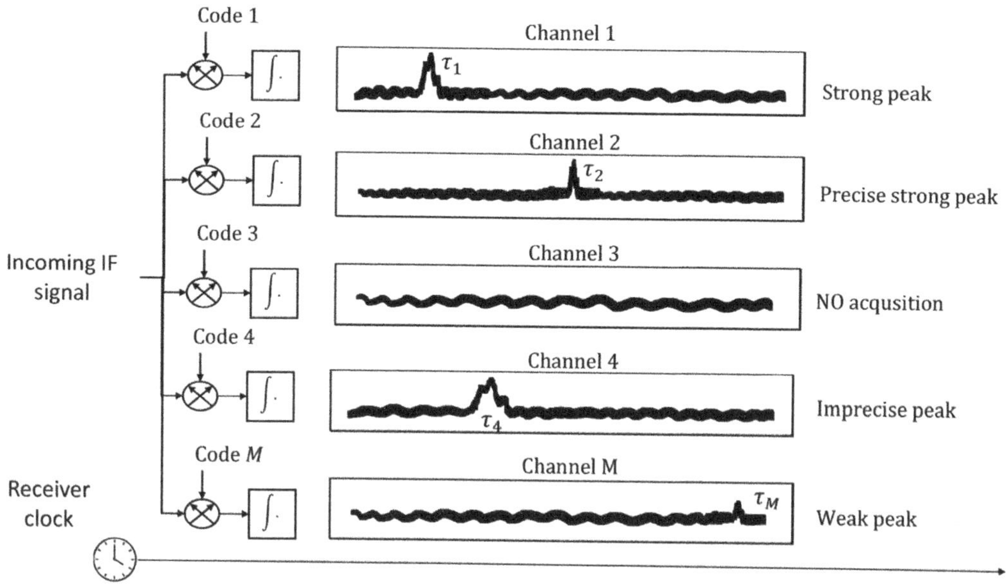

Figure 5.21 Illustration of code phase (code delay) for multiple channels.

5.6.3 Tracking Module

The tracking module achieves two important goals: (1) Refining the coarse code delay and IF carrier frequency/phase values estimated by the acquisition, and (2) As the anchors and the target are moving, the code phase and carrier frequency/phase are constantly changing over time. The tracking module keeps tracking the changes in the code phase and carrier frequency/phase. The tracking module does this using two control loops: one loop for code tracking and another loop for carrier tracking. Tracking the code phase means changing the internal code replica phase to continuously match the incoming code phase. This is done using a delay lock loop (DLL). Tracking the carrier phase means changing the internal carrier replica's phase and frequency to continuously match the incoming IF carrier signal phase and frequency. This is done using a phase lock loop (PLL). It is important to understand the relationship between the frequency and the phase. Phase is the integration of angular frequency. Recall angular frequency is given by:

$$\omega = 2\pi f \tag{5-28}$$

The units of ω are radians. Phase is an angle ranges from 0 to 2π in one cycle of a sinusoidal or periodic signal. Integrating the angular frequency ω gives the phase:

$$\phi = 2\pi \int f\,dt = \int \omega\,dt \tag{5-29}$$

Understanding this relationship is important to understand the phase lock loop (PLL). What PLL does is simply minimize the phase difference between the incoming IF signal carrier and the internal replica by changing the repla's phase (see Figure 5.22). Changing the replica's phase means a change in frequency as well. PLL works on the phase difference $\Delta\phi$. Keeping $\Delta\phi$ minimum (zero or constant) means that the frequency difference between the two signals $\Delta\omega$ is zero which is the desired goal of PLL, to keep the internal replica in synchronization with the

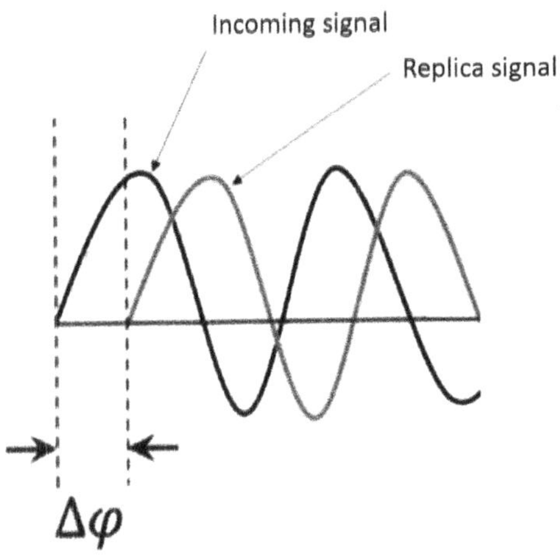

Figure 5.22 Illustration of the phase difference.

incoming signal carrier. By minimizing the phase difference between the two signals, PLL keeps the phase and the frequency matched.

5.6.3.1 Delay Lock Loop (DLL)

DLL tracks the code phase. To keep tracking the code phase, we generate three correlator variables: one correlator variable exactly at the phase estimated by the acquisition is called the *Prompt* correlator variable. Another correlator variable is generated at a slightly smaller or earlier code phase (minus half a chip, for example) called the *Early* correlator variable. Another correlator variable is generated at a slightly larger or earlier code phase (plus half a chip, for example) called the *Late* correlator variable. Assume the acquisition code delay is τ_D, the early, prompt, and late correlator variables are given by:

$$I_E = \int \hat{c}(t)c(t - \tau_D - 0.5\,T_{ch})\,dt \tag{5-30}$$

$$I_P = \int \hat{c}(t)c(t - \tau_D)\,dt \tag{5-31}$$

$$I_L = \int \hat{c}(t)c(t - \tau_D + 0.5\,T_{ch})\,dt \tag{5-32}$$

If τ_D is the right code delay, the I_P variable will be larger than I_L and I_E and that is what DLL is doing; keep changing code delay τ_D to keep I_P variable larger than I_L and I_E. This concept is illustrated in Figure 5.23.

To change τ_D, we use a discriminator function that takes I_E, I_P, and I_L and gives us a control signal that we can use to adjust τ_D to keep I_P variable larger than I_L and I_E. A straightforward simple code discriminator D_c can be defined as follows:

$$D_c = I_E - I_L \tag{5-33}$$

In practical implementation, much more complicated discriminators can be used like the ones in [7]. Keep in mind that the DLL must work on the signal after removing the carrier. The carrier is removed by multiplying the incoming signal with the internal carrier replica with the frequency that we estimated in the acquisition and refined by the PLL. Therefore, the code tracking loop will look like the one in Figure 5.24.

The DLL in Figure 5.24 works directly on the IF incoming signal. This is called the In-phase arm. This implementation is not perfect because it is sensitive to the phase of the carrier signal. To make the loop phase-insensitive, we process the two branches, I and Q, as we did in the acquisition. The I/Q branches are generated as follows:

$$I = A\cos(2\pi f_r t + \varphi_r)\cos(2\pi f_{nco} t + \varphi_{nco}) \tag{5-34}$$

$$Q = A\cos(2\pi f_r t + \varphi_r)\sin(2\pi f_{nco} t + \varphi_{nco}) \tag{5-35}$$

where f_{nco} is the internal oscillator carrier frequency.

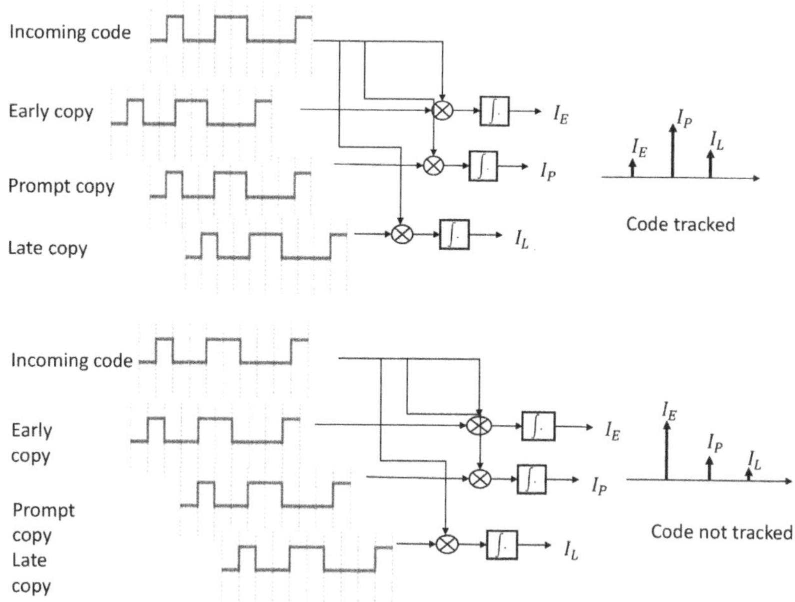

Figure 5.23 Illustration of tracking using early, prompt, and late correlators.

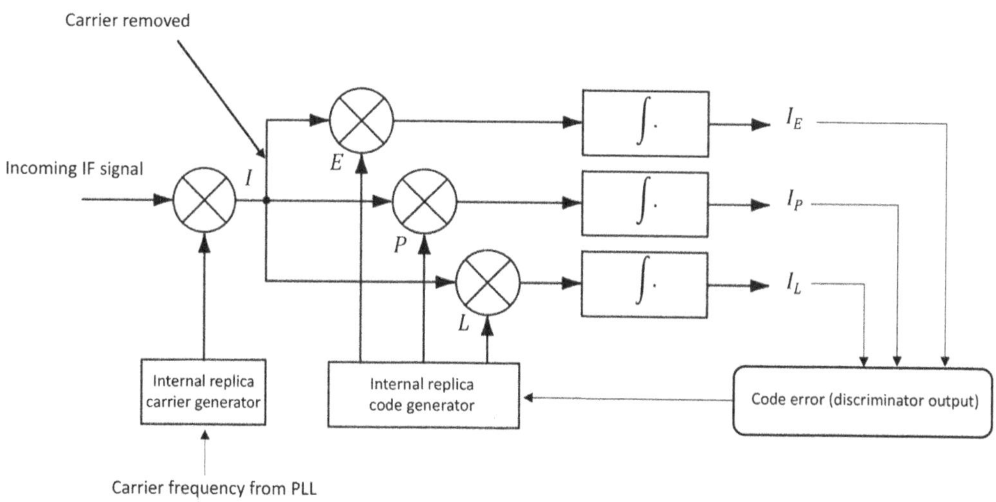

Figure 5.24 Basic code tracking loop.

We then implement the code tracking on the two branches I and Q as seen in Figure 5.25. Note that by perfectly removing the carrier and code, the remaining part of the signal becomes the navigation data, which (by using I/Q processing) will be completely carried out on the I-branch of the prompt copy, as seen in Figure 5.25. Given the early, late, and prompt correlators from the I and Q branches, a more sophisticated code discriminator can be defined as follows (see [5]):

$$D_c = I_P(I_E - I_L) + Q_P(Q_E - Q_L) \tag{5-36}$$

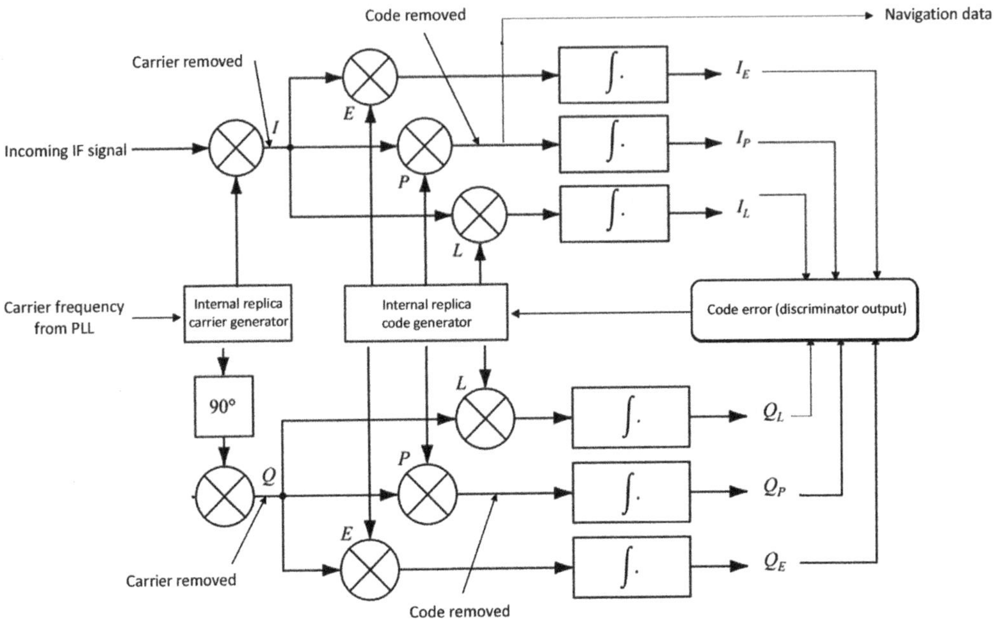

Figure 5.25 Code tracking loop.

5.6.3.2 Phase Lock Loop (PLL)

DL requires the perfect removal of the incoming IF carrier signal, which in turn requires the knowledge of the incoming IF signal carrier frequency. We have an initial value of the carrier frequency from the acquisition. The PLL will keep track of any change in the incoming IF signal carrier frequency using a simple control loop. This control loop is known as the Costas loop [4], [3], [7], which is shown in Figure 5.26. PLL tracks the IF carrier by first removing the code. Code is removed by multiplying the incoming IF signal with the internal code replica that is generated with code delay estimated by the acquisition and refined by the DLL. After removing the code, I and Q branches are generated using the internal oscillator (numerically controlled oscillator "NCO"). The oscillator takes its initial IF carrier frequency f_{nco} from the acquisition module. The I/Q form is necessary to get a stable error signal that represents the phase (and hence frequency) difference between

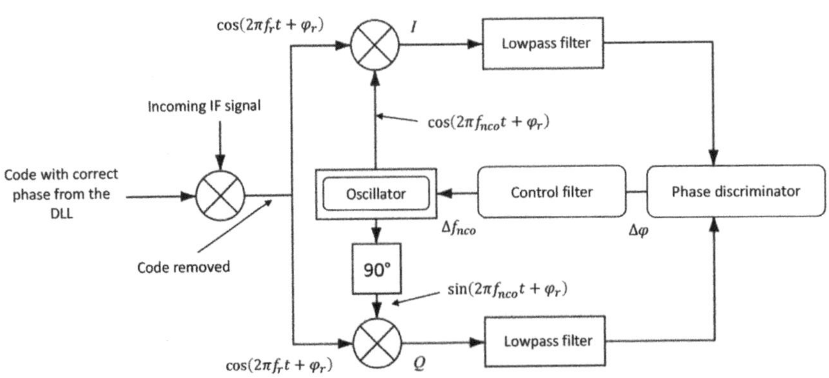

Figure 5.26 Costas phase lock loop (PLL) loop for carrier signal tracking.

the incoming IF carrier signal and the internal carrier replica. The PLL keeps all energy in the I (in-phase) arm. If we perfectly remove the code part, the I arm will be given by:

$$D^k(t)\cos(2\pi f_r t+\varphi_r)\cos(2\pi f_{nco}t+\varphi_{nco}) = \frac{1}{2}D^k(n)\cos(2\pi(f_r - f_{nco})+\varphi_r-\varphi_{nco}) + \frac{1}{2}D^k(n)\cos(2\pi(f_r+f_{nco})+\varphi_r+\varphi_{nco}) \tag{5-37}$$

where

$D^k(t)$ is the data part
$\cos(2\pi f_r t+\varphi_r)$ is the incoming signal
$\cos(2\pi f_{nco}t+\varphi_{nco})$ is the internal replica.

If our internal carrier frequency f_{nco} matches incoming signal carrier frequency f_r (which is the case as we get initial estimate of f_r from the acquisition module), then I arm will have a constant component and a high-frequency component as follows:

$$D^k(t)\cos(2\pi f_r t+\varphi_r)\cos(2\pi f_r t+\varphi_{nco}) = \frac{1}{2}D^k(t)\cos(\varphi_r-\varphi_{nco}) + \frac{1}{2}D^k(t)\cos(2\pi(2f_r)+\varphi_r+\varphi_{nco}) \tag{5-38}$$

The high frequency part $2\pi(2f_r)$ is removed by a low-pass filter. The remaining constant part is given by the following equation:

$$I^k = \frac{1}{2}D^k(t)\cos(\Delta\varphi) \tag{5-39}$$

Similarly, the Q arm will be given by:

$$Q^k = \frac{1}{2}D^k(t)\sin(\Delta\varphi) \tag{5-40}$$

where $\Delta\varphi$ is the phase difference between the incoming carrier and the internal carrier replica.

It is obvious that $\Delta\varphi$ is related to I and Q by the following form:

$$\frac{Q^k}{I^k} = \frac{\frac{1}{2}D^k(n)\sin(\Delta\varphi)}{\frac{1}{2}D^k(n)\cos(\Delta\varphi)} = \tan(\Delta\varphi) \tag{5-41}$$

$$\Delta\varphi = \tan^{-1}\left(\frac{Q^k}{I^k}\right) \tag{5-42}$$

The above phase difference $\Delta\varphi$ formula can be seen as a discriminator that tells us the phase difference between our internal IF carrier replica and the incoming IF carrier signal. Costas loop tries to keep the phase error $\Delta\varphi$ around zero by changing f_{nco} by Δf_{nco}. Changing the frequency means changing the phase (a step change in frequency means a ramp change in phase). If the phase difference is within a threshold, the PLL is considered phase-locked, and the observables are considered valid. The PLL function is to continuously keep the phase difference $\Delta\varphi$ near zero. While doing so, the PLL generates carrier-phase observables or measurements, which play a key role in precise positioning. The carrier-phase measurements are discussed in the next subsection.

5.6.3.3 PLL Control Filter

The details of the control filter that converts the phase discriminator $\Delta\varphi$ into a frequency change can be found in several references (see, e.g., [5], [4], [3]). As shown in Figure 5.26, Costas control loop has two main blocks, the control filter and oscillator block. The control filter diagram is shown in Figure 5.27.

The control filter converts the discriminator input $\Delta\varphi$ to a control input to the oscillator block. The transfer function of the control loop filter $F(s)$ is given by (see [7]):

$$F(s) = \frac{1}{s}\frac{\tau_2 s + 1}{\tau_1} \tag{5-43}$$

The transfer function of the oscillator block $N(s)$ can be written as follows [7]:

$$N(s) = \frac{K_o}{s} \tag{5-44}$$

The overall closed-loop transfer function of the loop is given by:

$$H(s) = \frac{\theta_o(s)}{\theta_i(s)} = \frac{K_d F(s) N(s)}{1 + K_d F(s) N(s)} \tag{5-45}$$

which is shown in Figure 5.27.

$H(s)$ is a second-order control system transfer function given by the following general formula:

$$H(s) = \frac{2\zeta\omega_n s + \omega_n^2}{s^2 + s\zeta\omega_n s + \omega_n^2} \tag{5-46}$$

where ω_n is the natural frequency given by the following equation:

$$\omega_n = \sqrt{(K_o K_d)/\tau_1} \tag{5-47}$$

and ζ is the damping ration defined as follows:

$$\zeta = (\tau_2 \omega_n)/2 \tag{5-48}$$

For a given noise bandwidth B_L, ω_n can be calculated as follows:

$$\omega_n = \frac{8\zeta B_L}{4\zeta^2 + 1} \tag{5-49}$$

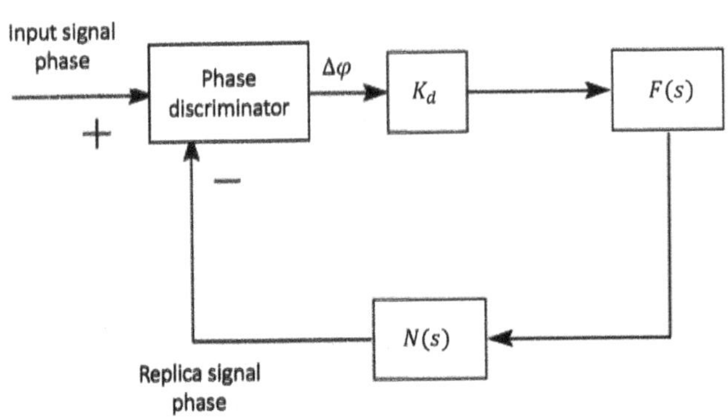

Figure 5.27 Basic PLL control filter.

The noise bandwidth B_L and damping ratio ζ depends on the characteristic of the phase error and the discriminator output signal characteristics. Given B_L and ζ, the parameters τ_1 and τ_2 can be calculated. Once τ_1 and τ_2 are calculated, the control input to the oscillator is calculated for a given phase error calculated by the discriminator function.

5.6.3.4 Carrier-phase Measurements

For any sinusoidal carrier signal, carrier phase is an angle that can be calculated by integrating the signal frequency overtime. We can also measure carrier phase in number of cycles where one cycle equals 2π. Assuming the NCO component of the PLL is running at frequency f_{nco}, the PLL would generate a carrier phase cycle count $\tilde{\varphi}$ given by the following equation:

$$\tilde{\varphi}(t_0) = f_{nco} t_0 \tag{5-50}$$

where t_0 is the time at which a phase lock is achieved.

Once the PLL is in phase-lock with the incoming IF signal, its NCO frequency would be matching the incoming IF signal carrier frequency f_r. The f_r can be used to estimate the Doppler shift as follows:

$$f_D = f_r - f_{IF} \tag{5-51}$$

Integrating the Doppler shift f_D measures, in number of carrier cycles, the change in the range between the receiver and the transmitter is calculated as follows:

$$\varphi(t) = \tilde{\varphi}(t_0) + \int_{t_0}^{t} f_D(t)dt \tag{5-52}$$

The carrier phase measurement $\varphi(t)$ has two parts, a known precise part which is $\int_{t_0}^{t} f_D(t)dt$ and an unknown part $\tilde{\varphi}(t_0)$. The unknown part $\tilde{\varphi}(t_0)$ has two parts, the number of cycles between the receiver and the transmitter (which is the true range between the receiver and the transmitter) and an unknown number N.

$$\varphi(t) = \rho(t) + N + \varepsilon \tag{5-53}$$

Since the PLL matches the fractional part of the phase, the unknown number of cycles N is an integer which is known as the integer ambiguity. Note that due to the cyclicity of angle values, the phase discriminator $\Delta\varphi$ is ambiguous. In other words, the PLL does not distinguish $\Delta\varphi$ from $\Delta\varphi + N2\pi$, where N is the Integer Ambiguity. The carrier-phase measurement $\varphi(t)$ can be used to provide precise range estimation because it is counted in carrier cycles which is in centimeter precision. The integer ambiguity N remain constant as long as the incoming IF signal is locked by the PLL.

5.6.3.5 Code and Carrier Tracking Combined

The DLL and PLL are combined to form the overall tracking module. A general block diagram that shows the DLL/PLL tracking modules combined is shown in Figure 5.28. The combined tracking module generates three fundamental pieces of information that are used to calculate the range and positioning of the receiver. These are the navigation data, code phase/delay, and carrier phase/frequency measurements. These three components (navigation data, code phase/delay, and carrier phase/frequency) can be used in different ways to calculate the receiver position and velocity. In more advanced configurations, we can also calculate the orientation of the target.

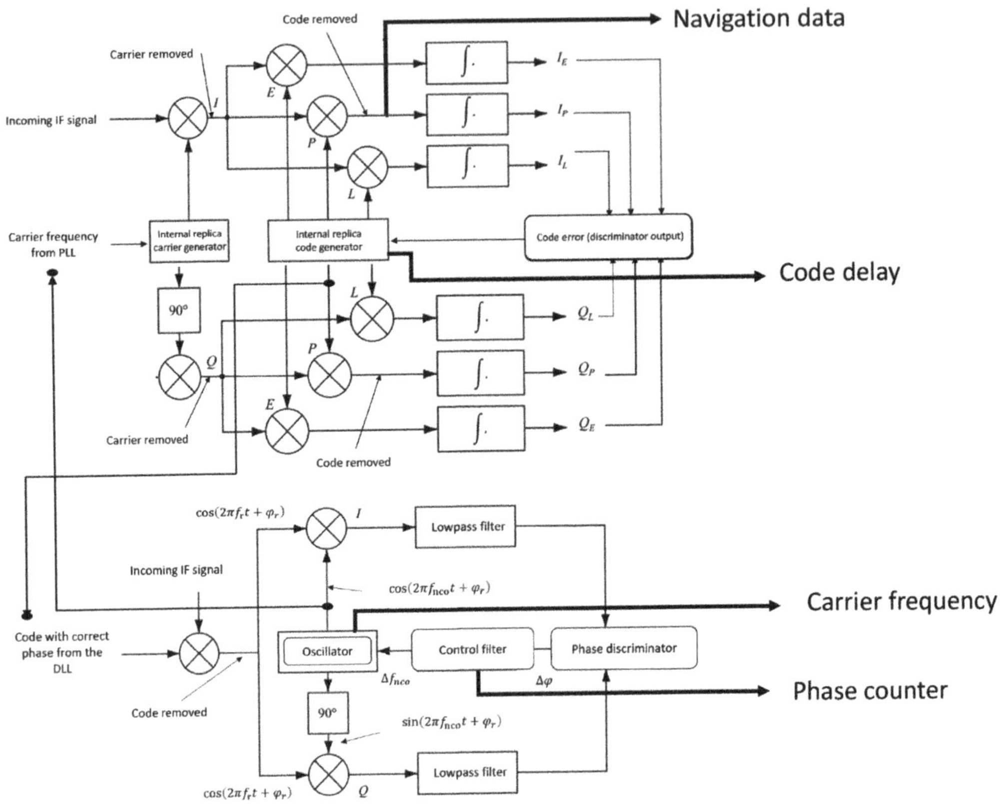

Figure 5.28 Basic tracking loops (PLL and DLL combined).

5.7 Radio Positioning Schemes

We explained how the anchors in radio positioning systems generate navigation radio signals and how receivers decode the navigation data, calculate the signal arrival time, and track the carrier phase and frequency. Given this capability, different time-based positioning schemes can be developed. The positioning schemes are based on how we use the above measurements. In the subsections below, we will explain different radio positioning schemes. It is important to note that the following radio positioning schemes can be applied to other wireless communication systems that use signals other than radio such as sound, infrared, light, or any wireless signals that can carry data, code, and a carrier component.

5.7.1 Time of Flight (ToF)

Positioning signals are always transmitted in frames that include data and code carried on a carrier. If the navigation data contains the time of the transmission of the frame, the receiver can calculate the time of the transmission of any code that is currently being processed T_t and we explained how the receiver can calculate the receiving time T_r using the matched filter (correlator) and tracking loops. As

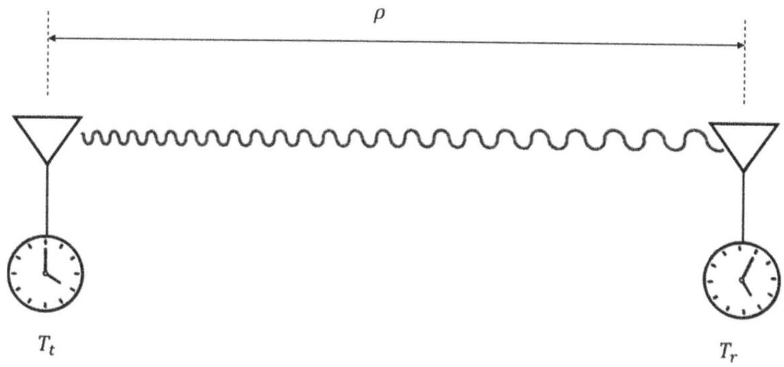

Figure 5.29 Radio ranging.

Figure 5.29 shows, the signal travels at the speed of light. By having the transmission and receiving time, we have the ToF of this code signal. Multiplying ToF will give us the distance as follows:

$$\rho = c(T_r - T_t) \tag{5-54}$$

5.7.1.1 ToF Measurement Model Equation

If the receiver receives waves from the other two anchors, as seen in Figure 5.30, the receiver can calculate the distance to the three anchors as follows:

$$\rho_1 = c(T_{r_1} - T_{t_1}) \tag{5-55}$$

$$\rho_2 = c(T_{r_2} - T_{t_2}) \tag{5-56}$$

$$\rho_3 = c(T_{r_3} - T_{t_3}) \tag{5-57}$$

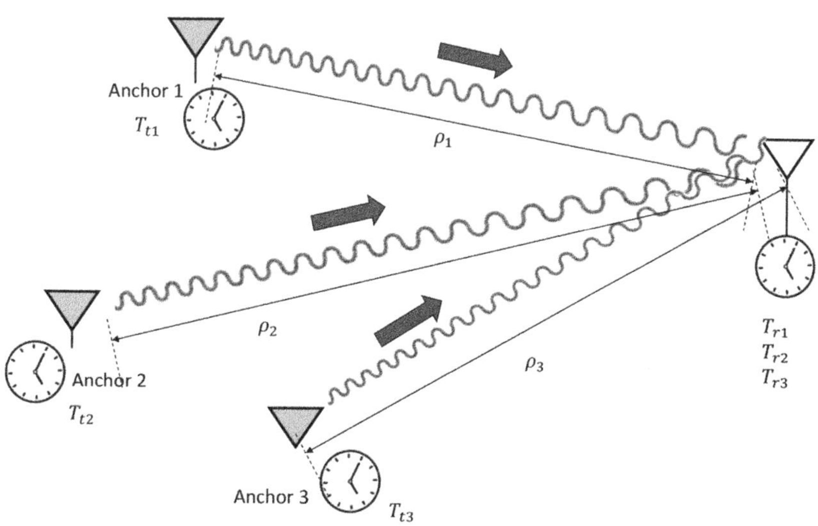

Figure 5.30 ToF positioning scheme.

If the positions of the three anchors are known ($\{x_i, y_i\}$, $i = 1, 2, 3$), the distances can be formulated as a function of the receiver position as follows:

$$\rho_1 = \sqrt{(x_1 - x)^2 + (y_1 - y)^2 + (z_1 - z)^2} \tag{5-58}$$

$$\rho_2 = \sqrt{(x_2 - x)^2 + (y_2 - y)^2 + (z_2 - z)^2} \tag{5-59}$$

$$\rho_3 = \sqrt{(x_3 - x)^2 + (y_3 - y)^2 + (z_3 - z)^2} \tag{5-60}$$

where

$(x_1\ y_1\ z_1)^T$, $(x_2\ y_2\ z_2)^T$, and $(x_3\ y_3\ z_3)^T$ are the positions of the three anchors
$(x, y, z)^T$ is the target's location.

5.7.1.2 Trilateration Positioning Algorithm

The trilateration algorithm is a nonlinear least-square algorithm that estimates the target location given range measurements from multiple known-location anchors. The nonlinear range equations described above can be linearized around an initial guess of the target location using first-order Tylor series as follows:

$$\rho_i = \bar{\rho}_i + \frac{\partial\sqrt{(x_i - x)^2 + (y_i - y)^2 + (z_i - z)^2}}{\partial x}(x - \bar{x}) + \frac{\partial\sqrt{(x_i - x)^2 + (y_i - y)^2 + (z_i - z)^2}}{\partial y}(y - \bar{y}) + \frac{\partial\sqrt{(x_i - x)^2 + (y_i - y)^2 + (z_i - z)^2}}{\partial z}(z - \bar{z}) \tag{5-61}$$

where

$(\bar{x}, \bar{y}, \bar{z})^T$ is our initial guess of the target location
$\bar{\rho}_i$ is our predicted range from the anchor i.

Based on our guess, the predicted range can be calculated as follows:

$$\bar{\rho}_i = \sqrt{(x_i - \bar{x})^2 + (y_i - \bar{y})^2 + (z_i - \bar{z})^2} \tag{5-62}$$

To simplify the notation, we will use the following assignments:

$$\delta\rho_i = \rho_i - \bar{\rho}_i \tag{5-63}$$

$$\delta x = (x - \bar{x}) \tag{5-64}$$

$$\delta y = (y - \bar{y}) \tag{5-65}$$

$$\delta z = (z - \bar{z}) \tag{5-66}$$

Doing the differentiation and putting the equation into a vector form will lead to the following error equation:

$$\delta\rho_i = \left(\frac{(x_i - \bar{x})}{\bar{\rho}_i}, \frac{(y_i - \bar{y})}{\bar{\rho}_i}, \frac{(z_i - \bar{z})}{\bar{\rho}_i}\right)\begin{pmatrix}\delta x \\ \delta y \\ \delta z\end{pmatrix} \tag{5-67}$$

Arranging multiple anchors' measurements into a matrix form, we get the following matrix equation:

$$\begin{pmatrix} \delta\rho_1 \\ \delta\rho_2 \\ \cdot \\ \cdot \\ \cdot \\ \delta\rho_M \end{pmatrix} = \begin{pmatrix} \frac{(x_1-\bar{x})}{\bar{\rho}_1} & \frac{(y_1-\bar{y})}{\bar{\rho}_1} & \frac{(z_i-\bar{z})}{\bar{\rho}_1} \\ \frac{(x_2-\bar{x})}{\bar{\rho}_2} & \frac{(y_2-\bar{y})}{\bar{\rho}_2} & \frac{(z_2-\bar{z})}{\bar{\rho}_2} \\ & \cdot & \\ & \cdot & \\ & \cdot & \\ \frac{(x_M-\bar{x})}{\bar{\rho}_M} & \frac{(y_M-\bar{y})}{\bar{\rho}_M} & \frac{(z_M-\bar{z})}{\bar{\rho}_M} \end{pmatrix} \begin{pmatrix} \delta x \\ \delta y \\ \delta z \end{pmatrix} \tag{5-68}$$

which is a linear model in the error vector and can be put in the following general linear measurement form:

$$\begin{pmatrix} \delta\rho_1 \\ \delta\rho_2 \\ \cdot \\ \cdot \\ \cdot \\ \delta\rho_M \end{pmatrix} = H \begin{pmatrix} \delta x \\ \delta y \\ \delta z \end{pmatrix} \tag{5-69}$$

From least-squares algorithm (review Chapter 2), the best estimate of the error vector $\begin{pmatrix} \delta x \\ \delta y \\ \delta z \end{pmatrix}^*$ is calculated as follows:

$$\begin{pmatrix} \delta x \\ \delta y \\ \delta z \end{pmatrix}^* = (H^T H)^{-1} H^T \begin{pmatrix} \delta\rho_1 \\ \delta\rho_2 \\ \cdot \\ \cdot \\ \cdot \\ \delta\rho_M \end{pmatrix} \tag{5-70}$$

We then update our initial guess as follows:

$$\begin{pmatrix} \bar{x} \\ \bar{y} \\ \bar{z} \end{pmatrix} = \begin{pmatrix} \bar{x} \\ \bar{y} \\ \bar{z} \end{pmatrix} + \begin{pmatrix} \delta x \\ \delta y \\ \delta z \end{pmatrix}^* \tag{5-71}$$

We repeat the process starting from the new guess $\begin{pmatrix} \bar{x} \\ \bar{y} \\ \bar{z} \end{pmatrix}$ and repeat the steps described above until the error $\begin{pmatrix} \delta x \\ \delta y \\ \delta z \end{pmatrix}$ becomes smaller than a threshold.

5.7.2 Time Difference of Arrival (TDoA)

ToF assumes perfect synchronization between all anchors' clocks and the receiver's clock which is very complex to achieve. Any small bias between the anchors' clock and the receiver's clock will be multiplied by the speed of light, leading to a large distance error. 1 microsecond (10^{-6}) error (bias or shift between an anchor's clock and receiver's clock) means 300 m distance error. Also, any drift in this bias will translate into an error in the Doppler frequency shift. It is extremely difficult to keep anchors and receivers' clocks perfectly synchronized because this will complicate the system design and reduce the system's scalability. Scalability means the ability of the system to handle a larger number of receivers or targets. A more scalable and manageable solution is to keep all anchors' clocks synchronized by a master clock and let the receivers perform self-synchronization and clock bias/drift correction, which is the technique adopted in GNSS technology, as we will discuss later.

5.7.2.1 Anchor-Initiated TDoA (AITDoA)

Anchor-initiated positioning mode refers to the mode where the user equipment or the target does not transmit any signals. Instead, the anchors transmit the reference positioning signal. If we can ensure that all anchors transmit a reference positioning signal frames at the same time, we can let the receiver calculate the differences of the arrival times of the signals from the anchors. This can be achieved by synchronizing all anchors' clocks with a master clock. In this configuration, the user equipment or the target is in listen-only mode. The time differences of arrival will be measured by the same clock (the user equipment receiver's clock); hence, there is no need to have the receiver's clock pre-synchronized with the anchor's clock. We will refer to this configuration by anchor-initiated TDoA (AITDoA). The configuration of AITDoA is illustrated in Figure 5.31.

5.7.2.2 User-Initiated TDoA (UITDoA)

User-initiated positioning mode refers to the mode where the target or the user equipment is the transmitter and the anchors are the receivers (or the listeners). In this case, there will be one reference positioning signal transmitted at the same time by the same clock. The user-transmitted reference positioning signal is received by the anchors at different times based on the distance between

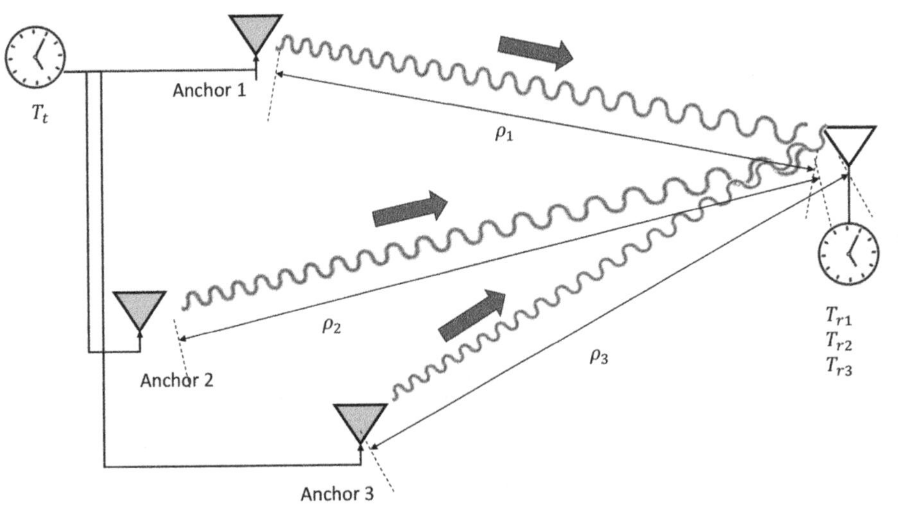

Figure 5.31 Anchor-initiated TDoA positioning.

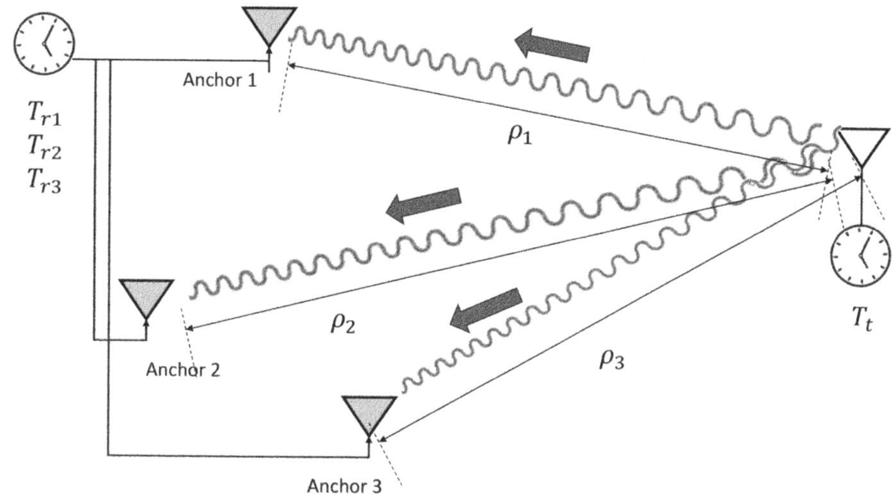

Figure 5.32 User-initiated TDoA.

the target's transmitter and the anchors. We will refer to this configuration by user-initiated TDoA (UITDoA). UITDoA configuration is shown in Figure 5.32. It is important to note that the AITDoA is more scalable than the UITDoA because in the AITDoA, as many users as we want can receive the reference positioning signal and calculate their position. In UITDoA mode, the number of users the system can handle depends on the anchors' processing capacity. The AITDoA mode is the mode adopted by the global satellite positioning systems.

5.7.2.3 TDoA Measurement Model and Positioning Algorithm

Based on the distances in the examples shown in Figures 5.31 and 5.32, the order of arrivals of the transmitted signals will be 3 (the closest anchor), 1 (second closest anchor), and 2 (the farthest anchor). So, we will have the following TDoA measurements, which can be linked to the range differences as follows:

$$\Delta\rho_{1,3} = \rho_1 - \rho_3 = c(T_{r_1} - T_{t_3}) = c\Delta T_{1,3} \tag{5-72}$$

$$\Delta\rho_{2,1} = \rho_2 - \rho_1 = c(T_{r_2} - T_{t_1}) = c\Delta T_{2,1} \tag{5-73}$$

$$\Delta\rho_{2,3} = \rho_2 - \rho_3 = c(T_{r_2} - T_{t_3}) = c\Delta T_{2,3} \tag{5-74}$$

We then have the following nonlinear measurement model:

$$\Delta\rho_{1,3} = \sqrt{(x_1 - x)^2 + (y_1 - y)^2 + (z_1 - z)^2} - \sqrt{(x_3 - x)^2 + (y_3 - y)^2 + (z_3 - z)^2} \tag{5-75}$$

$$\Delta\rho_{2,1} = \sqrt{(x_2 - x)^2 + (y_2 - y)^2 + (z_2 - z)^2} - \sqrt{(x_1 - x)^2 + (y_1 - y)^2 + (z_1 - z)^2} \tag{5-76}$$

$$\Delta\rho_{2,3} = \sqrt{(x_2 - x)^2 + (y_2 - y)^2 + (z_2 - z)^2} - \sqrt{(x_3 - x)^2 + (y_3 - y)^2 + (_3 - z)^2} \tag{5-77}$$

which is a nonlinear function in the target's location, and it can be solved using iterative least squares algorithm in the same way that we explained earlier in the trilateration algorithm. The derivation of the least squares algorithm that estimates the target's location given anchors' locations and TDoA measurements is left as an exercise (see the problems section).

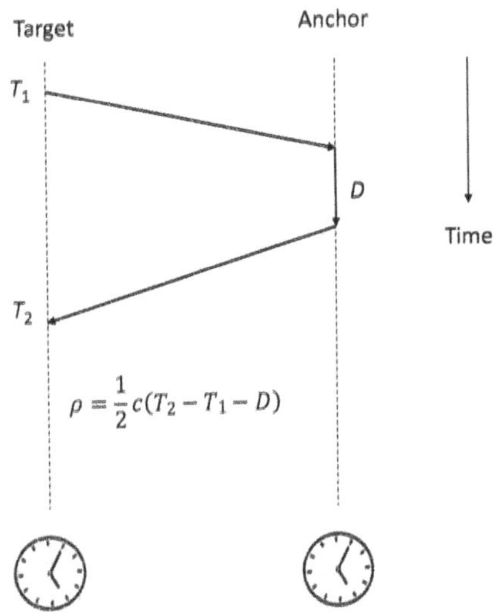

Figure 5.33 Round trip ranging.

5.7.3 Round Trip Delay (RTD)

Another time-based positioning scheme that does not need clock synchronization between target and anchors is round trip delay (RTD) scheme [8]. In RTD scheme, in one configuration (see Figure 5.33), target can send a ping message to an anchor, the anchor receives the ping message and transmits it back after embedding the delay the anchor took to respond to this return ping message. The target will receive this replay and decode the message to extract the delay T_D. The target knows the transmission time T_1 and receiving time T_2 using its own clock so there is no need for synchronization with another clock. Therefore, the distance between the target and the anchor can be calculated by the following equation:

$$\rho = \frac{1}{2}c\,(T_2 - T_1 - T_D) \tag{5-78}$$

The time $T_2 - T_1 - T_D$ is the round-trip delay the ping message took to reflect to the target. Given the RTD measurements from different anchors, the target can use the same trilateration positioning algorithm described in ToF positioning to estimate its position.

5.8 Dilution of Precision (DOP)

The geometric arrangement of the anchor nodes in radio positioning systems has a great influence on the accuracy of the positioning. It is easily understood that the range estimation accuracy affects the target's positioning accuracy. The geometric arrangement of the anchor nodes plays an important role in magnifying or minimizing the effect of range estimation error as seen in Figure 5.34.

A popular method to calculate the expected accuracy of radio positioning given a specific geometric anchor distribution and range estimation accuracy is the dilution of precision (DOP). DOP is a combined assessment of the uncertainty of the range estimation together with the effect of the geometric distributions of the anchors. In the trilateration algorithm, we showed that at convergence, the error in the target location is related to the error in range measurements according to the following equation:

$$\delta\vec{\rho} = H\delta\vec{x} + \vec{\varepsilon} \tag{5-79}$$

where $\vec{\varepsilon} \sim \mathcal{N}(0, \sigma^2)$ is combined target's receiver error due to range estimation error.

At convergence, $\delta\vec{x} \to 0$ given by the following least-square solution:

$$\delta\vec{x} = \left(H^T H\right)^{-1} H^T \delta\vec{\rho} \tag{5-80}$$

Taking the covariance of both sides of Eq. (5-89) gives the following:

$$E\left\langle \delta\vec{x} \right\rangle = E\left\langle \left(H^T H\right)^{-1} H^T \delta\vec{\rho} \right\rangle \tag{5-81}$$

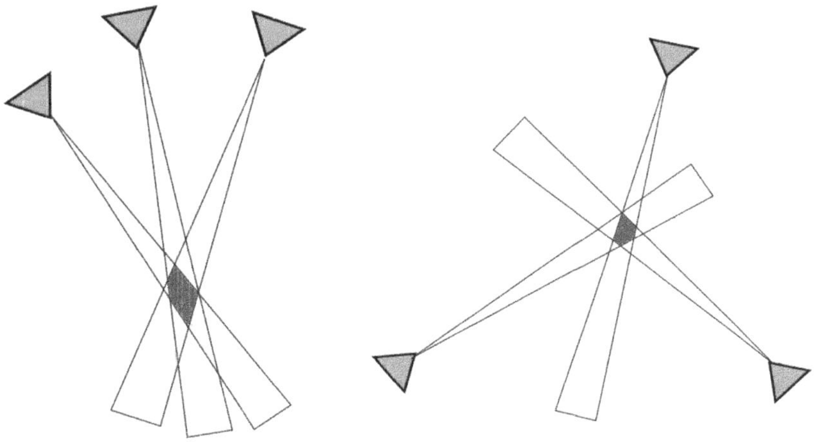

(a) (b)

Figure 5.34 Effect of geometric anchors' distribution on target's location accuracy (a) shows good geometry leading to small DOP (b) shows poor geometry leading to larger DOP.

The covariance of the target's location error $\delta\vec{x}$ defined as follows:

$$P = E\left\langle \delta x \delta x^T \right\rangle \tag{5-82}$$

$$E\left\langle \delta x \delta x^T \right\rangle = E\left\langle \left(H^T H\right)^{-1} H^T E\left\langle \delta\vec{\rho}\,\delta\vec{\rho}^{\,T} \right\rangle H \left(H^T H\right)^{-1} \right\rangle \tag{5-83}$$

$$E\left\langle \delta\vec{\rho}\,\delta\vec{\rho}^{\,T} \right\rangle = \sigma^2 I \tag{5-84}$$

$$E\left\langle \delta x \delta x^T \right\rangle = \left(H^T H\right)^{-1} \left(H^T H\right) \left(H^T H\right)^{-1} \sigma^2 I \tag{5-85}$$

$$E\left\langle \delta x \delta x^T \right\rangle = \left(H^T H\right)^{-1} \sigma^2 \tag{5-86}$$

Equation (5-86) shows that the covariance of the target's location error is related to the covariance of the ranging estimation error by a factor of $(H^T H)^{-1}$. The factor $(H^T H)^{-1}$ is called the geometric dilution of precision (DOP) defined as follows:

$$DOP = \left(H^T H\right)^{-1} \tag{5-87}$$

DOP indicates the effect of the geometric distribution of anchors on the reflection of range estimation error on the target's location error. The better the geometry, the smaller the DOP, the smaller the effect of range estimation error in target's localization, and the better the accuracy of the target's location estimation. The concept of the DOP is very important in GNSS technology as it is considered an independent assessment of the expected accuracy regardless of the range estimation accuracy.

5.9 GNSS Technology

One of the most popular radio positioning systems is global satellite navigation systems (GNSS). In this section, we will briefly discuss the basic components of GNSS technology and the different positioning configurations such as Single Point Positioning (SPP), Precise Point Positioning (PPP),

Differential GNSS (DGNSS), and real-time kinematics (RTK) GNSS. Several nations are developing their own GNSS. The first operational GNSS technology was the global positioning system (GPS) developed by the United States. In our discussion, we will always take the GPS as an example.

5.9.1 GNSS Overview

GNSS technology uses satellites as positioning anchors. GNSS consists of a constellation of satellites orbiting the Earth at specific orbits, as seen in Figure 5.1. Launching, monitoring, and controlling satellites in space and tracking satellites' positions with respect to Earth in real time is a complex independent field of its own, which is out of the scope of this book as many references explain this science in detail [3], [6], [4]. Satellites can be classified based on orbits according to their altitude (height) from Earth [3]. Orbits with a mean altitude below 1500 km are called low-Earth-orbit (LEO). Orbits with a mean altitude of around 20,000 km are called medium-Earth-orbit (MEO). GNSS satellites are MEO satellites orbiting the Earth at an altitude of around 20,000 km, which gives a revolution period of approximately 12 h. Utilizing MEO satellites as positioning anchors provides global coverage with clear line-of-sight in open-sky areas. In addition, the far distance between satellites and receivers makes the relative range error small. Note that any small error in the flight time of the electromagnetic wave transmitted by satellites is converted into a huge error in distance calculation because we multiply the flight time by the speed of light. For example, a microsecond (10^{-6}) error in time measurement will cause a 300 m error in distance estimation. An additional advantage of using satellites is that GNSS satellites orbit the Earth at a very high speed with respect to Earth. For example, GPS satellites orbit the Earth twice every 24 h at a speed of 3.9 km/s. This high speed causes a large Doppler shift proportional to the relative speed between the satellites and the receiver, which enables the GNSS receiver to estimate its speed with high accuracy.

Placing satellites at such high altitudes and at this high speed with respect to Earth imposed significant challenges on the signal propagation and processing of the electromagnetic waves transmitted by satellites. One challenge is the signal power. According to the Inverse-Square Law [1], an electromagnetic wave's power decreases with the square of the traveled distance. Satellites have limited payload, including the battery required to generate electromagnetic waves. This means that the GNSS receivers will receive a very weak signal from satellites. The average minimum power level of GNSS satellites is −160 decibel Watts (dBW). dBW is a unit that measures power on a log scale. If power in watts is P_W, then P_{dBW} unit is defined as $P_{dBW} = 10\log_{10} P_W$, which means that a −160 dBW is equivalent to 10^{-16} W. This low power is below the electronic circuits thermal noise (Johnson–Nyquist noise [9]) power level, which imposed a significant challenge on the GNSS receiver design to process such weak signal, which is discussed later in this chapter. Another challenge is the wave propagation. GNSS-transmitted waves travel through space, passing through different layers of the atmosphere. Each atmospheric layer has different effects on the wave propagation speed and direction. More details about orbital effects on signal's propagation will be covered later in the chapter.

The first operational GNSS was the United States' system known as the global positioning system (GPS) in 1994. In 1996, the Russian GNSS, known as GLONASS, was declared operational. The European version of GNSS is known as Galileo. The Chinese version of GNSS is called BeiDou. In addition to GNSS, there are several regional satellite navigation systems. Regional satellite navigation systems are either working independently or built to support and improve GNSS. Regional satellite systems utilize Geostationary satellites (GEO). GEO satellites orbit Earth in an orbit designed to keep the satellite's position constant relative to Earth so it is stationary with respect to Earth. Examples of regional GNSS systems include QZSS (Quasi-Zenith Satellite System), which was developed by Japan to enhance GPS accuracy in Japan. WAAS (Wide Area Augmentation System) is a U.S. regional system developed to improve GPS accuracy in North America. EGNOS (European Global Navigation Overlay System) is a regional GPS augmentation system that covers the European Union. GAGAN (GPS-Aided Geo-Augmented Navigation) is an Indian regional system to improve GPS accuracy in India. Most GNSS operational systems use CDMA communication protocol. Only the GLONASS system uses Frequency Division Multiple Access (FDMA) protocol on top of CDMA. FDMA protocol uses different frequencies for different satellites.

Satellite orbit design plays a key role in the robust, accurate operation of GNSS technology. Orbit models are mathematical models that describe the motion of an object in space. Orbit models have existed as far back as the first century AD [3]. The commonly used orbital models used in current GNSS technology are based on Kepler's laws [3] [5]. The fundamental three of Kepler's laws are described as follows: (1) A satellite is orbiting the Earth like a planet that is orbiting the Sun in an elliptical orbit where the Earth is located at one of the focuses of the ellipse as shown in Figure 5.35. Once launched at a specific velocity, the satellite will orbit the Earth in this elliptic orbit. The satellite reaches its maximum speed at the Perigee point and its minimum speed at the Apogee point. (2) A line that joins the Earth and the satellite spans equal areas in equal time intervals. (3) The sum of the Earth's and satellite's masses multiplied by the orbital period is proportional to the cube of the semimajor axis of the orbit.

Kepler's laws are the basis on which satellite position is calculated. The ephemeris data in the navigation message broadcasted by the satellite contains Kepler's parameters that are used by the receiver to calculate the satellite position at the time of the transmission of the signal. The calculation takes into consideration the effect of relativity, which is known as the Sagnac effect [10]. The Sagnac effect considers the small difference between the satellite positions at the time of the signal transmission and the time the receiver receives the signal. As the Earth and the satellites are rotating with respect to each other during the signal propagation, the receiver must correct the satellites' positions for these rotations. By tracking the satellites' positions at small intervals of time, satellites' velocities and speed can also be calculated. For more information about Kepler's model and satellite position calculation, readers are encouraged to check references [3, 10, 5]. To keep the

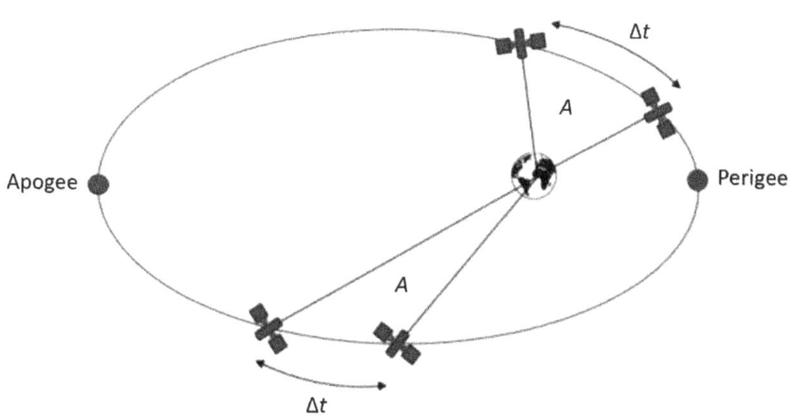

Figure 5.35 Satellite's orbit general shape.

parameters and models needed to calculate the satellites' positions up-to-date, ground control stations monitor the motion of the satellite and observe the appearance of the satellites around the Earth to constantly correct the model parameters. These corrections are transmitted to the satellites to be included in the navigation messages so the receivers can use these parameters to calculate the satellites' positions. This update procedure is also applied in other models such as the models used to calculate the ionospheric delays and satellite clock correction.

5.9.2 The Global Positioning System (GPS)

5.9.2.1 GPS System Overview

GPS is the first operational satellite radio navigation system. GPS became operational in 1993 initially for military applications. It became available to civilian and commercial applications in 1980 after the incident of the Korean Air Lines Flight 007 which was due to a navigation mistake. GPS and other GNSS realizations consist of three main segments: (1) space vehicles, (2) ground control, and (3) user equipment. A brief description of the three segments is given in the following subsections.

- **Space vehicles (SVs)**: These are the satellites that orbit the Earth and transmit navigation signals that include navigation data (like orbital information and satellite clock corrections), navigation codes, and carrier signals. The arrangement of SVs in space is called a constellation. GPS basic constellation consists of six orbits 55° with respect to the equator. Each orbit makes a plane that crosses the Earth and carries multiple SVs, as shown in Figure 5.36. The GPS constellation is designed to ensure that at least four satellites will be available anywhere on Earth. GPS constellation has limited coverage in the northern and southern poles.
- **Control segment**: The GPS control segment consists of globally distributed ground stations as seen in Figure 5.37 that monitor the status of SVs, calculate/update their orbital information, and correct their clocks to be synchronized with the global GPS timing system. The control segment sends commands and data to the satellites' constellation. The ground stations communicate this

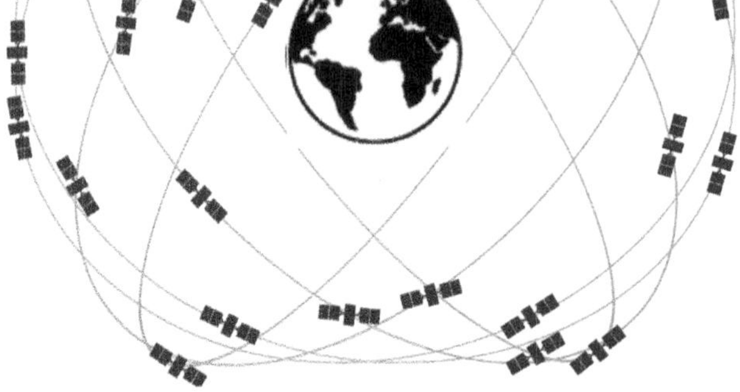

Figure 5.36 Illustration of GPS SVs constellation.

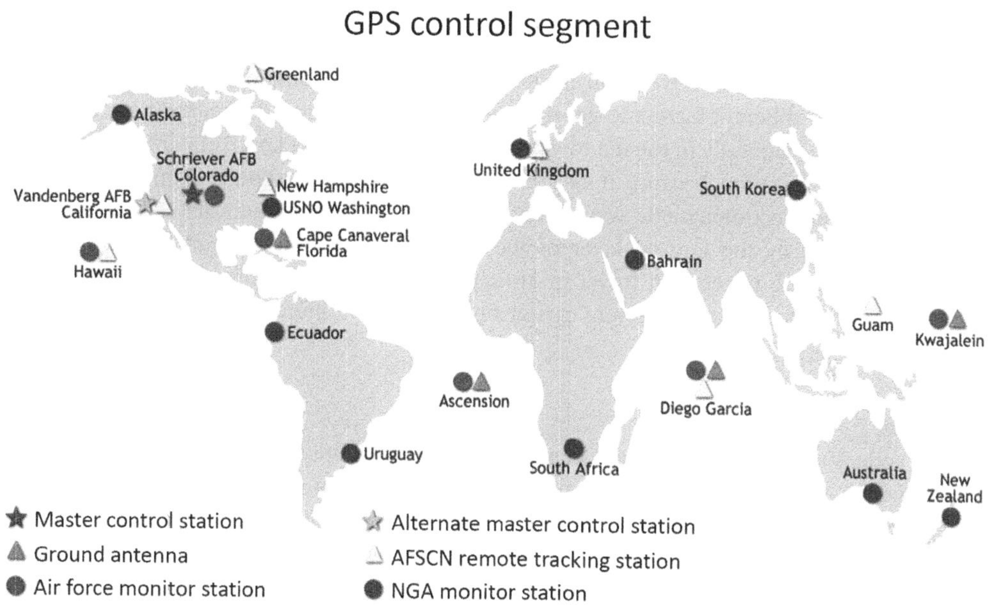

Figure 5.37 GPS control segment ground stations. *Source:* Global Positioning System (GPS)/https://www.gps.gov/systems/gps/control//last accessed on March 26, 2024.

information to the SVs, so the SVs broadcast it into the navigation data, which is in turn used by user receivers to calculate SVs positions, speed, clock corrections, satellite health status, and any other information that is important for navigation.

- **User segment**: The GPS user segment consists of the GPS receivers that receive GPS signals, decode navigation data, calculate the arrival times, apply clock corrections, ionospheric corrections, and tropospheric corrections, and run the state estimation algorithm that estimates the user's positioning from the radio signal measurements. GPS receivers have the same general design depicted in Figure 5.15. There are advanced types of GNSS receivers that can track multiple frequencies and multiple satellite constellations. More information about advanced GNSS receivers and modernized GNSS technologies can be found in several textbooks such as [11] and [12].

5.9.2.2 GPS Signals and Communication Protocol

GPS uses CDMA protocol where all satellites can transmit their signals over a common single carrier frequency. However, GPS uses multiple carrier frequencies (L_1, L_2, L_3, L_4, and L_5) and multiple codes (Coarse Acquisition "C/A code," encrypted precision code "P(Y)-code"). Different carrier frequencies and codes are used for different purposes. For example, P(Y)-code is a high-precision code signal that provides very precise correlation peaks, so it gives a better estimate of time of arrival and better positioning precision. P(Y) code is encrypted and used by authorized and military applications. In this chapter, we will explain the L_1 carrier and C/A code as examples to illustrate the main concepts.

GPS Carrier Signals

Among other frequencies, GPS satellites transmit the following two popular carrier frequencies:

$$L_1 = 154 \times 10.23\,\text{MHz} = 1{,}575.42\,\text{MHz} \tag{5-88}$$

$$L_2 = 120 \times 10.23\ \text{MHz} = 1{,}227.6\ \text{MHz} \tag{5-89}$$

In one configuration, internal GPS transmitters circuitry uses a common oscillator at 10.23 MHz and multiplies it by different factors to get the designed L_1 and L_2 carrier frequencies. These frequencies are selected carefully in the microwave spectrum. Microwave spectrum is suitable for satellite-Earth communication because it is less affected by the atmosphere. Atmospheres have different effects on electromagnetic waves. For example, at lower frequencies, electromagnetic waves are refracted by the ionized layer of the atmosphere. Microwaves are less affected by this phenomenon and, hence, can travel in an approximately straight line with less refraction. For more information about the different carrier frequencies for different GNSS, readers can refer to [3] and [6].

GPS C/A Code

GPS C/A code is also called pseudo-random code (or pseudo-random noise "PRN"). Pseudo-random, as discussed in Chapter 2, means the signal is random, but it repeats itself, and it is known by the designer. PRN is a predefined sequence of binary bits (zeros and ones or "−1" and "+1") that repeats itself at a desired frequency. Each GPS satellite has a unique PRN for the following purposes:

- Identifies the satellite.
- Spreads the spectrum of the overall signal (carrier, code, and data).
- Strong correlation enables detection of weak signals.
- Strong correlation enables precise estimation of signal arrival time.

C/A code has a length of 1,023 chips. GPS C/A code is transmitted at a chip rate of 1,023,000 chips/s, which means 1,000 PRN codes are transmitted each second. One PRN code takes 0.001 s, and one chip is approximately 1 μs. If we multiply single code chip duration by the speed of light (3×10^8) it gives 300 m. The complete PRN code has a distance length of approximately 300 km.

GPS Navigation Data

GPS navigation data [16, 17] is the message packets that carry the orbital information of the satellites, the time of transmission of the signal, satellite clock correction, atmospheric parameters, the status of the satellite, and other important information the receiver needs to calculate an accurate positioning. Navigation message has a much lower rate than code and carrier (to avoid affecting the correlation sign in the acquisition and tracking loops) and is transmitted as a stream of "−1" and "+1" like the code. The GPS navigation message is transmitted at a rate of 50 bits per second (bps). Each message contains 25 pages called frames. Each frame has 30 s of data (1,500 bits). It takes 12.5 min $(25 \times 30)/60$) to transmit a complete GPS navigation message. A more detailed demonstration of the 12.5-minute GPS navigation message is shown in Figure 5.38.

GPS L1 C/A Modulation

The three main components of GPS signal (carrier, code, and data) are combined into one signal using BPSK modulation as shown in Figure 5.39. Mathematically, GPS $L1$ signal for k^{th} satellite is given by:

$$s^k(t) = \sqrt{2P_{C/A}}\left(C^k(t) \oplus D^k(t)\right)\sin(2\pi f_{L1}t + \varphi) \tag{5-90}$$

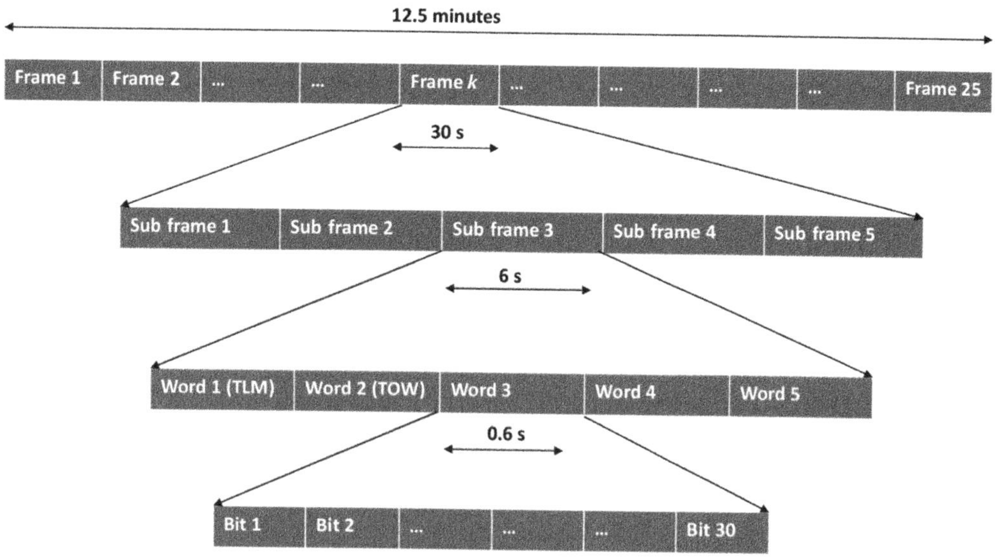

Figure 5.38 12.5-minute basic GPS navigation message.

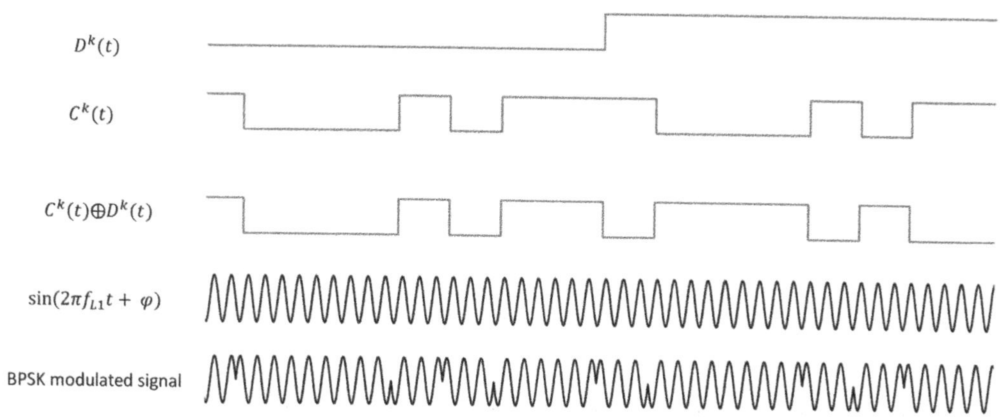

Figure 5.39 Illustration of GPS signal generation.

where $\sqrt{2P_{C/A}}$ represents the transmission power of the C/A signal. GPS $L1$ signal for all satellites is given by the following summation:

$$s(t) = \sum_{k=1}^{N} s^k(t) \tag{5-91}$$

where

N is the number of satellites
$s(t)$ is the combined signal that the receiver will receive from all satellites.

The receiver receives all electromagnetic waves passing through its antenna. To pick up a specific signal (for example, a signal from a specific GPS satellite on the $L1$ carrier), the receiver front-end filters the incoming waves and passes only the waves at and around the $L1$ frequency.

GPS L1 Signal Frequency Spectrum

The spectrum of the C/A code and the down converted carrier are shown in Figures 5.40 and 5.41. When we multiply the Carrier and the C/A code, the resulting signal is a spread spectrum centered around the $L1$ down-converted frequency as shown in Figure 5.42. The effect of the navigation data

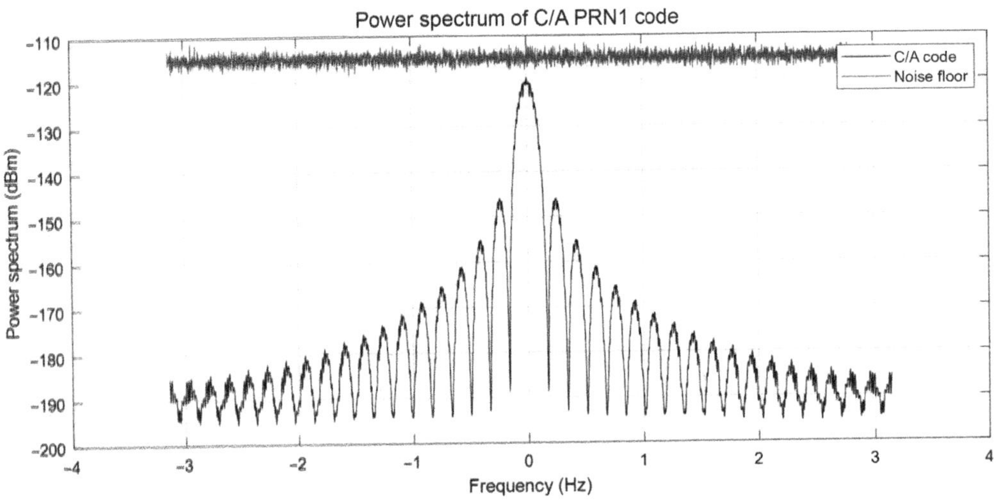

Figure 5.40 GPS C/A code spectrum.

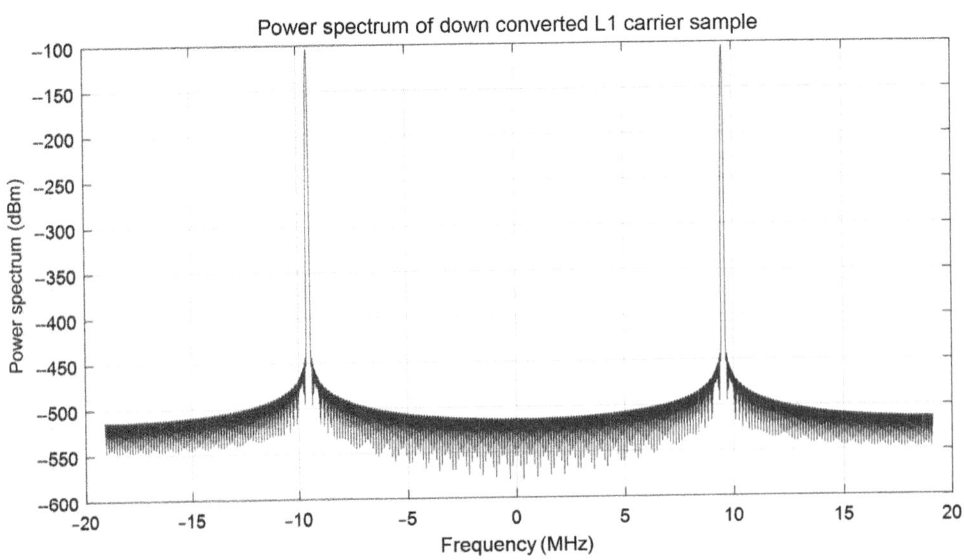

Figure 5.41 L1 Carrier after down conversion by the front-end.

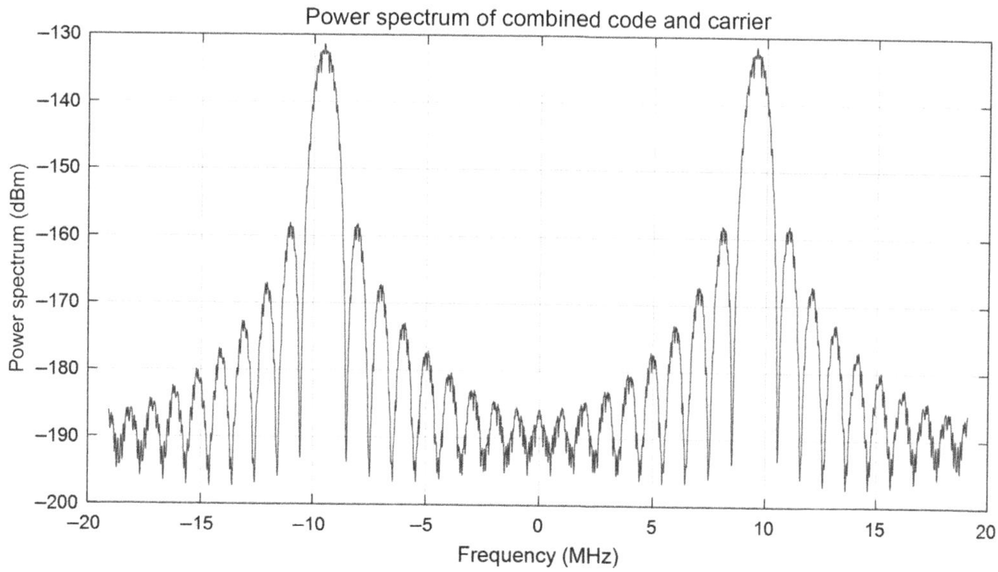

Figure 5.42 GPS spectrum of C/A code + $L1$ carrier. Note that the center frequency is the digital intermediate frequency generated by front-end component.

messages on the frequency spectrum of the signal is minimal since it has a very low frequency (50 Hz) compared to the code (in the range of 1 MHz) and the carrier (in the range of tens of MHz after down conversion by the front-end. Therefore, the overall signal spectrum (carrier, code, and data) will look like the C/A Code and Carrier spectrum in Figure 5.42. Note that the center frequency in Figure 5.42 is an intermediate frequency of the digital signal generated by the analog-to-digital converter of the receiver. Figures 5.40 and 5.42 have been generated by MATLAB using digital signal samples from [7, 13].

5.9.2.3 GPS Receiver Overview

The basic GPS receiver design for *C/A L1* signal is not different from the standard radio positioning receiver design explained in Section 5.6. The receiver has four main components: the front-end, acquisition, tracking, and positioning components. We will explain each component briefly in the following sections.

- **Front-end:** A sample front-end for $L1$ signal with specific design specifications is shown in Figure 5.43. The design parameters in this example are based on the front-end design in [7] and the MATLAB example in [13]. In the antenna, a bandpass filter (BPF) centered around $L1$ frequency with a 3 dBW bandwidth of 50 MHz is used to filter out all other signals of other frequencies. The 50 MHz bandwidth is large enough to pass the code, navigation data, and any Doppler shift that happens due to the relative speed between the satellite and the receiver. The signal is then amplified and mixed with a local oscillator to shift the high $L1$ frequency to a lower Intermediate Frequency (IF). In the example figure, the mixer's local oscillator frequency is 1,527.68 MHz, which means that the intermediate frequency equals the following value:

$$\bar{f}_{IF} = 1{,}575.42 - 1{,}527.68 = 47.74\ \text{MHz} \tag{5-92}$$

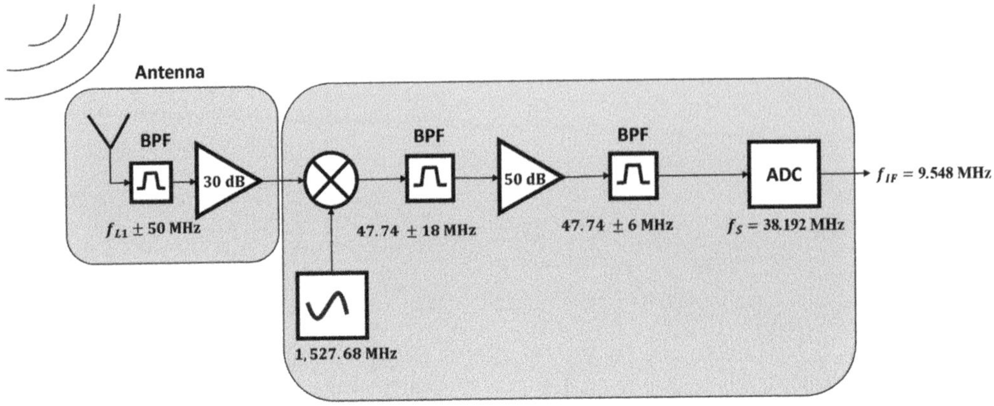

Figure 5.43 Example GPS L1 front-end.

The bar over $\bar{f}_{IF}$ refers to the fact that it is a temporary intermediate frequency. The final IF signal that will be processed is the digital output frequency, which is determined by the sampling frequency of the analog-to-digital converter (ADC). It is important to note that the mixing operation does not destroy the code and navigation data signal because the code has a frequency of approximately 1 MHz, and the data has a frequency of approximately 50 Hz. Both 1 MHz and 50 Hz are very low frequency compared to the IF, which is 47.74 MHz. Note that the bandpass filter has a bandwidth of 6 MHz, which guarantees that the code (1 MHz) and navigation data (50 Hz) will pass through this mixing and filtering process. In the given example, the sampling frequency is set to 38.192 MHz, which generates a final digital intermediate frequency of 9.548 MHz. Note that the sampling frequency affects the timing resolution in the receiver. The receiver calculates the signal receiving time with a precision of half of the sampling period. A sampling frequency of 38.192 MHz gives a timing precision as follows:

$$\Delta t = 0.5 \frac{1}{38.192\,\text{MHz}} \approx 1.31 \times 10^{-8}\,\text{s} \tag{5-93}$$

This timing precision gives a distance precision of the following value:

$$\Delta d = \Delta t \cdot c \approx 3.95\,\text{m} \tag{5-94}$$

The distance precision in Eq. (5-94) translates into position error range. Low-cost consumer-grade GPS receiver provides an accuracy in the range of 3–7 m. For more information about GPS accuracy, readers can refer to the report in [14], which includes a detailed GPS single point service (SPS) performance analysis.

- **Acquisition and tracking:** The acquisition and tracking module of GPS C/A L1 signal is not different from the basic acquisition and tracking modules described in Sections 5.6.2 and 5.6.3. Figure 5.44 shows a sample of a successful acquisition where a signal from a GPS satellite is detected at intermediate frequency 9.55 MHz and code phase at sample number 34,212 out of 38,192 samples per code.
- **Pseudo-range calculation:** In GPS, all satellites transmit a timed signal using a common clock time (the GPS time) every 6 seconds which is the length of the subframe. A frame has five subframes (Figure 5.38). Each subframe starts with a telemetric word (TLM) of 8 bits followed by a 17-bit time stamp called the time of the week (TOW). The TLM is used as a preamble to detect the

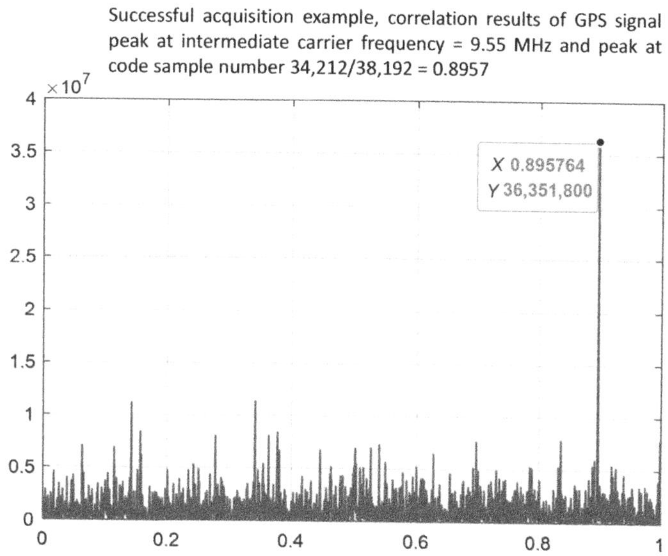

Figure 5.44 Example of a successful acquisition of a GPS satellite signal. The plot shows the code delay and carrier frequency (intermediate) at which the correlation peak is maximum.

occurrence of subframes. Each subframe has an ID. The receiver looks for the first subframe of each frame in each channel. To calculate the pseudo-range, the receiver calculates the arrival time only without caring about the transmission time of the subframe of each satellite channel since all subframes from all satellites/channels share the same TOW. The reason is that the satellites have been time-synchronized, and the orbits have been designed such that at any given time, the receiver will be processing the same subframe from all satellites (the word 'same subframe' here means subframes transmitted at the same TOW from all satellites). The "same subframe" from all satellites means all subframes currently processed by the receiver from all satellites have the same TOW. This feature is fundamental in GNSS technology, and it is guaranteed by the following two principles:

1) The timing between subframes is 6 s, which is long enough to guarantee that at any given time, all channels of the receiver are working on subframes that have the same TOW. Note that all satellites are time synchronized, and that is the reason they share the same TOW.
2) The orbital distances between the satellites and the receiver are small enough to guarantee that the receiver will always receive a subframe that has the same TOW from all satellites. In other words, there is no satellite that is too close to or too far from the receiver such that the receiver may receive multiple subframes from the close satellite while waiting for one subframe transmitted from a more distant satellite.

This careful configuration makes the GNSS, technically speaking, an Anchor-Initiated TDoA system thanks to the precise synchronization of satellite clocks and the orbit design of the satellites.

5.9.2.4 GPS Example from MATLAB Satellite Communication Toolbox

The MATLAB Satellite Communication Toolbox version R2023b includes a useful example project that demonstrates the acquisition and tracking of C/A code signals. The example can be downloaded and tested on MATLAB using the link in [13]. The project simulates four satellites, PRN 7, 11, 20, and 28 and it uses a sampling frequency of 38.192 MHz. The example generates and processes 0.4 s of data. The example accumulates a buffer of 1 ms long for processing at each iteration. This means that the number of samples per buffer equals 38,192, which is difficult to display in one plot. Figures 5.45–5.47 show a small portion of the generated (clean) waveforms and the received (noisy, delayed, and scaled) waveform. The example uses the MATLAB object GNSS Signal Acquirer to perform the acquisition. The acquisition results for a detected visible satellite are shown in Figure 5.48 and the acquisition results of an undetected (invisible) satellite are shown in Figure 5.49. Figure 5.48 shows a distinguished peak while Figure 5.49 does not which indicates that satellite number 7 is visible while satellite number 9 is not visible. The tracking loops (code and

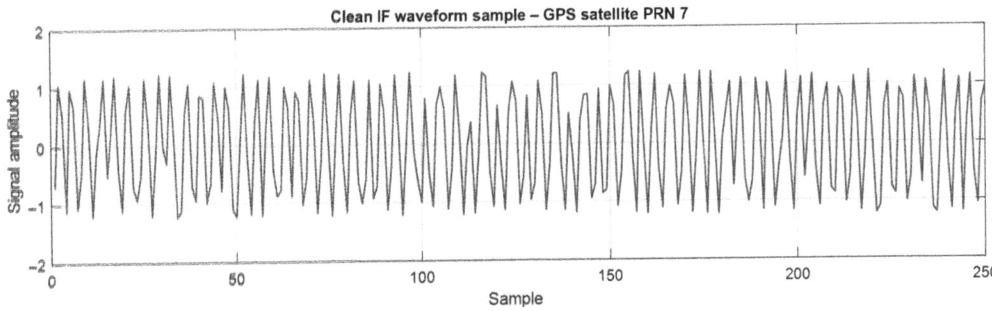

Figure 5.45 Generated IF waveform for GPS satellite 7.

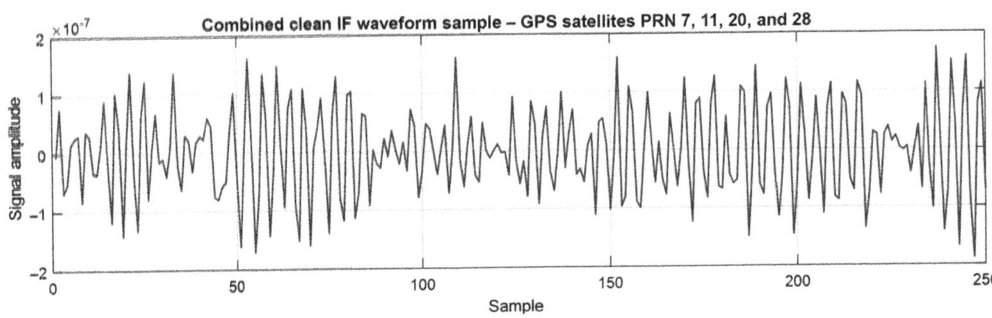

Figure 5.46 Combined generated IF waveforms for GPS satellites 7, 11, 20, and 28.

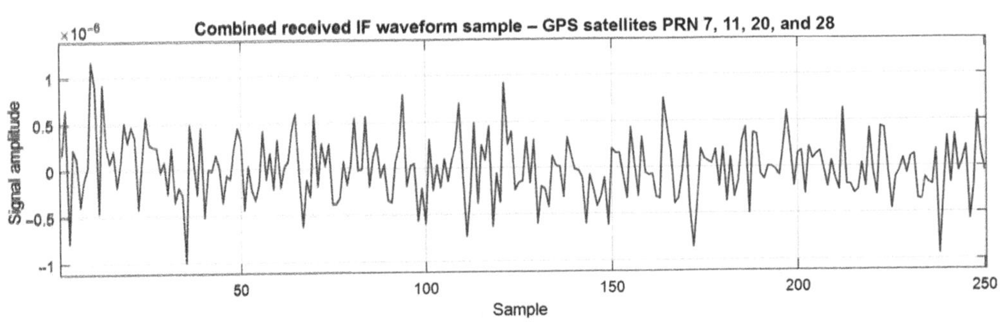

Figure 5.47 Sample of received IF waveform for GPS satellites 7, 11, 20, and 28.

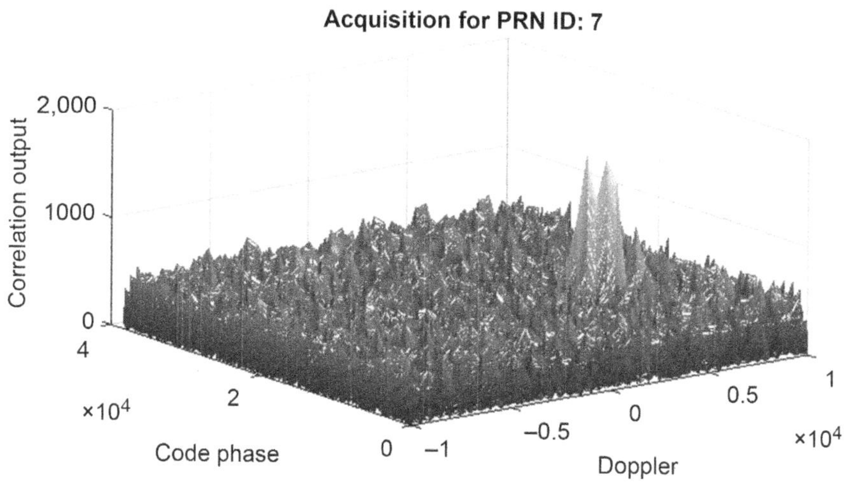

Figure 5.48 Detected acquisition from satellite PRN 7 (satellite 7 is visible).

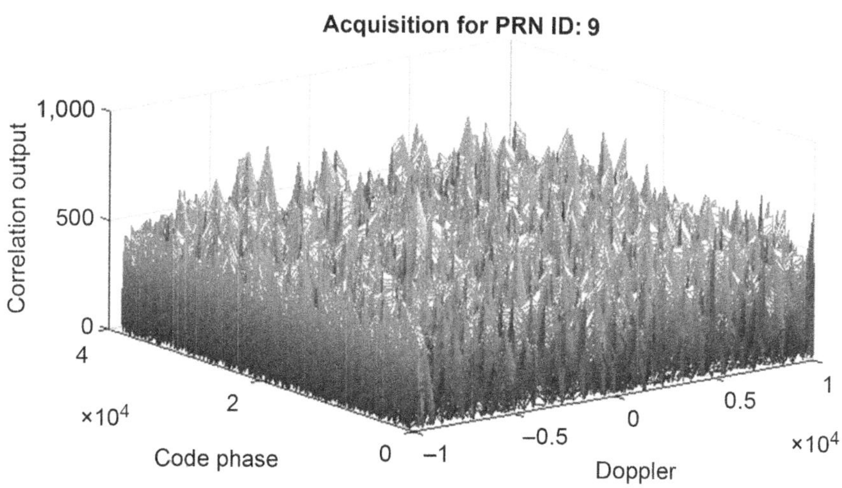

Figure 5.49 No acquisition detected from satellite PRN 9 (satellite 9 is not visible).

carrier tracking) start from the frequency and code delay estimated by the acquisition. Figure 5.50 shows the convergence of the PLL control loop that tracks the carrier frequency. A zero error indicates good tracking of the carrier frequency and phase. The code tracking result is shown in Figures 5.51 and 5.52. Figure 5.51 shows the code tracking error around zero and Figure 5.52 shows the estimated code delay. In this way, the receiver can remove the carrier and the code from the incoming signal, extract the navigation bits, decode the navigation message, and process the frames of the navigation message to calculate the satellite positions and other information like satellites' clock errors and any other correction information. As the receiver is perfectly locked with the satellite signals, the receiving time for each satellite's signal can be calculated precisely, and the pseudorange can be estimated.

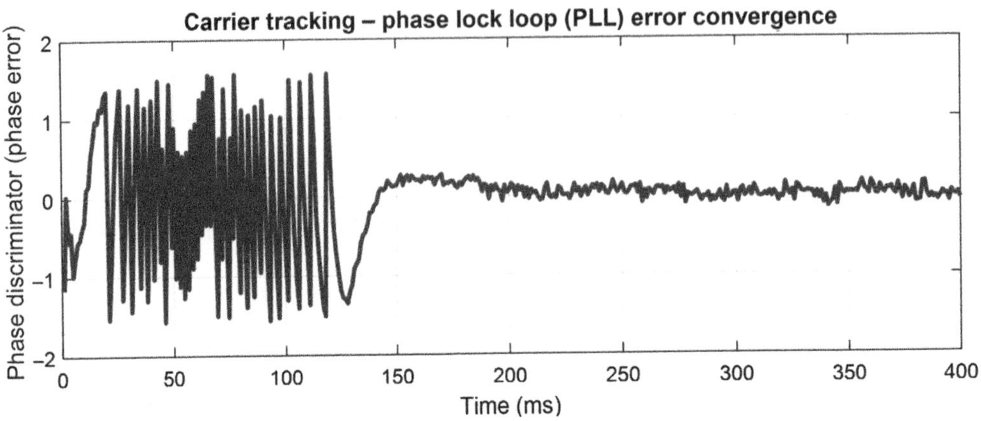

Figure 5.50 Error convergence in carrier tracking using phase lock loop.

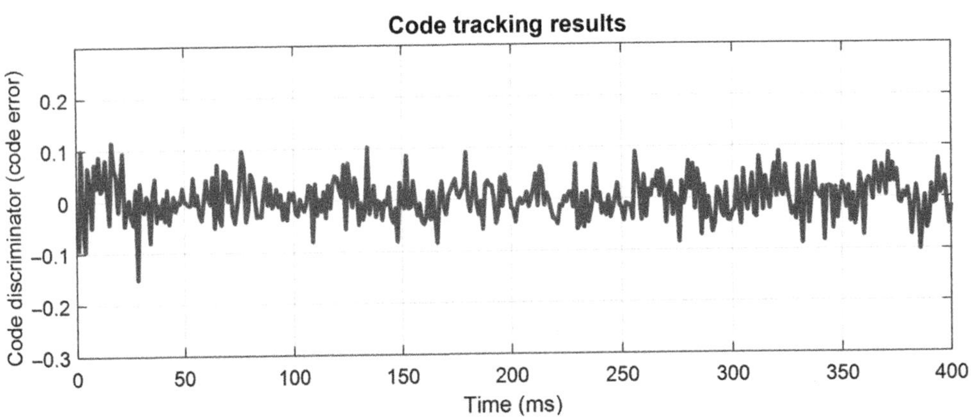

Figure 5.51 Code tracking error.

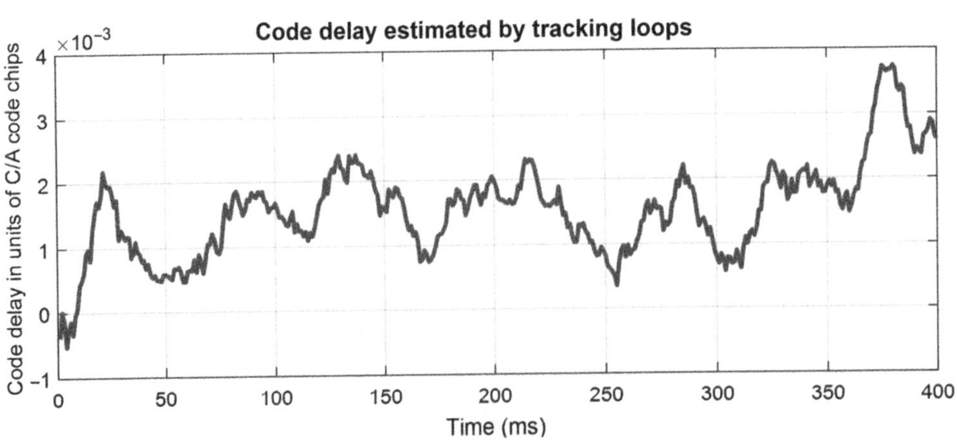

Figure 5.52 Code delay estimated by the tracking loop.

5.10 GNSS Positioning Methods

5.10.1 Single Point Positioning

The simplest GNSS positioning method is single-point positioning (SPP). In SPP, observations from one single receiver are used to calculate a position. SPP can be done using code-phase pseudo-range observations and carrier-phase observations. Assuming a GPS receiver, considering the common errors affecting the pseudo-range, the code-phase pseudo-range equation can be formulated as follows:

$$\tilde{\rho}^i = \|\vec{r} - \vec{s}^i\| + c\Delta t_r + c\Delta t^i + I_r^i + T_r^i + M_\rho^i + \varepsilon_\rho^i \tag{5-95}$$

and the carrier-phase pseudo-range equation can be formulated as follows:

$$\lambda\tilde{\varphi}^i = \|\vec{r} - \vec{s}^i\| + c\Delta t_r + c\Delta t^i + I_r^i + T_r^i + M_\rho^i + \varepsilon_\rho^i + \lambda N^i \tag{5-96}$$

where

$\tilde{\rho}^i$ is the code-phase pseudo-range measurement from i^{th} satellite
$\tilde{\varphi}^i$ is the carrier-phase measurement from i^{th} satellite
λ is the carrier signal wavelength
N^i is an Integer Ambiguity
$\vec{r} = (r_x, r_r, r_r)^T \in \mathbb{R}^3$ is the receiver position vector
$\vec{s}^i = \left(s_x^i, s_y^i, s_z^i\right)^T \in \mathbb{R}^3$ is the ith satellite's position vector
Δt_r is the receiver clock error
c is the speed of light
Δt^i is the satellite's clock error
I_r^i is the ionospheric error
T_r^i is the tropospheric error
M_ρ^i is the multipath error
ε_ρ^i is zero-mean random noise.

The quantity $\|\vec{r} - \vec{s}^i\|$ is the true range between the receiver and the ith satellite calculated by the following equation:

$$\|\vec{r} - \vec{s}^i\| = \sqrt{\left(r_x - s_x^i\right)^2 + \left(r_y - s_y^i\right)^2 + \left(r_z - s_z^i\right)^2} \tag{5-97}$$

The receiver clock error Δt_r multiplied by speed of light c can be replaced by a single variable called receiver bias $b_r = c\Delta t_r$. Assuming there is no multipath error, known satellite's clock error, known ionospheric and tropospheric delay errors, the pseudo-range equation reduces to the following form:

$$\tilde{\rho}^i = f\left(\vec{r}, b_r\right) + \varepsilon_\rho^i = \|\vec{r} - \vec{s}^i\| + b_r + \varepsilon_\rho^i \tag{5-98}$$

The unknowns in this equation are four unknowns, receiver position $\vec{r}$ and clock bias b_r. This explains why we need at least four satellites to get a solution from GPS. If M satellites are visible and tracked, and $M \geq 4$, pseudo-range measurements will be given as follows:

$$\vec{\rho} = \left(\tilde{\rho}^1, \tilde{\rho}^2, \tilde{\rho}^3, \ldots \tilde{\rho}^M\right)^T \tag{5-99}$$

$$\vec{\rho} = f\left(\vec{r}, b_r\right) + \vec{\varepsilon} = \begin{pmatrix} \|\vec{r} - \vec{s}^{\,1}\| \\ \|\vec{r} - \vec{s}^{\,2}\| \\ \|\vec{r} - \vec{s}^{\,3}\| \\ \vdots \\ \|\vec{r} - \vec{s}^{\,M}\| \end{pmatrix} + b_r + \varepsilon_{\rho}^{i} \tag{5-100}$$

We can combine the into one unknown vector $\vec{x} = (r_x\, r_r\, r_r\, b_r)^T$ so we get the following:

$$\vec{\rho} = f\left(\vec{x}\right) + \vec{\varepsilon} \tag{5-101}$$

Equation (5-100) is a nonlinear measurement model where $\vec{\varepsilon}$ is a zero-mean random noise. The estimation of $\vec{x}$ can be done iteratively using iterative least-square algorithm described in Chapter 3. To solve this problem using iterative least-square algorithm, we will need the Jacobian H matrix where the ith row of H is the derivative of $\tilde{\rho}^i$ with respect to r_x, r_r, r_r, b_r. Jacobian matrix is calculated as follows:

$$H = \frac{\partial f\left(\vec{x}\right)}{\partial \vec{x}} \tag{5-102}$$

$$H = \begin{pmatrix} \frac{(r_x - s_x^1)}{\|\vec{r} - \vec{s}^{\,1}\|} & \frac{\left(r_y - s_y^1\right)}{\|\vec{r} - \vec{s}^{\,1}\|} & \frac{(r_z - s_z^1)}{\|\vec{r} - \vec{s}^{\,1}\|} & 1 \\ \frac{(r_x - s_x^2)}{\|\vec{r} - \vec{s}^{\,2}\|} & \frac{\left(r_y - s_y^2\right)}{\|\vec{r} - \vec{s}^{\,2}\|} & \frac{(r_z - s_z^2)}{\|\vec{r} - \vec{s}^{\,2}\|} & 1 \\ & \vdots & & \\ \frac{(r_x - s_x^M)}{\|\vec{r} - \vec{s}^{\,M}\|} & \frac{\left(r_y - s_y^M\right)}{\|\vec{r} - \vec{s}^{\,M}\|} & \frac{(r_z - s_z^M)}{\|\vec{r} - \vec{s}^{\,M}\|} & 1 \end{pmatrix} \tag{5-103}$$

The vector $\frac{(r_x - s_x^i)}{\|\vec{r} - \vec{s}^{\,i}\|} \; \frac{\left(r_y - s_y^i\right)}{\|\vec{r} - \vec{s}^{\,i}\|} \; \frac{(r_z - s_z^i)}{\|\vec{r} - \vec{s}^{\,i}\|}$ can be defined as follows:

$$\vec{l}^{\,i} = \left(\frac{(r_x - s_x^i)}{\|\vec{r} - \vec{s}^{\,i}\|} \; \frac{\left(r_y - s_y^i\right)}{\|\vec{r} - \vec{s}^{\,i}\|} \; \frac{(r_z - s_z^i)}{\|\vec{r} - \vec{s}^{\,i}\|} \right)^T \tag{5-104}$$

where $\vec{l}^{\,i}$ is the line-of-sight vector between the ith satellite and receiver.

The Jacobian matrix can be written in a more compact form if we write as a function of $\vec{l}^{\,i}$ as follows:

$$H = \begin{pmatrix} \vec{l}^{\,1^T} & 1 \\ \vec{l}^{\,2^T} & 1 \\ \vdots & \vdots \\ \vec{l}^{\,M^T} & 1 \end{pmatrix} \tag{5-105}$$

The Jacobian relates the error in measurements $\vec{y}$ to the error in state $\vec{x}$ as follows:

$$\delta\vec{\rho} = H\delta\vec{x}$$

$$\begin{pmatrix} \delta\vec{\rho}^{\,1} \\ \delta\vec{\rho}^{\,2} \\ \vdots \\ \delta\vec{\rho}^{\,M} \end{pmatrix} = \begin{pmatrix} \vec{l}^{\,1T} & 1 \\ \vec{l}^{\,2T} & 1 \\ \vdots & \vdots \\ \vec{l}^{\,MT} & 1 \end{pmatrix} \begin{pmatrix} \delta r_x \\ \delta r_y \\ \delta r_z \\ \delta b_r \end{pmatrix} \tag{5-106}$$

Starting from an initial guess of the receiver position and clock bias $\vec{x}_0 = (r_{x,0}, r_{y,0}, r_{z,0}, b_{r,0})^T$ and applying the iterative least-square algorithm to get an improved estimate until convergence (review Chapter 3) as follows:

$$\vec{x}_{k+1} = \vec{x}_k + (H_k{}^T H_k)^{-1} H_k{}^T \left(\vec{y} - f\left(\vec{x}_k\right)\right) \tag{5-107}$$

where

k is the iteration number
H_k is the Jacobian evaluated at $\vec{x}_k$.

Equation (5-107) is implemented repeatedly for $k = 1, 2, ...$ In each iteration we start from the improved estimate until convergence. Convergence can be determined if the absolute difference between the old estimate and improved estimate $\left|\vec{x}_{k+1} - \vec{x}_k\right|$ or when the residual vector $\vec{e}_k = \left(\vec{y} - f\left(\vec{x}_k\right)\right)$ is extremely small. Note that in this calculation, the satellite positions are assumed to be in the ECEF (E) frame. Hence, the calculated receiver position will be in E frame. The calculated receiver position can then be converted to any coordinate such as geodetic or local-level coordinates as explained in Chapter 1. Given the Jacobian matrix H of Eq. (5-105), it is also possible to calculate the DOP for the GPS solution using the *DOP* equation given as follows:

$$DOP_E = \left(H^T H\right)^{-1} \tag{5-108}$$

The subscript E in DOP_E refers to the E frame. Note that *DOP* is a 4×4 matrix because we have four unknowns (three for position coordinates and 1 for receiver clock bias). Also, note that the first three diagonal elements of the *DOP* matrix calculated from the H matrix in Eq. (5-106) will be given in the E frame because the error state $\delta\vec{x}$ vector has the position represented in E frame coordinates. We can obtain DOP values in N or L frame either by applying frame transformations from E to N or L or by reformulating the positioning estimation problem using the N or L coordinates as states. The horizontal DOP (HDOP), vertical DOP (VDOP), position DOP (PDOP), and geometric DOP (GDOP) are given as follows:

$$HDOP = \sqrt{DOP(1,1)^2 + DOP(2,2)^2} \tag{5-109}$$

$$VHDOP = DOP(3,3) \tag{5-110}$$

$$VHDOP = \sqrt{DOP(1,1)^2 + DOP(2,2)^2 + DOP(3,3)^2} \tag{5-111}$$

$$GHDOP = \sqrt{DOP(1,1)^2 + DOP(2,2)^2 + DOP(3,3)^2 + DOP(4,4)^2} \tag{5-112}$$

5.10.2 Velocity Estimation

As discussed earlier in this chapter, the Doppler measurement is an indication of the relative speed between the receiver and the satellite. If the Doppler measurement and the satellite velocity are available, the receiver velocity can be calculated in a similar way like the position. We start from the range rate equation:

$$\dot{\rho}^i = \left(\vec{v} - \vec{v}^i\right) \cdot \vec{l}^i + c\delta\dot{t}_r - c\delta\dot{t}_s \tag{5-113}$$

where

$\dot{\rho}^i$ is the observed radial speed between i^{th} satellite and the receiver
$v = (v_x, v_y, v_z)^T \in \mathbb{R}^3$ is the receiver velocity
$\vec{v}^i = \left(v_x^i, v_y^i, v_z^i\right)^T \in \mathbb{R}^3$ is the ith satellite's velocity
$\delta\dot{t}_r$ is the receiver clock drift
$\delta\dot{t}_s$ is the receiver clock drift
c is the speed of light
$\vec{l}^i = \left(l_x^i, l_y^i, l_z^i\right)^T \in \mathbb{R}^3$ is the line of sight vector between the satellite and receiver defined in Eq. (5-104).

We know the satellite clock drift $\delta\dot{t}_s$ from the navigation message. The Doppler frequency f_D of ith satellite is related to the radial speed $\dot{\rho}^i$ by the following equation:

$$\dot{\rho}^i = \frac{-f_D{}^i c}{f_c} \tag{5-114}$$

where f_c is the carrier frequency.

We can embed $c\delta\dot{t}_r$ into one variable d_r so we get the following range rate equation:

$$\dot{\rho}^i = \left(\vec{v} - \vec{v}^i\right) \cdot \vec{l}^i + d_r \tag{5-115}$$

The only unknowns in Eq. (5-115) are $\vec{v} = \left(v_x, v_y, v_z\right)^T$ and d_r so if we have at least four satellites, we can apply the same iterative least-square algorithm that we used in the positioning to estimate the four unknowns. If M satellites are visible and tracked, and $M \geq 4$, range rate measurements will be given as follows:

$$\vec{\dot{\rho}} = \left(\dot{\rho}^1, \dot{\rho}^2, \dot{\rho}^3, \ldots \dot{\rho}^M\right)^T \tag{5-116}$$

$$\vec{\dot{\rho}} = f\left(\vec{v}, d_r\right) + \vec{\varepsilon} = \begin{pmatrix} \left(\vec{v} - \vec{v}^1\right) \cdot \vec{l}^1 \\ \left(\vec{v} - \vec{v}^2\right) \cdot \vec{l}^2 \\ \left(\vec{v} - \vec{v}^3\right) \cdot \vec{l}^3 \\ \vdots \\ \left(\vec{v} - \vec{v}^M\right) \cdot \vec{l}^M \end{pmatrix} + d_r + \varepsilon_\rho^i \tag{5-117}$$

We can combine the into one unknown vector $\vec{x} = \left(v_x\, v_y\, v_z\, d_r\right)^T$ so we get the following:

$$\vec{y} = f\left(\vec{x}\right) + \vec{\varepsilon} \tag{5-118}$$

Equation (5-118) is a nonlinear measurement model where $\vec{\varepsilon}$ is a zero-mean random noise. The estimation of $\vec{x}$ can be done iteratively using iterative least-square algorithm described in Chapter 3. To solve this problem using iterative least-square algorithm, we will need the Jacobian H matrix where the ith row of H is the derivative of $\dot{\vec{\rho}}^{\,i}$ with respect to v_x, v_r, v_r, d_r. Jacobian matrix is calculated as follows:

$$H = \begin{pmatrix} \vec{l}^{\,1T} & 1 \\ \vec{l}^{\,2T} & 1 \\ \vdots & \vdots \\ \vec{l}^{\,MT} & 1 \end{pmatrix} \tag{5-119}$$

The Jacobian relates the error in measurements $\vec{y}$ to the error in state $\vec{x}$ as follows:

$$\delta\vec{y} = H\delta\vec{x}$$

$$\begin{pmatrix} \delta\dot{\vec{\rho}}^{\,1} \\ \delta\dot{\vec{\rho}}^{\,2} \\ \vdots \\ \delta\dot{\vec{\rho}}^{\,M} \end{pmatrix} = \begin{pmatrix} \vec{l}^{\,1T} & 1 \\ \vec{l}^{\,2T} & 1 \\ \vdots & \vdots \\ \vec{l}^{\,MT} & 1 \end{pmatrix} \begin{pmatrix} \delta v_x \\ \delta v_y \\ \delta v_z \\ \delta d_r \end{pmatrix} \tag{5-120}$$

Starting from an initial guess of the receiver position and clock bias $\vec{x}_0 = \left(r_{x,0}, r_{y,0}, r_{z,0}, b_{r,0}\right)^T$ and applying the iterative least-square algorithm to get an improved estimate until convergence (review Chapter 3) as follows:

$$\vec{x}_{k+1} = \vec{x}_k + \left(H_k{}^T H_k\right)^{-1} H_k{}^T\left(\vec{y} - f\left(\vec{x}_k\right)\right) \tag{5-121}$$

where

k is the iteration number
H_k is the Jacobian evaluated at $\vec{x}_k$.

Equation (5-121) is implemented repeatedly for $k = 1, 2, \ldots$. In each iteration we start from the improved estimate until convergence. Convergence can be determined if the absolute difference between the old estimate and improved estimate $\left|\vec{x}_{k+1} - \vec{x}_k\right|$ or when the residual vector $\vec{e}_k = \left(\vec{y} - f\left(\vec{x}_k\right)\right)$ is extremely small. Note that in this calculation, the satellite velocities are assumed to be in the ECEF (E) frame. Hence, the calculated receiver velocity will be in E frame. The calculated receiver velocity can then be converted to any coordinate such local-level coordinates as explained in chapter 1.

5.10.3 Joint Position and Velocity Estimation

We can combine the position and velocity in one estimation problem as follows:

$$\underbrace{\begin{pmatrix} \delta\tilde{\rho}^1 \\ \vdots \\ \delta\tilde{\rho}^M \\ \delta\dot{\tilde{\rho}}^1 \\ \vdots \\ \delta\dot{\tilde{\rho}}^M \end{pmatrix}}_{\delta\vec{y}} = \underbrace{\begin{pmatrix} \vec{l}^{1T} & 1 & \vec{0}_{3\times 1} & 0 \\ \vdots & \vdots & \vdots & \vdots \\ \vec{l}^{MT} & 1 & \vec{0}_{3\times 1} & 0 \\ \vec{0}_{3\times 1} & 0 & \vec{l}^{1T} & 1 \\ \vdots & \vdots & \vdots & \vdots \\ \vec{0}_{3\times 1} & 0 & \vec{l}^{MT} & 1 \end{pmatrix}}_{H} \underbrace{\begin{pmatrix} \delta\vec{r} \\ \delta b_r \\ \delta\vec{v} \\ \delta d_r \end{pmatrix}}_{\delta\vec{x}} \tag{5-122}$$

Equation (5-122) can be solved using the same iterative least-squares algorithm that we used for position and velocity estimation. We start from initial guess of $\vec{x}_k$ (receiver's position, clock bias, velocity, and clock drift). From we calculate the Jacobian H_k and the residual $\delta\vec{y}$ (the difference between actual observations $\vec{y}$ and predicted observations $\left(\vec{x}_k\right)$) and then apply Eq. (5-121) to calculate $\vec{x}_{k+1}$. We repeat these steps until convergence.

5.10.4 Using Carrier-phase Observations

Carrier-phase observations $\tilde{\varphi}^i$ can be accurately obtained once the phase-lock loop locks with i^{th} satellite. Once the phase discriminator converges to near zero (see Figure 5.50), the number of carrier signal cycles can be tracked precisely by integrating the Doppler shift as illustrated in Eq. (5-57). The carrier-phase observations are more accurate than code-phase observations due to the high precision of the carrier cycle counter, which is a fraction of the wavelength (19 cm for the GPS L1 signal). From the moment of the lock, the distance between the satellite and the receiver can be measured as number of carrier cycles as follows:

$$\lambda\tilde{\varphi}^i = \|\vec{r} - \vec{s}^i\| + c\Delta t_r + c\Delta t^i + I_r^i + T_r^i + M_\rho^i + \lambda N^i + \varepsilon_\rho^i \tag{5-123}$$

where

$\tilde{\varphi}^i$ is the carrier-phase measurement from i^{th} satellite
λ is the carrier signal wavelength
N^i is the Integer Ambiguity.

To estimate the receiver's position $\vec{r}$, the value of the Integer Ambiguity N^i must be resolved. The Integer Ambiguity N^i remains constant as long as the PLL is locked-in with the ith satellite signal. N^i is estimated by including it as an unknown in the estimation algorithm, so the unknowns will include the receiver position, the receiver clock error, and the Integer Ambiguity. The terms $c\Delta t^i + I_r^i + T_r^i$ are known and the multipath error M_ρ^i is ignored. Hence, the linearized version of Eq. in (5-123) for multiple satellites after correcting for all known terms can be arranged in the following matrix form:

$$\lambda \begin{pmatrix} \delta\tilde{\varphi}^1 \\ \vdots \\ \delta\tilde{\varphi}^M \end{pmatrix} = \begin{pmatrix} \vec{l}^{1T} & c \\ \vdots & \vdots \\ \vec{l}^{MT} & c \end{pmatrix} \begin{pmatrix} \delta r_x \\ \delta r_y \\ \delta r_z \\ \delta\Delta t_r \end{pmatrix} + \lambda \begin{pmatrix} 1 & 0 & \dots & 0 \\ 0 & 1 & 0 & 0 \\ \vdots & 0 & \ddots & \vdots \\ 0 & \dots & 0 & 1 \end{pmatrix} \begin{pmatrix} \delta N_1 \\ \vdots \\ \delta N_M \end{pmatrix} + \vec{\varepsilon} \tag{5-124}$$

Equation (5-126) is in the following form:

$$y = A\vec{a} + B\vec{b} + \vec{\varepsilon} \tag{5-125}$$

where

$\vec{a}$ is a float vector

$\vec{b}$ is an integer vector.

The most used method to estimate the float vector and the integer vector simultaneously is the Least squares AMBiguity Decorrelation Adjustment (LAMBDA) method [15, 18, 19]. The LAMBDA method estimates the optimal integer ambiguities N^i using integer least square (ILS) optimization. The ILS optimization method is explained in Chapter 3 under the State Estimation section.

5.10.5 Carrier-Smoothed Positioning

The single-point positioning using code-phase pseudo-range observables is noisy due to the low precision of the code sequence, which makes the precision of the estimated pseudo-range low. The pseudo-range precision can be greatly enhanced by using the carrier phase changes. As explained earlier, tracking loops steer the internal carrier replica phase to match the incoming satellite signal carrier phase. This steering is done by constantly changing the internal carrier replica phase to match the incoming carrier signal phase. These carrier phase changes, if multiplied by the wavelength, give a more precise measure of the change in the range between the satellite and the receiver. Therefore, we can rely on the code-phase pseudo-range measurements in the first few epochs of processing and then update the pseudo-range using the carrier phase observable, which is smoother and more precise. This method is called carrier-smoothed positioning. A smoothed pseudo-range observable for satellite i can be estimated as follows (see [5]):

$$\tilde{\rho}^i_{r_1}(k) = W_\rho \tilde{\rho}^i_{r_1}(k) + W_\phi \left(\tilde{\rho}^i_{r_1}(k-1) + \lambda(\tilde{\phi}(k) - \tilde{\phi}(k-1)) \right) \tag{5-126}$$

where W_ρ and W_ϕ are weighting coefficients, so the smoothed pseudo-range is a combination of the code-phase pseudo-range given by $\tilde{\rho}^i_{r_1}(k)$ and the carrier-phase pseudo-range calculated by $\tilde{\rho}^i_{r_1}(k-1) + \lambda(\phi(k) - \phi(k-1))$.

The quantity $\lambda(\phi(k) - \phi(k-1))$ is the relative change in the range using carrier-phase change observables measured by $\phi(k) - \phi(k-1)$. The weighting coefficients are given by (see [5]):

$$\begin{cases} W_\rho = 1 - \dfrac{k-1}{N_s}, \quad W_\phi = \dfrac{k-1}{N_s} & (k = 1, 2, \cdots, N_s) \\ W_\rho = \dfrac{1}{N_s}, \quad W_\phi = 1 - \dfrac{1}{N_s} & (k > N_s), N_s = \tau_s / T_s \end{cases} \tag{5-127}$$

where τ_s is a smoothing time constant and is the sampling interval.

These weighting equations give more weight to the code-phase pseudo-range in the beginning and then shift the weight gradually to the carrier-phase pseudo-range. After τ_s seconds, the weight W_ρ will be 0.01 and W_ϕ will be 0.99 giving ultimate weight to carrier-phase pseudo-range.

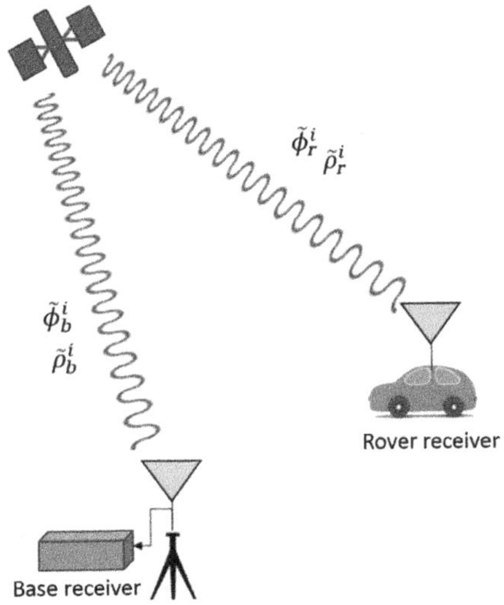

Figure 5.53 Single-difference DGNSS.

5.10.6 Differential Positioning

SSP and Carrier-smoothed positioning method still suffer from atmospheric errors, satellite clock error. A popular method to cancel or reduce these errors is differential GNSS (DGNSS). The DGNSS method uses a base-station receiver (base-receiver) as a reference and calculates other receivers' positions (rover-receivers) with respect to the base-receiver position. As the base-receiver position is known, the absolute position of the rover-receiver can be calculated. The benefit of DGNS is that we have two sets of observables from the same satellites (for DGNSS to work, the two receivers must see some common satellites), the base-receiver observables and the rover-receiver observables. If we take the difference between the two sets of observables, we can cancel out common errors, such as satellite clock errors, since they are common between the two receivers (the two receivers must at least see four common satellites). The difference also reduces atmospheric errors. As the ionospheric layer is higher in the sky, ionospheric errors are likely to be the same on the two receivers if the two receivers are located at a range of 100 km. Since the tropospheric layer is lower and more local to the receiver, it is less likely that the tropospheric error will be the same (unless the two receivers are very close to each other). Therefore, the differenced observations do not significantly reduce the atmospheric errors. It is important to note that differentiating the two sets of observations may increase the random noise. However, the benefit of canceling satellite clock errors and reducing ionospheric errors outweighs the increased random noise due to the differencing operation. DGNSS can be performed in three configurations: single difference, double difference, and epoch difference. Each configuration has its own advantages and limitations.

5.10.7 Single-Difference DGNSS

If a base station receiver "b" and a rover receiver "r" see the same i^{th} satellite (as seen in Figure 5.53), the code pseudo-range measurement equations for the base and rover receivers are given by:

$$\tilde{\rho}_b^i = \|\vec{r}_b - \vec{s}^i\| + c\Delta t_b + c\Delta t^i + I_b^i + T_b^i + M_b^i + \varepsilon_b^i \tag{5-128}$$

$$\tilde{\rho}_r^i = \|\vec{r} - \vec{s}^i\| + c\Delta t_r + c\Delta t^i + I_r^i + T_r^i + M_r^i + \varepsilon_r^i \tag{5-129}$$

The single difference code pseudo-range equation is defined as follows:

$$\Delta\tilde{\rho}_{b,r}^i = \tilde{\rho}_b^i - \tilde{\rho}_r^i \tag{5-130}$$

$$\Delta\tilde{\rho}_{b,r}^i = \|\vec{r}_b - \vec{s}^i\| - \|\vec{r} - \vec{s}^i\| + c(\Delta t_b - \Delta t_r) + (I_b^i - I_r^i) + (T_b^i - T_r^i) + (M_r^i - M_r^i) + \varepsilon_{r,b}^i \tag{5-131}$$

The term $p_b^i = \|\vec{r}_b - \vec{s}^i\|$ is known since the base station position $\vec{r}_b$ is known and the satellite position $\vec{s}^i$ is given in the ephemeris data. In Eq. (5-136), the ith satellite clock error Δt^i is removed.

In addition, the ionospheric delay error $\left(I_b^i - I_r^i\right)$ can be removed if the distance between the two receivers is relatively short (within 100 km), which makes the term $\left(I_b^i - I_r^i\right) \cong 0$. The same applies to the tropospheric delay error $\left(T_b^i - T_r^i\right) = 0$. In our solution, we ignore the multipath error term just like the case in single-point positioning algorithm so $\left(M_r^i - M_r^i\right) = 0$. These reductions will reduce the pseudo-range equation to the following form:

$$\Delta\tilde{\rho}_{b,r}^i \cong p_b^i - \| \vec{r} - \vec{s}^i \| + c(\Delta t_b - \Delta t_r) + \varepsilon_{r,b}^i \tag{5-132}$$

Equation (5-132) is a nonlinear equation that has five unknowns, the rover's receiver position, the rover's receiver clock bias Δt_r and the base station's receiver clock bias Δt_b. As the base station's receiver location is known, its clock bias Δt_b can be calculated using the standard single-point positioning algorithm described in Section 5.10.1. Therefore, Eq. (5-132) will be further reduced to the following form:

$$\| \vec{r} - \vec{s}^i \| + c\Delta t_r = p_b^i - \Delta\tilde{\rho}_{b,r}^i \tag{5-133}$$

Equation (5-133) has four unknowns, the rover's receiver position vector $\vec{r}$ and the rover's receiver clock bias Δt_r therefore, four satellites are sufficient to solve for the four unknowns. The solution of Eq. (5-133) is calculated using the same iterative nonlinear least-squares algorithm described in Section 5.10.1. The same differential positioning method can be used with carrier-phase measurements with one additional unknown, which is Integer Ambiguity. The carrier-phase differential measurement equation is given by:

$$\Delta\tilde{\phi}_{b,r}^i = \tilde{\phi}_b^i - \tilde{\phi}_r^i \tag{5-134}$$

where $\tilde{\phi}_b^i$ and $\tilde{\phi}_r^i$ are given by Eq. (5-96).

5.10.8 Double-Difference DGNSS

If we have two satellites i and j commonly visible by the two receivers, double differencing observations (see Figure 5.54) can be formulated as follows:

$$\Delta\Delta\tilde{\rho}_{1,2}^{i,j} = \Delta\tilde{\rho}_{1,2}^i - \Delta\tilde{\rho}_{1,2}^j = \left(\tilde{\rho}_1^i - \tilde{\rho}_2^i\right) - \left(\tilde{\rho}_1^j - \tilde{\rho}_2^j\right) \tag{5-135}$$

$$\Delta\Delta\tilde{\phi}_{1,2}^{i,j} = \Delta\tilde{\phi}_{1,2}^i - \Delta\tilde{\phi}_{1,2}^j = \left(\tilde{\phi}_1^i - \tilde{\phi}_2^i\right) - \left(\tilde{\phi}_1^j - \tilde{\phi}_2^j\right) \tag{5-136}$$

Using the double difference measurements, i^{th} and j^{th} satellite common errors (satellite clock error) will be removed, and the receivers' clock error will be removed, which is a major advantage of the doubled-difference technique. In addition, tropospheric and the ionospheric delay error are removed if the distance between the two receivers is relatively short within 100 km.

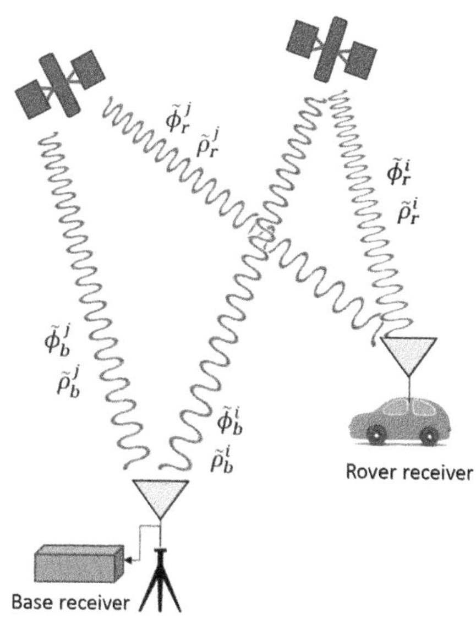

Figure 5.54 Double-difference DGNSS.

5.10.9 Triple-Difference DGNSS

In carrier-phase positioning, the Integer Ambiguity is a major source of error. If we take differences of double differences between sequential epochs (difference in time between double differences), we can remove this Integer Ambiguity. This is called triple differential DGNSS, and it works best if the rate of observables is high (so the effect of the propagation delays from one epoch to another is small). If the initial position of the receiver is known, this method works very well without the need for Integer Ambiguity resolution. The triple difference equation can be formulated as follows:

$$\Delta\Delta\Delta\tilde{\phi}_{1,2}^{i,j}(k) = \Delta\Delta\tilde{\phi}_{1,2}^{i,j}(k) - \Delta\Delta\tilde{\phi}_{1,2}^{i,j}(k-1) \tag{5-137}$$

5.10.10 Precise Point Positioning

Precise point positioning (PPP) can provide comparable accuracy to differential GNSS using a single receiver. DGNSS and RTK require a fixed-location base receiver and require that the rover-receiver be relatively close to the base station. PPP technique alleviates these requirements by using the following techniques:

- Using carrier-phase measurements and solving the integer ambiguity problem.
- Using dual frequency to accurately remove the ionospheric delay errors.
- Using more precise orbital parameters (better ephemeris information) and more accurate satellites' clock corrections from an external source. This external source can be the Internet, where the correction data, along with precise orbit data, is collected through a network of receivers and made available for users through the Internet. There are academic and commercial networks that provide precise satellite orbital and clock data. For example, the International GNSS Service (IGS) (https://igs.org/) provides precise orbital data and satellite clock correctios subdaily, daily, or weekly. The National Aeronautics and Space Administration (NASA) also provides precise satellite clock and orbit corrections, which can be found in https://www.gdgps.net/. There are commercial sources that gather precise satellites orbital and clock corrections data such as Trimble and NovAtel.

PPP is commonly used in postprocessing mode but can also be used in real time if the corrections, the precise orbital parameters, and satellites' clocks error information are available.

5.10.11 Real-Time Kinematics (RTK)

DGNSS methods can be applied in an offline or postprocessing mode where the observations of the base-receiver and the rover-receiver are collected and saved to a storage and processed later by a computer. RTK is simply DGNSS performed in real-time, where the information of the base-receiver is broadcast to rover-receivers in real-time. The communication link between the base receiver and the rover receiver is the key enabler of RTK. Obviously, the RTK-enabled rover receiver must be equipped with additional hardware and software to receive and decode the base-receiver information.

5.10.12 GNSS Assistance and Augmentation Systems

The idea of the GNSS assistance and augmentation systems is an extension of the control and monitoring stations where local stations are installed at known locations to estimate model parameters and corrections (such as orbital parameters and atmospheric delay parameters) and then broadcast

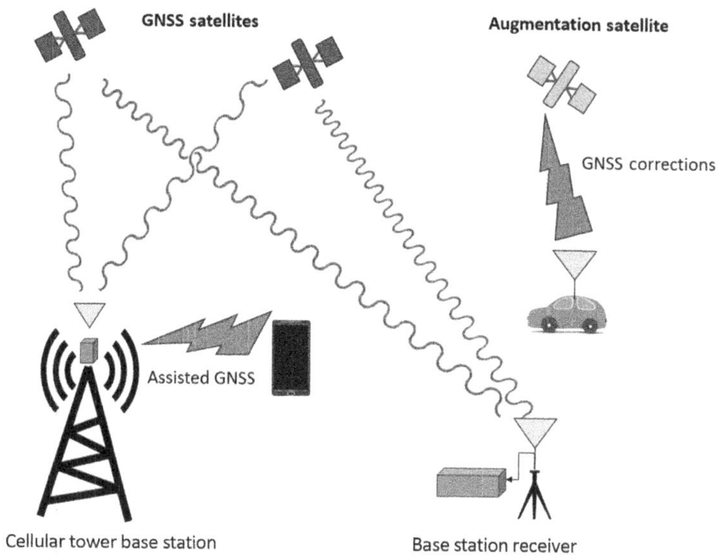

Figure 5.55 GNSS assistance and augmentation systems.

these parameters and corrections to the receivers to enhance the position calculations. Common assistance systems include cellular networks where cellular towers are used as GNSS base stations that estimate correction parameters as can be seen in Figure 5.55.

The correction parameters are then transmitted to smartphones, which can be used to assist in the position calculation. Augmentation systems are independent dedicated base stations installed either on the ground or in the sky as a network of geostationary satellites. These networks of base stations, which are installed at known locations, estimate and update correction parameters and broadcast them to receivers. To benefit from augmentation systems, receivers must have the capability to receive information from these augmentation base stations. Examples of common GNSS augmentation systems include the United States Wide Area Augmentation System (WAAS) and the European Geostationary Navigation Overlay Service (EGNOS).

5.11 GPS SPS MATLAB Project Sample

The projects' repository of the book includes a MATLAB project that demonstrates the single point positioning using GPS and GLONASS satellites. The project "*GnssPositioningEstimationProject*" includes datasets collected by a multi-constellation GNSS receiver that processes signals from GPS and GLONASS satellites. Readers are encouraged to run the demo scripts and investigate the results. The demo script is called "*demo_GPS_SPP*.m" and it loads the dataset and processes raw pseudo-range, estimates the receiver's position along with DOP values and compares the results against ground truth data. The pseudo-range measurements that come with the project have been corrected and cleaned to generate zero position errors. This is to give the readers a chance to add different noises and biases to the pseudo-range and test the effect of the added noise/biases on the positioning errors and the DOP values. Readers can even control how many satellites are to be processed at any given epoch by doing some simple modifications in the code. MATLAB 2023b environment will be needed to run the project scripts. Sample results of running the demo script with a Gaussian noise of 2.5 m standard deviation added to the pseudo-range measurements are shown in Figures 5.56–5.58. Note the relationship between the DOP values, the number of visible satellites, and the positioning errors. The higher the number of visible satellites, the smaller the DOP values and the smaller the errors. This shows how the number and geometry arrangements of the satellites affect the positioning accuracy. Note also how the vertical error component is larger than the horizontal (east and north). This is a common problem with GNSS technology since the visible satellites are the ones with high elevations with respect to the user's position on Earth.

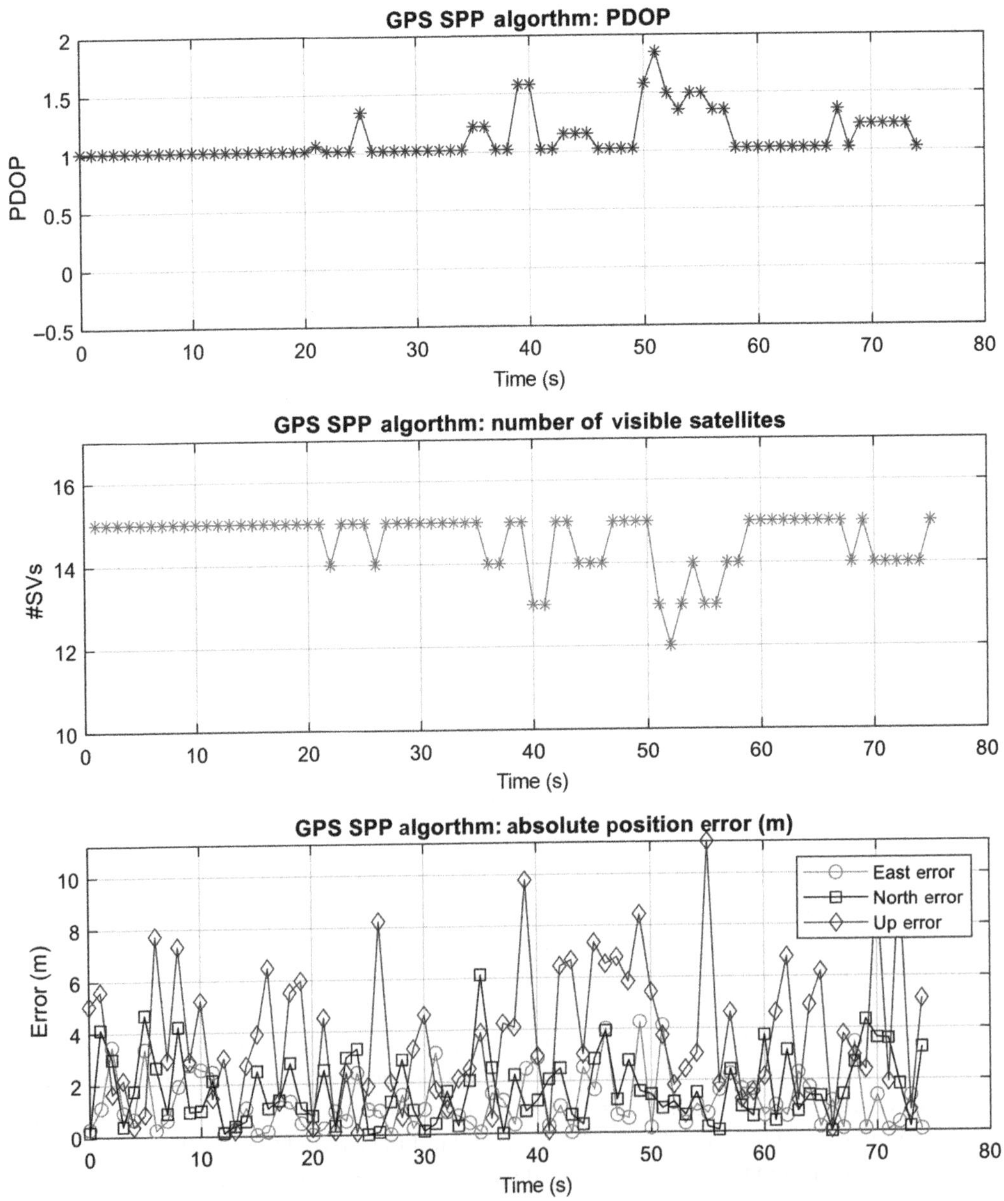

Figure 5.56 GPS SPS algorithm output: PDOP, number of visible satellites, and errors.

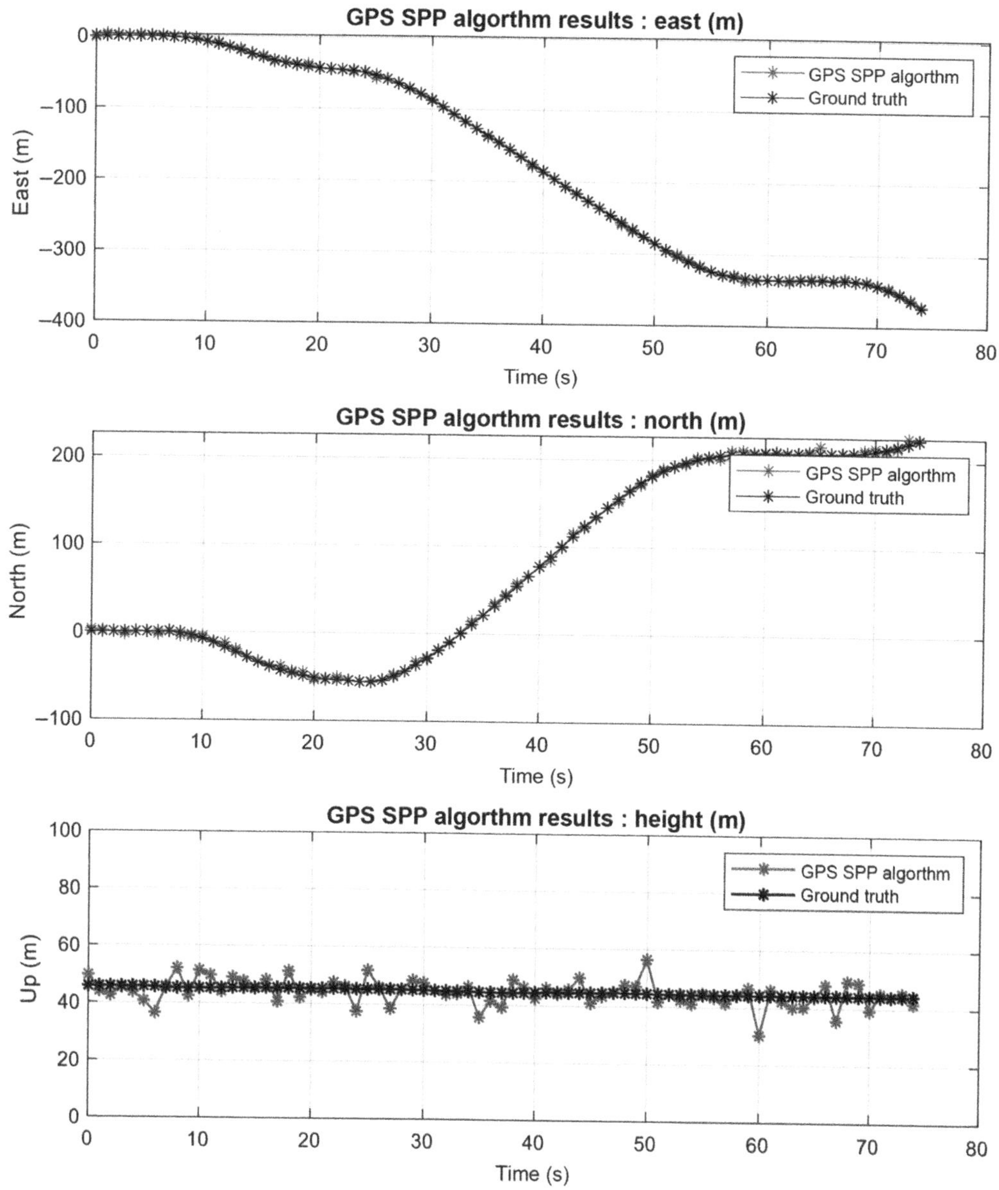

Figure 5.57 GPS SPS algorithm output: estimated vs. ground truth position.

Low-elevation satellites are usually not clearly visible to the receiver due to obstacles. Negative elevation satellites are impossible to see because they are on the other side of the Earth with respect to the current user's position.

5.12 Sagnac Effect

In all the positioning schemes that will be discussed in the next subsections, the location of the satellites is required to enable the localization of the receivers. As explained earlier, the navigation message includes the required orbital parameters to calculate the satellite positions at the time of

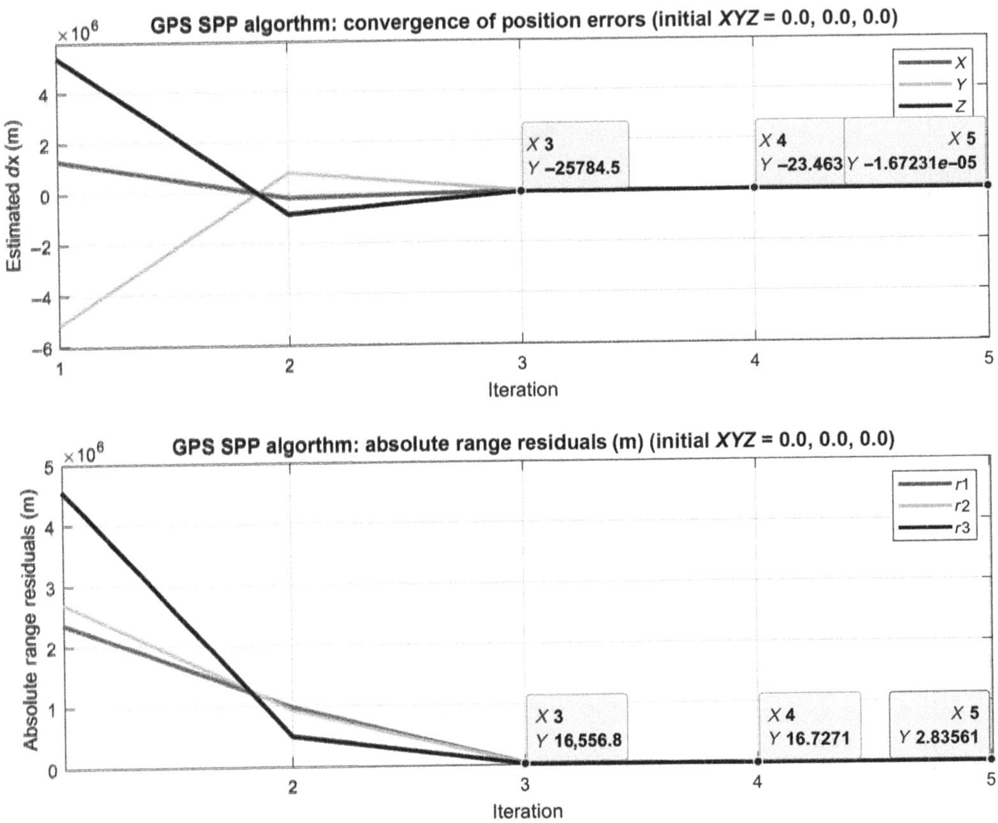

Figure 5.58 Convergence of first epoch of GPS SPS algorithm.

the transmission of the signal. However, at the time of the receiving of the signal, the satellite positions would have changed, which must be considered and corrected. This phenomenon is known as the Sagnac Effect due to the Earth's rotation during the travel of the signal from the satellite to the receiver. More information about the Sagnac Effect can be found in [10]. To correct for the Sagnac effect, the satellites' positions (in ECEF frame) are transformed by the following transformation matrix:

$$R_s = \begin{pmatrix} \cos(\omega_e \times \Delta T) & \sin(\omega_e \times \Delta T) & 0 \\ -\sin(\omega_e \times \Delta T) & \cos(\omega_e \times \Delta T) & 0 \\ 0 & 0 & 1 \end{pmatrix} \tag{5-138}$$

where

ΔT is the signal propagation time
ω_e is the Earth's rotation rate.

Some receivers and GNSS systems provide satellites' positions already corrected for the Sagnac effect. Users need to examine the datasheet of the GNSS equipment they are using to verify if the provided satellites' position coordinates are compensated for the Sagnac effect. The projects' repository includes a sample project that demonstrates the effect of the Sagnac effect on the receiver's positional errors. Sample results are shown in Figure 5.59.

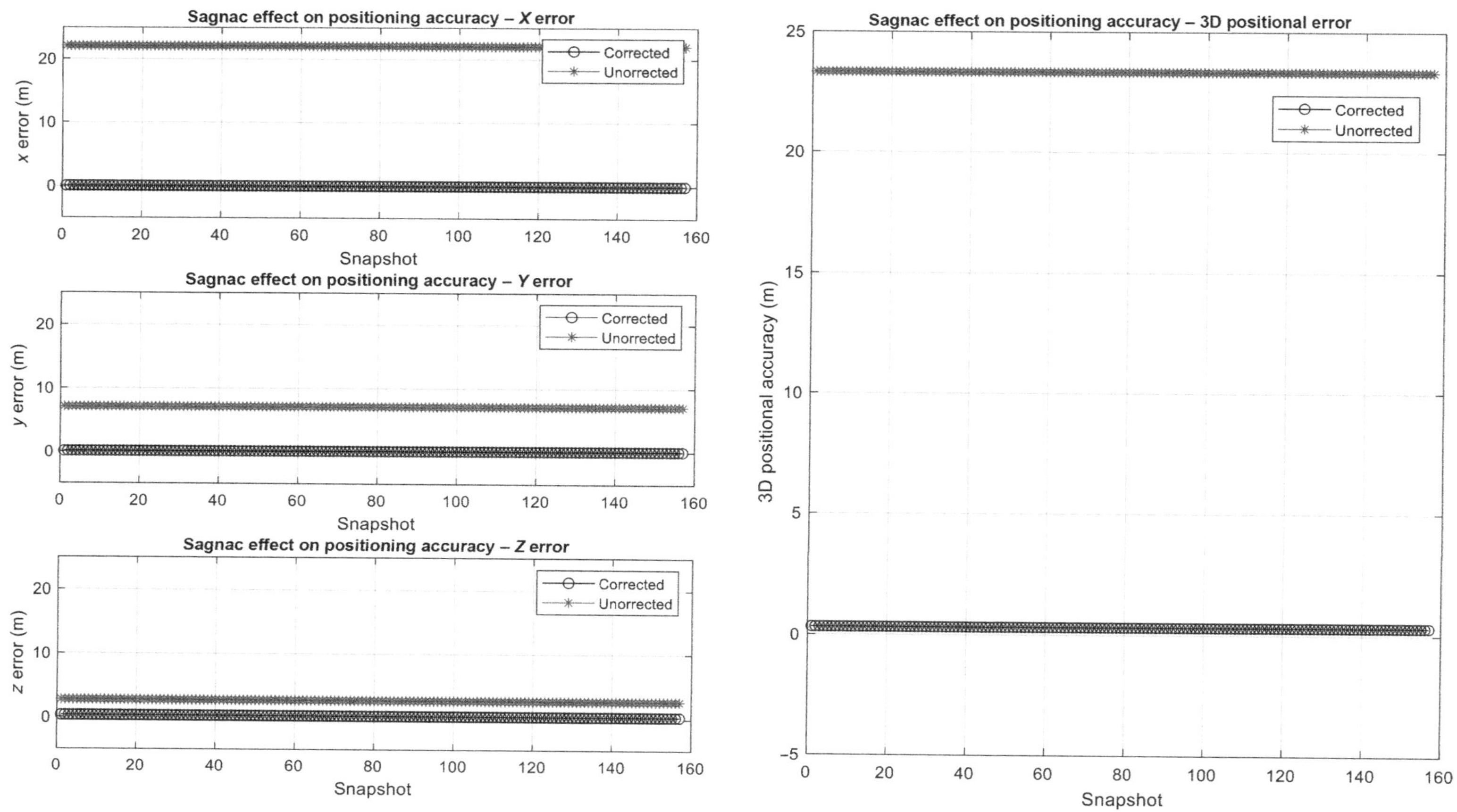

Figure 5.59 Sagnac effect on GNSS positional errors.

5.13 Problems

5.13.1 Problem 1

The Doppler measurement of GPS satellite signal received by an $L1$ GPS receiver is −2,740 Hz. Calculate the relative speed between this satellite and the receiver.

5.13.2 Problem 2

A user is located at $\vec{r} = (1{,}071{,}123.81, -4{,}450{,}502.41, 4{,}426{,}659.34)^T$ in ECEF coordinates $(X, Y, Z)^T$ in meters. If a GPS receiver at this location can acquire and track 4 GPS satellites, calculate the Jacobian matrix H given that the satellites' locations are as follows:

Satellite	X (m)	Y (m)	Z (m)
$\vec{s}^1$	16,426,866.75	−14,450,213.98	15,084,366.87
$\vec{s}^2$	1,497,442.01	−25,338,889.63	−7,379,759.82
$\vec{s}^3$	−856,246.57	−18,803,470.15	18,897,684.63
$\vec{s}^4$	−8,430,851.54	−16,640,365.30	19,414,393.48

5.13.3 Problem 3

In the previous problem, use the calculated H matrix to calculate the DOP.

5.13.4 Problem 4

In the previous problem, the position DOP vector $(DOP(1, 1), DOP(2, 2), DOP(3, 3))^T$ is given in E coordinates. Transform the position DOP to N frame.

5.13.5 Problem 5

Repeat Problems 2, 3, and 4 if the user location and satellite's locations are given as follows:

$$\vec{r} = (1{,}071{,}100.07, -4{,}450{,}517.42, 4{,}426{,}671.01)^T$$

Satellite	X (m)	Y (m)	Z (m)
$\vec{s}^1$	15,212,507.56	2,819,541.26	21,314,390.85
$\vec{s}^2$	16,436,228.29	−14,451,867.52	15,072,453.23
$\vec{s}^3$	1,499,962.97	−25,343,133.84	−7,364,700.03
$\vec{s}^4$	−847,259.38	−18,795,797.11	18,905,861.68
	−-8,426,567.18	−16,651,396.40	19,406,709.02

Comment on the values of the DOP in the two user's locations.

5.13.6 Problem 6

Given the following TDoA equations, assuming that the anchors' locations are known, derive the least-square solution to find the target's location given the TDoA measurements.

$$\Delta\rho_{1,3} = \sqrt{(x_1 - x)^2 + (y_1 - y)^2 + (z_1 - z)^2} - \sqrt{(x_3 - x)^2 + (y_3 - y)^2 + (z_3 - z)^2} \tag{5-139}$$

$$\Delta\rho_{2,1} = \sqrt{(x_2 - x)^2 + (y_2 - y)^2 + (z_2 - z)^2} - \sqrt{(x_1 - x)^2 + (y_1 - y)^2 + (z_1 - z)^2} \tag{5-140}$$

$$\Delta\rho_{2,3} = \sqrt{(x_2 - x)^2 + (y_2 - y)^2 + (z_2 - z)^2} - \sqrt{(x_3 - x)^2 + (y_3 - y)^2 + (z_3 - z)^2} \tag{5-141}$$

5.13.7 Problem 7

In TDoA localization configurations (passive or active see Figures 5.31 and 5.32), if we have measurements from a fourth anchor, derive an estimation algorithm that can simultaneously estimate the target's position and the transmission time.

References

1 A. A. Smith, "*Radio Frequency Principles and Applications: The Generation, Propagation, and Reception of Signals and Noise*," New York: John Wiley & Sons, 1998.

2 National Aeronautics and Space Administration (NASA), "Imagine The Universe," [Online]. Available: https://imagine.gsfc.nasa.gov/science/toolbox/emspectrum1.html. [Accessed 30th May 2023].

3 I. G. Petrovski and T. Tsujii, "*Digital Satellite Navigation and Geophysics*," Cambridge: Cambridge University Press, 2012.

4 T. Pany, "*Navigation Signal Processing for GNSS Software Receivers*," Norwood, MA: Artech House, 2010.

5 J. A. Farrell, "*Aided Navigation, GPS with High Rate Sensors*," New York: McGraw Hill, 2008.

6 P. D. Groves, "*Principles of GNSS, Inertial, and Multisensor Integrated Navigation Systems*," Boston, MA: Artech House, 2013.

7 K. Borre, D. M. Akos, N. Bertelsen, P. Rinder and S. H. Jensen, "*A Software-Defined GPS and Galileo Receiver, A Single-Frequency Approach*," Boston, MA: Birkhauser, 2007.

8 A. D. Angelis, A. Moschitta, P. Händel and P. Carbone, "Experimental Radio Indoor Positioning Systems Based on Round-Trip Time Measurement," in *Advances in Measurement Systems*, Edited by M. K. Sharma, IntechOpen, 2010, pp. 195–219.

9 P. Santagati and V. Beiu, "A Mathematical Model for the Analysis of the Johnson-Nyquist Thermal Noise on the Reliability in Nano-Communications," in *Bio-Inspired Models of Network, Information, and Computing Systems*, Springer, 2012, p. 49–59.

10 B. Bidikar, G. S. Rao and L. Ganesh, "Sagnac Effect and SET Error based Pseudorange Modeling for GPS Applications," in *4th International Conference on Recent Trends in Computer Science & Engineering*, Chennai, Tamil Nadu, India, 2016.

11 P. Misra and P. Enge, "*Global Positioning System: Signals, Measurements, and Performance*," Revised Second Edition, Lincoln, MA: Ganga-Jamuna Press, 2010.

12 B. Bhatta, "*Global Navigation Satellite Systems: New Technologies and Applications*," Second Edition, Boca Raton, FL: CRC Press, 2023.

13 Mathworks Inc., "GPS Receiver Acquisition and Tracking Using C/A-Code," Mathworks Inc., 2023. [Online]. Available: https://www.mathworks.com/help/satcom/ug/gps-receiver-acquisition-and-tracking-using-ca-code.html. [Accessed 24th January 2024].

14 FAA William J. Hughes Technical Center, "Global Positioning System Standard Positioning Service Performance Analysis Report," Federal Aviation Administration, Satellite Navigation Office (AJM-32), 2021.

15 P. Teunissen, "A New Method for Fast Carrier Phase Ambiguity Estimation," in *IEEE Position, Location and Navigation Symposium PLANS'94*, Las Vegas, NV, 1994.

16 E. S. Agency, "GPS Navigation Message," [Online]. Available: https://gssc.esa.int/navipedia/index.php/GPS_Navigation_Message. [Accessed 1st June 2023].

17 E. D. Kaplan and C. Hegarty, "*Understanding GPS/GNSS: Principles and Applications*," Third Edition, Boston, MA: Artech House, 2017.

18 J. J. van Aardt and Y. Pailhas, "Integer least squares techniques in remote sensing applications: Review and practical considerations," *ISPRS Journal of Photogrammetry and Remote Sensing*, vol. 58, no. 3–4, pp. 337–354, 2004. DOI: https://doi.org/10.1016/j.isprsjprs.2004.07.001.

19 A. N. Bishop, "Mixed-integer least squares for simultaneous estimation of integer and real parameters," *Journal of the Operational Research Society*, vol. 66, no. 1, pp. 51–60, 2015. DOI: https://doi.org/10.1057/jors.2014.31.

6

Active Ranging Sensors

6.1 Introduction

In the previous chapter, we studied radio positioning technology that uses electromagnetic waves to measure ranges and calculate objects' locations. The most common radio positioning example was global navigation satellite system (GNSS). In GNSS, communication is one way; the active transmitter is the satellite, and the receiver acts as a passive listener. Therefore, the GNSS receiver can be seen as a passive sensor that receives information from an external source. The benefit of this listen-only mode of GNSS receiver is that it does not depend on the environment and can support as many receivers as needed. Listen-only mode is also critical in military and security applications, so receivers cannot be detected or located because they do not transmit signals. Passive radio positioning systems do not provide a perception of the environment. Electromagnetic waves (and other waves like sound) can be used in another active configuration to provide a perception of the environment. This technology is called an active ranging sensor, where the sensor actively transmits or emits a signal and listens to the reflection to get a depth estimation with respect to objects in the environment. This signal could be an electromagnetic wave like the case in light detection and ranging (LiDAR [1] and [2]) sensors and radio detection and ranging (RADAR [3]) sensors, or it could be a sound wave (e.g., ultrasound wave) like sound navigation and ranging (SONAR [4]) sensors. Active-ranging sensors have the advantage of providing information about the environment, which is crucial for mapping and understanding the surrounding environment (Figure 6.1). That is the reason active ranging sensors are used in autonomous vehicles to provide obstacle detection, collision avoidance, and path planning. In addition to mapping and obstacle detection/collision avoidance, active ranging sensors can be used to provide positioning, as we will discuss in this chapter.

6.2 Principle of Operation of Active Ranging/Detection

Active ranging sensors are equipped with a transmitter part, receiver part, and a signal processing/estimation module, all in one package. In between, there are other components such as antennas, signal conditioning, and analog-to-digital converters. The main block diagram of active ranging sensors is shown in Figure 6.5. A brief explanation of each component is given in the next section.

Sensor Fusion Approaches for Positioning, Navigation, and Mapping: How Autonomous Vehicles and Robots Navigate in the Real World: With MATLAB Examples, First Edition. Mohamed M. Atia.
© 2025 The Institute of Electrical and Electronics Engineers, Inc. Published 2025 by John Wiley & Sons, Inc.

Figure 6.1 Active ranging sensor.

6.2.1 Ranging Sensor Components

Any ranging sensor has the following basic components:

- **Waveform generator:** It is the circuitry that generates an electrical current/voltage that matches the shape of the signal to be transmitted. The commonly used waveforms are pulses and continuous frequency-modulated waves. Pulsed signals can take different forms, as seen in Figure 6.2. Pulsed ranging systems transmit short pulses periodically at a period called pulse repetition interval (PRI). We will refer to this interval by T_{PRI}. When the pulse is transmitted, the sensor remains silent for a short period T_p. While the sensor is silent, it listens for a reflected echo of the transmitted signal. If an echo is detected, the sensor processor calculates the distance to a target by measuring the time delay between a transmitted pulse and the returning reflected echo. The distance can be calculated by the following simple equation:

$$d = \frac{sT}{2} \tag{6-1}$$

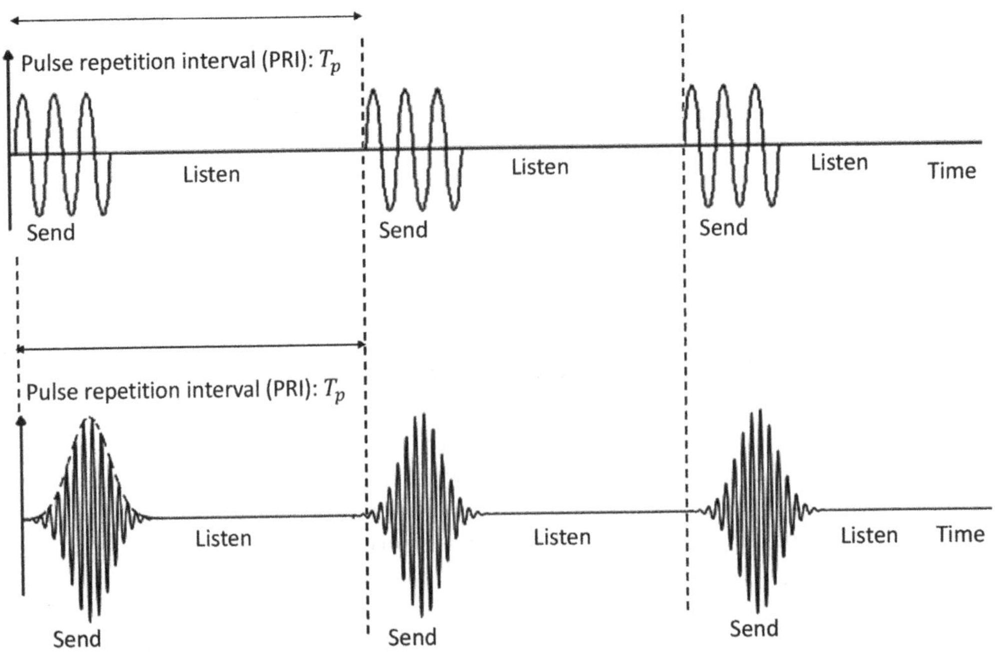

Figure 6.2 Sample pulse ranging waveform.

where

s is the speed of the signal
T is the round trip time.

Pulsed sensors must wait for the reflected echo of a target (if a target exists) before sending the next pulse. This sets a limit on the maximum range the sensor can sense. For example, a radar of maximum range $d_{max} = 150$ km will require a pulse repetition interval $T_{PRI} = 1$ ms. This can be calculated as follows [3]:

$$d_{max} = \frac{2\,T_{PRI}}{s} \tag{6-2}$$

where s is the speed of signal (e.g., speed of light in RADR/LiDAR or speed of sound in SONAR).

To avoid the limitations of the pulsed ranging technique, continuous waveforms can also be used. Continuous waveforms can be amplitude modulated or frequency modulated. The concept of the amplitude modulated continuous wave (AMCW [9]) ranging is illustrated in Figure 6.3. The distance equation for AMCW sensors is given by:

$$d = \frac{s}{2f}\frac{\Delta\varphi}{2\pi} \tag{6-3}$$

where

$\Delta\varphi$ is the phase difference between the transmitted chirp and the received echo
s is the signal speed
f is the frequency of the AMCW wave.

Due to the cyclicity of the phase shift measurement, $\Delta\varphi$, AMCW has limited range; otherwise, it may give ambiguous distance estimation. Modern ranging sensors including RADAR and LiDAR use frequency-modulated continuous wave (FMCW) [3] and [9] ranging technique. FMCW sensors constantly transmit a signal with linearly changing frequency as seen in Figure 6.4. This type of ranging system transmits a signal with a linearly varying frequency called chirps at a period called

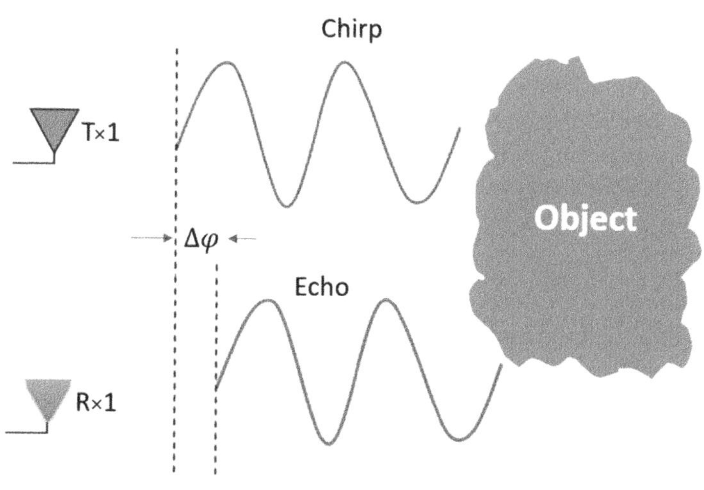

Figure 6.3 AMCW ranging.

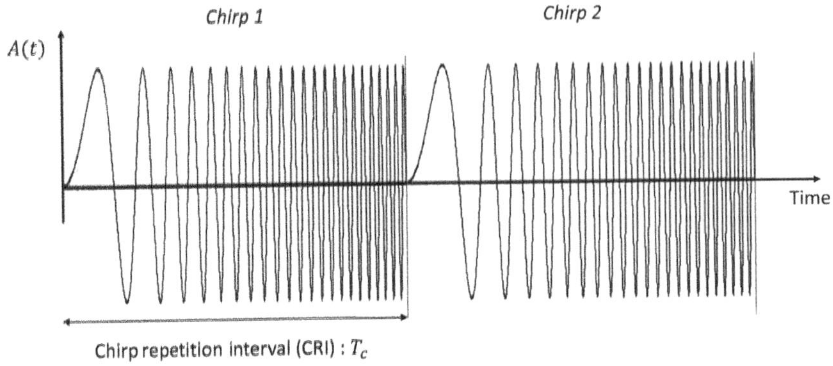

Figure 6.4 Sample FMCW ranging waveforms.

"chirp repetition interval (CRI) T_{CRI}" as seen in Figure 6.4. Simultaneously, the sensor listens to the reflected echo. If an echo is detected, the sensor uses the frequency shift between the transmitted chirp and the echo chirp to estimate the range and velocity of the targets. The transmitted signal is called a chirp, which is commonly used in the literature on ranging sensors. The signal processing and the advantages and limitations of both types of waveforms will be discussed in this chapter.

- **Transmitter**: This component converts electrical waveform into a physical wave (chirp). This wave could be electromagnetic waves (like RADAR and LiDAR) or sound waves (like SONAR). The characteristics of the ranging sensor, such as penetration properties, maximum range, range resolution, field of view, angular resolution, and accuracy, depend on the type of wave being used and the way the wave is transmitted. Generally, transmitters use multiple antennas or antenna arrays to illuminate a scene (flash sensing) or to transmit different beams in different directions (scan sensing).
- **Receiver**: The receiver component converts the reflected chirps (echo signals) into electrical voltage or current and may apply some preconditioning such as amplification, filtering (to select a specific frequency range), mixing with an internal signal replica, and analog-to-digital conversion. Like transmitters, receivers may have multiple antennas, which are used to estimate the angle of arrival of the reflected chirps (the echo).
- **Signal processor**: This component applies digital signal processing (DSP [5]) methods to detect targets, estimate ranges to targets, speed of the targets, and targets' angle or orientation with respect to the sensor. Usually, the transmitter and receiver are implemented in hardware, and the signal processing part is implemented using a microcontroller unit (MCU) or special digital signal processor (DSP) core. The details of the signal processing of both types of active ranging systems (pulsed systems and FMCW systems) will be explained later in this chapter.

6.2.2 Ranging Equation

Active ranging sensors measure, the round-trip time of the signal. The round-trip time $T = 2\tau$ where τ is the chirp propagation time (which is the same as the echo propagation time) as seen in Figure 6.5. To calculate the distance to the target, the following simple equation is used [3], [4]:

$$d = s\frac{T}{2} = s\,\tau \tag{6-4}$$

where s is the speed of the wave.

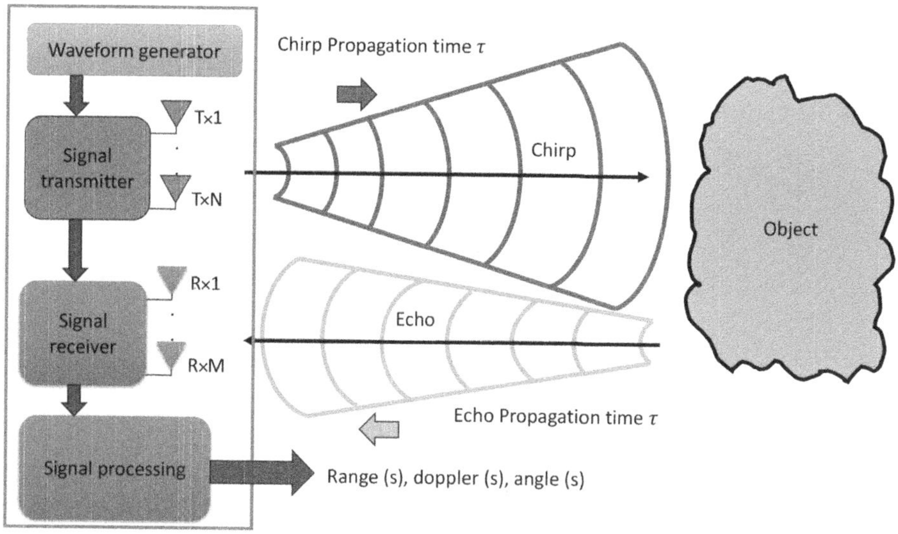

Figure 6.5 Basic block diagram of active ranging sensors.

If the wave is an electromagnetic wave, s will be the speed of light. If the wave is a sound (typically, an ultrasound wave is used), s will be the speed of sound. Although the principle of operation seems simple, the estimation of the range, the speed of the target, and the angle of the target from the reflected echo require complex signal processing, which will be explained in this chapter. We will also see how multiple targets (even the targets that are at the same distance from the sensor but at different angles or targets that lie on the same beam at different distances) can be detected.

6.2.3 Ranging Sensor's Specifications

Figure 6.6 shows some of the common specifications of active ranging sensors. Generally, any ranging sensor will have the following specifications:

- **Maximum range**: maximum detectable distance of a target.
- **Range resolution**: minimum detectable distance between two targets.
- **Field of view**: maximum angle the sensor can detect targets.
- **Angular resolution**: minimum distinguishable angle between two targets.
- **Maximum Doppler**: maximum radial speed the sensor can measure.
- **Doppler resolution**: minimum speed difference the sensor can distinguish.

In general, LiDAR has high angular and range resolution and high ranging accuracy. However, it requires more power and sophisticated optics components. LiDAR is more expensive and heavier than RADAR and SONAR. LiDAR is affected by fog and weather conditions. RADAR is less expensive and lighter. RADAR can see through fog and clouds and can penetrate objects depending on the wavelength. However, RADAR has low angular resolution and range accuracy compared to LiDAR. SONAR has the lowest maximum range, accuracy, and resolution. In addition, the speed of sound is significantly lower compared to the speed of light, which imposes limitations on the dynamics the sensors can have. However, SONAR is less expensive than LiDAR and RADAR. LiDAR is suitable for cars and large vehicles. RADAR is suitable for cars, drones, and small robotic platforms. SONAR is more suitable for low dynamics vehicles and small robotic platforms. LiDAR and RADAR can be used at far distances, so they can be used in space applications such as

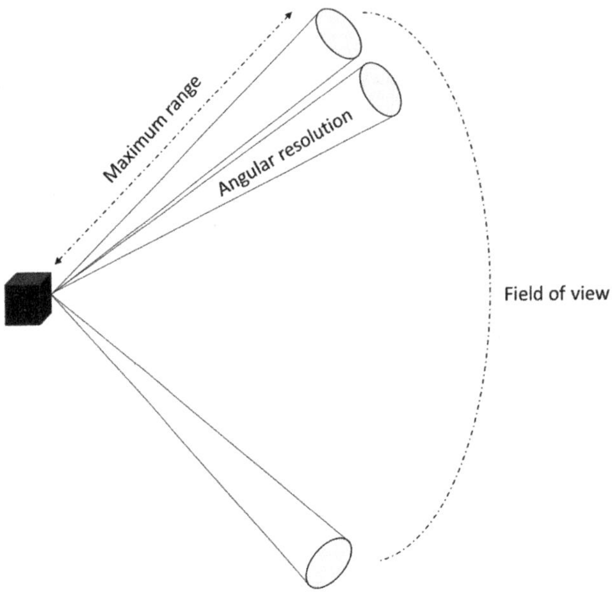

Figure 6.6 Ranging sensor specs.

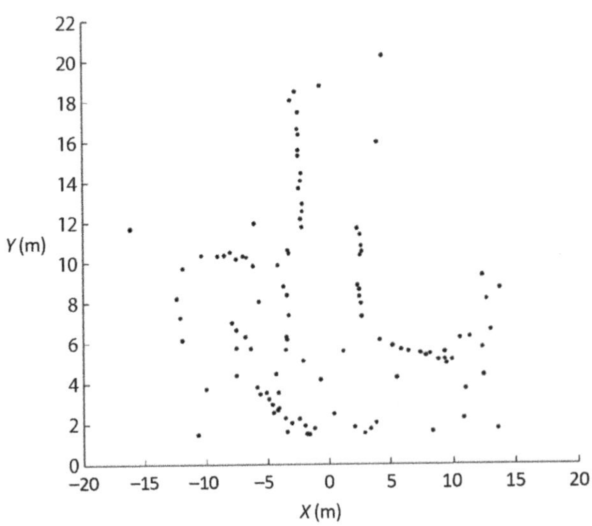

Figure 6.7 Raw FMCW mmWave radar scan.

monitoring and tracking satellites' motion. SONAR is suitable for small environments. Figure 6.7 shows a sample of an automotive-grade millimeter wave (mmWave) RADAR raw scan. Figure 6.8 shows multiple accumulated filtered (denoised) RADAR raw scans. Figure 6.9 shows a raw LiDAR scan. The figures show the superior accuracy of LiDAR in terms of angular resolution and range accuracy. The RADAR still has the advantage of penetrating objects and seeing through walls and undergrounds.

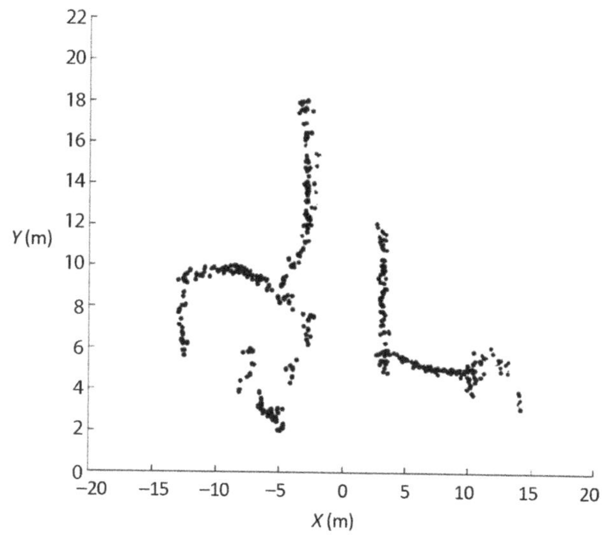

Figure 6.8 Accumulated/filtered scans FMCW mmWave radar scan.

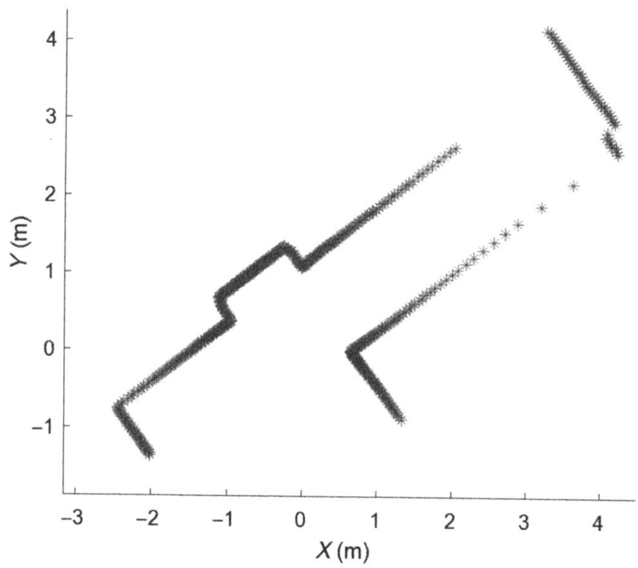

Figure 6.9 Raw LiDAR scan.

6.3 Sound, Radio, and Light Detection and Ranging

In this section, we discuss the different waves used in active-ranging sensors. Active ranging sensors can use any type of wave. The most used wave types are sound and electromagnetic waves. The fundamental difference between sound and electromagnetic waves is that sound propagates by moving particles in a physical matter like air, metal, or water, while electromagnetic waves are electromagnetic charges/fields that can travel through space. Sound waves travel by motion (compression and rarefaction of particles), generating a wave like the one shown in Figure 6.10. Depending on the sensor design, the beam of the sound wave propagates in 3D, making a shape like the one shown in Figure 6.11. The strength of the vibration will determine the maximum range the wave

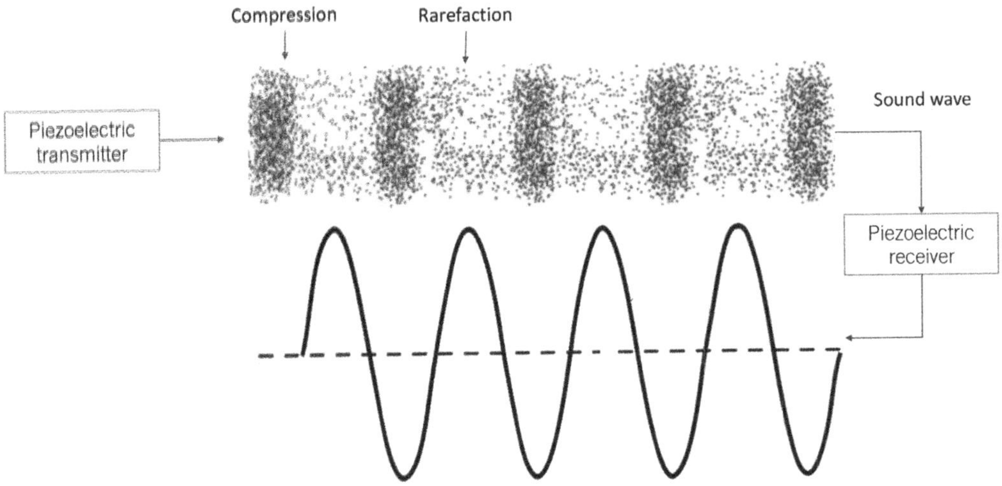

Figure 6.10 Illustration of Sound waves.

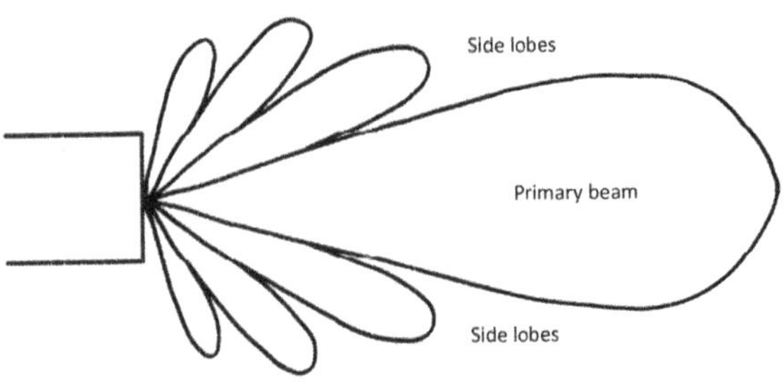

Figure 6.11 Sound wave beam.

can travel, and the shape of the beam will determine the angular resolution. SONAR systems use ultrasound waves. Ultrasound is a high-frequency (e.g., 1–5 MHz) sound wave generated using piezoelectric material (a material that vibrates when electric current/voltage is applied to it). Ultrasound waves cannot be heard by the human ear; however, they can be sensed by piezoelectric receivers that convert them back to electric current/voltage for processing.

The two common electromagnetic waves spectrums used in ranging and navigation are radio and light. When electromagnetic radio waves are generated using conventional radio circuits, the electromagnetic wave propagates in scattered beams in different directions like a cone (see Figure 6.8), and each direction may have a different phase due to scattering. Radio signals have the advantage of penetrating objects so they can see through clouds and rain. In addition, compared to light and laser waves, radio wave generation is more power efficient. However, the fading and scattering of the radio signal led to low angular resolution and more complicated signal processing. On the other side, light and laser (laser is an amplified/synchronized light wave) are narrow and precise (see Figure 6.12). The light spectrum has a shorter wavelength (see Figure 6.13), which improves the angular resolution. However, light sensors cannot see through objects, and they require more power. Raw light signals can be generated by a light-emitting diode (LED) and received by a photodetector (a material that generates current/voltage in response to light intensity such as a photodiode). To generate highly focused and synchronized electromagnetic beams with higher angular resolution (narrower beams), scientists invented a method called "amplification by stimulated emission of radiation (ASER) [1], [4]." When applied to visible or infrared electromagnetic waves, ASER technology is called LASER (it can be applied to the radio spectrum as well, and in this case, it is called RASER). LASER technology amplifies and synchronizes light photons to generate extremely focused narrow beams of light that are resistant to scattering over far distances. For example, LASER is used to track satellites in space. The details of ASER technology and

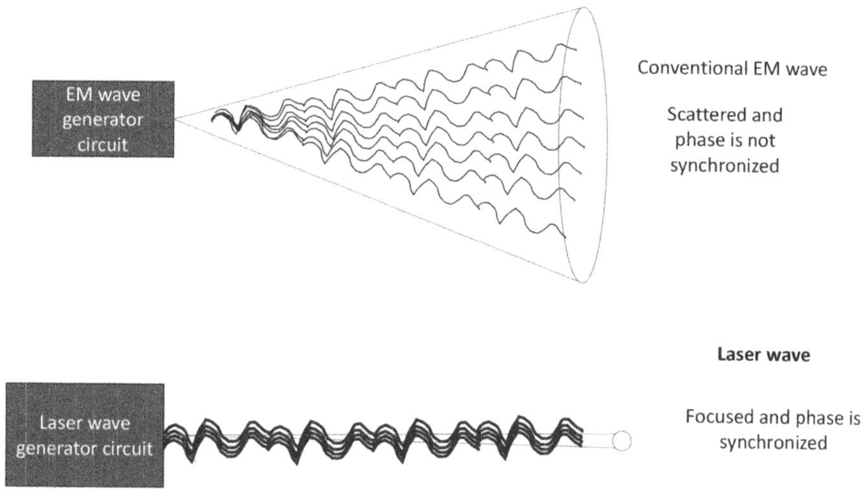

Figure 6.12 Laser wave vs. conventional electromagnetic wave.

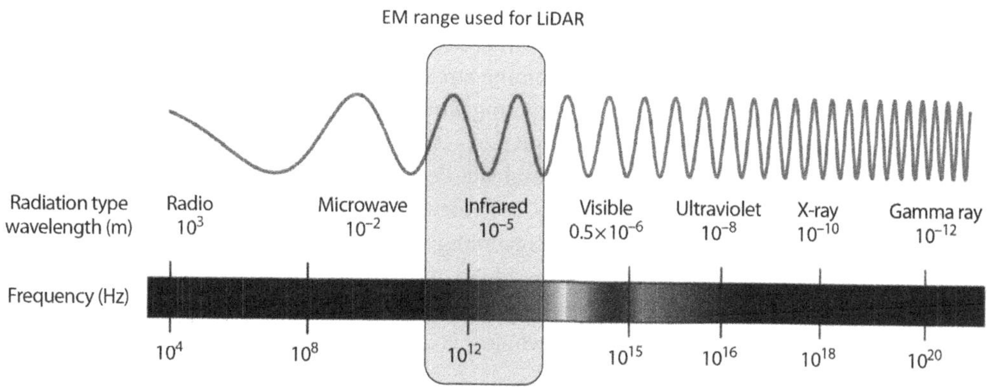

Figure 6.13 LiDAR electromagnetic region in the electromagnetic wave spectrum.

how LASER beams are physically generated are physics-intense subjects that have their own standalone research track. ASER technology can be applied to another electromagnetic wave spectrum. For example, MASER stands for microwave amplification by stimulated emission of radiation. LASER technology is commonly used to avoid interference with microwave and radio spectrums and to have better angular resolution (angular resolution enhanced with shorter wavelengths). LASER beam generation requires high energy, which raises the concern of power consumption in LiDAR-based ranging technologies. In modern LiDAR sensors, the core source of LASER wave is called a "LASER diode." A LASER diode is a complicated semiconductor device that is driven by a current and emits a LASER beam at a certain wavelength.

6.4 Signal Processing of Ranging Sensors

As discussed earlier in the previous section, ranging waveforms and ranging schemes are categorized into two categories: pulsed chirps and continuous wave chirps. The signal processing to estimate range, radial speed, and angle starts from the moment the wave hits the antenna (or antennas)

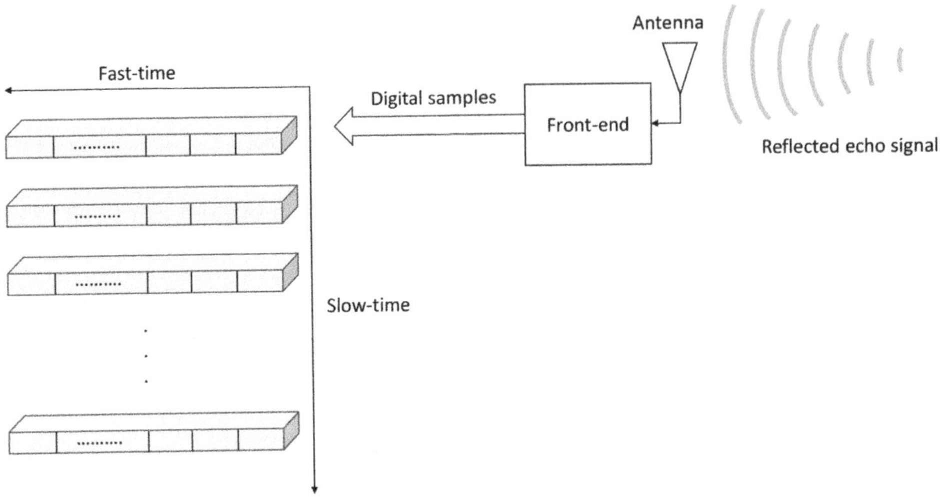

Figure 6.14 Processing of reflected echo – data array – single channel.

of the sensor. Like GPS receivers, active ranging sensors have an antenna and front-end that apply some filtering, conditioning, and preprocessing and convert the signal into a stream of digital samples. The sensor processor partitions this incoming stream of samples into windows and applies digital signal processing algorithms on the incoming window of data. Commonly, one single window of data is not enough for stable and accurate detection. Therefore, usually multiple windows are processed. Depending on the algorithm and the desired accuracy/resolution, the processor keeps a certain number of windows. In this way, the processor keeps a 2D array of samples as shown in Figure 6.14. In each processing cycle, new incoming windows of samples are pushed into the array and the oldest window of data is discarded. So, at any given time, the sensor processes a 2D array of data as seen in Figure 6.14. This is for one channel. Some sensors may have multiple processing channels. The time within a single window is called the fast time, and through windows, it is called the slow time.

6.4.1 Pulse-Based Ranging Signal Processing

6.4.1.1 Range Calculation

Assume the transmitted pulse has a waveform $p(t)$, if it hits a target within the maximum sensor range, a reflected echo signal $x(t)$ will be received. A reflected each that is delayed by T seconds $x(t)$ will be a noisy attenuated delayed copy of the pulse $p(t)$ as follows:

$$x(t) = \propto p(t-T) + n(t) \tag{6-5}$$

where

$\propto$ is an attenuation factor
T is the delay
$n(t)$ is a random noise commonly assumed white Gaussian noise.

Assuming we are processing a window of data with length N and period, T_w, we can detect the presence of the original pulse in this noisy attenuated delayed echo by applying cross correlation between $x(t)$ and a delayed version of $p(t)$ as follows:

$$r_{sx}(\tau) = \int_0^{T_w} x(t)p(t-\tau)dt \tag{6-6}$$

The signal $p(t-\tau)$ is a delayed version of the pulse shifted τ seconds. Note that this is a circular shift. In discrete form with a sampling period of T_s, the correlation is given by:

$$r_{sx}(nT_s) = \sum_{k=0}^{N} x(kT_s)s((k-n)T_s) \tag{6-7}$$

where

T_s is the sampling period
n is the correlation delay (in a number of samples).

This correlation operation multiplies the incoming signal with a delayed version of the pulse waveform delayed by n samples. This correlation will be maximum when $x(nT_s)$ is perfectly aligned and matched with $s((k-n)T_s)$, which will happen if $n = \frac{T}{T_S}$, which means that the correlation delay n (in samples) equals the true delay T/T_S (in samples). This operation is illustrated by an example simulated pulse in Figure 6.15, which shows the ideal transmitted pulse, the attenuated noisy delayed version, and the cross-correlation that shows a peak at the right delay (250 samples).

This cross-correlation operation is digitally implemented using a digital filter called a matched filter [5], [4]. A matched filter is a filter whose impulse response is a reversed version of the original signal to be detected. In discrete form, the impulse response of the matched filter is given by:

$$h_m[n] = s[-n] \tag{6-8}$$

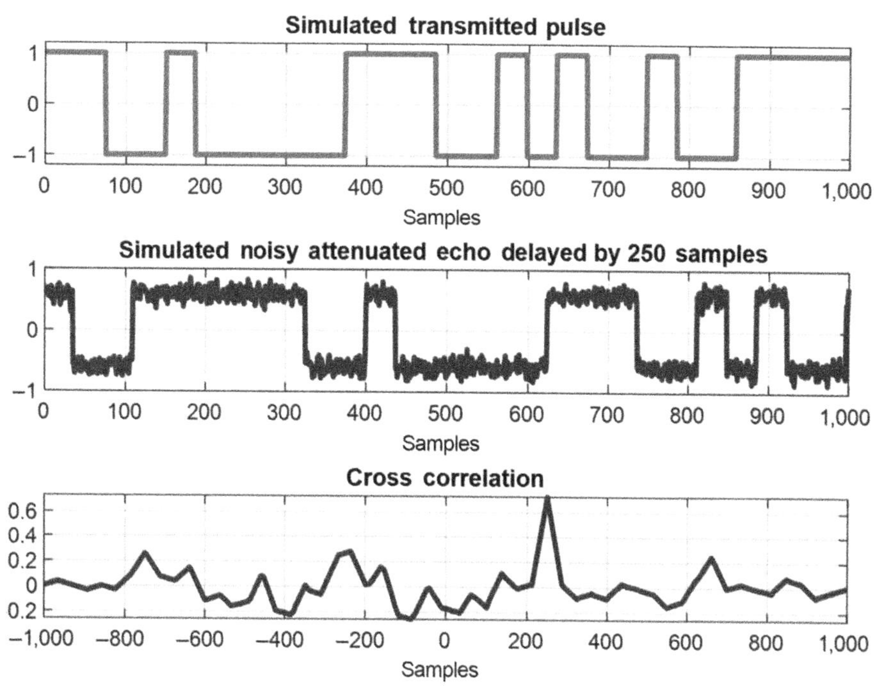

Figure 6.15 Cross-correlation illustration between pulse signal $s(t)$ and echo signal $x(t)$.

The incoming signal is convoluted with h_m as follows:

$$r_{sx}[n] = y[n] = \sum_{k=-0}^{N} x[k]h_m[n-k] \tag{6-9}$$

Note here $h_m[n-k]$ is a reversed delayed version of $s[k]$ (delayed by n samples). This convolution output $y[n]$ will be maximum when n equals the true delay T/T_s (in samples). Figure 6.16 shows

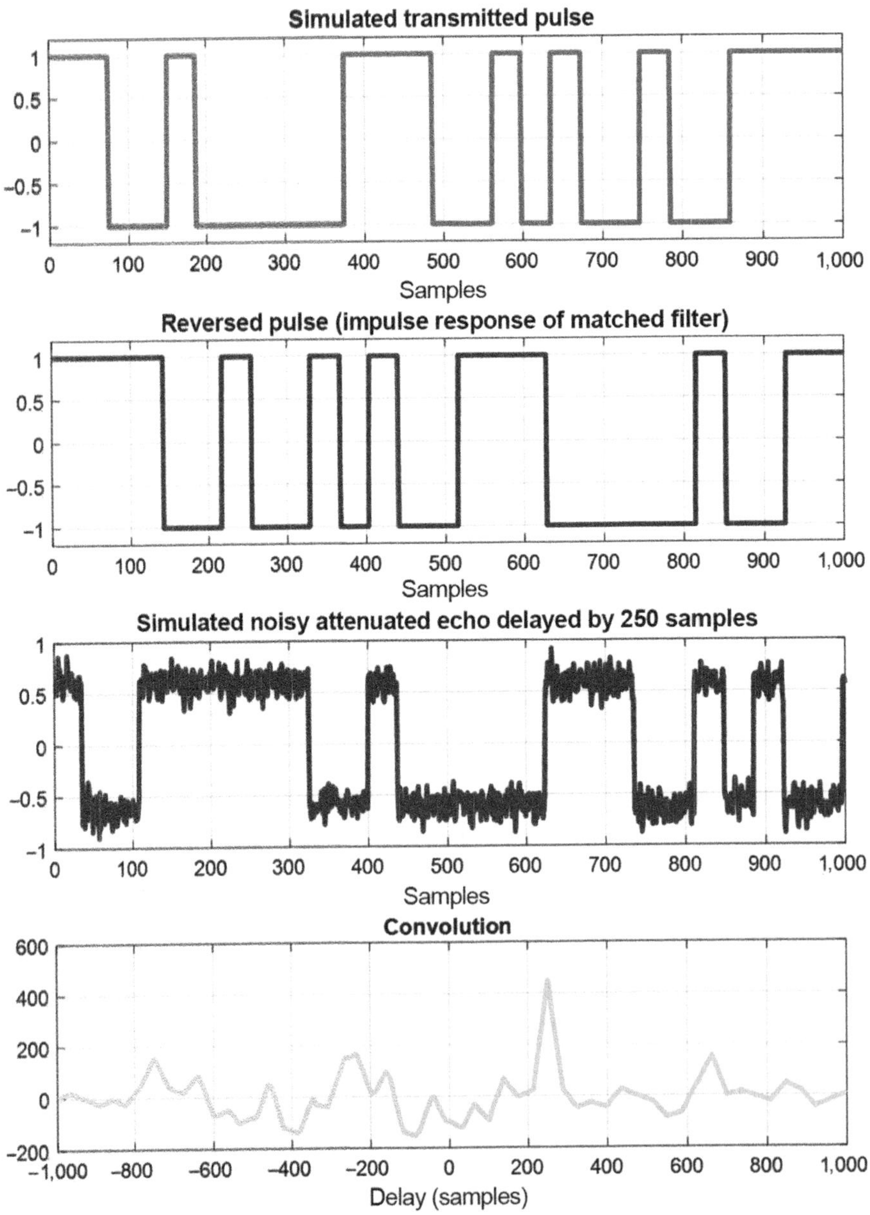

Figure 6.16 Illustration of how convolution is related to correlation.

how convolution with a matched filter is nothing but a correlation between the incoming signal and the reversed filter impulse response.

If there is a target at a distance $\frac{Ts}{2}$ where s is the signal speed, T is the round-trip time, the output of the matched filter should be maximum at $=T/T_s$. Convolution is multiplication in the frequency domain. So, practically, the samples of a window of data are collected into an array $x[k]$ and the matched filter impulse response is placed into another array $h_m[k]$ and discrete fast Fourier transform (FFT) is applied on each array and the resulting arrays are multiplied (element-by-element), and inverse-FFT is then applied to get $r_{xh}[n]$. The discrete FFT of the received signal $x[k]$, $n = 0, 1, \ldots, N$ is given by the following formula:

$$X[k] = \sum_{n=0}^{N-1} x[n]e^{-j2\pi\frac{k}{N}n} \; k = 0, 1, \ldots, N-1 \tag{6-10}$$

The discrete Fourier transform of the reversed internal replica of $h_m[k]$ is given by the following formula:

$$H[k] = \sum_{n=0}^{N-1} h_m[n]e^{-j2\pi\frac{k}{N}n} \; k = 0, 1, \ldots, N-1 \tag{6-11}$$

We then multiply $X[k]$ and $H[k]$:

$$R_{xh}[k] = X[k]H[k] \tag{6-12}$$

Inverse Fourier transform is applied on $Y[k]$ as follows:

$$r_{xh}[n] = \frac{1}{N}\sum_{k=0}^{N-1} R_{xh}[k]e^{+j2\pi\frac{k}{N}n} \tag{6-13}$$

A search in $r_{xh}[n]$ array is performed for a peak (peak detector). If a peak is detected at $n = n^*$, a target is announced present at the following distance d:

$$T = n^* * T_s \tag{6-14}$$

$$d = \frac{Ts}{2} \tag{6-15}$$

where s is the signal speed.

This processing is applied to a window of data. This is called fast-time processing. Due to noise, the correlation will have multiple small peaks and strong peak as Figure 6.17 shows. Note that the window length N is set as follows:

$$N = T_{PRI}/T_s \tag{6-16}$$

A single window may not be enough to detect a peak, especially if the target is at a far distance and the signal is significantly attenuated (low signal-to-noise ratio "SNR"). To overcome this problem, the same processing is repeated on all windows (rows of the 2D data array). By adding the matched filter output of all rows together, the effect of noise is minimized because noise is zero-mean random values that cancel each other by addition, and the peak becomes stronger as seen in Figure 6.18.

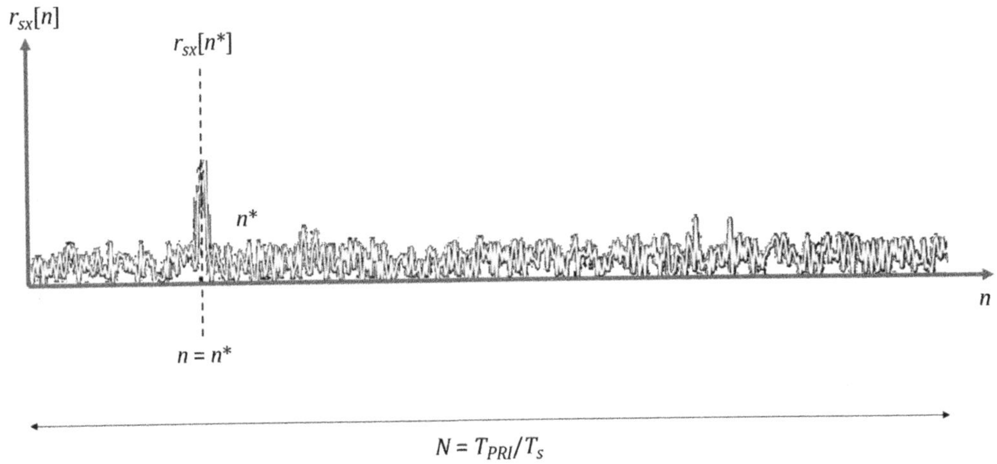

Figure 6.17 Fast-time window correlation result.

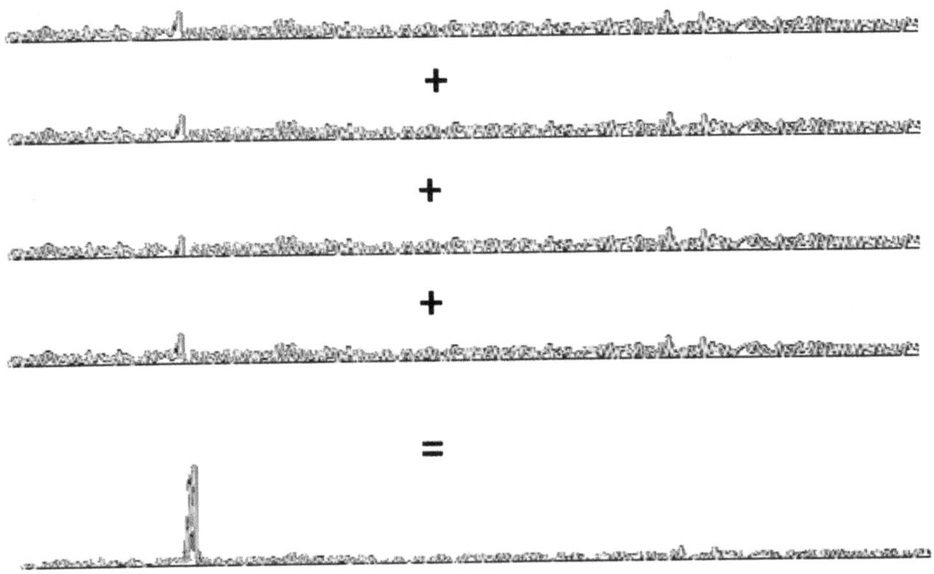

Figure 6.18 Adding multiple windows to cancel out noise.

Having to wait for an echo reflection before sending the next pulse (to avoid ambiguous reflections) imposes a limit on the maximum range given by:

$$d_{max} = \frac{s \cdot T_{PRI}}{2} \tag{6-17}$$

where s is the speed of the signal.

This is how the range is estimated. Let's discuss how the speed of the target (radial speed) is calculated. Radial speed is the relative speed in the direction of the line connecting the sensor to the target.

6.4.1.2 Radial Speed Calculation

If the target is stationary (or the relative speed between the sensor and the target is zero), the correlation peak will occur in the same place through all windows in the 2D array of data currently being processed. If the target is moving away from the sensor, the peak will appear at different times (samples) in the 2D data array as shown in Figure 6.19.

This change in the peak corresponds to a change in the range between the sensor and the target. The rate of change of the range between the sensor and the target is the radial speed that we want to measure. A simple way to calculate radial speed is to simply differentiate the range throughout consecutive pulses (rows in our 2D data array) and divide by T_{PRI}. However, this method is not accurate since the distance that the target moves from window to window is usually very small. Fortunately, there is a better way to measure the radial speed by measuring the phase difference of the received signal. Recall the Doppler effect that we discussed in radio positioning systems where a received signal will have a frequency that is different from the original transmitted signal frequency due to the relative motion between the transmitter and the receiver. This change in received frequency is called the Doppler shift. In practice, the Doppler shift is small, which makes it difficult to measure directly, and it is sensitive to noise. Fortunately, small frequency shifts cause considerable phase shifts that can be measured with better accuracy. To illustrate this concept, Figure 6.20 shows how frequency shift causes phase shift. Note that here we are talking about sinusoidal signals. That is the reason the pulsed ranging sensor's waveform consists of a pulse envelope and sinusoidal wave (see Figure 6.2). The sinusoidal wave makes it easy to measure phase changes as seen in Figure 6.20. We will assume that the transmitted pulse has a sinusoidal wave of frequency f_s, speed s and wavelength λ_s. There are different ways to measure the phase changes. One method is by using the Fourier transform. Applying the Fourier transform on the data window will give us a peak at the sinusoidal frequency. We can measure the phase at this peak (recall from Chapter 2, Fourier transform gives us the complex number that has an amplitude and a phase at each frequency). Phase change can be measured by differentiating the phases at the peak frequency of two consecutive windows of data. However, this method is sensitive to noise. To further improve the accuracy of phase change measurement, we apply Fourier transform on a column of data through all rows at sample $n = n^*$. As Figure 6.20 shows, a change in frequency causes an oscillatory change in phase through multiple rows of data. The frequency of this oscillation is linearly proportional to the target

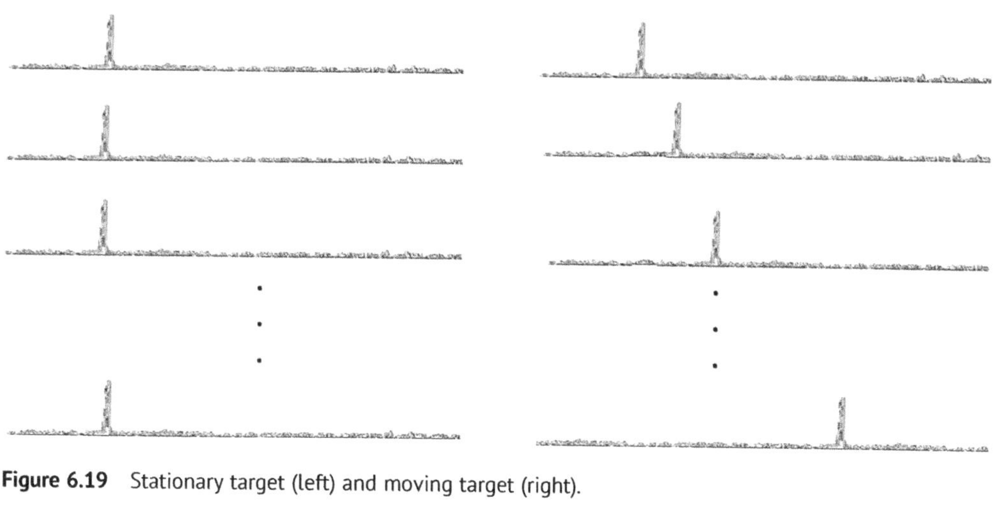

Figure 6.19 Stationary target (left) and moving target (right).

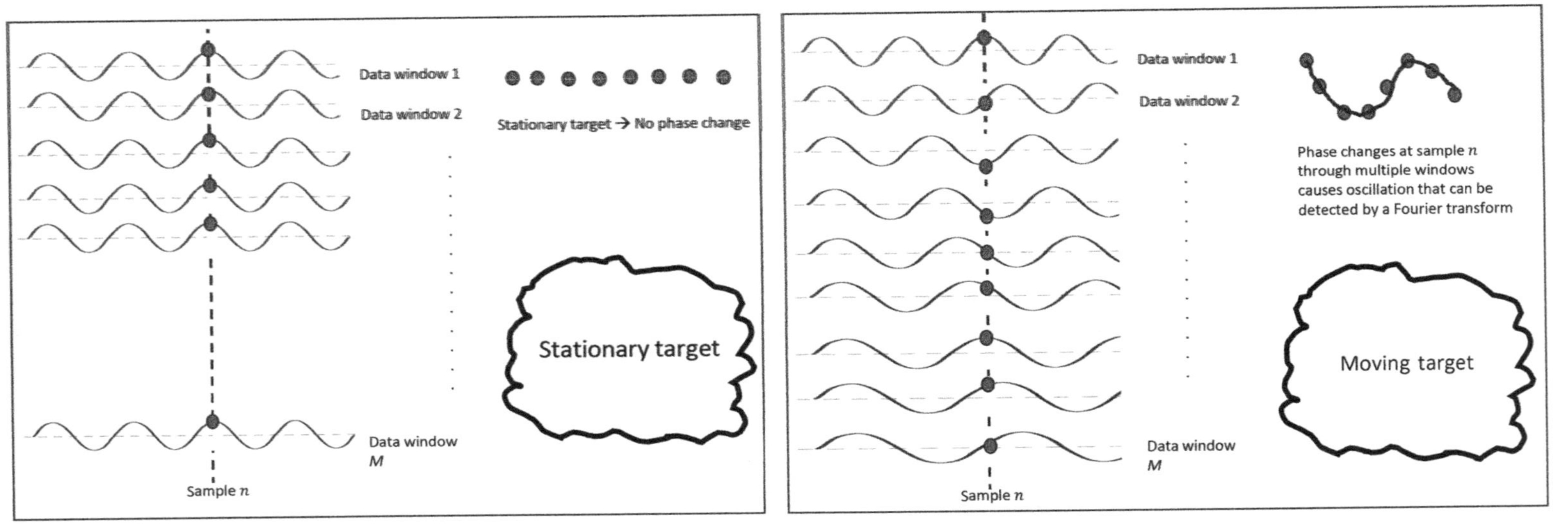

Figure 6.20 Phase changes due to frequency change.

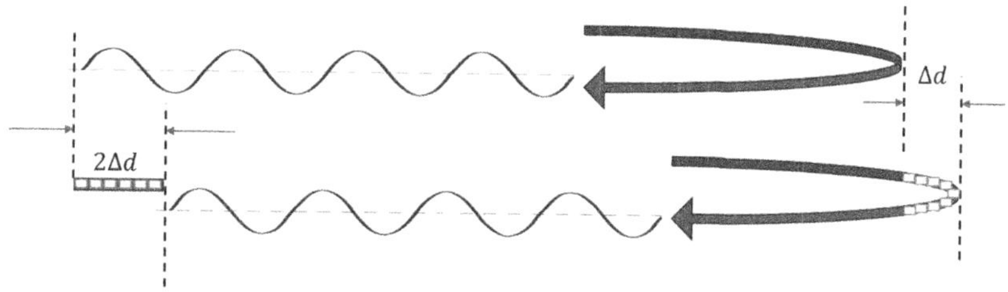

Figure 6.21 Phase change due to target motion.

speed. The maximum phase change from pulse to pulse should be less than π for this method to work. Otherwise, we will get ambiguous phase change measurements that lead to ambiguous speed estimation.

Now, given that we can measure the phase change between windows, let's try to relate the phase difference $\Delta\phi$ to the speed v of target. From one window to another, assume that the target moved a delta distance Δd. This will cause a phase change between the two windows given by:

$$\Delta\phi = 2\pi\frac{2\Delta d}{\lambda_s} \tag{6-18}$$

Note that we multiplied the delta distance Δd by two due to the round-trip delay as illustrated in Figure 6.21. If the target moved Δd, this would cause a phase change corresponding to $2\Delta d$.

At the same time, we know that within a T_{PRI} time, the target would have moved the following distance:

$$\Delta d = vT_{PRI} \tag{6-19}$$

Therefore, the target speed v is related to phase change $\Delta\phi$ by the following equation:

$$v = \frac{\lambda_s}{4\pi T_{PRI}}\Delta\phi \tag{6-20}$$

A simpler way to calculate the speed v is to use the frequency of the oscillation of the phase (calculated by Fourier transform) and relate it linearly directly to speed. The minimum frequency of the oscillation of the phase is zero, which corresponds to zero speed, and the maximum is $\frac{F_s}{2}$ which corresponds to the maximum speed where F_s is the sampling frequency (in our case $F_s = 1/T_{PRI}$). This method will only work if the change in range from a row to the next (or from a pulse to the next) is less than half of a wavelength (to guarantee that the phase change will be less than π). If the change in range is larger than half wavelength, the change in phase difference will flip its sign, which leads to ambiguous speed estimation. This puts a limit on the speed that can be sensed by the sensor to be as follows:

$$v_{max} = \frac{0.5\lambda_s}{T_{PRI}} \tag{6-21}$$

6.4.1.3 Angle Calculation

To calculate the angle of the target, multiple antennas are used, and the same idea of phase change is applied. To further illustrate the concept, Figure 6.22 shows two antenna configurations. As the target is usually at a distance that is much longer than a wavelength, we can assume that the two

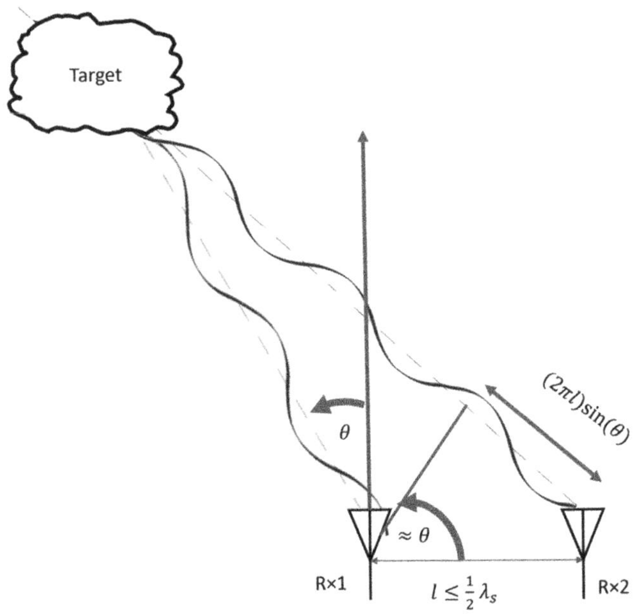

Figure 6.22 Target angle estimation from two antennas.

echo rays are parallel to each other. Depending on the angle of target θ, $R \times 2$ will receive the echo delayed by a distance $l \cdot \sin(\theta)$ where l is the distance between the two antennas. A distance difference $l \cdot \sin(\theta)$ between the two signals will cause a phase change given by the following equation:

$$\Delta\phi \approx \frac{2\pi l \cdot \sin(\theta)}{\lambda_s} \tag{6-22}$$

So, if we calculate the phase $\Delta\phi$ between two streams of two antennas, we can estimate the angle of the target using the following equation:

$$\theta = \sin^{-1}\left(\frac{\lambda_s \Delta\phi}{2\pi l}\right) \tag{6-23}$$

The phase difference between the signals of two antennas can be estimated using Fourier transform. To enhance the accuracy and the angular resolution of the sensor, more than two antennas can be used as illustrated in Figure 6.23. Phase change through multiple antennas can be calculated by applying the Fourier transform on the vector consisting of the data samples through the multiple antennas. The frequency of the oscillation of the phase is linearly proportional to the angle (a zero oscillation means zero angle and maximum oscillation means maximum angle). This will lead to a 3D array of data. This 3D array of data is called the sensor "data cube," which is the core data structure of the ranging sensor from which all parameters of interest are calculated including range, speed, and angle as seen in Figure 6.24 (The acronym FFT stands for Fast Fourier Transform.)

6.4.2 FMCW-Based Ranging Signal Processing

Pulsed ranging sensors have limitations on the range because the sensor needs to remain silent for the pulse repetition interval before sending another pulse. In addition, sending pulses requires high

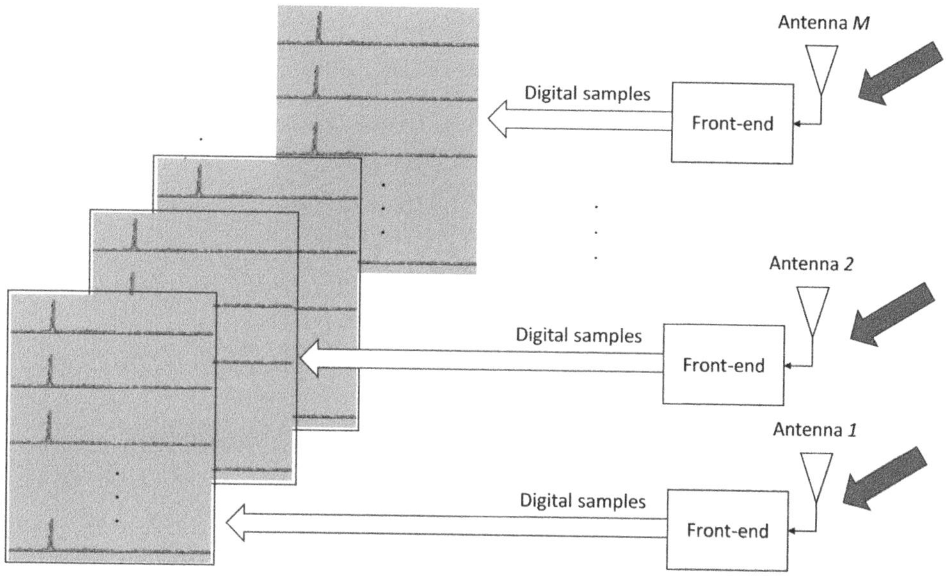

Figure 6.23 Using multiple antennas to measure angle of target.

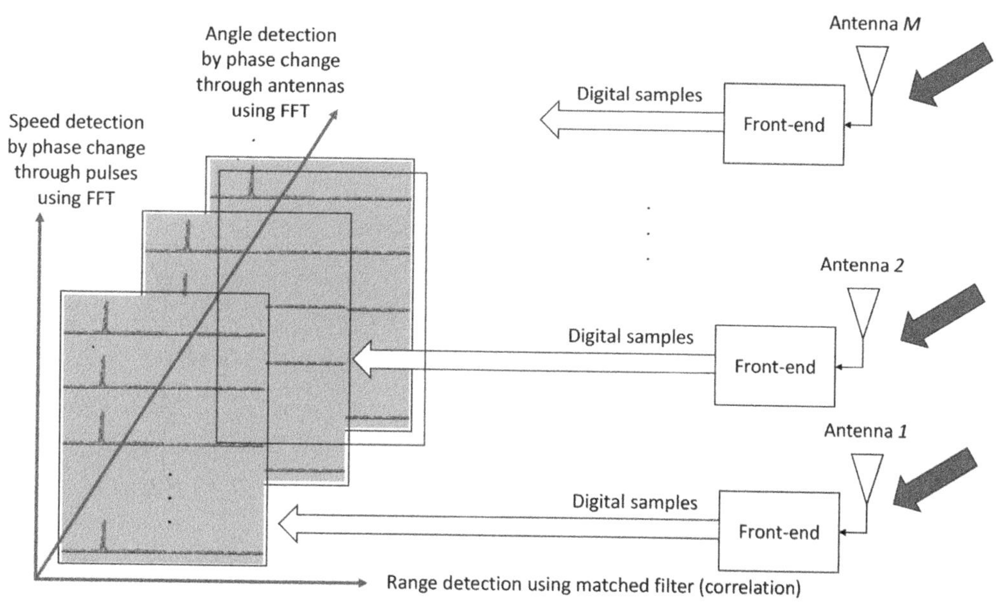

Figure 6.24 Ranging sensor data cube processing.

energy. An enhanced ranging technique is the frequency-modulated continuous wave (FMCW) ranging. As the name suggests, the sensor transmits a continuous wave called a frame. Each frame consists of multiple chirps. Each chirp is a sinusoidal signal that changes its frequency linearly with time. A sample FMCW waveform is shown in Figure 6.25. More information about waveform generation for FMCW radar can be found in [6]. The top plot shows the amplitude-time plot and the

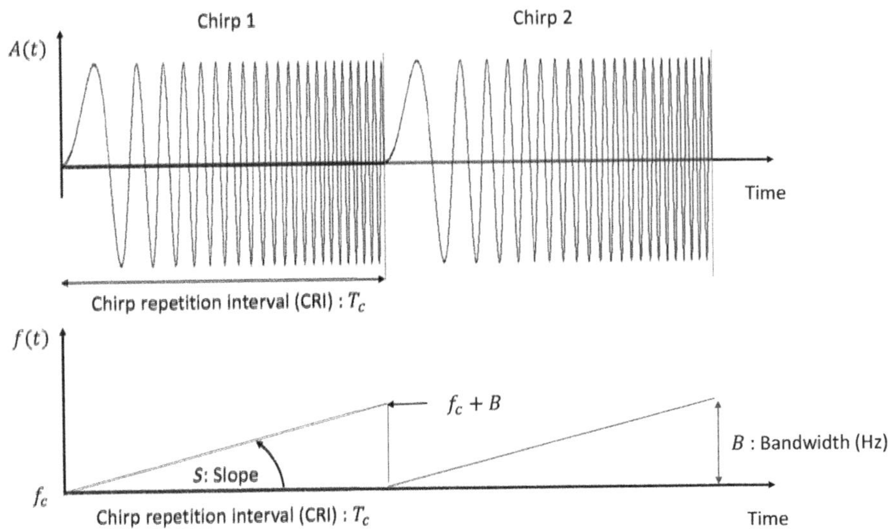

Figure 6.25 Example chirp and waveform for FMCW ranging.

bottom one shows the frequency-time plot. As seen in Figure 6.25, frequency starts at carrier frequency f_c and increases linearly with time to reach its maximum $f_c + B$ at time T_C where B is the bandwidth (Hz). We will discuss how this FMCW waveform can be used to measure range, speed, and angle. Figure 6.26 shows a simple block diagram of a single-channel (single antenna) FMCW ranging sensor.

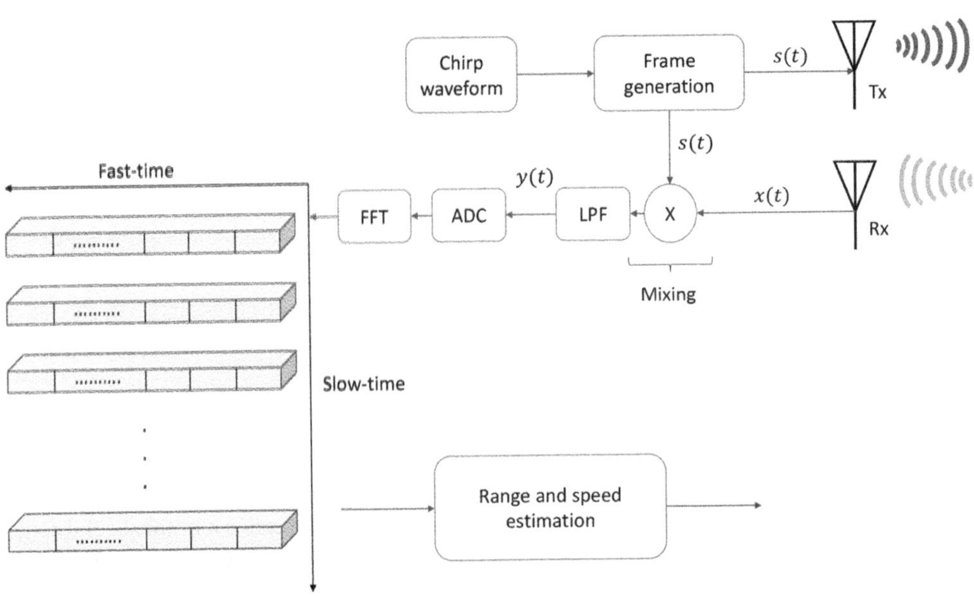

Figure 6.26 Single channel FMCW ranging sensor block diagram.

6.4.2.1 Range Calculation

Figure 6.26 shows a simplified block diagram of the signal processing sequence in the FMCW ranging sensor. The transmitted chirp signal is given by:

$$s(t) = A\cos(2\pi(f_c + St)t) \tag{6-24}$$

where S is the slope of the chirp.

The received signal (the echo) is given by:

$$x(t) = A'\cos(2\pi(f_c + St)(t - T)) \tag{6-25}$$

where T is the round trip time to the target.

The received echo is mixed with the transmitted signal as follows:

$$y(t) = A\cos(2\pi(f_c + St)t) \cdot A'\cos(2\pi(f_c + St)(t - T)) \tag{6-26}$$

Recall that for two sinusoidal signals, the following relationship holds:

$$\cos(a)\cos(b) = 0.5(\cos(a - b) + \cos(a + b)) \tag{6-27}$$

The high-frequency component ($\cos(a + b)$) can be easily removed by a low-pass filter. The remaining low-frequency part will be given by:

$$y(t) = 0.5AA'\cos[2\pi(ST)t] \tag{6-28}$$

which is a sinusoidal with the following intermediate frequency:

$$f_{IF} = ST \tag{6-29}$$

This intermediate frequency f_{IF} is easily detected by the first FFT as shown in Figure 6.26. This intermediate frequency is constant, and it is equal to the difference between the transmitted frequency and the received frequency as shown in Figure 6.27.

Given the speed of the signal s and the intermediate frequency f_{IF}, the distance to the target can be calculated by:

$$d = \frac{sT}{2} \tag{6-30}$$

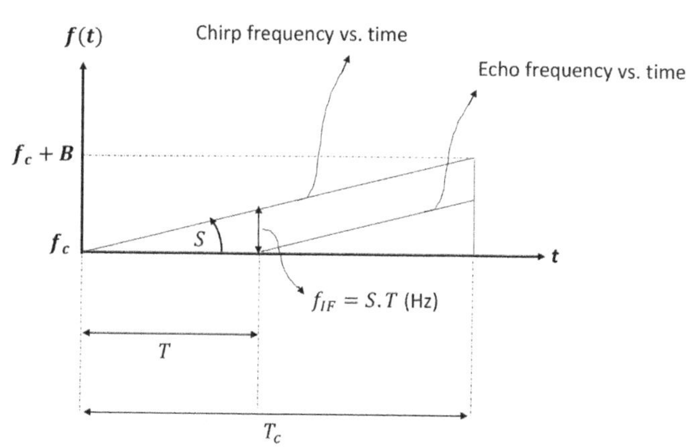

Figure 6.27 Frequency-time illustration of chirp/echo signals.

As Figure 6.27 shows, the delay T is related to intermediate frequency f_{IF} by:

$$f_{IF} = ST \tag{6-31}$$

which gives the final range equation as follows:

$$d = \frac{sf_{IF}}{2S} \tag{6-32}$$

where

s is the signal speed
S is the slope of the chirp frequency and both are known parameters of the sensor.

Once the intermediate frequency f_{IF} is estimated (by applying FFT and detecting a peak), we can calculate the distance to the target using Eq. (6-32). If we have two targets at a relative distance Δd from the sensor, they will appear as two peaks in the FFT. However, to get two distinguished peaks in the FFT corresponding to two targets separated by Δd distance from the sensor, the FFT needs to have enough samples, which depends on the chirp time T_c. Within chirp time T_c, the maximum detectable frequency difference is $\frac{1}{T_c}$ Hz. This gives the minimum detectable range resolution as follows:

$$\delta d = \frac{s\frac{1}{T_c}}{2S} \tag{6-33}$$

Given that the slope $= \frac{B}{T_c}$, the range resolution is given by:

$$d = \frac{s}{2B} \tag{6-34}$$

where B is the sensor bandwidth.

The maximum range will depend on the maximum detectable intermediate frequency f_{IFmax} within a chirp data. If the sampling frequency is F_s, F_s must be twice f_{IFmax}:

$$f_{IFmax} = \frac{F_S}{2} \tag{6-35}$$

which gives the maximum range by:

$$d_{max} = \frac{sf_{IFmax}}{2S} = \frac{sF_S}{4S} = \frac{sF_S T_c}{4B} \tag{6-36}$$

6.4.2.2 Radial Speed Estimation

Speed calculation is performed in the same way described above for the pulsed ranging sensor. We process multiple windows of data and apply a second FFT on the columns where we have a peak. If the target corresponding to this peak is moving, a phase oscillation will be detected through this column of data. The frequency of this phase oscillation is linearly proportional to the target speed. Assume this phase oscillation frequency is $f_\emptyset$ Hz, we know that if $f_\emptyset = 0$, speed $v = 0$ and if $f_\emptyset = maximum$, speed $v = maximum$. The maximum speed v that can be estimated equals to:

$$v_{max} = \frac{\lambda_c}{4T_c} \tag{6-37}$$

where λ_c is the carrier wavelength (corresponding to f_c, which is easily calculated by $\lambda_c = s/f_c$).

The maximum phase oscillation frequency $f_{\emptyset_{max}}$ is given by:

$$f_{\emptyset_{max}} = \frac{1}{2T_c} \tag{6-38}$$

Then, for any phase oscillation frequency $f_\emptyset$, the speed is given by:

$$v = \frac{\lambda_c}{2} f_\emptyset \tag{6-39}$$

Recall, for a sequence of length N, the frequency resolution is given by:

$$\delta f_N = \frac{F_S}{N} \tag{6-40}$$

The minimum detectable speed change (speed resolution) is given by:

$$\delta v = \frac{\lambda_c}{2NT_c} \tag{6-41}$$

where N is the number of chirps.

Recall, here we used the fact that an FFT on a sequence of length N can discriminate two frequencies if the difference is greater than $\frac{F_S}{N}$ Hz and the fact that the sampling frequency in the column dimension of the 2D data array is actually $\frac{1}{T_c}$ (the time from a window to the next is T_c).

6.4.2.3 Angle Estimation

Angle calculation is done in the same way as described in the pulsed ranging sensor by processing multiple frames from multiple antennas and measuring the phase oscillation frequency through the third dimension of the 3D data cube array.

6.5 MATLAB Implementation of FMCW Ranging Signal Processing

To further clarify FMCW Ranging signal processing details, the projects' repository of the book includes a MATLAB project "*RadarDSPProject.*" In this project, simulated FMCW radar signals are generated with a millimeter wave (mmWave) RADAR sensor specification that operates at 77 GHz carrier frequency. The readers can run the MATLAB script "*demo_RADAR.m*" and observe the output. The MATLAB script "*demo_RADAR.m*" shows the detailed steps to generate the chirp signal, simulate moving and static targets echo signals, the construction of the 2D data array, and the FFT processing of the 2D data array to estimate range and speed. The users can modify the script to simulate any number of targets at any speed/range configuration. A sample of the output of the 2D FFT processing is shown in Figure 6.28. The figure shows three targets: one target is stationary at a distance of 3 km, another target is at 1 km range and is moving away from the sensor at a speed of 50 m/s, and another target is at 2 km range and is moving toward the sensor at a speed of 50 m/s. The users can also add random noise and attenuation to the echo signals and test the effect of these noise/errors on the accuracy of the range/speed estimation.

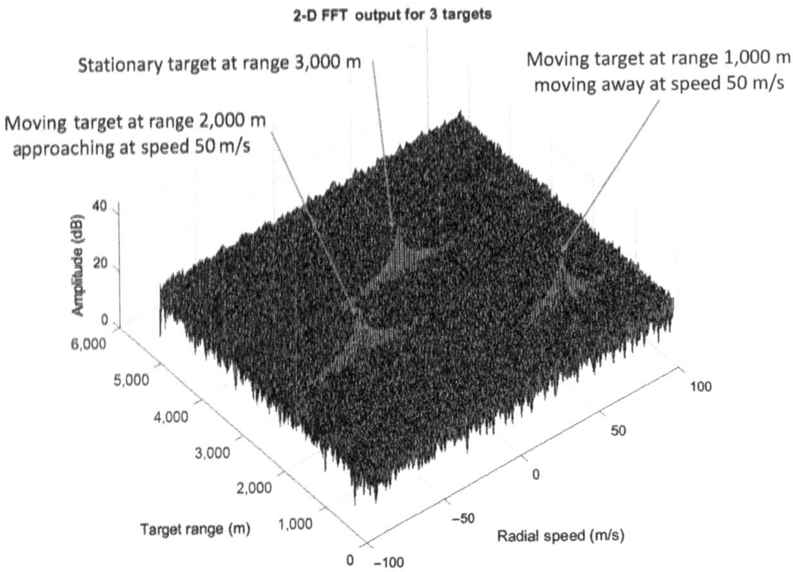

Figure 6.28 Sample of 2D FFT processing of a FMCW mmWave radar.

6.6 Range-Based Odometry

Range sensor scans provide a range associated with an angle. Range-based odometry is the estimation of the sensor pose changes (relative rotation matrix R and translation T) between two consecutive scans at two different times (epochs). We will illustrate the range-based odometry concepts using two consecutive scans i and j where scan i is called the reference scan and the scan j is called the current scan as seen in Figure 6.29. Point correspondences link each point in the current scan to the reference scan as seen in Figure 6.30. If point correspondences are available, the sensor's

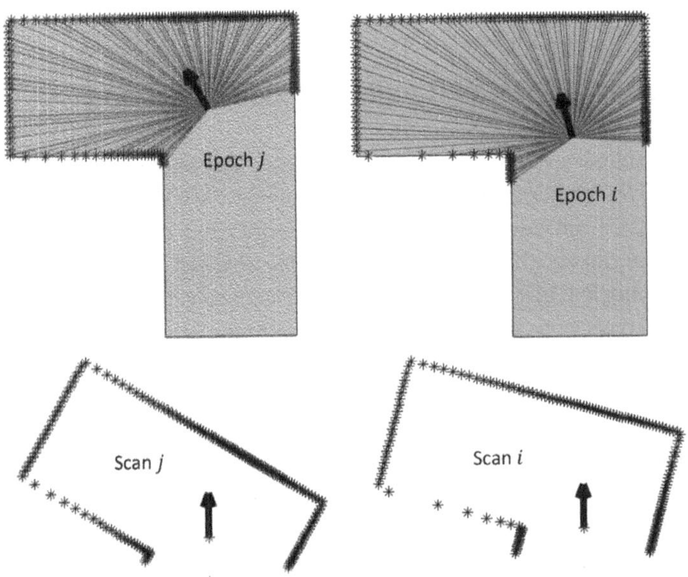

Figure 6.29 Two range measurement scans.

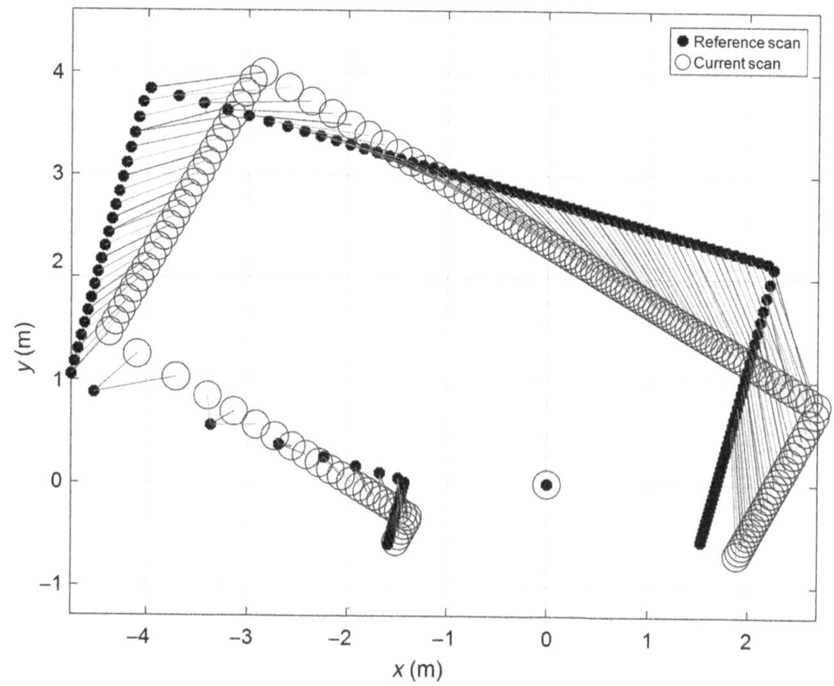

Figure 6.30 Point correspondences between two LiDAR scans.

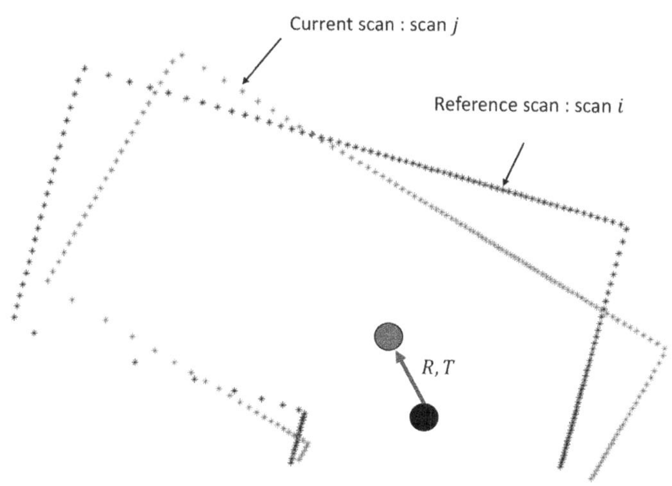

Figure 6.31 Range-based odometry (estimate of R and T from two scans).

relative motion (R, T) can be estimated using a scan-matching algorithm as illustrated in Figure 6.31. In the next subsections, we will explain the most used scan-matching algorithms using iterative closest point (ICP).

6.6.1 Range Scan Matching Algorithms

Scan-matching aims at finding the rotational and translational parameters (R, T) that transform common points of current scan (scan j) to current scan (scan i). The common points are the points that appear in both scans. When the sensor moves, some points disappear, some new points appear, and some common points remain in view of the sensor. Let us refer to these common points by $C \subset \{1, 2, ..., L\}$, where L is the number of beams per scan. Each LiDAR sensor has its own L based on the angular resolution and its field of view. Correspondence can be mathematically described as a mapping M as follows:

$$M \triangleq c(k), k \in C \tag{6-42}$$

The correspondence $M: c(k), k \in C$ maps each point in scan i (p_k^i) to a point in scan $\left(p_{c(k)}^j\right)$. Given some correspondences M, scan-matching problem can be mathematically formulated as follows:

$$R^*, T^* = \underset{R,T}{\operatorname{argmin}}\ J(R, T) \tag{6-43}$$

$$J(R, T) = \frac{1}{2}\sum_{k \in C}\left\|\left(p_k^i - \left(R\, p_{c(k)}^j + T\right)\right)\right\|^2 \tag{6-44}$$

If point correspondences are not known (which is usually the case), they can be estimated using the iterative closest point (ICP) algorithm, which will be explained shortly. Scan-matching can be done using the least-squares algorithm. We will assume that we have an accurate ranging sensor with high angular resolution (sharp beams) like LiDAR. However, the same concepts can be applied to any ranging sensor. Of course, the angular resolution and range accuracy will determine the accuracy of the range-based odometry algorithm output. In the following subsections, we will explain the ICP and scan-matching algorithms and how they can be used to estimate the trajectory of the sensor given consecutive scans.

6.6.2 Iterative Closest Point (ICP)

ICP solves the following problem: Given two range measurements scans as shown, estimate the point correspondence M and transformation (R, T) that minimizes the following cost function:

$$M \triangleq c(k), k \in C \tag{6-45}$$

$$M^*, R^*, T^* = \underset{M,R,T}{\operatorname{argmin}}\ J(R, T, M) \tag{6-46}$$

$$J(R, T, M) = \frac{1}{2}\sum_{k \in C}\left\|\left(p_k^i - \left(R\, p_{c(k)}^j + T\right)\right)\right\|^2 \tag{6-47}$$

where

p_k^i is the coordinate of point k of scan i

$p_{c(k)}^j$ is the correspondence of point k of scan i in scan j given by the corresponding function $M \triangleq c(k),\ k \in C$.

ICP algorithm is performed in two steps as follows.

6.6.2.1 Initialization

Initialize R, T to an initial value $\tilde{R}, \tilde{T}$. If no information is available, we usually set $\tilde{R} = I$ and $\tilde{T} = (0\,0\,0)^T$. For each point p_k^j in scan j, use the initial $\tilde{R}, \tilde{T}$ to find a transformed point $\tilde{p}_k^j = \tilde{R}p_k^j + \tilde{T}$. Then we search in scan i for the nearest neighbor point to transformed point $\tilde{p}_k^j$.

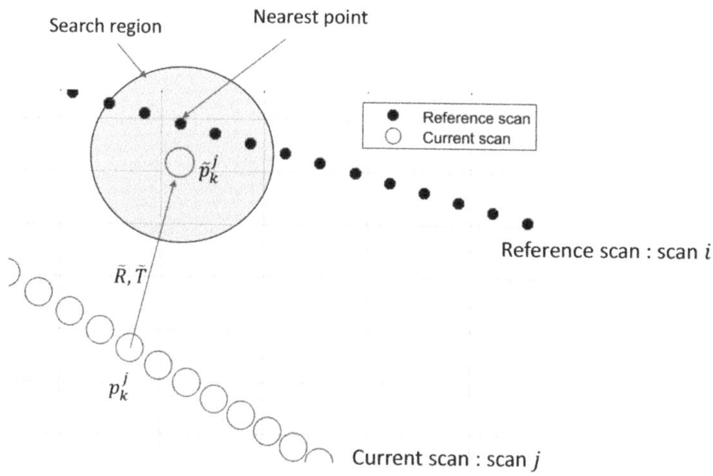

Figure 6.32 Finding correspondences using closest point search.

By having $\tilde{p}_k^j$ we reduce the search region that is needed to find the corresponding of point p_k^j in scan i as seen in Figure 6.32.

6.6.2.2 Correspondence Estimation

For each transformed point in the current in scan $\tilde{p}_k^j$, search in the neighborhood of $\tilde{p}_k^j$ for point p_l^i such that the Euclidean distance $\left\|\tilde{p}_k^j - p_k^i\right\|^2$ is minimum and set $c(k) = l$ and add $c(k)$ to the mapping M and k to the correspondence set C. To remove outliers, we may put a threshold on the distance $\left\|\tilde{p}_k^j - p_k^i\right\|^2$ to avoid corresponding a point in reference scan i to a wrong point in current scan j due to noise or due to the disappearance of this point from the field of view of the sensor. Closest neighbor search can be done in different ways including point-to-point search and point-to-line search.

6.6.2.3 Motion Estimation

Given the estimated correspondence or mapping $M \triangleq c(k), k \in C$, find the optimal rotation matrix and translation vector (R^*, T^*) that minimizes the following cost function:

$$R^*, T^* = \underset{R,T}{\operatorname{argmin}}\, J(R, T) \tag{6-48}$$

$$J(R, T) = \sum_{k \in C} \left\| \left(q_k - \left(R\, p_{c(k)} + T\right)\right) \right\|^2 \tag{6-49}$$

where

q_k is k^{th} point at scan i
$p_{c(k)}$ is the corresponding point in scan j.

We can obtain the optimal values R^*, T^* by finding the partial derivatives of $J(R, T)$ and set them to zero as follows:

$$\frac{\partial J(R, T)}{\partial T} = 0 \tag{6-50}$$

$$\frac{\partial J(R,T)}{\partial R} = 0 \tag{6-51}$$

Setting $\frac{\partial J(R,T)}{\partial T}$ to zero gives the following:

$$\sum_{k\in C}\left(R\,p_{c(k)} + T - q_k\right) = 0 \tag{6-52}$$

$$\sum_{k\in C}\left(R\,p_{c(k)}\right) + \sum_{k\in C}(T) - \sum_{k\in C}(q_k) = 0 \tag{6-53}$$

$$R\sum_{k\in C} p_{c(k)} + T\cdot|C| - \sum_{k\in C} q_k = 0 \tag{6-54}$$

$$T = \frac{1}{N}\sum_{k\in C} q_k - \frac{R}{N}\sum_{k\in C} p_{c(k)} \tag{6-55}$$

$$\mu_q = \frac{1}{N}\sum_{k\in C} q_k \tag{6-56}$$

$$\mu_p = \frac{1}{N}\sum_{k\in C} p_{c(k)} \tag{6-57}$$

$$T^* = \mu_q - R\mu_p \tag{6-58}$$

where $N = |C|$, μ_q is the center of mass of reference scan i μ_p is the center of mass of current scan j.

Now, let us find optimal value of R. The cost function now becomes as follows:

$$J(R) = \sum_{k\in C}\left\|\left(\left(R\,p_{c(k)} + \mu_q - R\mu_p\right) - q_k\right)\right\|^2 \tag{6-59}$$

$$J(R) = \sum_{k\in C}\left\|\left(R\left(p_{c(k)} - \mu_p\right) - \left(q_k - \mu_q\right)\right)\right\|^2 \tag{6-60}$$

$$J(R) = \sum_{k\in C}\left\|\left(Rp'_{c(k)} - q'_k\right)\right\|^2 \tag{6-61}$$

where $p'_{c(k)} = p_{c(k)} - \mu_p$ and $q'_k = q_k - \mu_q$ and we get the following cost function

$$J(R) = \sum_{k\in C}\left(Rp'_{c(k)}\right)^T\left(Rp'_{c(k)}\right) - 2(q'_k)^T Rp'_{c(k)} + \left(p'_{c(k)}\right)^T(q'_k) \tag{6-62}$$

$$J(R) = \sum_{k\in C}\left(p'_{c(k)}\right)^T R^T R\left(p'_{c(k)}\right) - 2(q'_k)^T Rp'_{c(k)} + \left(p'_{c(k)}\right)^T(q'_k) \tag{6-63}$$

As R is a rotation matrix, $R^T R = I$, which gives the following equation:

$$J(R) = \sum_{j=1}^{N}\left(p'_j\right)^T\left(p'_j\right) - 2\left(q'_j\right)^T Rp'_j + \left(p'_j\right)^T\left(q'_j\right) \tag{6-64}$$

$$J(R) = \sum_{k\in C}\left(p'_{c(k)}\right)^T\left(p'_{c(k)}\right) - 2(q'_k)^T Rp'_{c(k)} + \left(p'_{c(k)}\right)^T(q'_k) \tag{6-65}$$

This form of the cost function has only one term that depends on R, which is $\left(2(q'_k)^T R p'_{c(k)}\right)$. So, the cost function becomes:

$$J(R) = \sum_{k \in C} -2(q'_k)^T R p'_{c(k)} \tag{6-66}$$

which can be written in matrix form as follows:

$$J(R) = \mathrm{tr}\left(\begin{pmatrix} (q'_1)^T \\ (q'_2)^T \\ \vdots \\ (q'_N)^T \end{pmatrix} R\,(p'_1 p'_2 \dots p'_N)\right) \tag{6-67}$$

OR

$$J(R) = \mathrm{tr}\left(R\,(p'_1 p'_2 \dots p'_N)\begin{pmatrix} (q'_1)^T \\ (q'_2)^T \\ \vdots \\ (q'_N)^T \end{pmatrix}\right) \tag{6-68}$$

which can be written in the following compact form:

$$J(R) = \mathrm{tr}\left(Y^T R X\right) = \mathrm{tr}\left(R X Y^T\right) \tag{6-69}$$

OR

$$J(R) = \mathrm{tr}(R S) \tag{6-70}$$

where Y^T, X and S are defined as follows:

$$Y^{\top} = \begin{pmatrix} (q'_1)^T \\ (q'_2)^T \\ \vdots \\ (q'_N)^T \end{pmatrix} \tag{6-71}$$

$$X = (p'_1 p'_2 \dots p'_N) \tag{6-72}$$

$$S = X Y^{\top} \tag{6-73}$$

We will call the S matrix the information matrix. By replacing the information matrix S by its singular value decomposition (SVD),

$$S = U \Sigma V^T \tag{6-74}$$

we get the following cost function:

$$J(R) = \mathrm{tr}(R S) = \mathrm{tr}\left(R\, U \Sigma V^T\right) = tr\left(\Sigma V^T R U\right) \tag{6-75}$$

To minimize the trace of matrix term $\Sigma V^T R U$, we must guarantee that:

$$V^T R U = I \tag{6-76}$$

which gives the optimal value of R as follows:

$$R^* = VU^T \tag{6-77}$$

To guarantee that this is a valid rotation matrix, the determinant of VU^T must equal 1. If it equals −1, we calculate R as follows:

$$R^* = V\begin{pmatrix} 1 & & \\ & 1 & \\ & & |UV^T| \end{pmatrix}U^T \tag{6-78}$$

where $|UV^T|$ is the determinant of the matrix $|UV^T|$.

Once R^* and T^* are estimated, we repeat the algorithm again starting from the initialization step, where the values of $\tilde{R}$ and $\tilde{T}$ are set to the estimated R^* and T^* as follows:

$$\tilde{R} = R^* \tag{6-79}$$

$$\tilde{T} = T^* \tag{6-80}$$

We keep repeating the above steps until convergence. Convergence can be checked by testing the value of the cost function $J(R, T, M)$ reaches a threshold or does not change.

6.6.3 MATLAB Demonstration of ICP on LiDAR Scan Matching

The projects' repository includes a sample code that demonstrates the ICP algorithm applied to simulated LiDAR data. The MATLAB script "demo_LiDAR_ICP.m" under the folder "ICPLiDARProject" shows a step-by-step implementation of LiDAR scan matching using ICP algorithm. ICP is implemented in a separate script from [7]. An example of the ICP scan matching result is shown in Figure 6.33, where reference scan, current scan, ground-truth transformed scan, and ICP transformed scan are shown. The cost function value is shown in Figure 6.34. The convergence of the ICP

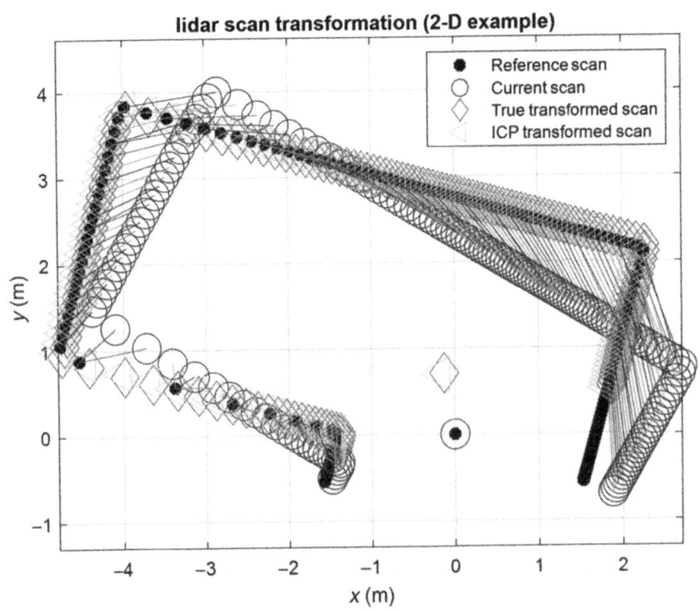

Figure 6.33 Two lidar scans transformation.

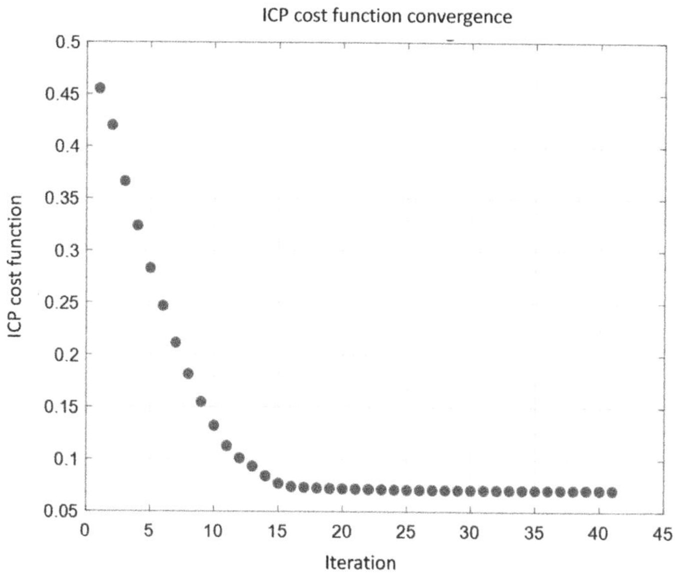

Figure 6.34 ICP cost function convergence.

cost function reflects the improvements in the rotation and translation estimation over the iterations. A convergence has been achieved after approximately 20 iterations. The simulation data in this example is noise-free.

6.6.4 Normal Distributions Transform (NDT)

Normal distributions transform (NDT) [8] is an alternative to ICP that is more robust and more efficient. This makes NDT more suitable for large and 3D point cloud data. NDT was first introduced by Biber and Straßer [9]. Given two scans k and $k+1$, NDT partitions the point cloud of scan k into cells and calculates a mean and covariance for each cell as follows:

$$\vec{\mu}_c = \frac{1}{M_c}\sum_{j\in c}\vec{x}_{j,k} \tag{6-81}$$

$$\Sigma_c = \frac{1}{M_c}\sum_{j\in c}\left(\vec{x}_{j,k} - \vec{\mu}_c\right)\left(\vec{x}_{j,k} - \vec{\mu}_c\right)^{\mathrm{T}} \tag{6-82}$$

where

$\vec{x}_{j,k}$ is a scan point in cell
$\vec{\mu}_c$ and Σ_c are the mean and covariance of the cell c at scan k.

The probability that any scan point $\vec{x}_{i,k}$ of belonging to cell c at scan k is given by the following Gaussian probability distribution:

$$p_c\left(\vec{x}_{i,k}\right) = \frac{1}{(2\pi)^{D/2}\sqrt{|\Sigma_c|}}e^{\left(-\frac{\left(\vec{x}_{i,k}-\vec{\mu}_c\right)^T(\Sigma_c)^{-1}\left(\vec{x}_{i,k}-\vec{\mu}_c\right)}{2}\right)} \tag{6-83}$$

For any point $\vec{x}_{i,k+1}$ in scan +1, it can be transformed back to its corresponding point in scan k using the transformation function $T\left(R, \vec{T}, \vec{x}_{i,k+1}\right)$ as follows:

$$\vec{\hat{x}}_{i,k} = T\left(R, \vec{T}, \vec{x}_{i,k+1}\right) = R\vec{x}_{i,k+1} + \vec{T} \quad (6\text{-}84)$$

The likelihood that the projected point $\vec{\hat{x}}_{i,k}$ belongs to cell c in scan k is given by:

$$p_c\left(\vec{\hat{x}}_{i,k}\right) \sim N(\vec{\mu}_c, \Sigma_c) \quad (6\text{-}85)$$

The right transformation should maximize the probability of all points as follows:

$$\Psi(R, T) = \prod_{i=1}^{n}\prod_{c} p_c\left(\vec{\hat{x}}_{i,k}\right) = \prod_{i=1}^{n}\prod_{c} p_c\left(T\left(R, \vec{T}, \vec{x}_{i,k+1}\right)\right) \quad (6\text{-}86)$$

Note that if $\vec{\hat{x}}_{i,k}$ do not belong to cell c, its probability $p_c\left(\vec{\hat{x}}_{i,k}\right)$ will be zero. Maximizing $\Psi(R, T)$ means minimizing the negative log-likelihood as follows:

$$J(R, T) = -\log\Psi(R, T) = -\sum_{i=1}^{n}\sum_{c}\log\left(p_c\left(T\left(R, \vec{T}, \vec{x}_{i,k+1}\right)\right)\right) \quad (6\text{-}87)$$

As $p\left(T\left(R, \vec{T}, \vec{x}_{i,k+1}\right)\right)$ is a Gaussian exponential function, the log of it is simple to obtain, which simplifies the cost function to the following summation form (ignoring the constant term in the Gaussian exponential formula $\left((2\pi)^{D/2}\sqrt{|\Sigma_c|}\right)$):

$$J(R, T) = \sum_{i=1}^{n}\sum_{c}\left(\left(T\left(R, \vec{T}, \vec{x}_{i,k+1}\right) - \vec{\mu}_c\right)^T (\Sigma_c)^{-1}\left(T\left(R, \vec{T}, \vec{x}_{i,k+1}\right) - \vec{\mu}_c\right)\right) \quad (6\text{-}88)$$

This cost function is a quadratic function of the rotation and translation parameters R and $\vec{T}$, which can be optimized using any optimization method. To optimize this cost function, we need to carefully consider the parameterization of the rotation parameter to facilitate the first and second derivatives of the cost function with respect to the rotation and translation parameters. As discussed earlier, using the rotation vector as a 3D parameterization for rotation and rotation error is a common approach. If the processing is only 2D, the parameterization of rotation is simplified to one angle (heading), which greatly simplifies the derivatives of the cost function. However, for the 3D case, exponential mapping of the rotation matrix, the utilization of rotation vector or quaternion parametrization is preferred to avoid the redundancy and singularities of the rotation matrix and to be able to correct rotation estimation by simple addition. The 2D NDT algorithm solved with Newton's optimization method can be found in [9]. A 3D implementation of NDT using Newton's optimization method can be found in [10]. Due to the clustering of the point cloud into cells and the probabilistic formulation of the problem, NDT is more robust and efficient compared to ICP.

6.6.5 Outlier Detection/Rejection

The sample data that we demonstrated above is clean as it is simulated data. In real-life scenarios, data always has noise and outliers. To avoid outliers, we apply the random sample consensus (RANSAC) algorithm. When we apply RANSAC algorithm on the ICP scan-matching problem, we begin with a random set of points and calculate M_c, R_c,T_c. We call this set (hypothetically) the inliers set. Then, for each other point in the scan, we use the calculated R_c,T_c to find its correspondence using the same nearest neighborhood search method described above. If this correspondence is below a certain small threshold, we consider the point as inlier and we add it to a set called consensus set. We then repeat the process again starting from a new random set of points and calculate and calculate M_c, R_c,T_c and check the inliers and make new consensus set and so on. After some iterations, we select the random set that has the largest number of inliers (the solution that has the largest consensus set). Once we get the best M_c, R_c,T_c that leads to the largest consensus set, can calculate the cost function as follows:

$$J(M_c, R_c, T_c) = \frac{1}{2}\sum_{k \in C}\left\| \left(p_k^i - \left(R\, p_{c(k)}^j + T\right)\right)\right\|^2 \tag{6-89}$$

RANSAC ICP scan matching is suboptimal in the sense that it is not guaranteed that the calculated parameters are the ultimate optimal parameters because the algorithm has a step that has some degree of randomness (the selection of the random set of points). However, RANSAC has proven improved performance and the ability to detect and reject outliers.

6.6.6 MATLAB Implementation of LiDAR Odometry

The projects' repository includes a sample code that demonstrates the range-based odometry using LiDAR data. The project "*LiDAROdometryProject*" main script code uses NDT algorithm to process consecutive LiDAR point cloud scans to estimate relative motions between scans. The algorithm further concatenates the estimated relative motions to estimate a full trajectory. In one test, the raw LiDAR scans are fed to the NDT algorithm scan matching algorithm without any initial guess of the transformation. In another testing scenario, we use inertial sensors to estimate an initial guess of the transformation and feed this initial guess to the NDT algorithm to see the effect of an initial guess on the accuracy of the estimation. Figure 6.35 shows the odometry results compared to the ground truth without providing any initial help to the NDT scan matching algorithm. As the figures show, there is a growing drift and some glitches. Figure 6.36 shows the inertial-only

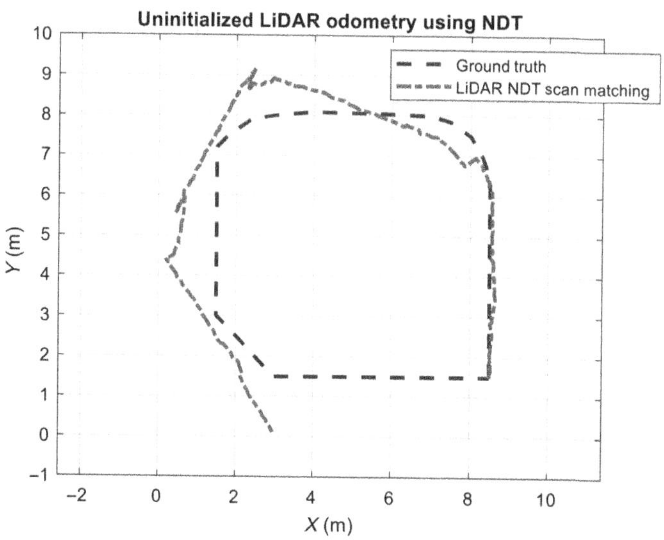

Figure 6.35 LiDAR odometry without initialization.

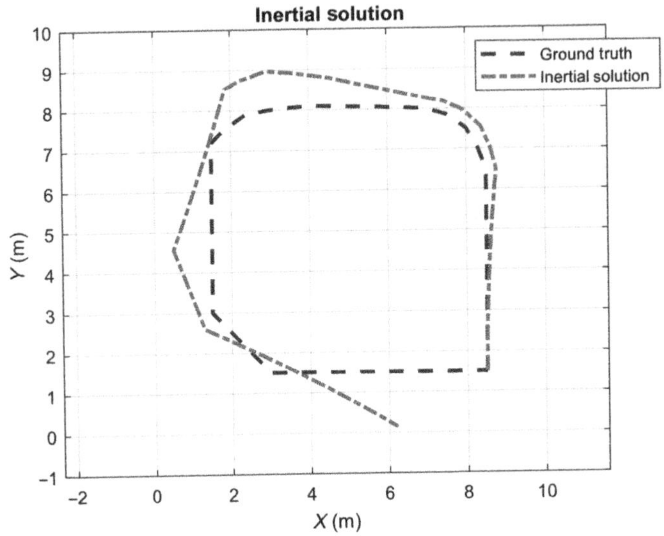

Figure 6.36 Inertial-only solution.

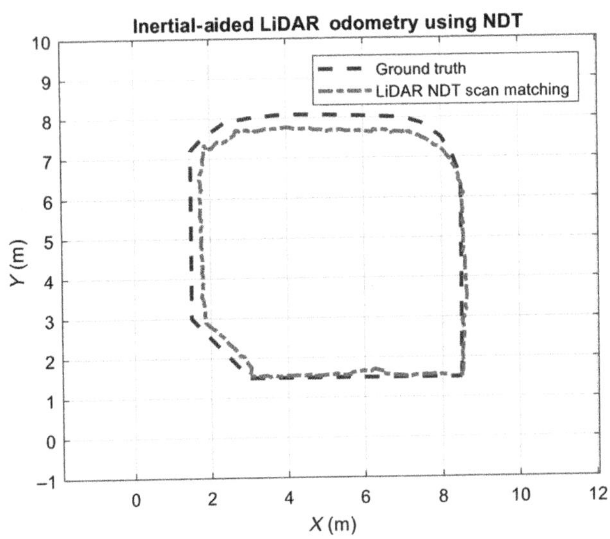

Figure 6.37 LiDAR odometry with initialization from inertial data.

solution, which is drifting constantly as expected. Figure 6.37 shows the results when initializing the NDT with an initial guess of the transformation from inertial sensors (rotation from gyroscope and translation from gyroscope/forward-speed measurements). As expected, providing an initial estimate to the NDT helps the algorithm to converge to the correct transformation which enhances the overall trajectory estimation accuracy. The project shows the benefits of fusing LiDAR and inertial sensors data which will be further discussed in future chapters.

6.7 Problems

6.7.1 Problem 1

Consider a LiDAR sensor that is tracked in 2D between time epochs k and $k+1$. At epoch k, the LiDAR sensor was initially horizontally aligned. By time epoch $k+1$, the sensor was rotated by an angle of $+15^{\circ}$ around its z-axis and translated by 0.1789 m in X direction and by -0.7028 m in Y direction. Calculate the 3D transformation T_{k+1}^{k} of the LiDAR sensor that can be used to transform scan points from epoch $k+1$ to epoch k.

6.7.2 Problem 2

In Problem 1, assume that at epoch k the position of the LiDAR sensor was (0, 0, 0) and the initial rotation angles were zeros. Calculate the 2D scan point coordinate q_k with respect to the LiDAR frame at epoch k given the following scan range/angle measurement:

$$r_k = 2.7439 \text{ m}$$

$$\psi_k = 38.5^{\circ}$$

6.7.3 Problem 3

In Problem 2, the point q_k was scanned in scan $k+1$ giving the following angle/range measurements:

$$r_{k+1} = 2.5275 \text{ m}$$

$$\psi_{k+1} = 8.5^{\circ}$$

A) Calculate p_{k+1}, the new coordinate of point q_k in scan $k+1$ with respect to the LiDAR frame at epoch $k+1$.
B) Using the transformation T_{k+1}^{k}, calculate $\hat{q}_k$, the reprojection of p_{k+1}.

Calculate the reprojection error $\|q_k - \hat{q}_k\|$.

6.7.4 Problem 4

A FMCW RADAR sensors with carrier frequency $F_c = 77$ GHz and bandwidth $B = 15.0$ MHz and chirp duration $T_c = 10$ μs. If the sampling rate $F_s = 1{,}024 \times 10^5$ sample/s, calculate the following:

A) Number of samples per chirp.
B) The maximum range this RADAR can sense.
C) The maximum speed this RADAR can measure.

Assume the speed of light is $s = 3 \times 10^8$ m/s

6.7.5 Problem 5

The RADAR in Problem 5 was used to scan an area and the absolute amplitude output from its 2D FFT is shown in Figure 6.38. Explain the results discussing how many targets are detected and their ranges, speeds, and their direction of motion with respect to the sensor (i.e., approaching or moving away from the RADAR).

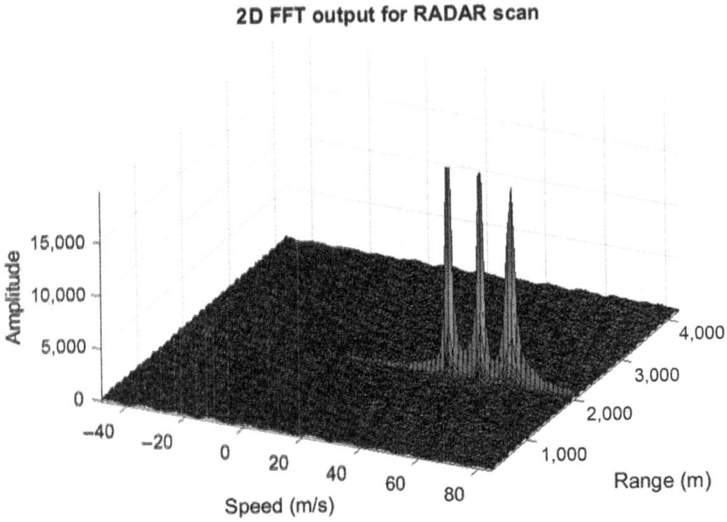

Figure 6.38 RADAR FFT2 processing output.

References

1 P. McManamon, "*LiDAR Technologies and Systems*," Washington, DC: SPIE PRESS, 2019.

2 W. Wei, "*ToF LiDAR for Autonomous Driving*," Bristol, UK: IOP Series in Advances in Optics, Photonics and Optoelectronics, 2023, DOI: https://doi.org/10.1088/978-0-7503-3723-6.

3 M. I. Skolink, "*Introduction to Radar Systems*," Boston, MA: McGraw-Hill Book Company, 1981.

4 G. Brooker, "*Sensors for Ranging and Imaging*," London, UK: SciTech Publishing, the Institution of Engineering and Technology, 2022.

5 J. G. Proakis and D. G. Manolakis, "*Digital Signal Processing, Principles, Algorithms, and Applications*," Prentice Hall, NJ: Pearson, 2007.

6 M. Kronauge, "*Waveform Design for Continuous Wave Radars*," Göttingen, Germany: Cuvillier Verlag, 2014.

7 J. Wilm, "An Implementation of Various ICP (Iterative Closest Point) Features," MathWorks Inc., Jan. 2013. [Online]. Available: https://www.mathworks.com/matlabcentral/fileexchange/27804-iterative-closest-point. [Accessed 8th July 2024].

8 K. Nagatani, T. Matsuzawa and K. Yoshida, "Scan-Point Planning and 3-D Map Building for a 3-D Laser Range Scanner in an Outdoor Environment," in *Field and Service Robotics*, Edited by B. Siciliano, O. Khatib, F. Groen, Cambridge, MA: Springer, 2010, pp. 207–217.

9 P. Biber and W. Straaer, "The Normal Distributions Transform: A New Approach to Laser Scan Matching," in *Proceedings 2003 IEEE/RSJ International Conference on Intelligent Robots and Systems (IROS 2003) (Cat. No.03CH37453)*, Las Vegas, NV, 2003.

10 M. Magnusson, "*The Three-Dimensional Normal-Distributions Transform - an Efficient Representation for Registration, Surface Analysis, and Loop Detection*," Örebro University, 2009.

7

Imaging Sensors

7.1 Introduction

In the previous chapters, we studied active ranging sensors, which provide in-depth information and are commonly used in mapping. Being an active sensor means the sensor requires more power, which may impose limitations on the practical implementation of the navigation systems. We have also seen how the global navigation satellite system (GNSS) receiver acts as a passive sensor that listens to satellites' signals, but it does not provide any information about the environment. Being a passive sensor, the GNSS receiver is generally more power efficient. Imaging sensors are electro-optical sensors that combine the advantages of being passive sensors and, at the same time, provide rich information about the environment. Conventional imaging sensors provide 2D images, which are projections of the physical 3D world onto a 2D plane. Imaging sensors can provide grayscale images or colored images. Most modern electronic imaging sensors provide colored images. This type of camera is called an RGB camera. RGB stands for red-green-blue color components. Each pixel in the image is a 3D vector that contains RGB components that determine the color of the pixel. The most challenging issue with RBG camera sensors is that they project the physical 3D world onto a 2D scaled domain where depth dimension is lost. Inferring position, depth, and environment information from a 2D-scaled domain is complex and more challenging. To overcome this challenge, some cameras combine the RGB passive imaging sensing and the active ranging feature by emitting a laser or an infrared wave and measuring the time-of-flight of the reflected waves, just like LiDAR and RADAR sensors. This type is called a depth camera or RGB-D camera, and they can provide depth information for each pixel in the image. However, due to power restrictions, the range of RGB-D cameras is always short, which makes them suitable for small environments such as rooms and offices. In this chapter, we will discuss how imaging sensors are used to provide 3D positioning and mapping.

7.2 How Imaging Sensors Work

We are surrounded by electromagnetic waves, and part of it is visible to our eyes. The visible range is called "light". Light hits objects and reflects to our eyes. Our eyes have narrow lenses where light beams go through and hit the retina. The retina is a thin layer of light-sensitive cells that converts light into signals and sends them through the optical nerves to the brain. The brain decodes the retina signals into color and brightness. Higher-level brain functions process color and brightness

Sensor Fusion Approaches for Positioning, Navigation, and Mapping: How Autonomous Vehicles and Robots Navigate in the Real World: With MATLAB Examples, First Edition. Mohamed M. Atia.
© 2025 The Institute of Electrical and Electronics Engineers, Inc. Published 2025 by John Wiley & Sons, Inc.

into higher-level information, such as regions, edges, and segments from which objects are detected and recognized. Computer vision and imaging sensors work similarly. Imaging sensors project the light reflected from the environment onto a light-sensitive material to form a 2D image. Imaging sensors transform the physical 3D metric world into a 2D scaled domain (pixels). Light beams reflected from objects travel in straight lines. When passing through a narrow transparent hole (a lens) and projected on an image plane that is coated with a light-sensing material, they form an image of the environment. In the old days, the sensing material was a "film," which is a chemical material that is sensitive to light. In modern digital cameras, the light-sensing material is an electronic material consisting of thousands or millions of small cells called pixels. The image sensor material can convert the physical light beam into a color code and a brightness level. Standard color cameras (RGB cameras) use color codes consisting of three color codes (numbers), red, green, and blue (RGB), from which any color can be represented. Some cameras produce grayscale images where only the brightness level is stored. Imaging sensors have a projection model that maps the physical 3D environment into a 2D image plane. The camera projection model is discussed in the next subsection. A simple illustration that shows how imaging sensors work is shown in Figure 7.1.

7.3 Camera Projection Model

There are several projection models that model the projection of the light beams from physical objects on the image plane. The simplest and most used camera model is the pinhole model [1, 2] illustrated in Figure 7.1. The view in Figure 7.1 is a side view to show the concept of focal length and virtual 2D image plane. The light beam travels from the object through the pinhole and hits the photo-sensitive material, which draws a flipped version of the object. The software in the camera reports the pixel coordinate in a virtual 2D image plane. The distance between

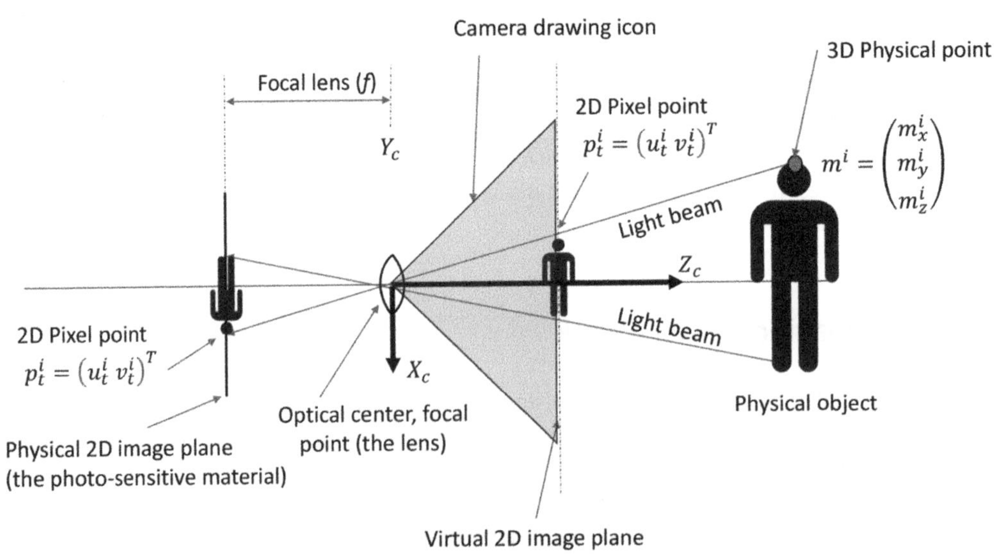

Figure 7.1 Imaging sensor: principle of operation.

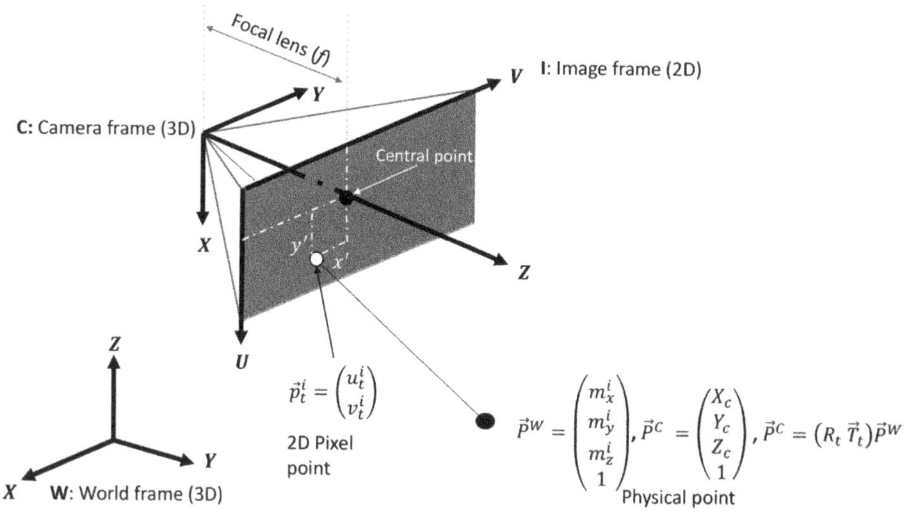

Figure 7.2 Basic coordinate frames used in camera projection.

the lens center and the photo-sensitive plane is called the focal length (f). In the drawings and diagrams, we use a pyramid shape as an icon to refer to the camera.

Basic coordinate frames used in camera projection are illustrated in Figure 7.2. Frame C defines the camera frame, frame W defines the word frame, and frame I defines the image frame. Figure 7.3 shows a top view of the pinhole projection where we have a physical point P projected to a pixel point $\vec{P}_t^i$. The t subscript refers to the fact that this 2D point is the projection at time t. The homogeneous coordinate of the point P in world and camera frames are $\vec{P}^W$ and $\vec{P}^C$, respectively. The relationship between P^C and P^W is given by the following transformation:

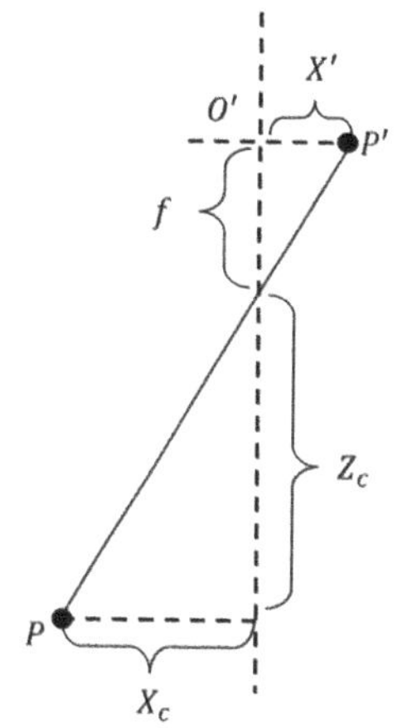

Figure 7.3 Top view of the projection diagram of pinhole projection.

$$\vec{P}^W = \begin{pmatrix} m_x^i \\ m_y^i \\ m_z^i \\ 1 \end{pmatrix} \quad (7\text{-}1)$$

$$\vec{P}^C = \begin{pmatrix} X_c \\ Y_c \\ Z_c \\ 1 \end{pmatrix} \quad (7\text{-}2)$$

$$\vec{P}^C = \left(R_t\ \vec{T}_t\right)\vec{P}^W \quad (7\text{-}3)$$

where $\left\langle R_t\ \vec{T}_t \right\rangle$ are the rotation and translation of the camera with respect to the world frame, which is simply the camera pose. $\left\langle R_t\ \vec{T}_t \right\rangle$ are called the extrinsic parameters of the camera. Now, we want to find the relationship between the point in camera frame $\vec{P}^C$ and its 2D image location $(x'\ y')^T$. The 2D image location $(x'\ y')^T$ is in metric units, not pixel units with respect to the central point. Figure 7.3 shows a top view of the projection as shown in Figure 7.2. Using triangle similarity properties as shown in Figure 7.3, we can derive the following relationships:

$$\frac{Z_c}{f} = -\frac{X_c}{x'} \tag{7-4}$$

where f is the focal length. Similarly, if we look at Figure 7.2 from the side, using the same triangle similarity rule, we can easily recognize (from the triangle similarity) the following relationship:

$$\frac{Z_c}{f} = -\frac{Y_c}{y'} \tag{7-5}$$

Given these equations, the 2D pixel point $(x'\ y')^T$ is defined as follows:

$$x' = f\frac{X_c}{Z_c} \tag{7-6}$$

$$y' = f\frac{Y_c}{Z_c} \tag{7-7}$$

In homogeneous coordinate, we can write the following equation:

$$\begin{pmatrix} x' \\ y' \\ 1 \end{pmatrix} = \frac{1}{Z_c}\begin{pmatrix} f & 0 & 0 \\ 0 & f & 0 \\ 0 & 0 & 1 \end{pmatrix}\begin{pmatrix} X_c \\ Z_c \\ Z_c \end{pmatrix} \tag{7-8}$$

$$Z_c\begin{pmatrix} x' \\ y' \\ 1 \end{pmatrix} = \begin{pmatrix} f & 0 & 0 \\ 0 & f & 0 \\ 0 & 0 & 1 \end{pmatrix}\begin{pmatrix} X_c \\ Z_c \\ Z_c \end{pmatrix} \tag{7-9}$$

The 2D image location $(x'\ y')^T$ is in metric units (not pixels) and it is described with respect to the center of the image plane. To convert $(x'\ y')^T$ into a pixel point $\vec{P}_t = \left(u_t^i\ v_t^i\right)^T$, we use the following equations:

$$u_t^i = \frac{x'}{p_x} + c_x = \frac{f}{p_x}\frac{X_c}{Z_c} + c_x \tag{7-10}$$

$$v_t^i = \frac{y'}{p_y} + c_y = \frac{f}{p_y}\frac{Y_c}{Z_c} + c_y \tag{7-11}$$

The homogeneous coordinate y_t^i of the pixel coordinate $\left(u_t^i\ v_t^i\right)^T$ is given by:

$$\vec{y}_t^i = \begin{pmatrix} u_t^i \\ v_t^i \\ 1 \end{pmatrix} \tag{7-12}$$

where p_x and p_y are size of the pixel in world metric units. Let us define the following parameters that convert from metric units to pixel units in image frame:

$$f_x = \frac{f}{p_x}$$

$$f_y = \frac{f}{p_y} \tag{7-13}$$

The relationship between the pixel coordinates y_t^i and physical point P^C is given by the following equation:

$$\vec{y}_t^{\,i} = \begin{pmatrix} u_t^i \\ v_t^i \\ 1 \end{pmatrix} = \frac{1}{Z_c}\begin{pmatrix} f_x & 0 & c_x \\ 0 & f_y & c_y \\ 0 & 0 & 1 \end{pmatrix}\begin{pmatrix} X_c \\ Z_c \\ Z_c \end{pmatrix} \triangleq \frac{1}{Z_c} K P^C \tag{7-14}$$

$$Z_c\begin{pmatrix} u \\ v \\ 1 \end{pmatrix} = K\begin{pmatrix} X_c \\ Y_c \\ Z_c \end{pmatrix} \tag{7-15}$$

$$\vec{y}_t^{\,i} = \begin{pmatrix} u_t^i \\ v_t^i \\ 1 \end{pmatrix} = K\begin{pmatrix} \frac{X_c}{Z_c} \\ \frac{Y_c}{Z_c} \\ 1 \end{pmatrix} \tag{7-16}$$

The matrix K is called the intrinsic parameters (or simply intrinsics) and the vector $\begin{pmatrix} \frac{X_c}{Z_c} & \frac{Y_c}{Z_c} & 1 \end{pmatrix}^T$ is called the normalized camera coordinate of the physical point where $Z_c = 1$. This relationship clearly explains how the depth dimension (the Z_c) is lost in the pixel frame. All 3D physical world points that lie in the projection line (see Figure 7.4) will project on the same pixel point in the 2D pixel image frame. All points on the projection line will have the same ratio of $\frac{X_c}{Z_c}$ and $\frac{Y_c}{Z_c}$ and hence will generate the same pixel coordinates y_t^i. This problem is referred to as the "scale ambiguity" and it is one of the major challenges in estimating 3D positioning and mapping information from 2D images.

Physical points are defined in the world frame $\vec{P}^W$ so they must be converted from the world frame to the camera frame before being converted to image pixel frame. The transformation from world frame to camera frame requires the knowledge of the camera's pose (position and orientation) with respect to the world frame. This transformation is described by a rotation matrix and

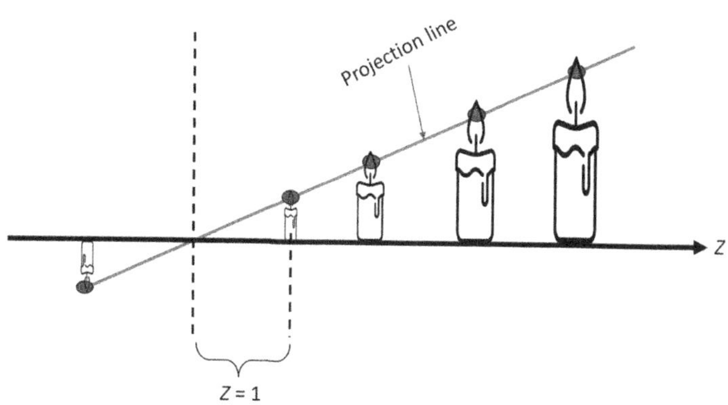

Figure 7.4 Scale ambiguity in camera projection.

translation vector $\left\langle R_t\, \vec{T}_t \right\rangle$ (the "extrinsic parameters"). Given a point in the 3D physical world $\vec{P}^{W}$, its projection on the 2D image pixel plane as follows:

$$\vec{y}_t^{\,i} = \begin{pmatrix} u_t^i \\ v_t^i \\ 1 \end{pmatrix} = \frac{1}{\lambda_t^i} K_{3\times 3} \left(R_t\, \vec{T}_t \right)_{3\times 4} \begin{pmatrix} m_x^i \\ m_y^i \\ m_z^i \\ 1 \end{pmatrix}_{4\times 1} \tag{7-17}$$

where λ_t^i is the depth of the point, which is the same as Z_c, the normal distance between the point and the XY plane of the camera frame. Sometimes λ_t^i is used to refer to the scale as well. The transformation can also be written as follows:

$$\lambda_t^i \vec{y}_t^{\,i} = Q\vec{P}^{W} \tag{7-18}$$

where Q is a matrix that combines the intrinsics and extrinsics of the camera. Pinhole projection is one of the most used models in computer vision. However, there are other camera models such as fish-eye camera model [3], spherical camera model [4], and omnidirectional camera model [5], and camera array models [6]. In addition, the projection model described above is a basic one for illustration purposes. In computer vision literature, there are advanced and more complicated projection models that include other parameters such as lens skew and distortion. Since the main scope of this book is navigation and mapping, we will discuss only the basic pinhole model, and we will discuss the monocular and stereo assemblies.

7.4 Estimating Camera Motion from Images

The key to understanding how camera motion (and map) can be estimated from images is the understanding of the camera projection model since it is the model that transforms 3D world points into the 2D image plane. Equally important are the key concepts of linear algebra related to matrices and vectors because there are several key algorithms and formulas that are based on proper manipulation of matrix properties like rank, singular value decomposition, eigenvalues, etc.

7.4.1 Problem Definition

Figure 7.5 shows camera poses at two-time instances k and $k+1$, while observing the same point m^j. At the two poses, the same point m^j are projected at different pixel coordinates $\vec{y}_k^{\,j}$ and $\vec{y}_{k+1}^{\,j}$ respectively. The problem of estimating camera motion from images can be described as follows: given the pixel coordinates $\vec{y}_k^{\,j}$ and $\vec{y}_{k+1}^{\,j}$ of the same world point m^j, and given the camera intrinsic parameters and assuming that at time k the camera frame is perfectly aligned with the world frame $\left(\left\langle R_k = I_{3\times 3}, \vec{T}_k = \vec{0}_{3\times 1} \right\rangle\right)$, we want to estimate how the camera rotated and translated from epoch k to epoch $k+1$. The rotation is described by the matrix $R_{k+1\rightarrow k}$ and translation is described by the vector $\vec{T}_{k+1\rightarrow k}$, respectively.

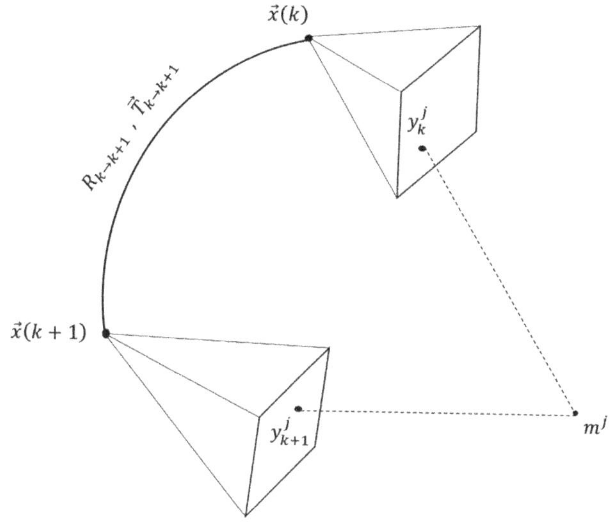

Figure 7.5 Camera motion estimation concept.

7.4.2 Principle of the Solution

We will assume that the camera motion from epoch k to $k+1$ is infinitesimally small. Let the world camera position at times k and $k+1$ be $\vec{x}(k)$ and $\vec{x}(k+1)$. The relationship between $\vec{x}(k)$ and $\vec{x}(k+1)$ is mathematically described as follows:

$$\vec{x}(k+1) = R_{k\to k+1}\vec{x}(k) + \vec{T}_{k\to k+1} \tag{7-19}$$

Given the homogeneous world location of the feature j given by $m^j = \left(m_x^j\ m_y^j\ m_z^j\ 1\right)^T$, the corresponding homogeneous pixel coordinates at time k and $k+1$ are give by the following equations:

$$\lambda_k \vec{y}_k^{\,j} = \lambda_k \begin{pmatrix} u_k \\ v_k \\ 1 \end{pmatrix} = K_{3\times 3}\left(R_{w\to k}\ \vec{T}_{w\to k}\right)_{3\times 4} \begin{pmatrix} m_x^j \\ m_y^j \\ m_z^j \\ 1 \end{pmatrix} \tag{7-20}$$

$$\lambda_{k+1} \vec{y}_{k+1}^{\,j} = \lambda_{k+1} \begin{pmatrix} u_{k+1} \\ v_{k+1} \\ 1 \end{pmatrix} = K_{3\times 3}\left(R_{k\to k+1}\ \vec{T}_{k\to k+1}\right)_{3\times 4} \begin{pmatrix} m_x^j \\ m_y^j \\ m_z^j \\ 1 \end{pmatrix} \tag{7-21}$$

where $\left(R_{w\to k}\ \vec{T}_{w\to k}\right)$ is the transformation from world to camera frame at time k. Assuming that the camera at time k is perfectly aligned with the world frame, $\left\langle R_{w\to k} = I_{3\times 3}, \vec{T}_{w\to k} = \vec{0}_{3\times 1}\right\rangle$, the equations simplify to the following form:

$$\lambda_k \vec{y}_k^{\,j} = \lambda_k \begin{pmatrix} u_k \\ v_k \\ 1 \end{pmatrix} = K_{3\times 3}\left(I_{3\times 3}\ \vec{0}_{3\times 1}\right)_{3\times 4} \begin{pmatrix} m_x^j \\ m_y^j \\ m_z^j \\ 1 \end{pmatrix} \tag{7-22}$$

$$\lambda_{k+1}\vec{y}^{j}_{k+1} = \lambda_{k+1}\begin{pmatrix} u_{k+1} \\ v_{k+1} \\ 1 \end{pmatrix} = K_{3\times 3}\left(R_{k\to k+1}\ \vec{T}_{k\to k+1}\right)_{3\times 4}\begin{pmatrix} m^j_x \\ m^j_y \\ m^j_z \\ 1 \end{pmatrix} \tag{7-23}$$

As the pixel coordinates $\vec{y}^{j}_{k}$ and $\vec{y}^{j}_{k+1}$ are known, and the intrinsics matrix $K_{3\times 3}$ is known, and the world point $m^j = \left(m^j_x\ m^j_y\ m^j_z\ 1\right)^T$ is static (did not move from time k to time $k+1$), the only unknowns in the above two equations are $\left\langle R_{k\to k+1}, \vec{T}_{k\to k+1}, m^j, \lambda_k, \lambda_{k+1}\right\rangle$. We can mathematically formulate the problem as a projection error minimization as follows:

$$J\left(R_{k+1\to k}, \vec{T}_{k+1\to k}, m^j, \lambda_k, \lambda_{k+1}\right) = \left\|\lambda_k\vec{y}^{j}_{k} - K_{3\times 3}\left(I_{3\times 3}\ \vec{0}_{3\times 1}\right)_{3\times 4} m^j{}_{4\times 1}\right\|^2 + \left\|\lambda_{k+1}\vec{y}^{j}_{k+1} - K_{3\times 3}\left(R_{k\to k+1}\ \vec{T}_{k\to k+1}\right)_{3\times 4}\left(I_{3\times 3}\ 0_{3\times 1}\right)_{3\times 4} m^j{}_{4\times 1}\right\|^2 \tag{7-24}$$

Because we assume that the motion is infinitesimally small, we can safely assume that $\lambda_k \cong \lambda_{k+1}$ which removes one of the unknowns and the cost function becomes a function in $\left(R_{k\to k+1}, \vec{T}_{k\to k+1}, m^j, \lambda_k\right)$. If at time k the world point was at a vertical distance from the camera XY plane that equals "1," this means that $X^j_c = m^j_z = 1$, which also means that $\lambda_k = 1$, which removes another unknown from the cost function, leaving it with $R_{k\to k+1}, \vec{T}_{k\to k+1}, m^j$ as unknowns. Although the solution principle is simple and clear, the mathematical solution and implementation are complex. One approach to simplify the problem is to remove m^j from the equations using some mathematical manipulation and, estimate the camera motion $R_{k\to k+1}, \vec{T}_{k\to k+1}$ and then use the estimated camera motion to estimate the feature point location m^j. This will take us to the concept of epipolar geometry and epipolar constraints, which is explained in the next subsection.

7.4.3 Epipolar Constraint

Figure 7.6 illustrates the concept of epipolar constraints. Camera motion between two frames (epochs) is illustrated in Figure 7.6. The projections of the 3D world point m^j on the two images

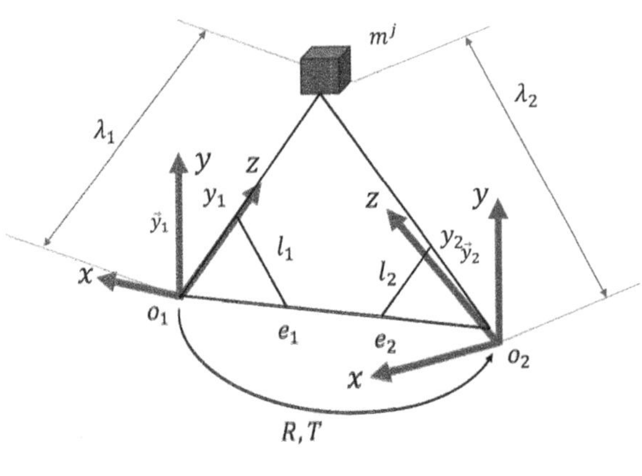

Figure 7.6 Epipolar geometry concept.

are denoted by $\vec{y}_1$ and $\vec{y}_2$. Note that the pixel projection coordinates are given in homogeneous form as follows:

$$\vec{y}_1 = \begin{pmatrix} u_1 & v_1 & 1 \end{pmatrix}^T \tag{7-25}$$

$$\vec{y}_2 = \begin{pmatrix} u_2 & v_2 & 1 \end{pmatrix}^T \tag{7-26}$$

For simplicity, we omitted the superscript j from the notation of $\vec{y}_1$ and $\vec{y}_2$ assuming, for now, we have only one feature point m^j in the environment. It is important to remember that R here is a rotation of the camera at epoch 1 with respect to camera at epoch 2. In other words, $R = R_{1\to 2}$ describes how camera frame 1 is rotated with respect to camera frame 2.

The optical centers of each camera pose are denoted by o_1 and o_2. The intersections of the line (o_1, o_2) with each image plane are called the epipoles e_1 and e_2. The intersections between the epipolar plane (o_1, o_2, m^j) and the image planes are called epipolar lines l_1 and l_2. There is one epipolar plane for each 3D world point. If the camera intrinsics are known, and the camera is initially perfectly aligned with the world frame, the homogeneous pixel coordinates of $\vec{y}_1$ and $\vec{y}_1$ are calculated as follows:

$$\lambda_1 \vec{y}_1 = K\left(I_{3\times 3}\begin{pmatrix} m_x^j \\ m_y^j \\ m_z^j \end{pmatrix} + \vec{0}_{3\times 1}\right) \tag{7-27}$$

$$\lambda_2 \vec{y}_2 = K\left(R\begin{pmatrix} m_x^j \\ m_y^j \\ m_z^j \end{pmatrix} + \vec{T}\right) \tag{7-28}$$

where λ_1, λ_2 are the depth of m^j in the two epochs, respectively. The camera motion between the two epochs is described by the rotation matrix R and translation vector $\vec{T}$. This will simplify the projection equations as follows:

$$\lambda_1 \vec{y}_1 = K\begin{pmatrix} m_x^j \\ m_y^j \\ m_z^j \end{pmatrix} \tag{7-29}$$

$$\lambda_2 \vec{y}_2 = K\left(R\begin{pmatrix} m_x^j \\ m_y^j \\ m_z^j \end{pmatrix} + \vec{T}\right) \tag{7-30}$$

Note that $m^j = \left(m_x^j\ m_y^j\ m_z^j\right)^T$ and $\vec{T}$ are vectors defined in the world frame and $\vec{y}_1$ and $\vec{y}_2$ are defined in the camera frame. Combining the two equations gives the following formula:

$$\begin{pmatrix} m_x^j \\ m_y^j \\ m_z^j \end{pmatrix} = \lambda_1 K^{-1}\vec{y}_1 \tag{7-31}$$

We can then substitute for $m^j = \left(m_x^j\ m_y^j\ m_z^j\right)^T$, which gives the following equation that does not have m^j variable:

$$\lambda_2 K^{-1}\vec{y}_2 = \lambda_1 R\left(K^{-1}\vec{y}_1\right) + \vec{T} \tag{7-32}$$

If we cross product the two sides by $\frac{\vec{T}}{\lambda_1}$, we get the following equation:

$$\frac{\lambda_2}{\lambda_1}\vec{T} \times K^{-1}\vec{y}_2 = \vec{T} \times R\left(K^{-1}\vec{y}_1\right) + 0 \tag{7-33}$$

If we dot-product both sides by the vector, $\left(K^{-1}\vec{y}_2\right)^T$, we get the following equation:

$$\frac{\lambda_2}{\lambda_1}\left(K^{-1}\vec{y}_2\right)^T\left(\vec{T} \times K^{-1}\vec{y}_2\right) = \left(K^{-1}\vec{y}_2\right)^T\left(\vec{T} \times R\left(K^{-1}\vec{y}_1\right)\right) \tag{7-34}$$

The cross product $\left(\vec{T} \times K^{-1}\vec{y}_2\right)$ will generate a vector perpendicular to both $\vec{T}$ and $\vec{y}_2$ which means that the dot product $\left(K^{-1}\vec{y}_2\right)^T\left(\vec{T} \times K^{-1}\vec{y}_2\right)$ equals zero because the angle between $K^{-1}\vec{y}_2$ and $\vec{T} \times K^{-1}\vec{y}_2$ is 90°. This gives the popular epipolar constraints equation defined as follows:

$$\left(K^{-1}\vec{y}_2\right)^T\left(\vec{T} \times R\left(K^{-1}\vec{y}_1\right)\right) = 0 \tag{7-35}$$

which can be rearranged as follows:

$$\left(\vec{y}_2\right)^T\left(K^{-1}\right)^T\left(\vec{T} \times R\left(K^{-1}\vec{y}_1\right)\right) = 0 \tag{7-36}$$

To further simplify the notation, let's define the following vectors:

$$\vec{y}_1' = K^{-1}\vec{y}_1 \tag{7-37}$$

$$\vec{y}_2' = K^{-1}\vec{y}_2 \tag{7-38}$$

$\vec{y}_1'$ and $\vec{y}_2'$ are the normalized world metric projection of the pixel-coordinates $\vec{y}_1$ and $\vec{y}_2$ respectively. Note that the z-components $\vec{y}_1'$ and $\vec{y}_2'$ will still be 1. If we use the skew matrix $\left(\vec{T}\times\right)$ multiplication instead of the cross product, this will simplify the notation to the following:

$$\left(\vec{y}_2'\right)^T\left(\vec{T}\times\right)R\vec{y}_1' = 0 \tag{7-39}$$

where the skew matrix form $\left(\vec{T}\times\right)$ is given by:

$$\left(\vec{T}\times\right) = \begin{pmatrix} 0 & -T_z & T_y \\ T_z & 0 & -T_x \\ -T_y & T_y & 0 \end{pmatrix} \tag{7-40}$$

7.4.4 The Essential Matrix

If we set E $= \left(\vec{T}\times\right)R$, we get the following equation:

$$\vec{y}_2'^T E\vec{y}_1' = 0 \tag{7-41}$$

which is one of the fundamental equations in photogrammetry. The matrix E is called the essential matrix and it combines both the rotational and translational motion of the camera. To estimate the

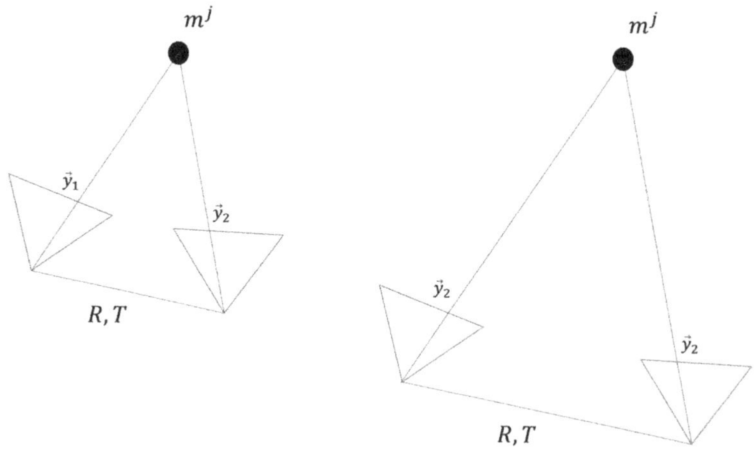

Figure 7.7 Depth/scale ambiguity. Both depths will generate the same pixel coordinates.

camera motion, the essential matrix E needs to be estimated. All camera motion estimation methods aim at estimating the essential matrix and extracting the camera motion from it. It is important to notice that the estimated translation will have a scale ambiguity depending on the depth of the observed point, as illustrated in Figure 7.7. Figure 7.7 shows two different points with different depths but they are located on the same projection line will give the same pixel coordinate as seen in Figure 7.7. Note that $\vec{y}_1'$ is aligned with $\overrightarrow{o_1 m^j}$, $\vec{T}$ is aligned with $\overrightarrow{o_2 o_1}$, and $R\left(\vec{y}_1'\right)$ is aligned with $\overrightarrow{o_2 m^j}$. Keeping this in mind, the Eq. (7-41) means that $\overrightarrow{o_1 m^j}$, $\overrightarrow{o_2 o_1}$, and $\overrightarrow{o_2 m^j}$ forms a plane which is consistent with the geometric illustration given in Figure 7.6. Forming a plane means that the parallelepiped volume formed by the three vectors $\overrightarrow{o_1 m^j}$, $\overrightarrow{o_2 o_1}$, and $\overrightarrow{o_2 m^j}$ is zero. Recall from Chapter 1 that the volume of an object formed by three vectors is the triple scalar product of the three vectors defined by $\vec{a} \cdot \left(\vec{b} \times \vec{c}\right)$, which proves the epipolar constraint formula in another way.

7.4.5 Estimating Camera Pose from the Essential Matrix

The essential matrix has some interesting properties that enable us to calculate the camera pose change or relative transformation (rotation matrix and translation vector). For example, for any essential matrix E, the single value decomposition must exist, and it satisfies the following condition:

$$E = U\Sigma V^{\top} \tag{7-42}$$

where $\Sigma = \text{diag}\{\sigma, \sigma, 0\}$, which means that E has three singular values, two must be equal and one must be zero. Another interesting property of E is that four solutions $(R_1\vec{T}_1, R_1\vec{T}_2, R_2\vec{T}_1, R_2, \vec{T}_2)$ can lead to the same essential matrix E as follows (see [7], [2], and [1]):

$$\left(\left(\overrightarrow{T_1} \times\right), R_1\right) = \left(U\begin{pmatrix} 0 & -1 & 0 \\ 1 & 0 & 0 \\ 0 & 0 & 1 \end{pmatrix}\Sigma U^{\top}, U\begin{pmatrix} 0 & 1 & 0 \\ -1 & 0 & 0 \\ 0 & 0 & 1 \end{pmatrix}V^{\top}\right) \tag{7-43}$$

$$\left(\left(\overrightarrow{T_2}\times\right),R_2\right)=\left(U\begin{pmatrix}0&1&0\\-1&0&0\\0&0&1\end{pmatrix}\Sigma U^{\top},U\begin{pmatrix}0&-1&0\\1&0&0\\0&0&1\end{pmatrix}V^{\top}\right)\tag{7-44}$$

To resolve this ambiguity, we always choose the rotation and translation that leads to reasonable results such as positive depth values of the metric coordinate of the physical world point. Due to numerical calculation issues and the assumptions, we have made in the derivations of the essential matrix, the estimated rotation R and translation $\vec{T}$ from the essential matrix are not perfectly accurate. However, they an be taken as a good initial estimate. We may perform more refining of the estimation solution by minimizing the pixel-coordinate reprojection error function defined as follows:

$$J\left(R,\vec{T}\right)=\left\|\lambda_j\vec{y}_j-K\left(R\vec{m}^j+\vec{T}\right)\right\|^2\tag{7-45}$$

In the coming sections, we will discuss how we can estimate the essential matrix from image sequences and how we can extract the translation and rotation from the essential matrix. Later, we will further demonstrate how the 3D physical world location of the feature point can be calculated.

7.4.6 The Eight-Point Algorithm

Given two image frames (image1 and image2), and given the homogeneous pixel coordinates of the same physical world point $m^j=\left(m_x^j\ m_y^j\ m_z^j\right)^T$ in the two images $\left\langle\vec{q}_j',\vec{p}_j'\right\rangle$ defined as follows (assuming we already know the intrinsic matrix K):

$$\vec{q}_j=\left(u_1^j,v_1^j,1\right)^T\in\mathbb{R}^3\tag{7-46}$$

$$\vec{p}_j=\left(u_2^j,v_2^j,1\right)^T\in\mathbb{R}^3\tag{7-47}$$

$$\vec{q}_j'=K^{-1}\vec{q}_j=\left(x_1^j,y_1^j,z_1^j\right)^T\in\mathbb{R}^3\tag{7-48}$$

$$\vec{p}_j'=K^{-1}\vec{p}_j=\left(x_2^j,y_2^j,z_2^j\right)^T\in\mathbb{R}^3\tag{7-49}$$

The epipolar constraint equation states that:

$$\left(\vec{p}_j'\right)^TE\vec{q}_j'=0\tag{7-50}$$

$$E=\left(\vec{T}\times\right)R\tag{7-51}$$

where $\vec{T}$ is the translation vector and R is the rotation matrix. Both $\vec{T}$ and R define the relative camera motion from image1 to image2. The essential matrix has nine elements as follows:

$$E=\begin{pmatrix}e_1&e_2&e_3\\e_4&e_5&e_6\\e_7&e_8&e_9\end{pmatrix}\tag{7-52}$$

which can be arranged in a 1D vector as follows:

$$\vec{e} = (e_1, e_2, e_3, e_4, e_5, e_6, e_7, e_8, e_9)^T \in \mathbb{R}^9 \tag{7-53}$$

We can also rearrange the vectors $\vec{p}'_j$ and $\vec{q}'_j$ as Kronecker product as follows:

$$\vec{a} = \vec{q}'_j \otimes \vec{p}'_j = \left(x_2^j x_1^j\ x_2^j y_1^j\ x_2^j z_1^j\ y_2^j x_1^j\ y_2^j y_1^j\ y_2^j z_1^j\ z_2^j x_1^j\ z_2^j y_1^j\ z_2^j z_1^j\right)^T \in \mathbb{R}^g \tag{7-54}$$

Note that both z_2^j and z_1^j always equal 1 for all homogeneous coordinates, so equation will be reduced to the following form:

$$\vec{a} = \vec{y}'_{1,j} \otimes \vec{y}'_{2,j} = \left(x_2^j x_1^j\ x_2^j y_1^j\ x_2^j\ y_2^j x_1^j\ y_2^j y_1^j\ y_2^j\ x_1^j\ y_1^j\ 1\right)^T \in \mathbb{R}^g \tag{7-55}$$

Using this vector form, the epipolar constraint equation can be rearranged in the following matrix form:

$$\left(\vec{y}'_{2,j}\right)^T E \vec{y}'_{1,j} = \vec{a}.\vec{e} = 0 \tag{7-56}$$

$$\left(x_2^j x_1^j, x_2^j y_1^j, x_2^j z_1^j, y_2^j x_1^j, y_2^j y_1^j, z_2^j z_1^j, z_2^j x_1^j, z_2^j y_1^j, z_2^j z_1^j\right) \begin{pmatrix} e_1 \\ e_2 \\ e_3 \\ e_4 \\ e_5 \\ e_6 \\ e_7 \\ e_8 \\ e_9 \end{pmatrix} = 0 \tag{7-57}$$

The reader can verify the above equation by doing the multiplication $\left(\vec{y}'_{2,j}\right)^T E \vec{y}'_{1,j}$ and verifying that it equals $\vec{a}.\vec{e}$ product. If we have multiple feature points, will have a matrix equation as follows:

$$\underbrace{\begin{pmatrix} x_2^1 x_1^1, x_2^1 y_1^1, x_2^1 z_1^1, y_2^1 x_1^1, y_2^1 y_1^1, z_2^1 z_1^1, z_2^1 x_1^1, z_2^1 y_1^1, z_2^1 z_1^1 \\ x_2^2 x_1^2, x_2^2 y_1^2, x_2^2 z_1^2, y_2^2 x_1^2, y_2^2 y_1^2, z_2^2 z_1^2, z_2^2 x_1^2, z_2^2 y_1^2, z_2^2 z_1^2 \\ \vdots \end{pmatrix}}_{A} \underbrace{\begin{pmatrix} e_1 \\ e_2 \\ e_3 \\ e_4 \\ e_5 \\ e_6 \\ e_7 \\ e_8 \\ e_9 \end{pmatrix}}_{\vec{e}} = 0 \tag{7-58}$$

which can be written in a shorter matrix form as follows:

$$A\vec{e} = 0 \tag{7-59}$$

This equation means that the vector $\vec{e}$ is a null vector for the matrix A ($\vec{e}$ is in the null space of A). To get a unique single solution (single null vector), the rank of the matrix A must be 8 (if the rank is 9, this means that null-space is empty; if rank is 8, this means that null-space has one single vector). That is the reason we need at least eight feature points to get a unique solution to this equation. That is the reason this algorithm is called the eight-point algorithm. It is nearly impossible to find null vector $\vec{e}$ that makes $A\vec{e}$ exactly zero due to random noise and the approximations we assumed in the derivation of the equation. Fortunately, the null vector of a matrix is the eigenvector corresponding to the eigenvalue zero. We can benefit from this property and calculate the eigen decomposition of the matrix A and chose the vector $\vec{e}$ to be the eigenvector corresponding to the smallest eigenvalue value. As the matrix A is not square, instead of eigen decomposition, we use singular value decomposition and choose the right singular vector corresponding to the smallest singular value. To get a unique solution for this equation, the points must be distributed evenly in the image and not concentrated in one place. If, for example, all points lie in a single plane, this will lead to matrix A with rank less than 8, which will lead to multiple solutions for $\vec{e}$. This case is called degenerate points, and it needs special handling to provide a solution.

Even after calculating a unique solution for $\vec{e}$, due to random noise and numerical errors, we cannot just convert $\vec{e}$ to a valid essential matrix E. Therefore, once the vector $\vec{e}$ is calculated, it is projected into the essential matrix space to satisfy the essential matrix properties. To project $\vec{e}$ to essential matrix space, we follow the following steps (see [7] and [1]):

1) Convert $\vec{e}$ to a 3-by-3 matrix E'
2) Calculate the singular value decomposition (SVD) of E'

$$E' = U \begin{pmatrix} s_1 & 0 & 0 \\ 0 & s_2 & 0 \\ 0 & 0 & s_3 \end{pmatrix} V^T \tag{7-60}$$

3) Calculate the essential matrix as follows:

$$\sigma = (s_1 + s_2)/2 \tag{7-61}$$

$$E = U \begin{pmatrix} \sigma & 0 & 0 \\ 0 & \sigma & 0 \\ 0 & 0 & 0 \end{pmatrix} V^T \tag{7-62}$$

As we can see, the SVD operation plays a key role in the calculations of E. SVD is explained in detail in Chapter 1. It is important to note that the calculated camera translation $\vec{T}$ vector is a unit vector $\| \vec{T} \| = 1$ because we marginalized (removed) the depth parameters (λ_1, λ_2) and normalized the physical world to be at a unity distance from the central point of the camera (the origin of the camera frame). The true translation in the physical world is a scaled version of the estimated translation vector as follows:

$$\vec{T}_{\text{true}} = \gamma \vec{T}_{\text{estimated}} \tag{7-63}$$

where γ is an unknown scale factor. If other sensors that can provide depth information or other aiding sensors that can estimate $\vec{T}_{\text{true}}$ such as inertial sensors, the scale factor γ can be estimated by dividing $\vec{T}_{\text{true}}/\vec{T}_{\text{estimated}}$.

7.4.7 MATLAB Implementation of Eight-Point Algorithm

To further clarify the eight-point algorithm and to get hands-on experience in calculating the rotation and translation of the camera from matched feature points in two images, a MATLAB implementation of the eight-point algorithm is given in the projects' repository. Part of the implementation of the algorithm is shown in Figure 7.8. The data of this project comes from MATLAB calibration example in [8]. The data in this example has nine camera poses. Our MATLAB project chose points 1 and 7 and applied the eight-point algorithm to estimate the relative rotation and translation between the two camera poses. The algorithm uses 54 matched feature points (q and p in the MATLAB script below). The camera poses from different views are shown in Figure 7.9. The relative rotation and translation of this example are as follows: roll angle is 2.3°, pitch angle is 4.3°, heading angle is −38.5°, and the translation vector $\vec{T} = (-0.072, -0.013, -0.997)^T$. By tracing the code in Figure 7.8, readers can inspect how the normalized features' coordinates are calculated from the pixel coordinates and the inverse of the intrinsic matrix K as described in Eqs. (7-48) and (7-49), how the matrix A in Eq. (7-59) is constructed, how the essential matrix vector $\vec{e}$ is estimated using singular value decomposition, how the essential matrix is refined using Eqs. (7-60)–(7-62), and how the rotation and translation matrix is calculated from the four possible solutions given in Eqs. (7-43) and (7-44).

```
%--> 8-points algorithms (inputs: matching points q,p and intrinsics K)

% Get normalized hmogeneous coordinates from pixel coordinates and K matrix
q_ = K\q;
p_ = K\p;

% Form the A matrix using the Kronecker product
A = zeros(length(p),9);
for i = 1:length(p)
    x1 = p_(1,i);
    y1 = p_(2,i);
    x2 = q_(1,i);
    Y2 = q_(2,i);
    A(i,:) = [x2*x1 x2*y1 x2 Y2*x1 Y2*y1 Y2 x1 y1 1];
end

%Find the essential matrix vector using SVD of A
[~, ~, V] = svd(A);
e = V(:,9);
F = reshape(e, 3, 3);

% Refine the essential matrix using SVD
[U, S, V] = svd(F,"econ");
E = U*diag([(S(1,1)+S(2,2))/2 (S(1,1)+S(2,2))/2 0])*V';

% Get the 4 possible solutions from the E matrix
P = get4PossibleSolutions(E);

% Select the solution that assures positive depth
P = SelectCorrectCameraTransformationMatrix(p,q,P,K);
R = P(:,1:3);
T = P(:,4);

% Display the orientation in Euler angles and the Translation vector
[roll,pitch,Azimuth] = dcm2angle(R,"XYZ");
fprintf('8-points algorithm: r,p,A = %f,%f,%f, T = [%f,%f,%f]\n',roll*180/pi,pitch*180/pi,Azimuth*180/pi,T);
```

Figure 7.8 Snapshot of the MATLAB code that implements the eight-point algorithm.

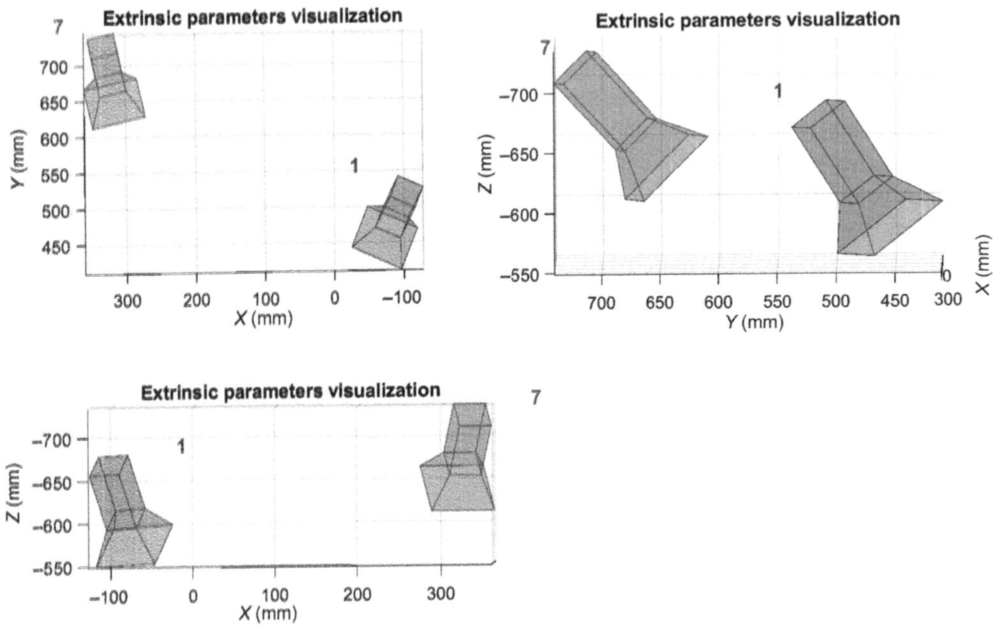

Figure 7.9 Camera poses in the eight-point algorithm MATLAB project.

7.5 Visual 3D Reconstruction: Estimating Physical Points' Location

Once we calculate the camera motion $\left\langle R, \vec{T} \right\rangle$, the 3D world coordinates of the physical point can be calculated. The projection equation for a point m^j from image1 to image2 is given by:

$$\lambda_2^j \vec{y}_2^{\prime j} = \lambda_1^j R \vec{y}_1^{\prime j} + \gamma \vec{T} \tag{7-64}$$

Cross product both sides by $\vec{y}_2^{\prime j}$, which means multiplying both sides by $\left(\vec{y}_2^{\prime j} \times \right)$,

$$\lambda_1^j \left(\vec{y}_2^{\prime j} \times \right) R \vec{y}_1^{\prime j} + \gamma \left(\vec{y}_2^{\prime j} \times \right) \vec{T} = 0, j = 1, \ldots, n \tag{7-65}$$

which can be rearranged in a matrix form as follows:

$$\begin{pmatrix} \left(\vec{y}_2^{\prime 1} \times \right) R \vec{y}_1^{\prime 1} & 0 & 0 \;\; 0 & 0 & \left(\vec{y}_2^{\prime 1} \times \right) \vec{T} \\ 0 & \left(\vec{y}_2^{\prime 2} \times \right) R \vec{y}_1^{\prime 2} & 0 \;\; 0 & 0 & \left(\vec{y}_2^{\prime 2} \times \right) \vec{T} \\ 0 & 0 & . \;\; 0 & 0 & . \\ 0 & 0 & 0 \;\; . & 0 & . \\ 0 & 0 & 0 \;\; 0 & \left(\vec{y}_2^{\prime n} \times \right) R \vec{y}_1^{\prime n} & \left(\vec{y}_2^{\prime n} \times \right) \vec{T} \end{pmatrix} \begin{pmatrix} \lambda_1^1 \\ \lambda_1^2 \\ . \\ . \\ \lambda_1^n \\ \gamma \end{pmatrix} = 0 \tag{7-66}$$

This matrix equation can be solved by finding the singular value decomposition and selecting the right singular vector corresponding to the smallest singular value. The more points observed in both image1 and image2, the better estimate of the depth and scale parameters. The depth of point j (λ_1^j) can the used to calculate the 3D coordinate of the point using the following equation:

$$\vec{m}^j = \lambda_1^j \vec{y}_1'^j \tag{7-67}$$

Note that the 3D point location is calculated with respect to the camera pose at epoch 1, which is assumed to be perfectly aligned with the world frame. In addition, recall that $\vec{y}_1'^j = K^{-1}\vec{y}_1^{\,j}$, where $\vec{y}_1$ is the pixel coordinate of the point m^j. This will give the full 3D coordinate of the point m^j in the camera frame in epoch 1, which is aligned with the world frame as assumed at the beginning of the section. It is important to remember that the estimated scale factor γ and depth λ_1^j are still scaled values (they do not reflect the true world. Instead, they are scaled versions of the true world). In subsequent epochs (e.g., $\left\langle R_{2\to3}, \vec{T}_{2\to3} \right\rangle$), if we encounter the same feature m^j again at epoch 3, we will have its world location (up to a scale γ). Knowing the 3D location of the feature m^j will facilitate the solution of $R_{2\to3}, \vec{T}_{2\to3}$ (because we already know the depth of m^j at epoch 2). This will make the solution $\left\langle R_{2\to3}, \vec{T}_{2\to3} \right\rangle$ (and more specifically, the translation $\vec{T}_{2\to3}$) scaled by the same unknown scale factor γ of the previous processing of epochs 1 and 2. When we know the camera pose $\left\langle R_{2\to3}, \vec{T}_{2\to3} \right\rangle$, a new feature m^l that is seen only in epochs 2 and 3 will be localized using the same unknown scale factor γ of the previous processing of epochs 1 and 2. This means that the same unknown scale factor will stay and propagate through epochs. However, due to the numerical errors and random noise, a drift will happen to this scale, and the solution will eventually drift. The main reason for this drift is that we process epochs pair by pair in an online fashion. We can apply batch optimization using bundle adjustment as we will see later in this chapter.

7.6 Optical Flow Estimation

Another important measurement that can be estimated from a sequence of images and used in positioning is the optical flow. Optical flow is the estimation of the velocity of a point in the image frame from a sequence of images. The goal is to predict the point location (in the image frame) in the next image frame to perform feature tracking. Feature tracking is important in building correct correspondences between features throughout the image sequence. In addition, in sensor fusion, the optical flow can be used as a measure of the speed of the camera if the scene is static.

7.6.1 Lucas–Kanade Method

One of the early and key methods that calculate the optical flow from image sequences is the Lucas–Kanade method [9]. The method assumes small displacement from image k to image $k+1$ will keep the intensity of the feature the same as shown in Figure 7.10. If we know the pixel velocity $\vec{v}_{j,k} = \left(v_{x,k}, v_{y,k}\right)^T$ of pixel point $q_{j,k}$, at image k, we expect that its intensity at pixel point $q'_{j,k+1}$ at image $k+1$ remains the same. This can be mathematically expressed as follows:

$$I\left(q_{j,k}\right) = I\left(q'_{j,k+1}\right) = I\left(q_{j,k} + \begin{pmatrix} \Delta x \\ \Delta y \end{pmatrix}\right) \tag{7-68}$$

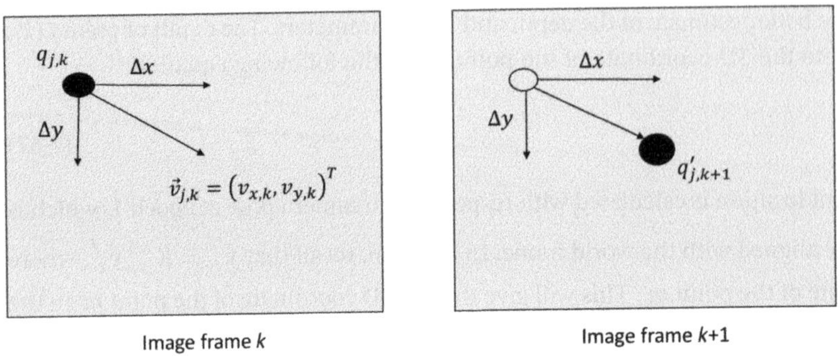

Figure 7.10 Illustration of optical flow.

Using Tylor series expansion, the intensity $I\left(q'_{j,k+1}\right)$ can be expressed as follows:

$$I\left(q'_{j,k+1}\right) = I\left(q_{j,k} + \begin{pmatrix}\Delta x\\ \Delta y\end{pmatrix}\right) \cong I\left(q_{j,k}\right) + \frac{\partial I\left(q_{j,k}\right)}{\partial x}\Delta x + \frac{\partial I\left(q_{j,k}\right)}{\partial y}\Delta y + \frac{\partial I\left(q_{j,k}\right)}{\partial t}\Delta T \tag{7-69}$$

as $I\left(q_{j,k}\right) = I\left(q'_{j,k+1}\right)$, we can easily observe that the following equation holds:

$$\frac{\partial I\left(q_{j,k}\right)}{\partial x}\Delta x + \frac{\partial I\left(q_{j,k}\right)}{\partial y}\Delta y + \frac{\partial I\left(q_{j,k}\right)}{\partial t}\Delta T = 0 \tag{7-70}$$

Noting that $v_{x,k} \cong \frac{\Delta x}{\Delta T}$ and $v_{y,k} \cong \frac{\Delta y}{\Delta T}$, the above equation can be written as follows:

$$\frac{\partial I\left(q_{j,k}\right)}{\partial x}v_{x,k} + \frac{\partial I\left(q_{j,k}\right)}{\partial y}v_{y,k} = -\frac{\partial I\left(q_{j,k}\right)}{\partial t} \tag{7-71}$$

Given a set of pixels $q_1, q_2, \ldots, q_n$ in a small window within image k, the following equation can be formed (we omitted the subscript k for simplicity of notation):

$$\begin{aligned}
\frac{\partial I(q_1)}{\partial x}v_x + \frac{\partial I(q_1)}{\partial y}v_y &= -\frac{\partial I(q_1)}{\partial t}\\
\frac{\partial I(q_2)}{\partial x}v_x + \frac{\partial I(q_2)}{\partial y}v_y &= -\frac{\partial I(q_2)}{\partial t}\\
&\vdots\\
\frac{\partial I(q_n)}{\partial x}v_x + \frac{\partial I(q_n)}{\partial y}v_y &= -\frac{\partial I(q_n)}{\partial t}
\end{aligned} \tag{7-72}$$

These equations can be written in matrix form as follows:

$$\begin{pmatrix}
\frac{\partial I(q_1)}{\partial x} & \frac{\partial I(q_1)}{\partial y}\\
\frac{\partial I(q_1)}{\partial x} & \frac{\partial I(q_2)}{\partial y}\\
\vdots & \vdots\\
\frac{\partial I(q_n)}{\partial x} & \frac{\partial I(q_n)}{\partial y}
\end{pmatrix}
\begin{pmatrix} v_x\\ v_y\end{pmatrix} =
\begin{pmatrix}
-\frac{\partial I(q_1)}{\partial t}\\
-\frac{\partial I(q_2)}{\partial t}\\
\vdots\\
-\frac{\partial I(q_n)}{\partial t}
\end{pmatrix} \tag{7-73}$$

which is a linear equation in the form $Av = b$. This equation can be solved using least-squares if A and b are known. We can calculate an approximate value of the partial derivatives $\frac{\partial I(q_1)}{\partial t}$, $\frac{\partial I(q_1)}{\partial x}$ and $\frac{\partial I(q_1)}{\partial y}$ as follows:

$$\frac{\partial I\left(q_j\right)}{\partial t} \cong \frac{I\left(q_{j,k+1}\right) - I\left(q_{j,k}\right)}{\Delta T} \tag{7-74}$$

$$\frac{\partial I\left(q_j\right)}{\partial x} = \frac{I\left(q_{j,k-1}\right) - I\left(q_{j,k-1} + (\Delta x\, 0)^T\right)}{\Delta x} \tag{7-75}$$

$$\frac{\partial I\left(q_j\right)}{\partial y} = \frac{I\left(q_{j,k-1}\right) - I\left(q_{j,k-1} + (0 \Delta y)^T\right)}{\Delta y} \tag{7-76}$$

As the derivatives are approximated by taking the difference of the intensity, the unit of the velocity will be pixels per frame interval, which can be converted into metric distance per unit time. When the scene is primarily static, and the camera mostly faces a plane (i.e., a downward-looking camera like a computer mouse or a camera mounted underneath an aerial vehicle), the optical flow of all pixels tends to be similar. Under this configuration, the average optical flow over all pixels can give a general estimate of the camera's velocity.

7.6.2 Relationship Between Optical Flow and Camera Motion

Given a feature point $P^j = \left(P^j_x\, P^j_y\, P^j_z\right)^T$ defined with respect to the camera frame, its 2D image metric coordinate $p^j = \left(p^j_x\, p^j_y\, 1\right)^T$ defined as follows:

$$p^j_x = \frac{f P^j_x}{P^j_z} \tag{7-77}$$

$$p^j_y = \frac{f P^j_y}{P^j_z} \tag{7-78}$$

where f is the focal length of the camera. When the camera moves at a 3D translational velocity $\vec{V}$ and 3D rotational velocity $\vec{\omega}$, the point p^j moves at a 2D velocity $v^j = \left(v^j_x\, v^j_y\right)^T$ defined as follows [9]:

$$v^j = \frac{f}{P^j_z}\begin{pmatrix} -1 & 0 & p^j_x \\ 0 & -1 & p^j_y \end{pmatrix}\vec{V} + \frac{1}{f}\begin{pmatrix} p^j_x p^j_y & -\left(f^2 + (p^j_x)^2\right) & f p^j_y \\ f^2 + \left(p^j_y\right)^2 & -p^j_x p^j_y & -f p^j_x \end{pmatrix}\vec{\omega} \tag{7-79}$$

The derivation of this formula is left as an exercise. The derivation starts from the assumption that the physical point P^j is moving and the camera is static (which is the same as if the camera is moving and the physical feature is static). The 3D feature velocity is related to the 3D camera motion as follows:

$$\left(\frac{dP^j_x}{dt}, \frac{dP^j_y}{dt}, \frac{dP^j_z}{dt}\right)^T = -\left(\left(\vec{\omega} \times \right)P^j + \vec{V}\right) \tag{7-80}$$

As we have the relationship between the 3D feature coordinate P^j and its corresponding 2D image coordinates p^j, we can derive the relationship between the optical flow (the time derivative of p^j) and the time derivative of P^j, which is linked to the camera motion (translational and rotational velocities) by equation (7-80). If we have several 3D features where their depths are known, their corresponding 2D image coordinates are known, and their optical flow is known, we can formulate the above equations into an optimization problem that can be solved using nonlinear optimization techniques to solve for camera translational and rotational (angular) velocities $\vec{V}$ and $\vec{\omega}$.

7.7 Stereo Camera System

The stereo camera system uses two cameras with known transformation to localize the features in 3D. Having the feature's 3D location resolves the depth and scale ambiguity problem. Below is an explanation of calculating depth and 3D location of features using a stereo camera system, followed by a discussion about depth, scale, and disparity values.

7.7.1 Estimating 3D Feature's Location from Stereo Vision

The stereo camera system consists of two camera sensors mounted at a known distance and commonly aligned with each other as seen in Figure 7.11. This means that the rotation and translation $\left\langle R, \vec{T} \right\rangle$ between the two camera sensors are known. If the two camera sensors see the same 3D world point, the 3D metric coordinate of the point can be estimated using the same method described in Section 7.5. In this case, the scale factor γ is known since $\vec{T}$ is fully known. We will assume that the world frame is aligned with camera 1 frame so the world coordinate of $\vec{m}^j$ is the same as its coordinate with respect to camera 1 frame. Therefore, the projection equation for point 3D point $\vec{m}^j = (X_j\, Y_j\, Z_j\, 1)^T$ from camera 1 to camera 2 is given by:

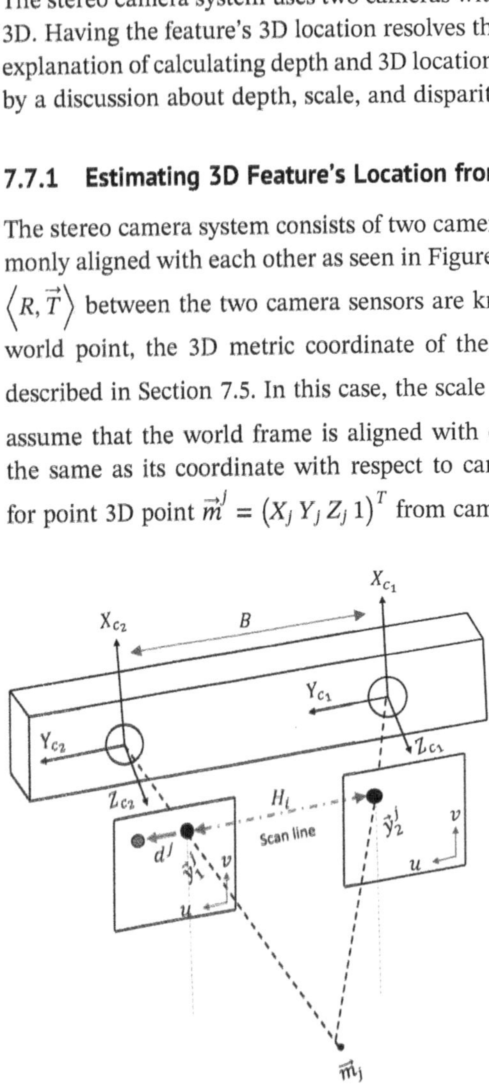

Figure 7.11 Stereo camera system.

$$\lambda_2^j \vec{y}_2^{\prime j} = \lambda_1^j R \vec{y}_1^{\prime j} + \vec{T} \tag{7-81}$$

where $\vec{y}_1 = \left(u_1^j, v_1^j\, 1\right)^T$ and $\vec{y}_2 = \left(u_2^j, v_2^j\, 1\right)^T$ are the pixel location of feature $\vec{m}^j$ in camera 1 and camera 2 image frames, respectively, $\vec{y}_1^{\prime j}$ and $\vec{y}_2^{\prime j}$ are the normalized world metric projection of the pixel-coordinates $\vec{y}_1$ and $\vec{y}_2$. Cross product both sides by $\vec{y}_2^{\prime j}$ by multiplying both sides by the skew matrix $\left(\vec{y}_2^{\prime j} \times\right)$ gives the following equation (note that cross product with itself gives zero):

$$0 = \lambda_1^j \left(\vec{y}_2^{\prime j} \times\right) R \vec{y}_1^{\prime j} + \left(\vec{y}_2^{\prime j} \times\right) \vec{T} \tag{7-82}$$

The only unknown in this equation is λ_1^j, which is the z-component of the 3D coordinate of the physical point $\vec{m}^j$. Substituting λ_1^j in the camera 1 projection equation gives the following equation:

$$\lambda_1^j \vec{y}_1 = K\vec{m}^j \tag{7-83}$$

The only unknown in Eq. (7-83) is the 3D point location, which can be calculated as follows:

$$\vec{m}^j = \lambda_1^j K^{-1} \vec{y}_1 \tag{7-84}$$

In stereo camera systems, the rotation matrix R is commonly I (the two camera frames are parallel) and the length of $\vec{T}$ is called the baseline B. If this is the case, and noting that the translation vector from camera 2 to camera 1 is $\vec{T} = (0 \quad B \quad 0)^T$, the 3D location of the feature can be calculated as follows:

$$Z_j \vec{y}_1 = K_1 \vec{m}^j \tag{7-85}$$

$$Z_j \vec{y}_2 = K_2 \vec{m}^j \tag{7-86}$$

where K_1 and K_2 are defined as follows:

$$K_1 = \begin{pmatrix} & 0 \\ K & 0 \\ & 0 \end{pmatrix} \tag{7-87}$$

$$K_2 = \begin{pmatrix} & 0 \\ K & B \\ & 0 \end{pmatrix} \tag{7-88}$$

where K is the intrinsics of camera 1 and camera 2, assuming that the two cameras are identical. The only unknowns in Eqs. (7-85) and (7-86) are the three coordinates of the feature point $\vec{m}^j = (X_j \; Y_j \; Z_j \; 1)^T$. By rearranging Eqs (7-85) and (7-86), the 3D feature's location can be calculated as follows:

$$\begin{pmatrix} -f_x & 0 & u_1 - c_x & 0 \\ 0 & -f_y & v_1 - c_y & 0 \\ -f_x & 0 & u_2 - c_x & 0 \\ 0 & -f_y & v_2 - c_y & B \end{pmatrix} \vec{m}^j = 0 \tag{7-89}$$

where f_x, f_y, c_x, and c_y are the metric calibration parameters of the cameras (entries of matrix K in metric units). $\vec{m}^j$ can be calculated by the smallest eigenvalue decomposition method noting that the resulting $\vec{m}^j$ will need a normalization to force the fourth component to be one to satisfy the 4D homogeneous coordinate condition.

7.7.2 Depth and Disparity Values

It can be noticed that the larger the baseline, the farther the 3D points that the system can localize accurately. However, if the baseline is too large, there is less possibility that the two cameras will see the same physical 3D point. If the baseline is small, there is a high possibility that the two cameras will see the same physical 3D point, but the range where the system can localize 3D points will be small. Figure 7.11 shows the scan line, which is the horizontal line connecting the pixel projection of the commonly seen feature $\vec{m}^j$ in the views of camera 1 and camera 2. Figure 7.11 also shows d^j which is the displacement that happened to the pixel location $\vec{y}_1^j$ due to the new pixel location $\vec{y}_2^j$ (within the same image or camera view) and H_i is the horizontal distance (through the scan line) between $\vec{y}_1^j$ and $\vec{y}_2^j$ (through the two images or cameras views). This displacement d^j is known as the disparity. The larger the depth, the smaller the disparity and the opposite is correct. From the triangle similarity rules, the disparity d_j is related to the base line B by the following equation:

$$\vec{m}^j = (X^j \; Y^j \; Z^j)^T \tag{7-90}$$

$$\frac{u_1^j}{f} = \frac{Y^j}{Z^j} \tag{7-91}$$

$$\frac{u_2^j}{f} = \frac{Y^j - B}{Z^j} \tag{7-92}$$

Combining the two equations gives the following useful relationship between the disparity d_j and the depth of the point $\vec{m}^j$:

$$u_1^j - u_2^j = d_j = \frac{Bf}{Z^j} \tag{7-93}$$

Note that Z^j is the depth of the point $\vec{m}^j$, which is the same for the cameras, which gives the following equality:

$$Z^j = \lambda_1^j = \lambda_2^j \tag{7-94}$$

Another approach to estimate the 3D point location in stereo camera configuration is by estimating the scale as follows: In the stereo camera configuration, the same feature $\vec{m}^i$ will be projected on the two cameras as follows:

$$\vec{y}^i_{k,l} = \begin{pmatrix} u^i_{k,l} \\ v^i_{k,l} \\ 1 \end{pmatrix} = \frac{1}{\lambda^i_{k,l}} K_{3\times 3} \left(R_{k,l} \; \vec{T}_{k,l} \right)_{3\times 4} \begin{pmatrix} m^i_x \\ m^i_y \\ m^i_z \\ 1 \end{pmatrix}_{4\times 1} \tag{7-95}$$

$$\vec{y}^i_{k,r} = \begin{pmatrix} u^i_{k,r} \\ v^i_{k,r} \\ 1 \end{pmatrix} = \frac{1}{\lambda^i_{k,r}} K_{3\times 3} \left(R_{k,r} \; \vec{T}_{k,r} \right)_{3\times 4} \begin{pmatrix} m^i_x \\ m^i_y \\ m^i_z \\ 1 \end{pmatrix}_{4\times 1} \tag{7-96}$$

where the subscript k, l and k, r refers to left and right camera image at epoch k, respectively. The commonly visible features $\vec{m}^i$ and the two views (left and right) can be used to estimate the scaled transformation between the left and the right camera $\left\langle \hat{R}_l^r \; \hat{\vec{T}}_l^r \right\rangle$, in the same way, we estimate the relative camera motion between two epochs k and $k+1$. Because we know the true values of $\left\langle R_l^r \vec{T}_l^r \right\rangle$, we can estimate the scale s as follows:

$$s = \frac{\hat{\vec{T}}_l^r}{\vec{T}_l^r} \tag{7-97}$$

Hence, we can calculate the depth and the world metric unscaled 3D location of the feature.

7.8 Camera Calibration

In the previous discussions, we assumed that the camera intrinsic matrix (K) is known. In most scenarios, intrinsics are provided by the manufacturer. However, it is important to understand how the camera calibration process is performed. Cameras can be calibrated in the lab using a special object called "calibration rig." Another approach is to just include the intrinsic matrix K in the unknowns of the problem of camera pose and 3D points estimation problem and try to estimate them all. The calibration procedure follows the same approach used in the camera pose and 3D points estimation, which starts with the projection equation:

$$\lambda_j \vec{y}_j = K\left(R\vec{m}^j + \vec{T}\right) \tag{7-98}$$

If we assume the camera frame is coincident with the world frame, we can set $R = I$ and $\vec{T} = 0$. This simplifies the equation to the following form:

$$\lambda_j \vec{y}_j = K\vec{m}^j \tag{7-99}$$

Cross product both sides by $\vec{y}_j$ gives the following equation:

$$0 = \left(\vec{y}_j \times\right) K\vec{m}^j \tag{7-100}$$

If the 3D physical point coordinate $\vec{m}^j$ is known and the pixel coordinate $\vec{y}_j$ is known, the only unknown in this equation will be the intrinsic matrix K, which can be estimated by a least-square algorithm. This can be done by separating the intrinsics in a separate vector as follows:

$$K = \begin{pmatrix} f_x & 0 & c_x \\ 0 & f_y & c_y \\ 0 & 0 & 1 \end{pmatrix} \tag{7-101}$$

$$\vec{k} = \begin{pmatrix} f_x & f_y & c_x & c_y \end{pmatrix}^T \tag{7-102}$$

and rearranging Eq. (7-100) into a linear system in the following form:

$$M\vec{k} = 0 \tag{7-103}$$

which is a simple linear system that can be solved using the least-squares algorithm or by applying singular value decomposition to the matrix A and choosing the vector $\vec{k}$ to be the right singular vector corresponding to the smallest singular value. Another approach is to optimize the following cost function:

$$J(K) = \left\| \lambda_j \vec{y}_j - K\vec{m}^j \right\|^2 \quad (7\text{-}104)$$

$$K^* = \arg\max_K J(K) \quad (7\text{-}105)$$

Note that the best solution K^* must satisfy the conditions of the intrinsic matrix. To implement this approach, we must have a set of known 3-D physical points whose locations are known with respect to the camera frame. This can be done by placing a calibration rig (like a checkerboard as seen in Figure 7.12) at a known distance from the camera. Knowing the dimensions of the checkerboard and its location with respect to the camera allows the calculation of the 3D coordinates of the corner features points of the checkerboard and their depths, assuming that the world frame is perfectly aligned with the camera frame. These corner points of the checkerboard are easy to detect using a corner detector.

One limitation of the above approach is that we must know the location of the calibration board with respect to the camera. We can develop a more general approach to calibrating the camera without knowing the location of the calibration rig with respect to the camera. In this approach, the world frame will originate at the first corner point in the checkerboard (axes X_b, Y_b, Z_b in Figure 7.12) and the camera poses with respect to this world frame (i.e., the extrinsic) will be included in the optimization problem and estimated together with the intrinsic parameters. To do so, we take multiple images of the calibration board from different angles and distances. Given a checker board with M corner feature, and given N images taken from different angles, the reprojection error of all features at all camera poses will be given by the following cost function:

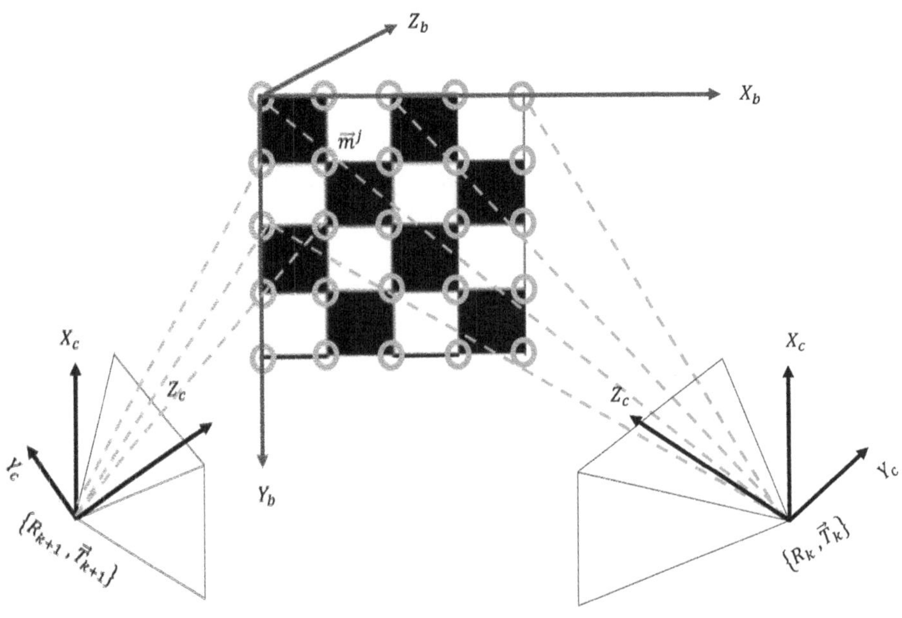

Figure 7.12 Camera calibration using checkerboard.

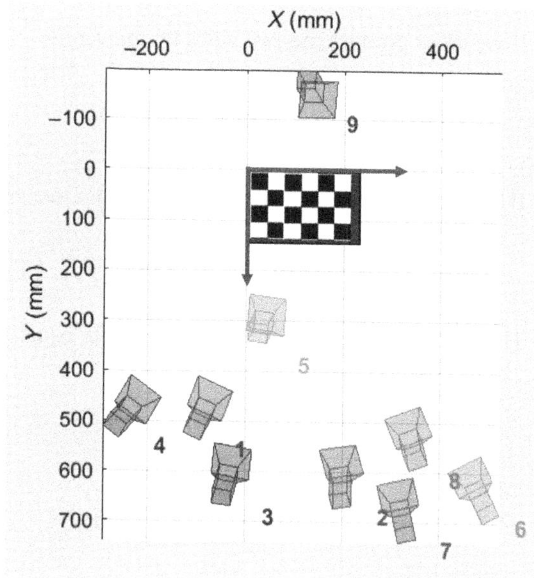

Figure 7.13 Visualization of camera calibration using MATLAB vision toolbox. *Source:* [8]/The MathWorks, Inc.

$$J\left(K, R_1 \ldots R_N, \vec{T}_1 \ldots \vec{T}_N, \lambda_{1,1} \ldots \lambda_{M,N}\right) = \sum_{k=1}^{N} \sum_{j=1}^{M} \left(\left\| \lambda_{j,k} \vec{y}_{j,k} - K\left(R_k \vec{m}^j + \vec{T}_k\right) \right\|^2 \right) \tag{7-106}$$

where $\left\{R_k, \vec{T}_k\right\}$ are the camera poses (transformation from world frame to camera frame) at image k,$\lambda_{j,k}$ is the depth of feature j at image k, $\vec{m}^j$ is the 3D location of feature j with respect to the world frame, and $\vec{y}_{j,k}$ is the pixel location of feature j at image k. The unknowns in Eq. (7-106) are the camera intrinsics K and camera poses $R_1 \ldots R_N, \vec{T}_1 \ldots \vec{T}_N$. There are available software tools that can perform the calibration automatically such as MATLAB camera calibration tool in [8]. These software tools include step-by-step instructions to calibrate a camera given a set of images taken by the camera to a checkerboard at different locations and orientations. Figure 7.13 shows the results of MATLAB camera calibration procedure applied on pre-prepared images in MATLAB computer vision toolbox. The figure shows the checkerboard and the camera poses (extrinsics). The tool generates the values of the camera intrinsics K and camera poses $R_1 \ldots R_N, \vec{T}_1 \ldots \vec{T}_N$. The optimization of the cost function J is performed by first parametrizing the problem using rotation vector instead of rotation matrices and applying any nonlinear optimization method such as Gauss–Newton method.

7.9 The Fundamental Matrix

Reviewing the projection equations of two views, we have the following equations:

$$\lambda_1 \vec{y}_1 = K\vec{m} \tag{7-107}$$

$$\lambda_2 \vec{y}_2 = K\left(R\vec{m} + \vec{T}\right) \tag{7-108}$$

$$\lambda_2 K^{-1} \vec{y}_2 = \lambda_1 R\left(K^{-1} \vec{y}_1\right) + \vec{T} \tag{7-109}$$

If we cross product the two sides by the vector $\frac{\vec{T}}{\lambda_1}$, we get

$$\frac{\lambda_2}{\lambda_1}\vec{T} \times K^{-1}\vec{y}_2 = \vec{T} \times R\left(K^{-1}\vec{y}_1\right) + 0 \tag{7-110}$$

If we dot-product both sides by $\left(K^{-1}\vec{y}_2\right)^T$, we get

$$\frac{\lambda_2}{\lambda_1}\left(K^{-1}\vec{y}_2\right)^T\left(\vec{T} \times K^{-1}\vec{y}_2\right) = \left(K^{-1}\vec{y}_2\right)^T\left(\vec{T} \times R\left(K^{-1}\vec{y}_1\right)\right) \tag{7-111}$$

This gives the popular epipolar constraints equation defined as follows:

$$\left(K^{-1}\vec{y}_2\right)^T\left(\vec{T} \times R\left(K^{-1}\vec{y}_1\right)\right) = 0 \tag{7-112}$$

which can be rearranged as follows:

$$\left(\vec{y}_2\right)^T(K^{-1})^T\left(\vec{T} \times R\left(K^{-1}\vec{y}_1\right)\right) = 0 \tag{7-113}$$

$$\left(\vec{y}_2\right)^T F\vec{y}_1 = 0 \tag{7-114}$$

where $F = (K^{-1})^T\left(\vec{T} \times \right)RK^{-1}$. This equation is like the epipolar constraint equation but with a different matrix F called the fundamental matrix. The fundamental matrix is the essential matrix shaped by the intrinsics matrix K as follows:

$$F = (K^{-1})^T EK^{-1} \tag{7-115}$$

The SVD of the fundamental matrix has three singular values, one of them is zero. The fundamental matrix is used to estimate camera pose and 3-D point coordinates in case of uncalibrated cameras (where K is unknown). However, this problem will require more than one view to be solved.

7.10 Enhanced Motion Estimation from More Than Two Images

If the 3D point feature $\vec{m}$ is observed in more than two images (m images), this can lead to a better estimate of camera pose and 3D coordinates of the point. So, in this subsection, we study the equations that model the projection through more than two images and the solution approach to estimate the camera pose and 3D coordinates of the feature points. The pixel coordinate $\vec{y}_i$ of the 3D feature point location $\vec{m}$ at image k is given by:

$$\lambda_k\vec{y}_k = \Pi_k\vec{m} \tag{7-116}$$

where Π_k matrix combines the calibration matrix K and the camera pose with respect to the world frame defined as follows:

$$\Pi_k = K\begin{pmatrix} R_0^k & \vec{T}_0^k \end{pmatrix} \tag{7-117}$$

where $R_0^k, \vec{T}_0^k$ are the rotation/translation between the camera frame at time i and the world frame. Combining all the images ($k = 1, 2 \ldots N$) in one equation gives us the following matrix equation:

$$\mathcal{I}\vec{\lambda} \equiv \begin{pmatrix} \vec{y}_1 & 0 & \cdots & 0 \\ 0 & \vec{y}_2 & 0 & 0 \\ \vdots & \vdots & \ddots & \vdots \\ 0 & 0 & \cdots & \vec{y}_N \end{pmatrix} \begin{pmatrix} \lambda_1 \\ \lambda_2 \\ \vdots \\ \lambda_N \end{pmatrix} = \begin{pmatrix} \Pi_1 \\ \Pi_2 \\ \vdots \\ \Pi_N \end{pmatrix} \vec{m} \equiv \Pi\vec{m} \tag{7-118}$$

$$\mathcal{I}\vec{\lambda} = \Pi\vec{m} \tag{7-119}$$

Recall the cross-product trick that we applied in the two-image projection equations. To apply a similar approach here, we will define the cross product in matrix form $\mathcal{I}^{\perp}$ as follows:

$$\mathcal{I}^{\perp} \equiv \begin{pmatrix} (\vec{y}_1 \times) & 0 & \cdots & 0 \\ 0 & (\vec{y}_2 \times) & \cdots & 0 \\ \vdots & \vdots & \ddots & \vdots \\ 0 & 0 & \cdots & (\vec{y}_N \times) \end{pmatrix} \in \mathbb{R}^{3N \times 3N} \tag{7-120}$$

where $\mathcal{I}^{\perp}\mathcal{I} = 0$. By multiply both sides by$\mathcal{I}^{\perp}$, we get the following equation:

$$\mathcal{I}^{\perp}\Pi\vec{m} = 0 \tag{7-121}$$

$$W_p\vec{m} = 0 \tag{7-122}$$

where $W_p \equiv \mathcal{I}^{\perp}\Pi$. It is clear that $\vec{m}$ is in the null space of the matrix W_p. In fact, if the feature point $\vec{m}$is visible in all images $k = 1, 2 \ldots N$, the matrix W_p will have a single null-vector so we can calculate $\vec{m}$ from this equation using the SVD method or by applying the singular value decomposition and choosing the 3D feature point location $\vec{m}$ to be the right singular vector corresponding to the smallest singular value. To calculate the camera poses, we use the cross-product trick by multiplying W_p by $\begin{pmatrix} (\vec{y}_1 \times) & \vec{y}_1 & 0 \\ 0 & 0 & 1 \end{pmatrix}$ to get the following matrix:

$$W_p \begin{pmatrix} (\vec{y}_1 \times) & \vec{y}_1 & 0 \\ 0 & 0 & 1 \end{pmatrix} = \begin{pmatrix} (\vec{y}_1 \times)(\vec{y}_1 \times) & 0 & 0 \\ (\vec{y}_2 \times) R_0^2 (\vec{y}_1 \times) & (\vec{y}_2 \times) R_0^2 \vec{y}_1 & (\vec{y}_2 \times) \vec{T}_0^2 \\ (\vec{y}_3 \times) R_0^3 (\vec{y}_1 \times) & (\vec{y}_3 \times) R_0^3 \vec{y}_1 & (\vec{y}_3 \times) \vec{T}_0^3 \\ \vdots & \vdots & \vdots \\ (\vec{y}_N \times) R_0^N (\vec{y}_1 \times) & (\vec{y}_N \times) R_0^N \vec{y}_1 & (\vec{y}_N \times) \vec{T}_0^N \end{pmatrix} \tag{7-123}$$

This matrix has entries like the epipolar constraint equation as follows:

$$M_p = \begin{pmatrix} \left(\vec{y}_2 \times\right) R_0^2 \vec{y}_1 & \left(\vec{y}_2 \times\right) \vec{T}_0^2 \\ \left(\vec{y}_3 \times\right) R_0^3 \vec{y}_1 & \left(\vec{y}_3 \times\right) \vec{T}_0^3 \\ \vdots & \vdots \\ \left(\vec{y}_N \times\right) R_0^N \vec{y}_1 & \left(\vec{y}_N \times\right) \vec{T}_0^N \end{pmatrix} \tag{7-124}$$

This matrix M_p is called the multiple-view matrix of the 3D feature point $\vec{m}$. We can show that the following equation holds:

$$\lambda_1 \left(\vec{y}_k \times\right) R_0^i \vec{y}_k + \left(\vec{y}_k \times\right) \vec{T}_0^i = 0, \quad k = 2, 3, \ldots . N \tag{7-125}$$

We can then estimate the depth parameter λ_1 as follows:

$$M_p \begin{pmatrix} \lambda_1 \\ 1 \end{pmatrix} = 0 \tag{7-126}$$

We can also see that the epipolar constraints for all points $\vec{y}_k$ against $\vec{y}_1$ holds:

$$\vec{y}_k^{\top} \left(\vec{T}_0^i \times\right) R_0^k \vec{y}_1 = 0, \quad k, l = 1, 2, 3 \ldots N \tag{7-127}$$

And the pairwise epipolar constraints for all pairs $\vec{y}_k$ and $\vec{y}_j$ are given as follows:

$$\vec{y}_k^{\top} \left(\vec{T}_l^k \times\right) R_l^k \vec{y}_l = 0, \quad k, l = 1, 2, 3 \ldots N \tag{7-128}$$

We can use the above epipolar constraints to estimate the essential matrix at any time. Another approach to solve for the camera poses is to benefit from the fact that the two columns of M_p matrix are linearly dependent as follows:

$$\left(\vec{y}_k \times\right) R_0^k \vec{y}_1 + \frac{1}{\lambda_1} \left(\vec{y}_k \times\right) \vec{T}_0^k = 0, \quad k = 2, 3 \ldots . N \tag{7-129}$$

If we have more than one features $\vec{m}^j$, where $j = 1, 2 \ldots M$, the above equation can be written in a matrix form as follows for feature $\vec{m}^j$:

$$\begin{pmatrix} \left(\vec{y}_{2,j} \times\right) R_0^{2j} \vec{y}_{1,j} \\ \left(\vec{y}_{3,j} \times\right) R_0^{3j} \vec{y}_{1,j} \\ \vdots \\ \left(\vec{y}_{N,j} \times\right) R_0^{Nj} \vec{y}_{1,j} \end{pmatrix} + \frac{1}{\lambda_1^j} \begin{pmatrix} \left(\vec{y}_{2,j} \times\right) \vec{T}_0^2 \\ \left(\vec{y}_{3,j} \times\right) \vec{T}_0^3 \\ \vdots \\ \left(\vec{y}_{N,j} \times\right) \vec{T}_0^N \end{pmatrix} = 0 \tag{7-130}$$

If we rearranged R_0^K into a vector form as follows:

$$\vec{r}_0^k = (r_{11}, r_{21}, r_{31}, r_{12}, r_{22}, r_{32}, r_{13}, r_{23}, r_{33})^T \in \mathbb{R}^9 \tag{7-131}$$

we can rewrite the above equation for time k and M feature points as follows:

$$\begin{pmatrix} \vec{y}_{1,1}^{\,T} \otimes \left(\vec{y}_{k,1}\times\right) & \frac{1}{\lambda_1^1}\left(\vec{y}_{k,1}\times\right) \\ \vec{y}_{1,2}^{\,T} \otimes \left(\vec{y}_{k,2}\times\right) & \frac{1}{\lambda_1^2}\left(\vec{y}_{k,2}\times\right) \\ \vdots & \vdots \\ \vec{y}_{1,M}^{\,T} \otimes \left(\vec{y}_{k,M}\times\right) & \frac{1}{\lambda_1^M}\left(\vec{y}_{k,M}\times\right) \end{pmatrix} \begin{pmatrix} \vec{r}_0^{\,k} \\ \vec{T}_0^{\,k} \end{pmatrix} = 0, k = 2, 3 \ldots N \tag{7-132}$$

where the operator $\otimes$ is the Kronecker product. Solving this matrix equation using the SVD method to calculate $R_0^k, \vec{T}_0^{\,k}$, which gives us the matrices Π_k. Similarly, if the camera poses are known, we can estimate the depths of all feature points at the first epoch using the following least-square solution formula:

$$\frac{1}{\lambda_1^j} = -\frac{\sum_{k=2}^{N}\left(\left(\vec{y}_k^{\,j}\times\right)\vec{T}_0^{\,k}\right)^{\mathrm{T}}\left(\vec{y}_{k,j}\times\right)R_0^k\vec{y}_{1,j}}{\sum_{k=2}^{N}\left\|\left(\vec{y}_{k,j}\times\right)\vec{T}_0^{\,k}\right\|^2}, \; j = 1, \ldots, M \tag{7-133}$$

We can solve iteratively for depths and camera poses recursively. This method gets a better estimate of the camera poses and 3-D point coordinates of the feature points.

7.11 Visual Odometry

We have seen how both camera's pose and 3-D point coordinates can be calculated from two or more image scans that observe common map feature points. The more common points that are visible and recognized by the image scans sequence, the more accurate the pose estimation. This process can be repeated through multiple scans and the estimated relative camera poses can be concatenated to estimate a full trajectory in a process known as visual odometry (VO) [7, 9]. The mathematics of visual odometry is described in the next subsection.

7.11.1 Problem Definition

If we have a single camera system (monocular), we will assume we are given an n-image sequence $I_{0:n} = \{I_0, \ldots, I_n\}$ and if we have stereo camera system (stereo), we will assume we are given n-image left-camera sequence $I_{l,0:n} = \{I_{l,0}, \ldots, I_{l,n}\}$ and n-image right-camera sequence $I_{r,0:n} = \{I_{r,0}, \ldots, I_{r,n}\}$, respectively. Given the cameras intrinsics K matrix, we want to calculate the camera trajectory $\{C_0, \ldots, C_n\}$, where C_k is a rotation/translation matrix that describes the camera frame position and orientation with respect to the world frame at time k. We will also assume that the world frame is the camera frame at time index $k = 0$. At time $k - 1$ and k, we have the following transformation matrices:

$$D_{k-1}^k = \begin{pmatrix} R_{k-1}^k & \vec{T}_{k-1}^{\,k} \\ 0 & 1 \end{pmatrix} \tag{7-134}$$

$$C_k = \begin{pmatrix} R_0^k & \vec{T}_0^k \\ 0 & 1 \end{pmatrix} \tag{7-135}$$

$$C_k = D_{k-1}^k C_{k-1} \tag{7-136}$$

where $R_{k-1}^k, \vec{T}_{k-1}^k$ are the relative rotation/translation from image frame $k-1$ to image frame k and $R_0^k, \vec{T}_0^k$ are the absolute rotation/translation of the camera with respect to the world frame (which is the camera frame at $k=0$). Keep in mind that C_k is a global pose description of the camera and D_{k-1}^k is relative pose between two images. Visual odometry aims at calculating all relative poses $D_0^1, D_1^2 \ldots D_{k-1}^k$ and concatenate them to calculate the camera pose at anytime C_k. We will assume that at each time instant, we have a set of features detected and matched between images. Given a 3D point m^j in the world frame, its 3D coordinates at the camera frame at time k is $\vec{X}_k^j = \begin{pmatrix} x^j & y^j & z^j \end{pmatrix}^T$ and its pixel projection on the image frame at time k is given by:

$$m^j = \vec{X}_0^j \tag{7-137}$$

$$\lambda_k \vec{y}_{k,j} = K\vec{X}_k^j \tag{7-138}$$

And the following transformations hold:

$$\vec{X}_k^j = R_0^k \vec{m}^j + \vec{T}_{0k}^k = C_k \vec{m}^j \tag{7-139}$$

$$\vec{X}_k^j = R_{k-1}^k \vec{X}_{k-1}^j + \vec{T}_{k-1,k} = D_{k-1}^k \vec{X}_{k-1}^j \tag{7-140}$$

As the intrinsic parameters matrix K is known, we will use the normalized metric coordinate of the pixel coordinate of the feature points as follows:

$$\vec{y}'_{k,j} = K^{-1} \vec{y}_{k,j} \tag{7-141}$$

7.11.2 Online Solution

In online visual odometry, we use pairs of epochs to estimate the camera motion and then concatenate the results through multiple epochs to estimate the whole trajectory. The epipolar equation between any two frames k and $k-1$ are given by:

$$\left(\vec{y}'_{k,j}\right)^T E_{k-1}^k \vec{y}'_{k-1,j} = 0 \tag{7-142}$$

from which we can estimate E_{k-1}^k from which we can estimate R_{k-1}^k and $\vec{T}_{k-1}^k$. If we have three image frames -2, $k-1$, and k, we will have E_{k-2}^{k-1} and E_{k-1}^k and we will have D_{k-2}^{k-1} and D_{k-1}^k. To concatenate D_{k-2}^{k-1} and D_{k-1}^k to C_{k-2} we use the following matrix multiplication equations:

$$C_{k-1} = D_{k-2}^{k-1} C_{k-2} \tag{7-143}$$

$$C_k = D_{k-1}^k C_{k-1} = D_{k-1}^k D_{k-2}^{k-1} C_{k-2} \tag{7-144}$$

This approach is called 2D to 2D visual odometry. If we have a stereo camera system, we can localize the features in 3D (with respect to the camera frames), and use the 3D locations of the features in epochs $k-1$, and k to find the R_{k-1}^k and $\vec{T}_{k-1}^k$ using the following cost function minimization:

$$J\left(R_{k-1}^{k}, \vec{T}_{k-1}^{k}\right) = \arg\min_{T_k} \frac{1}{2}\sum_i \| \vec{X}_k^j - \left(R_{k-1}^{k}\vec{X}_{k-1}^{i} + \vec{T}_{k-1}^{k}\right) \| \tag{7-145}$$

This optimization problem can be solved using the same least-square solution that we used in iterative closest point (ICP) algorithm (see Section 6.6 in Chapter 6 "Range-based Odometry"). The translation vector can be calculated as follows:

$$R_{k-1}^{k} = VU^{\top} \tag{7-146}$$

$$U\Sigma V^{T} = SVD\left(\left(\vec{X}_{k-1}^{M} - \vec{\mu}_{k-1}\right)\left(\vec{X}_{k}^{M} - \vec{\mu}_{k}\right)^{\top}\right) \tag{7-147}$$

where $\vec{X}_k^M$ is a matrix that contains all locations of all M features at epoch k and $\vec{\mu}_k$ is the mean or center of mass of all the M features at epoch k. To guarantee that R_{k-1}^{k} is a valid rotation matrix, the determinant of $V^{\top}$ must equal one. If it equals -1, we calculate R_{k-1}^{k} as follows:

$$R_{k-1}^{k} = V\begin{pmatrix} 1 & 0 & 0 \\ & 1 & 0 \\ 0 & 0 & \det\left(UV^{\top}\right) \end{pmatrix}U^{\top} \tag{7-148}$$

The translation vector $\vec{T}_{k-1}^{k}$ is then calculated as follows:

$$\vec{T}_{k-1}^{k} = \vec{\mu}_k - R_{k-1}^{k}\vec{\mu}_k \tag{7-149}$$

This solution is called 3D to 3D visual odometry and it requires a stereo system to obtain the 3D locations of the features so we can perform the above ICP step. Another approach to visual odometry is by using the pixel-coordinates reprojection error as a cost function. If at epoch $k-1$, the 3D location of the feature $\vec{X}_{k-1}^{i}$ is known (either from a stereo system or from a previous processing of pairs k–2 and k–1), the reprojection error is given by the following equation:

$$J\left(R_{k-1}^{k}, \vec{T}_{k-1}^{k}\right) = \arg\min_{T_k} \frac{1}{2}\sum_i \| \vec{y}_k^j - \vec{\hat{y}}_k^j \| \tag{7-150}$$

$$\vec{\hat{y}}_k^j = \frac{K}{\lambda_k^j}\left(R_{k-1}^{k}\vec{X}_{k-1}^{i} + \vec{T}_{k-1}^{k}\right) \tag{7-151}$$

where $\vec{\hat{y}}_k^j$ is the predicted pixel location of the feature at epoch k. Note that λ_k^j is the z-component of the vector $\left(R_{k-1}^{k}\vec{X}_{k-1}^{i} + \vec{T}_{k-1}^{k}\right)$ so for any transformation $R_{k-1}^{k}, \vec{T}_{k-1}^{k}$, we know the predicted λ_k^j value. This optimization function can be solved using a matrices manipulation approach or by non-linear optimization methods like the Gauss–Newton method.

7.11.2.1 Scale Normalization

If we processed pairs of monocular system epochs independently, we would have different scale every time. Assume the true translation in epoch pairs $\langle k-2, k-1\rangle$ and $\langle k-1, k\rangle$ are $\vec{T}_{k-2}^{k-1}$ and $\vec{T}_{k-1}^{k}$ and the estimated scaled translations are as follows:

$$\vec{\hat{T}}_{k-2}^{k-1} = s_1\vec{T}_{k-2}^{k-1} \tag{7-152}$$

$$\hat{\vec{T}}_{k-1}^{k} = s_2 \vec{T}_{k-1}^{k} \tag{7-153}$$

We cannot concatenate the pose of the two epochs before we make sure that both translations are scaled by the same scale as follows:

$$\hat{\vec{T}}_{k-2}^{k-1} = s_1 \vec{T}_{k-2}^{k-1} \tag{7-154}$$

$$\hat{\vec{T}}_{k-1}^{k} = s_1 \vec{T}_{k-1}^{k} \tag{7-155}$$

If we know the relative scale ration

$$s = \frac{s_1}{s_2} \tag{7-156}$$

we can rescale $\hat{\vec{T}}_{k-1}^{k}$ to make it $s_1 \vec{T}_{k-1}^{k}$ instead of $s_2 \vec{T}_{k-1}^{k}$ by multiplying $s_2 \vec{T}_{k-1}^{k}$ by s as follows:

$$s\left(s_2 \vec{T}_{k-1}^{k}\right) = \left(\frac{s_1}{s_2}\right) s_2 \vec{T}_{k-1}^{k} = s_1 \vec{T}_{k-1}^{k} \tag{7-157}$$

The relative scale s can be estimated from the relative distances between any two feature points i and j observed in the two epoch pairs $\langle k-2, k-1\rangle$ and $\langle k-1, k\rangle$ as follows (see [10]):

$$s = \frac{\| \vec{X}_{k-1}^{i} - \vec{X}_{k-1}^{j} \|}{\| \vec{X}_{k}^{i} - \vec{X}_{k}^{j} \|} \tag{7-158}$$

The calculated translation vectors $\vec{T}_{k-2}^{k-1}$ and $\vec{T}_{k-1}^{k}$ are then scaled by s before doing the concatenation. This step is important to unify the scale along the whole trajectory. In some scenarios (like the case in stereo system or an environment with known location features), the 3D world coordinates of the feature points $\vec{X}_{k-1}^{j}$ is known. In this case, we can use this additional information to refine the estimation of the relative rotation R_{k-1}^{k} and translation $\vec{T}_{k-1}^{k}$ as follows:

$$\left\langle \hat{R}_{k-1}^{k}, \hat{\vec{T}}_{k-1}^{k} \right\rangle = \underset{R_{k-1}^{k}, \vec{T}_{k-1}^{k}}{\arg\min} \sum_{j} \| \vec{y}_{k,j} - K\left(R_{k-1}^{k} \vec{X}_{k-1}^{j} + \vec{T}_{k-1}^{k}\right) \|^2 \tag{7-159}$$

The estimated camera poses can then be used to calculate the 3D world points coordinates. We can use this feature to perform visual simultaneous localization and mapping (V-SLAM). We will explain the SLAM problem and solution methods in more detail in the next chapter. SLAM is a key problem in robotics that has evolved significantly. Therefore, we will dedicate a separate chapter to explain this evolution and current state-of-art SMAL methods that use both ranging and imaging sensors to perform positioning, navigation, and mapping.

7.11.3 Bundle Adjustment

The online solution described above processes pairs of epochs at a time, which is not ideal. Estimating the camera pose and/or 3D world feature points from two-image sequences or three-image sequences or a few numbers of images where a certain feature is observed is called "local" or "online" processing and it provides simple and closed-form analytical solution. However, due to the limited information in the local online processing scheme, the effect of noise is strong, which leads to poor estimation accuracy especially if we have degenerative cases where the matrices are ill-formed (e.g., noninvertible, or unstable SVD). Local online estimation provides good starting initial condition for more advanced estimation schemes where a batch or a window of images is processed in one estimation problem. If more images can be processed at once, this will lead to a more robust and accurate estimation. Bundle Adjustment refers to the batch processing of a window or a buffer of images instead of pairs of images or a few numbers of images. In our discussion, we will refer to the 3-D to 2-D projection at time (or frame) k by the following form:

$$\vec{y}^{j}(k) = \pi_k\left(\vec{m}^{j}\right) \tag{7-160}$$

where $\vec{m}^{j}$ is the 3D world point and $\vec{y}^{j}$ is the pixel-coordinate of the point in the camera image frame and π_k is the projection matrix at frame k, which includes both the intrinsic matrix K, the extrinsics $R_0^k, \vec{T}_0^k$, and the depth Z_k^j defined as follows:

$$\pi_k\left(\vec{m}^{j}\right) = \frac{1}{Z_k^j} K \begin{pmatrix} R_0^k & \vec{T}_0^k \end{pmatrix} \vec{m}^{j} \tag{7-161}$$

Note here that $R_0^k, \vec{T}_0^k$ are absolute not relative, which means that $R_0^k, \vec{T}_0^k$ transform from the world point m^j (represented in world coordinate frame referred to as frame 0) to camera frame at time k. Due to sensor imperfection and noise, we get a noisy pixel measurement of $\vec{y}^{j}$, we will refer to it by $\vec{\tilde{y}}^{j}$. Given M feature points, and N images, the reprojection (between frame k and $k+1$) error function (a function in the camera pose $R_0^k, \vec{T}_0^k$ and the world points $\vec{m}^{j}$) is given by the following cost function equation:

$$J\left(R_1, \vec{T}_1 \ldots R_N, \vec{T}_N, \vec{m}^{1}, \ldots, \vec{m}^{M}\right) = \sum_{k=0}^{N} \sum_{j=1}^{M} c(k,j) \left\| \vec{\tilde{y}}^{j}(k) - \pi_k\left(\vec{m}^{j}\right) \right\|^2 \tag{7-162}$$

$$c(k,j) = \begin{cases} 1 & \text{if feature j appears in image frame } k \\ 0 & \text{otherwise} \end{cases} \tag{7-163}$$

This cost function formulates the visual SLAM problem. If we would like to estimate camera motion only (finding odometry solution), we can remove the 3D world feature point from the formulation. In this case, we will formulate the problem using relative poses $R_k^{k+1}, \vec{T}_k^{k+1}$ instead of absolute poses. In this case, the cost function will be defined as follows:

$$J\left(R_0^1, \vec{T}_0^1, R_1^2, \vec{T}_1^2 \ldots\ldots R_{N-1}^N, \vec{T}_{N-1}^N\right) = \sum_{k=0}^{N} \sum_{j=1}^{M} c(k,j) \left\| \vec{\tilde{y}}^{j}(k+1) - \pi_{k+1,k}\left(\vec{m}^{j}\right) \right\|^2 \tag{7-164}$$

$$c(k,j) = \begin{cases} 1 & \text{if feature } j \text{ appears in image frame } k \text{ and } k+1 \\ 0 & \text{otherwise} \end{cases} \tag{7-165}$$

$$\pi_{k+1,k}\left(\vec{m}^{j}\right) = \frac{1}{Z_{k+1}^j} K\left(R_k^{k+1}\left(K^{-1}\vec{\tilde{y}}^{j}(k)\right) + \vec{T}_k^{k+1}\right) \tag{7-166}$$

The operator $\pi_{k+1,k}\left(\vec{m}^{j}\right)$ is the pixel reprojection of the pixel point $\vec{\tilde{y}}^{j}(k+1)$ using the corresponding pixel point at frame k ($\vec{\tilde{y}}^{j}(k)$). This reprojection is done by converting the pixel point $\vec{\tilde{y}}^{j}(k)$ to a normalized metric 3D point in the camera frame k, which is obtained by the expression

$K^{-1}\vec{y}^{j}(k)$. This normalized metric 3D point in the camera frame k is transformed to camera frame at time $k+1$ using $\left\langle R_k^{k+1}, \vec{T}_k^{k+1} \right\rangle$ and then converted a to pixel point at time $k+1$ by multiplying by K and dividing by Z_{k+1}^j. This cost function still depends on the depths of the 3D world point, but these depths should not affect the minimization of the cost function. Minimizing cost function over camera relative poses $R_k^{k+1}, \vec{T}_k^{k+1}$ will generate all relative camera poses which then can be concatenated to establish a full trajectory of the camera.

Whether we optimize for camera poses only (visual Odometry) or poses and 3D environment feature points (visual SLAM), the optimization of the cost function $J(.)$ in Eq. (7-164) is not a simple task. $J(.)$ in Eq. (7-164) is a complex nonlinear optimization function that may have multiple local minima and ambiguous solutions where different combinations of camera-pose and world feature-points may lead to the same cost value. Starting from an arbitrary initial solution may get us stuck in a local minimum, which leads to a wrong solution. However, if we have a good initial solution, solving this optimization will be faster and more accurate. In practice, this initial solution is calculated from the local online processing (e.g., pairs of images) using the closed-form solution formulas as discussed earlier. Starting from this initial solution, optimization algorithms are applied iteratively to enhance this solution by optimizing or minimizing the cost function over a window of images.

Another key issue in bundle adjustment is the rotation parametrization. The cost function defined in (7-162) and (7-164) uses rotation matrices to describe the rotation. Optimization functions need the first and second derivative (Jacobian and Hessian) of the cost function with respect to the parameters of interest, including the rotation. Representing the rotation as a matrix complicates obtaining the Jacobian and/or the Hessian. In addition, the matrix representation is a Lie Group where only multiplication and inverse are defined. Lie Group is not a linear Euclidean space. We need to transform this matrix representation to a linear Euclidean space to apply iterative optimization algorithms. This can be done by representing the rotation matrix by the rotation vector. In this way, the derivative will be with respect to only three rotation parameters and the same for the rotation error. We discussed in Chapter 1 how to perform this mapping using exponential mapping.

This concludes the discussion about bundle adjustment. You can see $\vec{y}^{j}(k)$ is the actual measured quantity and $\pi_k\left(\vec{m}^{j}\right)$ as the predicted measurement and our objective is to bring the predicted measurement close to the observed measurement. This is the basic concept of estimation, optimization, and machine learning methods. We build a model where the parameters of the model are the quantities that we are interested in calculating and we keep changing these parameters until the model output fits the actual observed measurements. We then verify the accuracy of our model and our parameters by predicting future measurements and comparing them with actual observations. If the model is correct, and the parameters are accurate, our prediction should match the observed measurements. The trick is to change the parameters wisely and efficiently. Wisely means in the right direction. Efficiently means with minimum computation resources (processing and memory). Some of the common nonlinear optimization algorithms are given in the following list:

- Newton methods.
- Gauss–Newton algorithm.
- Levenberg–Marquardt algorithm.

These optimization methods are described in Chapter 3 "Sensor Fusion Methods and Algorithms"

7.11.4 Outlier Detection/Rejection

A final note about the accuracy is the need to detect and reject outlier measurements. In imaging sensor processing, in addition to random noises and potential glitches, outlier measurements can happen due to the mismatching between feature points, which leads to wrong correspondence. To detect and reject wrong correspondences, statistical algorithms such as RANSAC are commonly used. We already explained the RANSAC algorithm in Chapter 6. RANSAC in imaging sensors can be applied in the same way it is applied in ranging sensors. We iteratively take a random sample of features and process them to estimate rotation and translation. We then apply this estimated transformation to other features. The features that generate a small projection error under a threshold are considered inliers. The process is repeated and the transformation that has the maximum number of inliers is selected as the best solution.

7.11.5 MATLAB Visual Odometry Example

The vision toolbox in MATLAB R2023b comes with many built-in functions that implement most of the described algorithms in this chapter, which facilitate the study and understanding of visual navigation methods. A good example that demonstrates camera pose estimation, visual odometry, and bundle adjustment can be found in the MATLAB vision toolbox under the name "*monocular visual odometry*" which can be downloaded from [11]. Sample results of monocular visual odometry from this MATLAB example can be seen in Figure 7.14. Readers can download the example code and trace the steps of the processing. The example code uses key built-in MATLAB functions for important algorithms such as features detection "*detectSURFFeatures,*" features matching "*matchFeatures,*" "*estimateEssentialMatrix,*" "*bundleAdjustment,*" and "*triangulateMultiview.*" The example is a good practice to get familiar with several aspects of visual navigation.

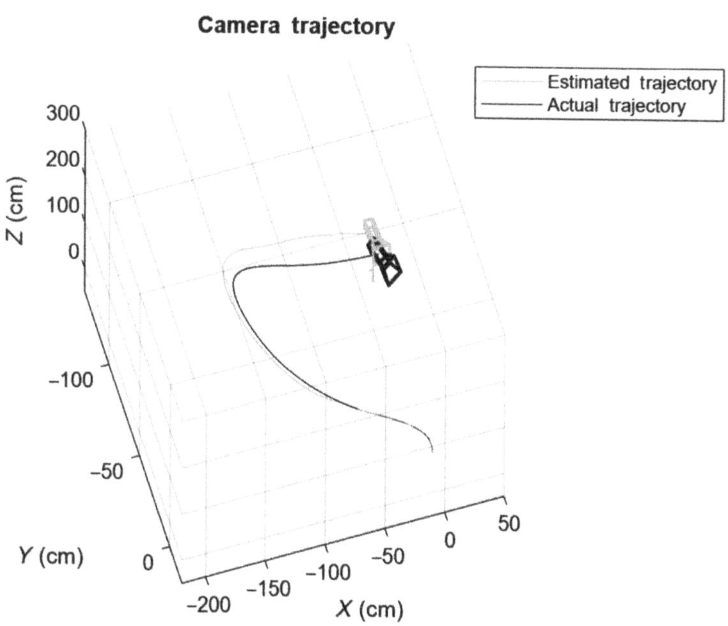

Figure 7.14 Sample results of MATLAB example "monocular visual odometry."Example code can be downloaded and tested on MATLAB environment. *Source:* The MathWorks, Inc. / https://www.mathworks.com/help/vision/ug/monocular-visual-odometry.html / Last accessed on March 29, 2024.

7.12 Problems

7.12.1 Problem 1

A pinhole camera with the following calibration parameters:

$$f_x = 714.18\ \text{(pixels)}$$

$$f_y = 710.38\ \text{(pixels)}$$

$$c_x = 563.65$$

$$c_y = 355.72$$

The following 3D world point was observed by the camera:

$$X = \begin{pmatrix} 29 \\ 116 \\ 0 \end{pmatrix}$$

If the camera pose is described by the following rotation and translation vectors:

$$\vec{\phi}_w^c = \begin{pmatrix} -0.6096 \\ -0.1789 \\ -0.3835 \end{pmatrix}$$

$$\vec{T}_w^c = \begin{pmatrix} -146.0516 \\ -26.8684 \\ 797.9027 \end{pmatrix}$$

A) Using the rotation vector $\vec{\phi}_w^c$ and Rodrigues' Formula, calculate the rotation matrix R_w^c, where w and c indicates that the transformation is done from world frame to camera frame.

B) Using the rotation matrix R_w^c and rotation vector $\vec{T}_w^c$, form the 4×4 transformation matrix T_w^c that transforms 4D homogeneous world point coordinates to 4D homogeneous camera frame coordinates.

C) Use T_w^c to transform the given 3D world point X from world frame to camera frame coordinate X^c.

Using the given calibration parameters, form the 3×3 calibration matrix K.

D) Calculate $\hat{p}$, the pixel coordinates of the world point X.

E) If the true pixel coordinate of the point is as follows:

$$p = \begin{pmatrix} 493.97 \\ 404.84 \end{pmatrix}\ \text{(pixels)}$$

Calculate the projection error $\|p - \hat{p}\|$

7.12.2 Problem 2

The absolute pose of the pinhole camera of Problem 1 was measured twice at time epochs k and $k+1$ given by the following rotations and translations:

$$R_w^{c_k} = \begin{pmatrix} 0.9145 & 0.4014 & -0.0513 \\ -0.2972 & 0.7523 & 0.5879 \\ 0.2745 & -0.5224 & 0.8073 \end{pmatrix}, \vec{T}_w^{c_k} = \begin{pmatrix} -146.0516 \\ -26.8684 \\ 797.9027 \end{pmatrix}$$

$$R_w^{c_{k+1}} = \begin{pmatrix} 0.9769 & -0.1434 & -0.1584 \\ 0.2125 & 0.7290 & 0.6507 \\ 0.0222 & -0.6693 & 0.7426 \end{pmatrix}, \vec{T}_w^{c_{k+1}} = \begin{pmatrix} -209.4357 \\ -59.4563 \\ 921.8198 \end{pmatrix}$$

where w refers to world frame and c_k refers to camera frame at time epoch k. Find the rotation $R_{c_k}^{c_{k+1}}$ and translation $\vec{T}_{c_k}^{c_{k+1}}$ that describes the relative transformation from camera frame at time epoch k to camera frame at epoch $k+1$.

7.12.3 Problem 3

Given the following essential matrix of two camera poses:

$$E = \begin{pmatrix} 1.6557 & 2.7745 & 0.3231 \\ -2.3728 & 2.3821 & -2.2493 \\ 0.9940 & 1.9652 & 0.1099 \end{pmatrix}$$

Calculate the four possible 3D rotations and translations for this essential matrix. You may use MATLAB or any math library to calculate the singular value decomposition of E.

7.12.4 Problem 4

Given the following fundamental matrix:

$$F = \begin{pmatrix} 0.0000034002 & 0.0000058933 & -0.0033300812 \\ -0.0000050263 & 0.0000051214 & -0.0024064228 \\ 0.0012185891 & -0.0020279018 & 0.9999887611 \end{pmatrix}$$

Use the calibration parameters given in Problem 1 to calculate the essential matrix corresponding to this fundamental matrix.

References

1 Y. Ma, S. Soatto, J. Košecká and S. S. Sastry, "*An Invitation to 3-D Vision, From Images to Geometric Models*," Springer - Science+Business Media, LLC, 2004.

2 R. Jain and R. Kasturi, "*Machine Vision*," McGraw Hill, Inc., 1995.

3 K. Wang, L. Zhao and R. Li, "Fisheye omnidirectional camera calibration — Pinhole or spherical model?," in *2014 IEEE International Conference on Robotics and Biomimetics (ROBIO 2014)*, Bali, Indonesia, 2014.

4 H. Li, H. Yu, J. Wang, W. Yang, L. Yu and S. Scherer, "ULSD: Unified Line Segment Detection Across Pinhole, Fisheye, and Spherical Cameras," *ISPRS Journal of Photogrammetry and Remote Sensing*, vol. 178, no. August 2021, pp. 187–202.
5 B. Chai, "On-Site Global Calibration of Mobile Vision Measurement System Based on Virtual Omnidirectional Camera Model," *Remote Sensing*, vol. 13, no. 10, p. 1982, 2021.
6 N. B. Monteiro and J. A. Gaspar, "Generalized Camera Array Model for Standard Plenoptic Cameras," in *Robot 2019: Fourth Iberian Robotics Conference*, 2019.
7 R. Hartley, "*Multiple View Geometry in Computer Vision*," New York: Cambridge University Press, 2003.
8 MATLAB, "Camera Calibration," MathWorks Inc., 2023. [Online]. Available: https://www.mathworks.com/help/vision/ug/camera-calibration.html [Accessed 4 August 2023].
9 A. Wedel and D. Cremers, "*Stereo Scene Flow for 3D Motion Analysis*," Springer-Verlag, London, 2011.
10 D. Scaramuzza and F. Fraundorfer, "Visual Odometry," *IEEE Robotics and Automation Magazine*, vol. 18, no. 4, pp. 80–92, 2011.
11 MATLAB, "Monocular Visual Odometry," MathWorks Inc., 2023. [Online]. Available: https://www.mathworks.com/help/vision/ug/monocular-visual-odometry.html. [Accessed 4 February 2024].

8

Mapping Algorithms

8.1 Introduction

A map can be defined as localized features that describe an environment. In the simplest form, the localized features could be simple points. Localized feature points are commonly known as point clouds. In more advanced forms, points can be integrated into higher-level features such as lines and planes. In more advanced maps, complex features can be integrated into polygons, shapes, and objects [1]. There are different forms of map representations and advanced 3D models of maps that are used in geographic information systems (GIS) [2, 3]. However, on the low level, they are all based on localized points. In this chapter, map-building algorithms that are concerned with localizing feature points (i.e., point clouds) in an environment are considered.

Mapping can be performed using active ranging sensors or passive imaging or a combination of both sensors that are carried by a moving platform. If the measurements and the pose (position and orientation) of the ranging sensor or the imaging sensor are accurately and precisely known, the location of the feature points in the environment can be accurately determined, and a map of the environment can be built. However, in practice, even if we can have a highly accurate and precise platform, ranging and imaging sensors have several challenges and measurement errors that make mapping challenging. Conventional mapping systems use a very accurate high-end positioning system, which usually consists of a high-end inertial measurement unit (IMU) and a carrier-phase. Differential global navigation satellite system (GNSS)-based integrated positioning system. The mapping platform can be any moving platform equipped with proper sensors and algorithms. This can be a ground vehicle, an aerial vehicle, a person, or a space vehicle. For example, satellites equipped with long-range RADAR (Radio Detection and Ranging) and LiDAR (Light Detection and Ranging) with accurate, precise orbital information can be used to build global maps of earth. A popular example of this type of mapping is Google Earth. If the pose of the mapping platform is not known or known with low accuracy, simultaneous localization and mapping (SLAM) [4] is performed. SLAM can also be applied if the sensor's positioning is accurately and precisely known to enhance the map's accuracy. In this chapter, we will discuss the two approaches, mapping and SLAM.

In this chapter, we will use a simple 2D mapping vehicle platform to explain the basic concepts of mapping and SLAM. This simple 2D mapping vehicle is illustrated in Figure 8.1. The vector $(x_k\ y_k\ \psi_k)^T$ is the vehicle's pose at time epoch k consists of x-coordinate x_k, y-coordinate y_k, and heading or yaw angle ψ_k. The vehicle is controlled by two control signals: the vehicle's forward speed v_k and the vehicle's turn rate ω_k at time epoch k. The vehicle is equipped with a ranging sensor that provides range r_k and angle θ_k measurements at time epoch k.

Sensor Fusion Approaches for Positioning, Navigation, and Mapping: How Autonomous Vehicles and Robots Navigate in the Real World: With MATLAB Examples, First Edition. Mohamed M. Atia.
© 2025 The Institute of Electrical and Electronics Engineers, Inc. Published 2025 by John Wiley & Sons, Inc.

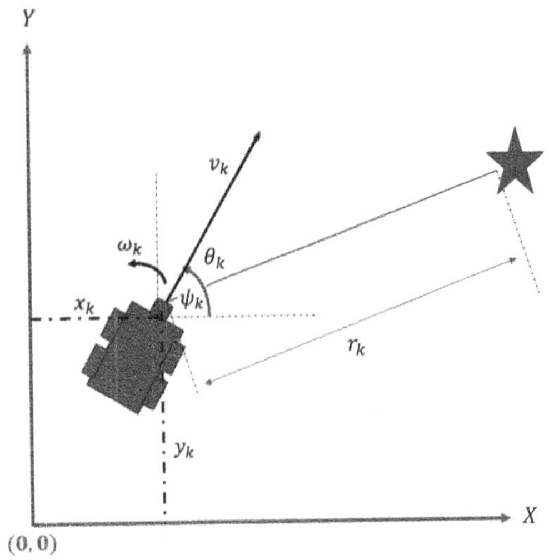

Figure 8.1 Simple 2D mapping vehicle.

8.2 Occupancy Grid Mapping

Occupancy grid is one of the simplest and most efficient mapping methods where the environment is partitioned into grid cells, and each cell is assigned a probability of being occupied by an object or obstacle. Occupancy grid mapping can be done using ranging sensors such as RADAR, LiDAR, and SONAR to provide physical range and orientation to target points in the environment. Given the pose of the mapping platform, range and orientation measurements can be converted into a 2D or 3D point cloud. Occupancy maps can also be built using imaging sensors if the scale ambiguity is resolved and the physical 2D or 3D location of the feature points is estimated. In both cases, the occupancy grid requires range and angle to the feature points, so in our discussion, we will consider a ranging sensor, but the same concepts apply with imaging sensors if range and angle to the feature point are calculated (e.g., from stereo images or from image sequence with known motion). For simplicity, we will assume a 2D ranging sensor and a flat horizontal environment like an office area or a room. Assume the area is partitioned into grid cells and each cell has a location $\vec{m}^j = (x^j\, y^j)^T$ with respect to a reference coordinate frame as shown in Figure 8.2. Note that $\vec{m}^j$ does not have time epoch subscript k as the map are assumed to be static. This reference coordinate frame is usually taken as the initial pose of the moving platform in the beginning of the motion as seen in Figure 8.2. Each cell has an occupancy probability $p\left(\vec{m}^j\right) \in [0, 1]$ where $p\left(\vec{m}^j\right) = 1$ means that we are certain that there is an object or obstacle at this cell and $p\left(\vec{m}^j\right) = 0$ means we are certain this cell is free. $p\left(\vec{m}^j\right) = 0.5$ means we are not sure about the status of cell $\vec{m}^j$. Any occupancy probability value between 0 and 1 expresses our belief about the occupancy status of this cell. The reason to model the occupancy as a probability is that the raging sensors always have random errors.

For any given angle/range measurement $\vec{y}_k = (\theta_k\, r_k)^T$ at time epoch k, the estimated location of the feature point $\vec{m}^j = (\hat{x}^j\, \hat{y}^j)^T$ is given as follows:

$$\hat{x}^j = x_k + r_k \cos(\theta_k + \psi_k) \tag{8-1}$$

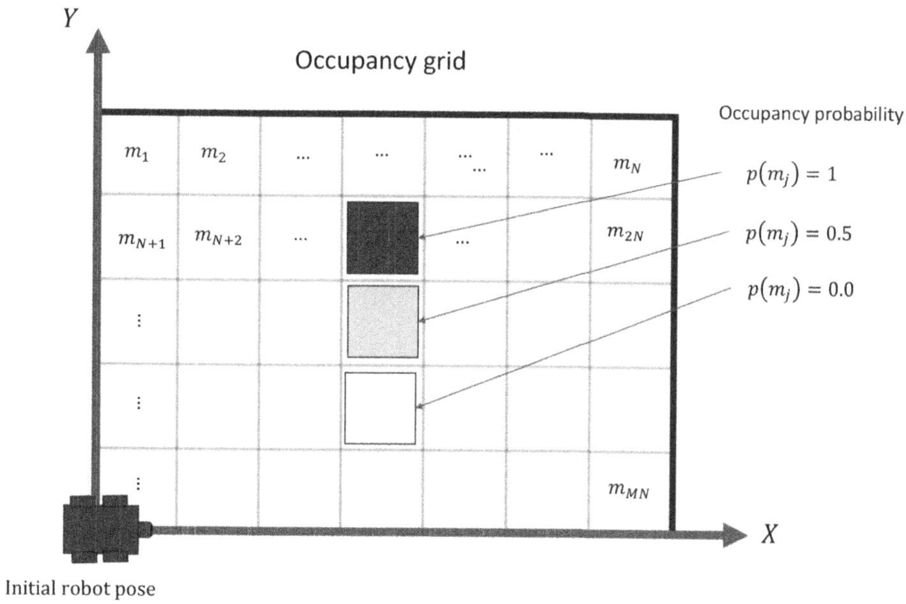

Figure 8.2 Occupancy grid and occupancy probability.

$$\hat{y}^j = y_k + r_k \sin(\theta_k + \psi_k) \tag{8-2}$$

Assuming the ranging sensor has a range/angular Gaussian errors δr and $\delta\theta$, there will be a region of uncertainty about the location of the feature point. This uncertainty region is illustrated in Figure 8.3. The probability of having an occupied cell at location $\vec{m}^j$ is as 2-D Gaussian distribution centered around the feature's location $\vec{\hat{m}}^j = (\hat{x}^j\ \hat{y}^j)^T$. Given angle/range measurement $\vec{y}_k = (\theta_k\ r_k)^T$ at epoch k with angular error $\delta\theta$ and range error δr, we can divide the environment into three regions, A, B,and C as shown in Figure 8.4. Figure 8.4 clearly shows the regions A, and B.

Figure 8.3 Relationship between occupancy probability and range/angle accuracy.

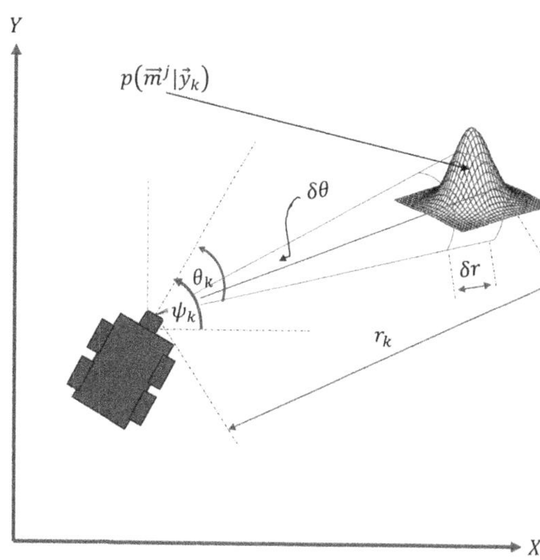

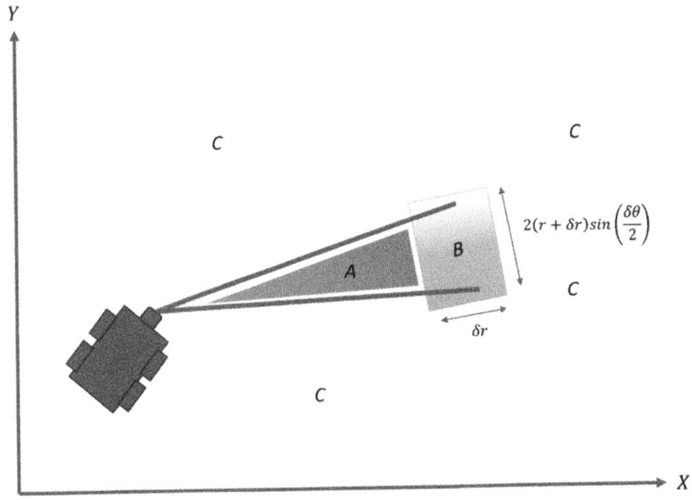

Figure 8.4 Measurement beam partitions the environment into three regions A, B, and C.

Region C is simply anywhere else (all cells outside the reach of the sensor beam). The occupancy likelihood (also known as the inverse measurement likelihood) of the cell $\vec{m}^j$ given the measurement $\vec{y}_k$ can be obtained by the following Gaussian probability equation:

$$p\left(\vec{m}^j \mid \vec{y}_k\right) = \frac{1}{\sqrt{(2\pi)^n}\sqrt{|\Sigma|}} e^{-\frac{1}{2}\left(\vec{m}^j - \hat{\vec{m}}^j\right)^T \Sigma^{-1}\left(\vec{m}^j - \hat{\vec{m}}^j\right)} \tag{8-3}$$

where Σ is covariance matrix, which depends on the ranging error δr and the angular resolution error $\delta\theta$ and n is the dimension of $\vec{m}^j$. Greater errors in range and angle measurements lead to greater covariance matrix Σ.

We will use the following expressions to describe the occupancy probability:

$p\left(\vec{m}^j\right)$: probability that cell $\vec{m}^j$ is occupaied

$p\left(\neg\vec{m}^j\right)$: probability that cell $\vec{m}^j$ is free or empty

Initially, the probability of occupancy $p\left(\vec{m}^j\right)$ equals probability of vacancy $p\left(\neg\vec{m}^j\right)$:

$$p\left(\vec{m}^j\right) = p\left(\neg\vec{m}^j\right) = 0.5 \tag{8-4}$$

From probability axioms, the following equation holds:

$$p\left(\vec{m}^j\right) + p\left(\neg\vec{m}^j\right) = 1 \tag{8-5}$$

Given range/angle measurement $\vec{y}_k = (r_k\ \theta_k)^T$ with range/angular errors δr and $\delta\theta$, we can construct the regions A, B,and C and calculate the occupancy likelihood of the cells of each region using Eq. (8-3). Now, we have two probabilities, the very initial occupancy probability $p\left(\vec{m}^j\right)$ and the occupancy likelihood $p\left(\vec{m}^j \mid \vec{y}_k\right)$. We want to combine these two pieces of information to estimate the posterior occupancy probability $p\left(\vec{m}^j \mid \vec{y}_{1:k}\right)$. The posterior probability $p\left(\vec{m}^j \mid \vec{y}_{1:k}\right)$ is the probability that cell $\vec{m}^j$ is occupied given all measurements from the beginning of the time to the current time epoch k. Using Bayes' rule, and keeping in mind that the measurements are independent, and the state at time epoch k depends only on the state at epoch $k-1$, the posterior $p\left(\vec{m}^j \mid \vec{y}_{1:k}\right)$ can be calculated as follows:

$$p\left(\vec{m}^j \mid \vec{y}_{1:k}\right) = \frac{p\left(\vec{y}_k \mid \vec{m}^j\right) p\left(\vec{m}^j \mid \vec{y}_{1:k-1}\right)}{p\left(\vec{y}_k\right)} \tag{8-6}$$

The measurement likelihood $p\left(\vec{y}_k \mid \vec{m}^j\right)$ can be calculated using Bayes' rule as follows:

$$p\left(\vec{y}_k \mid \vec{m}^j\right) = \frac{p\left(\vec{m}^j \mid \vec{y}_k\right) p\left(\vec{y}_k\right)}{p\left(\vec{m}^j\right)} \tag{8-7}$$

Combining Eqs. (8-6) and (8-7), we get the following equation:

$$p\left(\vec{m}^j \mid \vec{y}_{1:k}\right) = \frac{p\left(\vec{m}^j \mid \vec{y}_k\right) p\left(\vec{m}^j \mid \vec{y}_{1:k-1}\right)}{p\left(\vec{m}^j\right)} \tag{8-8}$$

Equation (8-8) calculates the posterior occupancy probability using the inverse measurement model $p\left(\vec{m}^j \mid \vec{y}_k\right)$ which is calculated by Eq. (8-3), the previous occupancy probability $p\left(\vec{m}^j \mid \vec{y}_{1:k-1}\right)$, and the very initial occupancy probability $p\left(\vec{m}^j\right)$. Due to the small values of the probability, this method is numerically unstable. A more numerically stable update formula that depends on ratios between probability is more feasible. We can derive a similar formula like (8-8) for the vacancy probability $p(\neg m_j \mid y_{1:k})$ as follows [2, 4]:

$$p\left(\neg\vec{m}^j \mid \vec{y}_{1:k}\right) = 1 - p\left(\vec{m}^j \mid \vec{y}_{1:k}\right) = \frac{p\left(\neg\vec{m}^j \mid \vec{y}_k\right) p\left(\neg\vec{m}^j \mid \vec{y}_{1:k-1}\right)}{p\left(\neg\vec{m}^j\right)} \tag{8-9}$$

By dividing (8-8) over (8-9), we get a ratio of occupancy over vacancy probabilities given by the following equation noting that $p\left(\neg\vec{m}^j \mid \vec{y}_{1:k}\right) = 1 - p\left(\vec{m}^j \mid \vec{y}_{1:k}\right)$ [1–5]:

$$\frac{p\left(\vec{m}^j \mid \vec{y}_{1:k}\right)}{1 - p\left(\vec{m}^j \mid \vec{y}_{1:k}\right)} = \frac{p\left(\vec{m}^j \mid \vec{y}_k\right) p\left(\neg\vec{m}^j\right) p\left(\vec{m}^j \mid \vec{y}_{1:k-1}\right)}{p\left(\neg\vec{m}^j \mid \vec{y}_k\right) p\left(\vec{m}^j\right) p\left(\neg\vec{m}^j \mid \vec{y}_{1:k-1}\right)} \tag{8-10}$$

Equation (8-10) can be rearranged as follows:

$$\frac{p\left(\vec{m}^j \mid \vec{y}_{1:k}\right)}{1 - p\left(\vec{m}^j \mid \vec{y}_{1:k}\right)} = \frac{p\left(\vec{m}^j \mid \vec{y}_k\right)}{1 - p\left(\vec{m}^j \mid \vec{y}_k\right)} \frac{1 - p\left(\vec{m}^j\right)}{p\left(\vec{m}^j\right)} \frac{p\left(\vec{m}^j \mid \vec{y}_{1:k-1}\right)}{1 - p\left(\vec{m}^j \mid \vec{y}_{1:k-1}\right)} \tag{8-11}$$

Defining the function $\text{logit}(x) = \frac{x}{1-x}$, we get the following recursive formula:

$$\text{logit}\left(p\left(\vec{m}^j \mid \vec{y}_{1:k}\right)\right) = \text{logit}\left(p\left(\vec{m}^j \mid \vec{y}_k\right)\right) + \text{logit}\left(p\left(\vec{m}^j \mid \vec{y}_{1:k-1}\right)\right) - \text{logit}\left(p\left(\vec{m}^j\right)\right) \quad (8\text{-}12)$$

Initially, $p\left(\vec{m}^j \mid \vec{y}_{1:k-1}\right) = p\left(\vec{m}^j\right)$. Once we have a measurement $\vec{y}_k$, the posterior probability $p\left(\vec{m}^j \mid \vec{y}_{1:k}\right)$ is calculated as a function in three terms, the inverse measurement likelihood $p\left(\vec{m}^j \mid \vec{y}_k\right)$, the prior occupancy probability $p\left(\vec{m}^j \mid \vec{y}_{1:k-1}\right)$, and the very initial probability $p(m_j)$. The inverse measurement likelihood $p\left(\vec{m}^j \mid \vec{y}_k\right)$ is the probability that the cell $\vec{m}^j$ is occupied given the measurement $\vec{y}_k$, which can be calculate using the following Gaussian equation:

$$p\left(\vec{m}^j \mid \vec{y}_k\right) = \begin{cases} 0 & \vec{m}^j \in \text{region } A \\ \frac{1}{\sqrt{(2\pi)^n}\sqrt{|\Sigma|}} e^{-\frac{1}{2}\left(\vec{m}^j - \hat{\vec{m}}^j\right)^T \Sigma^{-1}\left(\vec{m}^j - \hat{\vec{m}}^j\right)} & \vec{m}^j \in \text{region } B \\ 0.5 & \vec{m}^j \in \text{region } C \end{cases} \quad (8\text{-}13)$$

We have all variables to calculate the posterior occupancy probability $p\left(\vec{m}^j \mid \vec{y}_{1:k}\right)$ recursively using Eq. (8-12). The more the robot moves around in a certain area, the more the measurements become available and the more accurate the occupancy map will be in this portion. Although we described the occupancy grid mapping in 2D example, it can be applied and implemented in 3D case as well, where the grid will be 3D matrix, and the cell will be a cube instead of a square.

8.3 MATLAB Occupancy Grid Building Example

An example of 2D occupancy grid mapping is shown in Figures 8.5–8.7. These figures have been generated by a MATLAB project from the projects' repository of this book under the name "*Test-OccupancyMapProject*." The project uses simulated noisy LiDAR scans and the built-in MATLAB function *buildMap* that takes LiDAR scans and poses and builds an occupancy map. In this project, we used the ground truth poses as the objective is to test the mapping part only. Figure 8.5 shows the ground-truth robot poses, the true map of the area, and a sample of the LiDAR scan at a specific location/orientation. Figure 8.6 shows a sample of the LiDAR scan projected as a point cloud in the LiDAR frame, which demonstrates the noisy scans. Figure 8.7 shows the resulting occupancy map. The white area is free with a high probability that is near zero, the dark area is occupied with a high probability that approaches one, and the gray area is an unknown area, so it remains at the 0.5 probability level. Readers can test the demo project by changing the noise of the LiDAR scan and observing the effect on the resulting map.

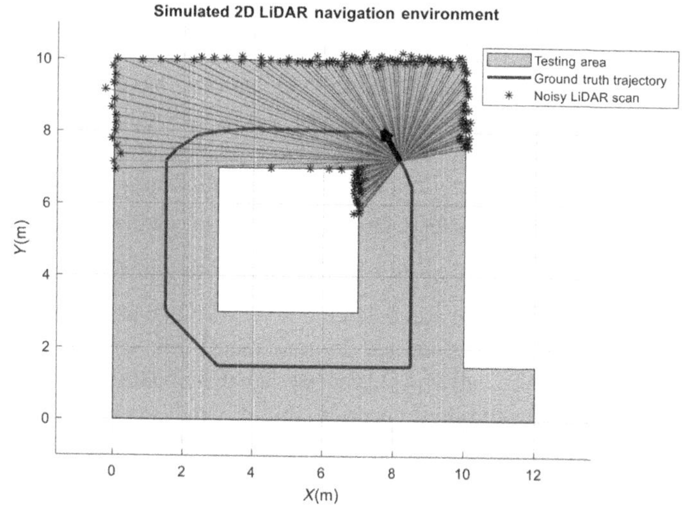

Figure 8.5 Example robot ground truth trajectory and map.

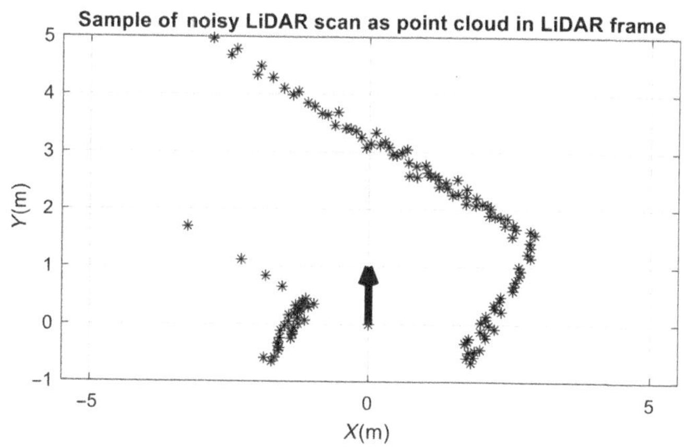

Figure 8.6 Sample of noisy LiDAR scan measurements.

Figure 8.7 LiDAR-based occupancy grid.

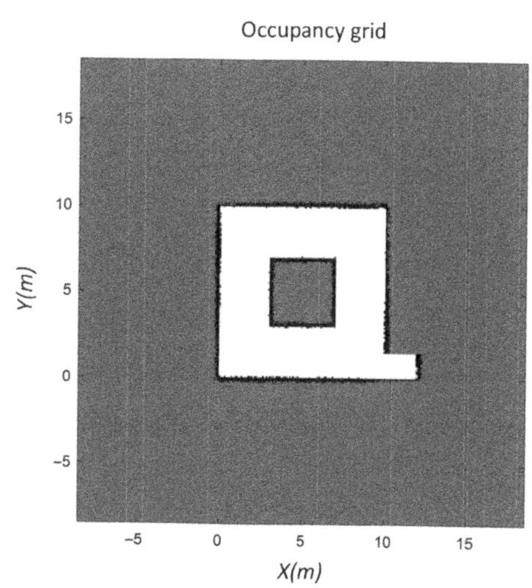

8.4 SLAM Problem and Solution Approaches

In the SLAM problem, we want to estimate both the pose of the moving platform that carries the sensors and the map of the environment. We will discuss four popular SLAM algorithms that are commonly used in the literature. The SLAM problem can be defined as follows: given the platform motion model, the sensor measurement model, static environment feature points, sensor measurements $\vec{y}_k$, and the control signal u_k that triggers the platform motion, we want to calculate $\vec{x}_k$, the platform pose, and a map $m = \left(\vec{m}^1\ \vec{m}^2 \ldots \vec{m}^M\right)^T$, where $\vec{m}^j$ is a feature point location. The SLAM problem can be approached in two modes: online mode and batch mode (also known as offline or postprocessing) mode. The online mode estimates the pose and the map recursively, considering the current and previous time epochs' data mainly. Full SLAM processes a batch of data of several time epochs at once. Online mode is fast but not the most accurate. Full or batch mode is more accurate as it performs global optimization, but it is slower and requires larger storage as it processes more data. In this section, we will discuss both types of SLAM algorithms and frameworks.

8.4.1 EKF-SLAM

EKF-SLAM is an online SLAM algorithm that uses the extended Kalman filtering technique to estimate both the location and the map. Therefore, before reading this subsection, it is recommended that you review the extended Kalman filter (EKF) basics from Chapter 3. The system state consists of the platform pose and the location of the environment feature points (i.e., the map). The platform pose includes the platform's location and orientation. The velocity of the platform can be added as well. Assuming a 2D environment, the state vector at epoch k is defined as $\vec{b}_k$ as follows:

$$\vec{b}_k = \begin{pmatrix} \vec{x}_k \\ m \end{pmatrix} \tag{8-14}$$

where $\vec{x}_k$ is the platform pose defined as follows:

$$\vec{x}_k = \begin{pmatrix} x_k \\ y_k \\ \psi_k \end{pmatrix} \tag{8-15}$$

and m is M features' locations defined as follows:

$$m = \begin{pmatrix} \vec{m}^1 \\ \vec{m}^2 \\ \vdots \\ \vec{m}^M \end{pmatrix} \tag{8-16}$$

The platform location is defined by the vector $(x_k\ y_k)^T$ and orientation is defined by the angle ψ_k, and the vector $\vec{m}^j = \left(m_x^j\ m_y^j\right)^T$ is the coordinate of map point $\vec{m}^j$. Note that the platform location is a function of time because it changes with time, while the map points are not function of time as the environment is assumed to be static. The state vector described above is modeled as a multi-variate Gaussian random variable with the following mean and covariance:

$$\vec{\mu}_k = \begin{pmatrix} \vec{\mu}_{\vec{x}_k} \\ \mu_m \end{pmatrix}, \Sigma_k = \begin{pmatrix} \Sigma_{\vec{x}_k} & \Sigma_{\vec{x}_{k,m}} \\ \Sigma_{\vec{x}_{k,m}} & \Sigma_m \end{pmatrix} \tag{8-17}$$

where $\vec{\mu}_{\vec{x}_k}$ is the mean of the platform's pose and $\vec{\mu}_m$ is the mean of the map features' location. Pose error is assumed to be zero-mean Gaussian as follows:

$$\varepsilon_t \sim \mathcal{N}(0, Q_k) \tag{8-18}$$

Given the 2-D platform example with a simple range/angle sensor assembly, this simple robot is controlled by forward speed v and angular speed (steering) ω. The control vector of this simple 2-D vehicle is given by:

$$\vec{u}_k = \begin{pmatrix} v_k \\ \omega_k \end{pmatrix} \tag{8-19}$$

where v_k is the vehicle's forward speed and ω_k is the vehicle's turn rate at time epoch k. Given the platform state at epoch $k-1$, the state at epoch k is predicted by the following system model (motion model):

$$\vec{x}_k = \begin{pmatrix} x_k \\ y_k \\ \psi_k \end{pmatrix} = f\left(\vec{x}_{k-1}, \vec{u}_k\right) + \varepsilon_k = \begin{pmatrix} x_{k-1} + v_k \cos\psi_{k-1}\Delta T \\ y_{k-1} + v_k \sin\psi_{k-1}\Delta T \\ \psi_{k-1} + \omega_k \Delta T \end{pmatrix} + \varepsilon_k \tag{8-20}$$

where ΔT is the epoch period (sampling time). The predictive probability distribution of $\vec{x}_k$ is given by:

$$p\left(\vec{x}_k\right) \sim \mathcal{N}\left(f\left(\vec{x}_{k-1}, u_k\right), Q_k\right) \tag{8-21}$$

When we get an angle/range measurement $\vec{y}_k^j = \left(\theta_k^j\, r_k^j\right)^T$ from a map feature point $\vec{m}^j = \left(m_x^j\, m_y^j\right)^T$, the relationship between this measurement, the feature location, and the platform pose is given by the following measurement model:

$$\vec{y}_k^j = \begin{pmatrix} \theta_k^j \\ r_k^j \end{pmatrix} = h\left(\vec{b}_k\right) = h\left(\vec{x}_k, \vec{m}^j\right) = \begin{pmatrix} \tan^{-1}\left(\dfrac{m_y^j - y_k}{m_x^j - x_k}\right) - \psi_k \\ \sqrt{\left(m_x^j - x_k\right)^2 + \left(m_y^j - y_k\right)^2} \end{pmatrix} + \delta_k \tag{8-22}$$

where δ_t is the measurement noise modeled as Gaussian as follows:

$$\delta_k \sim N(0, R_k) \tag{8-23}$$

The measurement likelihood of $\vec{m}^j$ is given by the following Gaussian equation:

$$p\left(\vec{y}_k^j \mid \vec{x}_k, \vec{m}^j\right) \sim \mathcal{N}\left(h\left(\vec{x}_k, \vec{m}^j\right), R_k\right) \tag{8-24}$$

Our prior knowledge about the platform's pose can be expressed in the following mean and covariance:

$$\vec{\mu}_{\vec{x}_0} = (0 \quad 0 \quad 0)^T, \Sigma_{\vec{x}_0} = \begin{pmatrix} \varepsilon & 0 & 0 \\ 0 & \varepsilon & 0 \\ 0 & 0 & \varepsilon \end{pmatrix} \tag{8-25}$$

where ε is a small value to express our confidence about the initial platform pose. Our prior knowledge about the map features is given by the following mean and covariance:

$$\mu_m = ([0 \quad 0] \ldots [0 \quad 0])^T$$
$$\Sigma_m = \begin{pmatrix} \beta & 0 & 0 \\ 0 & \ddots & 0 \\ 0 & 0 & \beta \end{pmatrix} \tag{8-26}$$

where β is a large value as we do not have any prior knowledge about the locations of the map features yet. In the very beginning, the robot scans the area and initializes the visible features' locations using the following equation:

$$\vec{\mu}_{\vec{m}^j} = \begin{pmatrix} x_k + r_k^j \cos\left(\theta_k^j + \psi_k\right) \\ y_k + r_k^j \sin\left(\theta_k^j + \psi_k\right) \end{pmatrix} \tag{8-27}$$

The above formula is called the inverse measurement model. We will refer to the inverse measurement model of feature j by $h^{j^{-1}}$ as follows:

$$\vec{\mu}_{\vec{m}^j} = h^{j^{-1}}\left(\vec{y}_k^j, \vec{x}_k\right) \tag{8-28}$$

When the robot moves, we perform the prediction step at which we predict the next pose of the robot using the motion model as follows:

$$\vec{\mu}_{\vec{x}_k} = f\left(\vec{\mu}_{\vec{x}_{k-1}}, \vec{u}_k\right) \tag{8-29}$$

The predicted covariance is given by the following formula (review the propagation of gaussian signal through a linear system in Chapter 3):

$$\Sigma_{\vec{x}_k} = F_k \Sigma_{\vec{x}_{k-1}} F_k{}^T + Q_k \tag{8-30}$$

where F_k is the Jacobian of the system model given by the following first-order derivative formula:

$$F_k = \frac{\partial}{\partial \vec{x}_k} f\left(\vec{x}_k, \vec{u}_k\right)\Bigg|_{\vec{x}_k = \vec{\mu}_{\vec{x}_k}} \tag{8-31}$$

$$= \begin{pmatrix} 1 & 0 & -v \sin \psi_k \Delta T \\ 0 & 1 & v \cos \psi_k \Delta T \\ 0 & 0 & 1 \end{pmatrix} \tag{8-32}$$

Once we have a measurement from feature $\vec{m}^j$, we perform an update to the whole state vector (which includes both the platform pose and the locations of the features) using the following update step:

$$H_k^j = \frac{\partial}{\partial\left(\vec{x}_k\,\vec{m}^j\right)^T} h\left(\vec{x}_k, \vec{m}^j\right)\Bigg|_{\vec{x}_k = \vec{\mu}_{\vec{x}_k}, \vec{m}^j = \vec{\mu}_{\vec{m}^j}} \tag{8-33}$$

$$K_k^j = \Sigma_k \left(H_k^j\right)^T \left(H_k^j \Sigma_k \left(H_k^j\right)^T + R_k\right)^{-1} \tag{8-34}$$

$$\mu_{\vec{b}_k} = \mu_{\vec{b}_k} + K_k^j\left(\vec{y}_k - h^j\left(\vec{\mu}_{\vec{x}_k}\right)\right) \tag{8-35}$$

$$\Sigma_k = \left(I - K_k H_k^j\right)\Sigma_k \tag{8-36}$$

In the update step, we calculate the design matrix H_k^j and Kalman Gain K_k^j, and then we updated the mean $\vec{\mu}_k$ and covariance Σ_k of the whole state that includes both the robot's pose and the map features. Given the measurement model in Eq. (8-22), the measurement design matrix H_k^j is the Jacobian of the measurement model calculated as follows:

$$\frac{\partial}{\partial x_t} h^j\left(\vec{x}_k, \vec{m}^j\right) = \begin{pmatrix} \frac{\Delta y_k^j}{\left(r_k^j\right)^2} & \frac{-\Delta x_k^j}{\left(r_k^j\right)^2} & -1 & 0\ldots0 & \frac{-\Delta y_k^j}{\left(r_k^j\right)^2} & \frac{\Delta x_k^j}{\left(r_k^j\right)^2} & 0\ldots0 \\ \frac{-\Delta x_k^j}{r_k^j} & \frac{-\Delta y_k^j}{r_k^j} & 0 & 0\ldots0 & \frac{\Delta x_k^j}{r_k^j} & \frac{\Delta y_k^j}{r_k^j} & 0\ldots0 \end{pmatrix} \tag{8-37}$$

$$r_k^j = \sqrt{\left(m_x^j - x_k\right)^2 + \left(m_y^j - y_k\right)^2} \tag{8-38}$$

$$\Delta x_k^j = \left(m_x^j - x_k\right) \tag{8-39}$$

$$\Delta y_k^j = \left(m_y^j - y_k\right) \tag{8-40}$$

If we have M features, the dimension of H_k^j will be 2 (3 + 2M). The three columns are the platform pose entries, and the 2M columns are the number of features times the feature coordinate dimension, which is 2 in our 2D example.

Once corrected, the corrected robot pose is used to scan the environment to localize new features that were not previously visible by the robot. As the robot moves and the EKF-SLAM algorithm continues, the robot's pose and existing features' locations will be corrected, and new features will be localized and so on until the whole environment is scanned and mapped. Keep in mind that we assume that all features are unique and can be identified through epochs without ambiguity, so the robot knows which feature is which through multiple scans and epochs. In practice, this assumption is not always available, and the robot must build correspondences between scans to track and identify a feature from one scan to another and through the entire map. In the case of range sensors like LiDAR, correspondence can be estimated using ICP or NDT algorithms (review Chapter 6 for more information about ICP and NDT algorithms).

Although EKF-SLAM is simple, it comes with several challenges. We will discuss a few of them here. One challenge is linearization. In this 2D example, the measurement model is simple and therefore, the linearization is relatively simple. In practice, the measurement model can be more complex, like the case of visual sensors, where measurements are pixel locations that are mapped to physical feature points coordinate through complex projection matrices. Another challenge is the distribution of map features. The measurement update will correct and improve both the robot pose and the map features' locations. To get the best accuracy, features must be distributed throughout the area and not concentrated in one place or one direction; otherwise, it will lead to numerical problems and matrix singularities. In addition, features must not be too close to the platform to avoid linearization problems. The diversity of features' location affects the accuracy of the EKF estimation accuracy in the same way the geometry of satellites affects the accuracy of GPS positioning. Another challenge with EKF-SLAM is the computational complexity. EKF-SLAM is efficient if the map is small. If the map is large, the matrix sizes will grow significantly, the linearization process will not be accurate, and the inverse operation will be inefficient, time consuming, and inaccurate, leading to significant errors. This is one of the major limitations of EKF-SLAM. More information about EKF-SLAM can be found in [5].

8.4.2 Rao–Blackwellized Fiter-SLAM

To avoid matrix inversion issues and linearization complexities and inaccuracies, particle filter-based SLAM (PF-SLAM) was proposed. Readers are encouraged to review the basics of particle filtering from Chapter 3. In particle filtering, the state $\vec{b}_k$ is represented as a probability distribution that is approximated by N particles that form a particle cloud $\left\{\vec{b}_{k-1}^{[1]}, \vec{b}_{k-1}^{[2]}, \ldots, \vec{b}_{k-1}^{[N]}\right\}$. We then apply the prediction step and the update step. In the prediction step, given the noisy input $\vec{u}_k$, each particle in the particle cloud goes through the system model and a new predicted particle $\left\{\vec{b}_k^{[1]}, \vec{b}_k^{[2]}, \ldots, \vec{b}_k^{[N]}\right\}$ is obtained. Since the map features are static, and their locations can be inferred from the platform's pose $\vec{x}_k$ using the inverse measurement model in Eq. (8-27), our state will keep only a cloud of poses $\left\{\vec{x}_k^{[1]}, \vec{x}_k^{[2]}, \ldots, \vec{x}_k^{[N]}\right\}$ and a set of map features $\vec{m} = \left(\vec{m}^1\ \vec{m}^2 \ldots \vec{m}^M\right)^T$. In other words, there is no particles for each map feature m^j. This approach is known as Rao-Blackwellized Particle Filter [6], and it has the advantage of saving huge space as representing each map features by a cloud of particles will consume large memory space. Given the noisy input $\vec{u}_k$, we apply the system model on the platform's pose as follows:

$$\vec{x}_k = f\left(\vec{x}_{k-1}, u_k\right) + \varepsilon_k \tag{8-41}$$

where ε_k is a zero-mean noise with covariance Q_k (system noise covariance). The new particle cloud approximates the predicted probability $p\left(\vec{x}_k | \vec{x}_{k-1}, \vec{u}_k\right)$. So far, the weight of all particles is equal and normalized. Once measurements $\vec{y}_k^j = \left(\theta_k^j\ r_k^j\right)^T$ become available, we check if the feature $\vec{m}^j$ has been previously seen. If it exists in the map features $\vec{m}$, this means that we have an estimated location of the feature $\vec{\hat{m}}^j$. We then use this estimated location $\vec{\hat{m}}^j$ to generate predicted measurements $\vec{\hat{y}}_k^j = \left(\hat{\theta}_k^j\ \hat{r}_k^j\right)^T$ using the measurement model $h\left(\vec{x}_k, \vec{\hat{m}}^j\right)$ in Eq. (8-22). The difference between the predicted measurements $\hat{\vec{y}}^j_k$ and the observed measurements $\vec{y}^j_k$ is then used to update the weight of the particles as follows:

$$w_k^{[i]} \propto \frac{1}{\sqrt{(2\pi)^2}\sqrt{|R|}} e^{-\frac{1}{2}\left(\vec{y}^j_k - \hat{\vec{y}}^j_k\right)^T R^{-1}\left(\vec{y}^j_k - \hat{\vec{y}}^j_k\right)} \tag{8-42}$$

where R is the measurement noise covariance. After the weights are updated, the resampling step is applied to generate a new particle cloud. In the new particle cloud, particles with larger weights are selected at higher probability and the weights of all particles are then normalized. The mean of the platform's pose is calculated as follows:

$$\vec{\mu}_{\vec{x}_k} = \frac{1}{N}\sum_{i=1}^{N} \vec{x}_k^{[i]} \tag{8-43}$$

If the feature $\vec{m}^j$ is a new feature, its estimated location $\vec{\mu}_{\vec{m}^j}$ is calculated using the current estimate pose $\vec{\mu}_{\vec{x}_k}$ and the inverse measurement model in Eq. (8-27) and the feature is added to the map $\vec{m}$.

8.4.3 Fast-SLAM

In the previous section, the new map features' locations are estimated using the estimated pose $\vec{\mu}_{\vec{x}_k}$ and the inverse measurement model in Eq. (8-27). Fast-SLAM algorithm (described in [7]) enhances the map features' location estimation by using an EKF. For each particle and for each map feature, an EKF is created. In this case, EKF becomes very efficient because the matrix's dimensions become small. In our 2D case, the dimension of the EKF matrices will be 2×2. The platform pose is estimated by particle filtering as described in Section 8.4.2. For each particle $\vec{x}_k^{[i]}$ for each map feature $\vec{m}^j$, we will have an EKF. The state vector $\vec{m}_i^j$ of this EKF has two components $\left(m_x^{i,j}, m_y^{i,j}\right)^T$. In the feature $\vec{m}_i^j = \left(m_x^{i,j}, m_y^{i,j}\right)^T$, the i refers to the particle number, and the j refers to the feature index in the map. Since the map features are static, the system model of the EKF of $\vec{m}_i^j$ associated with particle $\vec{x}_k^{[i]}$ is simply a random constant in the following form:

$$\begin{pmatrix} \dot{m}_x^{i,j} \\ \dot{m}_y^{i,j} \end{pmatrix} = \begin{pmatrix} 0 \\ 0 \end{pmatrix} + \vec{w} \tag{8-44}$$

where $\vec{w}$ is a zero-mean random noise. The measurement model $h\left(\vec{x}_k, \vec{m}^j\right)$ is given by Eq. (8-22). The prediction step is like Rao–Blackwellized filter described in Section 8.4.2 with only one modification. When a new feature $\vec{m}^j$ is detected by a new measurement $\vec{y}^j_k = \left(\theta^j_k\, r^j_k\right)^T$, for each particle $\vec{x}_k^{[i]}$ we initialize an EKF for this new feature. The state value of this EKF is simply the feature's location $\hat{\vec{m}}_i^j$ which is calculated using the particle' pose $\vec{x}_k^{[i]}$ and the inverse measurement model in Eq. (8-27). The initial covariance of the map feature $\hat{\vec{m}}_i^j$ is set to the following equations:

$$H^j_{k,i} = \frac{\partial}{\partial \vec{m}^j} h^j\left(\vec{x}_k, \vec{m}^j\right)\Bigg|_{\vec{x}_k = \vec{x}_k^{[i]}, \vec{m}^j = \hat{\vec{m}}^j_i} \tag{8-45}$$

$$\Sigma_{k,i}^{j} = \left(H_{k,i}^{j}\right) R_{k}^{j} \left(H_{k,i}^{j}\right)^{T} \tag{8-46}$$

If the feature is observed again in future epochs, we update the weights of the platform's pose particles using the same update step of Rao–Blackwellized filter described in Section 8.4.2. We also update the feature's location using EKF update step as follows:

$$H_{k,i}^{j} = \frac{\partial}{\partial \vec{m}^{j}} h^{j}\left(\vec{x}_{k}, \vec{m}^{j}\right)\bigg|_{\vec{x}_{k} = \vec{x}_{k}^{[i]}, \vec{m}^{j} = \vec{\hat{m}}_{i}^{j}} \tag{8-47}$$

$$K_{k,j}^{j} = \Sigma_{k-1,i}^{j}\left(H_{k,i}^{j}\right)^{T}\left(H_{k,i}^{j}\Sigma_{k-1,i}^{j}\left(H_{k,i}^{j}\right)^{T} + R_{k}^{j}\right)^{-1} \tag{8-48}$$

$$\vec{\hat{m}}_{i}^{j} = \vec{\hat{m}}_{i}^{j} + K_{k,j}^{j}\left(\vec{y}_{k}^{j} - h^{j}\left(\vec{x}_{k}^{[i]}, \vec{\hat{m}}_{i}^{j}\right)\right) \tag{8-49}$$

$$\Sigma_{k,i}^{j} = \left(I - K_{k,j}^{j}\Sigma_{k,i}^{j}\right)\Sigma_{k,i}^{j} \tag{8-50}$$

After the resampling step, the estimated map feature's location $\vec{m}^{j}$ is calculated by averaging $\vec{\hat{m}}_{i}^{j}$ over all the N particles all as follows:

$$\vec{\mu}_{\vec{m}^{j}} = \frac{1}{N}\sum_{i=1}^{N} \vec{\hat{m}}_{i}^{j} \tag{8-51}$$

The Fast-SLAM algorithm can be used to build occupancy maps. The steps are similar except in the map update. Once we have a corrected updated pose, we use the occupancy map building algorithm described in Section 8.2 to calculate the occupancy map probabilities. This Fast-SLAM occupancy mapping is called gmapping. A popular open-source software that implements gmapping is available in [8]. More information about PF-SLAM can be found in [9].

8.4.4 Graph-SLAM

EKF-SLAM, PF-SLAM, and Fast-SLAM are online techniques that process data recursively epoch by epoch. The online approach is efficient but lacks global optimization perspective. Online techniques can be used to provide fast but not perfect estimates. Online techniques can be improved by processing a buffer of data from time to time. Online techniques can be enhanced by applying batch processing from time to time, which is known as bundle adjustment approach. Graph-SLAM [10, 11] is a popular batch or full SLAM approach that incorporates the data of the whole trajectory into one optimization problem and solves it at once. This technique has better global insight into the problem, so it leads to better solutions compared to online techniques. Graph-SLAM can be seen as an extended bundle adjustment applied to a larger data buffer. In the next subsections, we will review the probabilistic formulation of the SLAM problem and the graph solution approach.

8.4.4.1 Probabilistic Formulation of SLAM Problem

The probabilistic formulation of the SLAM problem can be simplified to the conditional probability $p\left(\vec{b}_{0:k} \mid \vec{y}_{1:k}\right)$, where $\vec{b}_{0:k}$ includes all poses (the trajectory) and the map and $\vec{y}_{1:k}$ includes all measurements. The Bayes' rule splits this conditional probability into motion and measurement probabilities as follows:

$$p\left(\vec{b}_{0:k} \mid \vec{y}_{1:k}\right) \propto \underbrace{p\left(\vec{y}_{1:k} \mid \vec{b}_{0:k}\right)}_{\text{measurement likelihood}} \quad \underbrace{p\left(\vec{b}_{0:k}\right)}_{\text{motion system model}} \tag{8-52}$$

Since the measurements are independent (through epochs and within each epoch if there are multiple measurements per epoch), the product rule simplifies measurement likelihood to the following equation:

$$p\left(\vec{y}_{1:k} \mid \vec{b}_{0:k}\right) = \prod_{i=1}^{k} p\left(\vec{y}_i \mid \vec{b}_i\right) \tag{8-53}$$

$$p\left(\vec{y}_i \mid \vec{b}_i\right) = \prod_{l=1}^{M} p\left(\vec{y}_i^{\,l} \mid \vec{b}_i\right) \tag{8-54}$$

The subscript i in Eq. (8-53) refers to time epoch and superscript l in Eq. (8-54) refers to l^{th} measurement in epoch i and M is the number of measurements at epoch i. Given the Markov assumption, the product rule simplifies the predictive motion probability to the following equation:

$$p\left(\vec{b}_{0:k}\right) = \prod_{i=1}^{k} p\left(\vec{x}_i \mid \vec{x}_{i-1}, \vec{u}_i\right) \tag{8-55}$$

where $\vec{u}_i$ is the control input that triggers the platform motion. Note that in Eq. (8-55), we used the pose variable $\vec{x}_i$ not $\vec{b}_i$ since the map features are static and only the robot or the platform is moving. These simplifications lead to the following overall probability formula:

$$p\left(\vec{b}_{0:k} \mid \vec{y}_{1:k}\right) \propto p\left(\vec{b}_0\right) \prod_{i=1}^{k} p\left(\vec{x}_i \mid \vec{x}_{i-1}, \vec{u}_i\right) p\left(\vec{y}_i \mid \vec{x}_i\right) \tag{8-56}$$

$$p\left(\vec{b}_{0:k} \mid \vec{y}_{1:k}\right) \propto p\left(\vec{b}_0\right) \prod_{i=1}^{k} p\left(\vec{x}_i \mid \vec{x}_{i-1}, \vec{u}_i\right) \prod_{l=1}^{M} p\left(\vec{y}_i^{\,l} \mid \vec{b}_i\right) \tag{8-57}$$

Taking the log of the products of a series of probabilities converts the multiplications into sums as follows:

$$\ln p\left(\vec{b}_{0:k} \mid \vec{y}_{1:k}\right) = \ln\left(p\left(\vec{b}_0\right)\right) + \sum_{i=1}^{k}\left(\ln\left(p\left(\vec{x}_i \mid \vec{x}_{i-1}, \vec{u}_i\right)\right)\right) + \sum_{i=1}^{k}\sum_{l=1}^{M} \ln\left(p\left(\vec{y}_i^{\,l} \mid \vec{b}_i\right)\right) \tag{8-58}$$

A cost function $J\left(\vec{b}_{0:k}\right)$ can be defined by taking the negative of Eq. (8-58) as follows:

$$J\left(\vec{b}_{0:k}\right) = -\ln\left(p\left(\vec{b}_0\right)\right) - \sum_{i=1}^{k}\left(\ln\left(p\left(\vec{x}_i \mid \vec{x}_{i-1}, \vec{u}_i\right)\right)\right) - \sum_{i=1}^{k}\sum_{l=1}^{M} \ln\left(p\left(\vec{y}_i^{\,l} \mid \vec{b}_i\right)\right) \tag{8-59}$$

As all probabilities are assumed to be Gaussian as follows:

$$p\left(\vec{x}_i \mid \vec{x}_{i-1}, \vec{u}_i\right) \sim \mathcal{N}\left(f\left(\vec{x}_{i-1}, \vec{u}_i\right), Q\right) \tag{8-60}$$

$$pp\left(\vec{y}_i^{\,l} \mid \vec{b}_i\right) \sim \mathcal{N}\left(h^l\left(\vec{b}_i\right), R\right) \tag{8-61}$$

where $f\left(\vec{x}_{i-1}, \vec{u}_i\right)$ is the system model and $h^l\left(\vec{b}_i\right)$ is the measurement model of measurement l at epoch i. and thanks to the mathematical advantages of the exponential formula of the Gaussian probability function, taking the log of exponentials removes the log from the cost function as follows:

$$\begin{aligned} J\left(\vec{b}_{0:k}\right) &= \text{const.} + \left(\vec{b}_0 - \vec{\mu}_{\vec{b}_0}\right)^T \Sigma_0^{-1}\left(\vec{b}_0 - \vec{\mu}_{\vec{b}_0}\right) \\ &+ \sum_{i=1}^{k}\left(\vec{x}_i - f\left(\vec{x}_{i-1}, \vec{u}_i\right)\right)^T Q^{-1}\left(\vec{x}_i - f\left(\vec{x}_{i-1}, \vec{u}_i\right)\right) \\ &+ \sum_{i=1}^{k}\sum_{l=1}^{M}\left(\vec{y}_i^{\,l} - h^l\left(\vec{b}_i\right)\right)^T R^{-1}\left(\vec{y}_i^{\,l} - h^l\left(\vec{b}_i\right)\right) \end{aligned} \tag{8-62}$$

where $\vec{\mu}_{\vec{b}_0}$ is initial state (pose and map), Σ_0^{-1} is the information matrix that reflects our confidence in the initial state value, Q^{-1} is the information matrix of the motion model noise (system noise), R^{-1} is the information matrix of the measurement model noise (measurement noise). The information matrix is the inverse of the covariance. Optimizing the cost function $J\left(\vec{b}_{0:k}\right)$ aims at finding $\vec{b}_{0:k}$ (platform poses and map features' locations) that best explains the observations $\vec{y}_i^{\,l}$ and at the same time fits the system model. We can use conventional nonlinear optimization methods such as Gauss-Newton to solve this optimization problem. This approach will require the Jacobian and maybe the Hessian of the motion model and the measurement model. Numerical methods that do not require linearization can also be used. The major drawback of the conventional optimization approach is the size of the problem. If the map features are dense, which is the case in most scenarios, the unknown vector $\vec{b}_{0:k}$ will be too long and the matrices will be too complex to process, especially if matrix inversion is involved. Graph optimization was proposed to solve this complex optimization problem efficiently.

8.4.4.2 Graph Optimization

Graph-SLAM approach models the SLAM problem as a graph where nodes are the states (platform's poses and map features' locations) and links are system model and measurement models that link nodes together as seen in Figure 8.8. The graph-based approach builds a large information matrix and an information vector to represent the graph. The Graph-SLAM algorithms transform the SLAM problem into a graph like the one shown in Figure 8.8 and linearize the system models and measurement models around an initial solution $\vec{\mu}_{\vec{b}_0}$ which converts the graph into a linear estimation problem in the following linear form:

$$\mathrm{A}\vec{b}_{0:t} = \vec{B} \tag{8-63}$$

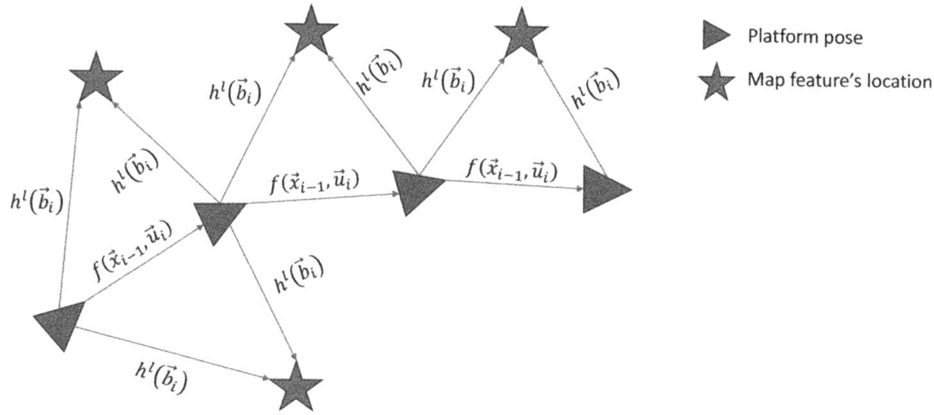

Figure 8.8 Graph modeling of the SLAM problem.

where A is called the information matrix and the vector $\vec{B}$ is called the information vector. The solution $\vec{b}_{0:t}$ that satisfies this model is given by the following matrix inversion formula:

$$\vec{b}_{0:t} = \mathrm{A}^{-1}\vec{B} \tag{8-64}$$

In the following discussion, we will see how to convert the cost function in Eq. (8-62) to the simple matrix form shown in Eq. (8-64), assuming that the platform is the 2D vehicle example shown in Figures 8.3 and 8.4. The first step is initialization to obtain an initial guess of the poses as follows:

$$\vec{\mu}_{\vec{x}_0} = (0 \quad 0 \quad 0)^T, \Sigma_{\vec{x}_0} = \begin{pmatrix} \varepsilon & 0 & 0 \\ 0 & \varepsilon & 0 \\ 0 & 0 & \varepsilon \end{pmatrix} \tag{8-65}$$

$$\vec{\mu}_{\vec{x}_i} = f\left(\vec{\mu}_{\vec{x}_{i-1}}, \vec{u}_k\right) = \begin{pmatrix} x_{i-1} + v_i \cos \psi_{i-1} \Delta T \\ y_{i-1} + v_i \sin \psi_{i-1} \Delta T \\ \psi_{i-1} + \omega_i \Delta T \end{pmatrix}, i = 1 \dots k \tag{8-66}$$

The initialization of poses $\vec{\mu}_{\vec{x}_{0:k}}$ together with the map features' measurements $\vec{y}_i^j = \left(\theta_i^j \, r_i^j\right)^T$ where θ_i^j is scan angle and r_i^j is range measurements, can be used to get an initialization of the map features' locations using the inverse measurement model as follows:

$$\vec{\mu}_{\vec{m}^j} = h^{j^{-1}}\left(\vec{y}_i^j, \vec{\mu}_{\vec{x}_i}\right) = \begin{pmatrix} x_i + r_i^j \cos\left(\theta_i^j + \psi_i\right) \\ y_i + r_i^j \sin\left(\theta_i^j + \psi_i\right) \end{pmatrix}, i = 1 \dots k \tag{8-67}$$

We have now an initial full-state vector (poses and map) $\vec{\mu}_{\vec{b}_{0:k}}$ for the entire trajectory ($i = 0, 1, 2 \dots k$) and all map features. The second step is to replace $f\left(\vec{x}_{i-1}, \vec{u}_i\right)$ and $h^l\left(\vec{b}_i\right)$ by their linearized approximation using Tylor series expansion as follows:

$$f\left(\vec{x}_{i-1}, \vec{u}_i\right) \cong f\left(\vec{\mu}_{\vec{x}_{i-1}}, \vec{u}_i\right) + \left.\frac{\partial f\left(\vec{x}_{i-1}, \vec{u}_i\right)}{\partial \vec{x}_{i-1}}\right|_{\vec{x}_{i-1} = \vec{\mu}_{\vec{x}_{i-1}}} \left(\vec{x}_{i-1} - \vec{\mu}_{\vec{x}_{i-1}}\right) \tag{8-68}$$

$$h^l\left(\vec{b}_i\right) \cong h^l\left(\vec{\mu}_{\vec{b}_i}\right) + \left.\frac{\partial h^l\left(\vec{b}_i\right)}{\partial \vec{b}_i}\right|_{\vec{b}_i = \vec{\mu}_{\vec{b}_i}} \left(\vec{b}_i - \vec{\mu}_{\vec{b}_i}\right) \tag{8-69}$$

To simplify the notation, we define the Jacobian matrices F_i and H_i^l as follows:

$$F_i = \left.\frac{\partial f\left(\vec{x}_{i-1}, \vec{u}_i\right)}{\partial \vec{x}_{i-1}}\right|_{\vec{x}_{i-1} = \vec{\mu}_{\vec{x}_{i-1}}} \tag{8-70}$$

$$H_i^l = \left.\frac{\partial h^l\left(\vec{b}_i\right)}{\partial \vec{b}_i}\right|_{\vec{b}_i = \vec{\mu}_{\vec{b}_i}} \tag{8-71}$$

Substituting $f\left(\vec{x}_{i-1}, \vec{u}_i\right)$ and $h^l\left(\vec{b}_i\right)$ by their linearized approximation by their linearized approximations, we get the following approximate cost function:

$$\begin{aligned} J\left(\vec{b}_{0:k}\right) \cong & \left(\vec{b}_0 - \vec{\mu}_{\vec{b}_0}\right)^T \Sigma_0^{-1} \left(\vec{b}_0 - \vec{\mu}_{\vec{b}_0}\right) \\ & + \sum_{i=1}^{k} \left(\vec{x}_i - f\left(\vec{\mu}_{\vec{x}_{i-1}}, \vec{u}_i\right) - F_i\left(\vec{x}_{i-1} - \vec{\mu}_{\vec{x}_{i-1}}\right)\right)^T Q^{-1} \left(\vec{x}_i - f\left(\vec{\mu}_{\vec{x}_{i-1}}, \vec{u}_i\right) - F_i\left(\vec{x}_{i-1} - \vec{\mu}_{\vec{x}_{i-1}}\right)\right) \\ & + \sum_{i=1}^{k}\sum_{l=1}^{M} \left(\vec{y}_i^l - h^l\left(\vec{\mu}_{\vec{b}_i}\right) - H_i^l\left(\vec{b}_i - \vec{\mu}_{\vec{b}_i}\right)\right)^T R^{-1} \left(\vec{y}_i^l - h^l\left(\vec{\mu}_{\vec{b}_i}\right) + H_i^l\left(\vec{b}_i - \vec{\mu}_{\vec{b}_i}\right)\right) \end{aligned} \tag{8-72}$$

This approximated version of the cost function $J\left(\vec{b}_{0:k}\right)$ has many advantages. First, it is quadratic in the state vector $\vec{b}_{0:k}$. After some rearrangements of terms and removing constant terms, the cost function can be formulated in the popular quadratic linear estimation form given by:

$$J\left(\vec{b}_{0:k}\right) = \text{const} - \left(\vec{b}_{0:k}\right)^T \mathrm{A}\left(\vec{b}_{0:k}\right) + 2\left(\vec{b}_{0:k}\right)^T \vec{B} \tag{8-73}$$

where A is an information matrix calculated as follows [10, 11]:

$$\mathrm{A} = \sum_{i=1}^{k} \begin{pmatrix} -(F_i)^T \\ I \end{pmatrix} Q^{-1}\left(-(F_i)^T I\right) + \sum_{i=1}^{k}\sum_{l=1}^{M} (H_i^l)^T R^{-1} (H_i^l) \tag{8-74}$$

and $\vec{B}$ is the information vector calculated as follows [10, 11]:

$$\vec{B} = \sum_{i=1}^{k} \begin{pmatrix} -(F_i)^T \\ 1 \end{pmatrix} Q^{-1}\left(f\left(\vec{\mu}_{\vec{x}_{i-1}}, \vec{u}_i\right) - (F_i)\vec{\mu}_{\vec{x}_{i-1}}\right) + \sum_{i=1}^{k}\sum_{l=1}^{M} (H_i^l)^T R^{-1}\left(\vec{y}_i^l - h^l\left(\vec{\mu}_{\vec{b}_i}\right) - (H_i^l)\vec{b}_{0:k}\right) \tag{8-75}$$

The optimal solution for $J\left(\vec{b}_{0:k}\right)$ is found by setting the derivative of Eq. (8-73) with respect to the unknown state vector $\vec{b}_{0:k}$ to zero which gives the following formula:

$$\frac{\partial J\left(\vec{b}_{0:k}\right)}{\partial \vec{b}_{0:k}} = -\mathrm{A}\vec{b}_{0:t} + \vec{B} = 0 \tag{8-76}$$

which leads to the following popular linear estimation formula:

$$\mathrm{A}\vec{b}_{0:k} = \vec{B} \tag{8-77}$$

which can be solved using the following inverse matrix formula:

$$\vec{b}_{0:k} = \mathrm{A}^{-1}\vec{B} \tag{8-78}$$

Although the solution formula seems simple, we still have a size problem. To overcome the problem of the large size of A and $\vec{B}$, Graph-SLAM reduces the size of the information matrix and information vector by removing the entries corresponding to the map features' locations using the marginalization rule. The basic idea is to remove the map features from the information matrix and information vector and solve only for the platform's poses and then construct the map using the optimized poses. We used this approach when we used the Rao–Blackwellized particle filter and Fast-SLAM algorithm in Sections 8.4.2 and 8.4.3.

8.4.4.3 Graph Reduction

The measurement model is the model that links poses nodes to map features' location nodes. To remove the map features' locations from the graph, we must replace the measurement model with a relationship that links the poses that see the same map features to each other. In other words, the measurement model that links some map features to some poses can be replaced by a relationship that links these poses that observe these common map features directly together. This step will remove the entries corresponding to the map features' locations from the information matrix and information vector, leading to much smaller matrix sizes that contain poses only as unknowns. This step is known as graph reduction, and it is performed by replacing each feature with links between the poses from which it was observed. This constraint is the key aspect that improves the pose estimation using the map features measurements without explicitly estimating the map features. The links between two poses and one feature observed by the two poses can be replaced by a link between the two poses in different ways. One way is by using the commonly seen features $\vec{m}_c$ to perform an ICP scan-matching to estimate the rotation and translation between the poses from which $\vec{m}_c$ were observed. These rotation and translation are treated as a measurement coming from an independent sensor, and it has its own measurement model given by:

$$\vec{y}_k = h\left(\vec{x}_k, \vec{x}_{k-1}\right) = \begin{pmatrix} R_k \\ \vec{T}_k \end{pmatrix} \tag{8-79}$$

which is a relationship between poses $\vec{x}_k, \vec{x}_{k-1}$ where R_k and $\vec{T}_k$ are the relative rotation and translation between poses $\vec{x}_k$ and $\vec{x}_{k-1}$. This new measurement model replaces the following two measurement models:

$$\vec{y}_k = h\left(\vec{x}_k, \vec{m}_c\right) \tag{8-80}$$

$$\vec{y}_{k-1} = h\left(\vec{x}_{k-1}, \vec{m}_c\right) \tag{8-81}$$

so we can remove the entries corresponding to $\vec{m}_c$ from the information matrix and information vector. Another approach to remove map features from the graph is probabilistic marginalization using the marginalization rule. This step is also known as factor graph optimization. For a given set of features $\vec{m}_c$ that are observed by poses $\vec{x}_{0:k}$, the full state (pose and map) probability is given by:

$$p\left(\vec{b}_{0:t} \mid \vec{y}_{1:k}, \vec{u}_{1:k}\right) = p\left(\vec{x}_{0:k} \mid \vec{y}_{1:k}, \vec{u}_{1:k}\right) p\left(\vec{m}_c \mid \vec{x}_{0:k}, \vec{y}_{1:k}, \vec{u}_{1:k}\right) \tag{8-82}$$

We can obtain a formula for $p\left(\vec{b}_{0:t} \mid \vec{y}_{1:k}, \vec{u}_{1:k}\right)$ that does not include map features $\vec{m}_c$ by marginalizing out the variable $\vec{m}_c$. This can be done by integrating $p\left(\vec{b}_{0:t} \mid \vec{y}_{1:k}, \vec{u}_{1:k}\right)$ over all possible values of $\vec{m}_c$ as follows:

$$p\left(\vec{b}_{0:t} \mid \vec{y}_{1:k}, \vec{u}_{1:k}\right) = \int p\left(\vec{x}_{0:k} \mid \vec{y}_{1:k}, \vec{u}_{1:k}\right) p\left(\vec{m}_c \mid \vec{x}_{0:k}, \vec{y}_{1:k}, \vec{u}_{1:k}\right) dm \tag{8-83}$$

Because we assume Gaussian distributions, this integration has a closed-form solution and the probability $p\left(\vec{b}_{0:t} \mid \vec{y}_{1:k}, \vec{u}_{1:k}\right)$ can be calculated such that the terms that have the measurement models $h\left(\vec{b}_i\right)$ is replaced by a term that has pose states $\vec{x}_i$ only. The marginalization rule states that if the probability distribution $p\left(\vec{x}, \vec{y}\right)$ over the random vectors $\vec{x}$ and $\vec{y}$ be a Gaussian represented in the information form:

$$\mathrm{A}_{xy} = \begin{pmatrix} \mathrm{A}_{xx} & \mathrm{A}_{xy} \\ \mathrm{A}_{yx} & \mathrm{A}_{yy} \end{pmatrix} \tag{8-84}$$

$$\vec{B} = \begin{pmatrix} \vec{B}_x \\ \vec{B}_y \end{pmatrix} \tag{8-85}$$

If A_{yy} is invertible, the marginal probability of $\vec{x}\left(p\left(\vec{x}\right)\right)$ is a Gaussian distribution whose information matrix and vector are defined as follows:

$$A_{xx} = \mathrm{A}_{xx} - \mathrm{A}_{xy}\mathrm{A}_{yy}^{-1}\mathrm{A}_{yx} \tag{8-86}$$

$$\vec{B}_x = \vec{B}_x - \mathrm{A}_{xy}\mathrm{A}_{yy}^{-1}\vec{B}_y \tag{8-87}$$

In our case, the map features' locations are the variable $\vec{y}$ and the poses that observe them are the variable $\vec{x}$. Hence, in the information matrix A, we replace the entries corresponding to each map feature's location by subtracting the quantity $(\mathrm{A}_{xy}\mathrm{A}_{yy}^{-1}\mathrm{A}_{yx})$ from the entries of the original matrix corresponding to pose $\vec{x}$. Similarly, we subtract the quantity $(\mathrm{A}_{xy}\mathrm{A}_{yy}^{-1}\vec{B}_y)$ from the entries of the original information vector corresponding to pose $\vec{x}$ and then we can remove the entries corresponding to map feature $\vec{y}$ from the original information matrix and information vector.

8.4.4.4 Trajectory Estimation

After removing all entries corresponding to map features' locations from the information matrix and vector (A and $\vec{B}$), we will have new information matrix and vector A$'$ and $\vec{B}'$, where the unknowns are the platform poses $\vec{x}_{0:k}$ only as follows:

$$\mathrm{A}'\vec{x}_{0:k} = \vec{B}' \tag{8-88}$$

which can be solved by matrix inversion as follows:

$$\vec{x}_{0:k} = (\mathrm{A}')^{-1}\vec{B}' \tag{8-89}$$

This step is like the variable elimination method in matrix inversion algorithms. The philosophy is that the information that two or more poses that see the same feature is used in enhancing the estimation of these poses. The reason for getting this enhancement is that if a previously seen feature is observed again, the difference between the predicted feature location and the observed feature location is used to correct the pose estimation. In return, the corrected pose is used to get a better estimate of the map features' locations. This is simply the core philosophy behind SLAM and Graph-SLAM, which is based on this fundamental concept or philosophy.

8.4.4.5 Map Estimation

Once we estimate the platform poses (the full trajectory), defined by $\vec{x}_{0:k}$, we can estimate the map features' locations using the Gaussian conditional probability rule that states that if the probability distribution $p\left(\vec{x}, \vec{y}\right)$ over the random vectors $\vec{x}$ and $\vec{y}$ is a Gaussian represented in the information form:

$$\mathrm{A}_{xy} = \begin{pmatrix} \mathrm{A}_{xx} & \mathrm{A}_{xy} \\ \mathrm{A}_{yx} & \mathrm{A}_{yy} \end{pmatrix} \tag{8-90}$$

$$\vec{B} = \begin{pmatrix} \vec{B}_x \\ \vec{B}_y \end{pmatrix} \tag{8-91}$$

the conditional $p\left(\vec{x}, \vec{y}\right)$ is a Gaussian with information matrix A_{yy} and information vector $\vec{B}_y + \mathrm{A}_{xy}\vec{B}_x$. We can Apply this rule to the original information matrix and vector to estimate the map features' locations. Assume that submatrix A_{jj} consists of the entries corresponding to the map feature $\vec{m}^j$ in the original information matrix A. Assume that $\mathrm{A}_{j,\vec{x}_k}$ is a submatrix of A_{jj} that consists of the entries corresponding to map feature $\vec{m}^j$ and the poses $\vec{x}_k$ from which this feature was observed, we would have the following linear estimation equation where unknowns are map feature location $\vec{m}^j$:

$$\mathrm{A}_{jj}\vec{m}^j = \vec{B}_j + \mathrm{A}_{j,\vec{x}_k}\vec{x}_k \tag{8-92}$$

which can be solved using the following matrix inverse equation:

$$\vec{m}^j = \mathrm{A}_{jj}^{-1}\left(\vec{B}_j + \mathrm{A}_{j,\vec{x}_k}\vec{x}_k\right) \tag{8-93}$$

Graph-SLAM implements other constraints such as loop closure and external GPS updates in the same way it implements system and measurement models. Loop closure happens when the platform visits a previously visited place. In Graph-SLAM, loop closure can be modeled as a constraint or connection between two poses in the graph. The transformation between these two poses will be used as a connection or link between the two poses in the graph. This constraint significantly improves the overall trajectory accuracy and, consequently, the map. External information, such as GPS locations, can also be added to the graph as a measurement link or measurement constraint. In a less common approach, we can also add the points correspondences as part of the optimization. Graph-SLAM is now the industry standard SLAM framework. The general steps of Graph-SLAM can be summarized in the following points:

1) Apply features extraction (range measurement or visual features detection)
2) Use odometry or ICP scan-matching to find an initial pose estimate.
3) Linearize motion and measurement models around the initial estimate.
4) Construct the full graph, including loop closure and external constraints.
5) Reduce the graph by eliminating map features.
6) Solve quadratically constrained optimization to optimize pose estimate.
7) Build the map using the optimized pose estimate.

More information and details about Graph-SLAM can be found in [10] and [11].

8.4.5 MATLAB LiDAR Graph-SLAM Example Project

MATLAB navigation toolbox has several built-in functions that can be used to build and test Graph-SLAM algorithms. The toolbox includes a graph optimization engine and several helper functions to store imaging and ranging sensor scans, relative poses, absolute poses, and point clouds. To demonstrate the concept of Graph-SLAM, a sample MATLAB code that performs Graph-SLAM is presented. The sample code is included in the projects' repository of this book under project "*LiDARGraphSLAMProject.*" The project uses the simulation environment that is used in the occupancy grid mapping project "*TestOccupancyMapProject.*" In this project, a vehicle with 2-D LiDAR, vertical gyroscope, and speedometer is simulated. The gyroscope and the speed data (we will refer to this by inertial data) are used to predict the relative motion change between epochs. The predicted relative motion is used as an initial relative pose to an NDT-based scan matching algorithm that uses LiDAR scans to estimate the relative pose change between two scans. Relative poses are added to a MATLAB point cloud object "pcviewset" of type "view set." View set is the object that MATLAB uses to build the graph and to apply graph optimization. The graph is built by adding a view set for each epoch. The view set is initialized by the first pose of the vehicle, which is taken as the world frame. Subsequent scans are added to the view set as connected nodes/views. Consecutive nodes/views are linked/connected by their relative poses. The relative poses are estimated with help from inertial data and from the NDT LiDAR scan matching algorithm. If there is a loop detected between view i and view j, an additional link (additional constraint) is added to the graph. This additional link/constraint holds the relative pose between view i and view j. After processing all epochs, a MATLAB pose graph optimization object is created and set to process the graph that has been built in the view set. The built-in MATLAB optimization function "*optimizePoseGraph*" is called with the desired optimization algorithm (e.g., Levenberg Marquardt) to optimize the graph and estimate the corrected poses. The optimized poses are then used along with the LiDAR point cloud to build an optimized map. Sample results of this LiDAR/Inertial Graph-SLAM project can be seen in Figures 8.9–8.11. Note the effect of loop closure on the estimated trajectory (Figure 8.10) and on the map accuracy (Figure 8.11).

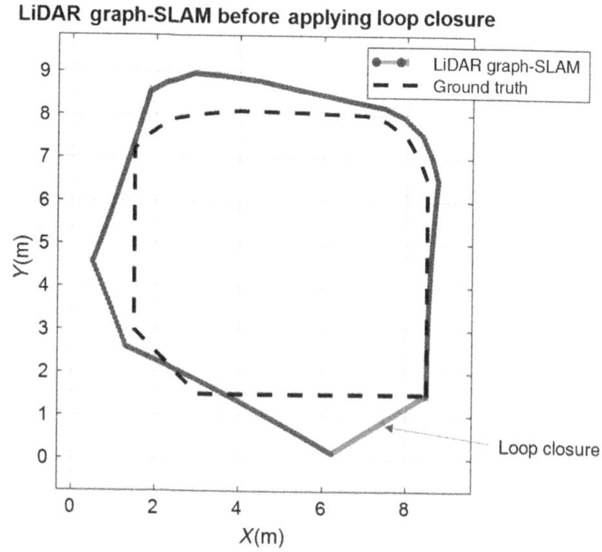

Figure 8.9 LiDAR Graph-SLAM before applying loop Closure. Illustration of the loop closure position.

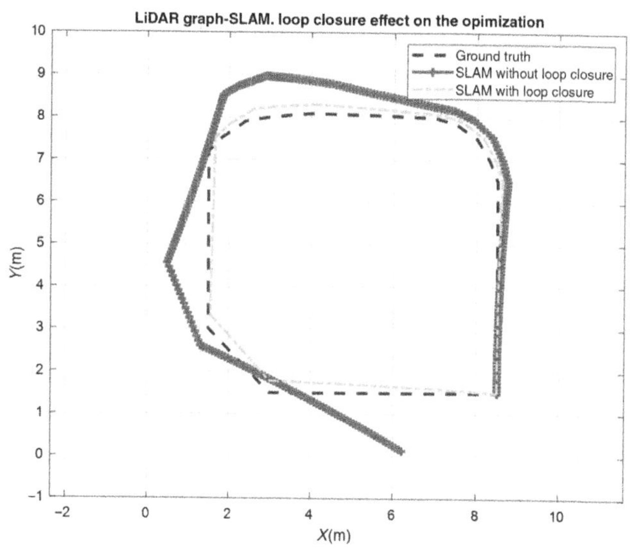

Figure 8.10 LiDAR Graph-SLAM. Effect of loop closure on the trajectory accuracy.

8.5 Visual SLAM

In the previous sections, we have seen how range sensors are used in mapping and SLAM. The main advantage of the range sensors is that the sensor provides physical metric measurements that, together with the platform pose, can be used to calculate the physical location of the map feature point. This makes the calculations more straightforward and the linearization of models easier, including the inverse measurement model. However, range sensors do not provide color

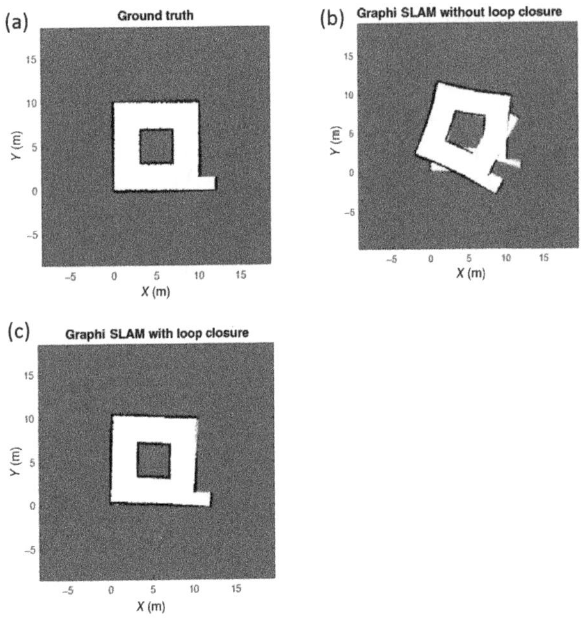

Figure 8.11 LiDAR Graph-SLAM. Effect of loop closure on the map accuracy. (a) Ground-truth solution, (b) Graph-SLAM without loop closure, (c) Graph-SLAM with loop closure.

information (although they do provide intensity measurement). In addition, finding the correspondence of feature points between different scans is challenging because the points are indistinguishable. Vision sensors have the advantage of providing richer and more distinguishable information about the environment, which makes feature matching and correspondence estimation more robust. In the next subsections, we will briefly discuss visual SLAM.

8.5.1 Measurement Model

The sensor measurement model is the model that relates physical features in the environment to system states. In imaging sensors, the measurements are a pixel location in the 2D image plane. We have seen in the previous chapter how the 3D environment is projected on a 2D image plane using the following projection model equation [12]:

$$y_t^i = \begin{pmatrix} u_t^i \\ v_t^i \\ 1 \end{pmatrix} = \frac{1}{\lambda_t^i} K_{3\times3} \left(R_t \ \vec{T}_t \right)_{3\times4} \begin{pmatrix} m_x^i \\ m_y^i \\ m_z^i \\ 1 \end{pmatrix}_{4\times1} \tag{8-94}$$

where $m^i = \left(m_x^i \ m_y^i \ m_z^i \right)^T$ is the 3D feature's location in the world frame, the vector, $\left(m_x^i \ m_y^i \ m_z^i \ 1 \right)^T$ is the homogeneous 4D location of the map feature point, $R_t, \vec{T}_t$ are the direction cosine matrix and translation vector that describes the orientation of the camera with respect to the world frame, $K_{3\times3}$ is the intrinsic calibration matrix of the camera that transforms a point from the 3D camera frame to the 2D image plane, λ_t^i is the depth of the feature's point (the vertical component of the point's coordinate in the camera frame), $p_t^i = \left(u_t^i \ v_t^i \right)^T$ is the 2D pixel location of the feature point, and $y_t^i = \left(u_t^i \ v_t^i \ 1 \right)^T$ is the homogeneous pixel coordinate. Note that we used the subscript t to indicate that the feature point's pixel location changes with time if the camera poses changes. This projection model is the measurement model that relates the 3D map feature point in the physical environment to a pixel measurement in a specific location in the 2D map image.

8.5.2 Visual Features

To track map feature points movement in the image plane, we need to identify the feature using a distinguishable description. Visual features are distinguishable descriptions that describe a map feature point using its visual projection (in pixels) in the image plane. This description may use the colors, intensities, shape, or texture features in the image to detect and identify a feature. There are several approaches to describe a point in a distinguishable way [13, 14]. For example, feature points can be described using corner detectors as seen in Figure 8.12. Figure 8.12 shows corners detected by applying the features-from-accelerated-segment test (FAST) corner detector (also known as kernel) [15] and [16]. The kernel is applied at all pixels in the image and calculating the correlation between the image pixel values (intensities or colors) and the kernel and then detecting the resulting peaks. If there are distinguishable peaks in the resulting image pixel matrix, this indicates the locations of the corners in the image as seen in Figure 8.12.

Corners are obviously not fully distinguishable globally in the whole image. However, it is distinguishable in the local region around the point. Therefore, if we are trying to match the feature in two or more image scans, we limit the search to a local region to avoid ambiguity as much as possible. This local region can be established in the same way we established the search region in the ICP scan-matching algorithm. Using an initial guess of the camera pose change from one scan to another, we can predict the region where the feature should appear in the next image scan if the point is still in the view of the camera. Corner is just an example of a feature detector. There are several feature types and feature detectors, such as edges and blobs. Detection of features is commonly performed by convoluting the image at different scales with different kernels and detecting the peaks in the resulting image. This convolution was previously done using conventional signal processing using man-crafted kernels. However, recently, this process is commonly done using conventional neural networks (CNN). Feature description and detection is a large topic in computer vision literature that is out of the scope of this book. More details about feature descriptors and detectors can be found in [13], [14], and [17]. In addition, there are software packages and libraries for visual feature detection such as OpenCV [18] and MATLAB computer vision toolbox [19] and [16].

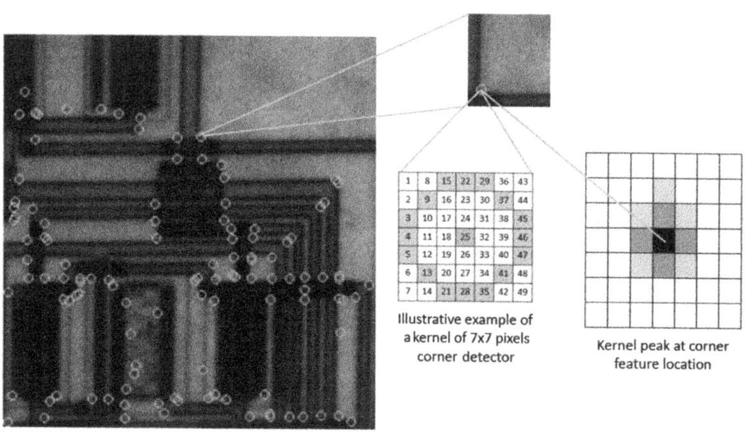

Figure 8.12 Illustration of visual FAST corner features detection. *Source:* The MathWorks, Inc.

Using the feature descriptors, we can perform feature matching between two or more images to find the correspondences of the map feature through multiple image scans. Given a reference scan, for each feature in the reference scan, feature matching applies a search and a compare process to find the correspondence location of the feature in the next or current image scan. This process is illustrated in Figure 8.13.

8.5.3 Visual Odometry

Given a feature point in the 3D physical environment, the camera sensor projects its location into a 2D pixel location. When the camera moves, the feature's pixel location moves on the image plane. By tracking the movement of the feature in the image plane, we can estimate the camera motion. The estimation of the relative pose changes of the camera is known as visual odometry. The concept of visual odometry is illustrated in Figure 8.14. The details of the visual odometry, camera motion

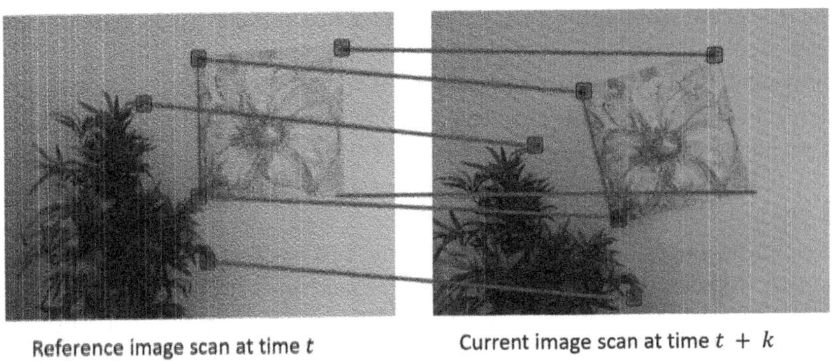

Figure 8.13 Features matching between current reference and current image scans.

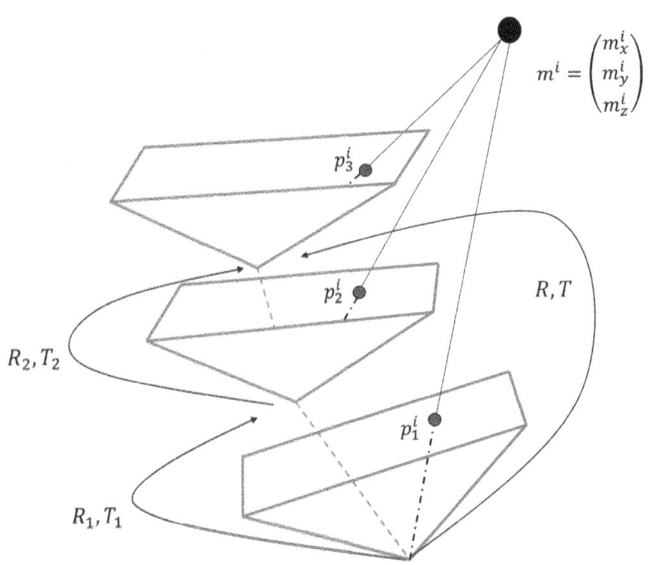

Figure 8.14 Visual odometry concept.

estimation from visual features, and related mathematical derivations are described in Chapter 7, "Imaging Sensors." Visual odometry calculates the relative poses between scans, which can be concatenated to form a full trajectory. For example, in Figure 8.14, if we have three epochs, and assuming the initial pose of the camera is aligned with the world frame ($R_0 = I, \vec{T}_0 = (0\,0\,0)^T$), visual odometry can calculate the following relatives pose:

$$R_{0\to1}, \vec{T}_{0\to1} \tag{8-95}$$

$$R_{1\to2}, \vec{T}_{1\to2} \tag{8-96}$$

The camera's pose at the end of the three epochs can be calculated by concatenation of $R_{0\to1}, \vec{T}_{0\to1}$ and $R_{1\to2}, \vec{T}_{1\to2}$ as follows:

$$\begin{pmatrix} R_{0\to2} & \vec{T}_{0\to2} \\ 0 & 1 \end{pmatrix} = \begin{pmatrix} R_{1\to2} & \vec{T}_{1\to2} \\ 0 & 1 \end{pmatrix} \begin{pmatrix} R_{0\to1} & \vec{T}_{0\to1} \\ 0 & 1 \end{pmatrix} \tag{8-97}$$

If we have more epochs, the camera's pose at any time k defined by the rotation and translation pair $\left\langle R_{0\to t}, \vec{T}_{0\to t} \right\rangle$ with respect to world frame is calculated recursively as follows:

$$\begin{pmatrix} R_{0\to t} & \vec{T}_{0\to t} \\ 0 & 1 \end{pmatrix} = \begin{pmatrix} R_{t-1\to t} & \vec{T}_{t-1\to t} \\ 0 & 1 \end{pmatrix} \begin{pmatrix} R_{t-2\to t-1} & \vec{T}_{t-2\to t-1} \\ 0 & 1 \end{pmatrix} \cdots \begin{pmatrix} R_{0\to1} & \vec{T}_{0\to1} \\ 0 & 1 \end{pmatrix} \tag{8-98}$$

The main issue with single camera (monocular) assembly is that the estimated translation vectors are scaled with an unknown scale factor.

We have discussed in Chapter 7 "Imaging Sensors" how visual odometry can be preformed in both online and batch mode. In Chapter 7, a MATLAB example project that performs visual odometry was demonstrated and can be found in [20]. The visual odometry solution is used in the same way we used the LiDAR scan-matching solution in LiDAR Graph-SLAM example in Section 8.4.5. The visual odometry solution is used to build a pose graph where connections between nodes are the relative pose transformation between epochs (image frames in this case).

8.5.4 Visual SLAM Approaches

Visual SLAM can be implemented using the same approaches that are used in the range-based SLAM. All popular SLAM algorithms such as EKF-SLAM, Fast-SLAM, and Graph-SLAM can be applied to visual sensors in the same way they are applied to range sensors. The main difference in visual SLAM is that the measurement model is different, and the scale ambiguity problem must be taken care of. Recall the measurement model of the 2D range sensor example given by the following equation:

$$y_t^i = \begin{pmatrix} \theta_t^i \\ r_t^i \end{pmatrix} = \begin{pmatrix} \tan^{-1}\left(\dfrac{m_y^i - x_{2,t}^r}{m_x^i - x_{1,t}^r}\right) - x_{3,t}^r \\ \sqrt{\left(m_x^i - x_{1,t}^r\right)^2 + \left(m_y^i - x_{2,t}^r\right)^2} \end{pmatrix} \tag{8-99}$$

where $(x^r_{1,t}, x^r_{2,t})$ is the platform's coordinate, $x^r_{3,t}$ is the platform's orientation, and $\left(m^i_x, m^i_y\right)$ is the world metric coordinate of the map feature point m^i. The linearization of this measurement model is given by:

$$\frac{\partial}{\partial x_t} h^i(x_t) = \begin{pmatrix} \frac{\Delta y^i_t}{(r^i_t)^2} & \frac{-\Delta x^i_t}{(r^i_t)^2} & -1 & 0\ldots0 & \frac{-\Delta y^i_t}{(r^i_t)^2} & \frac{\Delta x^i_t}{(r^i_t)^2} & 0\ldots0 \\ \frac{-\Delta x^i_t}{r^i_t} & \frac{-\Delta y^i_t}{r^i_t} & 0 & 0\ldots0 & \frac{\Delta x^i_t}{r^i_t} & \frac{\Delta y^i_t}{r^i_t} & 0\ldots0 \end{pmatrix} \tag{8-100}$$

where r^i_t, Δx^i_t, and Δy^i_t are calculated as follows:

$$r^i_t = \sqrt{\left(m^i_x - x^r_{1,t}\right)^2 + \left(m^i_y - x^r_{2,t}\right)^2} \tag{8-101}$$

$$\Delta x^i_t = \left(m^i_x - x^r_{1,t}\right) \tag{8-102}$$

$$\Delta y^i_t = \left(m^i_y - x^r_{2,t}\right) \tag{8-103}$$

The linearization of this 2D range/angle measurement model is simple. The situation is different in visual SLAM. The vision sensor measurement model is given by the following equation:

$$y^i_t = \begin{pmatrix} u^i_t \\ v^i_t \\ 1 \end{pmatrix} = \frac{1}{\lambda^i_t} K_{3\times3}\left(R_t\ \vec{T}_t\right)_{3\times4} \begin{pmatrix} m^i_x \\ m^i_y \\ m^i_z \\ 1 \end{pmatrix}_{4\times1} \tag{8-104}$$

In contrast to the range-based measurement model, the vision-based measurement model is more complex and cannot be linearized easily in its current form due to the matrix representation of the orientation. This is a clear difference between range-based SLAM and visual SLAM. Other differences can be mentioned. The following brief list discusses the main differences between visual SLAM and range-based SLAM, which can be described as follows:

- **The odometry method:** In visual SLAM, visual odometry explained in Chapter 7, "Imaging Sensors," is used instead of the range-based ICP or NDT scan-matching algorithm that is used in range-based SLAM. The main limitation of visual odometry compared to ICP range scan-matching is that the former gives a scaled solution if a single camera (monocular assembly) is used, while the latter gives a direct metric solution without any scale ambiguity. This issue can be resolved either by solving the scale ambiguity using known features in the scene, but this will require a higher-level understanding of the objects and their dimension in real life. Another solution to scale ambiguity is the utilization of a stereo camera assembly where camera poses and 3D map features' locations are estimated in metric units without scale ambiguity. The main drawback of this approach is the extra processing of two streams of images instead of one. In addition, stereo camera system performance deteriorates if the features are at a distance that is too long compared to the baseline distance.

- **Scale ambiguity:** the measurement model in range-based SLAM is relatively simple, especially the 2D model. In addition, the measurement model in range-based SLAM provides metric measurements without scale ambiguity. In visual SLAM, the measurement model provides pixel locations of map features, and due to scale ambiguity, multiple map features can lead to the same pixel locations, which leads to ambiguous estimation. This scale ambiguity issue can be solved using the same techniques described above.
- **The measurement model linearization**: Another challenge of the visual measurement model is linearization, as it is more complicated than the measurement model in range-based SLAM. It is difficult to linearize the measurement model in visual SLAM because the parameters we differentiate with respect to (the pose parameters) are embedded into a transformation matrix. The transformation matrix has several issues, such as singularities and redundant parameters (nine elements are used to describe rotation).
- **Addition of errors:** In EKF-SLAM, we need to correct the state by adding error vectors. In addition, in Graph-SLAMA, we build the information matrix and information vector by the addition of a linearized measurement model. As the measurement model includes rotation matrices that belong to $SO(2)$ or $SO(3)$ groups which are not Euclidean, we cannot simply add rotation matrices errors to correct a rotation matrix in EKF-SLAM or to build the information matrix and information vector in Graph-SLAM. To solve this problem, measurement models that utilize other rotation parametrization are used. The rotation parametrization is locally linear, and Euclidean is the rotation vector or the vector part of the quaternion vector. If the rotation vector or quaternion vector is used as an orientation parameter, linearization of the measurement model becomes easier, and the construction of the information matrix and information vector in Graph-SLAM becomes possible. The approach to implement this parametrization is to replace the rotation matrix by its corresponding rotation vector form or the quaternion form as follows:

$$R_t = I + \frac{\left(\vec{\phi}_t \times\right)}{\left\|\vec{\phi}_t\right\|} \sin\left(\left\|\vec{\phi}_t\right\|\right) + \frac{\left(\vec{\phi}_t \times\right)^2}{\left\|\vec{\phi}_t\right\|^2}\left(1 - \cos\left(\left\|\vec{\phi}_t\right\|\right)\right) \tag{8-105}$$

$$R_t = \begin{pmatrix} \left(a^2 + b^2 - c^2 - d^2\right) & 2(bc - ad) & 2(bd + ac) \\ 2(bc + ad) & \left(a^2 - b^2 + c^2 - d^2\right) & 2(cd - ab) \\ 2(bd - ac) & 2(cd + ab) & \left(a^2 - b^2 - c^2 + d^2\right) \end{pmatrix} \tag{8-106}$$

where $\vec{\phi}_{LB} = \left(\phi_{t,x} \quad \phi_{t,y} \quad \phi_{t,z}\right)^T$ is the rotation vector corresponding to the rotation matrix R_t and $q_t = (a, b, c, d)^T$ are the quaternion vector corresponding to the rotation matrix R_t. Given this parametrization of rotation, we can linearize the measurement model to be used in EKF-SLAM and Graph-SLAM. The same technique can be used for any 3D rotation parametrization either in visual SLAM or 3D range-based SLAM.

8.5.5 ORB Visual SLAM

There are several visual SLAM engines in the literature. One of the popular open-source visual SLAM engines is the ORB SLAM [21]. ORB stands for oriented FAST and rotated BRIEF. ORB is a feature descriptor that showed several advantages. The ORB features are known to be fast to compute and match. They also have good invariance to viewpoint, which allows feature match from wide baselines, and thereby boosting the accuracy of batch processing and bundle adjustment. The latest version of ORB SLAM can integrate inertial measurements. ORB visual SLAM uses the pinhole and fish-eye camera models, and it can work with monocular and stereo configurations. It can resolve the scale ambiguity with help from inertial sensors and it also applies loop closure to minimize drifts. ORB SLAM applies the concept of key frames to avoid wasting computational resources on feature-poor images, low-texture environments, or repeated image frames (if the vehicle is stationary). It also implements location recovery or re-localization if the matching between image frames is lost. ORB SLAM consists of two main blocks, front-end and back-end. The front-end processes incoming image frames, detects visual features, and establishes local maps. The back end establishes the optimization graph and solve the SLAM problem using optimization methods as described in Graph-SLAM. ORB works based on three main threads: Tracking, Mapping, and Loop Closure. The tracking thread processes the incoming image frames and performs feature extraction, estimates the initial pose, and identifies keyframes. The mapping threads use the keyframes and perform the key Graph-SLAM processing. The loop closure thread preforms loop detection and loop correction. Interested readers can download ORB SLAM from https://github.com/UZ-SLAMLab/ORB_SLAM3 and test it. ORB SLAM runs on Linux operating system and requires configurations. Therefore, testing ORB SLAM requires some knowledge of Linux operating system and C/C++ programming.

8.5.6 MATLAB Visual SLAM Sample Project

MATLAB computer vision toolbox has useful visual SLAM examples that further demonstrate the concepts of visual SLAM. A monocular visual SLAM example can be found in [22] and a stereo visual SLAM example can be found in [23]. In addition to the examples available by MATLAB, the projects' repository of this book includes a visual-inertial SLAM software that can be found under folder "*VisualInertialEKFSLAMProject.*" This implementation uses a multistate EKF to fuse data from inertial sensors and visual features from a single camera (monocular). The project utilizes speeded-up robust features (SURF) visual features and implements visual inertial odometry. The same algorithm can implement EKF-SLAM. The project comes with a dataset that is publicly available from the KITTI datasets website here [24]. Figure 8.15 shows a sample of an image with detected SURF features marked by + symbol. Figure 8.16 shows 2D and 3D views of the resulting trajectory, which demonstrate the ground-truth, IMU-stand-alone solution, and EKF fusion results. The IMU-stand-alone solution is drifting as can be seen in Figure 8.16. The fusion of the visual features measurements reduces the drifts; however, as both visual odometry and inertial odometry are incremental solutions (there is no absolute position fix update), the covariance is growing but at a slower pace because of the fusion. Figure 8.17 shows the covariance of the error to demonstrate the uncertainty of the estimation. The KITTI datasets come with a metric that assesses algorithms performance. The KITTI metric values of this visual inertial multistate EKF are given in Table 8.1. The mathematical details of the multistate-constrained EKF used in this MATLAB project are given in the reference [25].

Figure 8.15 Sample image frame from KITTI dataset with detected features highlighted.

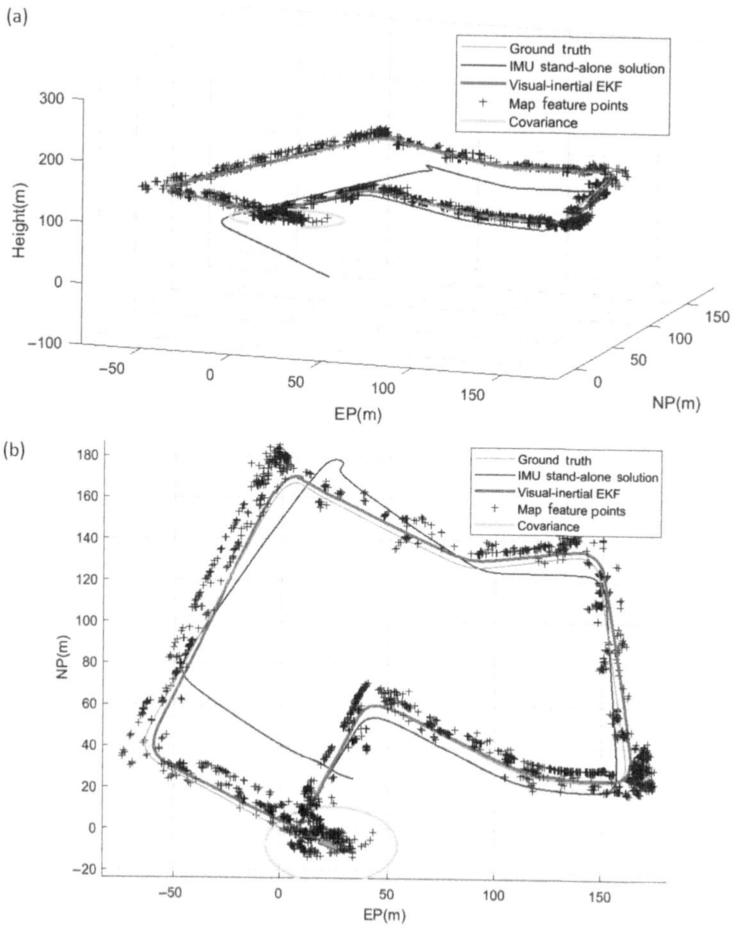

Figure 8.16 Sample results on a KITTI dataset using EKF visual/inertial SLAM Project. (a) 3D view: visual inertial multistate EKF MATLAB project. (b) 2D view: visual inertial multistate EKF MATLAB project.

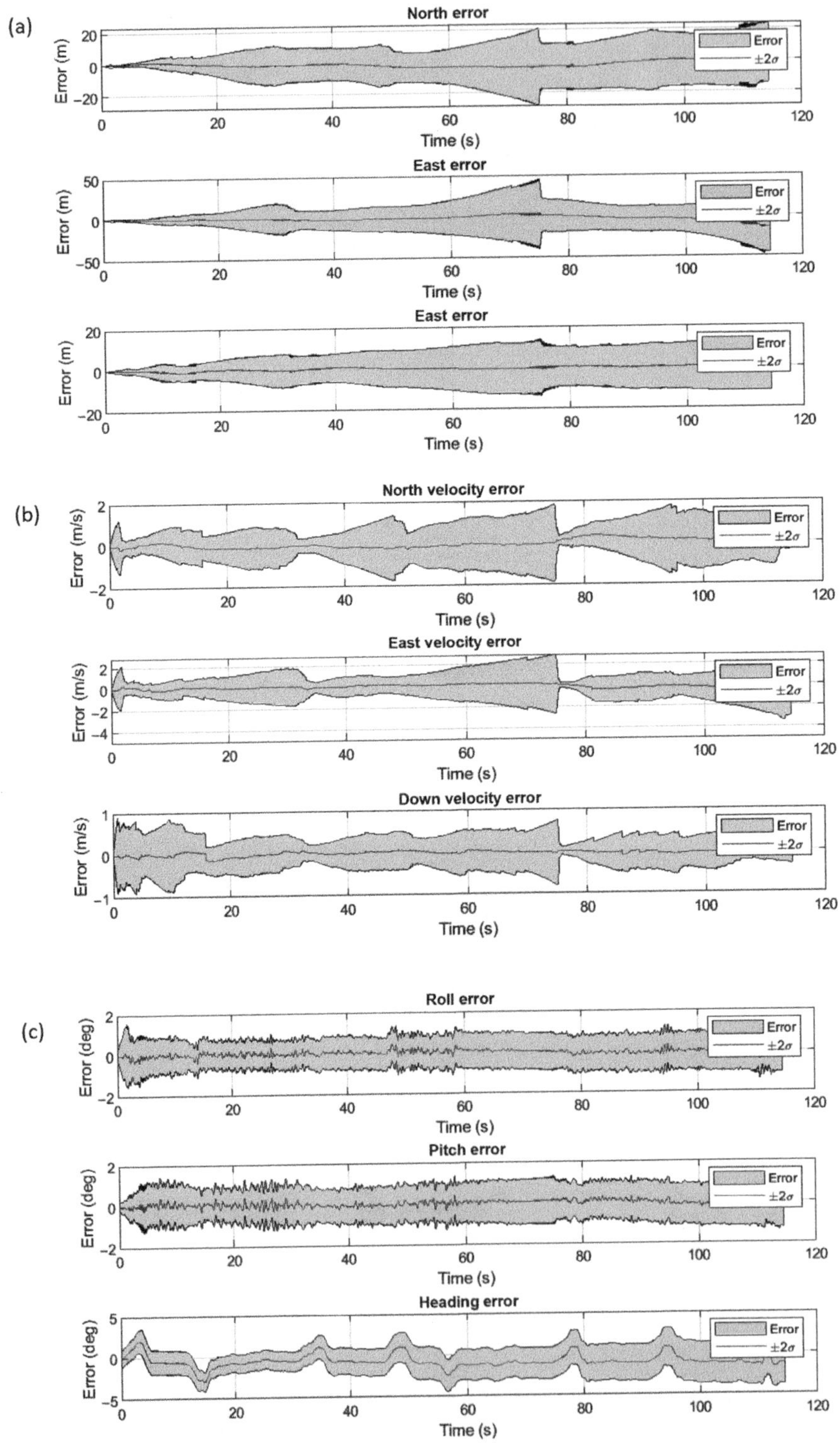

Figure 8.17 Errors of visual/inertial EKF SLAM MATLAB project. (a) Position error: visual inertial multistate EKF MATLAB project. (b) Velocity error: visual inertial multistate EKF MATLAB Project. (c) Orientation error: visual inertial multistate EKF MATLAB project.

Table 8.1 KITTI error metric for dataset sequence 07 under visual/inertial EKF.

Metric	Value
Trajectory length (m)	692.2427
Trajectory duration (s)	114
North translation error (m)	0.0261
East translation error (m)	0.0354
Alt translation error (m)	0.0134
Roll rotation error (deg)	0.0451
Pitch rotation error (deg)	0.0452
Heading rotation error (deg)	0.6901
Total number of map feature points	2321

References

1 M. McKenney and M. Schneider, Map Framework, "*A Formal Model of Maps as a Fundamental Data Type inInformation Systems*," Springer International Publishing AG, 2016.

2 P. McHaffie, S. Hwang and C. Follett, "*GIS: An Introduction to Mapping Technologies*," Boca Raton: CRC Press, 2018.

3 D. F. d. Lamotte, P. Leturmy, P. Souloumiac and A. F. d. Lamotte, "*Geological Objects and Structures in 3D: Observation, Interpretation and Building of 3D Models*," London: CRC Press, 2020.

4 J. Będkowski, "*Large-Scale Simultaneous Localization and Mapping*," Springer Nature Singapore Pte Ltd, 2022.

5 S. H. a. G. D. Zhan Wang, "*Simultaneous Localization and Mapping: Exactly Sparse Information Filters*," World Scientific Publishing Company, 2011.

6 K. P. Murphy, *Machine Learning: A Probabilistic Perspective*, London, England: The MIT Press, Cambridge, Massachusetts, 2012.

7 M. S. T. D. K. a. B. W. Montemerlo, "FastSLAM: a factored solution to the simultaneous localization and mapping problem," in *International AAAI Conference (AAAI/IAAI) 2002*, 2002.

8 ROS Gmapping, "gmapping," ROS, 1 Sept. 2023, [Online]. Available: https://wiki.ros.org/gmapping. [Accessed 1 September 2023].

9 C. Stachniss and W. Burgard, "Particle Filters for Robot Navigation," *Foundations and Trends® in Robotics*, vol. 3, pp. 211–282, 2014.

10 G. Grisetti, R. Kümmerle, C. Stachniss and W. Burgard, "A Tutorial on Graph-Based SLAM," *IEEE Intelligent Transportation Systems Magazine*, vol. 2, no. 4, pp. 31-43, 2010.

11 S. Thrun and M. Montemerlo, "The GraphSLAM Algorithm with Applications to Large-Scale Mapping of Urban Structures," *The International Journal of Robotics Research*, vol. 25, no. 5–6, p. 403, 2006.

12 Y. Ma, S. Soatto, J. Košecká and S. S. Sastry, "*An Invitation to 3-D Vision, from Images to Geometric Models*," Springer-Science+Business Media, LLC, 2004.

13 M. Weinmann, "Visual Features - From Early Concepts to Modern Computer Vision," in *Advanced Topics in Computer Vision, Advances in Computer Vision and Pattern Recognition*, Springer, 2013, pp. 1–34.

14 D. G. Lowe, "Distinctive Image Features from Scale-Invariant Keypoints," *International Journal of Computer Vision*, vol. 60, no. 2004, p. 91–110, 2004.

15 J. G. Proakis and D. G. Manolakis, "*Digital Signal Processing, Principles, Algorithms, and Applications*," New Jersey: Pearson, Prentice Hall, 2007.

16 MATLAB, "Local Feature Detection and Extraction," MathWorks Inc., [Online]. Available: https://www.mathworks.com/help/vision/ug/local-feature-detection-and-extraction.html. [Accessed 27 October 2024].

17 H. Bay, T. Tuytelaars and L. V. Gool, "SURF: Speeded Up Robust Features," *Lecture Notes in Computer Science*, Springer, vol. 3951, 2006.

18 S. Datta, "*Learning OpenCV 3 Application Development: Build, Create, and Deploy Your Own Computer Vision Applications with the Power of OpenCV*," Packt Publishing, Limited, 2016.

19 MATLAB, "Computer Vision Toolbox, Design and test computer vision, 3D vision, and video processing systems," MathWorks Inc., [Online]. Available: https://www.mathworks.com/products/computer-vision.html. [Accessed 27 October 2024].

20 MATLAB, "Monocular Visual Odometry," MathWorks Inc., 2023. [Online]. Available: https://www.mathworks.com/help/vision/ug/monocular-visual-odometry.html. [Accessed 4 February 2024].

21 C. Campos, R. Elvira, Juan J. G. Rodríguez, J. M. M. Montiel, J. D. Tardós "ORB-SLAM3: An Accurate Open-Source Library for Visual, Visual–Inertial, and Multimap SLAM," *IEEE Transactions on Robotics*, vol. 37, no. 6, pp. 1874–1890, 2021.

22 MATLAB, "Monocular Visual Simultaneous Localization and Mapping," MathWorks Inc., [Online]. Available: https://www.mathworks.com/help/vision/ug/monocular-visual-simultaneous-localization-and-mapping.html. [Accessed 27 October 2024].

23 MATLAB, "Stereo Visual SLAM for UAV Navigation in 3D Simulation," MathWorks Inc., [Online]. Available: Stereo Visual SLAM for UAV Navigation in 3D Simulation.

24 A. Geiger, P. Lenz, C. Stiller and R. Urtasun, "The KITTI Vision Benchmark Suite," Karisrushe Institute of Technology, [Online]. Available: https://www.cvlibs.net/datasets/kitti. [Accessed 27 October 2024].

25 S. Sheikhpour and M. M. Atia, "An Enhanced Visual-Inertial Navigation System Based on Multi-State Constraint Kalman Filter," 2020 IEEE 63rd International Midwest Symposium on Circuits and Systems (MWSCAS), Springfield, MA, USA, 2020, pp. 361–364, doi: https://doi.org/10.1109/MWSCAS48704.2020.9184501

9

Case Study #1: Wheeled Platforms

9.1 Introduction

In this chapter, we will apply the sensor fusion methods on wheeled platforms. Wheeled platforms (see Figure 9.1) include cars and wheeled robots. Wheeled platforms have two characteristics that make them distinguishable from other platforms. Firstly, motion is mostly constrained in the 2D plane. Secondly, ground speed can be accurately measured using wheel encoders. Wheel encoders are devices attached or embedded to wheels to count the number of times the wheel rotates per unit time. The knowledge about the diameter of the wheels is used to convert the number of wheel rotations per unit of time into speed. Differential encoders use multiple wheel encoders to measure steering orientation as well. The characteristics of wheeled platforms impose several constraints that lead to simplifications of the motion model and enhancement of the positioning, navigation, and mapping algorithms.

As discussed earlier in Chapter 2, we always need a motion prediction model (system model) and measurement model to perform sensor fusion. The motion model predicts the next platform's location, velocity, and orientation. This prediction of the navigation state is always noisy and erroneous due to sensor errors and random noise. Sensor errors are modeled and added to the motion model as part of the navigation state to correct these errors. Independent measurements are used to correct the state and compensate for sensor errors. To use the measurements to correct the states, we need a measurement model that relates the measurements to the system state. Sensor fusion is then done by predicting the motion using the motion model and then using the predicted state and the measurement model to predict the measurements. The predicted measurements are compared to the observed/sensed measurements, and the difference is used to correct the system state. This is the core philosophy of sensor fusion in general. In the chapter, we will discuss the application of this philosophy on wheeled platforms.

Sensor fusion algorithms differ in the details and design of the motion model, the measurement model, and the state correction algorithm. The design of accurate motion models and measurement models is key to performing robust and accurate sensor fusion. If we have multiple sensors, we divide them into motion prediction sensors and measurement sensors. As measurements are used to correct motion prediction, motion prediction must be performed at least at the same rate as the measurement rate or ideally higher. Therefore, motion prediction sensors are usually chosen to be the sensor with the highest rate. Also, motion prediction sensors should be self-contained and do not depend on any external factors (satellites, landmarks, visual features, light intensity, weather, etc.) to be able to provide continuous motion prediction under all circumstances. Given these constraints on the motion prediction sensors, inertial sensors, inertial measurement units (IMUs), and

Sensor Fusion Approaches for Positioning, Navigation, and Mapping: How Autonomous Vehicles and Robots Navigate in the Real World: With MATLAB Examples, First Edition. Mohamed M. Atia.
© 2025 The Institute of Electrical and Electronics Engineers, Inc. Published 2025 by John Wiley & Sons, Inc.

Figure 9.1 A wheeled robotic platform (*Source:* Indrobotics Inc.).

inertial navigation systems (INS) have unique characteristics that make them ideal choices to be used as a motion model. IMU provides high-rate measurements, lightweight, and low-power consumption at the same time. If other sources of internal odometry sensors are available (like wheel encoders in this case study), they can be used with orientation sensors to build a more simplified motion model. Therefore, in most sensor fusion systems, inertial and/or internal odometry sensors are used as motion models. In this chapter, we will mainly use IMU and internal odometry sensors as the main motion prediction sensor and INS as our main motion model. Other sensors/systems (radio positioning, RADAR, LiDAR, or Imaging sensors) will be used as external measurements. In this chapter, we will explain different configurations to fuse inertial sensors with other sensors/systems, such as radio positioning systems and ranging sensors.

In this chapter, we will discuss the fusion of radio positioning systems and LiDAR sensors. The integration of imaging and RADAR sensors will be explained in the next chapter where a 3D motion case study is considered. The most popular radio positioning systems used in modern technology are Global Navigation Satellite Systems (GNSS) such as GPS, GLONASS, Galileo, etc. (review Chapter 5 for more details about GNSS technology). In this chapter, we will assume that the radio positioning system is GPS, but the same concepts can be applied to any radio positioning system.

9.2 Reference Frames and the Lever Arm

In this section, we will explain a 2D platform, its motion model, and its linearization. A 2D motion model is used for illustration. It is suitable for most wheeled platforms that are moving mostly on flat ground. Wheeled platforms that are used on rough ground with uneven terrain will need a 3D model. 3D models are discussed in the next chapter when we consider the case of aerial vehicles. Figure 9.2 shows an illustration of the example 2D wheeled vehicle and the reference frames.

9.2.1 Reference Frames

Figure 9.2 also shows the coordinate reference frames we are going to use in our analysis which are described as follows:

- **Local-level frame (NED and ENU)**: Its origin is located a reference geodetic coordinate φ_o, λ_o, h_o (latitude, longitude, and altitude) which can be chosen to be the initial position of the vehicle.

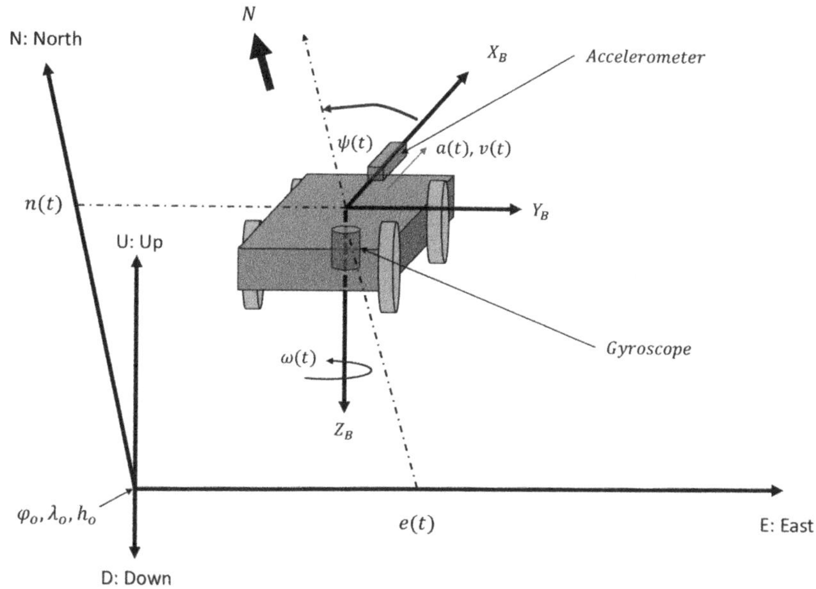

Figure 9.2 2D wheeled vehicle frames and sensors diagram.

The E-axis points to the East direction, the N-axis points to the north direction, and up-axis points to the upward vertical to the east-north plane. This configuration is called east-north-up (ENU) frame. In other configurations, a NED (north-east-down) configuration can be used, and, in this case, we would have a down (D-axis) instead of U-axis. D-axis would point downward vertical to the east-north plane. We will refer to NED frame as L frame and ENU frame as N frame. The geodetic coordinate φ_o, λ_o, h_o (latitude, longitude, and altitude) is the reference point where the vehicle's local position (east and north components) is referenced to.

- **Body-frame (X_B, Y_B, Z_B):** The origin of this frame is located at the center of mass of the vehicle and axes are defined as shown in Figure 9.2. Y_B-axis is pointing to forward motion direction of the vehicle, X_B axis is in the lateral direction of the vehicle, and z_B axis is vertical to X_B-Y_B plane and completes the right-hand rule. We will use the letter B to refer to the body frame.
- **Inertial sensor frame (X_i, X_i, Z_i):** This is the frame of the inertial sensors. Ideally, the inertial sensor frame should be located at or close to the center of mass of the vehicle. In this chapter, and in the following chapters, it is assumed that the inertial sensor frame is located at the center of mass and aligned with the body frame. We will refer to the inertial sensors frame by IMU frame.

9.2.2 Lever Arm Consideration

When the vehicle receives measurements from sensors other than the inertial sensors, like Global Positioning System (GPS), it is important to transform these measurements to the inertial sensors frame, which is assumed to be the vehicle body frame B. Figure 9.3 shows an illustration of the lever arm with GPS case. The GPS receiver calculates the position and velocity of the antenna. To fuse these measurements with inertial sensors, GPS measurements need to be transformed into the inertial sensors frame (IMU frame). For example, if GPS provides $\vec{p}_{GPS}$: north, east, up position measurements, we need to map to the IMU frame as follows:

$$\vec{p}_{IMU} = \vec{p}_{GPS} + \vec{L} \tag{9-1}$$

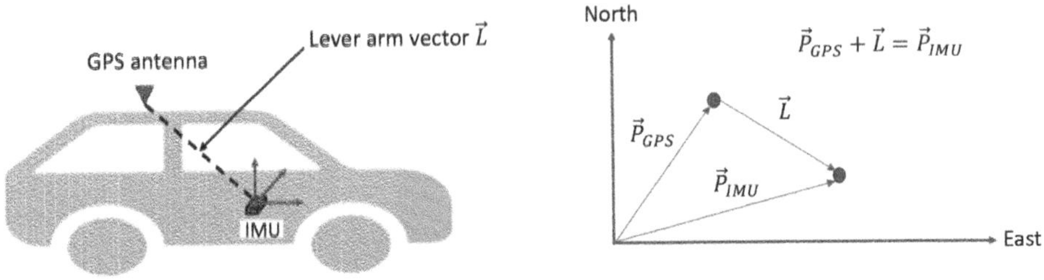

Figure 9.3 Lever arm illustration.

where $\vec{L}$ is the lever arm vector. In this example, the lever arm is simply a vector. In more complex cases where the other sensor provides raw measurements such as range/angle measurements (e.g., a LiDAR sensor), and we need to fuse these raw measurements with inertial sensors, the other sensor measurements must be transformed to the IMU frame using a transformation matrix that contains both orientation and translation mapping from the other sensor frame to the IMU frame. For simplicity, in this chapter, we will assume that the lever arm issue is already taken care of and other sensors' measurements are already transformed to the IMU frame so they are ready to be fused with IMU measurements to provide the vehicle's state.

9.3 2D System Model

The vehicle is equipped with a gyroscope that measures rotation around z_B-axis and an accelerometer that measures acceleration in y_B-axis. The navigation state of the vehicle can be described as follows:

- Azimuth angle $\psi(t)$
- North position $n(t)$
- East position $e(t)$
- North velocity $v_n(t)$
- East velocity $v_e(t)$

The east-north position components are Cartesian coordinates with respect to a reference location on Earth determined by unique reference latitude φ_o and longitude λ_o. The relationship between east-north coordinates and geodetic coordinates is given by the following equations:

$$e(t) = (\lambda(t) - \lambda_o)(R_N(t) + h(t))\cos(\varphi(t)) \tag{9-2}$$

$$n(t) = (\varphi(t) - \varphi_o)(R_M(t) + h(t)) \tag{9-3}$$

where $\varphi(t)$, $\lambda(t)$, $h(t)$ are the vehicle's coordinates in geodetic coordinate system (review Chapter 1 for details about coordinate systems). The control signals of the vehicle are the steering angular rate $\omega(t)$ (also known as turn rate), which is measured by the gyroscope, and the forward acceleration $a(t)$, which is measured by the accelerometer. As the gyroscope and the accelerometer measurements are contaminated with random noise and errors, we will model the error by a random bias as follows:

$$\hat{\omega}(t) = \omega(t) - b_\omega(t) + \sigma_\omega(t) \tag{9-4}$$

$$\hat{a}(t) = a(t) - b_a(t) + \sigma_a(t) \tag{9-5}$$

where $\hat{\omega}(t), \hat{a}(t)$ are the measured turn rate and forward acceleration, $\omega(t)$, $a(t)$ are the true measured turn rate and forward acceleration, $\sigma_\omega(t)$, $\sigma_a(t)$ are zero-mean Gaussian noise, and $b_\omega(t)$, $b_a(t)$ are bias errors modeled as random walk processes as follows:

$$\dot{b}_\omega(t) = w_\omega(t) \tag{9-6}$$

$$\dot{b}_a(t) = w_a(t) \tag{9-7}$$

where $w_\omega(t)$ and $w_a(t)$ are zero-mean Gaussian noises that reflect the quality of the gyroscope and accelerometer sensors. Note that we assumed a very simple model for the bias, which is a random constant. In practice, especially with low-quality sensors, more sophisticated time-varying random processes, such as Gauss–Markov processes, can be used to model the noise behavior of the sensors accurately. Given this information, we can model the motion of the vehicle as follows:

$$\psi(t) = \int (\omega(t) - b_\omega(t))\, dt \tag{9-8}$$

$$v_n(t) = \int (a(t) - b_a(t)) \cos \psi(t) dt \tag{9-9}$$

$$v_e(t) = \int (a(t) - b_a(t)) \sin \psi(t) dt \tag{9-10}$$

$$n(t) = \int v_n(t) dt \tag{9-11}$$

$$e(t) = \int v_e(t) dt \tag{9-12}$$

We can define the vehicle state vector $\vec{x}(t)$ and control vector $\vec{u}(t)$ as follows:

$$\vec{x}(t) = \begin{pmatrix} n(t) \\ e(t) \\ v_n(t) \\ v_e(t) \\ \psi(t) \\ b_a(t) \\ b_\omega(t) \end{pmatrix} \tag{9-13}$$

$$\vec{u}(t) = \begin{pmatrix} a(t) \\ \omega(t) \end{pmatrix} \tag{9-14}$$

The differential equation that describes how the vehicle's state vector is changing overtime (including the biases) is given by the following equations:

$$\dot{n}(t) = v_n(t) + w_n(t) \tag{9-15}$$

$$\dot{e}(t) = v_e(t) + w_e(t) \tag{9-16}$$

$$\dot{v}_n(t) = (a(t) - b_a(t)) \cos \psi(t) + w_{v_n}(t) \tag{9-17}$$

$$\dot{v}_e(t) = (a(t) - b_a(t))\ \sin\ \psi(t) + w_{v_e}(t) \tag{9-18}$$

$$\dot{\psi}(t) = (\omega(t) - b_\omega(t)) + w_\psi(t) \tag{9-19}$$

$$\dot{b}_\omega(t) = w_\omega(t) \tag{9-20}$$

$$\dot{b}_a(t) = w_a(t) \tag{9-21}$$

where $w_\psi(t), w_{v_n}(t), w_{v_e}(t), w_n(t), w_\omega$, w_a, and $w_e(t)$ are zero-mean Gaussian noise. Note that the above set of equations form the following general system model:

$$\dot{\vec{x}}(t) = f\left(\vec{x}(t), \vec{u}(t)\right) + \vec{w}(t) \tag{9-22}$$

where $\vec{x}(t)$ is the state vector, $\vec{u}(t)$ is the control signal vector, and $w(t)$ is the system noise vector defined as follows:

$$\vec{w}(t) = \begin{pmatrix} w_e(t) \\ w_n(t) \\ w_{v_n}(t) \\ w_{v_e}(t) \\ w_\psi(t) \\ w_a(t) \\ w_\omega(t) \end{pmatrix} \tag{9-23}$$

Assuming that all errors are independent, the covariance of the noise vector $\vec{w}(t)$ is given by the following matrix:

$$Q(t) = \langle w^T(t)\, w(t) \rangle = \begin{pmatrix} {q_e}^2 & 0 & 0 & 0 & 0 & 0 & 0 \\ 0 & {q_n}^2 & 0 & 0 & 0 & 0 & 0 \\ 0 & 0 & {q_{v_n}}^2 & 0 & 0 & 0 & 0 \\ 0 & 0 & 0 & {q_{v_e}}^2 & 0 & 0 & 0 \\ 0 & 0 & 0 & 0 & {q_\psi}^2 & 0 & 0 \\ 0 & 0 & 0 & 0 & 0 & {q_a}^2 & 0 \\ 0 & 0 & 0 & 0 & 0 & 0 & {q_\omega}^2 \end{pmatrix} \tag{9-24}$$

In practice, $Q(t)$ depends on the quality of the system prediction, which depends on the quality of the sensors and the accuracy of the motion model.

9.4 2D Error Model

We will need to linearize this motion model in later steps when we discuss the fusion of other sensor measurements. The linearized motion model is simply derived by applying the first-order Tylor series approximation to the motion. We obtain the linearized error model as follows:

$$\delta\dot{\vec{x}}(t) \simeq \frac{\partial f\left(\vec{x}(t), \vec{u}(t)\right)}{\partial \vec{x}} \delta\vec{x}(t) \tag{9-25}$$

which gives the following error model equations:

$$\delta\dot{n}(t) = \delta v_n(t) \tag{9-26}$$

$$\delta\dot{e}(t) = \delta v_e(t) \tag{9-27}$$

$$\delta\dot{v}_n(t) \simeq -(a(t) - b_a(t)) \sin \psi(t)\delta\psi(t) - \cos \psi(t)\delta b_a \tag{9-28}$$

$$\delta\dot{v}_e(t) \simeq (a(t) - b_a(t)) \cos \psi(t)\delta\psi(t) - \sin \psi(t)\delta b_a \tag{9-29}$$

$$\delta\dot{\psi}(t) \simeq -\delta b_\omega(t) \tag{9-30}$$

$$\delta\dot{b}_a(t) = 0 \tag{9-31}$$

$$\delta\dot{b}_\omega(t) = 0 \tag{9-32}$$

Rearranging these error equations into matrix form will give the following matrix equation:

$$\delta\dot{\vec{x}}(t) = F(t)\delta\vec{x}(t) \tag{9-33}$$

where $F(t)$ is the error model transition matrix defined as follows:

$$F(t) = \begin{pmatrix} 0 & 0 & 1 & 0 & 0 & 0 & 0 \\ 0 & 0 & 0 & 1 & 0 & 0 & 0 \\ 0 & 0 & 0 & 0 & -(a(t) - b_a(t)) \sin \psi(t) & -\cos\psi(t) & 0 \\ 0 & 0 & 0 & 0 & (a(t) - b_a(t)) \cos \psi(t) & -\sin\psi(t) & 0 \\ 0 & 0 & 0 & 0 & 0 & 0 & -1 \\ 0 & 0 & 0 & 0 & 0 & 0 & 0 \\ 0 & 0 & 0 & 0 & 0 & 0 & 0 \end{pmatrix} \tag{9-34}$$

The complete stochastic linearized motion error model is in the following form:

$$\delta\dot{\vec{x}}(t) = F(t)\delta\vec{x}(t) + G(t)\vec{w}_u(t) \tag{9-35}$$

where $\vec{w}_u$ is a unity-variance zero-mean Gaussian noise signal and $G(t)$ matrix is a system noise shaping matrix defined as follows:

$$G(t) = \begin{pmatrix} {q_e}^2 & 0 & 0 & 0 & 0 & 0 & 0 \\ 0 & {q_n}^2 & 0 & 0 & 0 & 0 & 0 \\ 0 & 0 & {q_{v_n}}^2 & 0 & 0 & 0 & 0 \\ 0 & 0 & 0 & {q_{v_e}}^2 & 0 & 0 & 0 \\ 0 & 0 & 0 & 0 & {q_\psi}^2 & 0 & 0 \\ 0 & 0 & 0 & 0 & 0 & {q_a}^2 & 0 \\ 0 & 0 & 0 & 0 & 0 & 0 & {q_\omega}^2 \end{pmatrix} \tag{9-36}$$

Assuming that the equations of motion are accurate, the accuracy of the motion model prediction depends on the quality of the sensors that measure the turn rate $\omega(t)$ and forward acceleration $a(t)$. As the sensors are always contaminated with stochastic errors, the motion model output will drift, leading to accumulated growing errors. In the remainder of this chapter, we will discuss how we can correct the errors and reduce drifts by fusing/integrating radio positioning systems and active ranging sensors (using LiDAR as an example) as external measurements. We will see how sensor fusion provides accurate estimation of the positioning, navigation and sensors' biases, which reflects on the accuracy of mapping as well.

9.5 Loosely Coupled GPS Measurement Model

In this section, we will assume that the external measurements are coming from a radio positioning system, which is GPS. GPS and GNSS technologies have been explained in Chapter 5, "Radio Positioning Systems." In loosely coupled mode, we use the position, velocity, (and maybe orientation) measurements that are independently calculated by GPS receivers where at least four satellites are visible, as illustrated in Figure 9.4. To fuse the measurements, we must have a measurement model that maps external measurements to system states. We will discuss the measurement model for different measurements in the next subsections.

9.5.1 Measurement Model Equation

The GPS receiver provides position and velocity measurements of the GPS antenna, which is located at a vector $\vec{L}$ with respect to the inertial sensors frame, which is assumed to be the vehicle body frame B. However, we will assume that the GPS positioning measurements are already given in the body frame B. This will simplify the notation as we do not have to model or consider the transformation between the GPS antenna location, the inertial sensor's frame, and the body frame. The transformation between the GPS antenna location and the origin of the body frame is called the lever arm, and it must be considered if the location of the GPS receiver's antenna is not collocated with the inertial sensors' body-frame origin. The general form of the model is given by:

$$\vec{y}(t) = h\left(\vec{x}(t)\right) + \vec{\varepsilon}(t) \tag{9-37}$$

where $\vec{\varepsilon}(t)$ is random noise and $\vec{y}(t)$ is GPS loosely coupled measurement. The measurement model can be linear or nonlinear in the system states depending on the measurement type. We will show an example for linear and nonlinear measurement model in the following subsections.

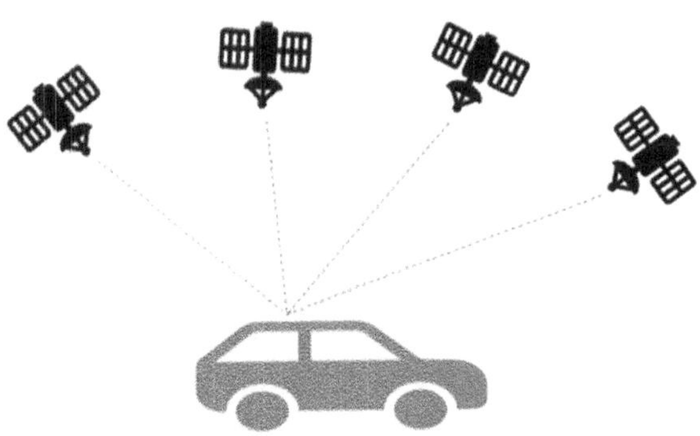

Figure 9.4 Minimum of four visible satellites = GPS available.

9.5.2 Linear Measurement Model

Assume the GPS receiver provides a measurement vector defined as follows:

$$\vec{y}(t) = \begin{pmatrix} \hat{n}(t) \\ \hat{e}(t) \\ \hat{v}_n(t) \\ \hat{v}_e(t) \end{pmatrix} \tag{9-38}$$

This measurement vector is related to the system state vector $\vec{x}(t)$ as follows:

$$\vec{y}(t) = \begin{pmatrix} 1 & 0 & 0 & 0 & 0 & 0 & 0 \\ 0 & 1 & 0 & 0 & 0 & 0 & 0 \\ 0 & 0 & 1 & 0 & 0 & 0 & 0 \\ 0 & 0 & 0 & 1 & 0 & 0 & 0 \end{pmatrix} \begin{pmatrix} n(t) \\ e(t) \\ v_n(t) \\ v_e(t) \\ \psi(t) \\ b_a(t) \\ b_\omega(t) \end{pmatrix} \tag{9-39}$$

which can be put in the following linear stochastic measurement model:

$$\vec{y}(t) = H(t)\vec{x}(t) + \vec{\varepsilon}(t) \tag{9-40}$$

where $H(t)$ is the design matrix defined as follows:

$$H(t) = \begin{pmatrix} 1 & 0 & 0 & 0 & 0 & 0 & 0 \\ 0 & 1 & 0 & 0 & 0 & 0 & 0 \\ 0 & 0 & 1 & 0 & 0 & 0 & 0 \\ 0 & 0 & 0 & 1 & 0 & 0 & 0 \end{pmatrix} \tag{9-41}$$

and $\vec{\varepsilon}(t)$ is zero-mean Gaussian noise that presents the quality of the GPS receiver measurements with the following covariance matrix:

$$R(t) = \left\langle \vec{\varepsilon}(t)^T \vec{\varepsilon}(t) \right\rangle = \begin{pmatrix} {r_n}^2 & 0 & 0 & 0 \\ 0 & {r_e}^2 & 0 & 0 \\ 0 & 0 & {r_{v_n}}^2 & 0 \\ 0 & 0 & 0 & {r_{v_e}}^2 \end{pmatrix} \tag{9-42}$$

This measurement model is linear, which facilitates the application of Kalman filtering. Given a linear measurement model, the update step of an extended Kalman Filter can be applied without the need for linearization.

9.5.3 Nonlinear Measurement Model

Assume the GPS provides a measurement vector defined as follows:

$$\vec{y}(t) = \begin{pmatrix} \hat{n}(t) \\ \hat{e}(t) \\ \hat{v}(t) \end{pmatrix} \tag{9-43}$$

where $\hat{v}(t)$ is the forward speed of the platform. This measurement vector is related to the system state $\vec{x}(t)$ as follows:

$$\vec{y}(t) = h\left(\vec{x}(t)\right) = \begin{pmatrix} n(t) \\ e(t) \\ \sqrt{(v_n(t))^2 + (v_e(t))^2} \end{pmatrix} \tag{9-44}$$

which is a nonlinear measurement model. To fuse these measurements using an EKF algorithm, we will need the linearized form of this measurement model. The linearization of this measurement model is given by the following derivative:

$$H(t) = \frac{\partial h\left(\vec{x}(t)\right)}{\partial \vec{x}} \tag{9-45}$$

Only the last element in the measurement vector is nonlinear. Given an operating state $(\overline{n}(t)\ \overline{e}(t)\ \overline{v}_n(t)\ \overline{v}_e(t)\ \overline{\psi}(t)\ )^T$, the linearized measurement model can be obtained by applying Tylor series expansion as follows:

$$v(t) = \overline{v}(t) + \frac{\overline{v}_n(t)}{\sqrt{(\overline{v}_n(t))^2 + (v_e(t))^2}}\delta v_n(t) + \frac{\overline{v}_e(t)}{\sqrt{(\overline{v}_n(t))^2 + (\overline{v}_e(t))^2}}\delta v_e(t) + H.O.T \tag{9-46}$$

Ignoring the higher-order terms (*H.O.T*), the linearized measurement error model will be given by:

$$\delta\vec{y}(t) = H(t)\delta\vec{x}(t) + \vec{\varepsilon}(t) \tag{9-47}$$

where *H*(*t*) is a design matrix given by the following matrix:

$$H(t) = \begin{pmatrix} 1 & 0 & 0 & 0 & 0 & 0 & 0 \\ 0 & 1 & 0 & 0 & 0 & 0 & 0 \\ 0 & 0 & \frac{\overline{v}_n(t)}{\sqrt{(\overline{v}_n(t))^2 + (v_e(t))^2}} & \frac{\overline{v}_e(t)}{\sqrt{(\overline{v}_n(t))^2 + (\overline{v}_e(t))^2}} & 0 & 0 & 0 \end{pmatrix} \tag{9-48}$$

and $\vec{\varepsilon}(t)$ is zero-mean Gaussian noise that presents the quality of the GPS receiver measurements with the following covariance matrix:

$$R(t) = \left\langle \vec{\varepsilon}(t)^T\, \vec{\varepsilon}(t) \right\rangle = \begin{pmatrix} {r_n}^2 & 0 & 0 \\ 0 & {r_e}^2 & 0 \\ 0 & 0 & {r_v}^2 \end{pmatrix} \tag{9-49}$$

9.5.4 Heading (Azimuth) Measurement

Some GPS receivers have a double antenna where the precise position of each antenna is calculated and used to calculate the heading of the platform. Figure 9.5 shows a simple illustration of double antenna GPS receiver. The heading angle can be calculated as follows:

$$\psi(t) = \tan^{-1}\left(\frac{n_1(t) - n_2(t)}{e_1(t) - e_2(t)}\right) \tag{9-50}$$

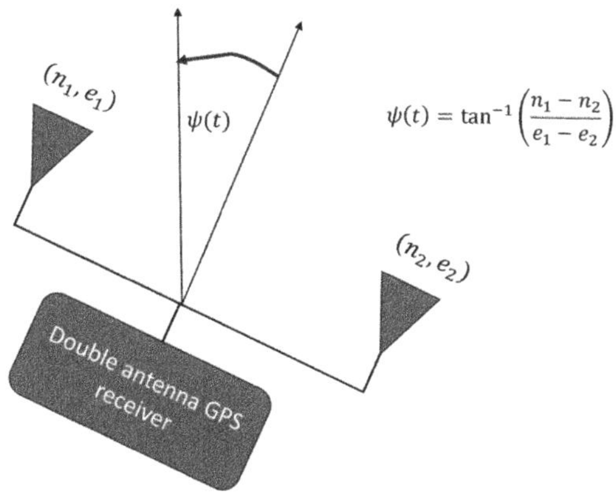

Figure 9.5 Double antenna GPS receiver can provide a heading angle.

This double antenna configuration assumes very precise and accurate positioning. If a heading angle is available, the measurement model can be augmented by adding an additional row to the measurement vector and design matrix. If we consider the linear measurement model where position and velocity updates are available, we will have the following measurement vector and model:

$$\vec{y}(t) = \begin{pmatrix} \hat{n}(t) \\ \hat{n}(t) \\ \hat{v}_n(t) \\ \hat{v}_e(t) \\ \hat{\psi}(t) \end{pmatrix} = \begin{pmatrix} 1 & 0 & 0 & 0 & 0 & 0 & 0 \\ 0 & 1 & 0 & 0 & 0 & 0 & 0 \\ 0 & 0 & 1 & 0 & 0 & 0 & 0 \\ 0 & 0 & 0 & 1 & 0 & 0 & 0 \\ 0 & 0 & 0 & 0 & 1 & 0 & 0 \end{pmatrix} \begin{pmatrix} n(t) \\ e(t) \\ v_n(t) \\ v_e(t) \\ \psi(t) \\ b_a(t) \\ b_\omega(t) \end{pmatrix} + \vec{\varepsilon}(t) \tag{9-51}$$

where $\vec{\varepsilon}(t)$ is zero-mean Gaussian noise that presents the quality of the GPS receiver measurements with the following covariance matrix:

$$R(t) = \left\langle \vec{\varepsilon}(t)^T \vec{\varepsilon}(t) \right\rangle = \begin{pmatrix} {r_n}^2 & 0 & 0 & 0 & 0 \\ 0 & {r_e}^2 & 0 & 0 & 0 \\ 0 & 0 & {r_{v_n}}^2 & 0 & 0 \\ 0 & 0 & 0 & {r_{v_e}}^2 & 0 \\ 0 & 0 & 0 & 0 & {r_\psi}^2 \end{pmatrix} \tag{9-52}$$

This measurement model is linear, which facilitates the application of Kalman filtering. Given a linear measurement model, the update step of an extended Kalman filter can be applied without linearization. If the measurement vector includes the position, speed, and heading, the linearized measurement error model is given by:

$$\delta\vec{y}(t) = H(t)\delta\vec{x}(t) + \vec{\varepsilon}(t) \tag{9-53}$$

where $H(t)$ is a design matrix given by the following matrix:

$$H(t) = \begin{pmatrix} 1 & 0 & 0 & 0 & 0 & 0 & 0 \\ 0 & 1 & 0 & 0 & 0 & 0 & 0 \\ 0 & 0 & \dfrac{\overline{v}_n(t)}{\sqrt{(\overline{v}_n(t))^2 + (v_e(t))^2}} & \dfrac{\overline{v}_e(t)}{\sqrt{(\overline{v}_n(t))^2 + (\overline{v}_e(t))^2}} & 0 & 0 & 0 \\ 0 & 0 & 0 & 0 & 1 & 0 & 0 \end{pmatrix} \tag{9-54}$$

and $\vec{\varepsilon}(t)$ is zero-mean Gaussian noise that presents the quality of the GPS receiver measurements with the following covariance matrix:

$$R(t) = \left\langle \vec{\varepsilon}(t)^T \vec{\varepsilon}(t) \right\rangle = \begin{pmatrix} r_n^2 & 0 & 0 & 0 \\ 0 & r_e^2 & 0 & 0 \\ 0 & 0 & r_v^2 & 0 \\ 0 & 0 & 0 & r_\psi^2 \end{pmatrix} \tag{9-55}$$

9.6 Tightly Coupled GPS Measurement Model

GPS loosely coupled fusion requires a position fix calculated independently by the GPS receiver, which needs at least four visible satellites. If less than four satellites are available, a loosely coupled fusion scheme does not work. This configuration is illustrated in Figure 9.6.

To overcome these limitations, a tightly coupled fusion scheme uses the range and Doppler measurements from individual satellites so it can work even with one visible satellite. Having range and Doppler measurements even from one satellite can help reduce the drifts. In the following sections, we will discuss the measurement model and fusion in a tightly coupled fusion scheme. As discussed earlier in Chapter 5 "Radio Positioning Systems," Doppler is a frequency shift that happens due to the relative radial speed between transmitter and receiver, and it is directly proportional to the rate of change of range (range rate or radial speed). Therefore, we will use the term Doppler to refer to range rate. The range and Doppler equations are more convenient when they are given in the ECEF frame. Therefore, we will formulate the measurement model and its linearization in the ECEF frame. Therefore, we will define the platform position and velocity in the ECEF frame. The correction of the platform state requires that the correction and the state be defined in the same frame (ECEF or NED). As the system model (the platform motion model) is given in the NED local frame, we will need to convert variables between the ECEF and NED frame. We will use $\vec{p}$ and $\vec{v}$ to refer to the platform position and velocity in the ECEF frame.

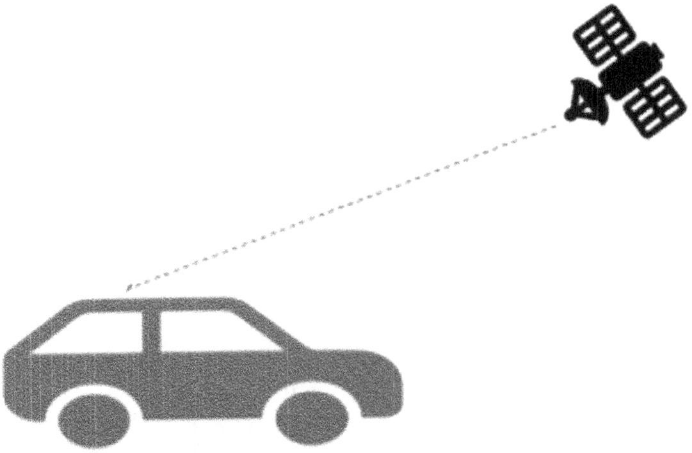

Figure 9.6 Illustration of limited GPS visibility (only one satellite is visible).

$$\vec{p} = \left(p_x, p_y, p_z\right)^T \tag{9-56}$$

$$\vec{v} = \left(v_x, v_y, v_z\right)^T \tag{9-57}$$

9.6.1 Range and Doppler Measurement Model

The measurement model in tightly coupled case is a model that relates range and range rate (doppler) measurements to the state vector. The general GPS range and Doppler equations are given by the following equation:

$$y_r^i(t) = \rho^i(t) = \|\vec{p}(t) - \vec{p}^i(t)\|_2 + c\Delta t_r(t) + c\delta t^i(t) + I_r^i(t) + T_r^i(t) + M_\rho^i + \nu_\rho^i(t) \tag{9-58}$$

$$y_d^i(t) = \dot{\rho}^i(t) = \left(\vec{v}(t) - \vec{v}^i(t)\right)^T . \vec{L}^i(t) + c\Delta\dot{t}_r(t) + \varepsilon_{\dot{\rho}}^i(t) \tag{9-59}$$

where c is the speed of light, $\rho^i, \dot{\rho}^i$ is pseudorange and Doppler measurements from i^{th} satellite, $\vec{p} \in \mathbb{R}^3$ is the vehicle's receiver's antenna position vector in the ECEF frame, $\vec{p}^i = \left(p_x^i, p_y^i, p_z^i\right)^T, \vec{v}^i = \left(v_x^i, v_y^i, v_z^i\right)^T$ are the ith satellite's position and velocity vector in the ECEF frame, $\vec{L}^i$ is the line-of-sight (LOS) vector between the receiver's antenna position and the satellite position, Δt_r and $\Delta\dot{t}_r$ are the receiver clock bias and drift, δt^i is the satellite's clock error, I_r^i is the ionospheric error, T_r^i is the tropospheric error, M_ρ^i is the multipath error, ν_ρ^i and $\varepsilon_{\dot{\rho}}^i$ are random noise, and $\| \vec{p} - \vec{p}^i \|_2$ is the true range between the receiver and the satellite calculated by the following equation:

$$\|\vec{p} - \vec{p}^i\|_2 = \sqrt{(p_x - p_x^i)^2 + \left(p_y - p_y^i\right)^2 + (p_z - p_z^i)^2} \tag{9-60}$$

where $\vec{p} = \left(p_x, p_y, p_z\right)^T$ and $\vec{p}^i = \left(p_x^i, p_y^i, p_z^i\right)^T$ are receiver's antenna position, and satellite position in ECEF frame, respectively. $\vec{L}^i$, the LOS vector between the receiver's antenna position and the ith satellite's antenna position is given by the following equation:

$$\vec{L}^i = \frac{\left((p_x - p_x^i), \left(p_y - p_y^i\right), (p_z - p_z^i)\right)^T}{\sqrt{(p_x - p_x^i)^2 + \left(p_y - p_y^i\right)^2 + (p_z - p_z^i)^2}} \tag{9-61}$$

As discussed in Chapter 5 "Radio Positioning Systems," satellite clock errors δt^i is calculated by the GPS ground control stations and transmitted to satellites, which in turn transmit it to receivers through the navigation message frames. Ionospheric and Tropospheric delays are estimated by well-known models so we will assume that they are known. We will also assume that there is no multipath error. The simplified range and Doppler measurement models are then given by the following equations:

$$\rho^i(t) = \sqrt{(p_x(t) - p_x^i(t))^2 + \left(p_y(t) - p_y^i(t)\right)^2 + (p_z(t) - p_z^i(t))^2} + b_r(t) + \nu_\rho^i(t) \tag{9-62}$$

$$b_r(t) = c\Delta t_r(t) \tag{9-63}$$

$$\dot{\rho}^i(t) = \left((v_x(t) - v_x^i(t)), \left(v_y(t) - v_y^i(t)\right), (v_z(t) - v_z^i(t))\right)^T . \vec{L}^i(t) + d_r(t) + \varepsilon_{\dot{\rho}}^i(t) \tag{9-64}$$

$$d_r(t) = \dot{b}_r(t) = c\Delta\dot{t}_r(t) \tag{9-65}$$

where $b_r(t)$ is the receiver clock bias converted into meters and $d_r(t)$ is the receiver's clock drift rate converted into meters per second.

9.6.2 Augmented State Error Model

From the measurement model, we notice that we have two new unknowns: the receiver clock bias $b_r(t)$ and the receiver clock rate $d_r(t)$. These new unknowns need to be added to the state vector and the system model. The following two equations give the system prediction model of these two new unknowns:

$$\dot{b}_r(t) = d_r(t) \tag{9-66}$$

$$\dot{d}_r(t) = w_r(t) \tag{9-67}$$

where $w_r(t)$ is a zero-mean Gaussian noise. Note that we modeled the clock drift rate as a random walk process. Other random processes, such as the Gauss–Markov random process, can be used. The linearized system model of these two new unknows are given by:

$$\delta\dot{b}_r(t) = \delta d_r(t) \tag{9-68}$$

$$\delta\dot{d}_r(t) = w_r(t) \tag{9-69}$$

Note that in all the above equations, all coordinates are given un ECEF frame, so we either perform the fusion in the ECEF frame and convert results back to the local NED frame or we move all equations to NED frame and perform the fusion directly in NED frame. Both approaches are correct. Given the two new unknowns $b_r(t)$ and $d_r(t)$, the augmented platform state vector (in NED frame) is given by the following vector:

$$\vec{x}(t) = \begin{pmatrix} n(t) \\ e(t) \\ v_n(t) \\ v_e(t) \\ \psi(t) \\ b_a(t) \\ b_\omega(t) \\ b_r(t) \\ d_r(t) \end{pmatrix} \tag{9-70}$$

9.6.3 GPS Linearized Range and Doppler Measurement Model

If we have M satellites, we will have a $2M$ measurement vector $\vec{y}(t)$ as follows:

$$\vec{y}(t) = \begin{pmatrix} \rho^1(t) \\ \rho^2(t) \\ . \\ . \\ \rho^M(t) \\ \dot{\rho}^1(t) \\ . \\ . \\ \dot{\rho}^M(t) \end{pmatrix} \tag{9-71}$$

where $\rho^i(t)$ and $\dot{\rho}^i(t)$ are defined by Eqs. (9-62) and (9-64). Applying first-order Tylor series linearizing, we get the following linearized measurement error model (for simplicity, we will omit the time variable from the notation):

$$\delta\rho^i \cong \left(\vec{L}^i(t)\right)^T \begin{pmatrix} \delta p_x \\ \delta p_y \\ \delta p_z \end{pmatrix} + \delta b_r \tag{9-72}$$

$$\delta\dot{\rho}^i \cong \left(\vec{L}^i(t)\right)^T \begin{pmatrix} \delta v_x \\ \delta v_y \\ \delta v_z \end{pmatrix} + \delta d_r(t) \tag{9-73}$$

Note here that we differentiated $\dot{\rho}^i$ with respect to the velocity state variables v_x, v_y, v_z although the $\dot{\rho}^i$ equation has the $\vec{L}^i(t)$ term which contains position state variables p_x, p_y, p_z. The $\vec{L}^i(t)$ formula contains the distance between the satellite and the receiver in the denominator. The differentiation will square this distance. The distance between the satellite and the receiver antenna is very large. Having the squared distance in the denominator will significantly attenuate the effect of positional errors on the Doppler error. That is the reason we ignored this part in the linearization of the Doppler measurement model. Rearranging this linearized measurement model into array form $\delta\vec{y}(t) = H(t)\delta\vec{x}(t)$ gives the following linearized measurement model:

$$\underbrace{\begin{pmatrix} \delta\rho^1 \\ \delta\rho^2 \\ \vdots \\ \delta\rho^M \\ \delta\dot{\rho}^1 \\ \delta\dot{\rho}^2 \\ \vdots \\ \delta\dot{\rho}^M \end{pmatrix}}_{\delta\vec{y}(t)} = \underbrace{\begin{pmatrix} \left(\vec{L}^1(t)\right)^T & 1 & 0_{3\times 1} & 0 \\ \vdots & \vdots & \vdots & \vdots \\ \left(\vec{L}^M(t)\right)^T & 1 & 0_{3\times 1} & 0 \\ 0_{3\times 1} & 0 & \left(\vec{L}^1(t)\right)^T & 1 \\ \vdots & \vdots & \vdots & \vdots \\ 0_{3\times 1} & 0 & \left(\vec{L}^M(t)\right)^T & 1 \end{pmatrix}}_{H(t)} \underbrace{\begin{pmatrix} \delta p_x \\ \delta p_x \\ \delta p_x \\ \delta b_r \\ \delta v_x \\ \delta v_x \\ \delta v_x \\ \delta d_r \end{pmatrix}}_{\delta\vec{x}} + \vec{\varepsilon}(t) \tag{9-74}$$

As the position and velocity vectors are in ECEF frame, we must convert them into local frame so we can integrate the platform's motion model and satellite measurements within the same coordinate frame. Therefore, we need to convert the error vector in the above model from the ECEF frame to the local frame. Given the current platform's position in geodetic coordinates (latitude φ, longitude λ) (review Chapter 1 for more details about Earth coordinate systems transformation), the transformation of a vector in NED frame to a vector in ECEF frame is given by:

$$T_E^L = \begin{pmatrix} -\sin\varphi\cos\lambda & -\sin\lambda & -\cos\varphi\cos\lambda \\ -\sin\varphi\sin\lambda & \cos\lambda & -\cos\varphi\sin\lambda \\ -\cos\varphi & 0 & \sin\varphi \end{pmatrix} \tag{9-75}$$

The linearized measurement error model in NED frame is given by:

$$\underbrace{\begin{pmatrix} \delta\rho^1 \\ \delta\rho^2 \\ \vdots \\ \delta\rho^M \\ \delta\dot{\rho}^1 \\ \delta\dot{\rho}^2 \\ \vdots \\ \delta\dot{\rho}^M \end{pmatrix}}_{\delta\vec{y}(t)} = H(t)\begin{pmatrix} \delta n \\ \delta e \\ \delta d \\ \delta b_r \\ \delta v_n \\ \delta v_e \\ \delta v_d \\ \delta d_r \end{pmatrix} + \vec{\varepsilon}(t) \tag{9-76}$$

where $H(t)$ is defined as follows:

$$H(t) = \begin{pmatrix} \left(\vec{L}^1(t)\right)^T & 1 & 0_{3\times1} & 0 \\ \vdots & \vdots & \vdots & \vdots \\ \left(\vec{L}^M(t)\right)^T & 1 & 0_{3\times1} & 0 \\ 0_{3\times1} & 0 & \left(\vec{L}^1(t)\right)^T & 1 \\ \vdots & \vdots & \vdots & \vdots \\ 0_{3\times1} & 0 & \left(\vec{L}^M(t)\right)^T & 1 \end{pmatrix}\begin{pmatrix} T_E^L & 0_{3\times1} & 0_{2\times2} & 0_{2\times2} \\ 0_{1\times3} & 1 & 0 & 0 \\ 0_{3\times1} & 0_{2\times2} & T_E^L & 0 \\ 0 & 0 & 0 & 1 \end{pmatrix} \tag{9-77}$$

where $(n, e, d)^T$ is the 3D platform position in NED frame and $(v_n, v_e, v_d)^T$ is the platform velocity in NED frame, and $\vec{v}$ is a zero-mean Gaussian noise with covariance R and $H(t)$ is the design matrix that will be used in the EKF fusion. As the motion is 2D, d will always be constant and v_d will always be near zero. Recall, the augmented platform state vector and error state vector are defined as follows:

$$\vec{x}(t) = \begin{pmatrix} n(t) \\ e(t) \\ v_n(t) \\ v_e(t) \\ \psi(t) \\ b_a(t) \\ b_\omega(t) \\ b_r(t) \\ d_r(t) \end{pmatrix}, \delta\vec{x}(t) = \begin{pmatrix} \delta n(t) \\ \delta e(t) \\ \delta v_n(t) \\ \delta v_e(t) \\ \delta\psi(t) \\ \delta b_a(t) \\ \delta b_\omega(t) \\ \delta b_r(t) \\ \delta d_r(t) \end{pmatrix} \tag{9-78}$$

Note that the state vector $\vec{x}(t)$ does not have vertical position and velocity (d and v_d). To remove the vertical position and velocity variables, the design matrix in (9-78) is multiplied by the following T matrix:

$$T = \begin{pmatrix} 1 & 0 & 0 & 0 & 0 & 0 & 0 & 0 & 0 \\ 0 & 1 & 0 & 0 & 0 & 0 & 0 & 0 & 0 \\ 0 & 0 & 0 & 0 & 0 & 0 & 0 & 0 & 0 \\ 0 & 0 & 0 & 0 & 0 & 0 & 1 & 0 & 0 \\ 0 & 0 & 1 & 0 & 0 & 0 & 0 & 0 & 0 \\ 0 & 0 & 0 & 1 & 0 & 0 & 0 & 0 & 0 \\ 0 & 0 & 0 & 0 & 0 & 0 & 0 & 0 & 0 \\ 0 & 0 & 0 & 0 & 0 & 0 & 0 & 0 & 1 \end{pmatrix} \tag{9-79}$$

This modified $H(t)$ will force the vertical position and velocity errors to be always zeros, which is consistent with the 2D motion assumption where the platform's height (altitude) and vertical velocity v_d are approximated by the following constants as follows:

$$d \cong d_0 \tag{9-80}$$

$$v_d \cong 0 \tag{9-81}$$

where d_0 is the reference height, which is forced to be fixed during the trajectory and can be obtained by GPS at the beginning of the platform motion. If the motion is mostly 2D, our measurement model will be accurate, which is the case in the satellite range/doppler measurements since the roads terrestrial changes are too small compared to the distance between the vehicle and the satellite. The same concept applies with Doppler. The vehicle's speed component in the vertical direction is too small compared to the doppler between the vehicle and the satellites which is in the range of kilometers per second.

9.7 GPS Fusion Using Extended Kalman Filter

In the previous sections, we defined the system state (see Eqs. (9-13) and (9-70)), derived the system model function $\dot{\vec{x}}(t) = f\left(\vec{x}(t), \vec{u}(t)\right)$ (see Eqs. (9-15)–(9-21) and (9-66) and (9-67)), and derived the measurement model $h\left(\vec{x}(t)\right)$ for different measurements (see Sections 9.5 and 9.6). We also have the linearized system model (see Eqs. (9-26)–(9-32) and (9-34)). To fuse GPS measurements with the vehicle's inertial sensors using extended Kalman filter (EKF), we initialize the state covariance matrix $P(t)$ that represents the system's confidence in the initial state estimation accuracy. Recall, state covariance matrix $P(t)$ is defined as follows:

$$P(t) = \left\langle \delta\vec{x}^{T}(t)\, \delta\vec{x}(t) \right\rangle \tag{9-82}$$

Initially, we initialize $P(t)$ according to our belief or confidence in the accuracy of the initial state values as follows:

$$P(0) = \begin{pmatrix} {\sigma_e}^2 & 0 & 0 & 0 & 0 & 0 & 0 \\ 0 & {\sigma_n}^2 & 0 & 0 & 0 & 0 & 0 \\ 0 & 0 & {\sigma_{v_n}}^2 & 0 & 0 & 0 & 0 \\ 0 & 0 & 0 & {\sigma_{v_e}}^2 & 0 & 0 & 0 \\ 0 & 0 & 0 & 0 & {\sigma_\psi}^2 & 0 & 0 \\ 0 & 0 & 0 & 0 & 0 & {\sigma_a}^2 & 0 \\ 0 & 0 & 0 & 0 & 0 & 0 & {\sigma_\omega}^2 \end{pmatrix} \tag{9-83}$$

By having the system model, transition matrix, measurement model, design matrix, and assuming noise parameters are given, we have all what we need to fuse GPS measurements with vehicle's inertial sensors using EKF. As we use digital computers to perform the fusion, we move all equations to the discrete time domain and use the subscript k to refer to the time index (or epoch) and the sampling period is ΔT. In our notation, we will use the "+" sign to refer to updated/corrected states (x_k^-) and the sign "−" to refer to the predicted state, F_k is the discrete form of $F(t)$, the error model transition matrix. Recall the EKF algorithm consists of two main steps; prediction and update as follows:

- **Prediction**

$$\vec{x}_k^- = f\left(\vec{x}_{k-1}^+, \vec{u}_{k-1}\right) \tag{9-84}$$

$$\emptyset_k \cong (I + \Delta T . F_k) \tag{9-85}$$

$$P_k^- = \emptyset_{k,k-1} P_k \emptyset_{k,k-1}{}^T + \Delta T^2 \left(G_k {G_k}^T\right) \tag{9-86}$$

where F_k is defined in (9-34) and G_k is defined in (9-36).

- **Update**

$$K_k = P_k^- H_k^T \left(H_k P_k^- H_k^T + R_k\right)^{-1} \tag{9-87}$$

$$P_k^+ = (I - K_k H_k) P_k^- \tag{9-88}$$

The matrix H_k depends on the measurements that we fuse (see (9-41), (9-54), and (9-77) for different measurement matrices). If the measurement model is linear, correction is performed as follows:

$$\vec{x}_k^+ = \vec{x}_k^- + K_k \left(\vec{y}_k - H_k \vec{x}_k^-\right) \tag{9-89}$$

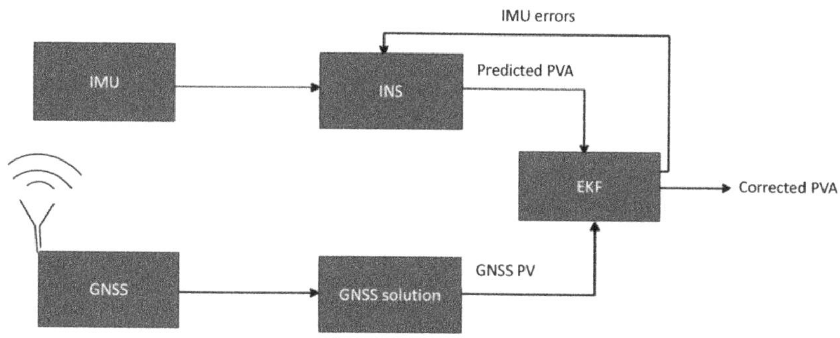

Figure 9.7 Loosely coupled EKF for GPS/IMU fusion.

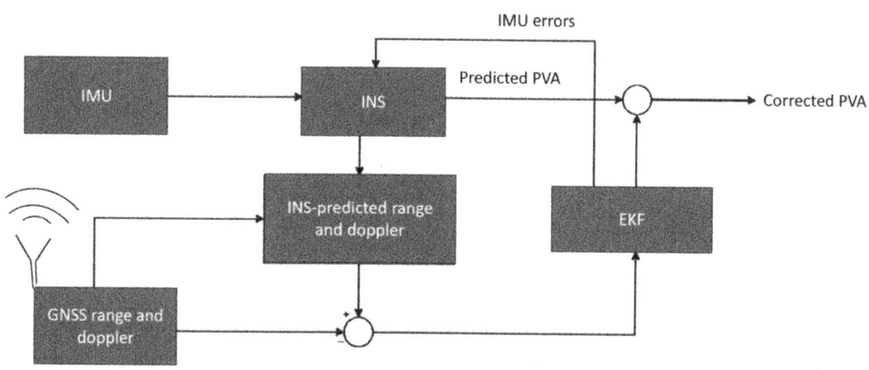

Figure 9.8 Tightly coupled EKF for GPS/IMU fusion.

If the measurement model is nonlinear, correction is performed as follows:

$$\vec{x}_k^{\,+} = \vec{x}_k^{\,-} + K_k\left(\vec{y}_k - h\left(\vec{x}_k^{\,-}\right)\right) \tag{9-90}$$

Loosely coupled and tightly coupled GPS/IMU fusion scheme using EKF is illustrated in Figures 9.7 and 9.8. To avoid linearization issues and matrix inversion in EKF, we can apply particle filtering as will be discussed in the next section.

9.8 GPS Fusion Using Particle Filter (PF)

PF has been discussed in Chapter 3 "Sensor Fusion Methods and Algorithms The main advantage of PF is that the linearization step is not required since PF works directly on the nonlinear models. In addition, although we adopt Gaussian distribution to model noise, the Gaussian assumption is not necessary as any probability distribution can be approximated by enough particles. The most significant limitation of PF is the computation time if the particle cloud size is large. As discussed in Chapter 3, PF is implemented in two steps: prediction and update. Both steps are described here again for convenience.

- **Initialization**
  A cloud of N particles with equal weights is initialized around an initial state $\vec{x}_0^{\,+}$. For simplicity, we will assume Gaussian distribution (which is not required). Therefore, the cloud of particles is initialized as follows:

$$\vec{x}_0^{\,+j} \sim \mathcal{N}\left(\vec{x}_0^{\,+}, \sigma_k^2\right), j = 1, 2 \ldots N \tag{9-91}$$

$$w_0^{+j} = 1/N \tag{9-92}$$

- **Prediction**
  All particles are advanced using the system model equation as follows:

$$\vec{x}_k^{\,-j} = f\left(\vec{x}_{k-1}^{\,+j}, \vec{u}_{k-1}\right) + \vec{w}_k \tag{9-93}$$

  Note that weights w_k^{-j} are still equal at this step.

- **Update**
  In the update step, the difference between the predicted measurements $h\left(\vec{x}_k^{\,-j}\right)$ and the observed measurements $\vec{y}_k$ are used to update the particles' weights as follows:

$$w_k^{+j} \propto \frac{1}{\sqrt{(2\pi)^m}\sqrt{|R_k|}} e^{-\frac{1}{2}\left(\vec{y}_k - h\left(\vec{x}_k^{\,-j}\right)\right)^T R^{-1}\left(\vec{y}_k - h\left(\vec{x}_k^{\,-j}\right)\right)} \tag{9-94}$$

  where w_k^{+j} is the posterior updated weight of particle j at epoch k. The measurement vector $\vec{y}_k$, measurement covariance matrix R_k and the measurement model $h\left(\vec{x}_k^{\,-j}\right)$ are all depending on the type of measurements we are fusing (GPS position, azimuth, range, or Doppler measurements).

- **Resampling**
  The particles are resampled such that each particle will be chosen a number of times that is proportional to its weight. A sample project with code that demonstrates PF fusion of inertial and GPS measurements in loosely coupled mode is presented at the end of this chapter.

9.9 GPS Deeply Coupled Fusion

A deeper level of integration between GPS and inertial/odometry measurements can be performed by fusing the platform motion model with the internal acquisition and tracking loops inside the GPS receiver to enhance the acquisition and tracking performance. This level of integration enhances signal re-acquisition in case of losing track of a satellite signal due to multipath or weak signals. For example, the platform speed information calculated by the system motion model can be used to expect the Doppler value, which can be used to narrow down the frequency search space in the acquisition module. The same approach can be used in the tracking loop. The platform's speed information can guide the tracking loop to keep the signal locked. The platform position calculated by the system model together with the satellite position can be used to estimate the distance between the satellite and the receiver and hence calculate the expected code phase and code delay, which narrows down the code delay search space and enhances the acquisition and tracking performance. The estimated distance can help in resolving the Integer Ambiguity as well. The fused corrected position obtained from vehicle system model in the short term can be used to enhance the acquisition peak accuracy and to mitigate multipath effects. The same can be used to enable acquisition and tracking under weaker noisier satellite signals, which enhances the sensitivity of the GPS (or GNSS) receiver.

9.10 Fusing Wheel Encoder Data

Wheeled platforms are commonly equipped with wheel encoders that count wheel revolutions using rotary wheel encoders. Rotary encoders are devices that use either mechanical brushes or light beams attached to a wheel, an electronic circuit that can use the mechanical brush movements or the light variations to count the number of revolutions a wheel rotates per unit time. Using the dimensions of the wheel, the revolutions per unit of time are converted into speed, angular velocity, and heading of the wheeled platform. In one possible fusion configuration, we can use wheel encoder data as measurements and fuse it with inertial data in the same way we fused the speed and heading measurements. In this case, the measurement vector will be given as follows:

$$\vec{y}(t) = \begin{pmatrix} \hat{v}(t) \\ \hat{\psi}(t) \end{pmatrix} \tag{9-95}$$

where $\hat{v}(t)$ is forward speed and $\hat{\psi}(t)$ is heading measurements. The measurement model relates the measurement vector to system state vector, so it does not need linearization. Regarding the heading $\hat{\psi}(t)$ it is directly mapped to the state $\psi(t)$. The forward speed $\hat{v}(t)$ is related to the state vector as follows:

$$\hat{v}(t) = \sqrt{(v_n(t))^2 + (v_e(t))^2} \tag{9-96}$$

which is a nonlinear relationship. Linearizing this equation and arranging both measurements into an error model in the form $\delta\vec{y}(t) = H(t)\delta\vec{x}(t) + \vec{\varepsilon}(t)$ will lead to the following design matrix $H(t)$:

$$H(t) = \begin{pmatrix} 0 & 0 & \dfrac{\overline{v}_n(t)}{\sqrt{(\overline{v}_n(t))^2 + (v_e(t))^2}} & \dfrac{\overline{v}_e(t)}{\sqrt{(\overline{v}_n(t))^2 + (\overline{v}_e(t))^2}} & 0 & 0 & 0 \\ 0 & 0 & 0 & 0 & 1 & 0 & 0 \end{pmatrix} \tag{9-97}$$

where $(\overline{n}(t)\ \overline{e}(t)\ \overline{v}_n(t)\ \overline{v}_e(t)\ \overline{\psi}(t))^T$ is an initial operating state of the system. To apply this measurement update in an EKF algorithm or a PF algorithm, we will use this design matrix and perform the update step as described in Section 9.7 or Section 9.8. It is important to note that in another configuration, the wheel encoders can be used as a system model to predict the vehicle's pose. In this configuration, other measurements (including inertial) can be used as external measurement updates. However, this configuration is less common.

9.11 Fusing LiDAR with 2D Inertial/Wheel Encoder Sensors

In clear weather conditions, LiDAR provides high-resolution, precise, and accurate ranging data. LiDAR data is commonly fused with internal odometry sensors (inertial or wheel encoders) to improve positioning and mapping accuracy. LiDAR can provide odometry data on its own using scan matching method such as iterative closest point (ICP) algorithm or the normal distribution transformation (NDT) algorithm (review Chapter 6 "Active Range Sensors"), which can be fused with internal odometry sensors to provide better odometry. This fusion can be referred to as loosely coupled fusion as the vehicle's pose (either relative or absolute) is calculated independently by LiDAR and fused with those calculated independently by internal odometry sensors such as wheel encoders or inertial sensors. Another approach is the tightly coupled approach, where LiDAR individual range/angle measurements are fused with inertial data. This section will describe both approaches, the loosely coupled and the tightly coupled. The fusion with LiDAR can be beneficial if GPS/GNSS is interrupted or blocked. In this section, we will use the same 2D system model and error model described in Sections 9.3 and 9.4. For more convenience, the 2D system model and error model are given again here. The state vector is given as follows:

$$\vec{x}(t) = \begin{pmatrix} n(t) \\ e(t) \\ v_n(t) \\ v_e(t) \\ \psi(t) \\ b_a(t) \\ b_\omega(t) \end{pmatrix} \tag{9-98}$$

And the system model is given by:

$$\dot{n}(t) = v_n(t) + w_n(t) \tag{9-99}$$

$$\dot{e}(t) = v_e(t) + w_e(t) \tag{9-100}$$

$$\dot{v}_n(t) = (a(t) - b_a(t)) \cos\psi(t) + w_{v_n}(t) \tag{9-101}$$

$$\dot{v}_e(t) = (a(t) - b_a(t)) \sin\psi(t) + w_{v_e}(t) \tag{9-102}$$

$$\dot{\psi}(t) = (\omega(t) - b_\omega(t)) + w_\psi(t) \tag{9-103}$$

$$\dot{b}_\omega(t) = w_\omega(t) \tag{9-104}$$

$$\dot{b}_a(t) = w_a(t) \tag{9-105}$$

where $w_\psi(t), w_{v_n}(t), w_{v_e}(t), w_n(t), w_\omega$, w_a and $w_e(t)$ are zero-mean Gaussian noise. The linearized system error model is given by $\delta\dot{\vec{x}}(t) = F(t)\delta\vec{x}(t) + G(t)\vec{w}_u(t)$ where $F(t)$ is the error model transition matrix defined as follows:

$$F(t) = \begin{pmatrix} 0 & 0 & 1 & 0 & 0 & 0 & 0 \\ 0 & 0 & 0 & 1 & 0 & 0 & 0 \\ 0 & 0 & 0 & 0 & -(a(t) - b_a(t)) \sin\psi(t) & -\cos\psi(t) & 0 \\ 0 & 0 & 0 & 0 & (a(t) - b_a(t)) \cos\psi(t) & -\sin\psi(t) & 0 \\ 0 & 0 & 0 & 0 & 0 & 0 & -1 \\ 0 & 0 & 0 & 0 & 0 & 0 & 0 \\ 0 & 0 & 0 & 0 & 0 & 0 & 0 \end{pmatrix} \tag{9-106}$$

and $G(t)$ matrix is a system noise shaping matrix defined as follows:

$$G(t) = \begin{pmatrix} {q_e}^2 & 0 & 0 & 0 & 0 & 0 & 0 \\ 0 & {q_n}^2 & 0 & 0 & 0 & 0 & 0 \\ 0 & 0 & {q_{v_n}}^2 & 0 & 0 & 0 & 0 \\ 0 & 0 & 0 & {q_{v_e}}^2 & 0 & 0 & 0 \\ 0 & 0 & 0 & 0 & {q_\psi}^2 & 0 & 0 \\ 0 & 0 & 0 & 0 & 0 & {q_a}^2 & 0 \\ 0 & 0 & 0 & 0 & 0 & 0 & {q_\omega}^2 \end{pmatrix} \tag{9-107}$$

Because we implement all algorithms in digital computers, we will use the discrete form of the vectors and models as follows:

$$\vec{x}_k = \begin{pmatrix} n_k \\ e_k \\ v_{n,k} \\ v_{e,k} \\ \psi_k \\ b_{a,k} \\ b_{\omega,k} \end{pmatrix} \tag{9-108}$$

$$\vec{u}_k = \begin{pmatrix} a_k \\ \omega_k \end{pmatrix} \tag{9-109}$$

Assuming ΔT is the sampling period, the discrete system model $\vec{x}_k = f\left(\vec{x}_{k-1}, \vec{u}_k\right)$ is approximated as follows:

$$n_k = n_{k-1} + (v_{n,k})\Delta T + w_{n,k} \tag{9-110}$$

$$e_k = e_{k-1} + (v_{e,k})\Delta T + w_{e,k} \tag{9-111}$$

$$v_{n,k} = v_{n,k-1} + ((a_k - b_{a,k})\cos\psi_k)\Delta T + w_{v_{n,k}} \tag{9-112}$$

$$v_{e,k} = v_{e,k-1} + ((a_k - b_{a,k})\sin\psi_k)\Delta T + w_{v_{e,k}} \tag{9-113}$$

$$\psi_k = \psi_{k-1} + (\omega_k - b_{\omega,k})\Delta T + w_{\psi,k} \tag{9-114}$$

$$b_{\omega,k} = b_{\omega,k-1} + w_{\omega,k} \tag{9-115}$$

$$b_{a,k} = b_{a,k-1} + w_{a,k} \tag{9-116}$$

The discrete linearized system error model is given by the following model:

$$\delta\vec{x}_k = \emptyset_k \vec{x}_{k-1} + G_k \vec{w}_{u,k} \tag{9-117}$$

where $\emptyset_k$ is approximated as follows:

$$\emptyset_k \cong (1 + \Delta T F_k) \tag{9-118}$$

$$F_k = \begin{pmatrix} 0 & 0 & 1 & 0 & 0 & 0 & 0 \\ 0 & 0 & 0 & 1 & 0 & 0 & 0 \\ 0 & 0 & 0 & 0 & -(a_k - b_{a,k}) \sin \psi_k & -\cos \psi_k & 0 \\ 0 & 0 & 0 & 0 & (a_k - b_{a,k}) \cos \psi_k & -\sin \psi_k & 0 \\ 0 & 0 & 0 & 0 & 0 & 0 & -1 \\ 0 & 0 & 0 & 0 & 0 & 0 & 0 \\ 0 & 0 & 0 & 0 & 0 & 0 & 0 \end{pmatrix} \tag{9-119}$$

We will discard the formula for the noise shaping matrix G_k as it can be seen as a tuning parameter that is tuned for specific sensors/systems to reflect the sensors' quality. In the tuning, the general rule is that high-end sensors have smaller G_k and low-quality sensors have larger G_k and it must be noted that what determines the weight of the system prediction is the ratio between system noise modeled by G_k matrix and measurement noise modeled by R_k matrix.

9.11.1 Loosely Coupled LiDAR Fusion

At time epoch k, LiDAR scans provide a vector of ranges $\vec{r}_k = \left(r_k^1, r_k^2 \ldots r_k^M\right)^T$ associated with angle vector $\vec{\theta}_k = \left(\theta_k^1, \theta_k^2 \ldots \theta_k^M\right)^T$. Given the vehicle's pose and LiDAR scan at epoch k, we can calculate the coordinate of the feature points (point cloud) in the local frame as follows:

$$\vec{p}^{\,i} = \begin{pmatrix} n^i \\ e^i \end{pmatrix} = \begin{pmatrix} n_k + r_k^i \sin\left(\theta_k^i - \psi_k\right) \\ e_k + r_k^i \cos\left(\theta_k^i - \psi_k\right) \end{pmatrix} \tag{9-120}$$

Note that the feature point coordinate $\vec{p}^{\,i}$ does not have the time subscript k as it is assumed to be static environment. Given two LiDAR scans $\left\{\vec{r}_k, \vec{\theta}_k\right\}$ and $\left\{\vec{r}_{k+1}, \vec{\theta}_{k+1}\right\}$, we have seen in Chapter 6 "Active Range Sensors" how ICP or NDT algorithms can be applied on LiDAR consecutive scans to estimate the pose change from epoch k to epoch $k+1$ by minimizing the following cost function:

$$J\left(M_c, R_{k+1}^k, \vec{T}_{k+1}^k\right) = \frac{1}{2}\sum\nolimits_{i\in C}\left\|\left(p_k^i - \left(R_{k+1}^k p_{k+1}^{c(i)} + \vec{T}_{k+1}^k\right)\right)\right\|^2 \tag{9-121}$$

$$M \triangleq c(i), i \in C \tag{9-122}$$

where M_c is the correspondence that map commonly visible points (C) between scan k and scan $k+1$, p_k^i is a feature point in scan k, $p_{k+1}^{c(i)}$ is its corresponding feature point in scan $k+1$. The optimization of this cost function provides an estimate of the platform relative rotation and translation. In 2D case, R_{k+1}^k and $\vec{T}_{k+1}^k$ give the heading angle change and position change as follows:

$$R_{k+1}^k = \begin{pmatrix} \cos \Delta\psi_{k,k+1} & \sin \Delta\psi_{k,k+1} \\ -\sin \Delta\psi_{k,k+1} & \cos \Delta\psi_{k,k+1} \end{pmatrix} \tag{9-123}$$

$$\vec{T}_{k+1}^k = \begin{pmatrix} \Delta n_{k,k+1} \\ \Delta e_{k,k+1} \end{pmatrix} \tag{9-124}$$

The transformation matrix given rotation and translation is given by:

$$\Delta T_{k+1}^{k} = \begin{pmatrix} R_{k+1}^{k} & \vec{T}_{k+1}^{k} \\ 0 & 1 \end{pmatrix} = \begin{pmatrix} \cos \Delta\psi_{k,k+1} & \sin \Delta\psi_{k,k+1} & \Delta n_{k,k+1} \\ -\sin \Delta\psi_{k,k+1} & \cos \Delta\psi_{k,k+1} & \Delta e_{k,k+1} \\ 0 & 0 & 1 \end{pmatrix} \tag{9-125}$$

The inverse transformation ΔT_{k}^{k+1} is given by:

$$\Delta T_{k}^{k+1} = \begin{pmatrix} \left(R_{k+1}^{k}\right)^{T} & -\vec{T}_{k+1}^{k} \\ 0 & 1 \end{pmatrix} \tag{9-126}$$

In the beginning, at epoch 0, the platform pose is defined as follows:

$$T_0 = \begin{pmatrix} 1 & 0 & 0 \\ 0 & 1 & 0 \\ 0 & 0 & 1 \end{pmatrix} \tag{9-127}$$

Let T_k be the platform pose at time k, given the relative transformation ΔT_{k}^{k+1}, the new platform pose T_{k+1} is given by the following concatenation equation:

$$T_{k+1} = T_k \Delta T_{k}^{k+1} \tag{9-128}$$

We can extract the following measurement vector from the matrix T_k as follows:

$$\vec{y}_k(t) = \begin{pmatrix} \hat{n}_k \\ \hat{n}_k \\ \hat{\psi}_k \end{pmatrix} = \begin{pmatrix} T_k(1,3) \\ T_k(2,3) \\ \tan^{-1} \dfrac{T_k(1,2)}{T_k(1,1)} \end{pmatrix} \tag{9-129}$$

This measurement vector can be used as measurement update to the platform state using either PF or EKF where the design matrix will be given by:

$$H(t) = \begin{pmatrix} 1 & 0 & 0 & 0 & 0 & 0 & 0 \\ 0 & 1 & 0 & 0 & 0 & 0 & 0 \\ 0 & 0 & 0 & 0 & 1 & 0 & 0 \end{pmatrix} \tag{9-130}$$

One possible additional fusion idea is to use the pose predicted by inertial sensors using the system model described in Section 9.3, as initial guess for the ICP or NDT algorithms to accelerate the convergence of scan matching algorithm. In this way, inertial sensors and system model improve the performance of the LiDAR scan matching, and the results of the LiDAR scan matching are fused with the pose predicted by the system model using either an EKF or a PF. The same idea can be applied in a different way where wheel encoders (if available) can be used to predict the platform pose, which can be used to boost the performance of the ICP or NDT scan matching algorithms.

It is important to highlight that using an initial pose guess estimated by inertial sensors or wheel odometer as a boost to the LiDAR scan matching algorithm whose output is used back as a measurement update to the system does not break the condition of independence between LiDAR measurements and inertial (or wheel encoders) system model output. The initial pose provided to the LiDAR ICP scan matching is just an initial guess. The final optimal pose estimated by the LiDAR scan matching is guided by the features' point measurements, which are independent of both inertial sensors and wheel encoder measurements.

Another point to highlight is that the fusion of inertial and wheel-encoders with LiDAR ICP or NDT scan matching integrates two sensors/systems of odometry nature. Neither LiDAR ICP nor NDT scan matching nor inertial and wheel odometry provide an absolute position, velocity, or orientation fix. They both update the current pose incrementally by accumulating relative pose changes from one epoch to the next. There is no absolute positioning fix like the GPS case. For this reason, we must expect that the fusion result will still have some drifts. The aim is to reduce this drift by integrating both sensors.

In practical applications, internal odometry sensors (inertial and/or wheel encoders) are usually of a higher rate than LiDAR. Therefore, the common scenario is to apply LiDAR pose updates at a lower rate than internal odometry sensors. The rate at which we apply LiDAR measurements and system prediction is a design choice that may differ from one vehicle to another or from one environment to another. It also depends on the sensors' quality and the accuracy of the system predictions. For example, if system prediction is accurate, we do not want to waste computing resources by applying the update step very frequently.

9.11.2 Tightly Coupled LiDAR Fusion

In this fusion scheme, we still consider two consecutive LiDAR scans, but we will directly use the raw LiDAR data (range and angles) in the fusion to correct the vehicle's state. We start the processing cycle from an initial state $\vec{x}_k^+$ that we trust. At time epoch k, LiDAR scans provide a vector of ranges $\vec{r}_k = \left(r_k^1, r_k^2 \ldots r_k^M\right)^T$ associated with angle vector $\vec{\theta}_k = \left(\theta_k^1, \theta_k^2 \ldots \theta_k^M\right)^T$. Given the vehicle's pose and LiDAR scans at epoch k, we can calculate the coordinate of the ith feature point of the scan as follows:

$$\vec{p}^{\,i} = \begin{pmatrix} n^i \\ e^i \end{pmatrix} = \begin{pmatrix} n_k^+ + r_k^i \sin\left(\theta_k^i - \psi_k\right) \\ e_k^+ + r_k^i \cos\left(\theta_k^i - \psi_k\right) \end{pmatrix} \tag{9-131}$$

Given two LiDAR scans $\left\{\vec{r}_k, \vec{\theta}_k\right\}$ and $\left\{\vec{r}_{k+1}, \vec{\theta}_{k+1}\right\}$, we apply ICP or NDT scan matching to find the correspondences $M \triangleq c(i)$, $i \in C$, where C is the indexes of feature points in scan k that are seen in both scan k and scan $k+1$ (commonly visible points). The locations of the common feature points already known from scan k and we know the predicted pose $\vec{x}_{k+1}^-$ from the system model $\vec{x}_{k+1}^- = f\left(\vec{x}_k^+, \vec{u}_{k+1}\right)$.

From the predicted state $\vec{x}_{k+1}^-$, we can calculate the predicted range/angle measurement using the following measurement model as follows:

$$\vec{y}_{k+1}^{\,c(i)} = h\left(\vec{x}_{k+1}^-, \vec{p}^{\,i}\right) = \begin{pmatrix} \bar{r}_{k+1}^{c(i)} \\ \bar{\theta}_k^{c(i)} \end{pmatrix} = \begin{pmatrix} \sqrt{\left(n_{k+1}^- - n^i\right)^2 + \left(e_{k+1}^- - e^i\right)^2} \\ \tan^{-1}\left(\dfrac{\left(n_{k+1}^- - n^i\right)}{\left(e_{k+1}^- - e^i\right)}\right) + \psi_{k+1}^+ \end{pmatrix} \tag{9-132}$$

At the same time, we have physical observed measurement of the ith feature point in scan k, which corresponds to feature $c(i)$ in scan $k+1$. These physical measurements form the observation vector $\vec{\hat{y}}_{k+1}^{\,c(i)}$ defined as follows:

$$\vec{\hat{y}}_{k+1}^{\,c(i)} = \begin{pmatrix} \hat{r}_{k+1}^{c(i)} \\ \hat{\theta}_k^{c(i)} \end{pmatrix} \tag{9-133}$$

Having the predicted measurement vector $\vec{y}_{k+1}^{c(i)}$ and physical observed measurements $\hat{\vec{y}}_{k+1}^{c(i)}$, we can then apply a measurement update to correct the state vector $\vec{x}_{k+1}^{-}$. This can be an update step in a PF or an EKF where the design matrix H_k is given by the following equation:

$$H_{k+1}^{i} = \left.\frac{\partial h\left(\vec{x}_{k+1}, \vec{p}^{\,i}\right)}{\partial \vec{x}_k}\right|_{\vec{x}_{k+1}^{-}} \tag{9-134}$$

$$H_{k+1}^{i} = \begin{pmatrix} \frac{\left(n_{k+1}^{-} - n^{i}\right)}{\sqrt{\left(n^{i} - n_{k+1}^{-}\right)^{2} + \left(e^{i} - e_{k+1}^{-}\right)^{2}}} & \frac{\left(e_{k+1}^{-} - e^{i}\right)}{\sqrt{\left(n^{i} - n_{k+1}^{-}\right)^{2} + \left(e^{i} - e_{k+1}^{-}\right)^{2}}} & 0 & 0 & 0 & 0 & 0 \\ \frac{-\left(e_{k+1}^{-} - e^{i}\right)}{\sqrt{\left(n^{i} - n_{k+1}^{-}\right)^{2} + \left(e^{i} - e_{k+1}^{-}\right)^{2}}} & \frac{\left(n_{k+1}^{-} - n^{i}\right)}{\sqrt{\left(n^{i} - n_{k+1}^{-}\right)^{2} + \left(e^{i} - e_{k+1}^{-}\right)^{2}}} & 0 & 0 & 1 & 0 & 0 \end{pmatrix} \tag{9-135}$$

The measurement noise covariance matrix in this case will be a 2×2 matrix because we have a 2×1 measurement vector as follows:

$$R_{k+1} = \begin{pmatrix} {\sigma_r}^2 & 0 \\ 0 & {\sigma_\theta}^2 \end{pmatrix} \tag{9-136}$$

As all feature points are independent, we can apply the updates iteratively feature by feature.

9.11.3 LiDAR Fusion Using EKF and PF

Regardless of the fusion mode (tightly or loosely coupled) we use the proper design matrix H_k^i to update the vehicle's state using the EKF update step as follows:

$$K_{k+1}^{i} = P_{k+1}^{-}\left(H_{k+1}^{i}\right)^{T}\left(H_{k+1}^{i} P_{k+1}^{-}\left(H_{k+1}^{i}\right)^{T} + R_{k+1}\right)^{-1} \tag{9-137}$$

$$P_{k+1}^{+} = \left(I - K_{k+1}^{i} H_{k+1}^{i}\right) P_{k+1}^{-} \tag{9-138}$$

$$\vec{x}_{k+1}^{+} = \vec{x}_{k}^{-} + K_{k+1}^{i}\left(\hat{\vec{y}}_{k+1}^{c(i)} - h\left(\vec{x}_{k+1}^{-}, \vec{p}^{\,i}\right)\right) \tag{9-139}$$

The prediction step of EKF will not change and it will be implemented exactly as described in Section 9.7. If PF is used, the difference between the predicted measurements and physical observed measurements are used to update the particle weight as follows:

$$p\left(\vec{x}_{k+1}^{+} \middle| \hat{\vec{y}}_{k+1}^{c(i)}\right) \triangleq w_k^{+j} \propto \frac{1}{\sqrt{(2\pi)^m}\sqrt{|R_k|}} e^{-\frac{1}{2}\left(\vec{y}_k - h\left(\vec{x}_k^{-j}\right)\right)^T R^{-1}\left(\vec{y}_k - h\left(\vec{x}_k^{-j}\right)\right)} \tag{9-140}$$

9.12 Graph SLAM Fusion of Inertial, Wheel-Odometry, and LiDAR

EKF and PF are online estimation methods. We have seen in Chapter 8 "Mapping Algorithms" how Graph SLAM can be used to perform full or batch estimation to simultaneously build a map of the environment and to optimize the platform's pose accordingly. In Graph SLAM derivation, we formulated the state estimation problem as a probability distribution of the state $\vec{x}_{0:N}$ given all control signals $\vec{u}_{0:N}$ and measurements $\vec{y}^{i}_{0:N}$ as follows:

$$J\left(\vec{x}_{0:k}\right) = -\ln\left(p\left(\vec{x}_0\right)\right) - \sum_{k=1}^{N}\left(\ln\left(p\left(\vec{x}^V_k \mid \vec{x}^V_{k-1}, \vec{u}_k\right)\right)\right) - \sum_{k=1}^{N}\sum_i \ln\left(p\left(\vec{y}^i_k \mid \vec{x}_k\right)\right) \tag{9-141}$$

where $\vec{x}_{0:N}$ is the full state that consists of both the platform pose $\vec{x}^V_{0:N}$ and maps features $\vec{x}^M$, $p\left(\vec{x}^V_k \mid \vec{x}^V_{k-1}, \vec{u}_k\right)$ is the probabilistic system model given by the following Gaussian distribution:

$$p\left(\vec{x}^V_k \mid \vec{x}^V_{k-1}, \vec{u}_k\right) \sim \mathcal{N}\left(f\left(\vec{x}^V_{k-1}, \vec{u}_k\right), Q_k\right) \tag{9-142}$$

where $f\left(\vec{x}^V_{k-1}, \vec{u}_k\right)$ is the system model, and $p\left(\vec{y}^i_k \mid \vec{x}_k\right)$ is the probabilistic measurement model given by the following Gaussian distribution:

$$p\left(\vec{y}^i_k \mid \vec{x}_k\right) \sim \mathcal{N}\left(h\left(\vec{x}^V_k\right), R_k\right) \tag{9-143}$$

where $h\left(\vec{x}^V_k\right)$ is the measurement model. Graph SLAM framework is technically fusing the system model information with the measurement model information in one estimation problem that estimates the entire trajectory and the map of the environment. Therefore, Graph SLAM can be seen as a batch sensor fusion framework. Batch fusion is the counterpart of online fusion methods such as EKF, PF and their variations. We have demonstrated the full derivations of Graph SLAM, where the sensor is a range/angle sensor that is applicable to LiDAR data. In our derivations, a single system model, the one that uses wheel odometry that provides forward speed v^w_k and turn rate ω^w_k to predict the platform pose as follows:

$$\begin{pmatrix} n_k \\ n_k \\ v_{n,k} \\ v_{e,k} \\ \psi_k \end{pmatrix} = \begin{pmatrix} n_{k-1} + (v_k \cos\psi_{k-1})\Delta t \\ e_{k-1} + (v_k \sin\psi_{k-1})\Delta t \\ v_k \cos\psi_{k-1} \\ v_k \sin\psi_{k-1} \\ \psi_{k-1} + \omega_k \Delta t \end{pmatrix} + \varepsilon_k \tag{9-144}$$

We will refer to this model by $f_w\left(\vec{x}^V_{k-1}, \vec{u}_k\right)$ and will refer to the inertial sensors-based system model by $f_I\left(\vec{x}^V_{k-1}, \vec{u}_k\right)$. If both inertial sensors and wheel encoder sensors are available, we will have two sets of control vectors as follows:

$$\vec{u}^I_k = \begin{pmatrix} a^I_k \\ \omega^I_k \end{pmatrix}, \vec{u}^w_k = \begin{pmatrix} v^w_k \\ \omega^w_k \end{pmatrix} \tag{9-145}$$

We can combine the two system models $f_I\left(\vec{x}^V_{k-1}, \vec{u}_k\right)$ and $f_w\left(\vec{x}^V_{k-1}, \vec{u}_k\right)$ into the Graph SLAM optimization function by defining a new joint probabilistic system model. The joint probabilistic system model is the product of two Gaussian distributions as follows:

$$p\left(\vec{x}^V_k \mid \vec{x}^V_{k-1}, \vec{u}_k\right) = p_I\left(\vec{x}^V_k \mid \vec{x}^V_{k-1}, \vec{u}_k\right) p_w\left(\vec{x}^V_k \mid \vec{x}^V_{k-1}, \vec{u}_k\right) \tag{9-146}$$

Incorporating this join system model probability in the Graph-SLAM cost function gives the following new cost function:

$$J\left(\vec{x}_{0:k}\right) = -\ln\left(p\left(\vec{x}_0\right)\right) - \sum_{k=1}^{N}\left(\ln\left(p_I\left(\vec{x}_k^V \mid \vec{x}_{k-1}^V, \vec{u}_k\right) p_w\left(\vec{x}_k^V \mid \vec{x}_{k-1}^V, \vec{u}_k\right)\right)\right) - \sum_{k=1}^{N}\sum_i \ln\left(p\left(\vec{y}_k^i \mid \vec{x}_k\right)\right) \tag{9-147}$$

From the log function properties, the cost function can be written as follows:

$$J\left(\vec{x}_{0:k}\right) = -\ln\left(p\left(\vec{x}_0\right)\right) - \sum_{k=1}^{N}\left(\ln\left(p_I\left(\vec{x}_k^V \mid \vec{x}_{k-1}^V, \vec{u}_k\right)\right)\right) - \sum_{k=1}^{N}\left(\ln\left(p_w\left(\vec{x}_k^V \mid \vec{x}_{k-1}^V, \vec{u}_k\right)\right)\right) - \sum_{k=1}^{N}\sum_i \ln\left(p\left(\vec{y}_k^i \mid \vec{x}_k\right)\right) \tag{9-148}$$

thanks to the mathematical advantages of the exponential formula of the Gaussian probability function, taking the log of exponentials removes the log from the cost function as follows:

$$\begin{aligned} J\left(\vec{x}_{0:k}\right) = &\text{const.} + \left(\vec{x}_0 - \vec{\mu}_0\right)^T \Sigma_0^{-1}\left(\vec{x}_0 - \vec{\mu}_0\right) \\ &+ \sum_{k=1}^{N}\left(\vec{x}_k^V - f_I\left(\vec{x}_{k-1}^V, \vec{u}_k\right)\right)^T R_I^{-1}\left(\vec{x}_k^V - f_I\left(\vec{x}_{k-1}^V, \vec{u}_k\right)\right) \\ &+ \sum_{k=1}^{N}\left(\vec{x}_k^V - f_w\left(\vec{x}_{k-1}^V, \vec{u}_k\right)\right)^T R_w^{-1}\left(\vec{x}_k^V - f_w\left(\vec{x}_{k-1}^V, \vec{u}_k\right)\right) \\ &+ \sum_{\tau=1}^{\tau=1}\sum_i \left(\vec{y}_\tau^i - h^i\left(\vec{x}_\tau\right)\right)^T Q^{-1}\left(\vec{y}_\tau^i - h^i\left(\vec{x}_\tau\right)\right) \end{aligned} \tag{9-149}$$

where $\vec{\mu}_0$ is initial state (pose and map). This cost function can be optimized using graph optimization by linearizing all models around an initial estimate of the states and building an information matrix and information vector. The effect of the joint system model is an additional term that will be added to the information matrix. The details of the derivation of the graph optimization method of the cost function $J\left(\vec{x}_{0:k}\right)$ is given in detail in Section 8.4.4 in Chapter 8 "Mapping Algorithms."

9.13 MATLAB Project Samples for 2D Wheeled Platforms

The projects' repository of this book includes projects that demonstrate the fusion of a ground-wheeled platform with GPS using EKF and PF. Project code and data are available under *"2D-EKF"* and *"2D-PF"* folders. Readers are encouraged to go through the MATLAB code implementation of EKF and PF given in these projects since they further clarify the details of the mathematical models and algorithmic steps described in this chapter. The 2D data simulates noisy, biased accelerometer and gyroscope data from a 2D vehicle, as shown in Figure 9.9. The accelerometer is contaminated by a bias of 0.5 m/s^2 and a Gaussian random noise of 60 $\mu g/\sqrt{Hz}$ and the gyroscope is contaminated by a bias of 0.5 rad/s and a Gaussian random noise of 0.01 deg/s/$\sqrt{Hz}$. These error specifications simulate a commercial IMU. A portion of the noisy, biased accelerometer and gyroscope data are shown in Figures 9.10 and 9.11.

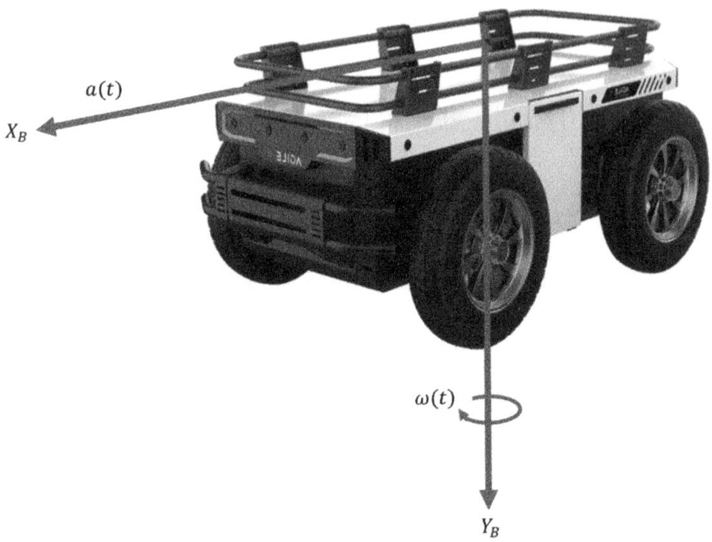

Figure 9.9 Illustration of accelerometer and gyroscope sensors on a 2D vehicle.

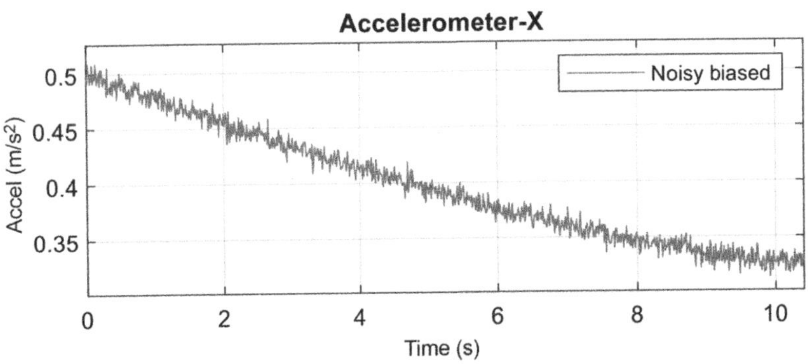

Figure 9.10 Noisy-biased accelerometer-X.

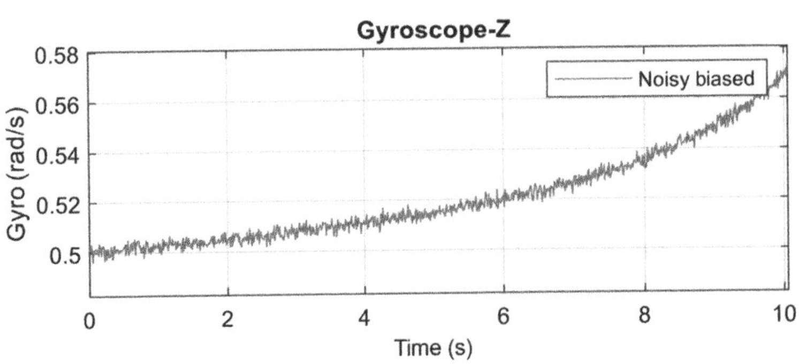

Figure 9.11 Noisy-biased gyroscope-Z.

The system model is simplified as follows:

$$\dot{n}(t) = v_f(t)\cos\psi(t) + w_n(t) \tag{9-150}$$

$$\dot{e}(t) = v_f(t)\sin\psi(t) + w_e(t) \tag{9-151}$$

$$\dot{v}_f(t) = (a(t) - b_a(t)) + w_{v_f}(t) \tag{9-152}$$

$$\dot{\psi}(t) = (\omega(t) - b_\omega(t)) + w_\psi(t) \tag{9-153}$$

$$\dot{b}_\omega(t) = w_\omega(t) \tag{9-154}$$

$$\dot{b}_a(t) = w_a(t) \tag{9-155}$$

where $v_f(t)$ is the forward speed, $\psi(t)$ is the heading (azimuth) angle, $n(t)$ and $e(t)$ are north and east position, $b_a(t)$ and $b_\omega(t)$ are biases of accelerometer-X and gyroscope-Z, respectively. This system model combines east and north velocities into one state $v_f(t)$, so it is more compact than the one described in Eqs. (9-15)–(9-21).

9.13.1 Extended Kalman Filter Sample Project

The linearized transition system matrix of this system model is given by:

$$F(t) = \begin{pmatrix} 0 & 0 & \cos\psi(t) & -v_f(t)\sin\psi(t) & 0 & 0 \\ 0 & 0 & \sin\psi(t) & v_f(t)\cos\psi(t) & 0 & 0 \\ 0 & 0 & 0 & 0 & -1 & 0 \\ 0 & 0 & 0 & 0 & 0 & -1 \\ 0 & 0 & 0 & 0 & 0 & 0 \\ 0 & 0 & 0 & 0 & 0 & 0 \end{pmatrix} \tag{9-156}$$

The measurements given in this example project are noisy north/east position and noisy heading angle measurements. These measurements simulate loosely coupled GPS position updates and magnetic sensor heading updates. The design matrix corresponding to these measurements is given as follows:

$$H = \begin{pmatrix} 1 & 0 & 0 & 0 & 0 & 0 & 0 \\ 0 & 1 & 0 & 0 & 0 & 0 & 0 \\ 0 & 0 & 0 & 0 & 0 & 0 & 0 \\ 0 & 0 & 0 & 1 & 0 & 0 & 0 \end{pmatrix} \tag{9-157}$$

Other matrices are parameters of EKF can be checked in the code of the project. If we applied the system prediction model only without any fusion, we would have a significantly drifting solution as shown in Figure 9.12.

The fusion results are shown in Figures 9.13–9.15 which demonstrate how EKF output is smooth and nondrifting. The ground truth is not visible in the figures as the EKF output is hiding it. To gain more insight into the internal workings and convergence of the EKF, Figures 9.16 and 9.17 show the convergence of the biases to the true values. To further inspect the performance of the filter, it is important to check the error covariance of the states to confirm the convergence of the filter and the consistency of the covariance values and the error values. Figures 9.18–9.20 show the pose errors along with the error envelope. The error envelope is shown as one sigma, which means the sigma is the square root of the covariance corresponding to the state (north/east position coordinates or heading angle). Note how the error covariance values start big and decrease with the

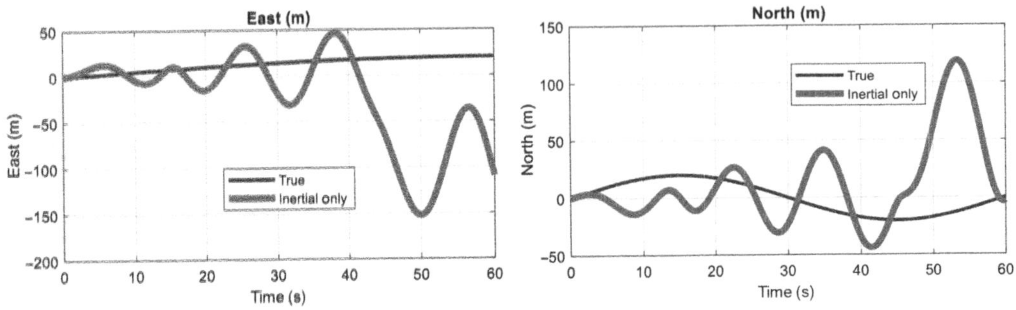

Figure 9.12 Inertial-only drifting solution (no fusion with GPS).

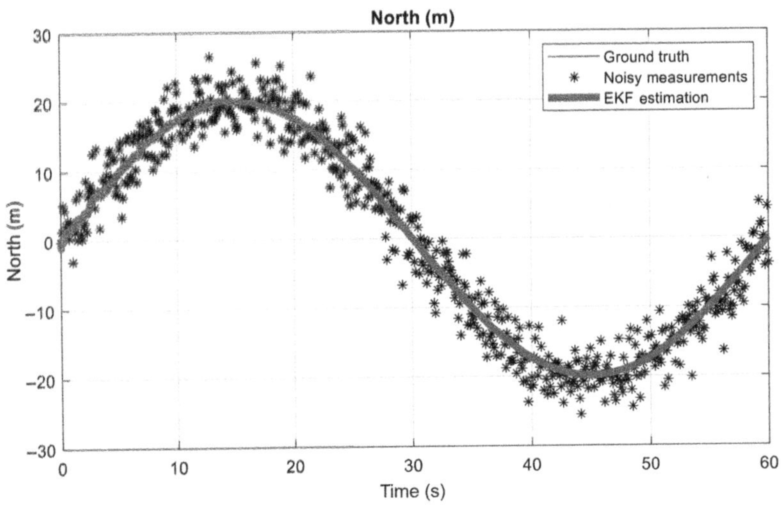

Figure 9.13 EKF estimated north position vs. noisy measurements.

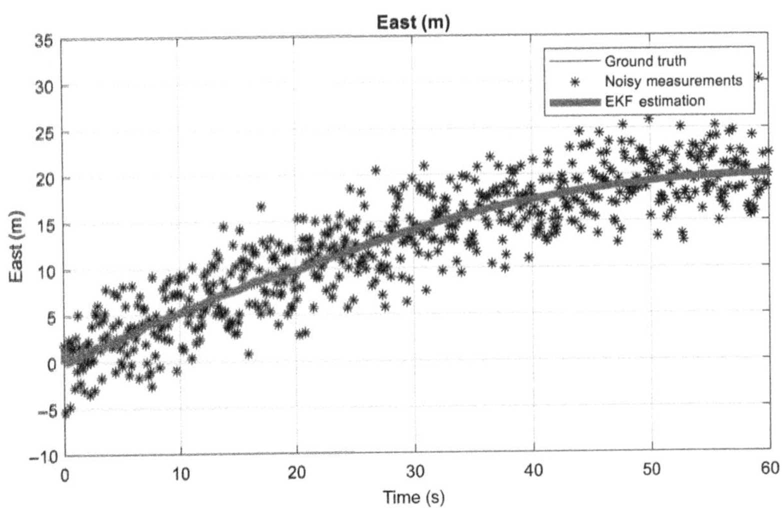

Figure 9.14 EKF estimated east position vs. noisy measurements.

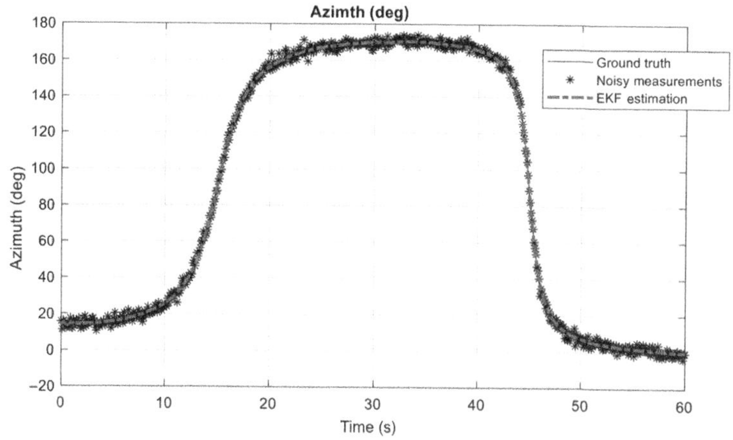

Figure 9.15 EKF estimated azimuth (heading) position vs. noisy measurements.

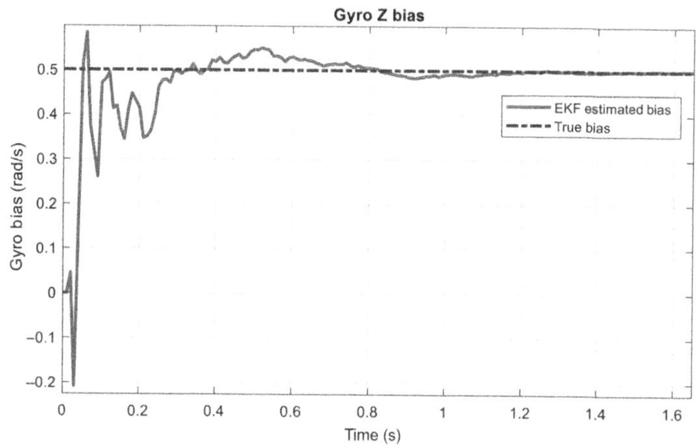

Figure 9.16 Gyroscope bias convergence.

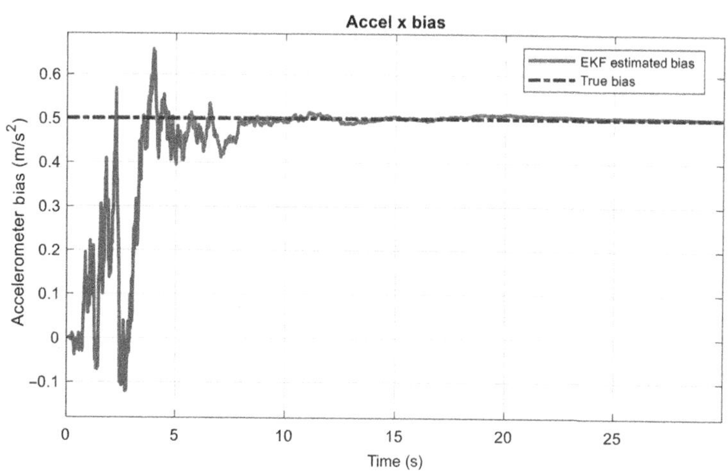

Figure 9.17 Accelerometer bias convergence.

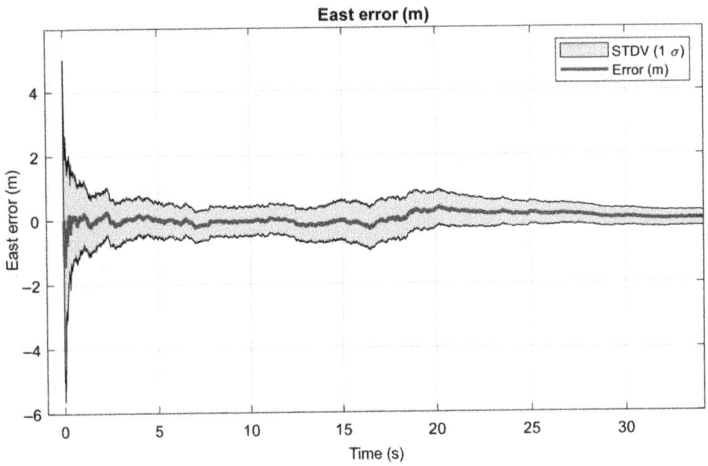

Figure 9.18 East position error plus error envelope.

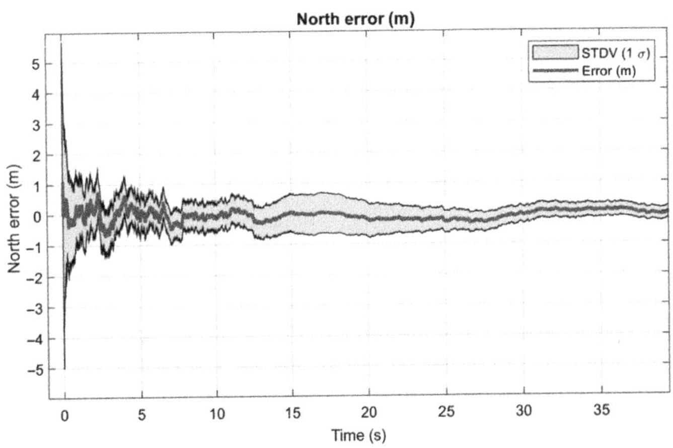

Figure 9.19 North position error + error covariance envelope.

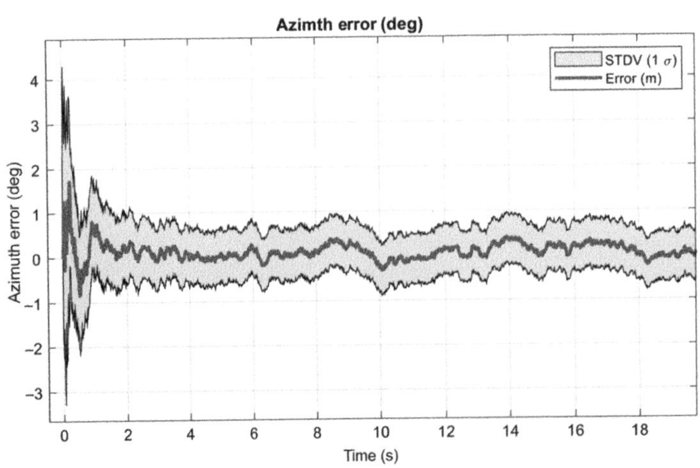

Figure 9.20 Azimuth error + error covariance envelope.

measurement's updates. This decrease in the covariance value is also synchronized with a decrease in the actual error value. The convergence of the biases and the values/consistency of the errors and error covariances indicate the correctness of the working of the filter. To further confirm the correct operation of the filter, we interrupt the measurements after convergence to check the accuracy after estimating the biases, which should be improved. Figures 9.21–9.23 show the pose errors under an outage from $t = 30$ s to $t = 50$ s. The pose error under outage is still bounded while the covariance increases to express the uncertainty in the estimation during this outage. During the outage, we apply system prediction only and benefit from the most recent estimated biases to clean incoming

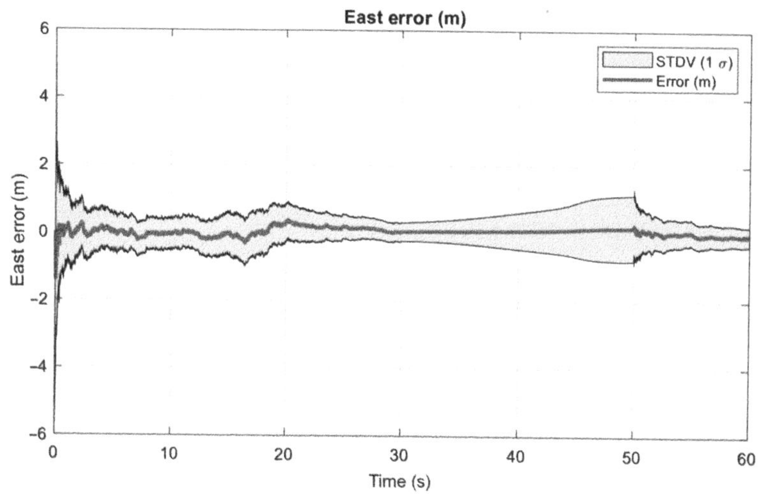

Figure 9.21 East position error with outage between 30 and 50 s.

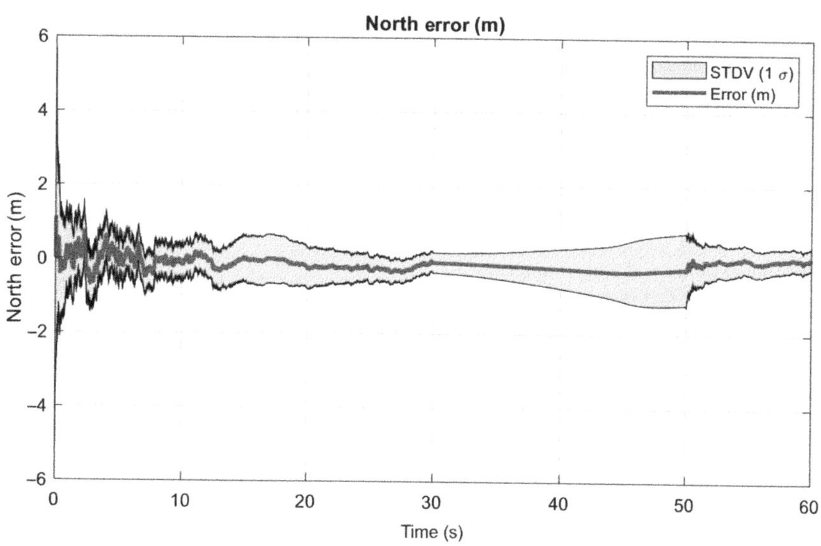

Figure 9.22 North position error with outage between 30 and 50 s.

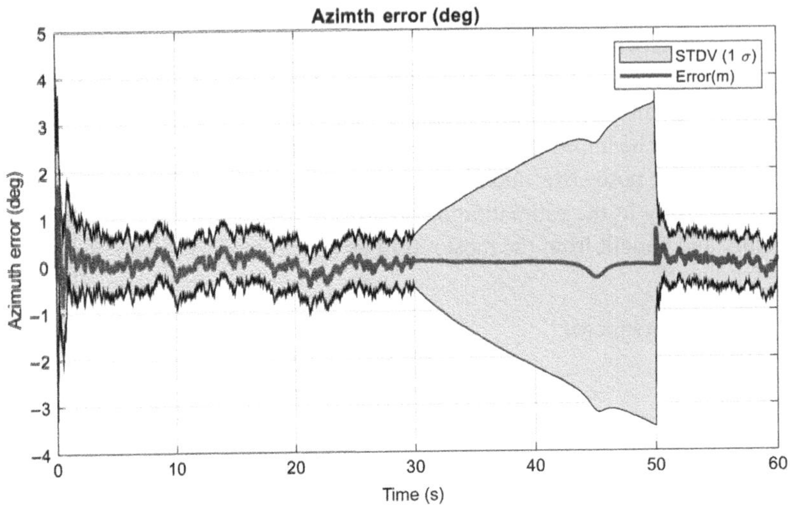

Figure 9.23 Azimuth error with outage between 30 and 50 s.

inertial measurements so the prediction is more accurate. This analysis confirms the correctness of our EKF's operation.

9.13.2 Particle Filter Sample Project

The particle filtering code can be found under *"2D-PF"* in the projects' repository. In this project, PF is applied to a similar but shorter dataset. The system model and the measurement models are the same models given in Eqs. (9-150)–(9-155). As PF approximates probability distributions by a cloud of particles, we display the cloud as an indication of the covariance of the estimation. Another difference between this PF code and the previous EKF sample is the measurement update rate. In this project, we apply measurement updates every 1 s to save computation time/resources. We also apply an outage from $t = 9$ s to $t = 14$ s. We also added biases of 0.8 rad/s for the gyroscope and 0.5 m/s^2 for the accelerometer. Figures 9.24 and 9.25 show the biases convergence. Figures 9.26 and 9.27 shows the positional error with the error envelope. Figures 9.28–9.30 show the pose particles. As the figures

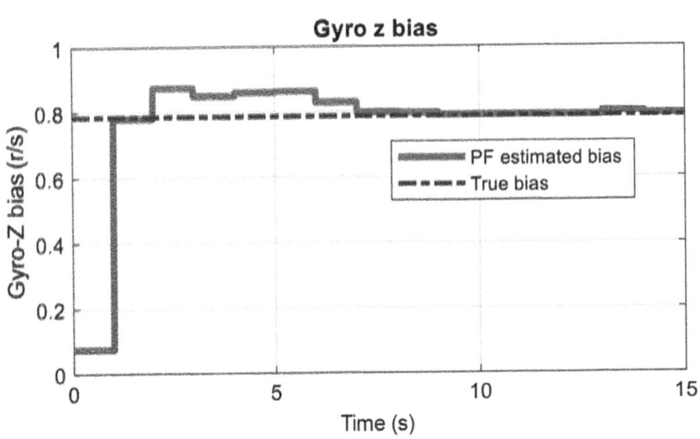

Figure 9.24 Gyroscope bias convergence under PF.

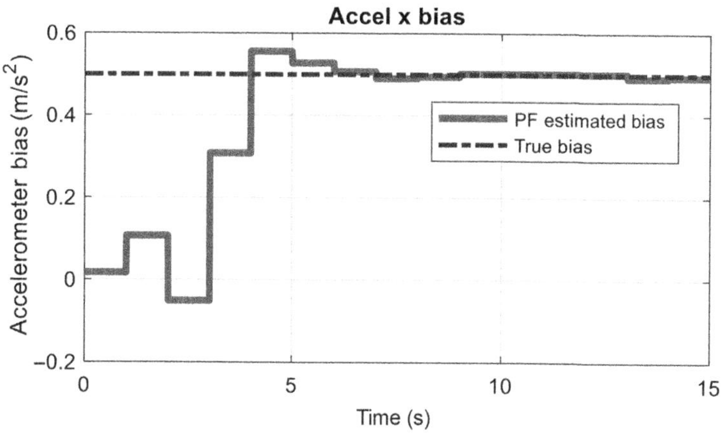

Figure 9.25 Accelerometer bias convergence under PF.

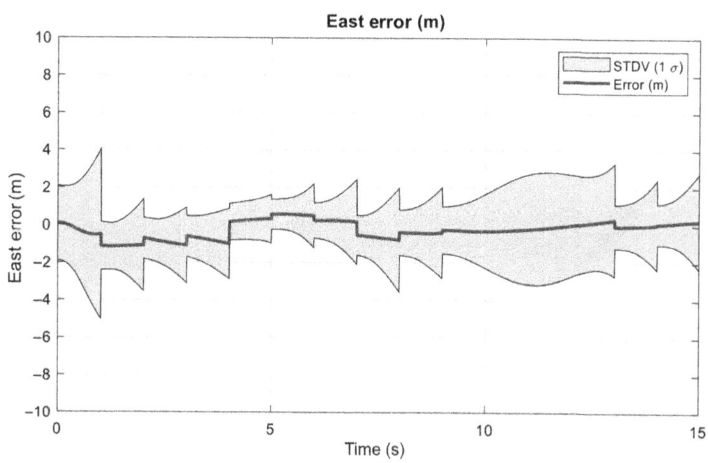

Figure 9.26 East position error + error envelope under PF with outage from second 9 to second 14.

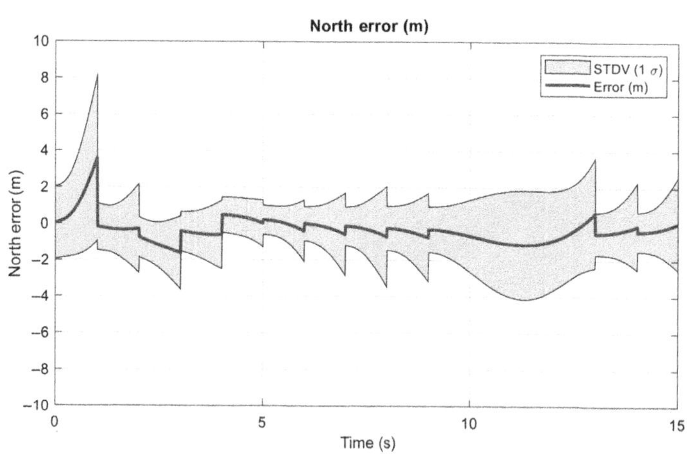

Figure 9.27 North position error + error envelope under PF with outage from second 9 to second 14.

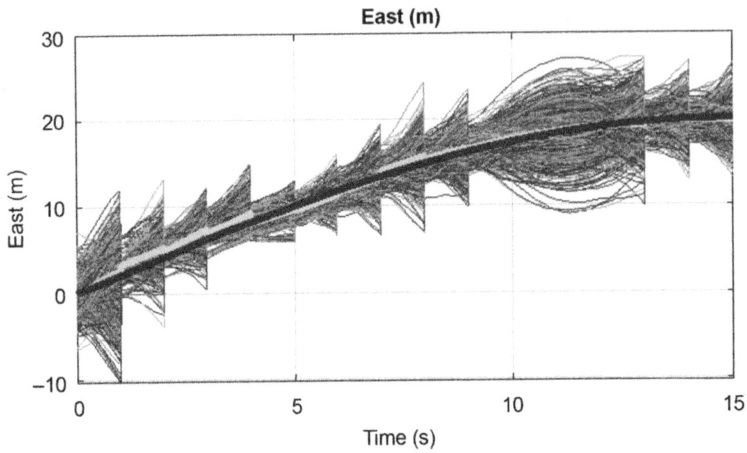

Figure 9.28 East position particles cloud with outage from second 9 to second 14.

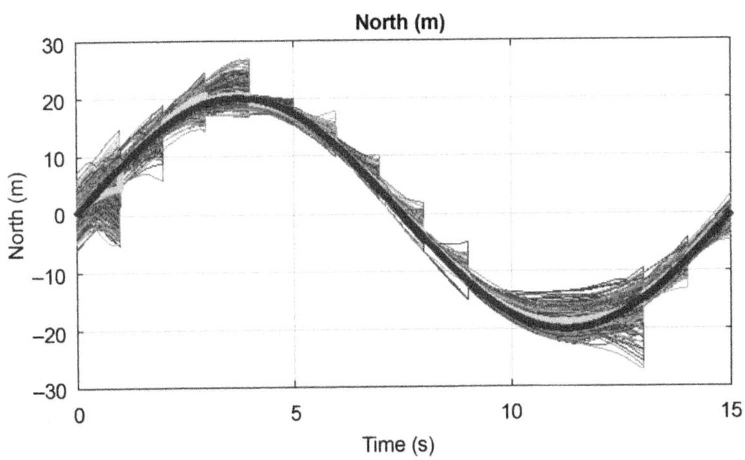

Figure 9.29 North position particles cloud with outage from second 9 to second 14.

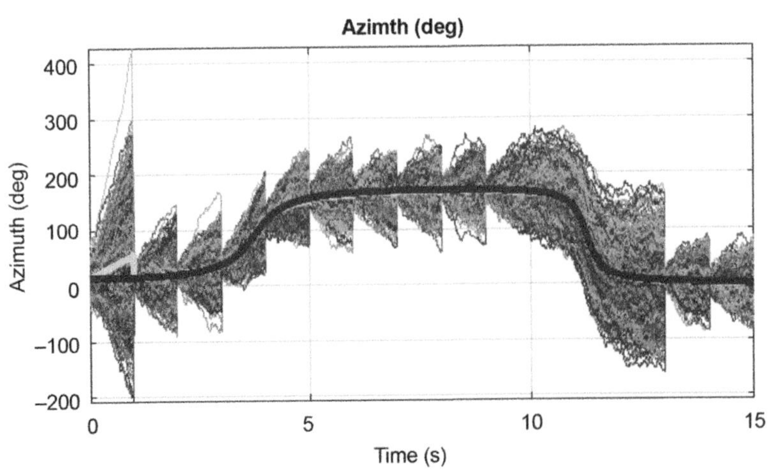

Figure 9.30 Azimuth particles cloud with outage from second 9 to second 14.

show, the biases are converging properly to the correct values, the error is reducing with the accumulation of measurements, and the performance during the outage is stable and accurate. The inspections of the figures show the correctness of the operation of the particle filter. Readers are encouraged to check the code to see more details about how the measurement update, weight updates, and resampling steps are performed in particle filtering.

10

Case Study #2: Aerial Vehicles

10.1 Introduction

In this chapter, we will apply the sensor fusion methods to aerial vehicles. Aerial platforms (see Figure 10.1) include manned and unmanned aerial vehicles (popularly known as drones). Unlike ground platforms, aerial platforms do not have horizontal motion constraints. Although aerial platforms may have a speedometer that works in a similar way to wheel encoders, they usually lack an accurate speed measurement due to several factors, such as wind disturbance. Large aerial platforms have some forward motion constraints, though. However, small-scale drones and vertical-take-off aerial platforms may exhibit aggressive 3D motion. These characteristics of aerial vehicles impose several challenges and make them more difficult than ground vehicles in terms of positioning, navigation, and mapping fusion algorithms. In this chapter, we will discuss some fusion configurations suitable for highly dynamic 3D aerial platforms. We will use a 3D motion model to accommodate highly dynamic motion. To support lightweight drones, we will consider lightweight power-effective sensors, including inertial sensors, imaging sensors, and RADAR sensors.

In this chapter, we will explain the details of the 3D system model, the measurement models, and the design matrix of each measurement type. By having the measurement vector and the design matrix, any fusion algorithm can be used. Therefore, in the measurement fusion part, we will limit our explanation to the measurement models and the design matrices as the steps to use the measurement models and design matrices in the fusion algorithms such as EKF, PF, or graph optimization have already been elaborated in detail in Chapter 9. Therefore, it is important that readers review chapter 9 before delving into this chapter.

10.2 3D Motion System Model

10.2.1 System Model

We discussed earlier in Chapters 1 and 4 the fundamentals of the kinematics of rigid bodies in 3D space, and we derived the differential equations that model the state of a rigid body in 3D space using the following general first-order differential equation (state space model) form:

$$\dot{\vec{x}}(t) = f\left(\vec{x}(t), \vec{u}(t)\right) + \vec{w}(t) \tag{10-1}$$

where $\vec{u}(t)$ is the control signal that triggers the transition of the state.

Sensor Fusion Approaches for Positioning, Navigation, and Mapping: How Autonomous Vehicles and Robots Navigate in the Real World: With MATLAB Examples, First Edition. Mohamed M. Atia.
© 2025 The Institute of Electrical and Electronics Engineers, Inc. Published 2025 by John Wiley & Sons, Inc.

Figure 10.1 Aerial platforms.

In our case, we will use inertial sensor measurements as the control signal that triggers the transition. This approach makes the developed fusion algorithms independent of any dynamic or control model and enables the developed fusion engine to work on any platform equipped with inertial sensors without interfering with its internal dynamic control systems, such as thrust and motors models. The control signal is then given by the following vector:

$$\vec{u} = \left(a_{SFx}^{B}\ a_{SFy}^{B}\ a_{SFz}^{B}\ \omega_{IBx}^{B}\ \omega_{IBY}^{B}\ \omega_{IBz}^{B}\right)^{T} \tag{10-2}$$

where

$\vec{a}_{SF}^{B} = \left(a_{SFx}^{B}\quad a_{SFy}^{B}\quad a_{SFz}^{B}\right)^{T}$ is accelerometer measurements (the acronym "*SF*" stands for specific force)

$\vec{\omega}_{IB}^{B} = \left(\omega_{IBx}^{B}\quad \omega_{IBY}^{B}\quad \omega_{IBz}^{B}\right)^{T}$ is the gyroscope measurements.

The subscript *IB* refers to the note that the gyroscope measures the rotation of the body frame with respect to the universal nonrotating inertial frame *I*. In both definitions, the superscript *B* means the measurements are presented in the body frame, which is assumed to be perfectly aligned with the inertial sensor frame. The state $\vec{x}(t)$ is the position, velocity, and attitude (orientation) of the object referenced to a specific coordinate reference frame in addition to the gyroscope and accelerometer biases. Details of the state vector will be discussed in the subsequent sections.

10.2.1.1 Coordinate Frames and State Vector

Recall from Chapter 4 "Inertial Sensors and Inertial Navigation," we used the following coordinate reference frames:

- ***B*: Body frame.** Origin is at the body's center of mass: x_B: longitudinal (forward) direction, y_B: transversal (lateral) direction, z_B: down (vertical) direction.
- ***L*: Local-level navigation frame.** Always leveled with respect to earth's ellipsoid surface. Origin is at the vehicle's location. *X*: True north direction, *Y*: East direction, *Z*: Down vertical direction
- ***E*: Earth-centered earth-fixed (ECEF).** Rotates with earth, origin: center of the earth. Z_E: extends through the North Pole. X_E: passes through the intersection of the equator and prime meridian, and Y_E: completes the right-hand rule in the equatorial plane.
- ***I*: Earth-centered inertial frame.** The *I* frame is nonrotating and it becomes aligned with ECEF at zero longitude.

Figure 10.2 illustrates the coordinate frames.

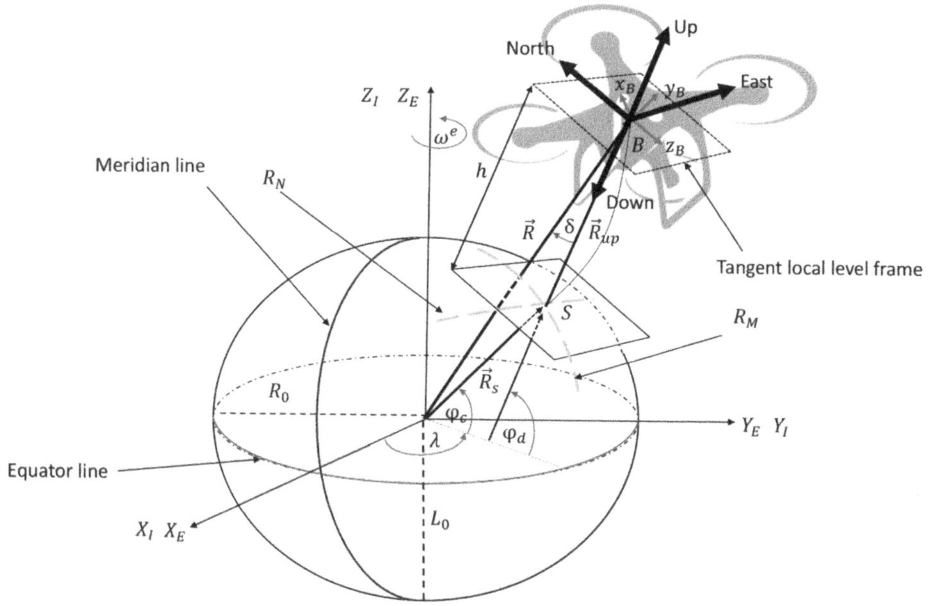

Figure 10.2 Coordinate frames illustration.

10.2.1.2 State Vector

The state vector can be defined as follows:

$$\vec{x}(t) =, \begin{pmatrix} \dot{\vec{r}}^L \\ \dot{\vec{v}}^L \\ \dot{\vec{q}}_B^L \\ \dot{\vec{b}}_g \\ \dot{\vec{b}}_a \end{pmatrix} \tag{10-3}$$

where

$\vec{r}^L$ and $\vec{v}^L$ are position and velocity of the vehicle in the local level frame (L or N)
$\vec{q}_B^L$ is the orientation quaternion
$\vec{b}_g$ and $\vec{b}_a$ are gyroscope and accelerometer biases.

However, due to the redundancy in the quaternion vector $\vec{q}_B^L$, which will lead to singularities in the covariance matrices, we will have to use only the vector part of $\vec{q}_B^L$. We will discuss this issue in Section 10.2.1.4.

10.2.1.3 System Model

In Chapter 4, we have demonstrated the derivation of the motion equations given the inertial sensors measurements, which are also known as the inertial navigation system (INS) mechanization equations that constitute the system model. The system model equations, including the stochastic random noise part, are given by:

$$\dot{\vec{q}}_B^L = \begin{pmatrix} \dot{q}_0 \\ \dot{q}_1 \\ \dot{q}_2 \\ \dot{q}_3 \end{pmatrix} = \frac{1}{2}\left(\vec{W}_{LB}^B \times\right)\vec{q}_B^L + \vec{w}_q, \vec{w}_q \sim \mathcal{N}\left(0, \sigma_q^2\right) \tag{10-4}$$

$$\dot{\vec{v}}^L = \begin{pmatrix} \dot{v}_n \\ \dot{v}_e \\ \dot{v}_d \end{pmatrix} = C_B^L \vec{a}_{SF}^B + \vec{g}^L - \left(\vec{\omega}_{EL}^L + 2\vec{\omega}_{IE}^L\right) \times \vec{v}^L \, \vec{w}_v, \vec{w}_v \sim \mathcal{N}\left(0, \sigma_v^2\right) \tag{10-5}$$

$$\dot{\vec{r}}^G = \begin{pmatrix} \dot{\varphi} \\ \dot{\lambda} \\ \dot{h} \end{pmatrix} = \begin{pmatrix} \frac{1}{(R_M + h)} & 0 & 0 \\ 0 & \frac{1}{(R_N + h)\cos(\varphi)} & 0 \\ 0 & 0 & -1 \end{pmatrix} \begin{pmatrix} v_n \\ v_e \\ v_d \end{pmatrix} + \vec{w}_r, \vec{w}_r \sim \mathcal{N}\left(0, \sigma_r^2\right) \tag{10-6}$$

$$\dot{\vec{r}}^L = \begin{pmatrix} \dot{N} \\ \dot{E} \\ \dot{D} \end{pmatrix} = \begin{pmatrix} v_n \\ v_e \\ v_d \end{pmatrix} \left(\omega_{EL}^L \times\right) r^L + \vec{w}_r, \ \vec{w}_r \sim \mathcal{N}\left(0, \sigma_r^2\right) \tag{10-7}$$

where

vector $\vec{r}^G$ is geodetic coordinates

vector $\vec{r}^L$ is position coordinates in L or N frame

R_m, R_n are the meridian and normal earth's curvature

$\vec{W}_{LB}^B = \begin{pmatrix} 0 & \omega_{LB_X}^B & \omega_{LB_Y}^B & \omega_{LB_Z}^B \end{pmatrix}^T$ is the homogeneous quaternion vector form of the rotation rate gyroscope measurements $\vec{\omega}_{LB}^B$

$\left(\left(\vec{W}_{LB}^B\right) \times\right)$ is the 4D skew matrix operator defined as follows:

$$\left(\left(\vec{W}_{LB}^B\right) \times\right) = \begin{pmatrix} 0 & -\omega_{LB_X}^B & -\omega_{LB_Y}^B & -\omega_{LB_Z}^B \\ \omega_{LB_X}^B & 0 & \omega_{LB_Z}^B & -\omega_{LB_Y}^B \\ \omega_{LB_Y}^B & -\omega_{LB_Z}^B & 0 & \omega_{LB_X}^B \\ \omega_{LB_Z}^B & \omega_{LB_Y}^B & -\omega_{LB_X}^B & 0 \end{pmatrix} \tag{10-8}$$

To get this value, we subtract the earth rate and transport rate $\vec{\omega}_{IL}^B$ from the raw gyroscope measurement $\vec{\omega}_{IB}^B$ as follows:

$$\vec{\omega}_{LB}^B = \vec{\omega}_{IB}^B - \vec{\omega}_{IL}^B \tag{10-9}$$

$\vec{\omega}_{IL}^B$ is the Earth rate plus transport rate calculated by the following equation:

$$\vec{\omega}_{IL}^B = C_L^B\left(\vec{\omega}_{IE}^L + \vec{\omega}_{EL}^L\right) \tag{10-10}$$

$$\vec{\omega}_{IE}^L = \begin{pmatrix} w^e\cos(\varphi) & 0 & -w^e\sin(\varphi) \end{pmatrix}^T \tag{10-11}$$

$$\vec{\omega}_{EL}^{L} = \begin{pmatrix} \frac{v_e}{(R_n + h)} & \frac{-v_n}{(R_m + h)} & \frac{-v_e}{(R_n + h)} \tan(\varphi) \end{pmatrix}^T \tag{10-12}$$

To compensate for the bias error, we modify $\vec{\omega}_{LB}^{B}$ as follows:

$$\vec{\omega}_{LB}^{B} = \vec{\omega}_{IB}^{B} - \vec{\omega}_{IL}^{B} - \vec{b}_g \tag{10-13}$$

where

$\vec{\omega}_{IB}^{B}$ is the raw gyroscope measurement
$\vec{w}_g$ is zero-mean random noise of gyroscope
$\vec{b}_g$ is the gyroscope bias vector
$\vec{\omega}_{IL}^{B}$ is the earth rotation plus transportation rates.

The accelerometer measurements are also compensated for biases as follows:

$$\vec{a}_{SF}^{B} = \vec{a}_{SF}^{B} - \vec{b}_a \tag{10-14}$$

where

$\vec{a}_{SF}^{B}$ is the raw accelerometer measurements
$\vec{b}_a$ is accelerometer bias
$\vec{w}_a$ is zero-mean random noise of accelerometer.

Note that we here considered the bias term only. In more sophisticated models, we may consider other inertial errors such as scale factor, nonorthogonality, and axes misalignments. To complete the system model, we will assume random walk process model for the biases as follows:

$$\dot{\vec{b}}_g = \vec{w}_b, \vec{w}_b \sim \mathcal{N}\left(0, \sigma_b^2\right) \tag{10-15}$$

$$\dot{\vec{b}}_a = \vec{w}_a, \vec{w}_a \sim \mathcal{N}\left(0, \sigma_a^2\right) \tag{10-16}$$

where $\vec{w}_b$ and $\vec{w}_a$ are zero-mean Gaussian noise.

Now, we put the system model in the stochastic nonlinear form $\dot{\vec{x}}(t) = f\left(\vec{x}(t), \vec{u}(t)\right) + \vec{w}(t)$ as follows:

$$\begin{pmatrix} \dot{\vec{r}}^{L} \\ \dot{\vec{v}}^{L} \\ \dot{\vec{q}}_{B}^{L} \\ \dot{\vec{b}}_g \\ \dot{\vec{b}}_a \end{pmatrix} = \begin{pmatrix} \vec{v}^{L} + \left(\vec{\omega}_{EL}^{L} \times\right) \vec{r}^{L} \\ C_B^L\left(\vec{a}_{SF}^{B}\right) + \vec{g}^{L} - \left(\vec{\omega}_{EL}^{L} + 2\vec{\omega}_{IE}^{L}\right) \times \vec{v}^{L} \\ 0.5\left(\left(\vec{W}_{LB}^{B}\right) \times\right) \vec{q}_{B}^{L} \\ 0_{1x3} \\ 0_{1x3} \end{pmatrix} + \begin{pmatrix} \vec{w}_q \\ \vec{w}_v \\ \vec{w}_r \\ \vec{w}_b \\ \vec{w}_a \end{pmatrix} \tag{10-17}$$

where $\vec{w}_q, \vec{w}_v, \vec{w}_r, \vec{w}_b, \vec{w}_a$ are zero-mean Gaussian random noise.

10.2.1.4 Minimal Coordinates Condition

In Chapter 4, we have seen how orientation (the transformation of the body frame B with respect to the local level navigation frame L can be described in several forms such as direction cosine matrix

$C_B^L \in \mathbb{R}^{3\times 3}$, quaternion vector $q_B^L \in \mathbb{R}^4$, rotation vector $\emptyset_B^L \in \mathbb{R}^3$, or Euler angles (roll: r, pitch: p, and azimuth: A). Because we will use this motion dynamic model in our fusion algorithms, we will need to linearize it to be in the following form:

$$\delta\dot{\vec{x}}(t) = F(t)\delta\vec{x}(t) + G(t)\vec{w}_u(t) \tag{10-18}$$

where $\vec{w}_u$ is a zero-mean unity variance Gaussian noise.

The fusion algorithm will estimate $\delta\vec{x}^*(t)$, the best estimate of the error state $\delta\vec{x}(t)$. We then perform state corrections using the following simple equation:

$$\vec{x}^+(t) = \vec{x}^-(t) + \delta\vec{x}^*(t) \tag{10-19}$$

To apply linear estimation techniques to this problem, we must ensure that the states are Euclidean so they can be corrected by the addition of small changes (errors) and that the state error covariance matrix is positive definite of rank n where n is the number of states. For these reasons, it is essential to ensure that the state vector $\vec{x}(t)$ has a minimum set of linearly independent parameters that can be corrected by addition of small changes. The position, velocity, and sensors' biases are Euclidean variables that have no issues related to these two conditions (closeness of addition and linear independence). The only part of the states that do not satisfy these conditions is the rotation part. As we discussed in Chapter 1, the rotation matrices belong to the Lie Group where Lie Algebra applies, and the set is closed under multiplication, not addition. Therefore, we must map the rotation matrix to a linear space to comply with the closeness of addition and linear independence conditions. We have seen in Chapter 1 how the exponential mapping maps small rotations from nine-element matrices into three-element rotation vectors that can be considered Euclidean within a small region (a manifold). We have also seen how rotation can be represented by a 4D quaternion vector. The advantage of the quaternion dynamic system model is that it is very simple, it does not have any sine or cosine terms, and it is numerically very stable compared to rotation matrices, Euler angles, and rotation vectors. However, quaternion has a small redundancy that causes linear dependency in any covariance matrix because the square of any element in the quaternion vector is a linear function of the square of the other three elements. In addition, the quaternion vector is not closed set under addition. To benefit from the numerical stability of quaternions and, at the same time, achieve the closeness of addition and linear independence conditions, we chose to parametrize the rotation using a quaternion vector $\vec{q}_B^L$ and we chose to model the rotation error using 3D rotation vector $\vec{\phi}_{LB}$. Direction cosine matrix C_B^L and quaternion vector $\vec{q}_B^L$ can be obtained from rotation vector as follows:

$$C_B^L = I + \frac{\left(\vec{\phi}_{LB}\times\right)}{\left\|\vec{\phi}_{LB}\right\|}\sin\left(\left\|\vec{\phi}_{LB}\right\|\right) + \frac{\left(\vec{\phi}_{LB}\times\right)^2}{\left\|\vec{\phi}_{LB}\right\|^2}\left(1-\cos\left(\left\|\vec{\phi}_{LB}\right\|\right)\right) \tag{10-20}$$

$$\vec{q}_B^L = \left(\cos\left(\frac{\left\|\vec{\phi}_{LB}\right\|}{2}\right), \phi_x \sin\left(\frac{\left\|\vec{\phi}_{LB}\right\|}{2}\right), \phi_y \sin\left(\frac{\left\|\vec{\phi}_{LB}\right\|}{2}\right), \phi_z \sin\left(\frac{\left\|\vec{\phi}_{LB}\right\|}{2}\right)\right)^T \tag{10-21}$$

To keep rotation parameters consistent, instead of using the rotation vector, we can use the vector part of the quaternion $\vec{q}_B^L(2:4)$ which is a scaled version of the unit rotation vector scaled by sine of the half of the rotation angle. Therefore, in the dynamic system prediction, we will use the full quaternion vector, but when we linearize the system for filtering, we will use the vector part of the quaternion, which is a scaled version of the rotation vector. To keep only the vector part of the quaternion $(q_1\ q_2\ q_3)^T$, the scalar part of the quaternion vector is replaced by the vector part as follows:

$$q_0 = \sqrt{1 - q_1^2 - q_2^2 - q_3^2} \tag{10-22}$$

or

$$q_0^2 = 1 - q_1^2 - q_2^2 - q_3^2 \tag{10-23}$$

We will do this replacement in the linearization process to get the linearized system model which is described in the next section. Given this minimal coordinate representation, the list of state variables is given as follows:

$$\vec{x}_{15\times 1} = \left(\left(\vec{r}^L\right)_{1\times 3} \quad \left(\vec{v}^L\right)_{1\times 3} \quad q_1 \quad q_3 \quad q_3 \quad \left(\vec{b}_g\right)_{1\times 3} \quad \left(\vec{b}_a\right)_{1\times 3} \right)^T \tag{10-24}$$

10.2.2 Linearized System Model

The system model equations are linearized using first-order Taylor approximation. In Chapter 4, we described a linearized error model for the INS. The described error model in Chapter 4 is more suitable for the E frame implementation, which can be used in L frame with some transformations. In this chapter, we derive another simplified error model that can be used in L frame. This simplified error model is the one we used in the sample projects provided in this book. To simplify the error model derivation, we can discard the terms that are multiplied by Earth rate and transportation rate as they are very small terms that are divided by Earth radius, which is a large value. We will ignore these terms only in the error model and not in the full state system model as we are keen to get the best accurate prediction of the state vector. We can tolerate the neglection if these terms are only in the error analysis as the errors should be small enough such that when divided by Earth radius, they become extremely insignificant. The first step to derive the error model is to replace rotation matrix C_B^L can be replaced by its quaternion equivalence as follows:

$$C_B^L = \begin{pmatrix} q_0^2 + q_1^2 - q_2^2 - q_3^2 & 2(q_1q_2 + q_0q_3) & 2(q_1q_3 - q_0q_2) \\ 2(q_1q_2 - q_0q_3) & q_0^2 - q_1^2 + q_2^2 - q_3^2 & 2(q_2q_3 + q_0q_1) \\ 2(q_1q_3 + q_0q_2) & 2(q_2q_3 - q_0q_1) & q_0^2 - q_1^2 - q_2^2 + q_3^2 \end{pmatrix} \tag{10-25}$$

by replacing C_B^L by its quaternion equivalent in the system model and ignoring the terms that are multiplied by Earth rate and transportation rate and the terms that are divided by the Earth radius, we obtain the following model (without showing the random noise part):

$$\dot{\vec{r}}^L \cong \vec{v}^L \tag{10-26}$$

$$\dot{v}_n \cong \left(-a_{SFx}^B(2q_2^2 + 2q_3^2 - 1) - a_{SFy}^B(2q_3q_0 - 2q_1q_2) + a_{SFz}^B(2q_2q_0 + 2q_1q_3)\right) \tag{10-27}$$

$$\dot{v}_e \cong \left(a_{SFx}^B(2q_3q_0 + 2q_1q_2) - a_{SFy}^B(2q_1^2 + 2q_3^2 - 1) + a_{SFz}^B(2q_1q_0 - 2q_2q_3)\right) \tag{10-28}$$

$$\dot{v}_d \cong \left(-a_{SFx}^B(2q_2q_0 - 2q_1q_3) + a_{SFy}^B(2q_1q_0 + 2q_2q_3) - a_{SFz}^B(2q_1^2 + 2q_2^2 - 1) + g\right) \tag{10-29}$$

$$\dot{q}_1 \cong \left(\left(\omega_{IBx}^B - b_{g_x}\right)\left(\frac{q_0}{2}\right) - \left(\omega_{IBy}^B - b_{g_y}\right)\left(\frac{q_3}{2}\right) + \left(\omega_{IBz}^B - b_{g_z}\right)\left(\frac{q_2}{2}\right)\right) \tag{10-30}$$

$$\dot{q}_2 \cong \left(\left(\omega_{IBx}^B - b_{g_x}\right)\left(\frac{q_3}{2}\right) + \left(\omega_{IBy}^B - b_{g_y}\right)\left(\frac{q_0}{2}\right) - \left(\omega_{IBz}^B - b_{g_z}\right)\left(\frac{q_1}{2}\right)\right) \tag{10-31}$$

$$\dot{q}_3 \cong \left(-\left(\omega_{IBx}^B - b_{g_x}\right)\left(\frac{q_2}{2}\right) + \left(\omega_{IBy}^B - b_{g_y}\right)\left(\frac{q_1}{2}\right) + \left(\omega_{IBy}^B - b_{g_y}\right)\left(\frac{q_0}{2}\right)\right) \tag{10-32}$$

$$\dot{\vec{b}}_g = 0 \tag{10-33}$$

$$\dot{\vec{b}}_a = 0 \tag{10-34}$$

We can notice that the left-hand side, the state vector, does not have the scalar quaternion element q_0 but the right-hand side does. We left q_0 in the right-hand side to keep the notation compact as much as possible. The second step is to do the linearization of Eqs. (10-26)–(10-34) by taking the derivative of the right-hand side with respect to the state vector. But before we do the linearization, we need to replace the occurrences of the quaternion scalar part q_0 with its equivalent $\sqrt{1-q_1^2-q_2^2-q_3^2}$ in both the orientation rate and the velocity rate equations so when we do the linearization, we do it with respect to q_1, q_2, and q_3 so we keep only the vector part $(q_1\ q_2\ q_3)^T$ in the states. As we can observe, the system model and the partial derivative process will get complex when we replace q_0 by $\sqrt{1-q_1^2-q_2^2-q_3^2}$. Fortunately, this linearization can be automatically done using symbolic math engines such as MATLAB symbolic engine. In MATLAB, we can define any symbolic expression and command MATLAB to perform a partial derivative with respect to any variable, and the results will be a symbolic expression. The detailed item-by-item approximated linearized error system model is given by the following equations:

$$\delta\dot{n} \cong v_n \tag{10-35}$$

$$\delta\dot{e} \cong v_e \tag{10-36}$$

$$\delta\dot{d} \cong v_d \tag{10-37}$$

$$\begin{aligned}\delta\dot{v}_n \cong &\left(a^B_{SFy}\left(2q_2 + 2\frac{q_1q_3}{q_0}\right) + a^B_{SFz}\left(2q_3 - 2\frac{q_1q_2}{q_0}\right)\right)\delta q_1 + \left(-a^B_{SFx}(4q_2) - a^B_{SFy}\left(2q_1 + 2\frac{q_2q_3}{q_0}\right)\right.\\ &\left. + a^B_{SFz}\left(2q_0 - 2\frac{q_2^2}{q_0}\right)\right)\delta q_2 + \left(-a^B_{SFx}(4q_3) - a^B_{SFy}\left(2q_0 - 2\frac{q_3^2}{q_0}\right) + a^B_{SFz}\left(2q_1 - 2\frac{q_2q_3}{q_0}\right)\right)\delta q_3\\ &+ (2q_2^2 + 2q_3^2 - 1)\,\delta b_{g_x} + (2q_3q_0 - 2q_1q_2)\delta b_{g_y} + (-2q_2q_0 - 2q_1q_3)\,\delta b_{g_z}\end{aligned} \tag{10-38}$$

$$\begin{aligned}\delta\dot{v}_e \cong &\left(a^B_{SFx}\left(2q_2 - 2\frac{q_1q_3}{q_0}\right) - a^B_{SFy}(4q_1) - a^B_{SFz}\left(2q_0 - 2\frac{q_1^2}{q_0}\right)\right)\delta q_1 + \left(a^B_{SFx}\left(2q_1 - 2\frac{q_2q_3}{q_0}\right)\right.\\ &\left. + a^B_{SFz}\left(2q_3 + 2\frac{q_1q_2}{q_0}\right)\right)\delta q_2 + \left(a^B_{SFx}\left(2q_0 - 2\frac{q_3^2}{q_0}\right) - a^B_{SFy}(4q_3) - a^B_{SFz}\left(2q_2 - 2\frac{q_1q_3}{q_0}\right)\right)\delta q_3\\ &+ (-2q_3q_0 - 2q_1q_2)\,\delta b_{g_x} + (2q_1^2 + 2q_3^2 - 1)\delta b_{g_y} + (2q_1q_0 - 2q_2q_3)\,\delta b_{g_z}\end{aligned} \tag{10-39}$$

$$\begin{aligned}\delta\dot{v}_d \cong &\left(-a^B_{SFx}\left(2q_3 + 2\frac{q_1q_2}{q_0}\right) - a^B_{SFy}\left(2q_0 - 2\frac{q_1^2}{q_0}\right) - a^B_{SFz}(4q_1)\right)\delta q_1 + \left(a^B_{SFx}\left(2q_0 - 2\frac{q_2^2}{q_0}\right)\right.\\ &\left. - a^B_{SFy}\left(2q_3 - 2\frac{q_1q_2}{q_0}\right) + a^B_{SFz}(4q_3)\right)\delta q_2 - \left(a^B_{SFx}\left(2q_1 + 2\frac{q_2q_3}{q_0}\right) + a^B_{SFy}\left(2q_2 + 2\frac{q_1q_3}{q_0}\right)\right)\delta q_3\\ &+ (2q_1q_3 - 2q_2q_0)\,\delta b_{g_x} + (2q_1q_0 + 2q_2q_3)\,\delta b_{g_y} + (-2q_1^2 - 2q_2^2 + 1)\delta b_{g_z}\end{aligned} \tag{10-40}$$

$$\begin{aligned}\delta\dot{q}_1 \cong &\left(-\frac{q_1}{2q_0}\left(\omega^B_{IBx} - b_{g_x}\right)\right)\delta q_1 + \left(-\frac{q_2}{2q_0}\left(\omega^B_{IBx} - b_{g_x}\right) + \frac{1}{2}\left(\omega^B_{IBz} - b_{g_z}\right)\right)\delta q_2\\ &+ \left(-\frac{q_3}{2q_0}\left(\omega^B_{IBx} - b_{g_x}\right) + \frac{1}{2}\left(\omega^B_{IBy} - b_{g_y}\right)\right)\delta q_3 - \left(\frac{q_0}{2}\right)\delta b_{g_x} + \left(\frac{q_3}{2}\right)\delta b_{g_y} - \left(\frac{q_2}{2}\right)\delta b_{g_z}\end{aligned} \tag{10-41}$$

$$\delta\dot{q}_2 \cong \left(-\frac{q_1}{2q_0}\left(\omega_{IBy}^B - b_{g_y}\right) + \frac{1}{2}b_{g_z}\right)\delta q_1 + \left(-\frac{q_2}{2q_0}\left(\omega_{IBy}^B - b_{g_y}\right)\right)\delta q_2 + \left(-\frac{q_3}{2q_0}\left(\omega_{IBy}^B - b_{g_y}\right) + \frac{1}{2}\left(\omega_{IBx}^B - b_{g_x}\right)\right)\delta q_3 - \left(\frac{q_3}{2}\right)\delta b_{g_x} - \left(\frac{q_0}{2}\right)\delta b_{g_y} + \left(\frac{q_1}{2}\right)\delta b_{g_z} \tag{10-42}$$

$$\delta\dot{q}_3 \cong \left(-\frac{q_1}{2q_0}\left(\omega_{IBz}^B - b_{g_z}\right) - \frac{1}{2}\left(\omega_{IBy}^B - b_{g_y}\right)\right)\delta q_1 + \left(-\frac{q_2}{2q_0}\left(\omega_{IBz}^B - b_{g_z}\right) + \frac{1}{2}\left(\omega_{IBx}^B - b_{g_x}\right)\right)\delta q_2 + \left(-\frac{q_3}{2q_0}\left(\omega_{IBz}^B - b_{g_z}\right)\right)\delta q_3 + \left(\frac{q_2}{2}\right)\delta b_{g_x} - \left(\frac{q_1}{2}\right)\delta b_{g_y} - \left(\frac{q_0}{2}\right)\delta b_{g_z} \tag{10-43}$$

$$\delta\dot{\vec{b}}_g = 0 \tag{10-44}$$

$$\delta\dot{\vec{b}}_a = 0 \tag{10-45}$$

Equations (10-35)–(10-45) form the complete error model after applying the approximations described above. The left-hand side contains all the error states, which are three position components, three velocity components, three orientation components, three gyroscope biases, and three accelerometer biases. This completes the 15-state error model. In addition, note that the right-hand side contains the scalar part of the quaternion q_0 which can be replaced by $\sqrt{1-q_1^2-q_2^2-q_3^2}$ so the whole error model is a function of the state vector described above. This linearized model gives us the transition matrix $F(t)$ where $\delta\dot{\vec{x}}(t) = F(t)\delta\vec{x}(t)$. Arranging Eqs. (10-35)–(10-45) into a matrix form where the time derivative of error vector is isolated in one side will give us the popular linearized error model $\delta\dot{\vec{x}}(t) = F(t)\delta\vec{x}(t)$. It is too complex to write down the full transition matrix $F(t)$ in one place here. However, $F(t)$ can be seen in detail in the project code that demonstrates the 3D IMU/GPS fusion that comes with this book. A brief explanation of this project sample is provided at the end of the chapter. To model the stochastic part, we add the noise shaping matrix term to get the full linearized error model in the form $\delta\dot{\vec{x}}(t) = F(t)\delta\vec{x}(t) + G(t)\vec{w}_u(t)$. Technically speaking, the $G(t)$ matrix is obtained by linearizing the system model with respect to noise terms as follows:

$$G(t) = \frac{\partial\left(\dot{\vec{x}} = f\left(\vec{x}, \vec{u}\right)\right)}{\partial \vec{w}} \tag{10-46}$$

In linear Gaussian systems, the function $\dot{\vec{x}} = f\left(\vec{x}, \vec{u}\right) = F\vec{x} + G\vec{w}_u$ so the derivative $\frac{\partial\left(\dot{\vec{x}} = f\left(\vec{x}, \vec{u}\right)\right)}{\partial \vec{w}}$ will lead to G directly. In nonlinear systems, the function $f\left(\vec{x}, \vec{u}\right)$ is not linear and the noise matrix G is not simply multiplied by the noise vector $\vec{w}$. Instead, the noise $\vec{w}$ is embedded within the states. For simplicity, we assume that the noise is Gaussian and linear as follows: $\dot{\vec{x}} = f\left(\vec{x}, \vec{u}\right) + G\vec{w}_u$, which is not very accurate, but it greatly simplifies the stochastic part of the linearization of nonlinear systems. In practice, we consider $G(t)$ as part of the system parameters that need to be tuned, keeping in consideration that $G(t)$ represents system prediction noise so if inertial sensors are accurate, $G(t)$ is set to small values and if the sensors are of low quality, $G(t)$ is set to large values. In practice, the exact values of $G(t)$ matrix is obtained by system tuning over several datasets. By having the transition matrix $F(t)$ and the noise shaping matrix $G(t)$, we are ready to use EKF to fuse inertial sensors with any other measurements. As we have demonstrated in the previous chapter, each measurement will have a measurement model that relates the observations to the internal system states. If the measurement model is linear, we directly apply the Kalman filter update step using the system design matrix H. If the measurement model is nonlinear, we linearize it to obtain the design matrix H and use it to apply EKF updates. In the next sections, we will discuss how to fuse a 3D inertial-based system model with GPS, imaging sensors, and radar in both loosely coupled and tightly coupled modes.

10.3 Fusion of GPS Measurements

Fusion of GPS measurements in the 3D case uses the same measurement models discussed in Chapter 9, which discuss the 2D wheeled platform case. For convenience, we briefly review GPS measurement models for loosely and tightly coupled modes in the next subsections.

10.3.1 Loosely Coupled Fusion

In the previous chapter, we discussed the GPS measurements in both loosely coupled mode and tightly coupled mode. Therefore, we assume that the reader is familiar with GPS measurements and measurement model derivations in each case. If this is not the case, readers are encouraged to review Sections 9.4 and 9.5 in Chapter 9. In the loosely coupled case, the GPS measurement model is as follows:

$$\vec{y}_{GPS}(t) = \begin{pmatrix} \hat{n}(t) \\ \hat{n}(t) \\ \hat{v}_n(t) \\ \hat{v}_e(t) \end{pmatrix} \tag{10-47}$$

This measurement model is linear in the form: $\vec{y}(t) = H(t)\vec{x}(t)$ where $H(t)$ is defined as follows:

$$H(t) = \begin{pmatrix} 1 & 0 & 0 & 0 & \\ 0 & 1 & 0 & 0 & \\ 0 & 0 & 1 & 0 & 0_{4\times 11} \\ 0 & 0 & 0 & 1 & \end{pmatrix} \tag{10-48}$$

10.3.2 Tightly Coupled Fusion

In case of tightly coupled, and noting that we are using NED coordinates for position and velocity, the GPS measurement model given M visible satellites is as follows:

$$\vec{y}_{GPS}(t) = \begin{pmatrix} \rho^1(t) \\ \rho^2(t) \\ \cdot \\ \cdot \\ \rho^M(t) \\ \dot{\rho}^1(t) \\ \cdot \\ \cdot \\ \dot{\rho}^M(t) \end{pmatrix} \tag{10-49}$$

where $\rho^i(t)$ and $\dot{\rho}^i(t)$ are range and range rate measurement of the ith satellite, which are defined as follows:

$$\rho^i(t) = \sqrt{(p_x(t) - p_x^i(t))^2 + \left(p_y(t) - p_y^i(t)\right)^2 + (p_z(t) - p_z^i(t))^2} + b_r(t) + v_\rho^i(t) \tag{10-50}$$

$$\dot{\rho}^i(t) = \left((v_x(t) - v_x^i(t)), \left(v_y(t) - v_y^i(t)\right), (v_z(t) - v_z^i(t))\right)^T . \vec{L}^i(t) + d_r(t) + \varepsilon_{\dot{\rho}}^i(t) \tag{10-51}$$

where

$\vec{p} = \left(p_x, p_y, p_z\right)^T$ and $\vec{p}^i = \left(p_x^i, p_y^i, p_z^i\right)^T$ are vehicle's position, and satellite position in ECEF frame respectively

$b_r(t)$ and $d_r(t)$ are the receiver clock bias and drift respectively.

This measurement model is nonlinear in the form: $\vec{y}(t) = h\left(\vec{x}(t)\right)$, which must be linearized around a given position/velocity to obtain the linear measurement error in the form $\delta\vec{y}(t) = H(t)\delta\vec{x}(t)$ and the design matrix $H(t)$. We have seen the linearization of this nonlinear measurement model in the previous chapter in Section 9.5.3. The linearized measurements model design matrix $H(t)$ is given by:

$$H(t) = \begin{pmatrix} \left(\vec{L}^1(t)\right)^T & 1 & 0_{3\times 1} & 0 \\ \vdots & \vdots & \vdots & \vdots \\ \left(\vec{L}^M(t)\right)^T & 1 & 0_{3\times 1} & 0 \\ 0_{3\times 1} & 0 & \left(\vec{L}^1(t)\right)^T & 1 \\ \vdots & \vdots & \vdots & \vdots \\ 0_{3\times 1} & 0 & \left(\vec{L}^M(t)\right)^T & 1 \end{pmatrix} \begin{pmatrix} T_E^L & 0_{3\times 1} & 0_{2\times 2} & 0_{2\times 2} \\ 0_{1\times 3} & 1 & 0 & 0 \\ 0_{3\times 1} & 0_{2\times 2} & T_E^L & 0 \\ 0 & 0 & 0 & 1 \end{pmatrix} \tag{10-52}$$

$$T_E^L = \begin{pmatrix} -\sin\varphi\cos\lambda & -\sin\lambda & -\cos\varphi\cos\lambda \\ -\sin\varphi\sin\lambda & \cos\lambda & -\cos\varphi\sin\lambda \\ -\cos\varphi & 0 & \sin\varphi \end{pmatrix} \tag{10-53}$$

$$\vec{L}^i = \frac{\left((p_x - p_x^i), \left(p_y - p_y^i\right), (p_z - p_z^i)\right)^T}{\sqrt{(p_x - p_x^i)^2 + \left(p_y - p_y^i\right)^2 + (p_z - p_z^i)^2}} \tag{10-54}$$

where

φ, λ are geodetic coordinates of the platform

$\vec{L}^i$ is the line of sight from the platform to the ith satellite's position.

Note that in case of tightly coupled GPS measurements, we need to augment the system model by the following two additional states:

$$\dot{b}_r(t) = d_r(t) + w_b(t) \tag{10-55}$$

$$\dot{d}_r(t) = 0 + w_d(t) \tag{10-56}$$

where $w_b(t)$ and $w_d(t)$ is zero-mean Gaussian noise that models the receiver's clock drift noise.

By having the system model and the measurement model for each measurement type, we can use any filtering technique, such as PF, EKF, or any of their variations, to perform the fusion. In the case of using EKF, we would need the linear (for loosely coupled mode) or the linearized (for tightly coupled mode) system and measurements model.

10.4 Fusion of Vision Measurements

Imaging sensors are suitable for aerial vehicles like drones as they are lightweight and have low power consumption compared to LiDAR. In the next subsections, we will discuss the measurement model of visual sensors and some possible fusion configurations. It is important to keep in mind that the position coordinates of the camera are usually described as a pose transformation matrix $\left\{R_k, \vec{T}_k\right\}$ at epoch k where R_k is a transformation matrix from world frame to camera frame, and $\vec{T}_k$ is a position vector that describes the world coordinate as seen from the camera frame. In the discussion about imaging sensors, we always considered the world frame as the frame of the camera at the first epoch when $k = 0$. We will refer to this frame by $c(0)$ (camera frame at $k = 0$). This means that the transformation from camera pose to world frame at epoch k is given by the following matrix:

$$T_{c(k)}^{c(0)} = \begin{pmatrix} (R_k)^T & -\vec{T}_k \\ 0 & 1 \end{pmatrix} \tag{10-57}$$

In many scenarios, we will need to generate results and perform fusion in the N or L frame coordinates. This can be done easily by knowing the transformation between the camera frame at epoch zero and the N or L frame. Let us define this transformation as $R_{c(0)}^{L}, \vec{T}_{c(0)}^{L}$ or $R_{c(0)}^{N}, \vec{T}_{c(0)}^{N}$ depending on which local frame we want to represent the results. Given the camera pose $\left\{R_k, \vec{T}_k\right\}$ with respect to its initial frame $c(0)$ and the transformations $R_{c(0)}^{L}, \vec{T}_{c(0)}^{L}$ or $R_{c(0)}^{N}, \vec{T}_{c(0)}^{N}$, the camera pose in N or L frame can be calculated as follows:

$$\begin{pmatrix} R_{c(k)}^{L} & \vec{T}_{c(k)}^{L} \\ 0 & 1 \end{pmatrix} = \begin{pmatrix} R_{c(0)}^{L} & \vec{T}_{c(0)}^{L} \\ 0 & 1 \end{pmatrix} \begin{pmatrix} (R_k)^T & -\vec{T}_k \\ 0 & 1 \end{pmatrix} \tag{10-58}$$

To avoid confusion, it is important to always remember which frame variables are defined.

10.4.1 Visual Measurements Model

In the following discussion, we will assume the word frame is the camera frame at the first epoch $k = 0$ referred to as $c(0)$. However, all variables can be represented in L or N frames if we know the transformations $R_{c(0)}^{L}, \vec{T}_{c(0)}^{L}$ or $R_{c(0)}^{N}, \vec{T}_{c(0)}^{N}$ as described in Eq. (10-60). We have seen in Chapter 7

"Imaging Sensors and Visual Odometry" how imaging sensors project 3D features on 2D image plan according to the following pinhole projection model:

$$\vec{y}_k^i = \begin{pmatrix} u_k^i \\ v_k^i \\ 1 \end{pmatrix} = \frac{1}{\lambda_k^i} K \begin{pmatrix} x_{c,k}^i \\ y_{c,k}^i \\ z_{c,k}^i \end{pmatrix} \tag{10-59}$$

$$\vec{y}_k^i = \begin{pmatrix} u_k^i \\ v_k^i \\ 1 \end{pmatrix} = \frac{1}{\lambda_k^i} K \left(R_k \; \vec{T}_k \right)_{3\times 4} \begin{pmatrix} m_x^i \\ m_y^i \\ m_z^i \\ 1 \end{pmatrix} \tag{10-60}$$

where

K is the camera intrinsic matrix
k is image frame number or time epoch
$\left(u_t^i \; v_t^i \right)^T$ is the pixel coordinate
$\vec{m}^i = \left(m_x^i \; m_y^i \; m_z^i \right)^T$ is the physical world metric 3D location of feature point i in the environment
$\vec{p}_k^i = \left(x_{c,k}^i \; z_{c,k}^i \; z_{c,k}^i \right)^T$ is the physical metric 3D feature's location with respect to the camera frame
$K_{3\times 3}$ is the camera intrinsic matrix
The rotation matrix R_k and the translation vector $\vec{T}_k$ form the transformation matrix $\left(R_k \; \vec{T}_k \right)_{3\times 4}$ from the world frame to the camera frame
λ_k^i is the physical feature's depth (also can be seen as the scale) with respect to the camera's frame origin (the physical metric z-component of the feature's location with respect to camera frame, which means $z_{c,k}^i = \lambda_k^i$).

This pinhole projection model can be seen as the measurement model of the camera defined as a function of the camera pose and feature's 3D location as follows:

$$\vec{y}_k^i = h\left(\vec{x}_k, \vec{m}^i \right) = h\left(R_k, \vec{T}_k, \vec{m}^i \right) \tag{10-61}$$

where $\vec{x}_k$ includes the camera pose.

In our analysis in Chapter 7, we used the rotation matrix and translation vector $\left\{ R_k, \vec{T}_k \right\}$ to describe the camera pose. However, like we did in the inertial mechanization equations, we can represent the camera's rotation using quaternion, rotation vector, or Euler angles. As we may need to linearize this measurement model to perform estimation and sensor fusion, we may need to use the vector part of the quaternion to describe the camera orientation, which is assumed to be aligned with the vehicle or platform orientation. In this case, the measurement model will be a function of quaternion vector and translation vector as follows:

$$\vec{y}_k^i = h\left(\vec{q}_k, \vec{T}_k, \vec{m}^i \right) \tag{10-62}$$

Assume $\vec{q}_k = \left(q_{0,k}, q_{1,k}, q_{2,k}, q_{3,k} \right)^T$, we replace the rotation matrix R_k by its quaternion equivalent to get the following measurement model:

$$h\left(\vec{q}_k, \vec{T}_k, \vec{m}^i \right) = \frac{1}{\lambda_k^i} K \left(\left(C\left(\vec{q}_k \right) \right)^T \vec{m}^i + \vec{T}_k \right) \tag{10-63}$$

where $C\left(\vec{q}_k\right)$ is the direction cosine matrix as a function of the quaternion vector $\vec{q}_k$ and is the transpose of R_k, which leads to the following measurement model:

$$C\left(\vec{q}_k\right) = \begin{pmatrix} q_{0,k}^2 + q_{1,k}^2 - q_{2,k}^2 - q_{3,k}^2 & 2\left(q_{1,k}q_{2,k} + q_{0,k}q_{3,k}\right) & 2\left(q_{1,k}q_{3,k} - q_{0,k}q_{2,k}\right) \\ 2\left(q_{1,k}q_{2,k} - q_{0,k}q_{3,k}\right) & q_{0,k}^2 - q_{1,k}^2 + q_{2,k}^2 - q_{3,k}^2 & 2\left(q_{2,k}q_{3,k} + q_{0,k}q_{1,k}\right) \\ 2\left(q_{1,k}q_{3,k} + q_{0,k}q_{2,k}\right) & 2\left(q_{2,k}q_{3,k} - q_{0,k}q_{1,k}\right) & q_{0,k}^2 - q_{1,k}^2 - q_{2,k}^2 + q_{3,k}^2 \end{pmatrix}^T \quad (10\text{-}64)$$

Note that $h\left(\vec{q}_k, \vec{T}_k, \vec{m}^i\right)$ still has an unknown depth parameter λ_k^i; however, we can replace it with a function of the minimal coordinate representation. We know from the camera projection model that λ_k^i is the third component of $\vec{p}_k^{\,i} = \left(x_{c,k}^i \; z_{c,k}^i \; z_{c,k}^i\right)^T$, the feature's metric location with respect to the camera which can be calculated as follows:

$$\lambda_k^i = 2\left(q_{1,k}q_{3,k} - q_{0,k}q_{2,k}\right)m_x^i + 2\left(q_{2,k}q_{3,k} + q_{0,k}q_{1,k}\right)m_y^i + \left(q_{0,k}^2 - q_{1,k}^2 - q_{2,k}^2 + q_{3,k}^2\right)m_z^i + T_{z,k} \quad (10\text{-}65)$$

By replacing λ_k^i by its equivalent equation, and the quaternion scalar part q_0 by its equivalent $\sqrt{1 - q_1^2 - q_2^2 - q_3^2}$, the measurement model will be a function of the minimal coordinate parameters only (camera pose and feature's location). Formulating the measurement model as a function of the minimal coordinate representation enables us to use the model in estimation and sensor fusion algorithms that require the linearization of the measurement model and the design matrix using the following partial derivative:

$$H_k^i = \frac{\partial h\left(\vec{q}_k, \vec{T}_k, \vec{m}^i\right)}{\partial\left(\vec{q}_k \; \vec{T}_k\right)} \quad (10\text{-}66)$$

The derivative in Eq. (10-66) can be obtained using MATLAB symbolic engine like the case with the linearization of the system model. To fuse image data, Like LiDAR scans, we can estimate the relative camera motion between epochs k and $k + 1$ defined as $\left\{R_k^{k+1} \; \vec{T}_k^{k+1}\right\}$, the depth λ_k^i and the feature's location with respect to the camera frame $\vec{p}_k^{\,i}$ using the methods discussed in Chapter 7. In addition, having the absolute pose of the camera $\left(R_k \; \vec{T}_k\right)_{3\times 4}$, we can calculate the feature's location with respect to the world frame m^i as follows:

$$\vec{m}^i = \vec{p}_k^{\,i}(R_k)^T - \vec{T}_k \quad (10\text{-}67)$$

It is important to realize that in case of a monocular camera, $\vec{T}_k$, $\vec{p}_k^{\,i}$, and $\vec{m}^i$ are scaled by an unknown scale. If a stereo camera system is available, we will be able to resolve the scale ambiguity, and we will be able to localize the point in the 3D world without scale ambiguity with respect to the camera frame and with respect to the world. The inertial-visual assembly is commonly used in smartphones and drones to build 3D models of physical objects. Whether we use the monocular system or a stereo system, there are different ways to fuse the visual measurements with inertial sensors in the 3D space. There are several similarities between visual measurements fusion and range measurements fusion. However, due to the 3D to 2D projection and the scale ambiguity problem, the fusion of visual measurements requires special processing steps. In the next subsections, we will discuss some ideas about fusing visual sensor information with inertial sensors.

10.4.2 Fusing Measurements from Known Location Landmarks

A simple and straightforward visual information fusion approach uses known-location visual landmarks. Once the platform's camera detects a known location landmark, it can use its known location as an update to correct the platform location. This approach is like a GPS trilateration algorithm, but instead of using satellites, we use known location landmarks in the environment, and instead of using radio measurements and ranging as a measurement model, we use visual pixels as measurements and the camera projection model as a measurement model. Like GPS measurements, we may fuse landmark visual measurements in loosely coupled mode or tightly coupled mode. We discuss the two modes in the next subsections.

10.4.2.1 Loosely Coupled Fusion

Given a landmark $\vec{m}^i$, its visual measurements model is given by the following equation:

$$\vec{y}_k^i = \begin{pmatrix} u_k^i \\ v_k^i \\ 1 \end{pmatrix} = h_k^i = h\left(\vec{q}_k, \vec{T}_k, \vec{m}^i\right) \tag{10-68}$$

where the terms of $h\left(\vec{q}_k, \vec{T}_k, \vec{m}^i\right)$ are defined in Eqs. (10-63) and (10-65).

Note that $\vec{y}_k^i$ and h_k^i are 3×1 vectors and the third component of $\vec{y}_k^i$ is "1." To avoid redundancy in the measurement vector and design matrices, which lead to singularity in the linearization, we take the first two elements of the two vectors. If the platform's camera system (assuming a monocular system) detects M landmarks, we would have the following measurement vector-matrix form:

$$\begin{pmatrix} \vec{y}_k^1(1:2)^T \\ \vdots \\ \vec{y}_k^M(1:2)^T \end{pmatrix} = \begin{pmatrix} h_k^1(1:2)^T \\ \vdots \\ h_k^M(1:2)^T \end{pmatrix} \tag{10-69}$$

The linearized measurement model will be given by the following equation:

$$\begin{pmatrix} \delta\vec{y}_k^1(1:2)^T \\ \vdots \\ \delta\vec{y}_k^M(1:2)^T \end{pmatrix} = \begin{pmatrix} \dfrac{\partial h_k^1(1:2)^T}{\partial\left(\vec{q}_k\ \vec{T}_k\right)} \\ \vdots \\ \dfrac{\partial h_k^M(1:2)^T}{\partial\left(\vec{q}_k\ \vec{T}_k\right)} \end{pmatrix} \begin{pmatrix} \delta\vec{q}_k \\ \delta\vec{T}_k \end{pmatrix} \tag{10-70}$$

which gives the popular linear measurement model in form $\delta\vec{y}_k = H_k \delta\vec{x}_k$, which can be solved iteratively using an iterative least-squares algorithm (review Chapter 3) to find the best state error and correct the state vector until convergence. The unknowns here are six parameters (three vector components of $\vec{q}_k$ and vector $\vec{T}_k$) so we should expect that we need at least six landmarks to be available. The quaternion vector $\vec{q}_k$ can be easily converted into rotation matrix R_k. The solution will give us an estimate of the camera platform pose (position $\vec{T}_k$ and orientation R_k) with respect to the world frame. If the world frame is defined as the initial camera frame at epoch $k = 0$, to fuse this pose estimate in a local level frame (L frame, for example), we would need to transform it from camera world frame to L frame as follows:

$$\begin{pmatrix} C_B^L & \vec{r}^L \\ 0 & 1 \end{pmatrix} = \begin{pmatrix} R_{c(0)}^L & \vec{T}_{c(0)}^L \\ 0 & 1 \end{pmatrix} \begin{pmatrix} (R_k)^T & -\vec{T}_k \\ 0 & 1 \end{pmatrix} \tag{10-71}$$

The rotation matrix C_B^L can be easily converted to a rotation vector or a quaternion vector $\vec{q}_B^L$. Now, we have a position and orientation update that we can use to apply either in an EKF or a PF or any of their variations. Recall, in a loosely coupled model, the platform state vector is given by:

$$\vec{x}_{15\times1} = \left(\left(\vec{r}^L\right)_{1\times3} \quad \left(\vec{v}^L\right)_{1\times3} \quad q_1 \quad q_3 \quad q_3 \quad \left(\vec{b}_g\right)_{1\times3} \quad \left(\vec{b}_a\right)_{1\times3} \right)^T \tag{10-72}$$

Given position and orientation update, the design matrix which relates measurements to the system state is given as follows:

$$H = \begin{pmatrix} 1 & 0 & 0 & 0 & 0 & 0 & 0 & 0 & 0 & \\ 0 & 1 & 0 & 0 & 0 & 0 & 0 & 0 & 0 & \\ 0 & 0 & 1 & 0 & 0 & 0 & 0 & 0 & 0 & \\ 0 & 0 & 0 & 0 & 0 & 0 & 1 & 0 & 0 & 0_{6\times6} \\ 0 & 0 & 0 & 0 & 0 & 0 & 0 & 1 & 0 & \\ 0 & 0 & 0 & 0 & 0 & 0 & 0 & 0 & 1 & \end{pmatrix} \tag{10-73}$$

This is a linear measurement model that can be applied easily in an EKF or a PF or any of their variations.

10.4.2.2 Tightly Coupled Fusion

If the number of visual features is less than six, we still can partially benefit from these partial measurements using the tightly coupled fusion mode. Given M visible landmarks, the linearized measurement model is given by the following matrix equation:

$$\begin{pmatrix} \delta\vec{y}_k^1(1:2)^T \\ \vdots \\ \delta\vec{y}_k^M(1:2)^T \end{pmatrix} = \begin{pmatrix} \dfrac{\partial h_k^1(1:2)^T}{\partial\left(\vec{x}_k\right)} \\ \vdots \\ \dfrac{\partial h_k^M(1:2)^T}{\partial\left(\vec{x}_k\right)} \end{pmatrix} \left(\delta\vec{x}_k\right) \tag{10-74}$$

The only difference between this measurement model and the model in the loosely coupled model is that the partial derivative is with respect to the platform 15-element state vector $\vec{x}_k$

and not only the six parameters camera pose $\left\{\vec{q}_k, \vec{T}_k\right\}$; therefore, the design matrix will be different. The design matrix of this measurement update is given by the following equation:

$$H_k = \begin{pmatrix} \dfrac{\partial h_k^1(1:2)^T}{\partial\left(\vec{x}_k\right)} \\ \vdots \\ \dfrac{\partial h_k^M(1:2)^T}{\partial\left(\vec{x}_k\right)} \end{pmatrix} \tag{10-75}$$

To calculate the design matrix, the derivative of the measurement model needs to be done, which is complex. However, it can be done with symbolic equation engines such as MATLAB symbolic engine. Once we calculate the design matrix H_k, we can apply an update in an EKF or a PF or any of their variations taking into consideration that the measurement model is nonlinear. One more important point to note is the frame where H_k is calculated. If the world frame is defined as the camera frame at epoch $k = 0$, and we want to do the fusion in L or N frame, we need to apply the proper transformation as described in Eq. (10-71). It is also important to note that due to scale ambiguity, using less than six features in a tightly coupled mode may affect the observability and the accuracy of the update.

10.4.3 Fusing Measurements from Optical Flow

The optical flow was discussed in Chapter 7, "Imaging Sensors and Visual Odometry." Optical flow is the velocity of pixels of the image described as pixels per unit time. Optical flow can be used to calculate the platform speed if the camera is configured under a constrained configuration such as a down-looking configuration and fixed altitude motion as seen in Figure 10.3. Under the configuration of Figure 10.3, and assuming the altitude is almost fixed, we can estimate the camera velocity (and speed) from the optical flow.

The derivation of this estimation starts from the camera projection model given as follows:

$$\vec{y}^{\,i} = \begin{pmatrix} u^i \\ v^i \\ 1 \end{pmatrix} = \frac{1}{z_c^i} K \begin{pmatrix} x_c^i \\ y_c^i \\ z_c^i \end{pmatrix} \tag{10-76}$$

where $\vec{p}_k^{\,i} = \left(x_{c,k}^i\ z_{c,k}^i\ z_{c,k}^i\right)^T$ is a feature metric location with respect to the camera frame.

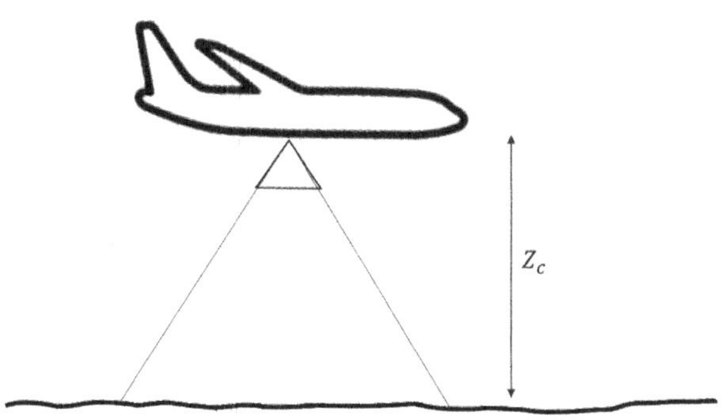

Figure 10.3 Down-looking camera attached to an aerial vehicle.

By differentiating both sides with respect to time, we get the following equation:

$$\begin{pmatrix} \frac{\partial u^i}{\partial t} \\ \frac{\partial v^i}{\partial t} \\ 1 \end{pmatrix} = \frac{K}{z_c^i} \begin{pmatrix} \frac{\partial x_c^i}{\partial t} \\ \frac{\partial y_c^i}{\partial t} \\ \frac{\partial z_c^i}{\partial t} \end{pmatrix} + \left(\frac{-K}{(z_c^i)^2} \right) \left(\frac{\partial z_c^i}{\partial t} \right) \begin{pmatrix} x_c^i \\ y_c^i \\ z_c^i \end{pmatrix} \tag{10-77}$$

where $\left(\frac{\partial u^i}{\partial t}, \frac{\partial v^i}{\partial t} \right)^T$ is the optical flow (pixels per second).

If the motion is mostly under constant altitude, the vertical velocity component will be approximately zero as follows:

$$\frac{\partial z_c^i}{\partial t} \cong 0 \tag{10-78}$$

Then the above equation can be simplified as follows:

$$\begin{pmatrix} \frac{\partial u^i}{\partial t} \\ \frac{\partial v^i}{\partial t} \\ 1 \end{pmatrix} = \frac{K}{z_c^i} \begin{pmatrix} \frac{\partial x_c^i}{\partial t} \\ \frac{\partial y_c^i}{\partial t} \\ 0 \end{pmatrix} \tag{10-79}$$

If we know the altitude (which can be known from a pressure sensor, a RADAR, or a sonar altimeter), and we know the optical flow $\left(\frac{\partial u^i}{\partial t}, \frac{\partial v^i}{\partial t} \right)^T$ using the Lucas–Kanade method describe in Chapter 7, we can calculate the camera velocity $\left(\frac{\partial x_c^i}{\partial t}, \frac{\partial y_c^i}{\partial t} \right)^T$ as follows:

$$\begin{pmatrix} \frac{\partial x_c^i}{\partial t} \\ \frac{\partial y_c^i}{\partial t} \\ 0 \end{pmatrix} = z_c^i K^{-1} \begin{pmatrix} \frac{\partial u^i}{\partial t} \\ \frac{\partial v^i}{\partial t} \\ 1 \end{pmatrix} \tag{10-80}$$

The camera speed (which is the same as the platform speed) is then calculated as follows:

$$\hat{v}(t) = \sqrt{\left(\frac{\partial x_c^i}{\partial t} \right)^2 + \left(\frac{\partial y_c^i}{\partial t} \right)^2} \tag{10-81}$$

This speed measurement can be fused with inertial sensors using its measurement model, which is defined as follows:

$$\hat{v}(t) = \left(\sqrt{(v_n(t))^2 + (v_e(t))^2} \right) \tag{10-82}$$

Which can be linearized around a given operating point $(\bar{v}_n(t), \bar{v}_e(t))^T$ as follows:

$$\delta\hat{v}(t) = \frac{\bar{v}_n(t)}{\sqrt{(\bar{v}_n(t))^2 + (v_e(t))^2}} \delta v_n(t) + \frac{\bar{v}_e(t)}{\sqrt{(\bar{v}_n(t))^2 + (\bar{v}_e(t))^2}} \delta v_e(t) \tag{10-83}$$

which gives the following design matrix:

$$H(t) = \begin{pmatrix} 0 & 0 & \frac{\bar{v}_n(t)}{\sqrt{(\bar{v}_n(t))^2 + (v_e(t))^2}} & \frac{\bar{v}_e(t)}{\sqrt{(\bar{v}_n(t))^2 + (\bar{v}_e(t))^2}} & \vec{0}_{1\times 11} \end{pmatrix} \tag{10-84}$$

which can be applied easily in an EKF or a PF or any of their variations, taking into consideration that the measurement model is nonlinear.

10.4.4 Fusing Measurements from Occasional Visual Features

Visual features are pixel descriptors of a distinguishable point in the environment, such as corners and blobs. We discussed visual feature types in Chapter 7. Visual features are the projection of a distinguishable point in the environment on the image pixel plan. Even if the locations of these distinguishable points in the environment are not known, they can still be fused with inertial sensors either in loosely coupled or tightly coupled mode. Both fusion modes will be discussed in the next sections.

10.4.4.1 Loosely Coupled Fusion

In loosely coupled mode, we use visual odometry to estimate the camera poses over a period, which then can be used for pose updates to reduce the drifts of inertial sensor prediction. Visual odometry was described in detail in Chapter 7. Visual odometry provides an estimate of the relative transformation from image $k-1$ to image k. This relative transformation can be arranged into a 4x4 matrix that includes rotation and translation as follows:

$$D_{k-1}^{k} = \begin{pmatrix} R_{k-1}^{k} & \vec{T}_{k-1}^{k} \\ 0 & 1 \end{pmatrix} \tag{10-85}$$

Given the absolute camera pose with respect to the world frame as follows:

$$C_{k-1} = \begin{pmatrix} R_{k-1} & \vec{T}_{k-1} \\ 0 & 1 \end{pmatrix} \tag{10-86}$$

The new absolute camera pose C_k can be calculated by concatenating the relative transformation to the existing camera pose as follows:

$$C_k = \begin{pmatrix} R_k & \vec{T}_k \\ 0 & 1 \end{pmatrix} = D_{k-1}^{k} C_{k-1} \tag{10-87}$$

The new camera pose $\left\{R_k, \vec{T}_k\right\}$ estimated from visual odometry can be used in a loosely coupled fusion mode to be fused with inertial sensors using the design matrix of Eq. (10-73). However, we need to resolve the scale ambiguity problem before applying this pose as an update to the inertial sensors system. The translation vector estimated by visual odometry defined by the variable $\left(\vec{T}_{k-1}^{k}\right)_{VO}$ is related to the true metric translation vector $\vec{T}_{k-1}^{k}$ by the following equation:

$$\left(\vec{T}_{k-1}^{k}\right)_{VO} = s\vec{T}_{k-1}^{k} \tag{10-88}$$

If we are using a stereo vision assembly, the scale s will be 1. If a monocular camera system is used, the scale s must be resolved. One way to resolve the scale is by using the inertial sensor's prediction of the translation vector from image $k-1$ to image k. The transition from image $k-1$ to image k is a short-term transition that makes the inertial prediction accurate, but this accuracy is conditioned on the relative motion and noise of inertial sensors. If a reliable motion between frame $k-1$ and k is detected and the noise is smoothed, the inertial prediction should be reliable. If the metric predicted translation from inertial sensors $\left(\vec{T}^{k}_{k-1}\right)_{IMU}$ is known, the scale can be estimated by dividing the translation vector estimated from visual odometry by the translation vector estimated from inertial prediction as follows:

$$s = \frac{\left(\vec{T}^{k}_{k-1}\right)_{VO}}{\left(\vec{T}^{k}_{k-1}\right)_{IMU}} \tag{10-89}$$

In this case, the platform position and quaternion that will be used as an update measurement are given by the following equations:

$$\vec{\hat{r}}^{L} = -\vec{T}_k \tag{10-90}$$

$$\vec{q}^{L}_{B} = q\left((R_k)^T\right) \tag{10-91}$$

where $q(R_k{}^T)$ is a conversion function that converts a direction cosine matrix $R_k{}^T$ into a quaternion vector $\vec{q}^{L}_{B}$.

Here we assumed that the world frame is already set to L frame. Like what we did in Section 10.4.2.1, we use the position vector $\vec{\hat{r}}^{L}$ and the vector part of the quaternion $(q_1\, q_2\, q_3)^T = \vec{q}^{L}_{B}(2:4)$ as updates where the design matrix is given by the following formula:

$$H = \begin{pmatrix} 1 & 0 & 0 & 0 & 0 & 0 & 0 & 0 & 0 & \\ 0 & 1 & 0 & 0 & 0 & 0 & 0 & 0 & 0 & \\ 0 & 0 & 1 & 0 & 0 & 0 & 0 & 0 & 0 & \\ 0 & 0 & 0 & 0 & 0 & 0 & 1 & 0 & 0 & 0_{6\times 6} \\ 0 & 0 & 0 & 0 & 0 & 0 & 0 & 1 & 0 & \\ 0 & 0 & 0 & 0 & 0 & 0 & 0 & 0 & 1 & \end{pmatrix} \tag{10-92}$$

This is a linear measurement model that can be applied easily in an EKF or a PF or any of their variations taking into consideration that the measurement model is linear.

10.4.4.2 Tightly Coupled Fusion

The tightly coupled approach fuses raw visual feature measurements directly with inertial sensors in one engine. In the processing of image frames $k-1$ and k, the pixel coordinates of a feature $\vec{P}^{i}_{k-1}$ (with respect to camera frame at epoch $k-1$) at epoch $k-1$ and epoch k are given by:

$$\begin{pmatrix} u^i_{k-1} \\ v^i_{k-1} \\ 1 \end{pmatrix} = \frac{K}{\lambda^i_{k-1}} \vec{P}^{i}_{k-1} \tag{10-93}$$

$$\begin{pmatrix} u^i_{k} \\ v^i_{k} \\ 1 \end{pmatrix} = \frac{K}{\lambda^i_{k}} \left(R^k_{k-1}\vec{P}^{i}_{k-1} + \vec{T}^{k}_{k-1}\right) \tag{10-94}$$

We have seen in Chapter 7 how the epipolar geometric constraints can be used to estimate the relative camera motion $\left\{R_{k-1}^{k}, \vec{T}_{k-1}^{k}\right\}$ and the 3D feature location $\vec{P}_{k-1}^{i}$. Given $\left(R_{k-1}^{k}, \vec{T}_{k-1}^{k}\right)$ and $\vec{P}_{k-1}^{i}$, we can also know the depth λ_{k-1}^{i} up to a scale if a monocular system is used. If we further process another image frame at next epoch $k+1$, and the feature i is still visible, and we know from inertial sensors prediction the predicted platform pose $\left\{R_{k-1}^{k+1}, \vec{T}_{k-1}^{k+1}\right\}$, we can predict the pixel location of feature i at epoch $k+1$ as follows:

$$\begin{pmatrix} \hat{u}_{k+1}^{i} \\ \hat{v}_{k+1}^{i} \\ 1 \end{pmatrix} = h_{k+1}^{i} = \frac{K}{\lambda_{k+1}^{i}}\left(R_{k-1}^{k+1}\vec{P}_{k-1}^{i} + \vec{T}_{k-1}^{k+1}\right) \tag{10-95}$$

If the motion from frames $k-1$, k, and $k+1$ are small, we can assume that the depth is the same in the three epochs $\lambda_{k-1}^{i} \cong \lambda_{k}^{i} \cong \lambda_{k+1}^{i}$. From the visual measurements, we have observed the pixel location of feature i at epoch $k+1$ defined as $\left(u_{k+1}^{i}, v_{k+1}^{i}\right)^{T}$. Now, we have an observed measurement and a predicted measurement of the feature i. The difference between the observed measurement and a predicted measurement is called the reprojection error defined as follows:

$$\begin{aligned} \vec{y} &= \begin{pmatrix} u_{k+1}^{i} \\ v_{k+1}^{i} \end{pmatrix} \\ \vec{\hat{y}} &= \begin{pmatrix} \hat{u}_{k+1}^{i} \\ \hat{v}_{k+1}^{i} \end{pmatrix} \\ \delta\vec{y} &= \begin{pmatrix} \delta u_{k+1}^{i} \\ \delta u_{k+1}^{i} \end{pmatrix} = \begin{pmatrix} u_{k+1}^{i} \\ v_{k+1}^{i} \end{pmatrix} - \begin{pmatrix} \hat{u}_{k+1}^{i} \\ \hat{v}_{k+1}^{i} \end{pmatrix} \end{aligned} \tag{10-96}$$

The predicted measurement is related to the camera pose by the measurement model given in Eq. (10-95). The reprojection error $\delta\vec{y}$ is related to the camera pose by the linearization of the measurement model in Eq. (10-95). We discussed the linearization of this measurement model in Sections 10.4.2.1 and 10.4.2.2. By replacing q_0 by its equivalent $\sqrt{1-q_1^2-q_2^2-q_3^2}$, the measurement model h_{k+1}^{i} will be a function of the platform state vector $\vec{x}$ (particularly the vector part of the quaternion $\vec{q}_k(2:4)$ and the position variable $\vec{r}^{L}$). The error $\left(\delta u_{k+1}^{i}, \delta v_{k+1}^{i}\right)^{T}$ between the predicted pixel location $\left(\hat{u}_{k+1}^{i}, \hat{v}_{k+1}^{i}\right)^{T}$ and the actual observed pixel coordinate $\left(u_{k+1}^{i}, v_{k+1}^{i}\right)^{T}$ can be used to correct the platform state vector $\vec{x}_{k+1}$. We can linearize h_{k+1}^{i} using first-order Tylor series expansion to obtain the design matrix H_{k+1}^{i}. Given M visible landmarks, the linearized measurement model is given by the following matrix equation:

$$\begin{pmatrix} \left(\delta u_{k+1}^{1}, \delta v_{k+1}^{1}\right) \\ \vdots \\ \left(\delta u_{k+1}^{M}, \delta v_{k+1}^{M}\right) \end{pmatrix} = H_{k+1}\left(\delta\vec{x}_{k+1}\right) \tag{10-97}$$

$$H_{k+1} = \begin{pmatrix} \dfrac{\partial h_{k+1}^{1}(1:2)^{T}}{\partial\left(\vec{x}_{k+1}\right)} \\ \vdots \\ \dfrac{\partial h_{k+1}^{M}(1:2)^{T}}{\partial\left(\vec{x}_{k+1}\right)} \end{pmatrix} \tag{10-98}$$

which can be applied in an EKF, PF, or any of their variations taking into consideration that the measurement model is nonlinear. It is important to note that if we are using a monocular camera system, we will need to resolve the scale ambiguity which can be resolved from the inertial sensors system prediction. We will continue to apply this update from this feature in future epochs if the feature is visible in these epochs.

10.4.5 Camera-IMU Calibration

In the previous discussions, we assumed that the camera and the IMU have the same coordinate frames, which is physically impossible. There must be a spatial transformation between the two frames, and it must be estimated to properly perform the integration of the measurements of the two sensors. We discussed the general idea of multisensor spatial calibration in Chapter 3. In this subsection, we briefly discuss the camera-IMU calibration. In this example, we will assume that the measurements of both sensors are time-synchronized.

Figure 10.4 shows an illustration of camera-IMU spatial calibration. Given clear and distinguishable visual features (like checkerboard, for example), a simple trajectory can be estimated by the camera. At the same time, the same trajectory can be estimated by the IMU. The pose of the IMU and the camera at epoch k are represented by $R_I(k)$, $r_I(k)$ and $R_c(k)$, $r_c(k)$ respectively where R refers

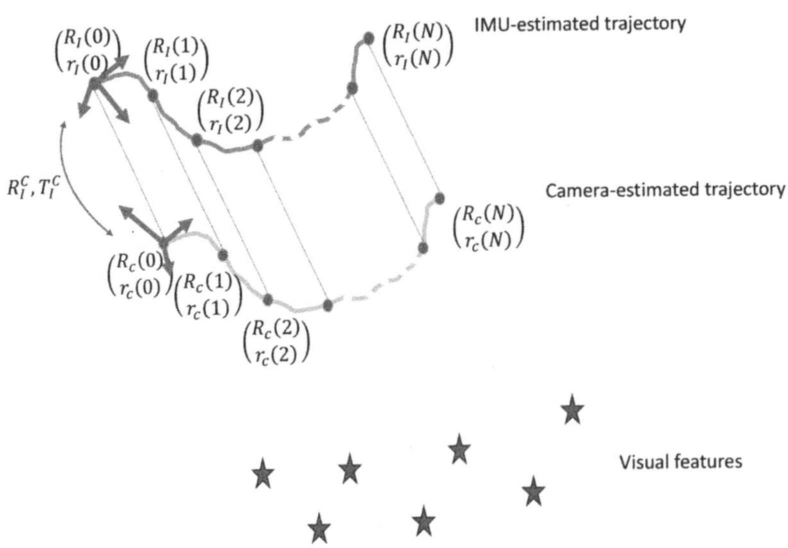

Figure 10.4 Illustration of camera-IMU calibration.

to the rotation matrix and r refers to the position. $R_c(k)$ refers to the rotation of the camera with respect to a world frame at time epoch k. The spatial transformation between the IMU and the camera is represented by R_I^C, T_I^C. Throughout the N points of the trajectory, the rotation transformation between the camera and the IMU is given by:

$$R_I(k) = R_C^I R_c(k) \tag{10-99}$$

The position of the IMU as a function of the position of the camera is given by:

$$r_I(k) = R_C^I r_c(k) + T_C^I \tag{10-100}$$

The world frame can be assumed to be the initial frame of the IMU or the camera. $R_I(k)$ and $r_I(k)$ are estimated using the IMU measurements, while $R_c(k)$ and $r_c(k)$ are estimated using the visual features. Given these measurements, the transformation between the camera and the IMU can be performed by minimizing the following cost functions:

$$\begin{aligned} J_1\left(R_C^I\right) &= \frac{1}{2}\sum_{j=1}^{N} \left\|R_I(k) - R_C^I R_c(k)\right\|^2 \\ J_2\left(R_C^I, T_C^I\right) &= \frac{1}{2}\sum_{j=1}^{N} \left\|r_I(k) - \left(R_C^I r_c(k) + T_C^I\right)\right\|^2 \end{aligned} \tag{10-101}$$

The cost function J_1 can be optimized first and then J_2 can be optimized. The optimization can be performed using least-squares algorithm or any other optimization method keeping in consideration that the rotation parameters must satisfy the minimal coordinate condition where the rotation variable is represented by a three-element vector, not a nine-element matrix. This three-element vector can be the rotation vector or the vector part of the quaternion. This calibration process is also known as inter-sensors extrinsic calibration. MATLAB navigation and computer vision toolboxes have examples that demonstrate the camera-IMU intersensors extrinsic calibration.

10.5 Fusion of RADAR Measurements

The RADAR sensor was discussed in Chapter 6. As a ranging sensor, RADAR provides a range and an angle to a target point in the environment, like LiDAR. However, RADAR angular resolution is much less than LiDAR. In addition, RADAR signals can travel through clouds and fog or even walls and objects depending on the signal power and frequency spectrum. RADAR measurements include the range, angle, and speed of target points.

Theoretically, as RADAR provides range and angle, it can be used to build a point cloud like LiDAR. Given two RADAR scans, point clouds can be matched using ICP or NDT scan matching algorithms and the relative motion of the sensor can be estimated. The estimated relative poses can then be concatenated to generate a full trajectory, which can be fused with inertial sensors in a loosely coupled fashion, just like LiDAR fusion discussed in chapters 6 and 9. In addition, theoretically, raw range/angle measurements can also be used as a tightly coupled measurement update. However, these fusion schemes do not usually provide accurate results because the RADAR point cloud is noisy and scattered. If RADAR range accuracy is improved and the angular resolution is increased, point-cloud fusion approaches, just like LiDAR, can be applied.

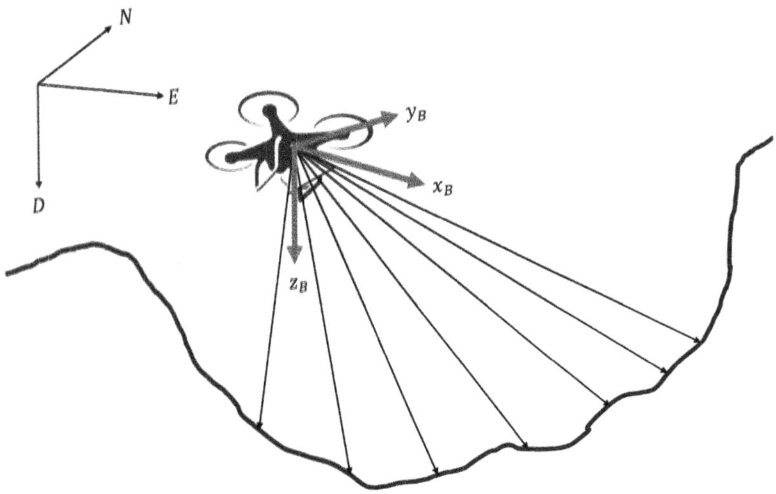

Figure 10.5 Illustration of radar measurements on unmanned aerial vehicle.

A better approach to fuse RADAR data is to benefit from the radial speed estimated by RADAR. If the environment is mostly static, the radial speed of RADAR measurements can be used to correct the velocity of the platform which, in turn, corrects other state variables. In this section, we discuss the fusion of RADAR radial speed measurements with inertial sensors. In our discussion, we will use the 3D inertial sensors system model described in Section 10.2.

Figure 10.5 shows an illustration of a drone with a RADAR sensor. Given the state of the drone $\vec{x}$ and M RADAR measurements $\{(r^1, B^1, E^1, s^1), \ldots (r^M, B^M, E^M, s^M)\}$ where r^i is range, B^i is the bearing angle, E^i is the elevation angle, and s^i is the target's radial speed for the ith beam. The location of the target point is given by the following equations:

$$x_B^i = r^i \cos(E^i) \cos(B^i) \tag{10-102}$$

$$y_B^i = r^i \cos(E^i) \sin(B^i) \tag{10-103}$$

$$z_B^i = r^i \sin(E^i) \tag{10-104}$$

The target's location with respect to the world frame (which is L frame in our case) is given by:

$$\vec{r}_i^{\,L} = \begin{pmatrix} n^i \\ e^i \\ d^i \end{pmatrix} = C_B^L \begin{pmatrix} x_B^i \\ y_B^i \\ z_B^i \end{pmatrix} \tag{10-105}$$

Assuming the target is a static point, its velocity is zero as follows:

$$\vec{v}_i^{\,L} = \begin{pmatrix} v_n^i \\ v_e^i \\ v_d^i \end{pmatrix} = \begin{pmatrix} 0 \\ 0 \\ 0 \end{pmatrix} \tag{10-106}$$

We can fuse radial speed measurements with inertial sensors in loosely coupled and tightly coupled modes. We will explain each mode in the next sections.

10.5.1 Loosely Coupled RADAR Fusion

The loosely coupled approach uses RADAR radial speed measurements to estimate the vehicle's velocity, and then the estimated vehicle's velocity is used as an update to the odometry navigation, whether it is inertial-based, wheel-encoders-based, or a combination of both. The radial speed of a target point i is related to the vehicle's own vehicle's speed by the following equation:

$$\hat{s}^i = \left(\vec{v}^L - \vec{v}_i^L\right)\vec{L}_i \tag{10-107}$$

where $\vec{L}_i$ is the line of sight (LOS) vector from the vehicle's position to the target point position, which is calculated as follows:

$$\vec{L}_i = \begin{pmatrix} \dfrac{(n-n^i)}{\left\|\vec{r}^L - \vec{r}_i^L\right\|} \\ \dfrac{(e-e^i)}{\left\|\vec{r}^L - \vec{r}_i^L\right\|} \\ \dfrac{(d-d^i)}{\left\|\vec{r}^L - \vec{r}_i^L\right\|} \end{pmatrix} \tag{10-108}$$

where

$\vec{r}_i^L = \left(n^i\, e^i\, d^i\right)^T$ is the target point location calculated using Eq. (10-105)
$\vec{r}^L = (n\, e\, d)^T$ is the current vehicle's position.

Assuming that the target points are static ($\vec{v}_i^L = 0$), and given an M target points, we will have the following linear measurement model in matrix-vector form:

$$\underbrace{\begin{pmatrix} \hat{s}^1 \\ \hat{s}^2 \\ \vdots \\ \hat{s}^M \end{pmatrix}}_{\vec{s}} = \underbrace{\begin{pmatrix} \vec{L}_1^{\,T} \\ \vec{L}_2^{\,T} \\ \vdots \\ \vec{L}_M^{\,T} \end{pmatrix}}_{A} \vec{v}^L \tag{10-109}$$

The Equation (10-109) is a linear regression model in the form $\vec{s} = A\vec{v}^L$, which can be solved using the least-square algorithm as follows:

$$\hat{\vec{v}}^L = \left(A^T A\right)^{-1} A^T \vec{s} \tag{10-110}$$

The estimated velocity $\hat{\vec{v}}^L$ is then used as a measurement update to correct the vehicle's state. The measurement model, in this case, is simple and linear. Assuming we have the state vector given in the Eq. (10-3), which is a 15-state vector, the measurement design matrix $H_{RADAR-LC}$ will be given as follows:

$$H_{RADAR\text{-}LC} = \begin{pmatrix} & 1 & 0 & 0 & \\ \vec{\mathbf{0}}_{3\times 3} & 0 & 1 & 0 & \vec{\mathbf{0}}_{3\times 11} \\ & 0 & 0 & 1 & \end{pmatrix} \tag{10-111}$$

The velocity measurement $\vec{v}^L$ and the design matrix H_{RADAR} can be used in any filtering algorithm such as PF, EKF, or any of their variations to be fused with inertial sensors.

10.5.2 Tightly Coupled RADAR Fusion

In the tightly coupled case, the measurements are the radial speeds of the target points detected by the RADAR, not the vehicle's velocity estimated by the RADAR. Assuming that the RADAR detects M static target points, the Eq. (10-107) will be a linear measurement model that relates RADAR radial speed to the internal state of the platform $\vec{x}$ as follows:

$$\underbrace{\begin{pmatrix} \hat{s}^1 \\ \hat{s}^2 \\ \vdots \\ \hat{s}^M \end{pmatrix}}_{\vec{y}} = \underbrace{\begin{pmatrix} & \vec{L}_1^{\,T} & \\ & \vec{L}_2^{\,T} & \\ \mathbf{0}_{M\times 3} & & \mathbf{0}_{M\times 9} \\ & \vdots & \\ & \vec{L}_M^{\,T} & \end{pmatrix}}_{H_{RADAR\text{-}TC}} \vec{x} \tag{10-112}$$

The state vector $\vec{x}$ is the one given in Eq. (10-3), which is a 15-state vector. A more elaboration of the $H_{RADAR\text{-}TC}$ design matrix is given by:

$$H_{RADAR\text{-}TC} = \begin{pmatrix} \vec{\mathbf{0}}_{1\times 3} & \dfrac{(n-n_1^L)}{\left\|\vec{r}^{\,L}-\vec{r}_1^{\,L}\right\|} & \dfrac{(e-e_1^L)}{\left\|\vec{r}^{\,L}-\vec{r}_1^{\,L}\right\|} & \dfrac{(d-d_1^L)}{\left\|\vec{r}^{\,L}-\vec{r}_1^{\,L}\right\|} & \vec{\mathbf{0}}_{1\times 9} \\ & & \vdots & & \\ \vec{\mathbf{0}}_{1\times 3} & \dfrac{(n-n_M^L)}{\left\|\vec{r}^{\,L}-\vec{r}_M^{\,L}\right\|} & \dfrac{(e-e_M^L)}{\left\|\vec{r}^{\,L}-\vec{r}_M^{\,L}\right\|} & \dfrac{(d-d_M^L)}{\left\|\vec{r}^{\,L}-\vec{r}_M^{\,L}\right\|} & \vec{\mathbf{0}}_{1\times 9} \end{pmatrix} \tag{10-113}$$

where

$\vec{r}_i^{\,L} = (n^i\, e^i\, d^i)^T$ is the target point location calculated using Eq. (10-105)

$\vec{r}^{\,L} = (n\, e\, d)^T$ is the current vehicle's position.

The RADAR radial speed measurement vector $\vec{y}$ and the design matrix $H_{RADAR\text{-}TC}$ can be applied in an EKF, a PF or any of their variations.

10.6 MATLAB Implementation of 3D IMU/GNSS Fusion Using EKF

The project repository has sample projects of 3D IMU/GNSS fusion using EKF. The projects implement the 3D INS system model described in Eqs. (10-4)–(10-7). The linearized error model in Eqs. (10-26)–(10-34) is calculated by the MATLAB symbolic engine. The orientation error is modeled by the 3D vector part of the quaternion vector. The projects include inertial and GNSS raw data. The projects implement both the loosely coupled fusion and the tightly coupled fusion. The following subsections demonstrate the plots resulting from running the provided codes on the provided data. The project includes two different datasets: a simulated dataset and a real dataset. The simulated dataset is useful in illustrating the biases estimation accuracy of the filter as the values of the true biases are known.

10.6.1 Loosely Coupled 3D IMU/GNSS Fusion

In the loosely coupled fusion, GNSS position and velocity are used as updates to the EKF. Figure 10.6 shows the noisy GNSS position/velocity measurements, the ground truth trajectory, and the EKF output. Figure 10.7 shows the EKF orientation estimation compared to the ground truth orientation. Figure 10.7 shows that the convergence of the heading and the pitch angles took longer than the convergence of the roll angle. This convergence is further illustrated in Figure 10.8. The filter fully converged after approximately 100 seconds when the airplane took the first 180° turn. Turning by 180° provided good visibility of the heading as seen in Figure 10.7, Figures 10.8 and 10.12. The positional errors are shown in Figure 10.9 along with the covariance estimated by the EKF. The covariance convergence is consistent with the error values, which confirms the correct operation of the filter. Figures 10.10 and 10.11 show the accelerometer and gyroscope biases convergence to the true values. The covariance value is also illustrated in Figures 10.10 and 10.11.

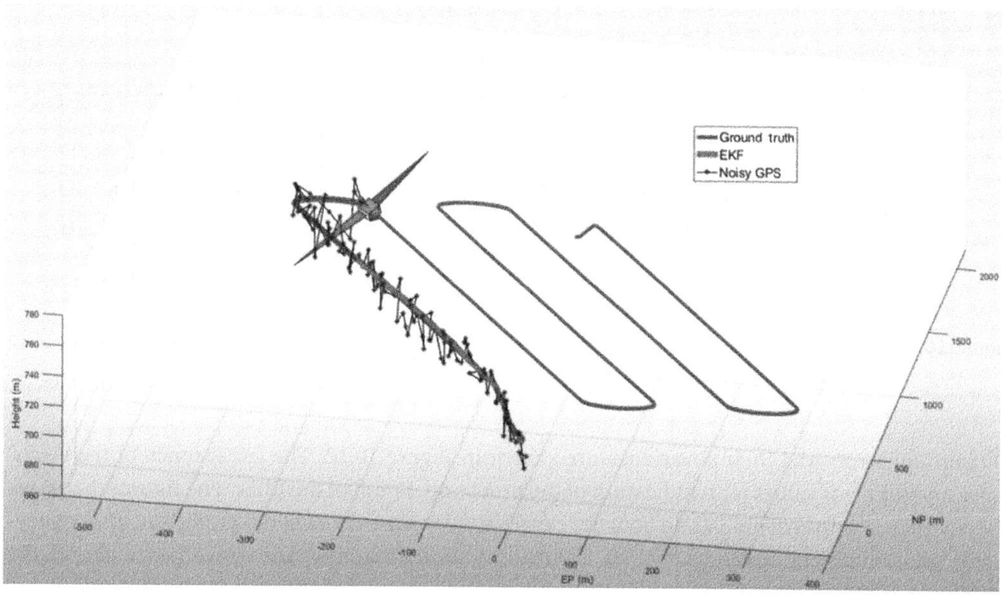

Figure 10.6 3D illustration of GPS/IMU loosely coupled fusion of aerial vehicle using EKF.

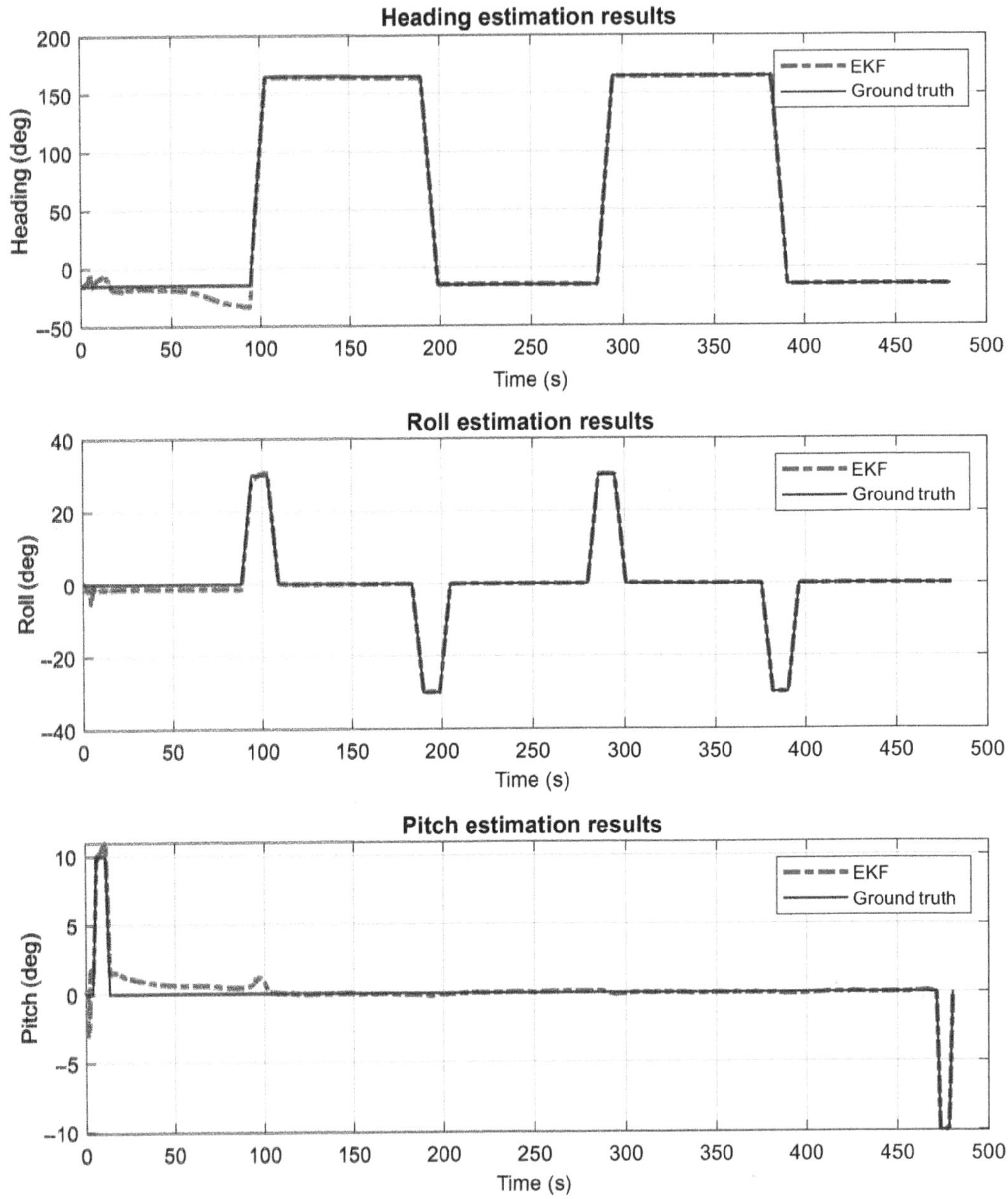

Figure 10.7 Orientation estimation in loosely coupled 3D IMU/GNSS fusion output.

Quaternion errors and their covariance are shown in Figure 10.12. The consistency of the covariance and the error values is an indication of the proper operation of the filter. The figures that show the covariance values demonstrate how the covariance increases during the prediction and reduces when a measurement update is applied. Figure 10.6 shows that after the convergence period, the EKF-estimated trajectory is smooth and consistent with the ground truth.

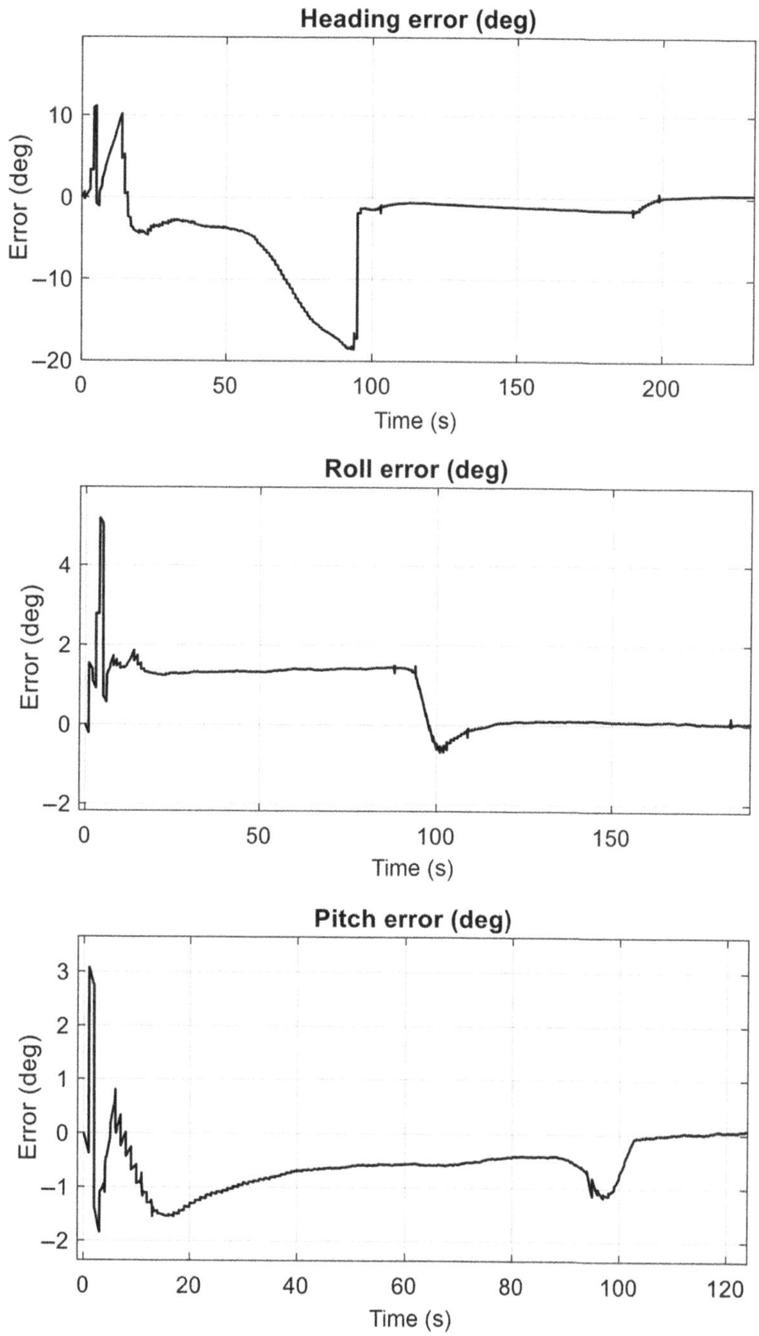

Figure 10.8 Orientation errors convergence in the first two minutes.

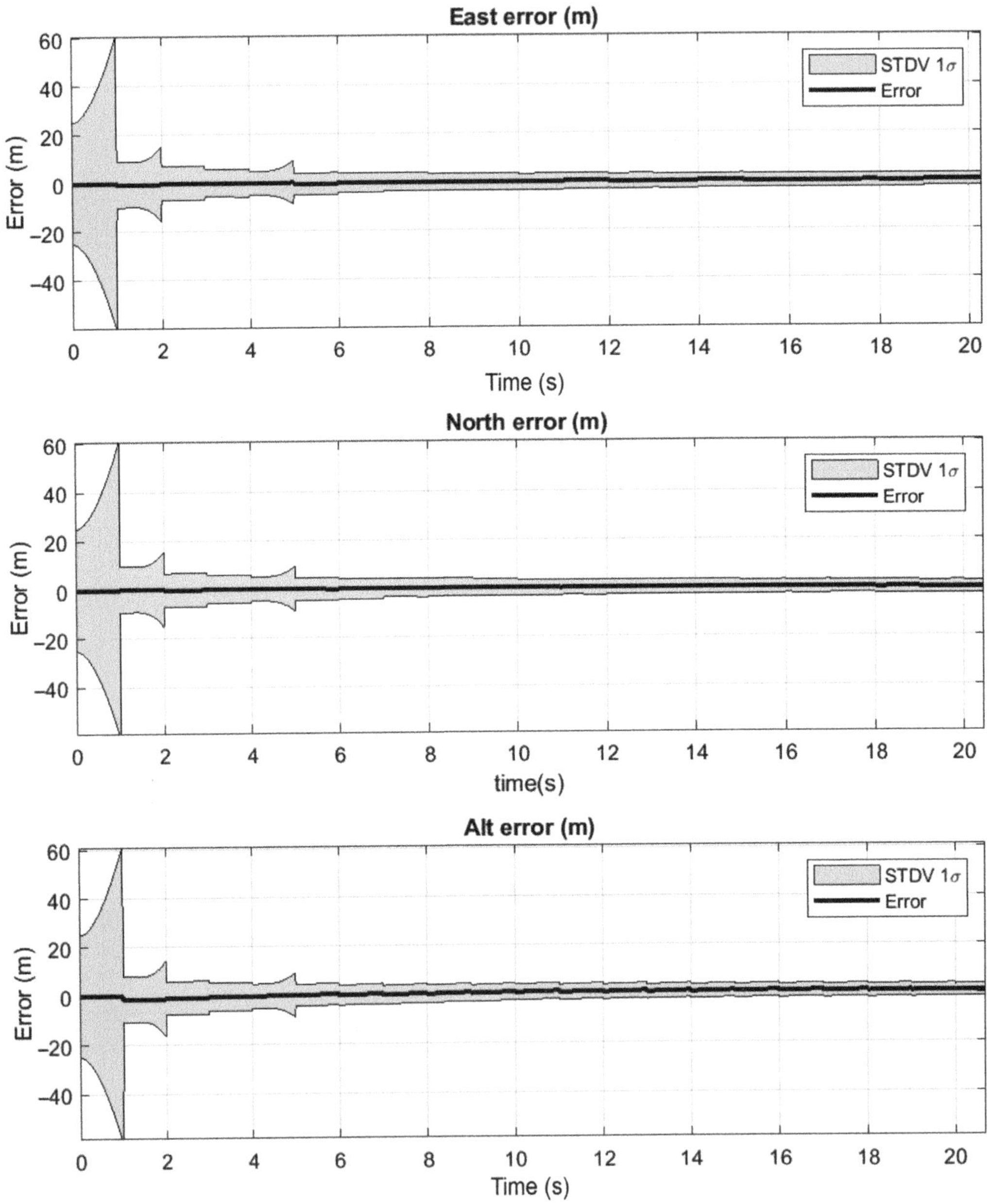

Figure 10.9 Positional errors and covariance convergence.

10.6.2 Tightly Coupled 3D IMU/GNSS Fusion

This project integrates GNSS pseudorange measurements with 3D inertial sensors in a tightly coupled mode using the system model described in Section 10.2 and the measurement model described in Section 10.3.2. The number of visible satellites throughout the trajectory is displayed in Figure 10.13. To test the fusion performance during a reduced number of visible satellites, we set

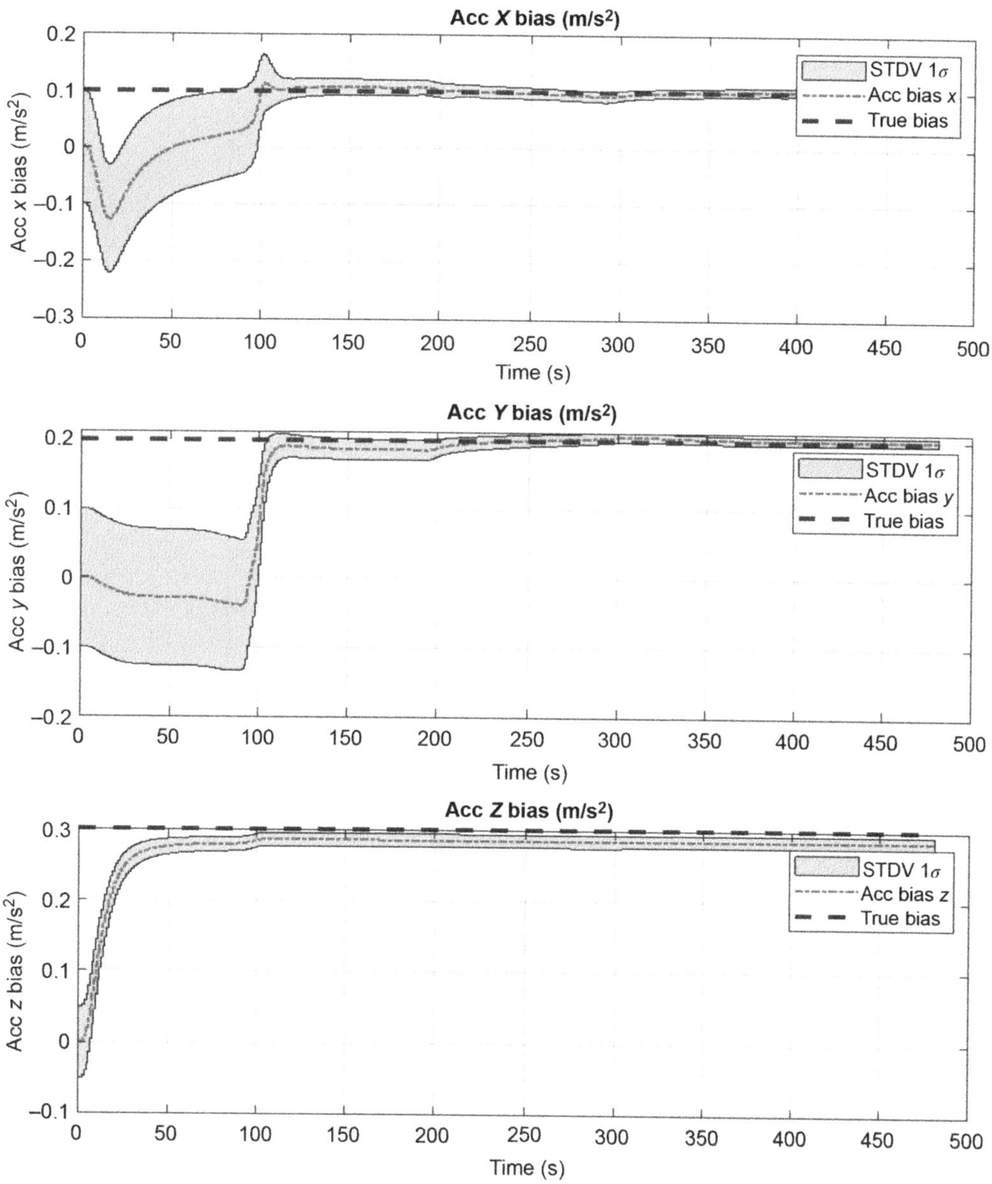

Figure 10.10 Accelerometer biases and covariance convergence.

the number of the satellites to 1 between the time $t = 140$ s and $t = 160$ s, as shown in Figure 10.14. The raw IMU data is shown in Figure 10.15. The raw IMU data is contaminated by known biases and Gaussian noises. GNSS pseudorange is corrected using the ground truth trajectory data and the positions of the satellites, and then a known receiver clock bias of 10 meters is added in addition to Gaussian random noise. Other information about the data can be found in the project folder. The estimation of 3D orientation angles compared to ground truth can be seen in Figure 10.16, and the

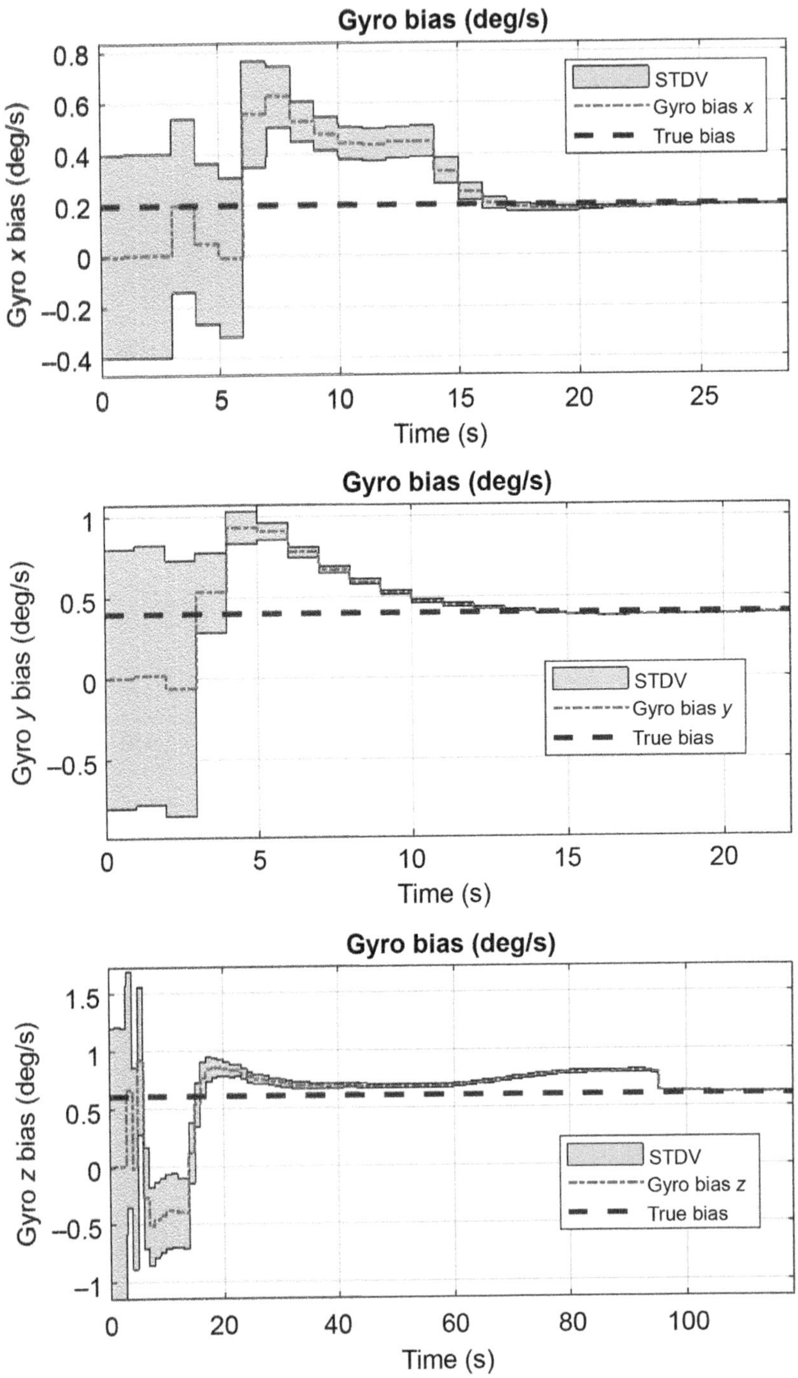

Figure 10.11 Gyroscope biases and covariance convergence.

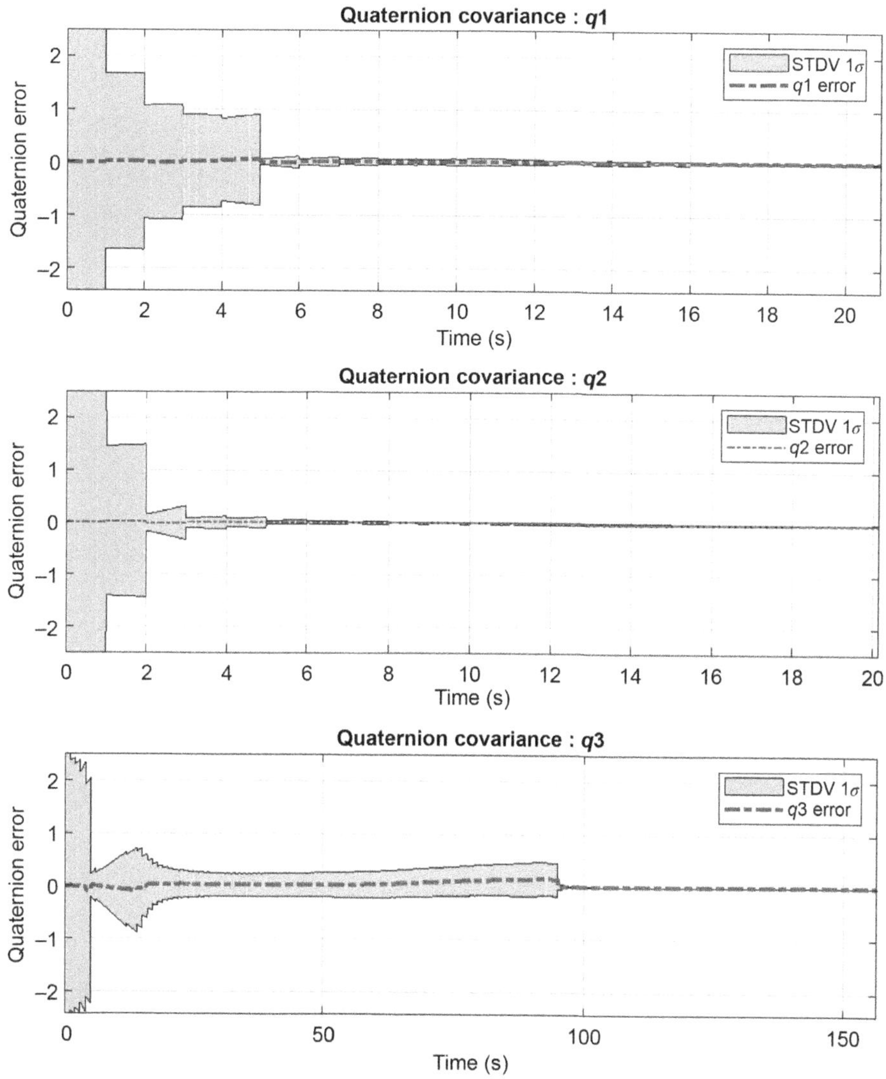

Figure 10.12 Quaternion errors and covariance convergence.

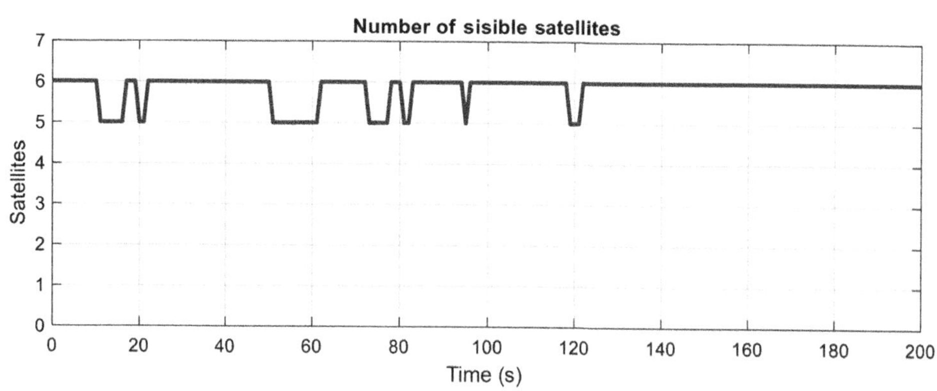

Figure 10.13 Number of visible satellites vs. time.

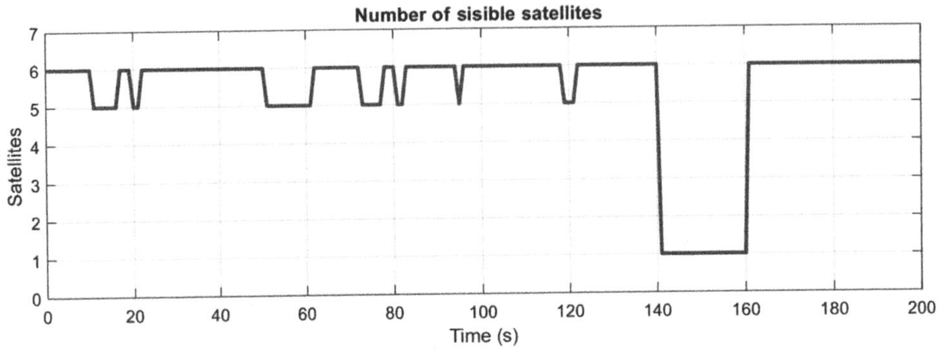

Figure 10.14 Reducing number of visible satellites between $t = 140$ s and $t = 160$ s.

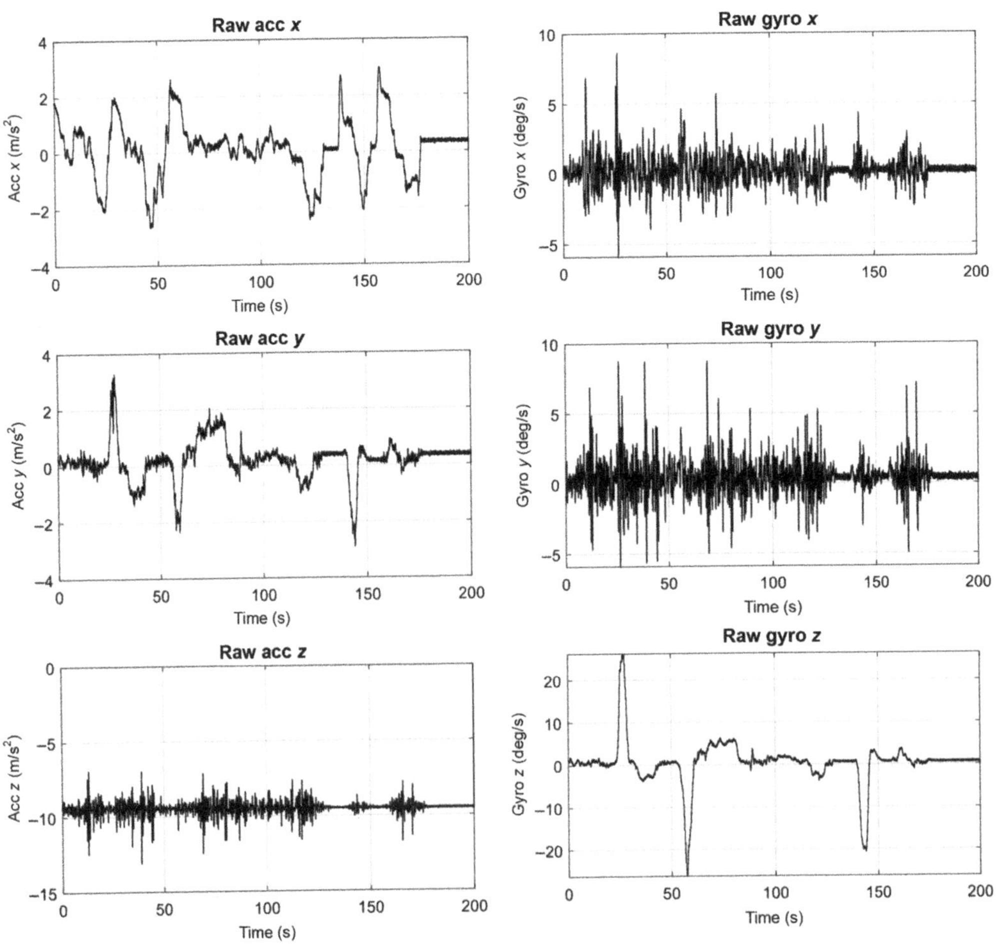

Figure 10.15 Raw 3D IMU data.

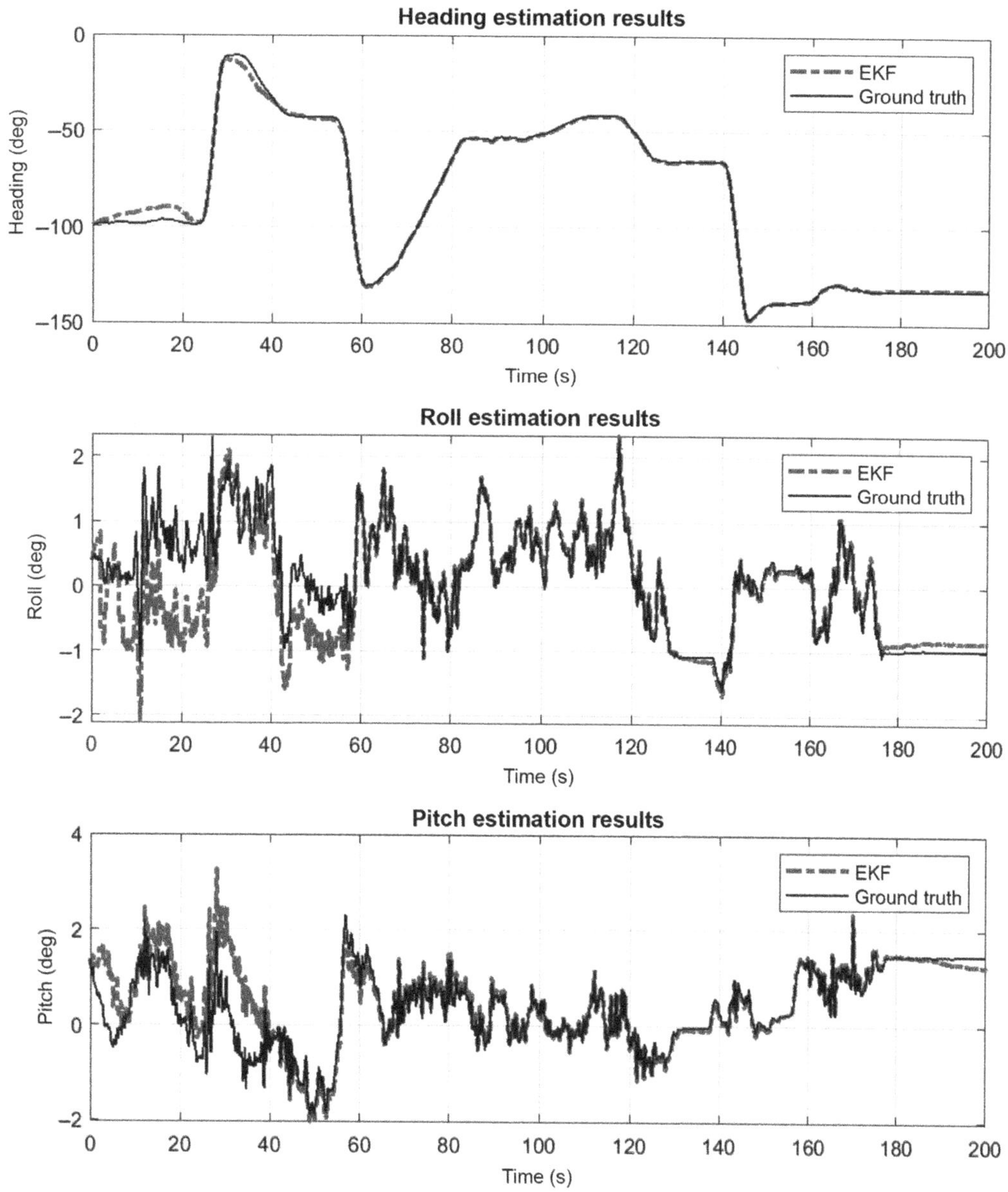

Figure 10.16 3D orientation, EKF vs. ground truth.

pose errors can be seen in Figures 10.16–10.18, which show that the filter converged approximately within a minute. The convergence is confirmed by Figures 10.19–10.22 which show the 3D biases of the accelerometers and gyroscopes to the true values in addition to the quaternion error convergence. The error figures also show the covariance estimated by the filter. Figure 10.23 shows the convergence of the receiver's clock bias. To test the filter performance during reduced satellite visibility, the number of satellites was reduced to 1 between times t = 140 s and t = 160 s. The

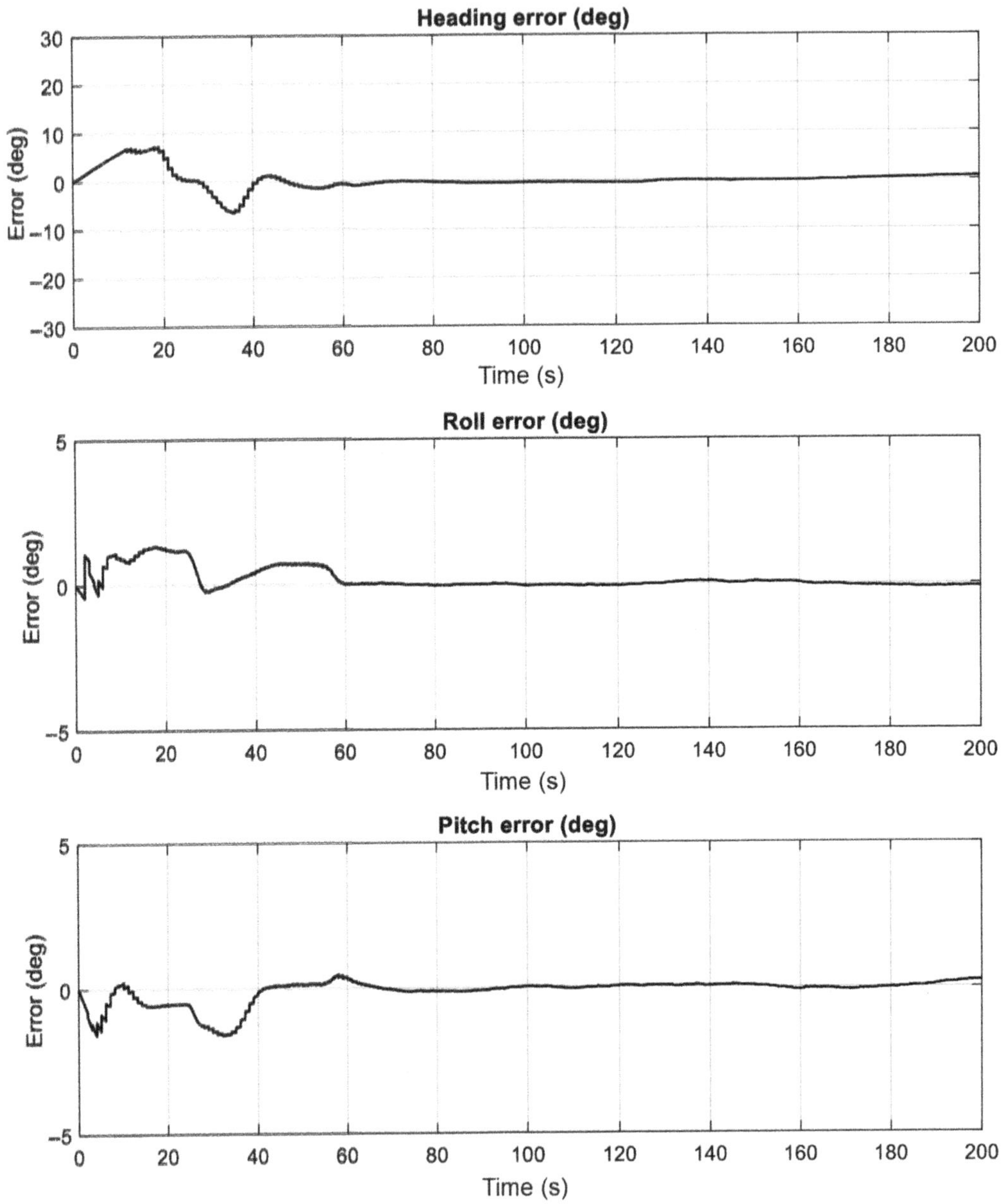

Figure 10.17 3D orientation errors.

covariances increased during these reduced satellites, as seen in Figure 10.24, but the error is still bounded and reduced, which confirms the correctness of the filter operation. Readers are encouraged to download the project code and inspect the code details to gain more detailed information about the implementation of the tightly coupled EKF fusion of GNSS and IMU sensors.

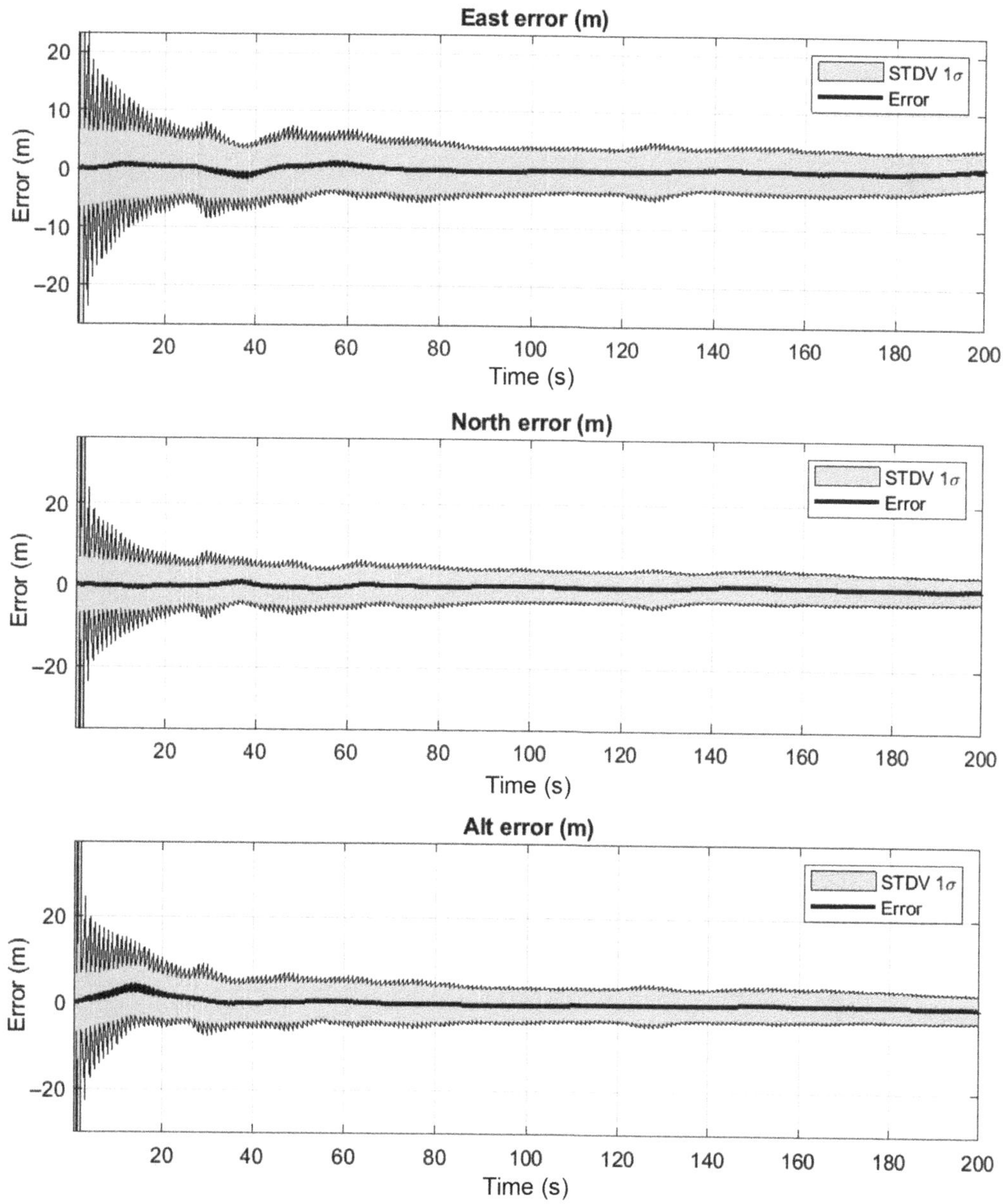

Figure 10.18 3D position errors.

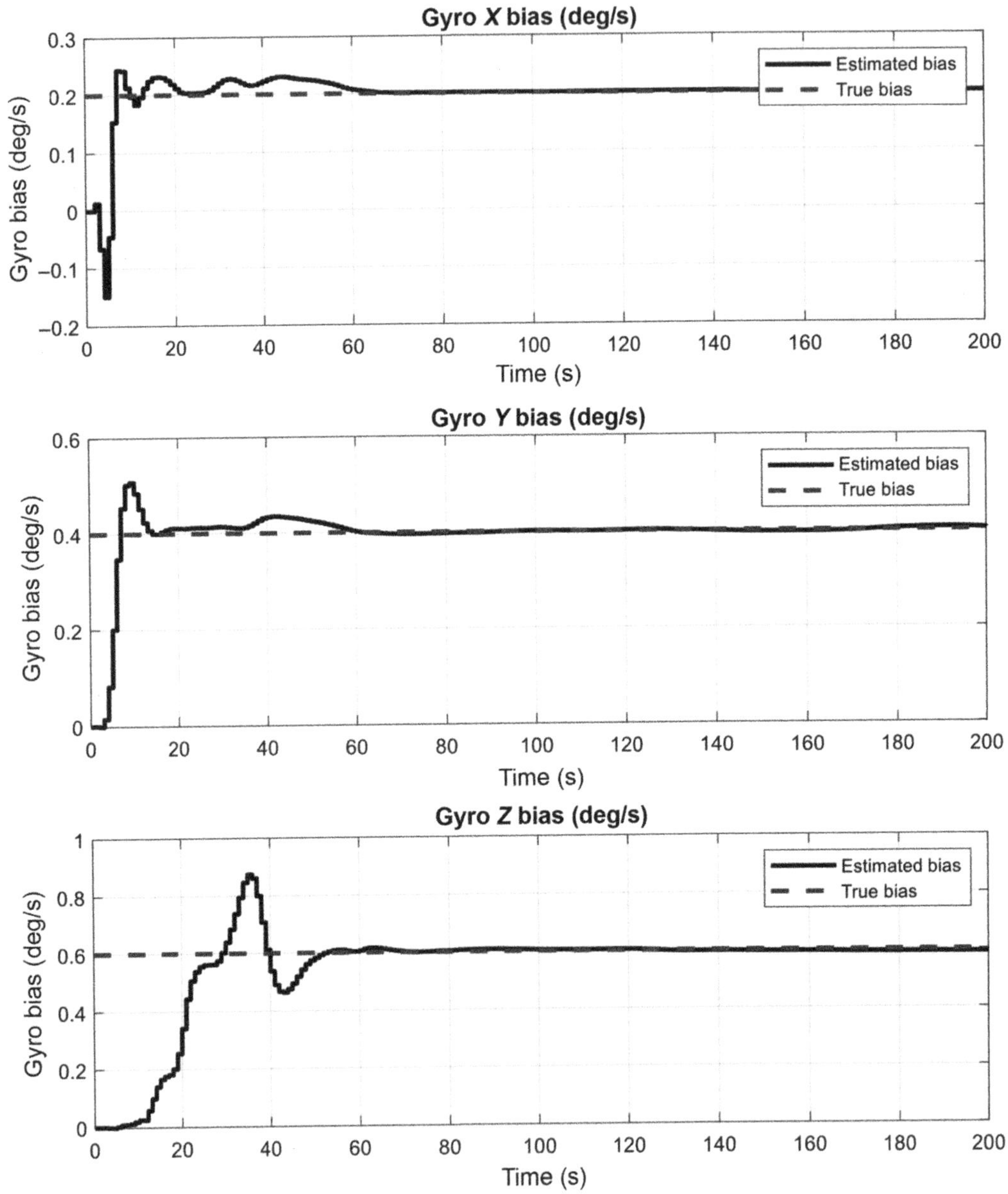

Figure 10.19 Gyroscope bias convergence.

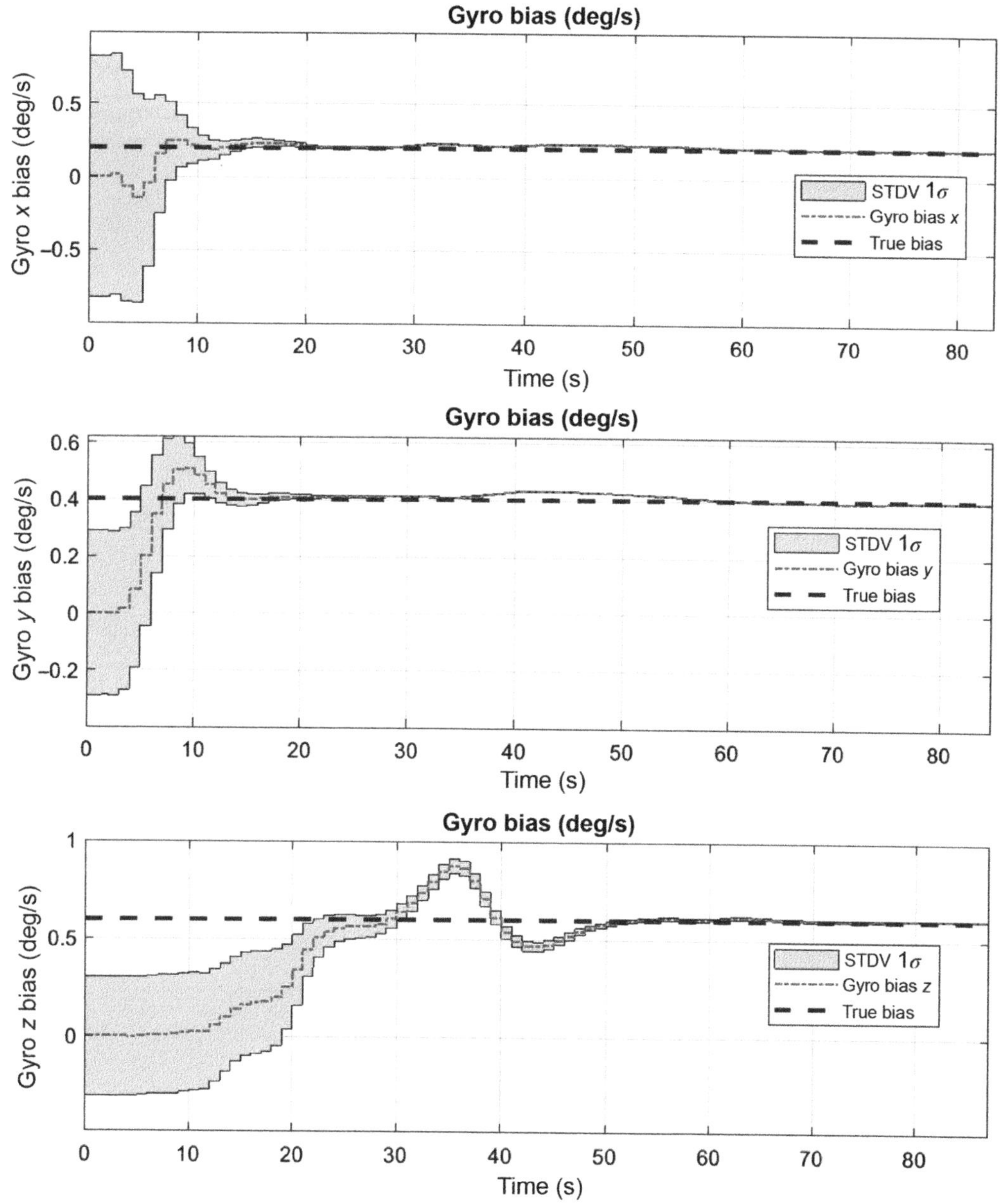

Figure 10.20 Gyroscope bias convergence with covariance.

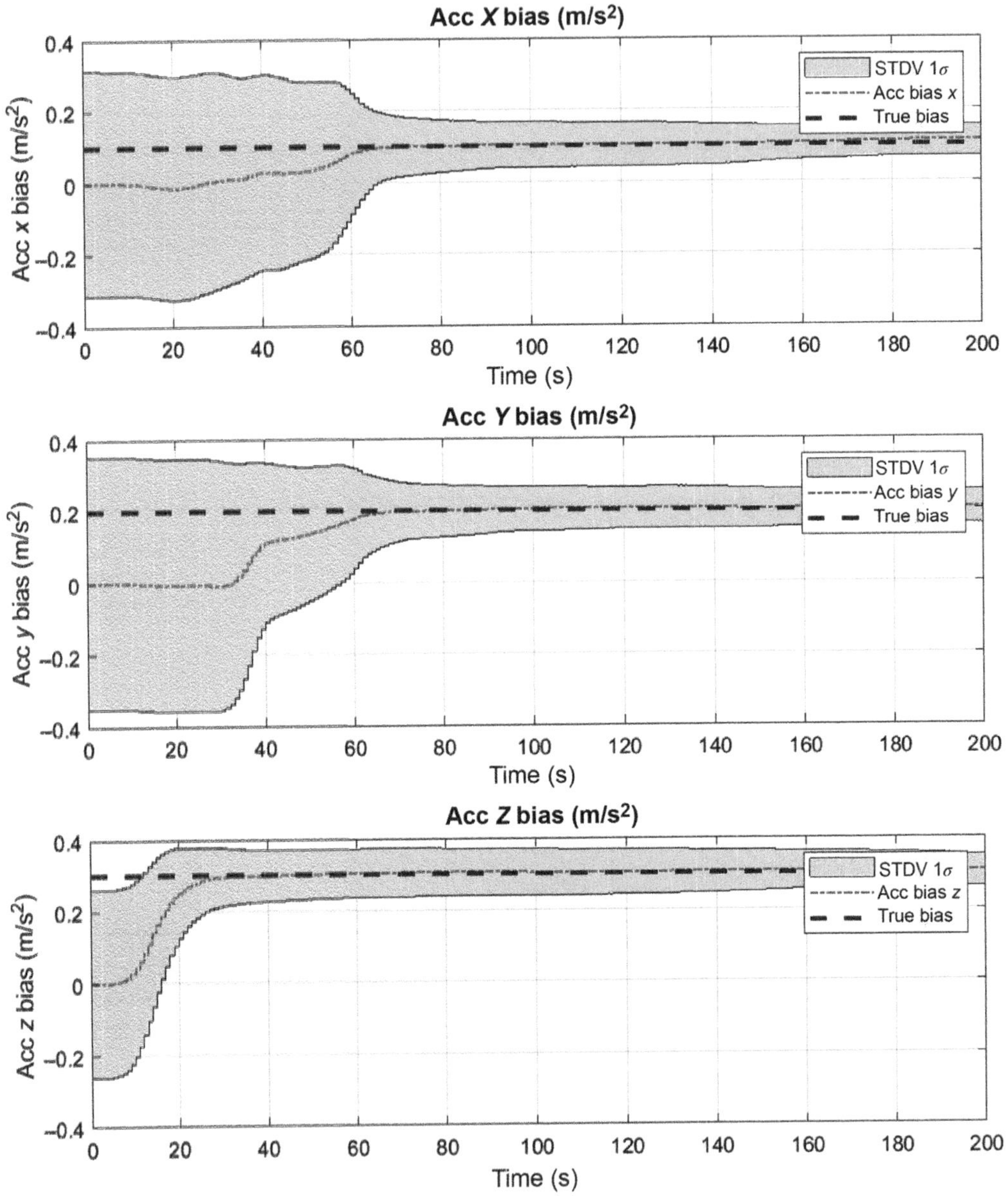

Figure 10.21 Accelerometer bias convergence with covariance.

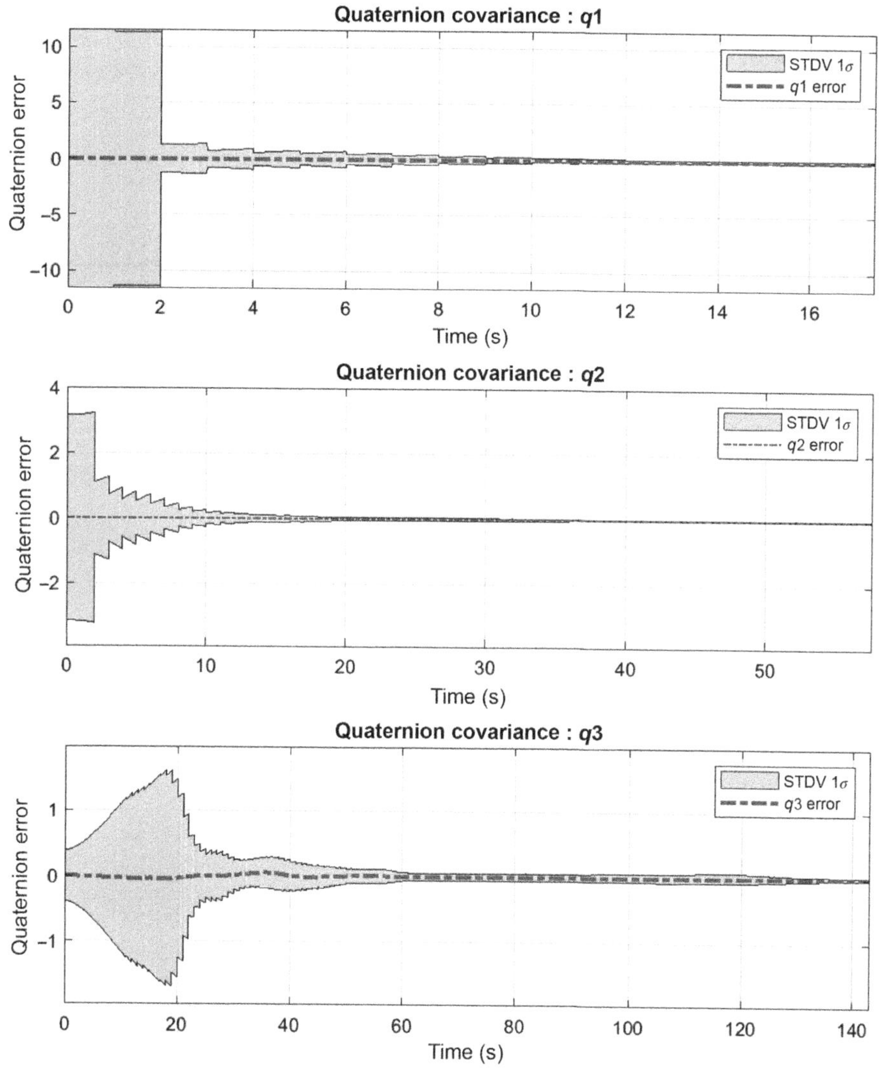

Figure 10.22 Quaternion error with covariance.

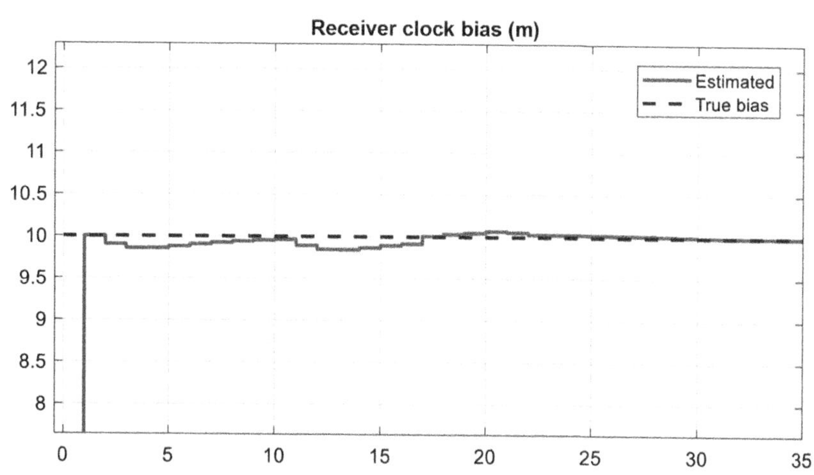

Figure 10.23 Receiver's clock bias convergence.

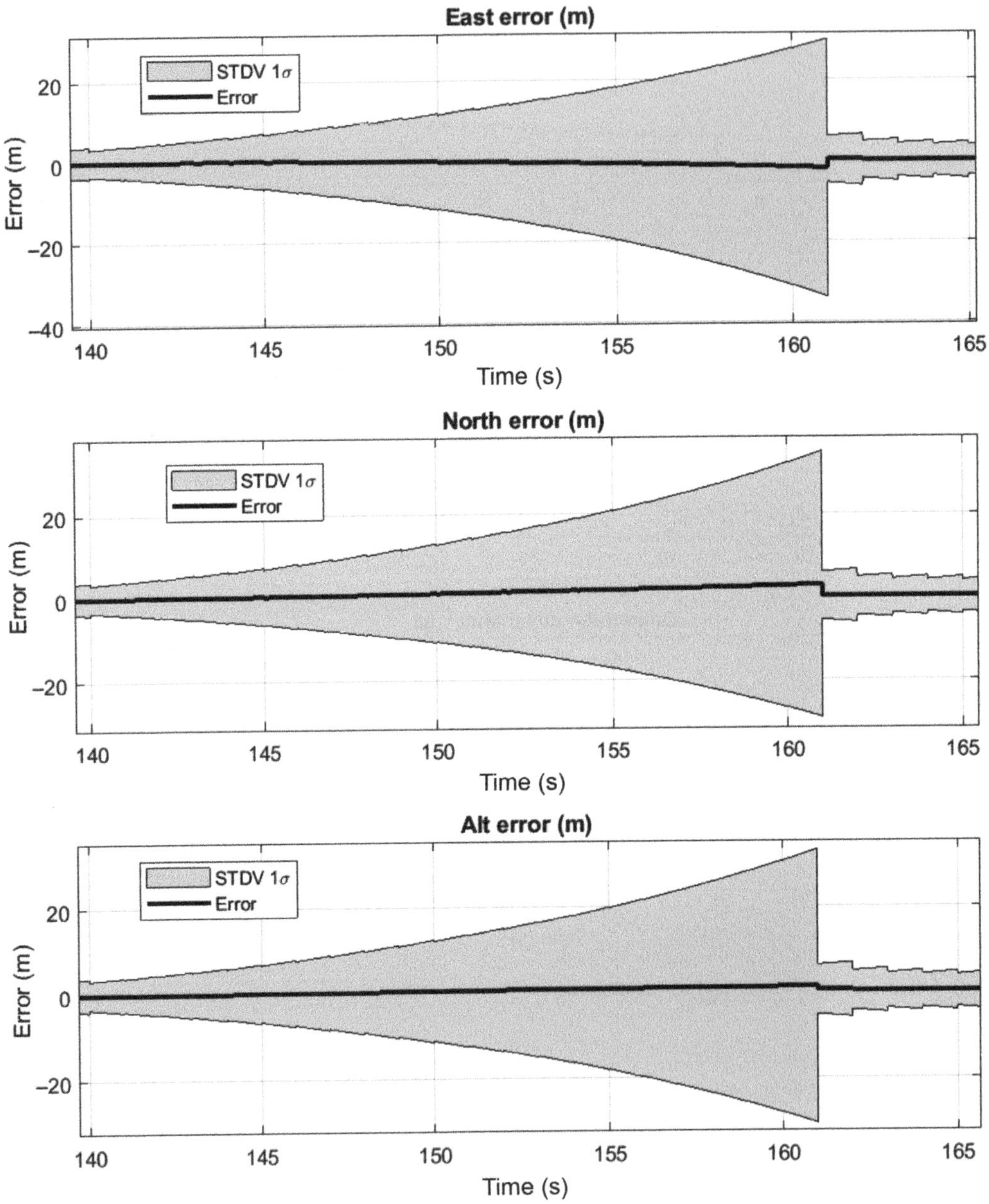

Figure 10.24 Error during reduced satellite visibility.

11

Case Study #3: AHRS and PDR

11.1 Introduction

Legged and humanoid platforms (see Figures 11.1 and 11.2) have recently gained increasing popularity. They are increasingly used in new applications, especially in factories and warehouses, to carry out labor-intensive, repetitive, routine tasks or safety/security-critical missions. Humanoid robots are promising technology that can handle the global labor shortage crisis. Legged and humanoid platforms differ from wheeled ground vehicles and aerial vehicles in the smoothness and constraints of the trajectory and motion patterns. For example, wheeled vehicles' motion is mostly concentrated on the body-forward axis and does not have significant motion in the lateral and vertical axes. Similarly, aerial vehicles, even small multi-rotor drones, exhibit constrained motion to a certain degree. They usually travel in a smooth trajectory and do not change state abruptly or very sharply for safety and stability purposes. The situation for legged and humanoid robotic platforms is different. Legged and humanoid platform motion has frequent stops and random movements in 3D. Therefore, the conventional inertial navigation system model described in Chapter 4 and further elaborated in Chapter 10 does not perform well under unconstrained motion. Fortunately, legged and humanoid platforms exhibit other constraints and patterns that can be used to provide better positioning and navigation and, consequently, mapping. This chapter will describe an enhanced inertial motion system model for legged and humanoid platforms. Having this enhanced inertial motion system model, integrating other sensors' measurements such as GPS, imaging, and ranging sensors is performed like other platforms using the proper measurement model.

One of the most important components in legged and humanoid platforms is the accurate calculation of orientation. The orientation of different parts of the body of the platform plays an important role in calculating other positioning and mapping states. For this reason, in this chapter, we will explain an important fusion scheme used to provide accurate orientation known as the attitude and heading reference system (AHRS). AHRS fuses inertial sensors (accelerometer and gyroscope) with a magnetometer and processes the sensor's measurements in a way that provides better orientation accuracy. AHRS are common in navigation and other industries that require precise orientation tracking. AHRS works best in static and low dynamics scenarios, which is the common scenario for legged and humanoid platforms. Having a good AHRS will help in performing motion prediction using step-detection mechanisms. The motion of legged and humanoid platforms is mostly driven by steps. Fortunately, inertial sensors (especially accelerometers) can easily detect steps with high accuracy using simple peak detection algorithms, even if the accelerometer is of very low quality. Detecting steps are used together with AHRS to predict the motion of legged

Sensor Fusion Approaches for Positioning, Navigation, and Mapping: How Autonomous Vehicles and Robots Navigate in the Real World: With MATLAB Examples, First Edition. Mohamed M. Atia.
© 2025 The Institute of Electrical and Electronics Engineers, Inc. Published 2025 by John Wiley & Sons, Inc.

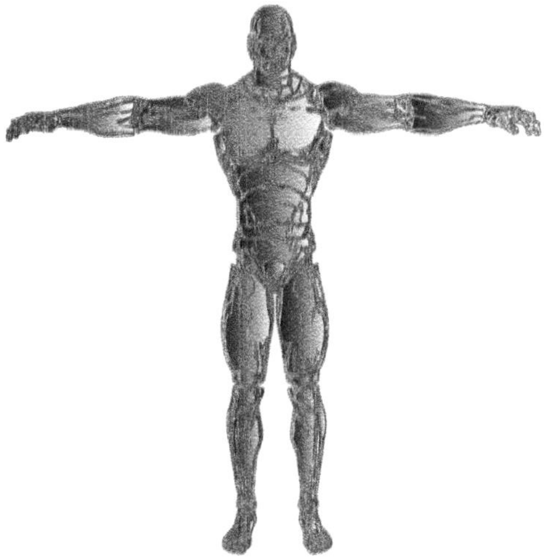

Figure 11.1 Humanoid robotic platform.

Figure 11.2 Legged robotic platform.
Source: Indrobotics Inc.

and humanoid platforms. If we have an estimate of the step length, we can easily project the step length in the orientation given by AHRS to calculate the next position. This process is called pedestrian dead reckoning (PDR). In this chapter, we will explain AHRS and PDR as they are key components in legged and humanoid platforms.

Other advanced applications, such as skeleton tracking and motion capture, can be developed using accurate AHRS and PDR. Skeleton tracking and motion capture applications apply additional constraints related to the dimensions of skeleton parts, dynamic constraints on body movements, and orientations of body parts with respect to each other. These constraints result in significant improvements. Skeleton tracking and motion capture are out of the scope of this book. However, this chapter will explain the main two components that are used in motion capture applications, AHRS and PDR.

11.2 Attitude and Heading Reference System (AHRS)

Due to inertial sensor errors (both systematic and stochastic), integrating raw angular rate data (gyroscope measurements) using the orientation rate equation will lead to significant drifts and errors in the resulting orientation angles. To generate an accurate and smooth orientation that does not drift over time, sensor fusion is applied to build the attitude and heading reference systems (AHRS). AHRS algorithms integrate accelerometer and magnetometer data with raw gyroscope measurements. We discussed the magnetometer sensor and its measurements in Chapter 4, and we demonstrated how magnetic field measurements can be used to estimate the heading of a platform. In this discussion, we will show how sensor fusion algorithms such as the extended Kalman filter (EKF) can be used to fuse measurements from gyroscopes, accelerometers, and magnetometers to provide accurate, smooth, and nondrifting orientation angle estimation. We will start by reviewing the dynamic system model for orientation calculation using inertial sensors. In our discussion, we will need to represent the rotation in different forms (rotation vector, directional cosine matrix, quaternion, and Euler angles). We will adopt the NED as the local (L) frame. We also assume the inertial sensors are attached and aligned with the body (B) frame (see Figure 11.3). In the equations, we will also refer to the inertial frame *I*. The details of the coordinate frames used in inertial sensors are given in Chapters 1 and 4.

The basic principle of operation of AHRS is that the accelerometer can provide nondrifting but noisy roll and pitch angles (tilt angles). The magnetometer (if tilt angles are known) can provide nondrifting but noisy heading angle estimation. Conversely, the system model integrates gyroscope measurements and provides smooth but drifting 3D orientation estimation. The complementary benefits of these sources of orientation estimation encourage the fusion of their measurements to provide smooth 3D, nondrifting orientation estimation. The following subsections will discuss the AHRS system model, tilt angle estimation from accelerometer measurements, and heading angle estimation from magnetic field sensors. Then, we will discuss the fusion of these measurements into an EKF to produce smooth and accurate 3D orientation estimation.

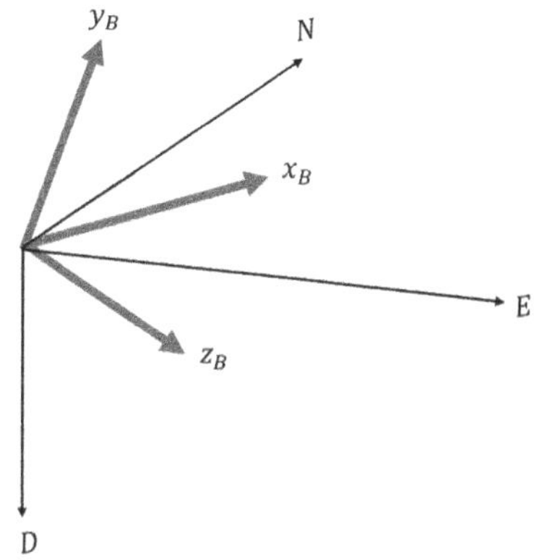

Figure 11.3 Illustration of orientation of body frame and the local frame.

11.2.1 AHRS System Model

Rotation can be represented by a direction cosine matrix C_B^L or quaternion vector $\vec{q}_B^L = (q_0\, q_1\, q_2\, q_3)^T$. As discussed in Chapter 1, the direction cosine matrix can be obtained from a quaternion vector as follows:

$$C_B^L(1,1) = \left(q_0^2 + q_1^2 - q_2^2 - q_3^2\right) \tag{11-1}$$

$$C_B^L(1,2) = 2(q_1q_2 - q_0q_3) \tag{11-2}$$

$$C_B^L(1,3) = 2(q_1q_3 + q_0q_2) \tag{11-3}$$

$$C_B^L(2,1) = 2(q_1q_2 + q_0q_3) \tag{11-4}$$

$$C_B^L(2,3) = 2(q_2q_3 - q_0q_1) \tag{11-5}$$

$$C_B^L(3,1) = 2(q_1q_3 - q_0q_2) \tag{11-6}$$

$$C_B^L(3,2) = 2(q_2q_3 + q_0q_1) \tag{11-7}$$

$$C_B^L(3,3) = \left(q_0^2 - q_1^2 - q_2^2 + q_3^2\right) \tag{11-8}$$

As discussed in Chapters 1 and 4, the orientation system model in the form $\dot{\vec{x}}(t) = f\left(\vec{x}(t), \vec{u}(t)\right) + \vec{w}(t)$ is given by the following differential equation:

$$\dot{\vec{q}}_B^L = \begin{pmatrix} \dot{q}_0 \\ \dot{q}_1 \\ \dot{q}_2 \\ \dot{q}_3 \end{pmatrix} = \frac{1}{2}\left(\vec{W}_{LB}^B \times\right)\vec{q}_B^L + \vec{w}_q, \vec{w}_q \sim \mathcal{N}\left(0, \sigma_q^2\right) \tag{11-9}$$

where

$\vec{W}^{B}_{LB} = \begin{pmatrix} 0 & \omega^{B}_{LB_X} & \omega^{B}_{LB_Y} & \omega^{B}_{LB_Z} \end{pmatrix}^{T}$ is the homogeneous quaternion vector form of $\vec{\omega}^{B}_{LB}$, the compensated rotation rate of the body frame with respect to the local frame L

$\left(\left(\vec{W}^{B}_{LB}\right)\times\right)$ is the 4D skew matrix operator defined as follows:

$$\left(\left(\vec{W}^{B}_{LB}\right)\times\right) = \begin{pmatrix} 0 & -\omega^{B}_{LB_X} & -\omega^{B}_{LB_Y} & -\omega^{B}_{LB_Z} \\ \omega^{B}_{LB_X} & 0 & \omega^{B}_{LB_Z} & -\omega^{B}_{LB_Y} \\ \omega^{B}_{LB_Y} & -\omega^{B}_{LB_Z} & 0 & \omega^{B}_{LB_X} \\ \omega^{B}_{LB_Z} & \omega^{B}_{LB_Y} & -\omega^{B}_{LB_X} & 0 \end{pmatrix} \tag{11-10}$$

To obtain $\vec{\omega}^{B}_{LB}$, the raw gyroscope measurements (referred as $\vec{\omega}^{B}_{IB}$) is compensated for the following components:

- Earth and transport rate $\vec{\omega}^{B}_{IL}$
- Gyroscope biases $\vec{b}_g$
- Random noise $\vec{w}_b \sim \mathcal{N}(0, \sigma_b^2)$

Therefore, $\vec{\omega}^{B}_{LB}$, the compensated rotation rate of the body frame with respect to the local frame L is given by:

$$\vec{\omega}^{B}_{LB} = \vec{\omega}^{B}_{IB} - \vec{b}_g - \vec{\omega}^{B}_{IL} \tag{11-11}$$

The Earth and translation rate $\vec{\omega}^{B}_{IL}$ is given by the following expressions:

$$\vec{\omega}^{B}_{IL} = C^{B}_{L}\left(\vec{\omega}^{L}_{IE} + \vec{\omega}^{L}_{EL}\right) \tag{11-12}$$

$$\vec{\omega}^{L}_{IE} = \left(w^{e}\cos(\varphi) \quad 0 \quad -w^{e}\sin(\varphi)\right)^{T} \tag{11-13}$$

$$\vec{\omega}^{L}_{EL} = \left(\frac{v_e}{(R_n + h)} \quad \frac{-v_n}{(R_m + h)} \quad \frac{-v_e}{(R_n + h)}\tan(\varphi)\right)^{T} \tag{11-14}$$

The state vector in this case is given by the following vector:

$$\vec{x} = \begin{pmatrix} q_1 \\ q_2 \\ q_3 \\ \vec{b}_g \\ \vec{b}_a \end{pmatrix} \tag{11-15}$$

Note that we added the accelerometer biases to the state vector as they will be needed when we fuse measurements from accelerometer. In addition, note that we used the vector part of the quaternion $(q_1\ q_2\ q_3)^T$ as this the minimal coordinate representation required to avoid redundancy, singularity, and to keep the closeness of addition and linear independence conditions, which are essential to apply any least square or linear estimation methods. In the following calculations, the scalar part of the quaternion component q_0 is replaced by the vector part as follows:

$$q_0 = \sqrt{1 - q_1^2 - q_2^2 - q_3^2} \tag{11-16}$$

To complete the final dynamic system model, gyroscope biases and scale factors are modeled as Gauss-Markov random processes. A stochastic signal x that is modeled as Gauss–Markov process has the following general differential equation:

$$\dot{x} = -\beta_x x + \sqrt{2\sigma_x^2 \beta_x} w_u \tag{11-17}$$

β_x and σ_x are the parameters of Gauss–Markov random process where β_x is the reciprocal of the time constant and σ_x is the standard deviation, and $w_u(t)$ is a zero-mean unit-variance Gaussian noise.

See Chapter 2 for more details about the Gauss–Markov random process. Given the minimal coordinate representation $\vec{x}$ that includes the vector part of the quaternion, biases, and scale factors, the system model is given by the following equations:

$$\dot{q}_1 \cong \left(\omega_{IBx}^B - b_{g_x}\right)\left(\frac{q_0}{2}\right) - \left(\omega_{IBy}^B - b_{g_y}\right)\left(\frac{q_3}{2}\right) + \left(\omega_{IBz}^B - b_{g_z}\right)\left(\frac{q_2}{2}\right) + \vec{w}_{q_1} \tag{11-18}$$

$$\dot{q}_2 \cong \left(\omega_{IBx}^B - b_{g_x}\right)\left(\frac{q_3}{2}\right) + \left(\omega_{IBy}^B - b_{g_y}\right)\left(\frac{q_0}{2}\right) - \left(\omega_{IBz}^B - b_{g_z}\right)\left(\frac{q_1}{2}\right) + \vec{w}_{q_2} \tag{11-19}$$

$$\dot{q}_3 \cong -\left(\omega_{IBx}^B - b_{g_x}\right)\left(\frac{q_2}{2}\right) + \left(\omega_{IBy}^B - b_{g_y}\right)\left(\frac{q_1}{2}\right) + \left(\omega_{IBz}^B - b_{g_y}\right)\left(\frac{q_0}{2}\right) + \vec{w}_{q_3} \tag{11-20}$$

$$\dot{\vec{b}}_g = -\vec{\beta}_b \odot \vec{b}_g + \sqrt{2\vec{\sigma}_b^2 \odot \vec{\beta}_b} \odot \vec{w}_u \tag{11-21}$$

$$\dot{\vec{b}}_a \cong -\vec{\beta}_{b_a} \odot \vec{b}_a + \sqrt{2\vec{\sigma}_{b_a}^2 \odot \vec{\beta}_{b_a}} \odot \vec{w}_u \tag{11-22}$$

where

$\odot$ is the element-by-element or component-wise vector multiplication
$\vec{w}_{q_1}$, $\vec{w}_{q_2}$, $\vec{w}_{q_3}$ are zero-mean Gaussian noise
$\vec{w}_u$ is unity-variance zero-mean Gaussian noise.

Equations (11-18)–(11-22) demonstrate the system model in the form $\dot{\vec{x}}(t) = f\left(\vec{x}(t), \vec{u}(t)\right) + \vec{w}(t)$. In the next subsection, we will derive the linearized error model of the above system model.

11.2.2 Linearized System Model

To obtain the linearized system model, we perform the partial derivative of the above system model equations with respect to the state vector $\vec{x}$. The partial differentiation may look complex, but it can be easily done using symbolic equation engines such as MATLAB symbolic engine. Doing the partial derivatives with respect to orientation, biases, and scale factors states gives the following linearized error model:

$$\begin{aligned}\delta\dot{q}_1 \cong &\left(-\frac{q_1}{2q_0}\left(\omega_{IBx}^B - b_{g_x}\right)\right)\delta q_1 + \left(-\frac{q_2}{2q_0}\left(\omega_{IBx}^B - b_{g_x}\right) + \frac{1}{2}\left(\omega_{IBz}^B - b_{g_z}\right)\right)\delta q_2 \\ &+ \left(-\frac{q_3}{2q_0}\left(\omega_{IBx}^B - b_{g_x}\right) + \frac{1}{2}\left(\omega_{IBy}^B - b_{g_y}\right)\right)\delta q_3 - \left(\frac{q_0}{2}\right)\delta b_{g_x} + \left(\frac{q_3}{2}\right)\delta b_{g_y} - \left(\frac{q_2}{2}\right)\delta b_{g_z} + \vec{w}_{q_1}\end{aligned} \tag{11-23}$$

$$\delta\dot{q}_2 \cong \left(-\frac{q_1}{2q_0}\left(\omega_{IBy}^B - b_{g_y}\right) + \frac{1}{2}b_{g_z}\right)\delta q_1 + \left(-\frac{q_2}{2q_0}\left(\omega_{IBy}^B - b_{g_y}\right)\right)\delta q_2$$
$$+ \left(-\frac{q_3}{2q_0}\left(\omega_{IBy}^B - b_{g_y}\right) + \frac{1}{2}\left(\omega_{IBx}^B - b_{g_x}\right)\right)\delta q_3 - \left(\frac{q_3}{2}\right)\delta b_{g_x} - \left(\frac{q_0}{2}\right)\delta b_{g_y} + \left(\frac{q_1}{2}\right)\delta b_{g_z} + \vec{w}_{q_2} \tag{11-24}$$

$$\delta\dot{q}_3 \cong \left(-\frac{q_1}{2q_0}\left(\omega_{IBz}^B - b_{g_z}\right) - \frac{1}{2}\left(\omega_{IBy}^B - b_{g_y}\right)\right)\delta q_1 + \left(-\frac{q_2}{2q_0}\left(\omega_{IBz}^B - b_{g_z}\right) + \frac{1}{2}\left(\omega_{IBx}^B - b_{g_x}\right)\right)\delta q_2$$
$$+ \left(-\frac{q_3}{2q_0}\left(\omega_{IBz}^B - b_{g_z}\right)\right)\delta q_3 + \left(\frac{q_2}{2}\right)\delta b_{g_x} - \left(\frac{q_1}{2}\right)\delta b_{g_y} - \left(\frac{q_0}{2}\right)\delta b_{g_z} - \omega_{IBx}^B \frac{q_2}{2}\delta s_{g_x} + \vec{w}_{q_3} \tag{11-25}$$

$$\delta\dot{\vec{b}}_g \cong -\vec{\beta}_{b_g} \odot \delta\vec{b}_g + \sqrt{2\vec{\sigma}_{b_g}^2 \odot \vec{\beta}_{b_g}} \odot \vec{w}_u \tag{11-26}$$

$$\delta\dot{\vec{b}}_a \cong -\vec{\beta}_{b_a} \odot \delta\vec{b}_a + \sqrt{2\vec{\sigma}_{b_a}^2 \odot \vec{\beta}_{b_a}} \odot \vec{w}_u \tag{11-27}$$

where

$\vec{w}_u$ is zero-mean unity-variance Gaussian noise
$\vec{w}_{q_1}, \vec{w}_{q_2}, \vec{w}_{q_3}$ are zero-mean Gaussian noise.

Note that left-hand side does not contain the scalar component of the quaternion q_0 which satisfies the minimal coordinate representation conditions (linear independence and closeness under addition). The right-hand side contains q_0 but it can be simply calculated using the vector part. This should not be a problem if we do not have the error of the scalar part of the quaternion δq_0 as part of the error state vector. Equations (11-23)–(11-27) constitute the error model in the form $\delta\dot{\vec{x}}\,(t) = F(t)\delta\vec{x}(t) + G(t)\vec{w}_u(t)$. The transition matrix $F(t)$ can be obtained by arranging Eq. (11-27) into vector-matrix form. The noise shaping matrix $G(t)$ is a tuning parameter that can be tuned given the sensors noise specifications as guidance as discussed in Chapter 10.

11.2.3 AHRS Loosely Coupled Fusion

In loosely coupled AHRS, the 3D orientation angles are independently calculated by accelerometers and magnetometer sensors and used as a measurement update to the fusion filter, whether it is a particle filter (PF), an EKF, or any of their variations. In this chapter, we will assume that the fusion algorithm is an EKF as can be seen in Figure 11.4. We will refer to the accelerometer measurements, which will be used to calculate the tilt angles in the loosely coupled mode by the variable $\vec{a}_{SF}^{B}$ and it goes under the same compensation as gyroscope measurement as follows:

$$\vec{a}_{SF}^{B} = \vec{\hat{a}}_{SF}^{B} - \vec{b}_a \tag{11-28}$$

where $\vec{\hat{a}}_{SF}^{B}$ is raw accelerometer measurements.

We will also refer to the magnetic field sensor measurements by the symbol $\vec{m}_q^L$. At stationary or very low-speed applications, accelerometers mainly measure gravity, which can be used to estimate tilt angles (roll and pitch), and magnetometers can be used to estimate the heading. In a local level

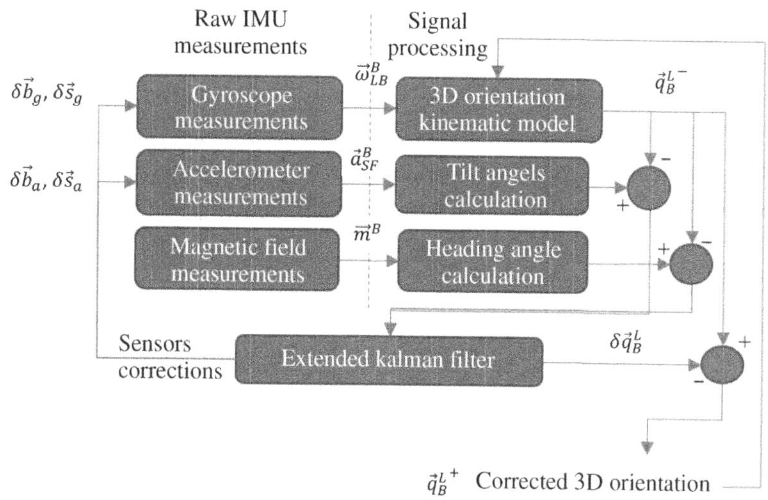

Figure 11.4 Loosely coupled AHRS EKF.

frame tangent to the Earth ellipsoid, the gravity vector has one nonzero component in the downward vertical direction. This gravity $\vec{g}^L$ is called a reference gravity vector. It can be given by:

$$\vec{g}^L = \begin{pmatrix} 0 \\ 0 \\ g_o \end{pmatrix} \tag{11-29}$$

where g_o is the standard gravity value at the current location.

The WGS84 standards provide accurate formulas to calculate g_o as a function of latitude and height. A perfectly leveled platform should ideally measure $\vec{g}^L$ (regardless of the heading). However, under low dynamics, a tilted body will measure the projection of $\vec{g}^L$ on the axes of the body frame. Therefore, the accelerometer sensor would read the following measurements:

$$\vec{a}^B_{SF} = C^B_L \vec{g}^L \tag{11-30}$$

where $\vec{a}^B_{SF}$ is the raw accelerometer data in body frame (subscript *SF* refers to specific force which is the quantity that the accelerometer measures).

As $\vec{g}^L$ is known and $\vec{a}^B_{SF}$ is measured by the accelerometer, tilt angles (roll and pitch angles) can be calculated as follows:

$$\hat{r} = a\ \tan 2\left(-a^B_{SFY}, -a^B_{SFZ}\right) \tag{11-31}$$

$$\hat{p} = a\ \tan 2\left(a^B_{SFX}, \sqrt{{a^B_{SFY}}^2 + {a^B_{SFZ}}^2}\right) \tag{11-32}$$

Heading angle is measured using Earth magnetic field sensors measurements. In arbitrary orientation, the magnetic field sensor will measure the earth's magnetic field projected on the body frame axes. We will refer to this measurement by the variable $\vec{m}^B = \left(m_x\ m_y\ m_z\right)^T$. A well-calibrated perfectly horizontal magnetometer sensor pointing to the earth's magnetic north will ideally measure the earth reference magnetic field $\vec{m}^L$. The reference earth magnetic field, like gravity, can be calculated from standard tables prepared by environment and physics scientists such as the one publicly available on the Natural Resources Canada portal (https://natural-resources.canada.ca/) or the National Centers for Environmental Information (https://www.ncei.noaa.gov/products/geomagnetic-data). If the magnetometer has a nonzero heading, $\vec{m}^L$ will be projected to body axes accordingly. To calculate heading from the magnetometer measurements, the following simple formula can be used:

$$\hat{A} = \tan^{-1}\left(\frac{-m_y\cos(r) + m_z\sin(r)}{m_x\cos(p)m_y\sin(r)\sin(p) + m_z\cos(r)\sin(p)}\right) + D(\varphi, \lambda) \tag{11-33}$$

where $D(\varphi, \lambda)$ is the declination angle, which is the deviation between magnetic north and geographic (true) north.

The declination angle is a function of the geodetic latitude φ and longitude λ and can also be obtained from standard tables like gravity and magnetic field values. This will complete the calculation of the three orientation angles: roll, pitch, and azimuth (heading).

If we have full orientation from accelerometers and magnetometer, we can convert the measurement vector $\vec{y} = (\hat{r}, \hat{p}, \hat{A})^T$ to a quaternion transformation $\vec{\hat{q}}_L^B = (\hat{q}_0\,\hat{q}_1\,\hat{q}_2\,\hat{q}_3)^T$ and feed the vector part of $\vec{\hat{q}}_L^B$ which is $(\hat{q}_1\,\hat{q}_2\,\hat{q}_3)^T$ to the EKF as a measurement update. The measurement model in this case is linear and the design matrix that relates the measurements to the state vector is given by the following matrix:

$$H = \begin{pmatrix} 1 & 0 & 0 & \\ 0 & 1 & 0 & \mathbf{0}_{3\times12} \\ 0 & 0 & 1 & \end{pmatrix} \tag{11-34}$$

and measurement noise covariance matrix is given by:

$$R = \begin{pmatrix} r_{q_1} & 0 & 0 \\ 0 & r_{q_2} & 0 \\ 0 & 0 & r_{q_3} \end{pmatrix} \tag{11-35}$$

This is a linear measurement model which can be applied easily in an EKF, a PF or any of their variations. The steps to apply EKF or PF are the same steps described in Chapter 3 and further elaborated on practical use cases in Chapters 9 and 10.

11.2.4 AHRS Tightly Coupled Fusion

In a tightly coupled approach, the raw measurements from accelerometers and magnetometer sensors are directly used as measurement updates to the fusion filter (EKF, PF, or any of their variations). In this chapter, we use an EKF as shown in Figure 11.5. The main advantage of tightly coupled mode is the ability of the fusion filter to continue working under partial updates. For example, if only a two-axis accelerometer is available, they can still be used as updates to a tightly coupled filter. Therefore, a tightly coupled approach provides better observability of accelerometer and magnetometer errors, which do not exist in loosely coupled mode. Raw acceleration and magnetic measurements from inertial and magnetic field sensors can be projected to the local frame L as follows:

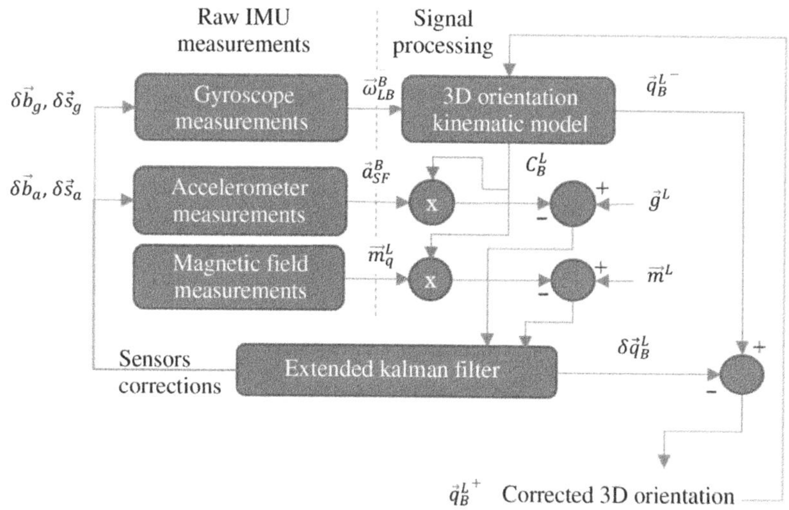

Figure 11.5 Tightly coupled AHRS EKF.

$$\hat{g}^L = C_B^L \vec{a}_{SF}^B \tag{11-36}$$

$$\vec{\hat{m}}^L = C_B^L \vec{m}^B \tag{11-37}$$

where $\hat{g}^L$ and $\vec{\hat{m}}^L$ are predicted measurements.

They are predicted based on our estimate of the platform state (particularly orientation represented in the direction cosine matrix form C_B^L). At the same time, given the current location of the platform, we can look up the standard reference values of the earth's magnetic field $\vec{m}^L$ and the gravity g^L. By comparing the predicted measurements with the reference measurements, we can correct the platform state using any fusion algorithm. In our case, we are using an EKF, so we will need the linearized measurement model.

The measurement model in our case can be written as follows:

$$\begin{pmatrix} \hat{g}^L \\ \vec{\hat{m}}^L \end{pmatrix} = \begin{pmatrix} C_B^L\left(\vec{\hat{a}}_{SF}^B - \vec{b}_a\right) \\ C_B^L \vec{m}^B \end{pmatrix} \tag{11-38}$$

This measurement model is a function of the platform state, but it uses the direction cosine matrix C_B^L which does not satisfy the minimal coordinate representation conditions. Minimal coordinate conditions are explained in Chapter 10. Therefore, we want to make Eq. (11-38) a function in the vector part of the quaternion $(q_1\ q_2\ q_3)^T$. We use the same trick we used earlier in Chapter 10 to enforce the minimal coordinate representation. We replace C_B^L by its quaternion equivalent as follows:

$$C_B^L = \begin{pmatrix} q_0^2 + q_1^2 - q_2^2 - q_3^2 & 2(q_1q_2 + q_0q_3) & 2(q_1q_3 - q_0q_2) \\ 2(q_1q_2 - q_0q_3) & q_0^2 - q_1^2 + q_2^2 - q_3^2 & 2(q_2q_3 + q_0q_1) \\ 2(q_1q_3 + q_0q_2) & 2(q_2q_3 - q_0q_1) & q_0^2 - q_1^2 - q_2^2 + q_3^2 \end{pmatrix} \tag{11-39}$$

$$q_0 = \sqrt{1 - q_1^2 - q_2^2 - q_3^2} \tag{11-40}$$

By replacing C_B^L by its quaternion equivalent, Eqs. (11-38)–(11-40) will form a nonlinear measurement model in the form $\vec{y} = h\left(\vec{x}\right)$, where $\vec{y}$ and $\vec{x}$ are defined as follows:

$$\vec{y} = \begin{pmatrix} \hat{g}^L \\ \vec{\hat{m}}^L \end{pmatrix}$$
$$\vec{x} = \begin{pmatrix} q_1 \\ q_2 \\ q_3 \\ \vec{b}_g \\ \vec{b}_a \end{pmatrix} \tag{11-41}$$

Equations (11-38)–(11-40) is a nonlinear measurement model that satisfies the minimal coordinate representation conditions. The last step to get the linearized measurement model is the partial derivative of the above measurement with respect to the variables of the state vector $\vec{x}$. The linearized measurement error model will give us the design matrix H defined as follows:

$$H = \begin{pmatrix} \dfrac{\partial \hat{g}^L}{\partial \vec{x}} \\ \dfrac{\partial \vec{\hat{m}}^L}{\partial \vec{x}} \end{pmatrix} \tag{11-42}$$

where the state vector is defined as follows:

$$H = \begin{pmatrix} \frac{\partial \hat{g}^L(1)}{\partial q_1} & \frac{\partial \hat{g}^L(1)}{\partial q_2} & \frac{\partial \hat{g}^L(1)}{\partial q_3} & 0 & 0 & 0 & \frac{\partial \hat{g}^L(1)}{\partial b_{a_x}} & \frac{\partial \hat{g}^L(1)}{\partial b_{a_y}} & \frac{\partial \hat{g}^L(1)}{\partial b_{a_z}} \\ \frac{\partial \hat{g}^L(2)}{\partial q_1} & \frac{\partial \hat{g}^L(2)}{\partial q_2} & \frac{\partial \hat{g}^L(2)}{\partial q_3} & 0 & 0 & 0 & \frac{\partial \hat{g}^L(2)}{\partial b_{a_x}} & \frac{\partial \hat{g}^L(2)}{\partial b_{a_y}} & \frac{\partial \hat{g}^L(2)}{\partial b_{a_z}} \\ \frac{\partial \hat{g}^L(3)}{\partial q_1} & \frac{\partial \hat{g}^L(3)}{\partial q_2} & \frac{\partial \hat{g}^L(3)}{\partial q_3} & 0 & 0 & 0 & \frac{\partial \hat{g}^L(3)}{\partial b_{a_x}} & \frac{\partial \hat{g}^L(3)}{\partial b_{a_y}} & \frac{\partial \hat{g}^L(3)}{\partial b_{a_z}} \\ \frac{\partial \vec{\hat{m}}^L(1)}{\partial q_1} & \frac{\partial \vec{\hat{m}}^L(1)}{\partial q_2} & \frac{\partial \vec{\hat{m}}^L(1)}{\partial q_3} & 0 & 0 & 0 & 0 & 0 & 0 \\ \frac{\partial \vec{\hat{m}}^L(2)}{\partial q_1} & \frac{\partial \vec{\hat{m}}^L(2)}{\partial q_2} & \frac{\partial \vec{\hat{m}}^L(2)}{\partial q_3} & 0 & 0 & 0 & 0 & 0 & 0 \\ \frac{\partial \vec{\hat{m}}^L(3)}{\partial q_1} & \frac{\partial \vec{\hat{m}}^L(3)}{\partial q_2} & \frac{\partial \vec{\hat{m}}^L(3)}{\partial q_3} & 0 & 0 & 0 & 0 & 0 & 0 \end{pmatrix} \tag{11-43}$$

Given the number of states and the matrix multiplication, differentiating the measurement model in Eqs. (11-38)–(11-40) with respect to the state vector $\vec{x}$ defined in Eq. (11-41) is long. A MATLAB implementation of AHRS using EKF is included in the project's repository. In the provided MATLAB code, the linearized measurement model is calculated using the MATLAB symbolic differentiation tool. The elements of the design matrix H are given as follows:

$$
\begin{aligned}
\frac{\partial \hat{g}^L(1)}{\partial q_1} &= \left(2q_2 + \frac{2q_1q_3}{q_0}\right)\left(a^B_{SFy} - b_{a_y}\right) + \left(2q_3 - \frac{2q_1q_2}{q_0}\right)\left(a^B_{SFz} - b_{a_z}\right)\\
\frac{\partial \hat{g}^L(1)}{\partial q_2} &= -4q_2\left(a^B_{SFx} - b_{a_x}\right) + \left(2q_1 + \frac{2q_2q_3}{q_0}\right)\left(a^B_{SFy} - b_{a_y}\right) + \left(2q_0 - \frac{2q_2{}^2}{q_0}\right)\left(a^B_{SFz} - b_{a_z}\right)\\
\frac{\partial \hat{g}^L(1)}{\partial q_3} &= -4q_3\left(a^B_{SFx} - b_{a_x}\right) - \left(2q_0 - \frac{2q_3{}^2}{q_0}\right)\left(a^B_{SFy} - b_{a_y}\right) + \left(2q_1 - \frac{2q_2q_3}{q_0}\right)\left(a^B_{SFz} - b_{a_z}\right)\\
\frac{\partial \hat{g}^L(1)}{\partial b_{a_x}} &= \left(2q_2{}^2 + 2q_3{}^2\right) - 1\\
\frac{\partial \hat{g}^L(1)}{\partial b_{a_y}} &= \left(2q_0q_3 - 2q_1q_2\right)\\
\frac{\partial \hat{g}^L(1)}{\partial b_{a_z}} &= \left(-2q_1q_3 - 2q_2q_0\right)
\end{aligned}
\tag{11-44}
$$

$$
\begin{aligned}
\frac{\partial \hat{g}^L(2)}{\partial q_1} &= \left(2q_2 - \frac{2q_1q_3}{q_0}\right)\left(a^B_{SFx} - b_{a_x}\right) - 4q_1\left(a^B_{SFy} - b_{a_y}\right) - \left(2q_0 - \frac{2q_1{}^2}{q_0}\right)\left(a^B_{SFz} - b_{a_z}\right)\\
\frac{\partial \hat{g}^L(2)}{\partial q_2} &= \left(2q_1 - \frac{2q_2q_3}{q_0}\right)\left(a^B_{SFx} - b_{a_x}\right) + \left(2q_3 + \frac{2q_1q_2}{q_0}\right)\left(a^B_{SFz} - b_{a_z}\right)\\
\frac{\partial \hat{g}^L(2)}{\partial q_3} &= \left(2q_0 - \frac{2q_3{}^2}{q_0}\right)\left(a^B_{SFx} - b_{a_x}\right) - 4q_3\left(a^B_{SFy} - b_{a_y}\right) + \left(2q_2 + \frac{2q_1q_3}{q_0}\right)\left(a^B_{SFz} - b_{a_z}\right)\\
\frac{\partial \hat{g}^L(2)}{\partial b_{a_x}} &= \left(-2q_1q_2 - 2q_3q_0\right)\\
\frac{\partial \hat{g}^L(2)}{\partial b_{a_y}} &= \left(2q_1{}^2 + 2q_3{}^2\right) - 1\\
\frac{\partial \hat{g}^L(2)}{\partial b_{a_z}} &= \left(2q_1q_0 - 2q_2q_3\right)
\end{aligned}
\tag{11-45}
$$

$$
\begin{aligned}
\frac{\partial \hat{g}^L(3)}{\partial q_1} &= \left(2q_3 + \frac{2q_1q_2}{q_0}\right)\left(a^B_{SFx} - b_{a_x}\right) + \left(2q_0 - \frac{2q_1{}^2}{q_0}\right)\left(a^B_{SFy} - b_{a_y}\right) - 4q_1\left(a^B_{SFz} - b_{a_z}\right)\\
\frac{\partial \hat{g}^L(3)}{\partial q_2} &= -\left(2q_0 - \frac{2q_2{}^2}{q_0}\right)\left(a^B_{SFx} - b_{a_x}\right) + \left(2q_3 - \frac{2q_1q_2}{q_0}\right)\left(a^B_{SFy} - b_{a_y}\right) - 4q_2\left(a^B_{SFz} - b_{a_z}\right)\\
\frac{\partial \hat{g}^L(3)}{\partial q_3} &= \left(2q_1 + \frac{2q_2q_3}{q_0}\right)\left(a^B_{SFx} - b_{a_x}\right) + \left(2q_2 - \frac{2q_1q_3}{q_0}\right)\left(a^B_{SFy} - b_{a_y}\right)\\
\frac{\partial \hat{g}^L(3)}{\partial b_{a_x}} &= \left(2q_2q_0 - 2q_1q_3\right)\\
\frac{\partial \hat{g}^L(3)}{\partial b_{a_y}} &= \left(-2q_2q_3 - 2q_1q_0\right)\\
\frac{\partial \hat{g}^L(3)}{\partial b_{a_z}} &= \left(2q_1{}^2 + 2q_2{}^2\right) - 1
\end{aligned}
\tag{11-46}
$$

The elements of the fourth to sixth rows of the design matrix are given as follows:

$$\begin{aligned}
\frac{\partial \vec{\hat{m}}^{L}(1)}{\partial q_1} &= \left(2q_2 + \frac{2q_1q_3}{q_0}\right)\left(m_x^B\right) + \left(2q_3 - \frac{2q_1q_2}{q_0}\right)\left(m_z^B\right) \\
\frac{\partial \vec{\hat{m}}^{L}(1)}{\partial q_2} &= -4q_2\left(m_x^B\right) + \left(2q_1 + \frac{2q_2q_3}{q_0}\right)\left(m_y^B\right) + \left(2q_0 - \frac{2{q_2}^2}{q_0}\right)\left(m_z^B\right) \\
\frac{\partial \vec{\hat{m}}^{L}(1)}{\partial q_3} &= -4q_3\left(m_x^B\right) - \left(2q_0 - \frac{2{q_3}^2}{q_0}\right)\left(m_y^B\right) + \left(2q_1 - \frac{2q_2q_3}{q_0}\right)\left(m_{gz}^B\right)
\end{aligned} \tag{11-47}$$

$$\begin{aligned}
\frac{\partial \vec{\hat{m}}^{L}(2)}{\partial q_1} &= \left(2q_2 - \frac{2q_1q_3}{q_0}\right)\left(m_x^B\right) - 4q_1\left(m_y^B\right) - \left(2q_0 - \frac{2{q_1}^2}{q_0}\right)\left(m_{gz}^B\right) \\
\frac{\partial \vec{\hat{m}}^{L}(2)}{\partial q_2} &= \left(2q_1 - \frac{2q_2q_3}{q_0}\right)\left(m_x^B\right) + \left(2q_3 + \frac{2q_1q_2}{q_0}\right)\left(m_{gz}^B\right) \\
\frac{\partial \vec{\hat{m}}^{L}(2)}{\partial q_3} &= \left(2q_0 - \frac{2{q_3}^2}{q_0}\right)\left(m_x^B\right) - 4q_3\left(m_y^B\right) + \left(2q_2 + \frac{2q_1q_3}{q_0}\right)\left(m_z^B\right)
\end{aligned} \tag{11-48}$$

$$\begin{aligned}
\frac{\partial \vec{\hat{m}}^{L}(3)}{\partial q_1} &= \left(2q_3 + \frac{2q_1q_2}{q_0}\right)\left(m_x^B\right) + \left(2q_0 - \frac{2{q_1}^2}{q_0}\right)\left(m_y^B\right) - 4q_1\left(m_z^B\right) \\
\frac{\partial \vec{\hat{m}}^{L}(3)}{\partial q_2} &= -\left(2q_0 + \frac{2{q_2}^2}{q_0}\right)\left(m_x^B\right) + \left(2q_3 - \frac{2q_1q_2}{q_0}\right)\left(m_y^B\right) - 4q_2\left(m_z^B\right) \\
\frac{\partial \vec{\hat{m}}^{L}(3)}{\partial q_3} &= \left(2q_1 + \frac{2q_2q_3}{q_0}\right)\left(m_x^B\right) + \left(2q_2 - \frac{2q_1q_3}{q_0}\right)\left(m_y^B\right)
\end{aligned} \tag{11-49}$$

To complete the measurement model, we define the measurement noise covariance matrices as follows:

$$R = \begin{pmatrix} R_g & 0 \\ 0 & R_m \end{pmatrix} \tag{11-50}$$

$$R_m = \begin{pmatrix} r_{mx} & 0 & 0 \\ 0 & r_{my} & 0 \\ 0 & 0 & r_{mz} \end{pmatrix} \tag{11-51}$$

$$R_g = \begin{pmatrix} r_{gx} & 0 & 0 \\ 0 & r_{gy} & 0 \\ 0 & 0 & r_{gz} \end{pmatrix} \tag{11-52}$$

The measurement noise matrices in Eqs. (11-50)–(11-52) are part of the parameters of the fusion algorithm to be tuned for best performance.

11.3 Pedestrian Dead Reckoning (PDR)

Inertial navigation can be described as odometry navigation, where an incremental measurement is integrated over time to calculate the location of an object. The standard inertial navigation equations (i.e., mechanization equations) described in Chapter 4 are suitable for constrained motion such as wheeled platforms, ships, or aerial vehicles. Standard mechanization motion equations are not suitable for some other motion dynamics and platforms due to the highly unconstrained stochastic motion patterns, such as running or climbing a stair, which is common in legged and humanoid platforms. These platforms cannot be precisely located using standard inertial motion equations. Therefore, other inertial-based processing models can be used for the legged and humanoid platforms. In this section, we will discuss inertial-based pedestrian dead-reckoning (PDR). Instead of the standard inertial system model where we integrate raw accelerometer and gyroscope data, we apply pattern recognition to detect the steps of the platform and the direction of motion (i.e., orientation or attitude). In the previous sections, we already discussed how to estimate the platform orientation precisely and smoothly using AHRS. In the next section, we will discuss how steps can be detected and how PDR works.

11.3.1 PDR System Model

PDR navigation systems use inertial sensors (mainly accelerometers) for step counting as a measure of forward motion instead of direct integration of acceleration data like in the conventional inertial system model. A step is the movement of one foot while the other foot remains stationary, whereas a stride is the successive movement of both feet. The person's step length can be projected in the direction of motion to estimate the person's new position. To be feasible for pedestrian use, sensors must be small, lightweight, low power, and low cost. Thus, micro-electromechanical systems (MEMS) inertial sensors are used. In the literature, some PDR algorithms use shoe-mounted sensors to benefit from zero velocity updates that happen when the person's foot touches the ground during a walking step. These zero velocity updates are used as an external measurement update that greatly helps in correcting inertial drifts and estimating inertial sensor errors. The general steps of PDR algorithm are as follows:

- Heading angle estimation
- Step detection
- Step length estimation
- Position update

A general block diagram of PDR is shown in Figure 11.6. The following subsections will explain each block of Figure 11.6. Pedestrian walking patterns generate peaks in acceleration, as shown in Figure 11.7. Therefore, step detection can be implemented using a peak detector applied to the accelerometer data. To avoid the effects of sensor tilt and body swing, instead of using a specific acceleration axis or component, the magnitude of the acceleration can be used. The peak detection algorithm can be applied to the magnitude of the acceleration as follows:

$$a_r = \sqrt{\left(a_{SFX}^B\right)^2 + \left(a_{SFY}^B\right)^2 + \left(a_{SFZ}^B\right)^2} \tag{11-53}$$

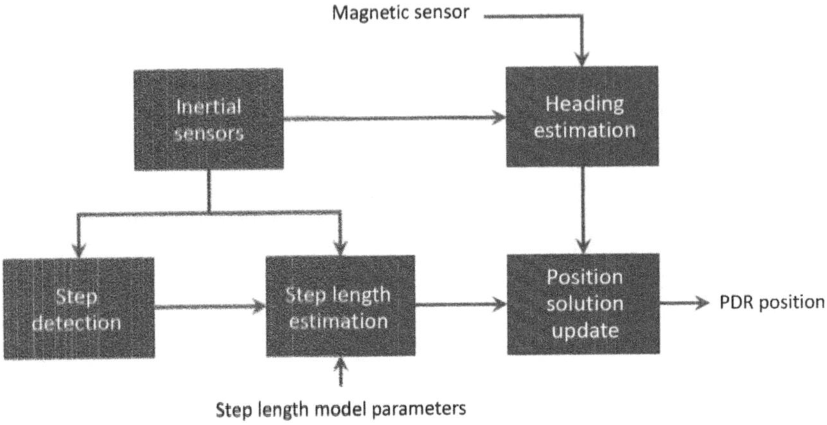

Figure 11.6 PDR-general block diagram.

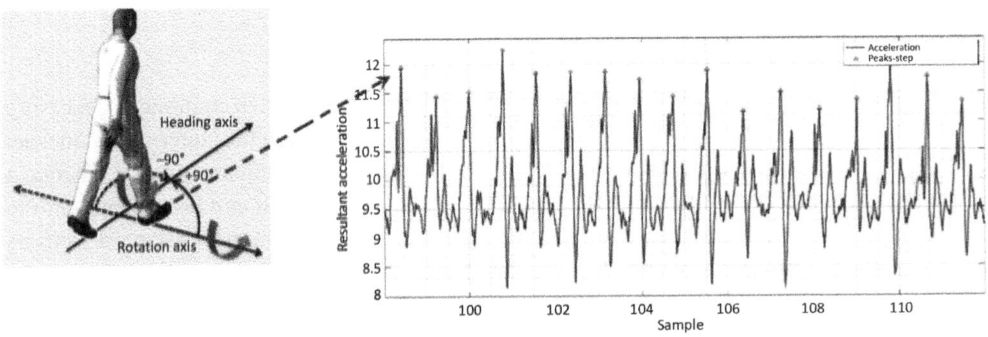

Figure 11.7 Peak-step detection using foot-mounted acceleration data.

In another configuration, a peak detector can be applied to the vertical component of the accelerometer measurements calculated by the following equations:

$$\begin{aligned} a_v &= \vec{a}^{\,L}_{SF}[3] \\ \vec{a}^{\,L}_{SF} &= C^L_B \vec{a}^{\,B}_{SF} \end{aligned} \tag{11-54}$$

where

$\vec{a}^{\,L}_{SF}[3]$ is the third component of the vector $\vec{a}^{\,L}_{SF}$
C^L_B is the direction cosine matrix calculated from an AHRS module.

Applying a peak detector to $a_r(t)$ or $a_v(t)$ will detect the steps of the platform as indicated by the small red triangles in Figure 11.7. The small red triangles on a peak correspond to step detection. Having the step detected and the heading estimated using the AHRS module, the next step is to estimate the step length. Step length estimation can be performed in different ways. The simplest one is to assume a fixed step length. This simple approach works well if the platform has a fixed known step length. If human or animal platforms are considered, gait analysis can be used to estimate the step length based on the skeleton dimensions and body mechanics. Different approaches, such as artificial neural networks, can be used to estimate step length. Step length can also be estimated if a precise positioning system like GPS is available for a portion of the trajectory. If the distance traveled by the platform is known or can be calculated using a GPS or a visual odometry device, and the number of steps is calculated, the step length for this platform can be calibrated. Once a peak is detected, a step forward is applied to update the pedestrian's position. The pedestrian's step length l_k and the heading ψ_k calculated from the AHRS are used to calculate the 2D pedestrian's position $(n\ e)^T$ using the following model:

$$n_k = n_{k-1} + l_k \cos(\psi_k) \tag{11-55}$$

$$e_k = e_{k-1} + l_k \sin(\psi_k) \tag{11-56}$$

11.3.2 Linearized System Model

We need the linearized error system model to fuse other sensors with PDR. We will assume that the AHRS is an independent component that has its own fusion modules. Starting from the prediction equations of the PDR system, we have:

$$n_k - n_{k-1} = l_k \cos(\psi_k) \tag{11-57}$$

$$\frac{(n_k - n_{k-1})}{T_k} = \left(\frac{l_k}{T_k}\right) \cos(\psi_k) \tag{11-58}$$

$$e_k - e_{k-1} = l_k \cos(\psi_k) \tag{11-59}$$

$$\frac{(e_k - e_{k-1})}{T_k} = \left(\frac{l_k}{T_k}\right) \sin(\psi_k) \tag{11-60}$$

where T_k is the step duration.

We can approximate the time derivative of the position vector $(n\ e)^T$ and the moving speed v_k of the platform as follows:

$$\dot{n}(t) \cong \frac{(n_k - n_{k-1})}{T_k} \tag{11-61}$$

$$\dot{e}(t) \cong \frac{(e_k - e_{k-1})}{T_k} \tag{11-62}$$

$$v_k \cong \frac{l_k}{T_k} \tag{11-63}$$

We can assume that due to the AHRS errors and step length estimation errors ψ_k and v_k will be contaminated by bias errors as follows:

$$\hat{v}(t) = v(t) - b_v(t) \tag{11-64}$$

$$\hat{\psi}(t) = \psi(t) - b_\psi(t) \tag{11-65}$$

where $b_v(t)$, $b_\psi(t)$ are biases in speed and azimuth input signals.

We can assume a random walk model for the biases as follows:

$$\dot{b}_v(t) = w_v(t) \tag{11-66}$$

$$\dot{b}_\psi(t) = w_\psi(t) \tag{11-67}$$

where $w_v(t)$ and $w_\psi(t)$ are zero-mean Gaussian noise.

Given this set of approximations and assumptions, the PDR state vector $\vec{x}(t)$ and control vector $\vec{u}(t)$ can be defined as follows:

$$\vec{x}(t) = \begin{pmatrix} e(t) \\ n(t) \\ b_v(t) \\ b_\psi(t) \end{pmatrix}, \quad \vec{u}(t) = \begin{pmatrix} \hat{v}(t) \\ \hat{\psi}(t) \end{pmatrix} \tag{11-68}$$

And the overall system model in the form $\dot{\vec{x}}(t) = f\left(\vec{x}(t), \vec{u}(t)\right) + \vec{w}(t)$ can be formulated as follows:

$$\dot{\vec{x}}(t) = \begin{pmatrix} \dot{e}(t) \\ \dot{n}(t) \\ \dot{b}_v(t) \\ \dot{b}_\psi(t) \end{pmatrix} = \begin{pmatrix} \hat{v}(t)\sin\hat{\psi}(t) \\ \hat{v}(t)\cos\hat{\psi}(t) \\ 0 \\ 0 \end{pmatrix} + \begin{pmatrix} w_e(t) \\ w_n(t) \\ w_v(t) \\ w_\psi(t) \end{pmatrix} \tag{11-69}$$

where $w_e(t)$, $w_n(t)$, $w_v(t)$, $w_\psi(t)$ are zero-mean random noise for speed and azimuth control signals.

Therefore, the linearized error model of this PDR system is given by the partial derivative of the system model with respect to the state vector variables, which gives the following system model:

$$\delta\dot{\vec{x}}(t) = \begin{pmatrix} \delta\dot{e}(t) \\ \delta\dot{n}(t) \\ \delta\dot{b}_v(t) \\ \delta\dot{b}_\psi(t) \end{pmatrix} = \begin{pmatrix} -\sin\hat{\psi}(t)\delta b_v(t) - \hat{v}(t)\cos\hat{\psi}(t)\delta b_\psi(t) \\ -\cos\hat{\psi}(t)\delta b_v(t) + \hat{v}(t)\sin\hat{\psi}(t)\delta b_\psi(t) \\ 0 \\ 0 \end{pmatrix} + \begin{pmatrix} w_e(t) \\ w_n(t) \\ w_v(t) \\ w_\psi(t) \end{pmatrix} \tag{11-70}$$

In matrix form, the linearized system model in the form $\delta\dot{x}(t) = F(t)\delta x(t) + G(t)w(t)$ is given by the following equation:

$$\begin{pmatrix} \delta\dot{e}(t) \\ \delta\dot{n}(t) \\ \delta\dot{b}_v(t) \\ \delta\dot{b}_\psi(t) \end{pmatrix} = \begin{pmatrix} 0 & 0 & -\sin\hat{\psi}(t) & \hat{v}(t)\cos\hat{\psi}(t) \\ 0 & 0 & -\cos\hat{\psi}(t) & \hat{v}(t)\sin\hat{\psi}(t) \\ 0 & 0 & 0 & 0 \\ 0 & 0 & 0 & 0 \end{pmatrix} \begin{pmatrix} \delta e(t) \\ \delta n(t) \\ \delta b_v(t) \\ \delta b_\psi(t) \end{pmatrix} + \begin{pmatrix} \delta_e & 0 & 0 & 0 \\ 0 & \delta_n & 0 & 0 \\ 0 & 0 & \delta_v & 0 \\ 0 & 0 & 0 & \delta_\psi \end{pmatrix} \vec{w}_u(t) \tag{11-71}$$

where

$\vec{w}_u(t)$ is a unity-variance zero-mean Gaussian noise
δ_v and δ_ψ are noise standard deviations of speed and azimuth measurements.

From Eq. (11-71), the transition matrix $F(t)$ is given by:

$$F(t) = \begin{pmatrix} 0 & 0 & -\sin\hat{\psi}(t) & \hat{v}(t)\cos\hat{\psi}(t) \\ 0 & 0 & -\cos\hat{\psi}(t) & \hat{v}(t)\sin\hat{\psi}(t) \\ 0 & 0 & 0 & 0 \\ 0 & 0 & 0 & 0 \end{pmatrix} \tag{11-72}$$

Equations (11-69) and (11-71) form a basic simple PDR system model with a limited set of state variables. When we fuse other sensors/systems measurements with PDR, we may find out that we need to model more states. In this case, we can augment the state vector and state model accordingly like the case where we added two additional variables in the GPS tightly coupled fusion in Chapters 9 and 10 to model the receiver's clock bias and drift. Once we have the system model, the measurement model, transition matrix $F(t)$ and noise shaping matrix $G(t)$, we can fuse the PDR system with other sensors using EKF or PF or any of their variations as described in Chapters 9 and 10. To fuse a specific sensor measurement, we will need the measurement model of the sensor to apply the sensor's measurement update step. If EKF is used, and the measurement model is nonlinear, we will need to linearize the measurement model, which we have demonstrated for different sensors in Chapters 9 and 10, like GPS, RADAR, LiDAR, and imaging sensors.

11.4 MATLAB Project Sample of an AHRS Using EKF

The projects' repository includes a sample MATLAB software that implements AHRS using EKF in both loosely coupled and tightly coupled. The code demonstrates the detailed implementation steps and displays the raw inertial and magnetic sensor data, orientation errors, and error covariances of the filter. Figure 11.8 shows the raw inertial data (gyroscope and accelerometer). Calibrated magnetic sensor measurements are displayed in Figure 11.9. If the raw gyroscope measurements are processed using Eq. (11-9), the solution will be smooth but drifting as seen in Figure 11.10.

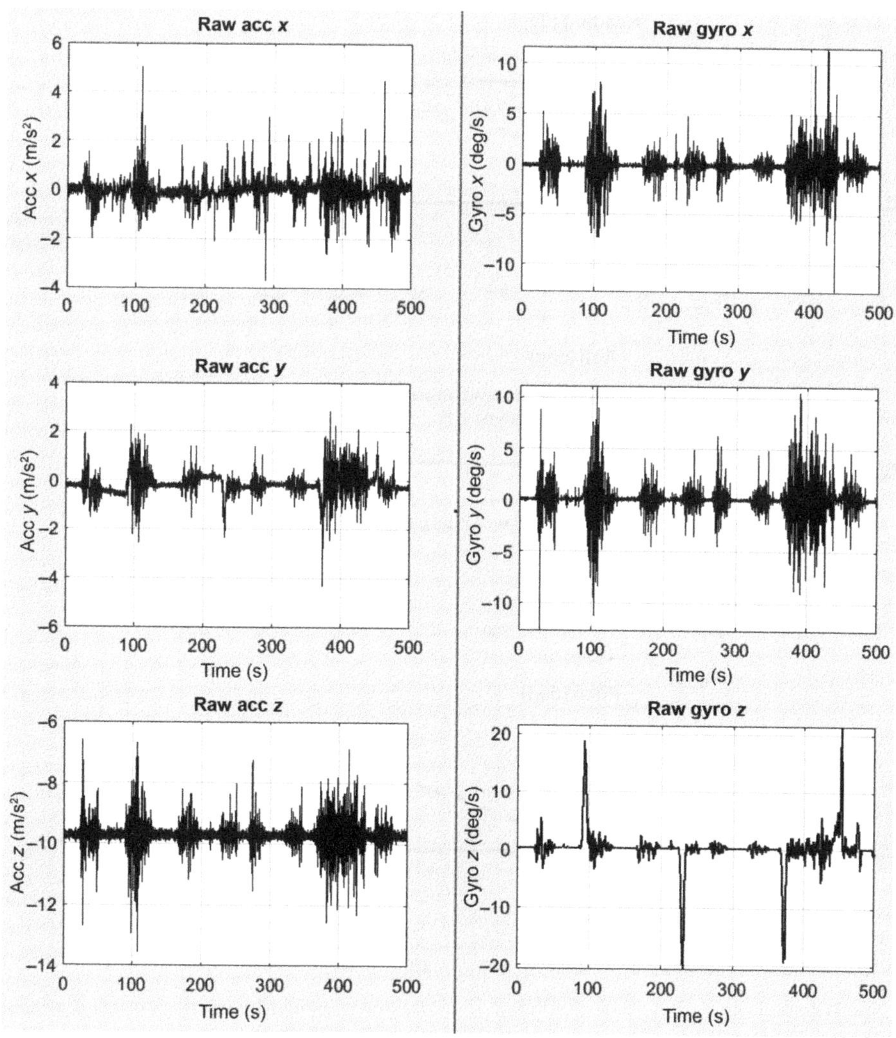

Figure 11.8 Raw inertial sensors data.

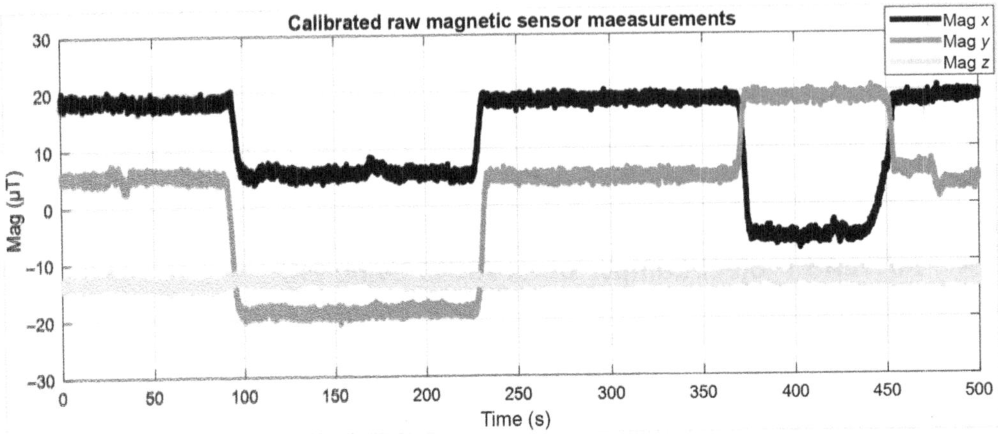

Figure 11.9 Calibrated raw magnetic sensor measurements.

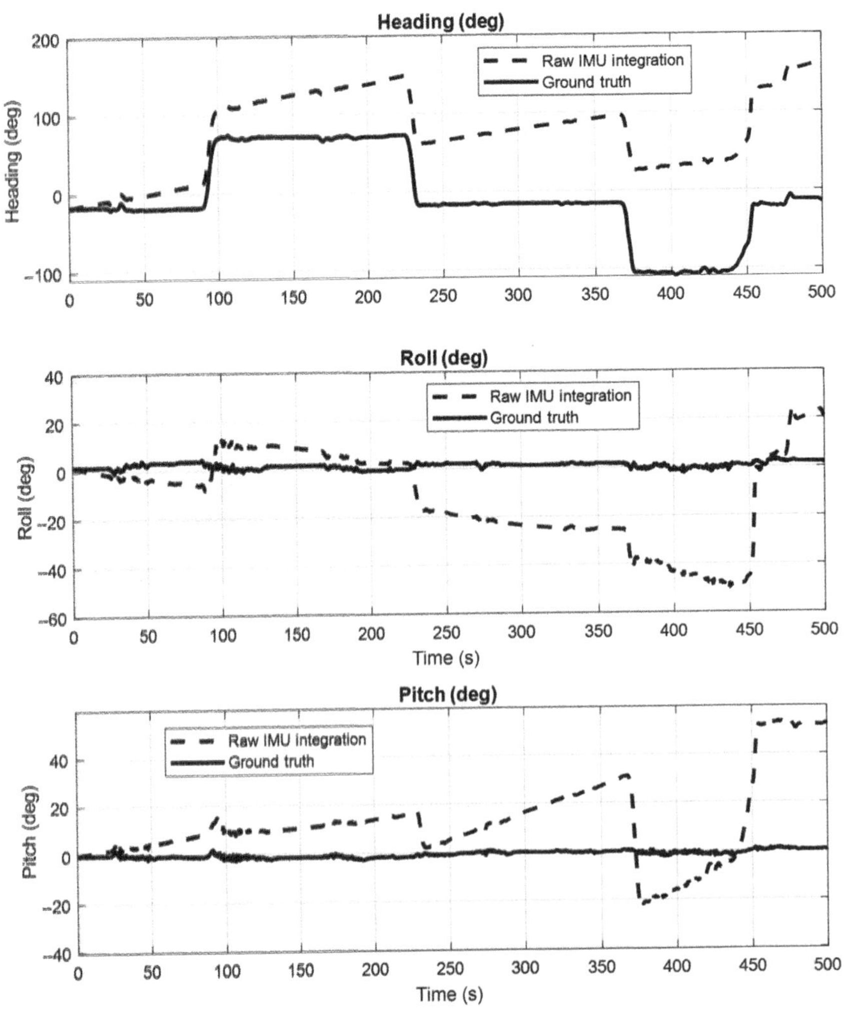

Figure 11.10 Orientation drifts of raw inertial sensors integration (INS).

The EKF-AHRS fusion results are displayed in Figure 11.11, which shows a smooth and nondrifting solution compared to roll/pitch calculated from accelerometer measurements and heading calculated from the magnetometer. The orientation errors are shown in both Euler angle form (Figure 11.12) and quaternion form (Figure 11.13). Figure 11.13 shows the EKF estimated covariance of the solution, which starts with large values and converges to small values as the measurements' updates accumulate. The same pattern is shown in Figure 11.14 and Figure 11.15 which displays the gyroscope biases convergence with their covariance. Readers are encouraged to inspect the code to get a closer look at the implementation details of the EKF-based AHRS module.

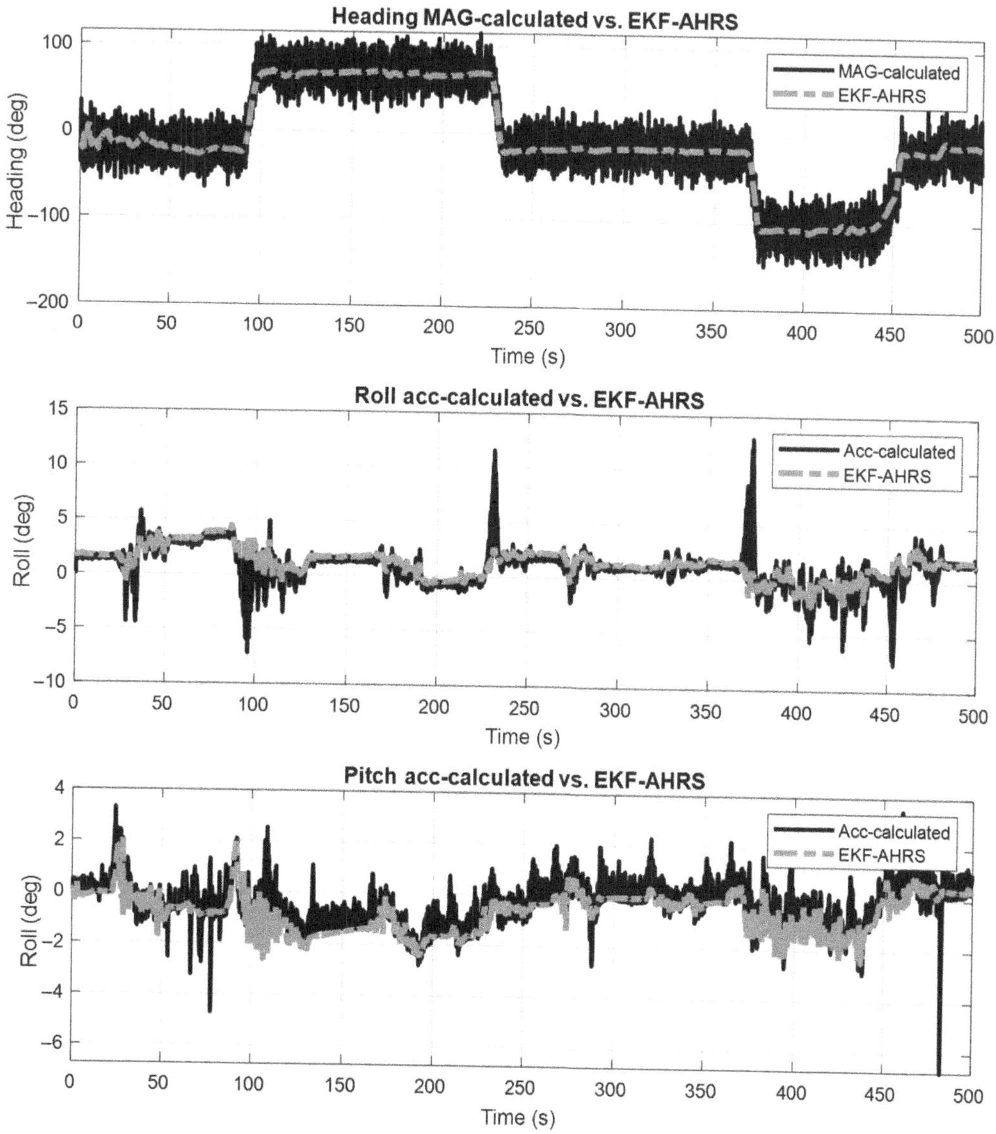

Figure 11.11 AHRS output compared to magnetometer/accelerometer-based orientation results.

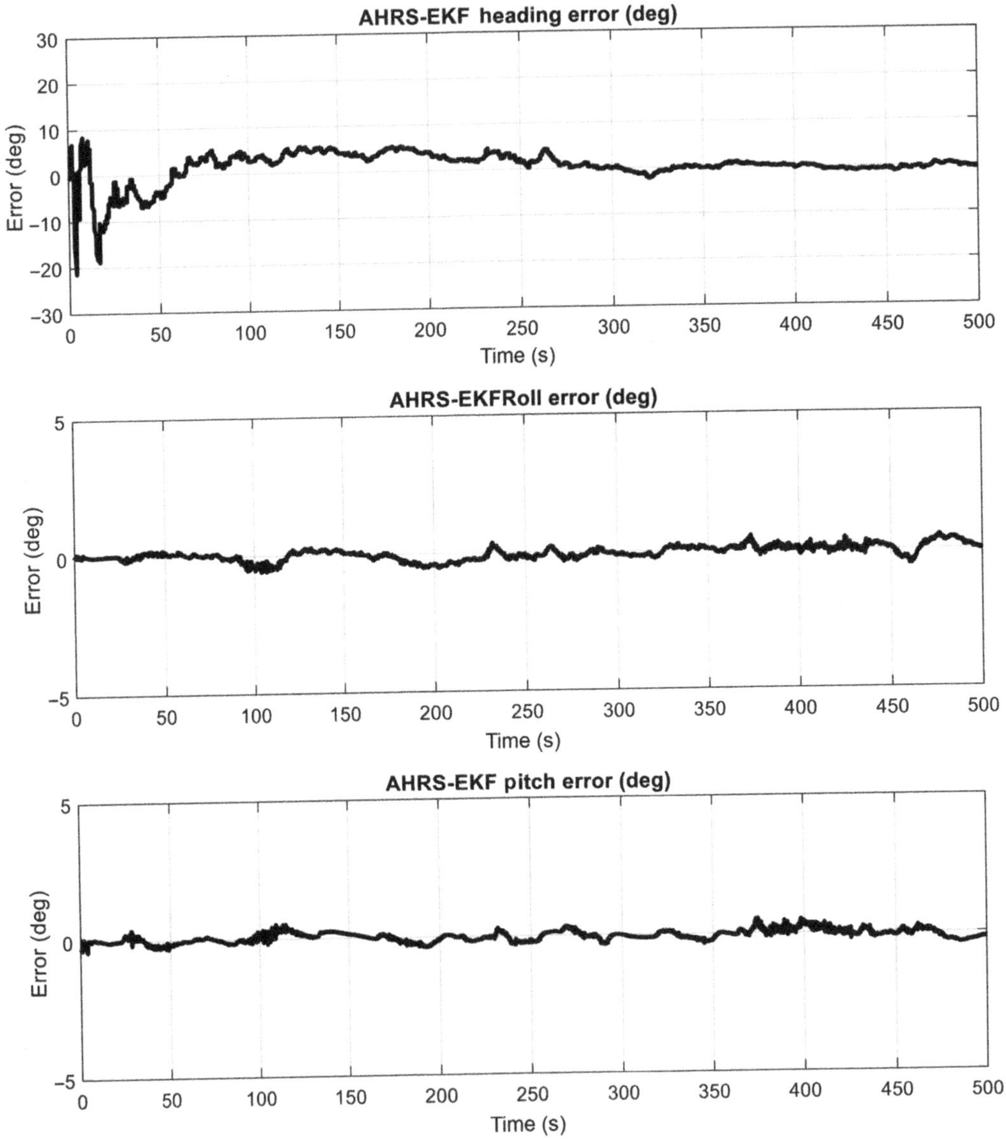

Figure 11.12 Orientation error after applying AHRS-EKF fusion.

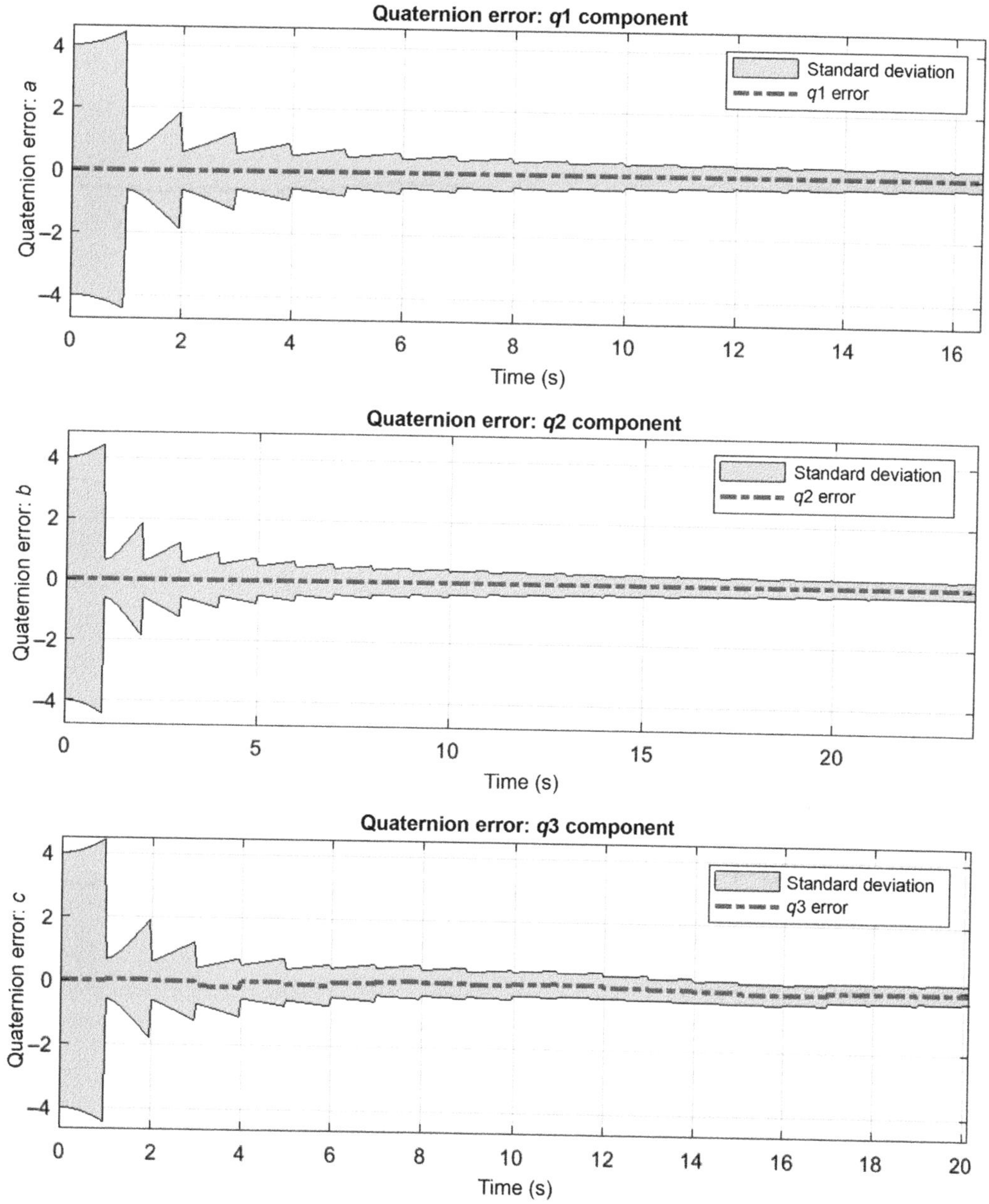

Figure 11.13 Orientation error with error envelope ($\pm 2\sigma$).

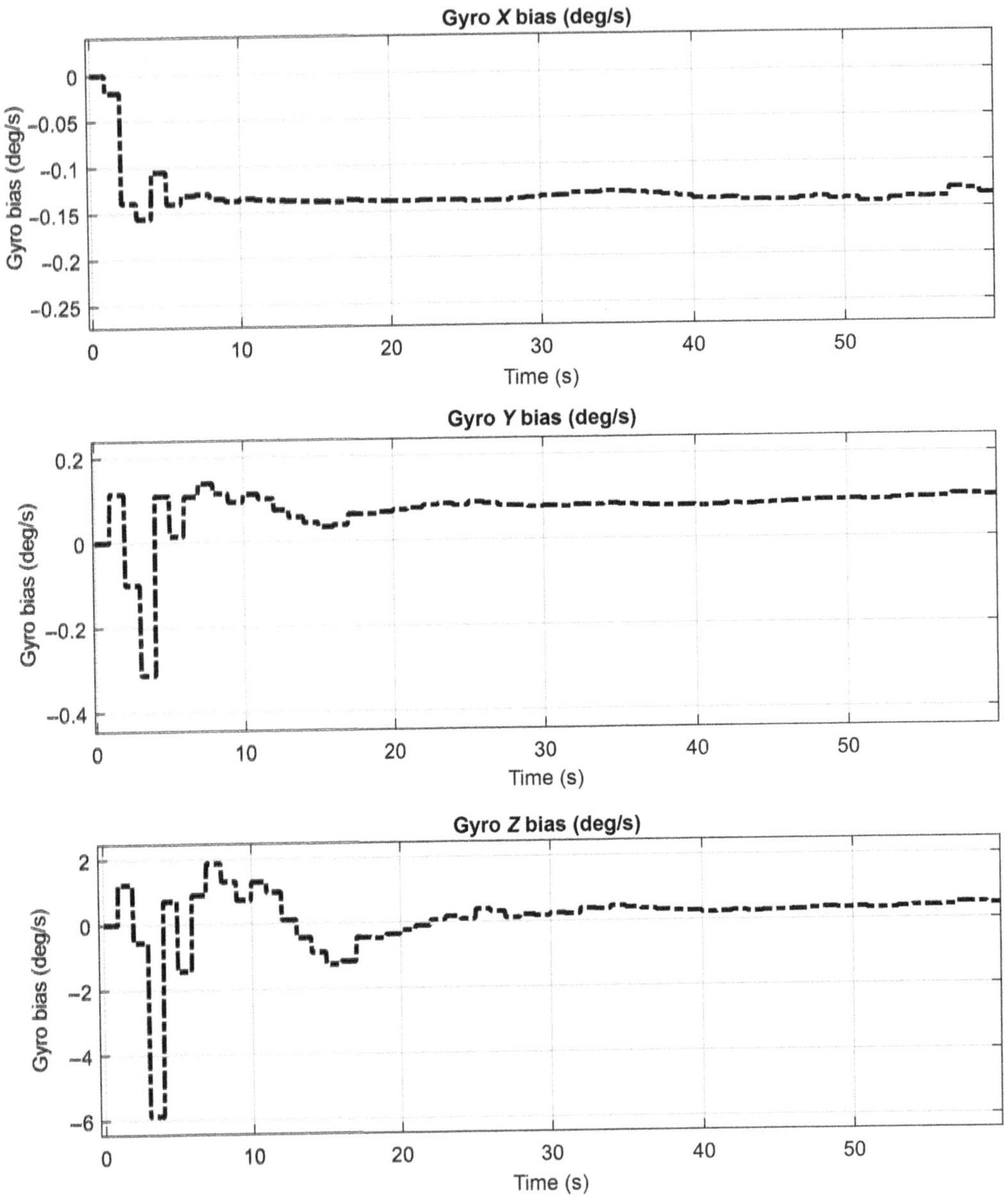

Figure 11.14 Gyroscope biases.

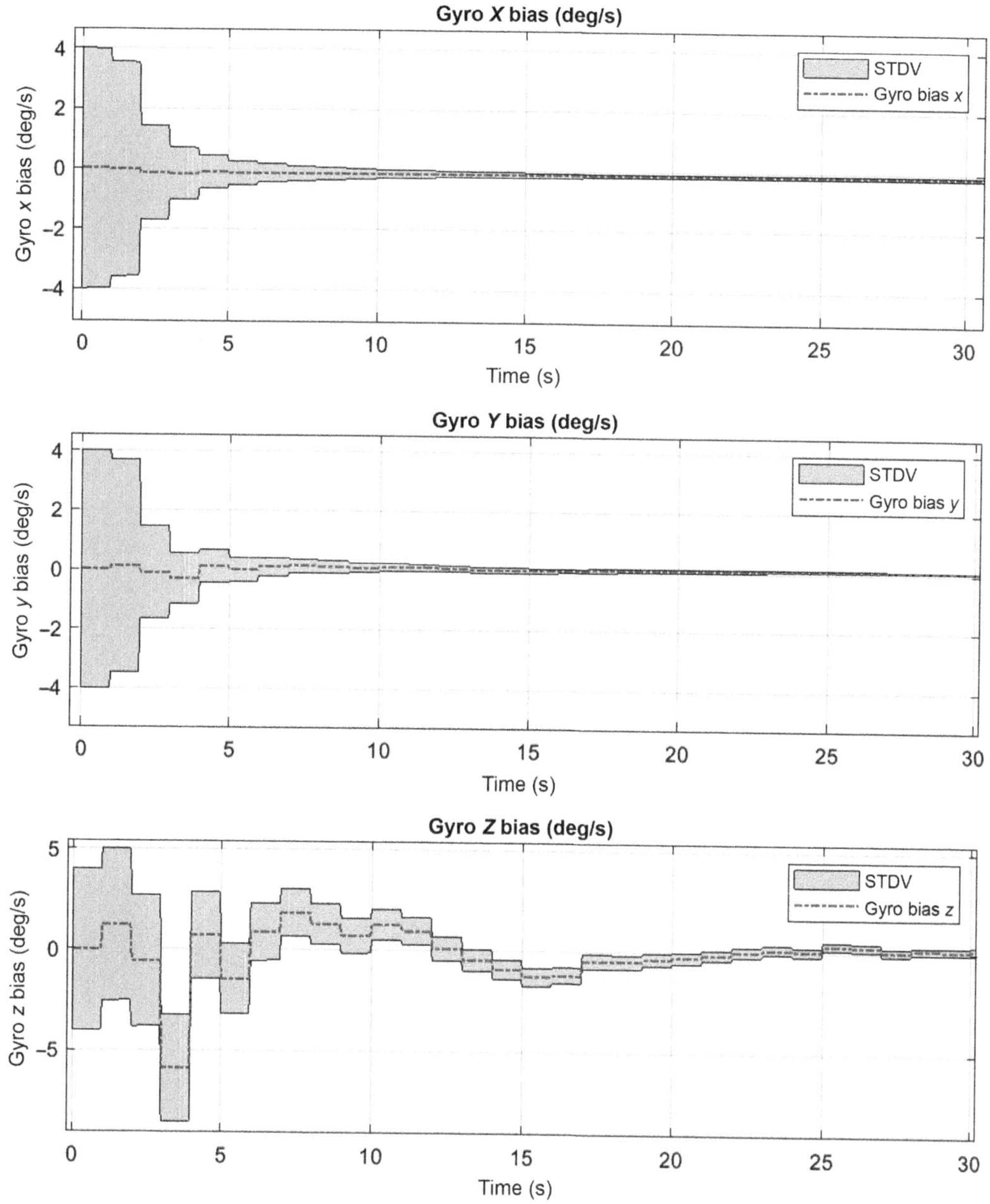

Figure 11.15 Gyroscope biases with error envelope ($\pm 2\sigma$).

12

Learning-Based Fusion Methods

12.1 Introduction

Machine learning (ML) and artificial intelligence (AI) are developing rapidly, and they are finding their way into several fields, including navigation and mapping applications. A brief introduction to AI and ML and the differences between them were presented in Chapter 3. Chapter 3 explained the logical and mathematical background of common AI and ML methods such as artificial neural networks (ANN), deep learning (DL), convolutional neural networks (CNN), recurrent artificial neural networks (RANN), recurrent convolutional neural network (RCNN), genetic algorithms (GA), fuzzy systems (FS), decision trees (DT), and reinforcement learning (RL). In this chapter, we will discuss the application of AI and ML in the field of multi-sensor positioning, navigation, and mapping. AI/ML models can be thought of as advanced, flexible computational tools that provide highly cognitive, learning and reasoning abilities beyond conventional signal processing and estimation algorithms. Applying ML and AI in navigation and mapping is currently mostly limited to higher-level tasks such as object recognition and semantic segmentation. Low-level positioning and mapping estimation are still dominated by conventional filtering and optimization techniques. However, recently, researchers have started to explore the application of AI and ML in low-level sensor fusion, where raw multi-sensor measurements are processed to provide better positioning and mapping accuracy.

Low-level sensor fusion is still dominated by conventional filtering and optimization methods where the fusion problem is formulated as a cost function (also known as loss function) that is minimized using least-squares or gradient-descent methods. Under assumptions such as linearity of system and measurement models and Gaussian distribution of stochastic noise, applying Bayesian rules or setting the first derivative of the cost function with respect to the parameter of interest to zero usually leads to a closed-form analytical solution. If an analytical solution is too complex to derive, the cost function is linearized and minimized iteratively using iterative optimization algorithms such as Newton–Gauss or Levenberg–Marquardt optimization. Conventional filtering methods assume linear or at least linearizable models, and they have little or no room for incorporating higher-level reasoning. In addition, the conventional methods depend on models that have been mostly derived under simplification assumptions and approximations that are not always satisfied and sometimes the cost function is not differentiable.

Note that both conventional filtering/optimization methods and ML models, especially ANN, CNN, RCNN, and DL, are optimized using a gradient-descent approach. They are both optimizing a cost or a loss function. Backpropagation itself, which is the core learning algorithm in neural

Sensor Fusion Approaches for Positioning, Navigation, and Mapping: How Autonomous Vehicles and Robots Navigate in the Real World: With MATLAB Examples, First Edition. Mohamed M. Atia.
© 2025 The Institute of Electrical and Electronics Engineers, Inc. Published 2025 by John Wiley & Sons, Inc.

networks, is based on gradient-descent optimization (review Chapter 3 for a detailed derivation of the backpropagation algorithm). The main distinction between conventional fusion methods and AI/ML methods is in the architecture of the models. AI and ML models' architectures are far more complex compared to conventional filtering and optimization techniques. The complexity of the architectures of AI and ML enables them to incorporate higher-level reasoning and cognitive capabilities, which are difficult and sometimes impossible to include in conventional signal processing and estimation algorithms. This advantage comes at the expense of computational cost, which is a critical factor in navigation systems where power consumption is a major concern. These limitations of conventional fusion methods left several cases and scenarios difficult to solve. This opened the discussion about benefiting from ML and AI tools in addressing these limitations. This chapter will discuss state-of-the-art approaches and designs that utilize ML and AI tools and models in low-level sensor fusion for enhanced positioning, navigation, and mapping.

12.2 Learning-Based Approaches

A general block diagram of a low-level conventional sensor fusion framework is shown Figure 12.1. The diagram covers almost any sensor fusion configuration. Given any number of sensors, designers, and engineers can design a fusion system by implementing the following steps:

1) Defining a system state $\vec{x}$.
2) Choosing one or more sensors as control inputs $\vec{u}_k$.
3) Defining a system prediction model $\vec{x}_{k+1} = f\left(\vec{x}_k, \vec{u}_k\right)$.

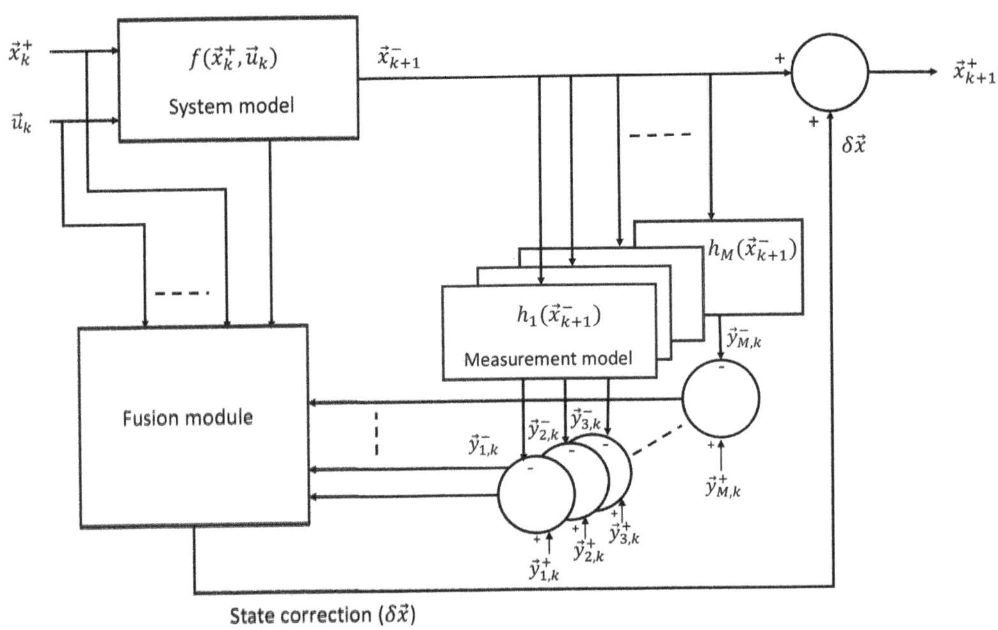

Figure 12.1 General conventional sensor fusion engine diagram.

4) Choosing the sensors/systems that will be used as measurements $\vec{y}_{i,k}$.
5) Defining a measurement model $\vec{y}_{i,k} = h_i\left(\vec{x}_k\right)$ for each measurement.
6) Linearizing the system and measurement models (if needed).
7) Choosing a fusion algorithm.
8) Collecting time-synchronized data.
9) Testing the fusion system on the collected data.
10) Tuning the fusion algorithm parameters such as noise covariance matrices.

In Figure 12.1, the superscript "−" means a predicted state value or a predicted measurement value and the superscript "+" means a corrected state value or an actual observed measurement value. The key blocks in this diagram are the system model, the measurement model, and the fusion module. The methods and algorithms we discussed so far in this book that address these three blocks were conventional filter-based or optimization-based methods. There are different approaches where AI and ML can be applied to improve low-level sensor fusion.

One approach to developing learning-based fusion is to utilize AI and ML to augment or replace one or more of the blocks of the conventional fusion framework depicted in Figure 12.1. If the analytical derivation of the system model or the measurement model is complex and inaccurate, or if the conventional fusion algorithm fails under challenging scenarios, ML and AI can be used to build alternative models. For example, conventional inertial mechanization equations fail drastically when applied to pedestrian motion due to the complex, rapid dynamics of pedestrian motion patterns. An alternative system model that is based on AI and/or ML can replace conventional inertial mechanization equations to achieve better pedestrian tracking. In a similar fashion, AI and ML models can augment or replace the conventional fusion module or measurement models.

In another approach, AI and ML models can be used to detect additional constraints that can be applied to conventional fusion frameworks to improve accuracy. AI and ML models can be trained to recognize objects, detect patterns, and predict behaviors. These abilities can provide additional constraints to conventional fusion algorithms. For example, an AI/ML-based computer vision module that can recognize lanes, road geometry, and/or drivable regions can be used to enhance the accuracy of conventional fusion algorithms by limiting the positioning estimation to the lane boundaries, road segments, or drivable regions. Another example is the utilization of semantic segmentation. SLAM (simultaneous localization and mapping) algorithms assume a static environment. Semantic segmentation can detect dynamic objects, which improves conventional SLAM algorithms by rejecting features that belong to a dynamic object.

A different approach utilizes ML and AI models to replace the conventional system/measurement model and the fusion module entirely and perform the fusion directly on the raw sensor data in one block that consists entirely of AI and/or ML models. This approach is called an end-to-end approach, and it is illustrated in Figure 12.2. One of the main advantages of the end-to-end approach is that it does not need separate models for prediction or measurements. In addition, it can automatically learn complex calibration parameters from data, such as camera intrinsics or inertial sensor errors. The end-to-end approach has been recently investigated and applied in optical flow, depth estimation, odometry/SLAM, and lane-level navigation using multisensors, including inertial sensors, imaging sensors, active ranging sensors, and wireless positioning systems.

Regardless of the approach we adopt to develop learning-based fusion frameworks, we need to be aware of the potential challenges and risks when we apply AI and ML to solve a specific problem. First, conventional system models and measurement models are usually computationally efficient, explainable, and linearizable. In case of system failures, it is always possible to trace and understand

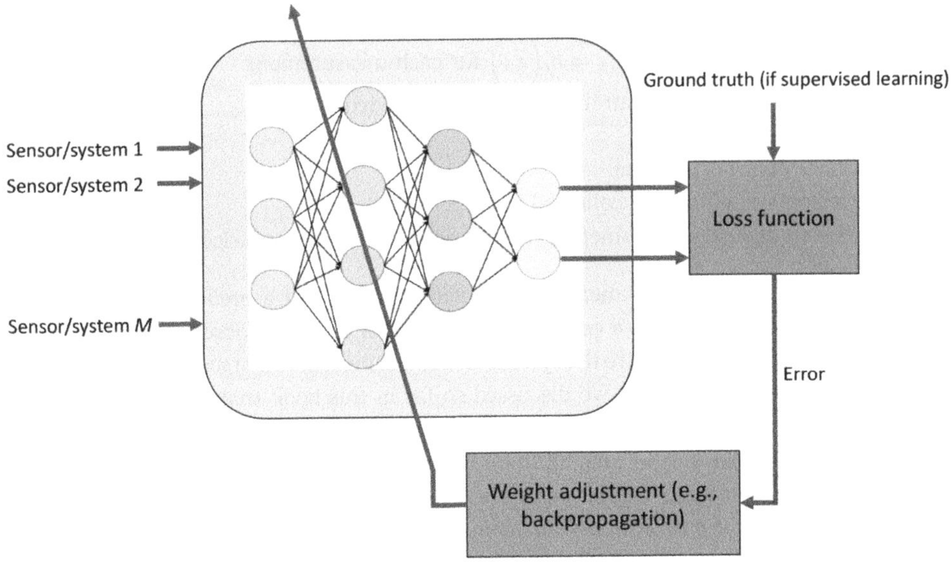

Figure 12.2 End-to-end learning-based fusion approach.

the causes of the failures and to fix them. On the other side, ML/AI models are generally more complicated, time-consuming to train, computationally expensive, difficult to linearize, and difficult to explain. Second, the accuracy of the ML/AI models depends on the training data. Trained models may perform very well on similar data but fail drastically if tested on different data. This is known as overfitting, which affects the generalizability of ML/AI models. So far, the only viable solution to enhance the generalizability of ML/AI is to train the models on large, diverse data sets. However, in many applications, it may be impossible to obtain or generate data sets that cover all possible scenarios. In this chapter, we will briefly discuss some ideas where ML and AI can be applied to develop better low-level sensor fusion methods for positioning, navigation, and mapping. It is recommended that the reader reviews Chapter 3 before continuing in the remainder of this chapter. Chapter 3 explains conventional filtering/optimization and AI/ML methods and tools.

12.3 Formulation of Loss Function

In all ML and AI models, a loss function is needed to guide the optimization of the weights and the parameters of the learning framework. In Chapter 3, we have shown how the backpropagation mechanism that is used to train neural networks starts with the derivative of the loss function and the derivative of the neuron activation function. Therefore, the loss function plays a vital role in the success of the model training and performance since it is the primary guidance for learning/adjusting the AI/ML parameters (e.g., neural network weights) to achieve the desired objective. The loss function is commonly a squared error between a predicted value and a desired or a target value. For example, in a positioning problem, a loss function could be the error between an estimated pose (translation and rotation) and a ground-truth pose. However, in more complex cases which involve multiple sensors, various methods and formulas can be used. More details about the loss function are discussed in the next sub-sections.

12.3.1 Direct Loss Functions

The loss function is commonly a Euclidean distance between a predicted output y and a desired output d using the squared norm $\|.\|^2$ operator as follows:

$$e = 0.5(d - y)^2 \tag{12-1}$$

The overall error over multiple n epochs (e.g., an entire trajectory) is given as a sum of the squared error as follows:

$$e = 0.5\sum_{i=1}^{n}(d_i - y_i)^2 \tag{12-2}$$

As poses consist of rotation and translation, we must represent the rotation in the minimal coordinate form (review Chapter 10 for the definition of minimal coordinates). Therefore, the direction cosine matrix, which belongs to the $SO(3)$ group, for example, is not suitable for the cost function in Eq. (12-2) because it cannot be used to calculate an Euclidean error. The $SO(3)$ group is a Lie Group where only multiplication and inversion operators form a continuous space. Euclidean distance is not defined in Lie Group. However, as we showed in Chapter 1, the $SO(3)$ group forms a manifold where infinitesimally small rotations can be approximated using a tangent space. This tangent space is Euclidean, where Euclidean distance is defined and, hence, a squared error can be calculated. The tangent space can be obtained from the $SO(3)$ using the exponential mapping. The exponential mapping maps the direction cosine matrix $C_B^L(\Delta t)$ with respect to an identity matrix (zero-rotation) over small time Δt to a 3D rotation vector $\vec{\phi}(t)$ as follows:

$$C_B^L(t) = e^{\left(\vec{\phi}(t)\times\right)} \tag{12-3}$$

Using the rotation vector $\vec{\phi}(t)$, we can calculate an Euclidean distance (error) between two rotations assuming that the error is small enough such that the local Euclidean assumption on the manifold holds. In ANN-based ML systems that estimate a pose, the network output can be designed to directly output 6D pose vector consisting of a 3D rotation vector $\vec{\hat{\phi}}$ and a 3D translation vector $\vec{\hat{T}}$ as pose $\left(\vec{\hat{\phi}}\ \vec{\hat{T}}\right)^T$ which can be directly compared to a ground truth pose $\left(\vec{\phi}\ \vec{T}\right)^T$ to calculate an error as follows:

$$e = 0.5\sum_{i=1}^{n}\gamma\left\|\vec{\phi} - \vec{\hat{\phi}}\right\|^2 + \beta\left\|\vec{T} - \vec{\hat{T}}\right\|^2 \tag{12-4}$$

where γ and β are weighting factors.

In the above case, we have a one-to-one mapping between the predicted output and the desired ground truth. In some scenarios, like comparing entire trajectories against a ground truth trajectory at once, we may not have this one-to-one link between estimated trajectory epochs and ground truth trajectory epochs. In this case, an alignment process is needed to align the two trajectories before we can calculate an overall error. After the alignment of the two trajectories, the error can be calculated either as absolute pose error (APE) or relative pose error (RPE). For more information about APE and RPE methods, readers can refer to [1].

12.3.2 Constraints Loss Function

The above loss function is simple and direct, which facilitates the backpropagation learning in neural networks. Recall from Chapter 3 that in the neural network training, the first layer of the backpropagation requires the derivative of the loss function with respect to the output of the network. This derivative is evaluated at the current output value of the network, whatever this value is. That is the reason we must initialize the weights, so we have a prediction output to calculate the derivative (Jacobians) at a specific value. If the network output is defined as y and the desired output is d, the simple loss function $J = \frac{1}{2}(d-y)^2$ makes the derivative of J with respect to the output y easy (just $(d-y)$). If the loss function is more sophisticated than this simple direct squared error formula, the derivative of the loss function must be calculated to run the backpropagation algorithm.

More sophisticated loss functions are defined in the form of constraints. Constraints are derived from the problem domain, and they can be used to formulate strong loss functions, especially in visual and range-aided navigation systems. These constraints could be geometric or photometric constraints like the case in visual and range sensors. For example, visual image sequence processing includes key parameters such as depth, disparity, and optical flow. These parameters constitute a set of constraints that are commonly used to build strong loss functions in ML-based fusion systems for positioning, navigation, and mapping. These parameters are used to impose geometric, photometric, spatial, and temporal constraints. Including these constraints in the loss functions, in addition to the obvious measures like relative and absolute pose error, has a great influence on the convergence and success of ML and AI models' training.

One of the most used loss functions in LiDAR and Camera processing is the reprojection and image wrapping losses. We have discussed the reprojection error in imaging sensor fusion in conventional filtering and optimization techniques. The reprojection error is commonly used in camera-based systems. For example, a neural network architecture can be designed to output a set of features $\vec{P}_k^j$ with correspondences and a relative pose prediction $\left(\vec{\hat{\phi}}_{k,k+1}\ \vec{\hat{T}}_{k,k+1}\right)^T$ from time epoch k to epoch $k+1$. In addition to the simple one-to-one direct loss function that compares the output pose against the ground-truth pose, features reprojection error can be added to strengthen the loss function. The reprojection error for a feature $\vec{P}_k^j = \left(P_{x,k}^j\ P_{y,k}^j\ P_{z,k}^j\right)^T$ defined with respect to the camera frame at epoch k is calculated as follows:

$$e_j = \left\| \vec{p}_k^j - \frac{K}{d_k^j}\left(\vec{P}_k^j R\left(\vec{\hat{\phi}}_{k,k+1}\right) + \vec{\hat{T}}_{k,k+1}\right)\right\|^2 \tag{12-5}$$

where

K is the camera intrinsics

$\vec{p}_k^j$ is the homogeneous pixel location of the feature $\vec{P}_k^j$ on the image plane at epoch k

$R\left(\vec{\hat{\phi}}_{k,k+1}\right)$ is the rotation matrix corresponding to the rotation vector $\vec{\hat{\phi}}_{k,k+1}$.

For M features, the overall reprojection error is given by:

$$J = \sum_{j=1}^{M} \left\| \vec{p}_k^j - \frac{K}{d_k^j}\left(\vec{P}_k^j R_k^{k+1} + \vec{\hat{T}}_{k,k+1}\right)\right\|^2 \tag{12-6}$$

However, we must be able to calculate the derivative of the cost function with respect to the output parameters. If it is difficult to calculate the derivative analytically, we may need to linearize the cost function numerically.

The image warping loss function can be applied in unsupervised learning when the network output is a disparity or an optical flow output (review Chapter 7 for the definitions of disparity and optical flow) and we have a stereo camera system or a sequence of monocular images. In this case, the network processes two images, a source and a target. The target image could be the left image I_L in a stereo system or an image I_k in a monocular image sequence. The source image can be the right image I_R in the stereo system or an image I_{k+1} in the monocular image sequence. Let us say the predicted disparity for pixel or feature i is d_i and the predicted optical flow is v_i. d_i and/or v_i can be used to perform an image wrapping on the source image and compare the wrapped image against the target image. Image wrapping means applying a transformation (either translation or rotation) to the pixels on the image to form a different image. This method is very popular in unsupervised depth estimation fusion systems where we have a stereo camera system or in optical flow estimation systems where we have a monocular image sequence. From Chapter 7, we know that for a stereo system, for each pixel p_i the following wrapping applies:

$$I_L = I_R(p_i + d_i) \tag{12-7}$$

where

d_i is the pixel disparity
p_i is the image coordinate of pixel or feature i.

Also, in a stereo system, the relationship between disparity d_i and depth Z_i is given by the following equation:

$$d_i = Bf/Z_i \tag{12-8}$$

Therefore, in ML-based depth estimation systems, if the network predicts the depth $\hat{Z}_i$ or disparity $\hat{d}_i$ of a feature i, the following loss function term can be applied:

$$J(\hat{Z}_i) = \frac{1}{2}\sum_{i\in\Omega}\|I_L(p_i) - I_R(p_i + Bf/\hat{Z}_i)\|^2 \tag{12-9}$$

$$J\left(\hat{d}_i\right) = \frac{1}{2}\sum_{i\in\Omega}\left\|I_L(p_i) - I_R\left(p_i + \hat{d}_i\right)\right\|^2 \tag{12-10}$$

where

B is the stereo baseline
f is the focal length of the cameras
Ω refers to the index of all possible pixels or features within the image.

If we have a sequence of images and the ML-based model predicts the optical flow $\hat{v}_i$, the following loss function term can be found:

$$J = \sum_{i\in\Omega}\|I_k(p_i) - I_{k+1}(p_i + \hat{v}_i\Delta T)\|^2 \tag{12-11}$$

where ΔT is the time between image I_k and image I_{k+1} of the image sequence.

It is important to note that such loss functions are nonlinear, and its first derivative is difficult to obtain analytically. Therefore, to get the backpropagation algorithm to work, the derivative of the loss function must be approximated numerically. For example, if the predicted output is disparity $\hat{Z}_i$, the output layer of the network will need the following derivative:

$$\frac{\partial}{\partial \hat{Z}_i} J(\hat{Z}_i) = \left(I_L(p_i) - I_R(p_i + Bf/\hat{Z}_i)\right) \frac{\partial I_R(p_i + Bf/\hat{Z}_i)}{\partial \hat{Z}_i} \tag{12-12}$$

The derivative $\frac{\partial I_R(p_i + Bf/\hat{Z}_i)}{\partial \hat{Z}_i}$ can be approximated numerically using Tylor series expansion as follows:

$$\frac{\partial I_R(p_i + Bf/\hat{Z}_i)}{\partial \hat{Z}_i} \cong \frac{I_R\left(p_i + \dfrac{Bf}{(\hat{Z}_i + \Delta\hat{Z}_i)}\right) - I_R(p_i + Bf/\hat{Z}_i)}{\Delta\hat{Z}_i} \tag{12-13}$$

In this approximation, $\Delta\hat{Z}_i$ is the difference between the estimated depth at learning iteration $n-1$ and iteration n. This approximation will be valid only for small $\Delta\hat{Z}_i$, which is a challenge that must be considered when training the network. In addition to these photometric constraints, in vision-based fusion, one of the common loss measures is the structural similarity index, which measures the similarity between two images. Another loss function term that is always used is smoothness. In many scenarios in positioning and mapping estimation, the network output must be smooth. A smoothing loss can be added to the overall loss function to smooth the output. This smoothing can be added as a gradient (differential) term of the output to be smoothed out. When the network minimizes the gradient of a variable, it will consequently enhance its smoothness. There are several cost functions related to depth, disparity, and optical flow in several references such as [2], [3], and [4].

12.4 Recent Trends and Examples

This section is a summary of state-of-the-art approaches and examples of AI/ML-enhanced sensor fusion methods for positioning, navigation, and mapping. The general idea of each approach is briefly explained, followed by an example realization of the concept from the literature.

12.4.1 Handling Dynamic Objects in SLAM

One of the main challenges in SLAM is dynamic objects. Conventional SLAM frameworks assume that the environment is static. If accurate information about the dynamics of objects in the environment is available, SLAM algorithms can be modified to handle dynamic features, but this will complicate SLAM frameworks. A more straightforward solution is excluding dynamic features from the processing. AI/ML-based machine vision models apply deep learning and high-level cognitive and reasoning rules to recognize objects, classify their types, and detect the dynamic status of visual features (static or moving). This capability can be used to enable SLAM algorithms to work only on the features that belong to static regions, such as buildings, walls, bridges, and trees, and exclude dynamic features, such as cars and people. An example of this idea can be found in [5] where LiDAR is fused with a camera and an inertial sensor. In this fusion scheme, semantic segmentation of the scene is performed using a CNN framework. Semantic segmentation assigns a class to each pixel that describes the type of the object/region the pixel belongs to and the dynamic status of the pixel. LiDAR point cloud is then projected on the 2D image plane. Consequently, we can determine the dynamic status and the object type of each point in the LiDAR scan. We can then perform LiDAR and/or visual SLAM using only static features. To perform this type of fusion, we must precisely know the transformation between the camera frame and the LiDAR frame to perform the LiDAR projection on the image plane accurately. Estimating the transformation between the camera frame and the LiDAR frame is called camera-LiDAR calibration, and it is an active research problem. Some solutions to the camera-LiDAR calibration can be found in [6]. Multisensor calibration has also been addressed by deep learning using CNN, as can be seen in [7].

12.4.2 Semantic SLAM

Conventional SLAM algorithms work on a geometric level where pixels, points, and features are used to build maps and to localize the platform within the map. Conventional SLAM approaches suffer from errors resulting from inaccurate feature matching/tracking and wrong loop closures. Also, plain 3D point cloud maps have little information, which leads to a limited understanding of the environment. SLAM frameworks can benefit from the higher-level semantic information that AI and ML models can provide, such as object recognition, semantic segmentation, and layout estimation. This semantic information is more distinguishable than pixels and features, which can reduce matching errors and wrong loop closures. Integrating semantic information into the SLAM framework provides a better understanding of the environment. This approach is called Semantic SLAM. There are different approaches to Semantic SLAM. The simplest approach uses conventional SLAM to build a map represented by 3D points and uses Semantic Segmentation and Object Recognition to detect objects in two separate threads and then project the objects and semantic information on the map. Other advanced methods use objects as features and apply conventional SLAM steps to the object-based features. In this approach, new feature extraction, matching, and tracking modules are replaced by object recognition, matching, and tracking. Object matching and tracking are more complicated than feature matching and tracking. However, object-based matching and tracking are less prone to errors and more robust. The current trend is to combine conventional feature-based SLAM frameworks with AI/ML-based semantic segmentation and object recognition and try to enhance both sides by exchanging information between them, as seen in Figure 12.3.

12.4.3 Improved Relocalization and Loop Closure using Deep CNN

One of the key challenges in vision-based positioning and mapping is the relocalization problem and loop closure detection. Both relocalization and loop closure detection aim at localizing the platform within an existing map. The conventional approach is to detect some features and match the detected features with the existing map. As the locations of the matched map features are known, triangulation can be used to estimate the platform pose. This relocalization process, including feature detection, matching, and pose estimation, can be automated in one framework using deep CNN. Deep CNN networks can be trained to estimate poses by matching input images with a database that includes absolute poses and their corresponding images in an environment. With proper architecture, good training, and proper loss function formulation, the trained CNN model can learn feature detection, correlation, and pose interpolation. An example realization of this idea can be found in [8], where the authors developed and trained a CNN-based framework and they called it PoseNet. PoseNet is a deep CNN that performs feature detection, matching, and pose estimation directly by formulating the problem as a regression problem with a training dataset containing individual images taken at different known poses along a trajectory. The main limitation of the designed network is that the relocalization accuracy was limited to the places used in the training data set. The authors in [8] used GoogLeNet as a starting architecture, which is a 22-layer neural network that can classify and identify over 1,000 types of objects from images. The pooling layers

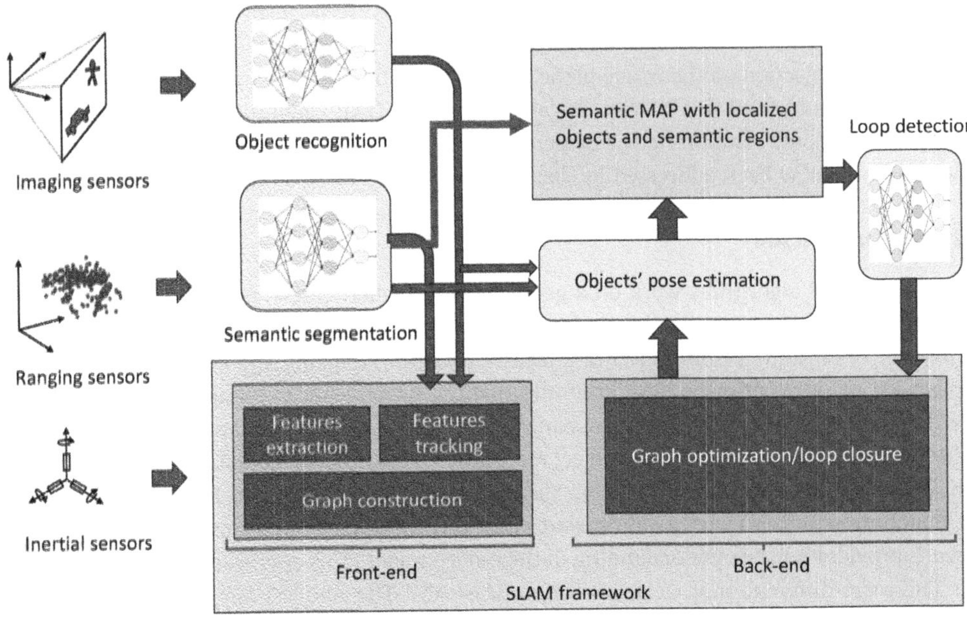

Figure 12.3 General illustrative diagram of semantic SLAM.

were replaced by a regression pooling layer, and the fully connected layer was modified to generate a camera pose as an output. The camera pose output is two vectors, a position vector and a quaternion vector, that are always normalized to guarantee a valid rotation is generated. The modified network is retrained over a dataset of images and their corresponding poses. The loss function was defined as follows:

$$J(I) = \left\| \vec{r} - \hat{\vec{r}} \right\|^2 + \beta \left\| \vec{q}_B^{\,L} - \frac{\hat{\vec{q}}_L^{\,B}}{\left\| \hat{\vec{q}}_L^{\,B} \right\|} \right\|^2 \tag{12-14}$$

where

I is the input image

$\left\{ \hat{\vec{r}}, \hat{\vec{q}}_B^{\,L} \right\}$ is the network output position and quaternion vectors

$\left\{ \vec{r}, \vec{q}_B^{\,L} \right\}$ is the ground-truth pose

β is a weighting factor

B and L refer to the body and local frames, respectively.

The GoogLeNet architecture is explained in [9].

12.4.4 Improved RADAR Scan Matching using ANN Models

One of the fundamental problems of RADAR is the scattered, sparse, noisy measurements due to the weak angular resolution of radio beams and radio signal refraction and reflection. This causes the point cloud estimated from RADAR measurements to be noisy and inaccurate. Consequently, any ICP scan-matching would fail to accurately estimate relative poses from consecutive RADAR scans. This is an active research area due to the weather resistance, lower cost, and lower power of

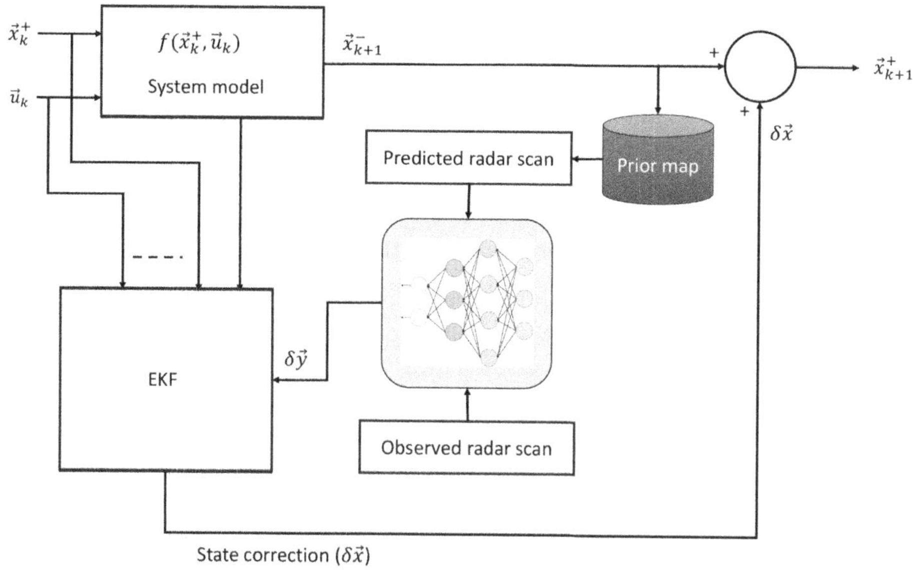

Figure 12.4 Deep ANN replaces ICP/NDT acting as a measurement model to an EKF.

RADAR compared to LiDAR technology. A promising solution to this problem is the replacement of ICP and NDT algorithms with a deep neural network that can learn the relative pose from two consecutive RADAR scans. The estimated relative pose can then be applied as a measurement update to a conventional filtering algorithm such as EKF. In this case, the ML model replaced the measurement model in the conventional fusion framework as seen in Figure 12.4. Figure 12.4 shows a scenario where we have a prior map of RADAR scans. Given a predicted pose and the map, we extract a predicted RADAR scan from the map. The predicted RADAR scan is matched with the actual observed RADAR scan by the vehicle, and a relative pose $\delta\vec{y}$ is estimated by the neural network. The estimated relative pose acts as a measurement update to an EKF that corrects the system state accordingly. The ML model is trained by providing several scan pairs and their corresponding relative poses. The loss function will be the relative pose error (RPE). A simplified diagram of the network training is shown in Figure 12.5.

Figure 12.5 Network training for RADAR scan matching.

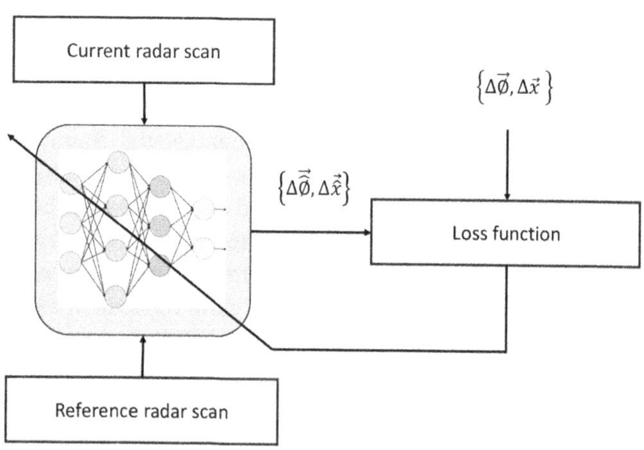

The above scheme can be further improved with LiDAR measurements, which are highly accurate and have a high angular resolution. A potential approach to apply ML to solve this problem is using training a network that accepts noisy RADAR scans and clean LiDAR scans and estimates the relative pose. This scenario is useful if the prior map is LiDAR map. Realization of this idea is demonstrated in [10]. The authors in [10] modified the architecture of the U-Net CNN, which is described in [11].

12.4.5 Wireless Fingerprinting for Indoor Positioning Using ANN Models

Radio positioning systems like GPS do not work well in indoor environments due to signal blockage. Even if local wireless networks such as Wi-Fi or Bluetooth are used, multipath and signal fading make it extremely difficult to precisely measure ranges and estimate positioning information from time-of-flight measurements. A popular trend in indoor positioning is radio fingerprinting. Radio fingerprinting consists of registering the radio signal characteristics (the radio fingerprints) (such as signal strength or channel state information "e.g., fading, scattering, and signal strength") at known reference locations and then use this pre-built radio-map to localize any radio-enabled device by matching what user equipment measures with the fingerprints stored in the radio-map database. This matching process is challenging due to the noisy nature of signal strength and channel state information. Therefore, ML models such as radial-basis function (RBF) ANN and Gaussian process regression (GPR) have been used to build fingerprinting radio maps and to match user equipment measurements with the radio map to localize users within the network. Examples of the realization of this idea can be seen in [12] and [13].

12.4.6 Fusing Inertial Sensors with Indoor Wireless Measurements Using ANN

Inertial sensors are fused to overcome the noisy wireless measurements (signal strength and channel state information) in indoor areas. If the platform is a legged or humanoid platform, Pedestrian Dead Reckoning (PDR) is used as a system model. Classical fusion algorithms usually fail to fuse signal strength measurements with PDR prediction due to the large non-Gaussian noise and large nonlinearity in the prediction and measurements models. In an effort to overcome these limitations, ANN was proposed. Given a training dataset that contains a large volume of ground-truth data (signal strength-location pairs), an ANN can be trained on pairs of signal strength patterns as inputs and locations as output. An example realization work of this approach was reported in [14]. Similar work for indoor positioning can be found in [15], where the authors aim at enhancing the accuracy of PDR framework. In this work, GPS data is used as ground truth to train a deep neural network that performs PDR by accepting raw inertial data and generating relative pose changes. Using raw GPS location data for training may not be easy as GPS is always contaminated by different noises and errors. Standard GPS has a 3- to 5-meter error margin, which is larger than human steps. Therefore, some refinement of the training data must be performed, such as using differential GPS (DGPS), RTK, or trajectory smoothing algorithms. One advantage of this learning-based fusion is that the trained network learns the stride length automatically from the training data. Stride length is usually fixed to a predefined value in conventional PDR algorithms, which makes it difficult for the algorithm to fit different platforms with different stride lengths. Learning the stride length from data is more convenient and potentially leads to better accuracy. If training can be done online, the system will be more adaptable to user changes and immune to outdated data problems.

12.4.7 Enhanced System Models using Deep CNN and Recurrent ANN

One of the sources of errors in conventional fusion algorithms is the accuracy of the system and measurement models, especially for constraint-free 3D motion like multi-rotor drones or legged/humanoid platforms. Dynamic models (like motion prediction) need special attention if we want to replace them with a neural network. Dynamic models have memory and feedback loops where they use past states (and maybe past control signals as well) to calculate the current state. The type of neural network that can remember and store state information is called a recurring artificial neural network (RANN). One of the very popular types of RANNs is the long-short-term-memory (LSTM) neural network. In many works, LSTM has been used to build motion prediction models using raw inertial, speedometer, images, and range data as input and relative poses as output. In some other hybrid configurations, the network can take the output of other subsystems as input, where these subsystems are conventional fusion modules such as EKF. An example realization of this approach in autonomous underwater vehicles (AUV) can be seen in [16], where authors use Attitude and Heading Reference System (AHRS) and Doppler velocity log (DVL) as input to an LSTM neural network to predict relative poses. The estimated relative poses are then used as an update to a conventional EKF. Similar work but for ground vehicles, in [17], a deep ANN was used to complement the kinematic model (system model) that takes raw inertial data as input and estimates position, velocity, and attitude (PVA). In this work, the deep neural network takes raw inertial sensor data as input and generates relative position and velocity changes as output. The learning-based estimated relative position and velocity are then fed as an update to an EKF that fuses the conventional inertial navigation system with the network output. Although some improvements are reported in these works, a major concern is that predictions and measurements are not exactly independent in these works since both the conventional inertial navigation systems and the deep neural network take the same input (raw inertial data) as input which may affect the assumption of the independence of measurements.

12.4.8 Improved Optical Flow using Deep CNN

Optical flow is one of the challenging problems in vision-based positioning, navigation, and mapping. In Chapter 7, we discussed how conventional methods, such as the Lucas-Kanade method, are applied to the image to estimate the optical flow of each pixel. The main challenge of conventional methods is the inaccuracies of feature detection and correlation. The Lucas–Kanade method can be applied to estimate the velocity of each pixel by applying a derivative operator on a window of pixels around the pixel of interest. This window-based derivative operator is an ideal use case for CNN. CNN has shown success in learning feature detection and correlation from image pairs. Given reliable ground-truth training data, optical flow can be learned from image pairs using a deep CNN framework. An example realization of this work can be found in [18], where researchers developed FlowNet, a deep CNN that estimates optical flow from pairs of raw images. The CNN architecture takes two consecutive images and the convolutional layers act as the windowed derivative operator. The subsequent layers generate the final output, which is the optical flow for each pixel in the input images. The network can be designed to work on raw pixels (dense optical flow estimation) or work on features. If the points of interest are features, the network can be trained to solve the feature matching if accurate ground-truth data is available.

12.4.9 Visual Odometry Using Deep Recurrent CNN Models

Authors in [19] proposed that the optical flow estimated by the CNN in [18] can be used to estimate the relative poses (odometry) between the two input images using Recurrent CNN particularly LSTM neural networks to model the dynamics involved in estimating the relative poses given pairs of image frames. The architecture presented in [19] is called DeepVO. When using CNN and LSTM to perform visual odometry, CNN can be seen as an automatic features detector, extractor, descriptor, and correlator. The LSTM network can be thought of as a dynamic system that learns relative poses from image sequences using the detected features. As there is a functional relationship between feature pixel locations in the image pairs and relative pose changes, the hypothesis that a neural network can learn this functional relationship is valid, given the right architecture and the right loss function. What verifies the hypothesis is the experimental testing by training the network structure on reliable datasets and watching the error after each learning iteration. The ultimate verification of this learning hypothesis is the testing of independent data set that was unseen by the network during the training. The general architecture of RCNN-based visual odometry is shown in Figure 12.6.

12.4.10 Visual Inertial Odometry using Deep CNN and LSTM Models

The integration of inertial sensors increases the stability and accuracy of visual odometry. To fuse inertial sensor data with image information, a more advanced neural network architecture can be designed like the work demonstrated in [20]. The work in [20] is building on top of the FlowNet work in [18] where the correlation version of the FlowNet (named FlowNetC) is used. What distinguishes the work presented in [20] is the way the ground-truth training data is generated. Obtaining ground-truth data is one of the key challenges in developing ML-based visual-inertial fusion. An additional advantage of the work in [20] is how the training data is obtained. The training data is obtained by fusing GPS (or GNSS) with an inertial sensors measurement unit (IMU) using EKF. During the reliable availability of GNSS, the IMU/GNSS EKF provides smooth, non-drifting, accurate positioning. The obtained poses are time synchronized with images, which form a ground-truth training data set for the system. The input to the network framework is a historical window of inertial sensors and a pair of images, and the output is the relative pose between the image pairs. Another advantage of the work in [20] is that it is an end-to-end approach that does not need pre-calibration of sensors or complex linearizable models. The overall architecture of this learning framework is shown in Figure 12.7.

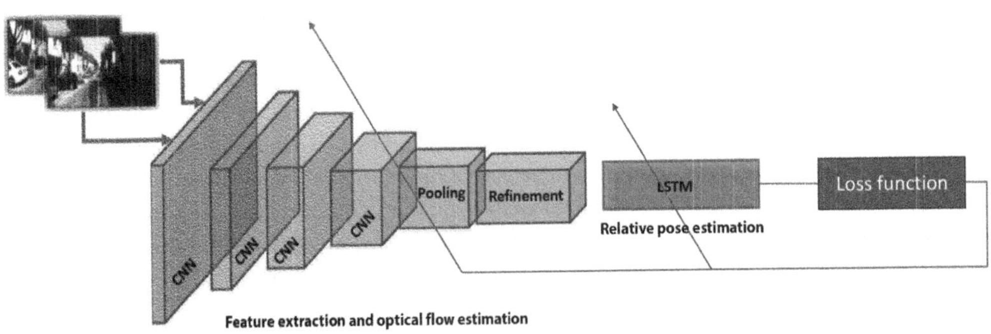

Figure 12.6 General architecture for RCNN-based visual odometry.

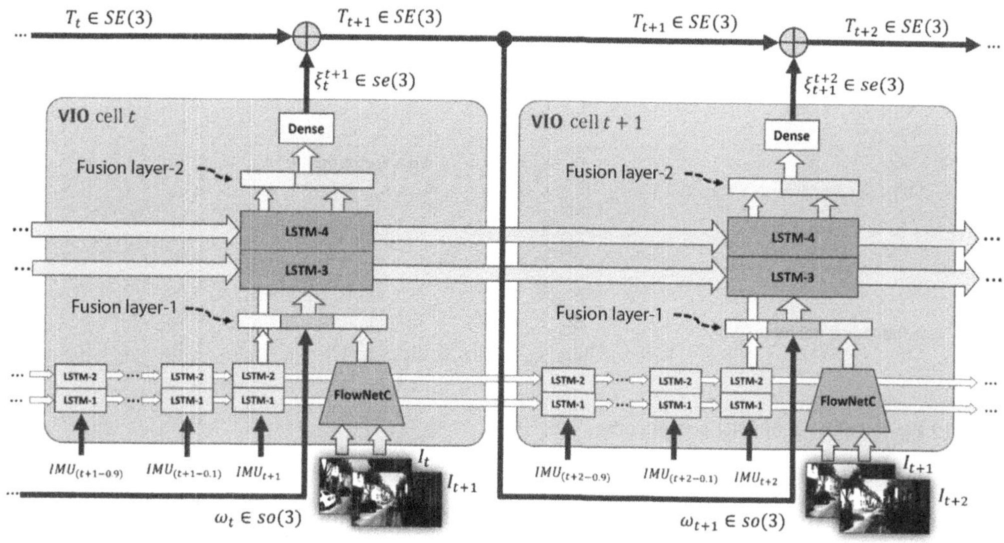

Figure 12.7 Visual inertial odometry learning-based framework.

12.4.11 Enhanced Odometry Using CNN-Based Fusion of LiDAR and Camera

LiDAR and camera can be fused to estimate better positioning and mapping. In the previous chapters, we discussed multiple scenarios where conventional filtering and optimization methods are applied to fuse inertial sensors with LiDAR and cameras for better odometry and mapping. An end-to-end deep learning approach can be designed to perform the fusion of inertial, camera, and LiDAR raw measurements. The loss function in this fusion is formulated as the error between the relative poses estimated by the network and the ground-truth poses dataset. A sample realization of this approach can be seen in [21].

12.4.12 Correcting Inertial Sensors Measurements Using DT Regression Models

Denoising inertial sensors is challenging, especially low-cost types that involve complex stochastic errors. Conventional signal processing and stochastic process modeling have several limitations when applied to clean the measurements of low-cost inertial sensors. Examples of these limitations include the computational constraints and the availability of ground-truth raw inertial data. AI and ML models have been explored as denoising tools to clean erroneous inertial sensors. An example realization of this approach can be found in [22], where an adaptive neuro-fuzzy-inference system (ANFIS) is developed and trained to denoise inertial sensors within a GPS/IMU fusion framework. Another example where the Decision Tree regression model is trained to clean erroneous inertial sensors can be found in [23]. The ground-truth data in [23] is generated by applying inverse kinematics on poses that are obtained by fusing GPS and inertial sensors using a conventional EKF. The application of inverse kinematics eliminates the need for ground-truth raw data from high-end expensive inertial sensors.

12.4.13 Depth Estimation Using Stereo Images and CNN Autoencoders

In Chapter 7, we discussed the conventional methods of depth estimation using stereo images. The conventional methods depend on feature detection and matching, which is vulnerable to errors and

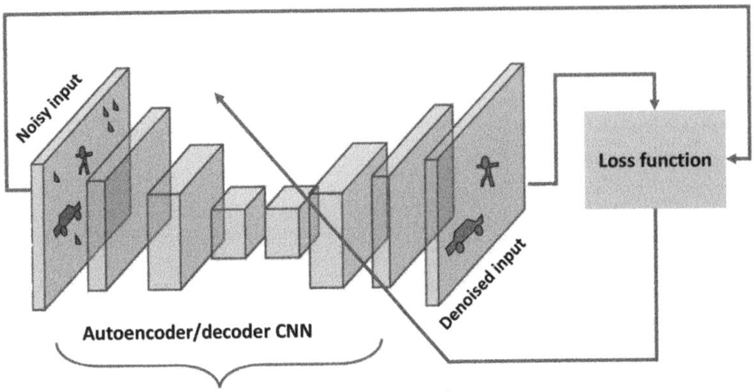

Figure 12.8 Illustration of CNN autoencoder.

mismatching. Predefined human-engineered features are one limiting factor. In addition, conventional methods require accurate camera calibration. As discussed in Chapter 7, feature detection is basically a convolution operation, which is the ideal scenario for CNN. Given the flexibility of CNN, it can learn new complex features autonomously. Therefore, researchers explored the utilization of CNN to develop an ML-based depth estimation model. Depth estimation from stereo images can be formulated as an unsupervised learning scenario that can be solved using CNN Autoencoders. CNN Autoencoders is a special type of deep neural network that is optimized to code the input data in a reduced size. Decoders CNN are used to decode this code to generate a copy of the input data. Autoencoders/decoders can be used for dimensionality reduction and noise cancellation applications. Extracting features from images is technically a dimensional reduction of the input image that can be learned by CNN autoencoders. The general architecture of CNN autoencoder is shown in Figure 12.8.

The autoencoder/decoder CNN architecture can be used to learn the depth from stereo images. The basic idea is to take the left image as an input and a wrapped (transformed) version of the right image as output and tune the network until the error between the left image and the wrapped version of the right image is minimized. The wrapping of the right image is performed using the depth estimated from autoencoder/decoder CNN as follows: if the network predicts depth $\hat{Z}_i$ for pixel p_i, the following loss function term can be applied:

$$J(\hat{Z}_i) = \frac{1}{2}\sum_{i\in\Omega}\left\|I_L(p_i) - I_R\left(p_i + Bf/\hat{Z}_i\right)\right\|^2 \qquad (12\text{-}15)$$

where I_L and I_R are left and right images, respectively.

By minimizing the error between the left image and the wrapped right image, we technically correct the depth estimation. This architecture is shown in Figure 12.9.

12.4.14 Deep Learning-Based SLAM

Researchers in [4] combined the optical flow, pose estimation, and depth estimation architecture described above to build a complete visual SLAM solution. The mapping module of the framework in [4] is an autoencoder/decoder CNN architecture that estimates the depth from images. The tracking module is based on the FlowNet and deep VO architecture where CNN and LSTM are

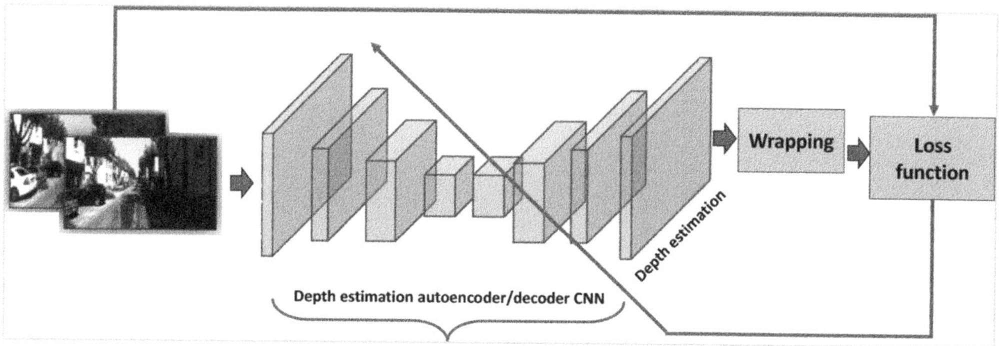

Figure 12.9 Depth estimation using autoencoder/decoder CNN architecture.

combined to estimate relative poses. A third CNN is trained to detect previously seen scenes to detect loops. The outputs from the three modules (mapping, tracking, and loop detection) are fed to the conventional graph building and graph optimization modules. The loss function in this framework included terms for photometric spatial constraints, pose consistency constraints, and temporal transformation constraints. Photometric spatial constraints use the predicted depth to warp the right image and compare the wrapped right image to the left image. Poses consistency constraints impose that the relative transformation of the left and the right cameras must be the same. The temporal transformation constraints use the predicted relative pose to transform the image at epoch k to the image at epoch $k + 1$ and compare the transformed image against the true image. Additional geometric constraints, such as point-cloud scan matching (like ICP or NDT), were also applied to the predicted depth map. In these constraints, the depth output is used to build a 3D point cloud at epoch k and epoch $k + 1$ and an ICP is applied to the two point clouds.

12.4.15 Dense 3D Depth Estimation Using CNN-Based Camera and LiDAR Fusion

Another approach where ML is applied in LiDAR and camera fusion is the estimation of dense, colored, and accurate 3D point clouds with accurate depth. LiDAR provides accurate depth measurements, but it does not provide colour information, and its point cloud is sparse due to the growing separation between light beams, especially with far distances. On the other hand, the camera provides a rich and dense projection of the environment, but it lacks depth information. Even with a stereo configuration, depth estimation may fail if the scene has repeated features or poor texture. This complementary characteristic is illustrated in Figure 12.10. Given their complementary characteristics, ML models can be applied to fuse cameras and LiDAR to obtain dense, colored, and accurate 3D point clouds with accurate depth is important. A sample realization of the idea of fusing LiDAR and a camera can be found in [3], where the authors developed a deep CNN framework called SLFNet. The developed CNN framework uses stereo camera measurements to train a deep CNN framework that fuses LiDAR and monocular camera images to generate dense 3D point clouds with accurate depth estimation. The CNN architecture used in [3] is a modified version of ResNet [24] that has several convolution layers that perform complex feature detection (encoding) at multiscales of the input image. Besides the complex structure of the SLFNet architecture, the loss function is very sophisticated and considers several spatial, geometric, photometric, and temporal constraints.

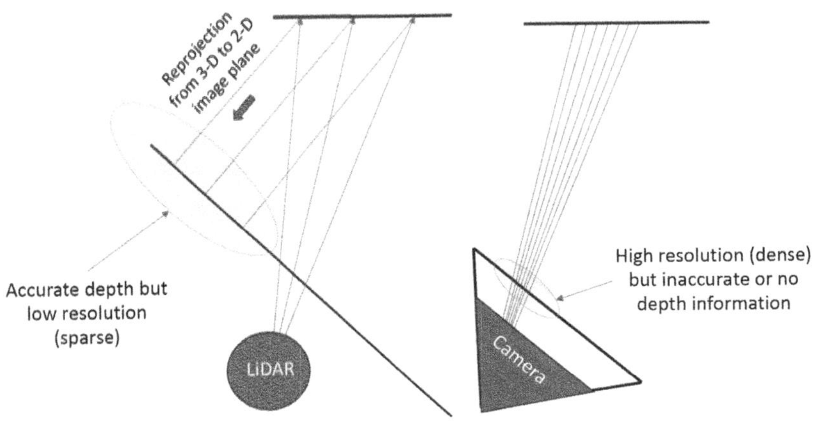

Figure 12.10 LiDAR vs. camera depth characteristics.

12.4.16 Learning Noise Parameters Using Genetic Algorithms

Fusion algorithms depend heavily on hyper-parameters such as the noise covariance parameters in EKF, PF, and their variations. Tuning these parameters for best performance is a cumbersome process that is usually performed by experts using a kind of trial-and-error approach guided by noise specifications provided by the sensors' manufacturers, where noise is assessed and reported in the form of power spectral density using statistical tools such as Allan variance. The overall tuning process is illustrated in Figure 12.11.

A clear example of this problem is Kalman filter tuning. Throughout the book, we have demonstrated Kalman filter fusion examples without discussing how parameters such as system noise covariance matrix Q, measurement noise covariance matrix R and initial error covariance matrix P_0 are determined (review Kalman filtering in Chapter 3). The parameters set $\{Q, R, P_0\}$ are key parameters that have a huge effect on the filter performance. In practice, noise covariance matrices are initialized by nominal values guided by the noise specifications of the sensors/systems provided by datasheets prepared by the sensor manufacturer based on laboratory tests. Since sensors come from different manufacturers, their standards and units of the noise parameters are not normalized, which causes inconsistency in Kalman filtering. Due to the lack of a closed-form formula to calculate the parameters $\{Q, R, P_0\}$, heuristic AI-based approaches such as genetic algorithms (GA) can be beneficial. An example realization of this approach can be found in [25] where the author applied GA as a heuristic stochastic search engine supported by the design of experiment (DoE [25])

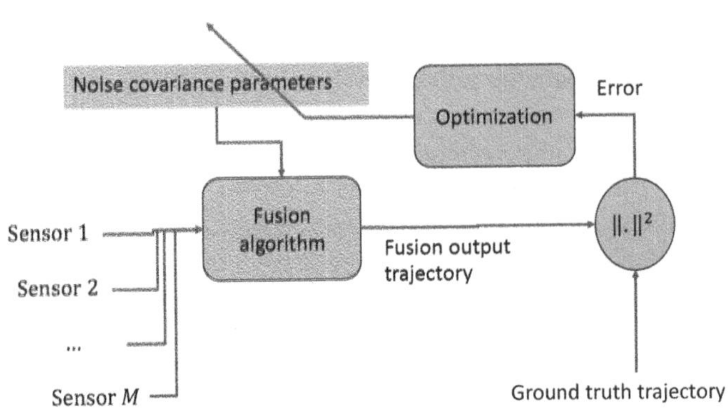

Figure 12.11 Noise covariance parameters tuning.

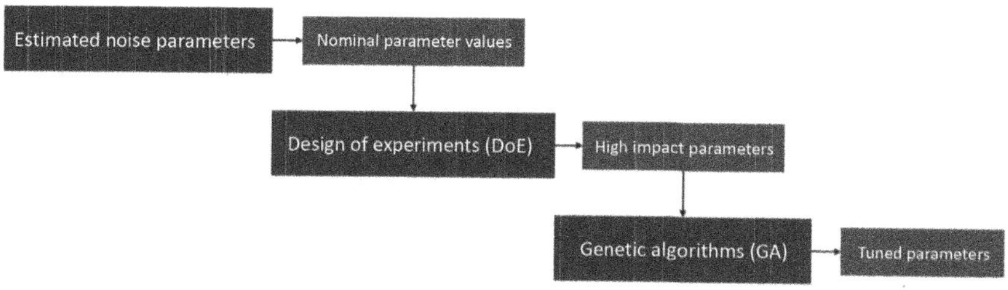

Figure 12.12 GA and DoE optimization of EKF parameters.

approach to assess the impact of each parameter that guides the GA search. The overall block diagram that shows the steps of the technique is shown in Figure 12.12.

12.4.17 Real-Time IMU Correction Using ML Models

In conventional filtering techniques, sensor errors (such as inertial sensor biases) are estimated by the estimation filters (EKF or PF). However, to estimate sensor errors, accurate models of sensor errors must be available. Biases, for example, are overly simplified error parameters as there are several other errors that need to be considered, especially for low-cost MEMS-based inertial measurement units (IMUs). However, considering all errors analytically will complicate the system models and increase the tuning and design of the system. In addition, not all errors can be modeled using linear stochastic systems. This problem becomes significant if low-grade MEMS IMUs are used. Even under EKF and accurately estimated biases, the navigation accuracy deteriorates quickly when external measurements are unavailable for small periods. This scenario can happen if, for example, GPS is blocked or light intensity deteriorates for a short period of time. As smooth and accurate pose estimation is usually available during the availability of external measurements, an ML can be applied to design a learning-based module to clean erroneous inertial sensors during short outages of external measurements. The overall diagram of this ML-aided fusion configuration is given in Figure 12.13. Example application of this scheme can be found in [23]. The authors in

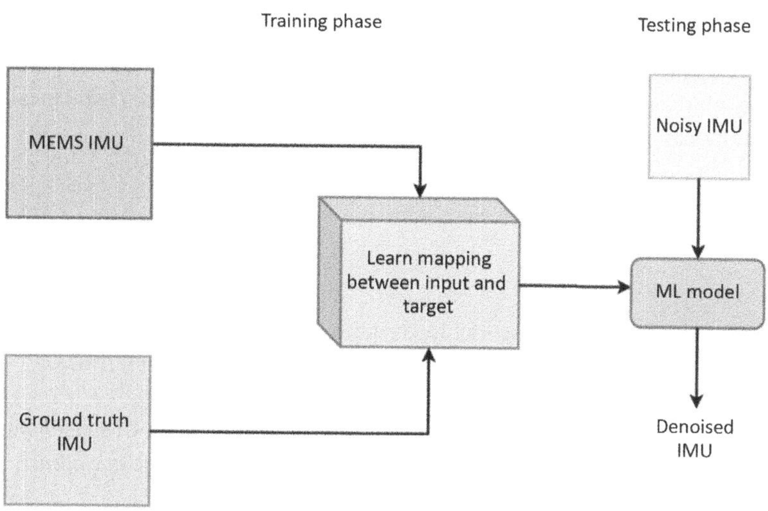

Figure 12.13 General diagram of ML-aided IMU correction configuration.

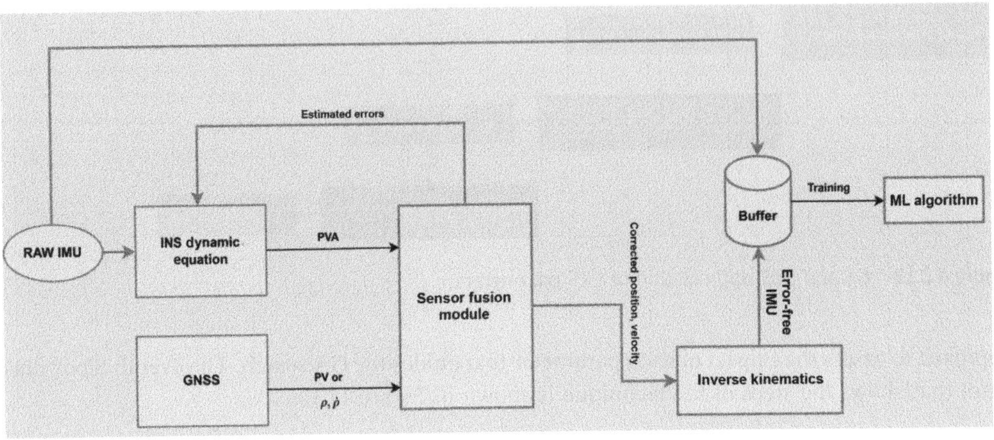

Figure 12.14 Learning IMU errors using inverse kinematics and ML algorithm.

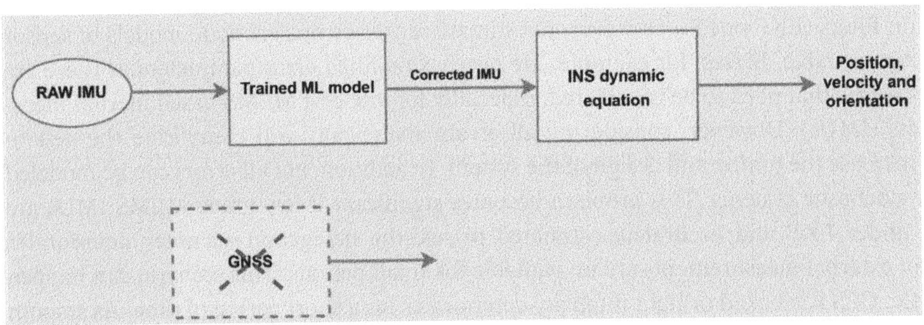

Figure 12.15 Testing the developed IMU correction ML algorithm during the GNSS outage.

[23] developed a technique that does not need high-end super-accurate IMUs to obtain ground-truth data. Instead, they applied inverse kinematics and smoothing techniques on the poses of well-filtered portions of the trajectory during the availability of reliable GPS and used the resulting smoothed raw IMU as ground truth for training. The training and testing phases of the work in [23] are illustrated in Figure 12.14 and Figure 12.15. The term PVA in Figure 12.14 stands for Position, Velocity, and Attitude. The approach works in tightly coupled fusion mode where the GNSS measurements are the pseudorange and pseudorange rate $\rho, \dot{\rho}$.

References

1 Z. Zhang and D. Scaramuzza, "A Tutorial on Quantitative Trajectory Evaluation for Visual-Inertial Odometry," in *2018 IEEE/RSJ International Conference on Intelligent Robots and Systems (IROS)*, Madrid, Spain, 2018.

2 L. Wang, Y. Guo, Y. Wang, Z. Liang, Z. Lin, J. Yang and W. An, "Parallax Attention for Unsupervised Stereo Correspondence Learning," *IEEE Transactions on Pattern Analysis and Machine Intelligence*, vol. 44, no. 4, p. 2108, 2022.

3 Y. Zhang, L. Wang, K. L. Fu and Y. Guo, "SLFNet: A Stereo and LiDAR Fusion Network for Depth Completion," *IEEE Robotics and Automation Letters*, vol. 7, no. 4, p. 10605, 2022.
4 R. Li, S. Wang and D. Gu, "DeepSLAM: A Robust Monocular SLAM System With Unsupervised Deep Learning," *IEEE Transactions on Industrial Electronics*, vol. 68, no. 4, p. 3577, 2021.
5 N. Abdelaziz and A. El-Rabbany, "Deep Learning-Aided Inertial/Visual/LiDAR Integration for GNSS-Challenging Environments," *Sensors (Basel)*, vol. 23, no. 13, p. 6019, 2023.
6 D. Lee, "Efficient Camera–LiDAR Calibration Using Accumulated LiDAR Frames," *IEEE Access*, vol. 10, pp. 132349–132362, 2022.
7 N. Schneider, F. Piewak, C. Stiller and U. Franke, "RegNet: Multimodal Sensor Registration Using Deep Neural Networks," in *2017 IEEE Intelligent Vehicles Symposium (IV)*, Los Angeles, CA, USA, 11–14 June 2017.
8 A. Kendall, M. Grimes and R. Cipolla, "PoseNet: A Convolutional Network for Real-Time 6-DOF Camera Relocalization," in *2015 IEEE International Conference on Computer Vision (ICCV)*, Santiago, Chile, 2015.
9 C. Szegedy, W. Liu, Y. Jia, P. Sermanet, S. Reed, D. Anguelov, D. Erhan, V. Vanhoucke and A. Rabinovich, "Going Deeper with Convolutions," in *2015 IEEE Conference on Computer Vision and Pattern Recognition (CVPR)*, Boston, MA, USA, 2015.
10 H. Yin, R. Chen, Y. Wang and R. Xiong, "RaLL: End-to-End Radar Localization on Lidar Map Using Differentiable Measurement Model," *IEEE Transactions on Intelligent Transportation Systems*, vol. 23, no. 7, p. 6737, 2022.
11 O. Ronneberger, P. Fischer and T. Brox, "U-Net: Convolutional Networks for Biomedical Image Segmentation," in *International Conference on Medical Image Computing and Computer-Assisted Intervention*, Springer, Cham, Switzerland, 2015.
12 N. Lv, F. Wen, Y. Chen and Z. Wang, "A Deep Learning-Based End-to-End Algorithm for 5G Positioning," *IEEE Sensors Letters*, vol. 6, no. 4, pp. 1–4, 2022.
13 M. Atia, A. Noureldin and M. Korenberg, "Dynamic Online-Calibrated Radio Maps for Indoor Positioning in Wireless Local Area Networks," *IEEE Transactions on Mobile Computing*, vol. 12, no. 9, p. 1774, 2013.
14 C. Kjellson, M. Larsson, K. Astrom and M. Oskarsson, "Accurate Indoor Positioning Based on Learned Absolute and Relative Models," in *International Conference on Indoor Positioning and Indoor Navigation (IPIN)*, Lloret de Mar, Spain, 2021.
15 K.-S. Kim and Y. Shin, "Deep Learning-Based PDR Scheme That Fuses Smartphone Sensors and GPS Location Changes," *IEEE Access*, vol. 9, no. 2021, p. 158616, 2021.
16 X. Zhang, B. He, G. Li, X. Mu, Y. Zhou and T. Mang, "Navnet: AUV Navigation Through Deep," *IEEE Access*, vol. 8, no. 2020, p. 59845, 2020.
17 Y. Jin, W.-A. Zhang, H. Sun and L. Yu, "Learning-Aided Inertial Odometry With Nonlinear State Estimator on Manifold," *IEEE Transactions on Intelligent Transportation Systems*, vol. 24, no. 9, p. 9792, 2023.
18 A. Dosovitskiy, P. Fischer, E. Ilg, P. Häusser, C. Hazirbas, V. Golkov, P. Smagt, D. Cremers and T. Brox, "FlowNet: Learning Optical Flow with Convolutional Networks," in *2015 IEEE International Conference on Computer Vision*, Santiago, Chile, 2015.
19 S. Wang, R. Clark, H. Wen and N. Trigoni, "DeepVO: Towards End-to-end Visual Odometry with Deep Recurrent Convolutional Neural Networks," in *2017 IEEE International Conference on Robotics and Automation (ICRA)*, Singapore, 2017.
20 S. Sheikhpour and M. Atia, "Calibration-free Visual-Inertial Fusion with Deep Convolutional Recurrent Neural Networks," in *Proceedings of the 32nd International Technical Meeting of the Satellite Division of The Institute of Navigation (ION GNSS+ 2019)*, Miami, FL, USA, 2019.

21 D. Zhang, T. Peng and J. S. Loomis, "Optimized Deep Learning for LiDAR and Visual Odometry Fusion in Autonomous Driving," *IEEE Sensors Journal*, vol. 23, no. 1, p. 29594, 2023.

22 A. E. Mahdi, A. Azouz, A. E. Abdalla and A. Abosekeen, "A Machine Learning Approach for an Improved Inertial Navigation System Solution," *Sensors (MDPI)*, vol. 22, p. 1687, 2022.

23 R. K. R. Damagatla and M. Atia, "Improving EKF-Based IMU/GNSS Fusion Using Machine Learning for IMU Denoising," *IEEE Access*, vol. 12, pp. 114358–114369, 2024. DOI: https://doi.org/10.1109/ACCESS.2024.3440314.

24 K. He, X. Zhang, S. Ren and J. Sun, "Deep Residual Learning for Image Recognition," in *IEEE Conference on Computer Vision and Pattern Recognition*, Las Vegas, NV, USA, 2016.

25 A. Zhang and M. Atia, "An Efficient Tuning Framework for Kalman Filter Parameter Optimization using Design of Experiments and Genetic Algorithms," *Navigation – Journal of The Institute of Navigation*, vol. 67, no. 4, pp. 775–793, 2020.

Index

Sensor Fusion Approaches for Positioning, Navigation, and Mapping: How Autonomous Vehicles and Robots Navigate in the Real World: With MATLAB Examples, First Edition. Mohamed M. Atia.
© 2025 The Institute of Electrical and Electronics Engineers, Inc. Published 2025 by John Wiley & Sons, Inc.

Printed and bound by CPI Group (UK) Ltd, Croydon, CR0 4YY
17/09/2026
14964259-0007